RELATION ALGEBRAS BY GAMES

STUDIES IN LOGIC

AND

THE FOUNDATIONS OF MATHEMATICS

VOLUME 147

Honorary Editor:

P. SUPPES

Editors:

S. ABRAMSKY, *London*
S. ARTEMOV, *Moscow*
D.M. GABBAY, *London*
A. KECHRIS, *Pasadena*
A. PILLAY, *Urbana*
R.A. SHORE, *Ithaca*

ELSEVIER
AMSTERDAM • BOSTON • LONDON • NEW YORK • OXFORD • PARIS
SAN DIEGO • SAN FRANCISCO • SINGAPORE • SYDNEY • TOKYO

RELATION ALGEBRAS BY GAMES

ROBIN HIRSCH
Department of Computer Science, University College
Gower Street, London, WC1E 6BT
United Kingdom

IAN HODKINSON
Department of Computing, Imperial College
180 Queen's Gate, London, SW7 2BZ
United Kingdom

2002

ELSEVIER
AMSTERDAM • BOSTON • LONDON • NEW YORK • OXFORD • PARIS
SAN DIEGO • SAN FRANCISCO • SINGAPORE • SYDNEY • TOKYO

ELSEVIER SCIENCE B.V.
Sara Burgerhartstraat 25
P.O. Box 211, 1000 AE Amsterdam, The Netherlands

© 2002 Elsevier Science B.V. All rights reserved.

This work is protected under copyright by Elsevier Science, and the following terms and conditions apply to its use:

Photocopying
Single photocopies of single chapters may be made for personal use as allowed by national copyright laws. Permission of the Publisher and payment of a fee is required for all other photocopying, including multiple or systematic copying, copying for advertising or promotional purposes, resale, and all forms of document delivery. Special rates are available for educational institutions that wish to make photocopies for non-profit educational classroom use.

Permissions may be sought directly from Elsevier Science via their homepage (http://www.elsevier.com) by selecting 'Customer Support' and then 'Permissions'. Alternatively you can send an e-mail to: permissions@elsevier.com, or fax to: (+44) 1865 853333.

In the USA, users may clear permissions and make payments through the Copyright Clearance Center, Inc., 222 Rosewood Drive, Danvers, MA 01923, USA; phone: (+1) (978) 7508400, fax: (+1) (978) 7504744, and in the UK through the Copyright Licensing Agency Rapid Clearance Service (CLARCS), 90 Tottenham Court Road, London W1P 0LP, UK; phone: (+44) 207 631 5555; fax: (+44) 207 631 5500. Other countries may have a local reprographic rights agency for payments.

Derivative Works
Tables of contents may be reproduced for internal circulation, but permission of Elsevier Science is required for external resale or distribution of such material.
Permission of the Publisher is required for all other derivative works, including compilations and translations.

Electronic Storage or Usage
Permission of the Publisher is required to store or use electronically any material contained in this work, including any chapter or part of a chapter.

Except as outlined above, no part of this work may be reproduced, stored in a retrieval system or transmitted in any form or by any means, electronic, mechanical, photocopying, recording or otherwise, without prior written permission of the Publisher.
Address permissions requests to: Elsevier Global Rights Department, at the fax and e-mail addresses noted above.

Notice
No responsibility is assumed by the Publisher for any injury and/or damage to persons or property as a matter of products liability, negligence or otherwise, or from any use or operation of any methods, products, instructions or ideas contained in the material herein. Because of rapid advances in the medical sciences, in particular, independent verification of diagnoses and drug dosages should be made.

First edition 2002

Library of Congress Cataloging in Publication Data
A catalog record from the Library of Congress has been applied for.

ISBN: 0 444 50932 1
ISSN: 0049-237X

♾ The paper used in this publication meets the requirements of ANSI/NISO Z39.48-1992 (Permanence of Paper).
Printed in The Netherlands.

Preface

Relation algebras are algebras arising from the study of binary relations. They form a part of the field of algebraic logic, and have applications in proof theory, modal logic, and computer science. This book uses combinatorial games to develop some of the theory of relation algebras, focusing on the fundamental notion of representation. Games allow an intuitive and appealing approach to the subject, and permit substantial advances to be made. The introduction explains our perspective on the material. We hope that the book will be used by graduate students and researchers interested in relation algebras and games.

The book proper is divided into six parts. The lengthy first part presents some necessary background material, including the formal definitions of relation algebras, cylindric algebras, their basic properties, and some connections between them. Examples are given. Part I ends with a short survey of other work beyond the scope of the book. In part II we introduce the games, and use them to axiomatise various classes of algebras. Part III discusses approximations to representability, using relational bases, hyperbases, relation algebra reducts, and relativised representations. In part IV we present some constructions of relation algebras, including Monk algebras and the 'rainbow construction', and use them to show that various classes of representable algebras are non-finitely axiomatisable or even non-elementary. Part V continues this work, showing that the representability problem for finite relation algebras is undecidable, and then in contrast proving some finite base property results. The book ends in part VI with a condensed summary of the work done, and a list of problems.

What prerequisites are there? The book is generally self-contained on relation algebras and on games, and introductory text is scattered throughout. Some familiarity with elementary aspects of first-order logic and set theory is assumed, though many of the definitions are given. Chapter 2 summarises the necessary universal algebra and model theory, and more specific model-theoretic ideas are explained as they arise. But we do not believe that the book should be read in total isolation, and we take it that the reader has access to standard texts that will provide a more detailed treatment of relevant material where required.

There are more than 400 exercises, ranging from elementary drill to exercises proving standard theorems in non-standard ways, weak versions of later results, or results beyond those in the text, and occasionally solving open problems. Exercises are usually placed at the end of a section. Exercise 4 at the end of section 5.2 is referred to within section 5.2 as 'exercise 4' or 'exercise 4 below', and outside section 5.2 as 'exercise 5.2(4)'.

The bibliography includes, in square brackets at the end of each reference, the numbers of the pages in the book on which the reference is cited. We first saw this device in [Hod93], and have found it enormously useful.

Many thanks are owed to the following people who have contributed to this book by pointing out errors, discussing the ideas, suggesting improvements, helping with historical information, or sending sections for inclusion in the text: Ágnes Kurucz, Maarten Marx, Szabolcs Mikulás, Mark Reynolds, Gabor Sági, András Simon, Yde Venema, and Michael Zakharyaschev; we owe special thanks to Hajnal Andréka, Steven Givant, Roger Maddux, Istvan Németi, and Tarek Sayed Ahmed, who sent us many pages of detailed comments. (This should not be taken as implying that they endorse the way the book has turned out.) Section 17.6 is joint work with Mikulás and we thank him for granting permission to include it. Model theory and games have heavily influenced the book, and we are very happy and grateful that Wilfrid Hodges has contributed a foreword. We also thank Dov Gabbay and Jane Spurr for handling some contractual matters, Mike Nugent for his careful copy-editing of the book, and Andy Deelen and the staff of Elsevier Science for bringing the book to press.

Parts of chapters 2 and 8 are modified versions of parts of [HirHod97b] and [HirHod97a] and appear by permission of the Association for Symbolic Logic. Parts of chapters 7, 15, and 16 are modified versions of parts of [Hir00, HirHod01a, HirHod00], respectively, and appear by permission of Elsevier Science. Part of chapter 14 is a modified version of part of [HirHod02a] and part of chapter 18 is a modified version of part of [HirHod01b] and they appear by permission of the American Mathematical Society. Parts of chapter 19 are modified versions of parts of [Hodk02, AndHod$^+$99], and appear by permission of Kluwer Academic Publishers and the Association for Symbolic Logic, respectively.

Foreword

There is something strange about the relations between logic and games in the first half of the twentieth century. A number of mathematicians (Zermelo, von Neumann and Julia Robinson for starters) were making deep contributions to both subjects, and yet nobody at that time saw any systematic connections between the two. You should find this puzzling after reading just a few pages of this book of Robin Hirsch and Ian Hodkinson. As they illustrate in a profusion of ways, games lie very close to some of the most general and fundamental notions of model theory, such as axiomatisability, step-by-step construction and even satisfaction of formulas. Of course many of the devices of model theory were yet to be discovered in 1950, but I can't help wondering if the idea of applying games in logic might have speeded up the discovery.

This is idle speculation, but here is one possible reason why it took time for logic and games to come together. The notion of a game has to do with people acting together, setting themselves and each other tasks. As a result, game-theoretic versions of mathematical ideas often have a direct intuitive appeal when compared with more formalistic treatments. In the period 1900–1950 logic was fighting to establish itself as a serious branch of mathematics, and if you want your mathematics to be serious you don't start by talking about people setting up competitions or exercise sessions. Today logic has won its battle for recognition, and Hirsch and Hodkinson can afford to make intuitiveness one of their chief aims.

Although the idea behind them is quite straightforward, relation algebras were a minority topic until recently. I used to feel that they have an uncomfortably ambiguous place on the boundary between syntax and model theory, and I think many of my contemporaries shared that impression. But it often happens that an unexpected viewpoint yields dividends, and today relation algebras take their proper place as an important tool of theoretical computer science among other things. Basic open questions about them, some old and some new, have given Hirsch and Hodkinson a perfect opportunity to show the power and elegance of game-theoretic techniques in model theory. You can read this book to equip yourself for research in relation algebras, or just to enjoy the many clear descriptions of how to analyse a problem and solve it.

Wilfrid Hodges   May 2002

Contents

Preface   v

Foreword   vii

1 Introduction   1
- 1.1 History 1
- 1.2 To the games 9
- 1.3 Non-finite axiomatisability 13
- 1.4 Approximations to representability 15
- 1.5 Constructions of algebras 17
- 1.6 Some remarks on methods 19
- 1.7 Summary of contents 20

I Algebras of Relations   23

2 Preliminaries   25
- 2.1 Foundations 25
- 2.2 Model theory 31
  - 2.2.1 Syntax 31
  - 2.2.2 Semantics — structures 33
  - 2.2.3 Models, validity 34
  - 2.2.4 Homomorphisms, embeddings, substructures 35
  - 2.2.5 Generating sets 37
  - 2.2.6 Compactness, Löwenheim–Skolem–Tarski theorems . . . 37
  - 2.2.7 Relativisation, interpretations, second-order logic 37
- 2.3 Boolean algebras 38
  - 2.3.1 Definition and examples 38
  - 2.3.2 Atoms 40
  - 2.3.3 Dense sets 41

		2.3.4	Ideals, filters, ultrafilters	41
		2.3.5	Representations of boolean algebras	42
		2.3.6	Canonical extensions	43
		2.3.7	Infinite sums and products	44
		2.3.8	Complete representations	46
		2.3.9	Completions of boolean algebras	48
	2.4	Products and ultraproducts		56
		2.4.1	Products	56
		2.4.2	Ultraproducts, ultrapowers	56
	2.5	Boolean algebras with operators		60
		2.5.1	Definitions	60
		2.5.2	Homomorphisms and ideals	61
		2.5.3	Completely additive and conjugated algebras	63
		2.5.4	Completions of BAOs	65
	2.6	Varieties and quasi-varieties of BAOs		67
		2.6.1	Basic concepts	67
		2.6.2	HSP notation and Birkhoff's theorem	68
		2.6.3	Subdirect products	69
		2.6.4	Discriminator varieties	70
	2.7	Aspects of duality for BAOs		77
		2.7.1	Atom structures of BAOs	77
		2.7.2	Complex algebras	79
		2.7.3	Canonical (perfect) extensions of BAOs	80
		2.7.4	Axiomatising the atom structures of a variety	84
		2.7.5	Recovering a variety from its atom structures?	89
		2.7.6	Sahlqvist varieties	93
3	**Binary relations and relation algebra**			**99**
	3.1	Algebraic logic		99
	3.2	Binary relations		101
		3.2.1	Proper relation algebras	101
		3.2.2	Square proper relation algebras	102
	3.3	Relation algebras		105
		3.3.1	Definition of relation algebras	105
		3.3.2	Peircean law	107
		3.3.3	**RA** is a completely additive variety of BAOs	109
		3.3.4	**RA** is a canonical variety	110
		3.3.5	**RA** is a discriminator variety	110
		3.3.6	Atom structures of relation algebras	111
		3.3.7	Consistent and forbidden triples of atoms	115
	3.4	Representations of relation algebras		118
		3.4.1	The class **RRA**	118

Contents   xi

	3.4.2		Model-theoretic view of representations	119
	3.4.3		Saturation	121
	3.4.4		**RRA** is a canonical variety	123

4 Examples of relation algebras   **133**
- 4.1 Set algebras . . . 133
- 4.2 Group relation algebras . . . 134
- 4.3 n-variable logic . . . 136
- 4.4 Examples . . . 137
- 4.5 The Lyndon algebras . . . 144

5 Relativisation and cylindric algebras   **151**
- 5.1 Relativisation . . . 152
  - 5.1.1 Relativised representations . . . 152
  - 5.1.2 Non-associative algebras . . . 155
  - 5.1.3 Weakly associative algebras . . . 157
  - 5.1.4 Semi-associative algebras . . . 157
  - 5.1.5 Basic facts about **NA**, **WA**, **SA** . . . 158
- 5.2 Weakly representable relation algebras . . . 163
- 5.3 Cylindric algebras . . . 166
- 5.4 Substitutions in cylindric algebras . . . 170
  - 5.4.1 Basic facts about substitutions . . . 170
  - 5.4.2 More valid substitution-cylindrification identities . . . 172
- 5.5 Relativised cylindric algebras . . . 180
- 5.6 Relation algebra reducts of cylindric algebras . . . 186
  - 5.6.1 Neat reducts and relation algebra reducts . . . 186
  - 5.6.2 Relation algebra reducts and canonical extensions . . . 188
  - 5.6.3 Relation algebra reducts are relation algebras . . . 189
  - 5.6.4 The classes $\mathbf{S\mathfrak{N}r_\beta CA_\alpha}$ and $\mathbf{S\mathfrak{R}aCA}_n$ . . . 190
- 5.7 Relation algebra reducts of other cylindric-type algebras . . . 194

6 Other approaches to algebras of relations   **199**
- 6.1 Diagonal-free algebras . . . 199
- 6.2 Polyadic algebra . . . 201
- 6.3 Pinter's substitution algebras . . . 204
- 6.4 Finitisation problem . . . 205
  - 6.4.1 Reducts, subreducts, generalised subreducts . . . 205
  - 6.4.2 Expansions . . . 206
  - 6.4.3 Special conditions for representability . . . 208
- 6.5 Decidability . . . 210
- 6.6 Amalgamation . . . 210
- 6.7 Technical innovations . . . 211

6.8	Applications		212

II Games   213

7 Games and networks   217
7.1	Networks		217
7.2	Refining networks		222
7.3	All weakly associative algebras have relativised representations		225
7.4	Games on relation algebra networks		233
7.5	Strategies		236
7.6	Games and representations of relation algebras		239
7.7	Networks for cylindric algebras		248
7.8	Games for cylindric algebra networks		249
7.9	Games for temporal constraint handling		252
7.10	Summary of chapter		258

8 Axiomatising representable relation algebras and cylindric algebras   261
8.1	The relation algebra case		262
8.2	An axiomatisation using 'Q-operators'		264
	8.2.1	The new function symbols	264
	8.2.2	Equations using these function symbols	265
	8.2.3	Proof that the equations characterise representability	266
	8.2.4	The Jónsson Q-operators	267
8.3	Axiomatising $\mathbf{RCA}_d$ for $3 \leq d < \omega$		269
8.4	Axiomatising $\mathbf{RCA}_\alpha$ for infinite α		271

9 Axiomatising pseudo-elementary classes   273
9.1	Introduction		273
9.2	Pseudo-elementary classes		277
9.3	Examples		278
9.4	Model theory of pseudo-elementary classes		284
	9.4.1	Alternative single-sorted view	284
	9.4.2	Equivalence of sorted and unsorted approaches	285
	9.4.3	Survey of known results	287
9.5	More explicit axioms		292
	9.5.1	The game	292
	9.5.2	The game characterises K	294
	9.5.3	Short games	296
	9.5.4	Axioms for the short games	298
	9.5.5	The axioms define K	301
	9.5.6	Varieties and equations	302

Contents xiii

- 9.6 Axiomatising pseudo-elementary classes 302
- 9.7 Generalised Q-operators 306

10 Game trees 309
- 10.1 Trees, and games on them 310
- 10.2 Strategies 314
- 10.3 Examples 318
  - 10.3.1 The game $G_n(I_a, \mathcal{A})$ 319
- 10.4 Formulas expressing a winning strategy 321
- 10.5 Games and non-finite axiomatisability 325
  - 10.5.1 Ultraproducts and games 326
  - 10.5.2 Countable, elementary subalgebra 329
  - 10.5.3 Non-finite axiomatisability 331

11 Atomic networks 335
- 11.1 Introduction 335
- 11.2 Atomic networks and games 338
- 11.3 Alternative views of the game 340
  - 11.3.1 Relation to the game G_n of chapter 7 340
  - 11.3.2 Lyndon conditions 341
  - 11.3.3 Game tree view 343
- 11.4 Atomic games and complete representations 347
- 11.5 Axioms for complete representability? 349

III Approximations 353

12 Relational, cylindric, and hyperbases 363
- 12.1 Hypernetworks 363
  - 12.1.1 Definition of hypernetworks 363
  - 12.1.2 Comparing and altering hypernetworks 365
- 12.2 Relational bases and hyperbases 366
  - 12.2.1 Relational bases 367
  - 12.2.2 Hyperbases 368
- 12.3 Elementary properties of bases 369
  - 12.3.1 Symmetric bases 369
  - 12.3.2 Interpolation in hyperbases 370
  - 12.3.3 From hyperbasis to cylindric algebra 371
  - 12.3.4 Reducing the dimension of a relational basis 372
  - 12.3.5 Reducing the dimension of a hyperbasis 373
- 12.4 Games 376
  - 12.4.1 Game for relational bases 376

    12.4.2 Game for hyperbases 378
    12.4.3 Expressing the games by game trees 380
12.5 The variety $\mathbf{RA}_n$ 384
12.6 Maddux's bases 388
    12.6.1 Relational and cylindric bases 389
    12.6.2 Comparing cylindric bases with hyperbases 390
12.7 Cylindric bases and homogeneous representations 395

13 Approximations to RRA   399

13.1 Representation theory 399
    13.1.1 Relativised semantics for $\mathcal{L}(\mathcal{A})$ 400
    13.1.2 Square relativised representations 401
    13.1.3 Flat relativised representations 402
    13.1.4 Smooth relativised representations 403
    13.1.5 Links between the notions 405
    13.1.6 Elementary view 405
13.2 From relativised representations to relation algebra reducts 408
13.3 From reducts to relational bases 412
13.4 From reducts to hyperbases 416
    13.4.1 Preliminary results on substitutions 416
    13.4.2 Finding the hyperbasis 417
13.5 From bases to relativised representations 421
13.6 From smooth to hyperbasis 427
13.7 Summary and discussion 428
    13.7.1 Atomic non-associative algebras 428
    13.7.2 Arbitrary non-associative algebras 430
    13.7.3 Three-dimensional version of theorem 13.46 433
    13.7.4 Finite versions of theorem 13.46 (first part) 435
    13.7.5 Finite versions of theorem 13.46 (second part) 435
13.8 Equational axioms for $\mathbf{RA}_n$ and $\mathbf{S\mathfrak{R}aCA}_n$ 437

IV   Constructing Relation Algebras   439

14 Strongly representable relation algebra atom structures   445

14.1 Introduction 445
14.2 **SRAS** is not an elementary class 447
    14.2.1 Graphs and colourings 447
    14.2.2 The construction 449
    14.2.3 **SRAS** is not elementary 453
14.3 Consequences of the theorem 454
    14.3.1 Closure properties 454

Contents   xv

	14.3.2 Related classes 455
14.4	Maddux's construction 459
	14.4.1 The atom structures 459
	14.4.2 $\mathcal{X}_q$ is strongly representable 461

15 Non-finite axiomatisability of S$\mathfrak{Ra}$CA$_{n+1}$ over S$\mathfrak{Ra}$CA$_n$ · 463
- 15.1 Outline of chapter 463
- 15.2 The algebras $\mathfrak{A}(n,r)$ and C_r 466
- 15.3 $\mathfrak{A}(n,r) \in \mathbf{S\mathfrak{Ra}CA}_n$ 467
- 15.4 $\mathfrak{A}(n,r) \notin \mathbf{S\mathfrak{Ra}CA}_{n+1}$ 469
- 15.5 $\exists$ can win $G_r^{m,n+1}(\mathfrak{A}(n,r),\Lambda)$ 476
- 15.6 Non-finite axiomatisability 484
- 15.7 Proof theory 486

16 The rainbow construction for relation algebras · 491
- 16.1 Ehrenfeucht–Fraïssé 'forth' games 492
  - 16.1.1 The standard Ehrenfeucht–Fraïssé game 492
  - 16.1.2 The modified Ehrenfeucht–Fraïssé game 493
- 16.2 The rainbow algebra $\mathcal{A}_{A,B}$ 494
- 16.3 How $\forall$ can win $G(\mathcal{A}_{A,B})$ 496
- 16.4 How $\exists$ can win $G(\mathcal{A}_{A,B})$ 500
- 16.5 Modifications to the rainbow algebra 509

17 Applying the rainbow construction · 513
- 17.1 Non-finite axiomatisability of **RRA** 514
- 17.2 Complete representations 515
- 17.3 There is no n-variable equational axiomatisation of **RRA** 517
- 17.4 **RA**$_{n+1}$ is not finitely based over **RA**$_n$ 520
- 17.5 Infinite-dimensional bases and relativised representations 525
- 17.6 Weakly representable relation algebras 528
- 17.7 Completions 531
  - 17.7.1 The example 531
  - 17.7.2 Corollaries and problems 535

V Decidability   537

18 Undecidability of the representation problem for finite algebras · 539
- 18.1 Introduction 539
- 18.2 The tiling problem 541
- 18.3 The definition of $RA(\tau)$ 543
- 18.4 Games 546

18.5 Winning $\exists$-strategy implies tiling 546
18.6 $RA(\tau) \in \mathbf{S\Re aCA_5}$ implies tiling 548
18.7 Tiling implies winning $\exists$-strategy 551
 18.7.1 $\exists$'s strategy for non-tile edges 554
 18.7.2 Tile edges 558
 18.7.3 Attached and linked tile edges 558
 18.7.4 Inductive conditions T1, T2, T3 on N 559
 18.7.5 Tiling functions and coordinates for $\forall$'s tile edges 560
 18.7.6 Tiling functions for $\exists$'s new tile edges 560
 18.7.7 Coordinates for $\exists$'s new tile edges 563
 18.7.8 Conditions T1, T2 hold for M 566
 18.7.9 $\exists$'s strategy for tile edges, T3, and consistency 568
18.8 Conclusion 569
18.9 Weak representability is undecidable 571
18.10 Undecidability of equational theories 577

19 Finite base property   581
19.1 Introduction 581
19.2 Guarded fragments 586
 19.2.1 Loosely guarded fragment 586
 19.2.2 Packed fragment 587
 19.2.3 Clique-guarded fragment 588
 19.2.4 Finite model property 589
19.3 The finite base property 590
19.4 Finite base property for $\mathbf{WA}$ 595
19.5 Finite algebra on finite base property for $\mathbf{RA}_n$ 601
19.6 The finite algebra on finite base property for $\mathbf{S\Re aCA}_n$? 602

VI   Epilogue   607

20 Brief summary   609
20.1 Basic definitions 609
20.2 Games for representability 611
20.3 Relativised representations, bases, reducts 613
 20.3.1 Relativised representations 613
 20.3.2 Relational bases and hyperbases 615
 20.3.3 Relation algebra reducts 616
 20.3.4 Equivalences between the notions 618
20.4 The rainbow construction 618
20.5 Atom structures 620
20.6 Decidability 621

20.7	Summary of relations between the classes	621
20.8	Summary of properties of classes	622

21 Problems   625

Bibliography   629

Symbol index   655

Subject index   667

Chapter 1

Introduction

This book is entitled 'Relation algebras by games'. It is intended both for beginners and for specialists in the field of algebraic logic. In it, we study binary relations using some model-theoretic methods, such as games, together with some combinatorial constructions. These tools have answered some long-standing questions about relation algebras, and the results presented here may be of interest to workers in this area. They may also find (and in fact already have found) application in other fields concerned with relations, such as modal logic and computer science.

In this introduction, we will try to explain the meanings of the words in the book's title, and how they are connected. In doing so, we will discuss in outline the contents of the book, and indicate some of the connections to other fields. We end with some remarks on the approach taken in the book, and a summary of what is covered and what is not.

1.1 History

In 1860 Augustus de Morgan published [Mor60], thereby launching an investigation into the logic of relations. This developed into the subject now called *Algebraic Logic*, though in the 19th century it was simply thought of as mathematical logic. This work, along with Frege's quantifier logic, became the foundation of modern logic and model theory.

De Morgan wanted to unveil the laws of rational thought. He was particularly interested in discovering the principles of everyday thinking and of mathematical argument. Two authors influenced him greatly: Aristotle and Boole. Aristotle's syllogism had held sway for 2000 years. Indeed, Kant [Kant92, page 10] had argued that

> Since Aristotle's time Logic has not gained much in extent, as indeed nature forbids it should. ... Aristotle has omitted no essential point of the understanding; we have only become more accurate, methodical, and orderly.

But De Morgan was among a number of philosophers who found the Aristotelian syllogism inadequate. He wrote [Mor66, page 29]:

> Observing that every inference was frequently declared to be reducible to syllogism, with no exception unless in the case of mere transformation, as in the deduction of 'No X is Y' from 'No Y is X', I gave a challenge in my work on formal logic to deduce syllogistically from 'Every man is an animal' that 'every head of a man is the head of an animal'. From the total absence of attempt to answer this challenge, I conclude that no one has succeeded in whose way it has fallen.

Aristotle's system was unable to deal with relations between two objects (like the relation 'belongs to', which relates a head to a man), and indeed Aristotle did not believe that such relations were suitable for formalisation (see [Ari63, chapter 7]). De Morgan criticised this limitation in the Aristotelian syllogism and sought a framework for reasoning about these higher order relations. A *unary* relation, like *red* or *old*, describes a single object: $red(x)$ is true just when x is red. A *binary* relation such as *father* or *older-than* describes a pair of objects: $older\text{-}than(x,y)$ is true when x is older than y. Higher-order relations such as *between* can describe triples, and in general, sequences of objects of any fixed length. We'll see later that binary and higher order relations are far harder to handle than unary relations.

The other decisive influence was George Boole, who had put forward a highly successful calculus of propositions [Boo51]. De Morgan wrote [Mor66, page 255]:

> When the ideas thrown out by Mr Boole shall have borne their full fruit, algebra, though only founded on ideas of number in the first instance, will appear like a sectional model of the whole form of thought. Its forms, considered apart from their matter, will be seen to contain all the forms of thought in general. The anti-mathematical logician says that it makes thought a branch of algebra, instead of algebra a branch of thought. It *makes* nothing; it *finds*: and it finds the laws of thought symbolized in the forms of algebra.

From unary to binary relations Criticising the limitations of Aristotle's syllogisms, De Morgan turned to the logic of binary relations. This research developed, particularly with the works of Peirce and Schröder, into algebras of binary relations, rather like boolean algebra. Eventually it led to what we now call relation al-

1.1. History

gebra. The reader may consult Maddux's interesting historical survey [Madd91d], which also discusses many of the mathematical details.

By analogy with boolean algebra, what is wanted is a set of algebraic axioms that identify precisely the true laws of binary relations. Of course, one has to specify the desired algebraic operations on these relations. We won't need the details at this stage, but in order that we know what we're talking about, let us briefly examine the options.

The chief operations on binary relations considered since De Morgan's time have been the boolean operations $+,\cdot,-,0,1$, and identity, conversion, and composition. To a first approximation,[1] they are as follows. Let r,s be binary relations on a set X. For objects x,y in X, we write $r(x,y)$ to denote that (x,y) stands in the relation r, and similarly for s. So, for example, if X is a set of people then $sister(x,y)$ means that x is a sister of y. Now, the *boolean join* or *sum* of r and s, written $r+s$, is the binary relation on X given by

$$(r+s)(x,y) \iff r(x,y) \text{ or } s(x,y),$$

where here and below, x and y range over elements of X. Their *boolean meet* or *product* $r\cdot s$ is the binary relation on X given by

$$(r\cdot s)(x,y) \iff r(x,y) \text{ and } s(x,y),$$

the *boolean complement* $-r$ of r is given by

$$-r(x,y) \iff r(x,y) \text{ does not hold},$$

and the *zero* and *universal* relations are given by

$$0(x,y) \text{ never holds, and } 1(x,y) \text{ always holds}.$$

The *identity* relation 1' on X is given by

$$1'(x,y) \iff x = y,$$

the *converse* $\check{r}$ of r is the relation on X given by

$$\check{r}(x,y) \iff r(y,x),$$

and the *composition* or *relative product* $r;s$ of r and s is given by

$$r;s(x,y) \iff \text{for some } z \in X, r(x,z) \text{ and } s(z,y).$$

---

[1] For technical reasons, the definitions of $-$ and 1 used with higher-order relations and employed later in the book will sometimes be a little different. Furthermore, different notations will be used for the concrete and abstract operations.

For example, *father* ; (*father* + *mother*) is the relation *grandfather.*

The choice of these as basic operations leads to the relation algebra approach to higher-order relations. An alternative approach, using Frege's logic or Peirce's algebraic logic with its explicit quantification (or 'cylindrification') instead of composition and conversion, leads to first-order logic (or 'cylindric algebra'), and became dominant in the twentieth century. Nonetheless, there are connections between the two views: it can be seen that composition gives a qualified (relativised) form of existential quantification, and indeed, [AneHou91] argues that both can be considered foundations of modern logic. We will describe some of the connections between them later.

Around the end of the nineteenth century, Peirce and Schröder worked towards identifying the true laws sought after by De Morgan, inasmuch as they studied the algebraic laws governing binary relations. They discovered a large number of true equations about binary relations — and even some true statements involving quantification over relations — but there seemed no end in sight, and in fact Peirce remarked that it appeared that one must create an ad-hoc argument to show validity of the statement in each case:

> The logic of relatives is highly multiform; it is characterised by innumerable immediate conclusions from the same set of premises. ... The effect of these peculiarities is that this algebra cannot be subjected to hard and fast rules like those of the Boolian calculus; and all that can be done in this place is to give a general idea of the way of working with it. [Pei33, 3.342]

Tarski's school The modern stage of the subject began with Tarski and his school at Berkeley in the 1940s. Tarski laid down the first axioms for relational algebra in the 1940s;[2] in their modern form, they consist of the axioms for boolean algebra, plus half a dozen additional axioms governing the specifically relational operations — composition, conversion, identity. The axioms are all equations, and we will

---

[2] An axiomatisation of the class of simple relation algebras occurs in [Tar41]. Tarski says on pages 86–87 of that paper (vol. 2, pp. 584–585 in [Tar86]): 'This metalogical theorem suggests still another way of constructing the calculus of relations. For it shows that we may confine ourselves, in developing this calculus, to sentences which have the form of equations ... , thus dispensing with the concepts and theorems of the sentential calculus. For this purpose we should have to put all our axioms into the form of equations, and to give rules which would permit us to derive new equations from given ones. Though this plan has not been worked out in detail, the realisation of it presents no essential difficulty.' An equational axiomatisation — precisely the standard one now used — was worked out by Tarski in the 1943–45 period, given in his seminar on relation algebras (which Jónsson attended) and published in [ChiTar51]. (In fact, it occurs in Chin's 1948 doctoral thesis.) There, Tarski writes: 'The first part of this work consists of some of the material given by A. Tarski in his seminar on relation algebras at the University of California, Berkeley, in 1945.'
We thank Steven Givant for providing this information.

1.1. History

quote them in definition 3.8. To illustrate, one axiom says that for all binary relations x,y, we have $(x;y)^\smile = \breve{y};\breve{x}$. This tells us for example that $(parent;sibling)^\smile$ is the same relation as $sibling;child$.

It was hoped that these axioms would be sufficient to characterise all true properties of binary relations: about a slightly earlier axiomatisation, Tarski had asked:

> Is it the case that every sentence of the calculus of [binary] relations which is true in every domain of individuals is derivable from the axioms ... ? This problem presents some difficulties and remains open. I can only say that I am practically sure that I can prove ... all of the hundreds of theorems to be found in Schröder's *Algebra und Logik der Relativ*. [Tar41]

Algebras and representations Suppose that the answer to Tarski's problem were negative: the axioms were not strong enough to capture all true properties of binary relations. Since it is easy to check that the axioms do make valid statements about relations, we imagine that there must exist 'situations' in which the axioms are true but some valid property of binary relations is not. Such a situation obviously cannot be one of real-world relations. So what could it be? In other words, if the Tarski axioms are not describing real binary relations, what are they describing?

If we recall De Morgan's hope that

> [Algebra's] forms, *considered apart from their matter,* will be seen to contain all the forms of thought in general [our italics]

we obtain, perhaps, a different slant on the same problem. How can we abstract away from the concrete subject matter of specific relations to arrive at their essential forms?

To answer these questions, we have to introduce the notion of *algebras* and their *representations*. The reader could be forgiven for thinking that what is meant by 'algebra' is simply a formal or abstract style of calculation, and that 'boolean algebra' means calculation according to the rules laid down by Mr Boole. Though it does have that meaning, nowadays it also has another. An *algebra* is a domain or set of elements together with some functions defined on this domain and taking values in it. In a *boolean-type algebra*, the functions are called $+, \cdot, -, 0, 1$, the first two being binary (taking two arguments), '−' being unary (taking one argument), and $0, 1$ nullary (they take no arguments at all and have a fixed value in the domain). For the algebra to be a *boolean algebra*, these functions must obey a few specified laws or axioms. For example, '$x + y = y + x$' must hold, for any x, y in the algebra's domain. We will not go into the axioms or their history now (see definition 2.3 later on), but we should mention that the first axiomatisation of boolean algebras was given by Huntington [Hun04].

6   Chapter 1. Introduction

One view of the objects in a boolean algebra is that they are intended to be unary relations, and the operations $+, -, \cdot$ on them are intended to reflect real-world manipulations of them ($+$ is intended to formalise 'or', $\cdot$, 'and', and $-$, 'not'). The axioms defining boolean algebras do express valid properties of unary relations. For example, an object is red or green if and only if it is green or red — $x + y = y + x$ is valid. However, it is critical to realise that this relational view is not part of the *definition* of a boolean algebra. For example, it is possible to make a boolean algebra whose domain consists of the four objects 0, 1, *red, green,* by defining $-red = green, -green = red, red + green = 1, red \cdot green = 0, -1 = 0$, and so on, stating the values of the functions on all possible arguments. Here, *red, green,* 0, and 1 are merely abstract entities. They are not necessarily unary relations on a set of objects, nor have we provided an 'interpretation' of them as unary relations by giving a list of objects and laying down which of them are red and which are green. In this case, it's easy to do this: we can take some objects, such as *nose, idea,* and *prairie,* and stipulate that they lie in the relations *red, green,* 0, 1 as follows:

0	(none)		
red	nose		
green		idea	prairie
1	nose	idea	prairie

The operations 'and', 'or', 'not' (corresponding to $\cdot, +, -$) now have natural interpretations on these objects — we know by the meanings of the words that all three objects are 'red or green', only the nose is 'not green', etc.

Given a complete knowledge of the abstract algebra functions on *red, green,* 0, and 1, we can check whether they match the natural meanings of the operations as applied to the specified relations on the objects *nose, idea, prairie.* For example, because in the algebra we have defined $-red = green$, we need to confirm that the objects that are not red are precisely those that are green — and this is clearly true. For $-1 = 0$, we confirm that the objects satisfying 0 (i.e., none of them) are precisely those objects not satisfying 1 — true again. Because $red + green = 1$ in the algebra, we need to confirm that all objects are red or green — and this is true too. And so on.

What we have done is define a second boolean-type algebra, with domain consisting of the four unary relations $\emptyset$, {nose}, {idea, prairie}, {nose, idea, prairie} on the three-element set {nose, idea, prairie}, with functions defined 'naturally', and check that this algebra is isomorphic to the first. Put another way, the nose, idea, and prairie, with their allotted colours, form a kind of 'model' of our boolean algebra, and we will call it a *representation* of the algebra. Most workers in algebraic logic take the former, 'isomorphism' view; we prefer the latter.

Now, there is more than one boolean algebra — any boolean-type algebra satisfying the axioms in the standard definition of 'boolean algebra' is by definition

1.1. History

a boolean algebra. There are finite and infinite boolean algebras, and they are of rich diversity. As a small illustration, we may define a partial ordering on the elements of a boolean algebra by $x \leq y \iff x + y = y$. In a representation, this corresponds to saying that 'every object satisfying x also satisfies y'. A smallest non-zero element in this ordering is called an *atom*. A boolean algebra in which any element except 0 lies above an atom is said to be *atomic*. Any finite boolean algebra is atomic, but there are infinite boolean algebras that have no atoms at all.

Because the axioms defining boolean algebras are true of real relations, any boolean-type algebra that is 'like' a genuine collection of unary relations, in the sense that it has a representation, is in fact a boolean algebra. But the wide variety of possible boolean algebras mean that it is not immediate that every boolean algebra has a representation. Indeed, it was not until 1936 that it was proved, by Stone, that this is so: any boolean algebra is *representable*. The significance of this is that the axioms for boolean algebras exactly capture the true properties of unary relations: all and only those properties that are true in every domain of individuals endowed with unary relations are derivable from the axioms. This is a great (perhaps the greatest) success for the algebraic viewpoint.

Relation algebras So much for unary relations. What about binary relations? A *relation-type algebra* is an algebra whose functions are called $+, \cdot$ (binary), 0, 1, 1' (nullary), ˘ (unary), and ; (binary). Generalising from boolean algebra, a *representation* of a relation-type algebra is a domain of individuals endowed with binary relations corresponding to the elements of the algebra; as in the boolean case, the natural interpretations of the relational operations on these relations, as we defined them before,[3] should exactly match the abstract definitions of the operations on the algebra elements. Any relation-type algebra satisfying the axioms given by Tarski is called a *relation algebra*. To answer our earlier question, this is what the axioms are describing: a 'situation' in which the axioms hold is a relation algebra.

As with boolean algebras, any relation-type algebra that is representable (i.e., has a representation) is easily seen to be a relation algebra. But being abstract, a relation algebra is only restricted by the relation algebra axioms and not by properties of real-world relations, and it could in principle be 'pathological' in that some true properties of binary relations are false in it. In particular, if Tarski's axioms were not strong enough to derive all true properties of binary relations that can be expressed in the language of relation algebras, then the negation of some such property would be consistent with them and there would exist a relation algebra in which this property failed. Such an algebra could not be representable. So Tarski's problem, to show 'that every sentence of the calculus of [binary] relations which is true in every domain of individuals is derivable from the axioms', would be solved by (indeed, is equivalent to) showing that this never happens — that *any relation*

---

[3] Except that the definitions of − and 1 are altered. See footnote 1.

algebra is representable.

The crisis Unfortunately, in [Lyn50] Lyndon was to find an example of a non-representable relation algebra (with 2^{56} elements). In 1961, he produced examples of relation algebras arising from projective geometry, some of which were representable as 'true' algebras of binary relations and some of which were not [Lyn61]. Monk soon used these [Mon64] to prove that no finite set of equations or indeed first-order axioms would capture all of the true properties of binary relations.

At this point, we perceive something of a crisis in the development of the subject. As it seemed, the hopes to produce a simple, elegant, or at least finite set of algebraic properties that captured exactly the true properties of binary relations (the representable relation algebras) were not to be realised. This impasse is still creating employment today, and most of the current book is concerned with it.

Circumventing the crisis Several different stratagems to get round the obstruction were evolved. One, promulgated by Tarski especially, was to find elegant sufficient conditions for representability. This approach has continued to the present, and now forms an extensive field. Other approaches were also developed — for example, seeking positive results by varying the signature of relation algebras (removing or adding operations). We will survey some of this work in section 6.4.

Lyndon's stratagem Persevering in the attempt to obtain both necessary and sufficient conditions for representability brings us to another stratagem which is more relevant here. It brings us back to the work of Lyndon. In [Lyn50], we see an early example of a *step-by-step construction,* to attempt to build a representation of an (atomic) relation algebra. The objects of the representation, and the relations they satisfy, are introduced and defined one by one in a process of potentially infinite length. This allows fine control over the properties of the relations at every step. We find it helpful to view the process as a two-player game, which we will describe a little later.

Further, Lyndon described first-order axioms, which we like to call the *Lyndon conditions,* and he showed that these conditions fully axiomatise the real binary relations, at least for finite relation algebras. What is so important about the Lyndon conditions for us is their meaning in terms of the step-by-step construction. This is the aspect of Lyndon's stratagem that distinguishes it from Tarski's, above; the game-theoretic tools used later can be seen as mere variations on it. *The Lyndon conditions are essentially the literal translation into the language of relation algebras of the statement that the proposed step-by-step construction of a representation can be carried through successfully.*

The question arises as to whether such a statement can be regarded as a satisfactory solution to the problem raised in De Morgan's project and indeed whether it is truly algebraic in nature. Although this question is imprecise and cannot be given a definitive answer, by De Morgan or anyone else, it has been of some importance historically. There was a problem posed in the monograph of Henkin, Monk, and Tarski [HenMon⁺71, p461]: to find a 'simple intrinsic characterisation' of the true properties of n-ary relations. Though the question was formally about cylindric algebras, we will take representable relation algebras (binary relations) to be included too. This problem seems to be regarded as still open (see, e.g., [AndNém⁺01, problem 1.12]), in spite of the existence of several axiomatisations, including one by Lyndon in a later paper [Lyn56], game-theoretic ones by us (chapter 8), one for cylindric algebras by Monk [Mon69], and others. The precise objections are hard to pin down, but broadly it seems that these axiomatisations are regarded as unsatisfactory in some way: they are too complicated — or perhaps too trivial, just paraphrasing the original problem without providing any new 'algebraic insight'.

We will have more to say about this in the next section, but we believe that such a dispute can miss the point. In our view, the acid test of success of a research viewpoint should be what further work it stimulates and what problems can be solved by adopting it. The step-by-step technique of building representations, especially when viewed as a game, is extremely potent. Not only does it allow the construction of axiomatisations for relation algebras and other kinds of algebra, but close examination of the way that games can be played on given algebras will elicit very fine and detailed information about their structure. The use of games may seem just a presentational matter but it is important, because games make life so much easier and so permit us to go much further in this analysis. This is no longer controversial: the step-by-step approach, whether by games or otherwise, is now widely accepted and has been used by — hardly exaggerating — many workers, from Andréka to Zakharyaschev.

1.2 To the games

Let us now jump forward to a modern perspective on the use of games to axiomatise the representable relation algebras, and other kinds of algebras of relations characterised by other notions of representation. The methodology has four ingredients:

1. Building a 'representation' of an algebra by a 'step by step' construction.
2. Doing this by a game played on the algebra. Following Keisler and Hodges, we call the players $\forall$ (male) and $\exists$ (female), and we view $\exists$ as a doctoral student in the Faculty of Representability of Algebras and $\forall$ as examiner of her dissertation on the algebra in question.[4] During play, $\exists$ tries to build a

---

[4] Keisler named his players $\forall$ and $\exists$ in [Kei65], but did not elaborate on pronunciation. We tend

representation by a sequence of approximations in response to prompts by $\forall$. The game has infinitely many rounds, and in each round, $\forall$ challenges her to refine her current approximation in one way or another. To do this, she can introduce new objects and define or refine which relations (corresponding to the algebra elements) hold on them.

The rules of the game, determining what $\forall$ can ask and how $\exists$ may legally respond, are fixed according to the kind of algebra and notion of representation we have in mind. Consider the representable relation algebras, for example. The game is played on a fixed relation algebra whose representability is at issue. Let us illustrate the kind of moves that can be made. (The formal rules adopted in chapter 8 are not quite these, but in effect they do allow these actions to happen.) $\forall$ may ask $\exists$ at some point whether a particular pair of objects satisfies a given relation or not — and she must tell him. For algebra elements a, b, if $\exists$ has stated that $a;b$ holds on the pair of objects (x, y), then $\forall$ can demand that she include some object z with (x, z) satisfying a and (z, y) satisfying b. $\exists$ loses the play (or match) of the game if she gives blatantly inconsistent responses, such as stating that some pair of objects are related by 0. As is often the way in examinations, she wins if she doesn't lose at any stage.

As one would expect, a *strategy* for one of the players is just a set of rules telling that player what move to make at every stage. The strategy is *winning* if its owner wins any play in which s/he uses the strategy. We are actually interested in whether or not $\exists$ has a winning strategy in the game, not in whether she wins a particular play. This is because for many algebras, $\exists$ has a winning strategy in the game if and only if the algebra is representable. (Mostly the games are *determined,* meaning that one of the players has a winning strategy; so assuming that $\forall$ is a thorough examiner, this presents no problem for Hodges' metaphor.)

3. Approximating the infinite game by finite ones. In these, $\exists$'s examination is curtailed after some arbitrarily large finite number of rounds; she is told just before the start how many rounds there will be. The hope is that this does not lead to a drop in standards: that if she can pass any such examination, she

---

to pronounce the names as 'A' and 'E', but the reader has a wide choice of more or less romantic options. Hodges [Hod85] used the $\forall$-$\exists$ notation and called the players Abelard and Eloïse, saying that Abelard was a 12th-century Parisian logician who used to play games with Eloïse, the niece of a canon of Notre Dame. Finite model theorists will call them Spoiler and Duplicator, though in the current context nothing much is being duplicated. Another possibility is Player I and Player II, though some may prefer Gurevich–Harrington's Mr ε and Mr 1 − ε, for ε < 2 [GurHar82]. Németi calls $\exists$ 'us' and $\forall$ 'the enemy'. Many other options are used, depending on context: van Benthem has used Verifier ($\exists$) and Falsifier ($\forall$), for example. The male-female distinction, which we think originated in [Hod85], is very useful in reducing the use of proper names in the text. The 'examination' metaphor is from [Hod97].

1.2. To the games

could pass the full exam of infinite length. For finite algebras, and for those games when ∃ is never asked to choose an element of the algebra, this hope is realised. But in general it is too naïve, because in a finite game, however long, ∃ may be able to prevaricate and not be caught out in the available time. Still, ∃'s surviving all finite examinations does entitle her to an M.Phil. or similar degree, with the meaning that some algebra similar (elementarily equivalent) to the original will be representable. Whether employers would understand such a qualification is doubtful, but it is often sufficient for the original algebra to be representable.

4. Writing out axioms expressing that the finite games can be won. We write a single axiom for the game of each finite length. This is where it helps to have finite-length games — an axiom for an infinite-length game would likely be infinitary. The process is technical but not otherwise difficult.

These ideas have an interesting history (see chapter 9). Here, we only note that most of them can be found in [Lyn50].

Formulating a general game We have tried to get a feel for how games can be *used* to determine representability of algebras, but we have not really pinned down how to *devise* a game suitable for axiomatising a given kind (or class) of algebras. In practice it is not difficult to do this in an *ad hoc* fashion for each kind of algebra and representation encountered, and for some purposes it is necessary to do so in order to get one's hands on the specific rules of the game. But for axiomatising classes, it can be misleading and repetitive.

An alternative is to follow the spirit of Lyndon's stratagem and develop a general 'meta-style' reasoning about the games themselves. If our notion of representation is such that there exist first-order axioms defining when a given structure is actually a representation of a given algebra — in technical terms, if the class of algebras in hand is pseudo-elementary — then we can use the axioms to define a game. Hence, we can synthesise axioms for the representable algebras directly (and constructively) from the defining axioms. This method is developed in chapter 9, and it applies to a wide range of algebras. The construction is related to Henkin's completeness proof in first order logic, and to model-theoretic forcing as described in [Hod85], for example.

What would the founders have said? Let us now return to the original 19th century objective to find algebraic forms of the laws of thought, and to the related problem of [HenMon+71] to find 'simple, intrinsic' characterisations for various classes of algebras in algebraic logic. We have seen that games can be used to provide explicit axioms for a wide range of these classes, 'automatically' and directly from their definition. On the basis that a general procedure is worth two *ad hoc* ones, there is at least something to be said for this. The axioms do presumably constitute an intrinsic characterisation in the sense that they are evaluated solely in the algebra, but are they simple and enlightening?

First, simplicity. Well, one person's complications are another's non-trivial theorems. Substantial effort has gone into proving negative results about what kinds of axiomatisation are possible for certain classes. We have mentioned the key theorem of Monk [Mon64] that the representable relation algebras are not finitely axiomatisable; the situation for higher-order relations is similar, as he showed later [Mon69]. Many similar results exist: to give just two, the representable relation algebras cannot be axiomatised by equations using only k variables, for any finite k [Jón91], nor by Sahlqvist equations (a result of Venema [Ven97b], using the work of section 17.7). The lesson of these many years' work is presumably that *any* axiomatisation of them is going to be complicated. The axioms obtained by games can be written fairly simply, by an inductive definition. If we eliminate the induction, to really 'see' the axioms, they do become much more complicated in appearance. But their *meaning* is always clear — it comes from the games.

Second, what about the 'quality' of game-axiomatisations? A sceptic might argue that the axioms produced by games do not give much 'algebraic insight', adding up to little more than the original statement that the algebra is representable, and that inasmuch as they are 'really' speaking about representations, they are not actually intrinsic at all. In contrast, elegant finite axiomatisations of the classes $\mathbf{D}_n, \mathbf{G}_n$ are known [AndTho88, And01], and a similar situation applies to the class **WA** of weakly associative algebras [Madd82].[5] The axiomatisations obtained for these classes by playing games get nowhere near that ideal.

Three responses can be made to this. First, it is true that axioms expressing winning strategies in games arise very directly from the definition of the class being axiomatised. But this is not the end of the matter. Representable relation algebras can be characterised in more than one way: simply by means of the notion of representation, or alternatively, by work of Monk, using 'relation algebra reducts of ω-dimensional cylindric algebras'. Games will synthesise axioms by either of these characterisations, and others. So we have a question for our sceptic: which of these axiomatisations has the *least* algebraic insight? The difference between them is simply because of the differing characterisations of the class. Arguably, the axioms obtained by games carry about the same insight as the chosen characterisation, and that is where attention should be directed.

Second, we admit that the basic game-theoretic axiomatisation method is not of itself much help in telling whether a given class is finitely axiomatisable, or providing a finite set of axioms when it is. Of course, many (most?) classes of algebras of interest in algebraic logic are not finitely axiomatisable, and here game-theoretic axiomatisations come into their own. But we can surely agree that for

---

[5]These classes will be discussed in chapter 5.

many classes, winning strategies in games will serve to build 'representations' of algebras. In a number of cases, algebraic ingenuity shows that the conditions for such strategies to exist are finitely axiomatisable. (For example, the characterising game for the class **WA** of Maddux is such that a winning strategy for ∃ in the game curtailed to two rounds — a finitely axiomatisable property — is already enough to guarantee a winning strategy for her in the full, infinite game.) But from then on, the proof of completeness of the axioms is essentially the standard game-theoretic one — in each case it was presented as a 'step by step' argument and this can be recast as a game. Indeed, it is quite likely that many elegant axiomatisations were discovered by considering what conditions were needed to carry through some or other step-by-step construction. Looked at in this way, the distinction between 'natural' algebraic axioms and axioms obtained from games becomes somewhat artificial.

So we do not admit that axioms obtained by games impart no 'algebraic insight'. Our third point also contributes to this view. The existence of a winning strategy in a game can often be expressed in a different way, by inventing some 'algebraic device' to represent it. The simplest examples, hardly a shift at all, are bisimulations and back-and-forth systems; but there are more serious transformations than this. We have in mind Fraïssé's characterisation of homogeneity by amalgamation [Fra54], the related cylindric and relational bases of Maddux [Madd83, Madd89b], the 'IRR theories' of Gabbay [GabHod$^+$94, definition 6.2.3], and the 'mosaic' method of Németi [Ném86]. If such a device can be found, it can itself be used as an alternative definition of the class in question, and perhaps to produce new axioms for it. It can also be used in other ways: see, for example, the decidability and complexity results of [Ném86, Ném95, Ném96, VenMar98] and the finite base property results of chapter 19. There is some 'algebraic insight' in all this, and we feel much more remains to be discovered, though it is at some remove from the games we started with.

But it is important to note that games yield their own kind of insight. We believe, and we hope this book will show, that games themselves constitute a 'simple intrinsic characterisation' of representable algebras. We will see that representability of an algebra is often quite easy to test by games, and that games can shed quite considerable light on the nature of an algebra.

1.3 Non-finite axiomatisability

Games are a potent tool in proving non-finite axiomatisability results such as those of Monk. For a given relation algebra, we said that if ∃ has a winning strategy in every finite-length game of the kind described before, then the algebra is representable, and conversely. So if she can only win the games up to a certain length, say with 16 rounds, but not longer ones, then the algebra will not be representable.

If this algebra is replaced by another, on which ∃ can win games of up to 32 rounds but no more, then the second algebra, while still not representable, would seem to be 'more representable' than the first. Recalling that each finite-length game yields an axiom stating that ∃ has a winning strategy in it, all the axioms together axiomatising the representable relation algebras, it is clear that the second algebra satisfies more of the axioms than the first.

Now for each finite m, suppose that we could find a relation algebra $\mathcal{A}_m$ such that ∃ has a winning strategy in the game with m rounds played on $\mathcal{A}_m$, but not in all the finite-length games played on it. $\mathcal{A}_m$ would satisfy the first m axioms, but not all of them. Given such relation algebras, the sequence $\mathcal{A}_1, \mathcal{A}_2, \ldots$ therefore approaches representability by satisfying more and more of the axioms, but no algebra in the sequence satisfies all of them and is representable. It can now be shown using standard algebraic (ultraproducts) or model-theoretic (compactness) techniques that there is no finite set of axioms defining when a relation algebra is representable. So if we could find relation algebras $\mathcal{A}_m$ as described, we would have a proof of the crisis-precipitating theorem of Monk [Mon64].

Monk also used ultraproducts in his proof, but his argument was not based on games and he used a sequence of algebras taken from Lyndon's work with projective planes. Our point here is that the games have told us what is required of the algebras: for $\mathcal{A}_m$, 'all' we need to ensure is that ∃ can last for m rounds of the game but not for much longer. This is important in other situations, because if we can devise a game characterising a class of algebras in the same way as for representable relation algebras, and then construct a sequence of algebras witnessing ∃'s ability to survive any given finite number of rounds but not forever, we will have a non-finite axiomatisability result for the class in our hands.

For some kinds of representation, the fact that ∃ has a winning strategy in all finite-length games on an algebra is not enough to ensure its representability: she must be able to win the full infinite game. In such cases, if we can find a single algebra on which ∃ can win every finite-length game but not the infinite-length one, it follows that the class of representable algebras is not first-order definable at all (non-elementary). The chief example of such a kind of representation is one that respects all existing meets and joins in the algebra, not just the finite meets and joins which are necessarily respected since they are definable in the language of relation algebras. These representations were called 'strong' by Scott, and here we call them *complete representations*. It turns out that the appropriate game for complete representations of relation algebras is almost exactly the one obtained from Lyndon's original work in [Lyn50]. He assumed implicitly that any representation of an atomic relation algebra must be complete; this turned out not to be so, but the error has been very fruitful. See chapters 11 and 17.

It goes without saying that knowing what is required of the algebras $\mathcal{A}_m$ does not of itself establish their existence. We will examine specific constructions in section 1.5 below.

1.4 Approximations to representability

Earlier, we considered some stratagems to survive the non-finite axiomatisability results of Monk. There is another one as well. It is associated with Monk and Maddux, and is related to dynamic logic, studied intensively in Amsterdam and elsewhere. We see it primarily as *trying to approximate representability,* typically by weakening the notion of representation. We have found games a great help here: paying close attention to the ways that games are played on algebras, and what is required for $\exists$ to have a winning strategy, can provide considerable insights into the 'fine structure' of algebras. Much of the book is concerned with this, so let us now discuss it.

We have just seen one kind of approximation to representability, when $\exists$'s examination was stopped after a finite number of rounds, to cut costs. Another is when the examination can go on forever — the remuneration of examiners being low, this is not a problem — but the examination room contains only a small whiteboard. More precisely, the *size* of $\exists$'s approximation to a representation is finitely bounded, say by a natural number n. If $\forall$'s challenge in some round would force her to add a new object and take her approximation to the representation above the size limit n, then the rules say that he must first remove an object from the approximation. The effect is that it is easier for $\exists$ to win.

We will see in chapter 13 that such games characterise a class of relation algebras defined by Maddux [Madd83] and called $\mathbf{RA}_n$. A notion of representation can be given for them: a non-classical 'relativised' representation, which nonetheless appears classical when examined by a movable 'window' that can only show n points at once.

A strengthened game allows $\forall$ to demand that any two approximations played during the game be amalgamated, provided the amalgam's size does not exceed n. This makes it considerably harder for $\exists$ to win, though still not as hard as it was for the original notion of representation. The corresponding class of algebras is called $\mathbf{S\mathfrak{R}aCA}_n$, and though it is defined quite differently, using the connection of binary to n-ary relations (formalised by 'relation algebra reducts'), there is a similar notion of (relativised) representation for it. This will also be discussed in chapter 13.

Having found an appropriate game for these classes, we are in a position to attempt non-finite axiomatisability results in the way outlined in section 1.3. In chapter 15 we will show that for each $n > 4$, $\mathbf{S\mathfrak{R}aCA}_n$ is not finitely axiomatisable, even within $\mathbf{S\mathfrak{R}aCA}_{n-1}$. In chapter 17 we will do the same for $\mathbf{RA}_n$. In chapter 18, we show that the problem of whether a given finite relation algebra is in $\mathbf{S\mathfrak{R}aCA}_n$ or not is undecidable (for $\mathbf{RA}_n$, it is decidable).

Proof theory Historically, one of the motivations for defining $\mathbf{RA}_n$ and $\mathbf{S\mathfrak{R}aCA}_n$ was connected to proof theory of first-order logic. The reader may already have wondered why the negative result of Monk was so terminal. After all, we may well choose to reason about binary relations using classical first-order logic. There are Hilbert systems with finite-schema axiomatisations in the textbooks, and a completeness theorem. The properties expressible in the language of relation algebras can all be expressed by first-order sentences using only three variables. So all the true properties of binary relations written in the language of relation algebras can be established by proofs in first-order logic. This approach began with Peirce; for more information, see [Tar41, TarGiv87]. A related approach to axiomatising the representable relation algebras occurs in [Ven92].

This is correct, but such an approach takes us outside the language of relation algebras and the relational calculus, and so at the least lacks some elegance. More to the point here, one may ask about the resources needed in the first-order proofs. For example, the number of variables required in a proof may be much larger than the number used in the theorem that is being proved. Algebraic logic is a useful tool for analysing the number of variables needed in proofs. It transpires that the true properties of binary relations (expressed in the language of relation algebras) that have a first-order proof using n variables are precisely those properties that are valid in all algebras in $\mathbf{S\Re aCA}_n$ [HenMon$^+$85]. A similar result for a different proof theory holds for $\mathbf{RA}_n$ [Madd83]. The fact that $\mathbf{S\Re aCA}_{n+1} \subset \mathbf{S\Re aCA}_n$ (for all $n \geq 4$) shows that there are such properties that really need $n+1$ variables to prove, and results related to the non-finite axiomatisability of $\mathbf{S\Re aCA}_{n+1}$ over $\mathbf{S\Re aCA}_n$ show that proof theory with n variables cannot be strengthened by finitely many axiom schemata using fewer than n variables to match the strength of $(n+1)$-variable proof theory. See chapter 15.

Moreover, it can be shown that $\mathbf{RA}_4$ and $\mathbf{S\Re aCA}_4$ are precisely the class of all relation algebras: the half-dozen axioms of Tarski defining relation algebras succeed in identifying precisely those true properties of binary relations that are provable with four variables. Thus, the relation algebra axioms are in some sense optimal — the true properties provable with five variables cannot be finitely axiomatised.

Finite base property The final chapter of the book contains some positive results on finite relativised representations. There are simple examples of finite representable relation algebras without finite representations, but we will prove that with a relativised notion of representation they do have finite representations. The same goes even for non-representable algebras such as those in the $\mathbf{RA}_n$ and Maddux's class $\mathbf{WA}$. A little can be said about $\mathbf{S\Re aCA}_n$, too, but there are theoretical limits here. Our argument uses the fact that the 'loosely guarded fragment' of first-order logic has the finite model property. This fragment was introduced in [Ben97]; it generalised the 'guarded fragment' of [AndBen$^+$98], whose introduction was motivated by results in relativised cylindric algebras. That is the extent of this

1.5. Constructions of algebras

book's contribution to decidability results, as the area is well-covered elsewhere. However, we prove some undecidability results in chapter 18.

1.5 Constructions of algebras

We should say a little about how to construct the sequence of relation algebras $\mathcal{A}_m$ for non-finite axiomatisability proofs introduced in section 1.3. It would be very interesting to synthesise it (where possible) straight from the definition of the representation in the way that we do when axiomatising the class of algebras, but this is beyond our present means.

1. Monk algebras We will give two kinds of construction. The first is of variants of what are sometimes called 'Monk algebras'. These are certain kinds of relation algebra whose non-representability is shown by Ramsey-theoretic arguments.

The Monk-style algebras of chapter 14 are quite easily understood. They are atomic relation algebras based on atoms whose structure arises from graphs. It turns out that such an algebra is representable if and only if the graph has infinite chromatic number; the left-to-right direction is shown using Ramsey's theorem. These algebras will be used in conjunction with graphs constructed probabilistically by Erdös to show that the class of 'atom structures of representable complex relation algebras' is not elementary.

A different kind of Monk-style algebra will be used in chapter 15. Here, $\mathcal{A}_m$ will be a finite relation algebra with m special elements (for some finite m), and we arrange the definition of composition so that any two distinct objects in any representation of $\mathcal{A}_m$ are related by one of them but no three objects are related by the same one. There is a finite bound, say m', on the number of objects in a representation of such an algebra: simply choose m' such that any colouring of the edges of a complete graph of size larger than m' using only m colours has a monochromatic triangle. We also ensure somehow that $\mathcal{A}_m$'s representations, if any, have more than m' objects. So $\mathcal{A}_m$ cannot be representable. Moreover, $\forall$ can demonstrate this by winning the game of infinite length where the approximations to the putative representation are of bounded size, but amalgamations are allowed (see above). But he will not be able to demonstrate it in the game with only m rounds, so $\exists$ has a winning strategy in this game. This serves to show that the classes $\mathbf{S\mathfrak{R}aCA}_n$ for $n \geq 5$ are not finitely axiomatisable. See chapter 15 for stronger results along these lines.

2. Rainbow algebras The second construction, which Yde Venema has called the 'rainbow construction', creates algebras $\mathcal{A}_{A,E}$ that are superficially similar to Monk algebras. Precursors to this construction can be found in [And94, Madd91c]. $\mathcal{A}_{A,E}$ contains a device A 'belonging to $\forall$', that allows him to (force $\exists$ to) create a large region of (her approximation to) the representation, during play of the game to build a representation of $\mathcal{A}_{A,E}$. The algebra also contains a device E 'belonging to $\exists$', which she must use to structure (or colour) the region. The key question in deciding if $\exists$ has a winning strategy in the game is whether or not $\forall$'s device can make the region grow so large or complicated that $\exists$'s device cannot manage to colour it.

The two devices A, E are in fact two first-order structures in a binary relational language: typically they are undirected graphs. Their relative strength is measured by playing a certain Ehrenfeucht–Fraïssé forth-only game from A to E. Roughly, $\forall$ places a pebble on an element of A, $\exists$ responds by placing a corresponding pebble on an element of E, $\forall$ puts another pebble on an element of A, and so on. $\exists$ has to ensure that the pebble positions always define a partial homomorphism from A to E. We can show that $\exists$ has a winning strategy in this 'forth' game from A to E of length t if and only if she has one in the original game on $\mathcal{A}_{A,E}$ of length $t+1$. This remains true for many variants of the algebra game — for example, $\exists$ has a winning strategy in the game on $\mathcal{A}_{A,E}$ of length t where the approximation is finitely bounded by $n > 4$ if and only if she has a winning strategy in the forth game on A, E with $t-1$ rounds and $n-2$ pebbles. The proof relies on the fact that in the game, the objects in the supposed representation are built by $\exists$ in a certain order, one by one.

The key difference of rainbow algebras from Monk algebras is this. In Monk algebras, the two devices ($\forall$'s and $\exists$'s) operate using cardinality, $\forall$ trying to make the special region too large for $\exists$ to colour; it may be clear that $\exists$ has no winning strategy for the infinite-length game, but not clear exactly which finite-length games she can win. Also, the two devices are interlinked, in that $\forall$'s device has to operate through $\exists$'s to some extent. In the rainbow algebras, $\forall$'s device A is completely independent of E. A and E are arbitrary structures for a binary relational language, and we have a precise characterisation, in terms of forth games on the structures, of the algebraic games for which $\exists$ has a winning strategy. This gives us much more flexibility in designing the algebras and control over the games. We can build algebras $\mathcal{A}_{A,E}$ with any two devices A, E, having any relative strength we choose and not necessarily relying on cardinalities or Ramsey theory. It is much easier to dream up first-order structures A, E such that $\exists$ has a winning strategy in (say) the forth game with three rounds but not four, than to think straight off of a relation algebra where $\forall$ can win the length-4 game but not the length-5 one: see figure 1.1.

Another possibility is to take A and E to be finite linear orders with A longer than E. $\forall$ can win a finite-length forth game from A to E with only two pebbles. So, taking $n = 5$ above, he can win the size-5 approximation game on the algebra $\mathcal{A}_{A,E}$, and we can calculate how many rounds he needs to win.

We will build the rainbow algebras $\mathcal{A}_{A,E}$ in chapter 16, and apply them in subsequent chapters to prove non-finite axiomatisability theorems including Monk's

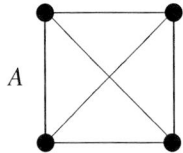 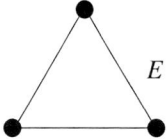

Figure 1.1: ∃ can win the 'forth' game from A to E with three rounds but not four

original result and the result for **RA**$_n$ mentioned previously, but also theorems on 'weakly representable relation algebras' and completions of relation algebras. A variant of the construction will be used in chapter 18 to show that it is undecidable whether a finite relation algebra is representable, 'weakly representable', or in $\mathbf{S\Re aCA}_n$ for finite $n \geq 5$.

1.6 Some remarks on methods

We use quite a lot of model theory in this book: compactness, saturation, interpretations, games and forcing, amalgamation. (We do not assume that the reader is an expert in model theory. Much of the time we merely use methods with a model-theoretic feel, and the book is self-contained on most of the model-theoretic results that are used explicitly.) The model theory used to prove the finite model property for the loosely guarded fragment came into being in the 1990s, but much of what we use dates from the 1970s or before. Model theory has undergone several revolutions since then, and it would be of interest to apply the more recent and powerful work such as stability theory to algebraic logic.

Our use of model theory in this field is unusual: most treatments of algebraic logic rely instead on universal algebra. We do use universal algebra, but we also believe (though others do not) that model-theoretic methods can simplify many matters. One of the problems in the presentation of an algebraic approach to relations is that many of the results and proofs are highly technical, sometimes made harder by an elaborate notation. In this book we attempt, not to avoid these technicalities, but to help the reader through them by presenting the material in a model-theoretic way, and by using games. In the model-theoretic framework, a representation of a relation algebra is simply a model of a certain first-order theory, rather than an isomorphism from a relation algebra to a field of binary relations. In the games, ∃ tries to build a representation of an algebra while ∀ tries to make this difficult. We believe that this makes some of the proofs more transparent.

1.7 Summary of contents

In outline, subsequent chapters of the book cover the following:

2. A rundown of the necessary model theory and universal algebra.

3. Basic facts about binary relations, relation algebras and their representations. Saturation is used to prove that the class **RRA** of representable relation algebras is a canonical variety.

4. Examples of relation algebras, including set algebras, group relation algebras, McKenzie's 4-atom non-representable algebra, and Lyndon algebras.

5. Relativised and weak representations of relation algebras. Cylindric algebras and relation algebra reducts.

6. A survey chapter that outlines some other approaches to algebras of relations, including diagonal-free algebras and polyadic algebras.

7. The game for representability of a weakly associative algebra, relation algebra and cylindric algebra. Applications in computer science.

8. Using the game to axiomatise the representable relation algebras and cylindric algebras.

9. A general approach to axiomatising pseudo-elementary classes of algebras, using games to obtain explicit axioms.

10. A general approach to games, via 'game trees'.

11. Games for complete representations of atomic relation algebras. The Lyndon conditions.

12. Bases for atomic relation algebras. Relational bases, cylindric bases, and hyperbases; the variety $\mathbf{RA}_n$.

13. Approximations to representable relation algebras. Special relativised representations are introduced: square, flat, and smooth. The chapter presents a number of approximations to **RRA** and gives three equivalent characterisations of these approximations: by bases, by relativised representations, and by relation algebra reducts of cylindric-type algebras.

14. Variants of Monk algebras, which are used with a graph-theoretic construction of Erdös to prove that the strongly representable relation algebra atom structures do not form an elementary class. It follows that **RRA** is not closed under completions: there are two atomic relation algebras with the same atom structure, one being representable, the other, not. It then follows that **RRA** is not Sahlqvist-axiomatisable.

1.7. Summary of contents

15. Other variants of Monk algebras, used to prove non-finite axiomatisability of $\mathbf{S\mathfrak{R}aCA}_{n+1}$ over $\mathbf{S\mathfrak{R}aCA}_n$ for $n \geq 4$, and of $\mathbf{S\mathfrak{N}r}_m\mathbf{CA}_{n+1}$ over $\mathbf{S\mathfrak{N}r}_m\mathbf{CA}_n$ for $3 \leq m < n < \omega$.

16. Rainbow algebras.

17. Applications of rainbow algebras:

    (a) The theorem of Monk that the representable relation algebras are not finitely axiomatisable.

    (b) The completely representable relation algebras do not form an elementary class.

    (c) The representable relation algebras cannot be defined by any equational theory using only a finite number of variables.

    (d) $\mathbf{RA}_{n+1}$ is not finitely axiomatisable over $\mathbf{RA}_n$ for finite $n \geq 4$.

    (e) The class of weakly representable relation algebras is not finitely axiomatisable.

    (f) A second proof that the class of representable relation algebras is not closed under completions. This proof also covers $\mathbf{RA}_n$ and $\mathbf{S\mathfrak{R}aCA}_n$ for finite $n \geq 6$; the case $n = 5$ is open.

18. Using a variant of the rainbow construction, encoding tiling problems, we show that it is undecidable whether a given finite relation algebra is representable, weakly representable, or in $\mathbf{S\mathfrak{R}aCA}_n$ for $n \geq 5$.

19. Using the fact that the loosely guarded fragment of first-order logic has the finite model property, we prove various kinds of finite base property for $\mathbf{WA}$ and $\mathbf{RA}_n$ (finite $n \geq 4$). Related but (necessarily) weaker results are proved for $\mathbf{S\mathfrak{R}aCA}_n$. Other results are covered in exercises.

20. A short summary of the key results in this book.

21. Some open problems.

What is not covered

We have not tried to cover the entire field of relation algebra, much less that of algebraic logic, and detailed treatment of some important issues has been left out.

- Decidability questions have already been mentioned as an area we soft-pedal on: see, e.g., [AndGiv[+]97] instead.

- Cylindric algebras are mostly only covered insofar as they relate to relation algebras, and even there we have not told the whole story (for example, the results of Monk and Maddux that $S\mathfrak{R}a\mathbf{CA}_4$ is precisely the class of relation algebras are not proved in full here). Attractive material such as twisting has been omitted, though we do tacitly use splitting [AndMad$^+$91] once or twice. Many of our results, such as the rainbow construction, generalise to cylindric algebras — see, e.g., [HirHod97a, Hodk97, HodMik00] — but the generalisations will mostly not be covered here.

- Similarly, diagonal-free and polyadic algebras are not extensively covered, nor their connections to cylindric algebras.

- Issues relating to functional atoms, rectangular density, tabulation, and fork algebras are not covered in detail.

- We cover relativised representations of relation algebras, but not so much those of cylindric-type algebras such as **Crs, D, G**, Resek–Thompson theorems, nor algebraic relativisations of relation and other algebras.

- Proof theory of first-order logic, an important application of algebraic techniques, is not much discussed here; see [HenMon$^+$85, TarGiv87] for more.

- Connections to modal logic and computer science, and on the other side, abstract algebraic logic, are also not discussed very much.

Some of these will be discussed briefly in chapters 5 and 6, where we give some citations to the literature for readers wishing to pursue them. Our choice of material has been governed by

1. not making the book too long — both the reader and the authors have much else to do,

2. not covering work well-covered elsewhere,

3. as befits the title, concentrating on games and the role they can play in elucidating and deepening the study of relation algebras.

Part I

Algebras of Relations

Chapter 2

Preliminaries

We do not want to assume any previous knowledge about algebraic logic, although a certain familiarity with mathematical concepts and notation is needed, and a background in formal logic will help. We have included this chapter in order to cover notions from outside algebraic logic that will be useful in later chapters. Most of them are from universal algebra — equational varieties and Birkhoff's theorem, subdirectly irreducibles, conjugates, and discriminators. We also cover boolean algebras with operators and associated constructions such as atom structures, canonical embedding algebras, and completions. Some model theory will also be used in the book, and most of the later chapters are heavily influenced by model-theoretic ideas, but we use few specific model-theoretic theorems, so we confine ourselves here to the basic concepts and introduce the remaining ones, such as saturation, at the time of use. The chapter is intended to be selective and expository, and not a full or rigorous development of either universal algebra or model theory. References to universal algebra, which has historically been closely related to algebraic logic, include [Grät79, BurSan81, McKeMcN$^+$87]. Standard texts on model theory, not so often seen as related to algebraic logic, include [ChaKei90, Hod93].

The reader who wishes to get straight into algebraic logic and the representation theory might be advised to run quickly through this preliminary chapter, or perhaps skip it altogether, and then refer back to it as the need arises.

2.1 Foundations

We assume a basic knowledge of sets, relations and functions, and other basic mathematical notions. In particular, the following concepts and notations will arise. Many of them are entire fields with many books devoted to them; we only want to give a handy reference for items used later.

Iff We often use 'iff' to abbreviate 'if and only if'; this useful word was invented by Halmos. '$\iff$' means the same; '$\Rightarrow$' means 'implies', and '$\Leftarrow$' means 'is implied by'.

Sets We work implicitly in ZFC — Zermelo-Fraenkel set theory with the axiom of choice. For sets X, Y, $X \times Y$ denotes the set of all ordered pairs (x,y) ($x \in X, y \in Y$). (Formally, $(x,y) = \{\{x\}, \{x,y\}\}$.) For a whole number $n \geq 1$, X^n denotes

$$\overbrace{X \times X \times \cdots \times X}^{n \text{ times}};$$

we think of this as the set of sequences $\{(x_0, \ldots, x_{n-1}) : x_0, \ldots, x_{n-1} \in X\}$. For a set X, $\bigcup X$ and $\bigcup_{x \in X} x$ denote $\{y : y \in x \text{ for some } x \in X\}$. For finite sets X, we simplify the notation, so that $x \cup y$ denotes $\bigcup \{x, y\}$, etc. Similar conventions apply to $\cap$ when $X \neq \emptyset$. The *disjoint union* of sets X_i ($i \in I$) is formally defined to be $\bigcup \{X_i \times \{i\} : i \in I\}$, though normally we think of it as the union of pairwise disjoint copies of the X_i. In the text, $\subseteq$ will denote set or class inclusion (see any standard text on set theory for information about classes), or substructure or subalgebra (see section 2.2.4), as appropriate. $\subset$ will be used to indicate that the inclusion is proper: $X \subset Y$ iff $X \subseteq Y$ and $X \neq Y$. $\wp(X)$ denotes the *power set* (set of all subsets) of X. We write $X \setminus Y$ for the set $\{x \in X : x \notin Y\}$.

Binary relations These are the main topic of the book; here we confine ourselves to their basic aspects. A *binary relation* R on a set X is a subset of $X \times X$. Generally, X will be non-empty, but sometimes it is useful to allow empty X. We write $R(x,y)$ or xRy as alternative notations for $(x,y) \in R$. The *domain* and *range* of R are $\{x : (x,y) \in R\}$ and $\{y : (x,y) \in R\}$, respectively. R is said to be

reflexive, if xRx for all $x \in X$,

irreflexive, if xRx for no $x \in X$,

symmetric, if xRy implies yRx,

antisymmetric, if xRy and yRx imply $x = y$,

transitive, if xRy and yRz imply xRz.

The *reflexive* (or *transitive*) *closure* of R (with respect to X) is the smallest reflexive (respectively, transitive) binary relation on X containing R. The reflexive closure of R with respect to X is $R \cup \{(x,x) : x \in X\}$. Its transitive closure is the set of

2.1. Foundations

all pairs (x,y) such that there exist finite $n > 0$ and $x_0, \ldots, x_n \in X$ with $x_0 = x$, $x_n = y$, $x_0 R x_1$, $x_1 R x_2$, $\ldots$, $x_{n-1} R x_n$. That is, it is the intersection of all transitive relations on X containing R — or, the smallest such relation. Similarly, the *reflexive transitive closure* of R with respect to X is the smallest reflexive transitive relation containing R.

Equivalence relations An *equivalence relation* on a set X is a reflexive, symmetric, transitive binary relation E on X. An equivalence class of E is a subset of X of the form $x/E \stackrel{\text{def}}{=} \{y \in X : yEx\}$ for some $x \in X$. We write X/E for the set of equivalence classes of E. The equivalence classes of E partition X: i.e., every element of X lies in exactly one of them.

The *equivalence relation generated by a binary relation R* on X is the smallest equivalence relation on X that contains R.

Orderings A *pre-order* is a set X endowed with a reflexive transitive binary relation $\leq$. A *partial order* or *partial ordering* is an antisymmetric pre-order. A partial order $(X, \leq)$ is *well-founded* if every non-empty subset $Y \subseteq X$ has a minimal element (there is $y \in Y$ such that if $z \in Y$ and $z \leq y$ then $z = y$), or equivalently, there is no infinite descending sequence of distinct elements $x_0 \geq x_1 \geq \cdots$ in X. $(X, \leq)$ is a *linear order* if $x \leq y$ or $y \leq x$ for all $x, y \in X$, and a *well-order* if it is well-founded and linear.

An *irreflexive partial order* on X is an irreflexive binary relation on X, usually written $<$, whose reflexive closure is a partial order on X. Similar definitions are made for *irreflexive linear order*, etc. If $(X, \leq)$ is a given partial order, $x < y$ denotes the binary relation on X given by $x < y$ iff $x \leq y$ and $x \neq y$. Then $(X, <)$ is an irreflexive partial order. Conversely, for an irreflexive partial order $(X, <)$, $x \leq y$ denotes the binary relation on X given by $x \leq y$ iff $x < y$ or $x = y$; then, $(X, \leq)$ is a partial order.

Two linear orders are said to be *order-isomorphic*, and to have the *same order type*, if there is an order-preserving bijection between them. (See 'Functions' below for 'bijection', and section 2.2.4 for isomorphisms.) Similar definitions are made for partial orders, etc.

Numbers $\mathbb{N}$ denotes the set $\{0, 1, 2, \ldots\}$ of natural numbers. $\mathbb{Z}$ denotes the set $\{\ldots, -2, -1, 0, 1, 2, \ldots\}$ of integers. $\mathbb{Q}$ denotes the set $\{p/q : p, q \in \mathbb{Z}, q \neq 0\}$ of rational numbers (fractions). $\mathbb{R}$ denotes the set of real numbers. These four sets are ordered by $<, \leq$ in the usual way, and we may write, e.g., $(\mathbb{Z}, <)$ for the ordered integers.

Ordinals and cardinals An *ordinal* is a transitive set (i.e., any member of it is also a subset of it) that is well-ordered by $\in$. Every well-ordered set is order-isomorphic to a unique ordinal. For ordinals α, β, $\alpha < \beta$ means $\alpha \in \beta$. An ordinal is therefore the set of all smaller ordinals, so for a finite ordinal n we have $n = \{0, 1, \ldots, n-1\}$. A *successor ordinal* is one with a $<$-maximal element; a *limit ordinal* is a non-zero ordinal with no $<$-maximal element. All ordinals are either 0, successor, or limit. The first limit ordinal is $\omega = \{0, 1, \ldots\}$, though we sometimes also write $\mathbb{N}$ for this set.

A *cardinal* is an ordinal not in bijection with any smaller ordinal (i.e., an 'initial ordinal'), and the *cardinality* $|X|$ of a set X is the unique cardinal in bijection with X (see 'Functions' below for an explanation of 'bijection'). Cardinals are ordinals and are therefore ordered by $<$ (i.e., $\in$). The first few cardinals are $0 = \emptyset, 1, 2, \ldots, \omega$ (the first infinite ordinal), ω_1 (the first uncountable ordinal). A set will be said to be *countable* if it has cardinality $\leq \omega$, *uncountable* otherwise, and *countably infinite* if it has cardinality ω. A *cofinite* subset X of a set Y is one such that $Y \setminus X$ is finite.

For ordinals α, β, $\alpha + \beta$ denotes the unique ordinal with the same order type as a copy of α followed by a copy of β (formally, $\alpha \times \{0\} \cup \beta \times \{1\}$, lexicographically ordered). For example, $1 + \omega = \omega$. Similarly, $\alpha \cdot \beta$ denotes the unique ordinal with the order type as β copies of α laid end to end: for example, $\alpha \cdot 2 = \alpha + \alpha$. For cardinals κ, λ, sum and product have different definitions: $\kappa + \lambda$ denotes the cardinality of the ordinal $\kappa + \lambda$, and similarly for $\cdot$, which for cardinals is more usually written $\times$. (Which definition of $+$ is intended in a given context must be derived from the context.) For cardinals κ, λ of which at least one is infinite, $\kappa + \lambda$ and $\kappa \times \lambda$ are both equal to the larger of κ, λ. Ordinal sum and product are not commutative operations (i.e., $\alpha + \beta \neq \beta + \alpha$ and $\alpha \cdot \beta \neq \beta \cdot \alpha$ in general), but cardinal sum and product are commutative.

On the whole we tend to write α, β, i, j, k for ordinals, δ for a limit ordinal, m, n for finite ordinals, and κ, λ for cardinals.

Functions If X, Y are sets, a *map* or *function* $f : X \to Y$ is a subset of $X \times Y$ such that for all $x \in X$ there is a unique $y \in Y$ with $(x, y) \in f$. This unique y is written as $f(x)$, and we may write $f : x \mapsto y$. If X has the form S^n, we may say that f is an *n-ary* function and write $f((s_0, \ldots, s_{n-1}))$ as $f(s_0, \ldots, s_{n-1})$. Binary (i.e., 2-ary) functions are sometimes written in infix form: e.g., $a + b$ rather than $+(a, b)$. A binary function $* : X \times X \to Y$ is *associative* if $(a * b) * c = a * (b * c)$ for all $a, b, c \in X$; in this case we often omit parentheses and write $a * b * c$, etc. $*$ is *commutative* if $a * b = b * a$ for all $a, b \in X$, and *idempotent* if $a * a = a$ for all $a \in X$.

If $X' \subseteq X$, the *restriction* $f\!\upharpoonright_{X'}$ of f to X' is $\{(x, y) \in f : x \in X'\}$; it is a map $: X' \to Y$. The *range* of f is $\mathrm{rng}(f) \stackrel{\mathrm{def}}{=} \{f(x) : x \in X\}$. For $X' \subseteq X$, $Y' \subseteq Y$, we sometimes write $f[X']$ for $\mathrm{rng}(f\!\upharpoonright_{X'})$, and $f^{-1}[Y']$ for $\{x \in X : f(x) \in Y'\}$. For $y \in Y$, $f^{-1}[\{y\}]$ is often abbreviated to $f^{-1}[y]$.

If $f : X^n \to X$ and $X' \subseteq X$, we say that X' is *closed under* f if $f\!\upharpoonright_{X'^n} : X'^n \to X'$.

2.1. Foundations

The *closure* of X' under a set of functions $f_i : X^{n_i} \to X$ (for $i \in I$) is the smallest subset of X that contains X' and is closed under every f_i.

f is said to be *injective*, or *one-one*, if for all $x, x' \in X$, if $f(x) = f(x')$ then $x = x'$. f is said to be *surjective* or *onto* if $\mathrm{rng}(f) = Y$. f is said to be *bijective*, or a *bijection*, if it is injective and surjective. The *inverse* of a bijection $f : X \to Y$ is $f^{-1} \stackrel{\text{def}}{=} \{(y,x) : (x,y) \in f\}$; it is a bijection : $Y \to X$. The *identity map* on a set X is $\{(x,x) : x \in X\}$, sometimes written Id_X. When $X \subseteq Y$, Id_X is actually a map : $X \to Y$, called the *inclusion map*. A function $f : X \to X$ is said to be an *involution* if $f(f(x)) = x$ for all $x \in X$; such a function is necessarily bijective.

A *partial function* on X is a function $f : D \to Y$ for some (unique) $D \subseteq X$. We write $\mathrm{dom}(f)$ (the *domain* of f) for this D. We will often write '$f : X \to Y$ is a partial function' in this situation. In the case where $D = X$, we say that f is *total*. For (possibly partial) functions f, g, $f \circ g$ is the function defined by $(f \circ g)(x) = f(g(x))$.

If X, Y are sets, $^Y X$ denotes the set of all functions $f : Y \to X$. For a finite ordinal n, $^n X$ is naturally isomorphic to X^n, and we sometimes blur the distinction between them. For an ordinal α, we write $^{<\alpha} X$ for $\bigcup_{\beta < \alpha} {}^\beta X$, and $^{\leq \alpha} X$ for $\bigcup_{\beta \leq \alpha} {}^\beta X$.

Relations For an ordinal α, an *α-ary relation* on a set X is a subset of $^\alpha X$. For finite n, an n-ary relation on X is usually regarded as a subset of X^n.

Tuples $\bar{x}$ denotes a usually finite sequence or *tuple* $(x_0, x_1, \ldots, x_{n-1})$ for some $n < \omega$ (or more generally, any ordinal n; but infinite sequences are not usually called tuples). The x_i could be variables or elements of some structure, for example. We write $|\bar{x}|$ for the length n of the tuple. An *n-tuple* is a tuple of length n. When we write $\bar{x}$ for a tuple, it will frequently be implicit that its ith entry is x_i, for each $i < |\bar{x}|$. If S is a set, we abuse notation by writing $\bar{x} \in S$ to mean $x_i \in S$ for each $i < |\bar{x}|$. We sometimes write $\bar{x} \in {}^n S$ (and occasionally $\bar{x} \in S^n$) to denote that $\bar{x}$ is an n-tuple of elements of S; this comes down to regarding $\bar{x}$ as a function : $n \to S$, and we will pass between the sequence and function views of tuples whenever convenient. Thus, $\mathrm{rng}(\bar{x})$ denotes the set $\{x_0, \ldots, x_{|\bar{x}|-1}\}$. If $f : X \to Y$ is a map, and $\bar{x} = (x_0, \ldots, x_{n-1}) \in {}^n X$, we write $f(\bar{x})$ for the n-tuple $(f(x_0), \ldots, f(x_{n-1}))$ of elements of Y. Formally, this is $f \circ \bar{x}$. Similarly, if $m < \omega$ and $\theta : m \to n$ is a map, $\bar{x} \circ \theta$ is the m-tuple $(x_{\theta(0)}, \ldots, x_{\theta(m-1)})$. If $\bar{x} = (x_0, \ldots, x_{n-1})$, $\bar{y} = (y_0, \ldots, y_{m-1})$ are tuples, we write $\bar{x}\bar{y}$ for the $(n+m)$-tuple $(x_0, \ldots, x_{n-1}, y_0, \ldots, y_{m-1})$.

For a sequence $\bar{x} = (x_i)_{i < \alpha} \in {}^\alpha S$, for some ordinal α, an *initial segment* of $\bar{x}$ is a sequence of the form $(x_i)_{i < \beta}$ for some $\beta \leq \alpha$. A *final segment* of $\bar{x}$ is a sequence of the form $(x_\beta, x_{\beta+1}, \ldots, x_i, \ldots)_{i < \alpha}$ for some $\beta \leq \alpha$.

Graphs A *(directed) graph* is a set X (of 'nodes' or 'vertices') endowed with a binary relation E, the edge relation. A pair (x, y) of elements of X is said to be an *edge* if xEy holds. A graph with set of nodes X is a *subgraph* of a second graph with set of nodes Y if $X \subseteq Y$ and each pair of nodes of X is an edge of the first graph iff it is an edge of the second (this notion is sometimes called *induced subgraph*.) A directed graph is said to be *complete* if (x,y) is an edge for all nodes x,y. A graph is said to be *loop-free* if E is irreflexive, and *undirected* if E is symmetric and irreflexive (see figure 1.1 for an example). So by default we take undirected to imply loop-free, though we may remind the reader of it by writing 'undirected loop-free'. An undirected graph is *complete* if (x,y) is an edge for all distinct nodes x,y. A *clique* in an undirected graph with set of nodes X is a set $C \subseteq X$ such that each pair of distinct nodes of C is an edge. A *path* from node x to node y in a graph is a sequence $x = x_0, x_1, \ldots, x_n = y$ of nodes, for some $n < \omega$, such that $x_i E x_{i+1}$ for all $i < n$. The *length* of the path is n. The graph is *connected* if there is a path between any two nodes. The *degree* of a node x in an undirected graph is the cardinality of the set of nodes y such that xEy.

A *(directed) hypergraph* is a set X of nodes together with a subset $H \subseteq {}^{\leq n}X$ for some arbitrary natural number n. A *hyperedge* is an element of H. A *labelled (hyper)graph* has additionally a map from the set of (hyper)edges to some set (of 'labels').

Groups A *group* is a non-empty set G endowed with a binary function $* : G \times G \to G$ that is associative (i.e., $x * (y * z) = (x * y) * z$ for all $x, y, z \in G$), an identity element $e \in G$ such that $x * e = e * x = x$ for all $x \in G$, and a unary function $-^{-1}$ such that $x * x^{-1} = x^{-1} * x = e$ for all $x \in G$. G is *abelian* (or *commutative*) if $x * y = y * x$ for all $x, y \in G$. A *subgroup* of G is a subset of G containing e and closed under $*, ^{-1}$; it is a group in its own right with functions $*, ^{-1}$ defined by restriction from G. The smallest subgroup is just $\{e\}$.

Permutation groups Examples of groups include the set $\text{Sym}(X)$ of all *permutations* of a non-empty set X (bijections from X to itself), where e is the identity map on X, f^{-1} is the inverse of f, and $f * g(x) \stackrel{\text{def}}{=} g(f(x))$. Any subgroup of $\text{Sym}(X)$ (for some X) is called a group of permutations, or a *permutation group*, acting on X. Any group is isomorphic to a permutation group (Cayley's theorem); see section 2.2.4 for the definition of 'isomorphism'.

Given a permutation group G acting on X, an *orbit* of G is a subset of X of the form $\{g(x) : g \in G\}$ for some $x \in X$. The action of G on X is said to be *transitive* if for all $x, y \in X$ there is $g \in G$ with $g(x) = y$. That is, there is only one orbit of G.

Topological spaces A *topological space* is a pair $(X, \mathcal{T})$ where $X \neq \emptyset$ and $\mathcal{T} \subseteq \wp X$ contains $\emptyset, X$, and is closed under arbitrary unions and finite intersections. Such a $\mathcal{T}$ is said to be a *topology* on X. A subset Y of X is said to be *open* if $Y \in \mathcal{T}$, *closed* if $X \setminus Y \in \mathcal{T}$, and *clopen* if it is closed and open. An *open cover* of X is a set $\mathcal{U} \subseteq \mathcal{T}$ with $\bigcup \mathcal{U} = X$. $(X, \mathcal{T})$ is *compact* if every open cover of X contains a finite open cover of X. $(X, \mathcal{T})$ is *discrete* if $\mathcal{T} = \wp(X)$, and *totally disconnected* if for any distinct $y, y' \in X$ there is a clopen set $Y \subseteq X$ with $y \in Y$, $y' \notin Y$.

Ramsey's theorem This useful combinatorial result states that if $n < \omega$, S is a finite set, and $f : [\omega]^n \to S$ is a map, then there is infinite $H \subseteq \omega$ such that $f \restriction_{[H]^n}$ is constant (i.e., $f(x) = f(y)$ for all $x, y \in [H]^n$). Here, for a set X, $[X]^n$ denotes $\{Y \subseteq X : |Y| = n\}$. The finite Ramsey theorem says that given n, S as above and a number $k < \omega$, there is $m < \omega$ such that any $f : [m]^n \to S$ is constant on $[H]^n$ for some $H \subseteq m$ with $|H| \geq k$. For proofs, see, e.g., [ChaKei90, 3.3.7], [Hod93, 11.1.3].

2.2 Model theory

We assume some familiarity with basic first-order logic, but for convenience we list the main ideas we will need. Readers wanting more details may refer to [ChaKei90, Hod93], for example. Anyone familiar with basic model theory could probably skip this section permanently if they are willing to work out at sight notations such as $\bar{a} \in M$, x^s for sorted variables, $\exists_{i \in I} x_i \varphi$.

2.2.1 Syntax

A *signature* (or similarity type, or vocabulary) is a set L of relation symbols, function symbols, and constants, each relation and function symbol having an associated type (either relation or function) and a finite *arity*, its number of arguments or places. A nullary relation or function symbol has arity 0 (e.g., a constant may be thought of as a nullary function symbol); unary, 1 (e.g., $P(x)$); binary, 2 ($x \leq y$); ternary, 3 ($f(a,b,c)$); and n-ary, n ($R(x_0, \ldots, x_{n-1})$). A signature is said to be *relational* if it has no function or constant symbols, and *functional* if it has no relation symbols (note that it can have constants, though). In *many-sorted first-order logic*, each place of the relation and function symbols has an associated sort, and each function symbol and constant has a sort associated with its value. We generally write sorts in boldface as $\mathbf{s_1, s_2, \ldots}$, and display the associated sorts of a relation symbol R as, e.g., $R(\mathbf{s_1, \ldots, s_n})$, of a function symbol f as $f : \mathbf{s_1} \times \cdots \times \mathbf{s_n} \to \mathbf{s}$, and of a constant as $c^\mathbf{s}$.

The *first-order language* associated with a signature L is made by forming *terms* and *formulas* from the L-symbols in the usual way (see [Hod93, ChaKei90] for more details), using a countably infinite set of variables (e.g., $\{x_0, x_1, \ldots\}$), the boolean connectives $\wedge, \neg$, and the quantifier $\exists$. In brief, any variable or constant in L is an L-term, and if $f \in L$ is an n-ary function symbol and $t_1, \ldots, t_n$ are

L-terms then $f(t_1,\ldots,t_n)$ is an L-term. For any n-ary relation symbol $R \in L$ and L-terms $t,u,t_1,\ldots,t_n$, $t=u$ and $R(t_1,\ldots,t_n)$ are (atomic) L-formulas, and if φ,ψ are L-formulas and x is a variable then $\neg \varphi$, $\varphi \wedge \psi$, and $\exists x \varphi$ are L-formulas. We generally regard $\top, \bot, \vee, \rightarrow, \leftrightarrow, \forall$ as the usual abbreviations. Equality ($=$) is regarded as a logical symbol and will always be included. In many-sorted logic, each variable has an associated sort; we write x^s to indicate that variable x has sort $\mathbf{s}$. We generally write L for the language as well as the signature, and $|L|$ denotes the cardinality of the set of L-formulas. Equivalently, $|L|$ is the least infinite cardinal at least as large as the cardinality of the signature L. If $n < \omega$, we write L^n for the sublanguage of L consisting of formulas using only the variables $x_0,\ldots,x_{n-1}$. The infinitary language $L_{\infty\omega}$ is obtained by additionally allowing the conjunction $\bigwedge \Phi$ of an arbitrary set Φ of formulas to be a formula. For $n < \omega$, $L_{\infty\omega}^n$ is the fragment of $L_{\infty\omega}$ consisting of formulas using only the variables $x_0,\ldots,x_{n-1}$, and $L_{\infty\omega}^\omega$ is the fragment consisting of formulas using finitely many variables (i.e., $\bigcup_{n<\omega} L_{\infty\omega}^n$).

The *subformulas* of a formula φ are those formulas constructed along the way to building φ: if φ is atomic, its only subformula is itself; the subformulas of $\neg \varphi$ are $\neg \varphi$ and the subformulas of φ; the subformulas of $\varphi \wedge \psi$ are $\varphi \wedge \psi$ and the subformulas of φ and of ψ; and the subformulas of $\exists x \varphi$ are $\exists x \varphi$ and the subformulas of φ. An occurrence of a variable x in a formula φ is *bound* if it is in a subformula of φ of the form $\exists x \psi$; all other occurrences are *free*, and the *free variables* of φ are those variables with free occurrences in φ. Unless otherwise stated, when we write a formula as $\varphi(\bar{x})$ or $\varphi(x_0,\ldots,x_{n-1})$ we mean that the free variables of φ are all in the tuple $\bar{x}$ or set $\{x_0,\ldots,x_{n-1}\}$ of variables and that the x_i are distinct. Not all the x_i need actually occur free in φ. A similar convention applies to terms. This notation is useful but can be dangerous: note for example that any given formula has infinitely many representations of the form $\varphi(\bar{x})$, depending on the choice of $\bar{x}$. Where problems may arise, we will be more careful over the notation.

In model theory, we are interested in semantics and not in fine details of syntax. For example, we omit parentheses where possible, so that in an expression such as $\varphi \wedge \psi \wedge \chi$, the parentheses can be added arbitrarily (as far as semantics is concerned). For formulas $\varphi_0,\ldots,\varphi_{n-1}$, $\bigwedge_{i<n}\varphi_i$ denotes $\varphi_0 \wedge \ldots \wedge \varphi_{n-1}$, and if $\Phi = \{\varphi_0,\ldots,\varphi_{n-1}\}$, $\bigwedge \Phi$ denotes $\bigwedge_{i<n}\varphi_i$ (semantically, the order of the conjuncts is immaterial). $\bigvee_{i<n}\varphi_i$, $\bigvee \Phi$ are similarly defined. $\bigwedge \emptyset$ is $\top$ and $\bigvee \emptyset$ is $\bot$. $\forall \bar{x} \varphi$ will abbreviate $\forall x_0 \ldots x_{n-1} \varphi$, which in turn abbreviates $\forall x_0 \ldots \forall x_{n-1} \varphi$, where $\bar{x} = (x_0,\ldots,x_{n-1})$. For a finite set $I = \{i_0,\ldots,i_{n-1}\}$ and variables x_i ($i \in I$), $\forall_{i \in I} x_i \varphi$ will denote $\forall x_{i_0} \ldots x_{i_{n-1}} \varphi$. This notation hides the order of the quantifiers; when the order is significant we will be more careful about notation. Notations $\exists \bar{x} \varphi$, $\exists_{i \in I} x_i \varphi$ are defined similarly.

An *equation* is an atomic formula of the form $t = u$, where t,u are terms. A *boolean combination* of formulas is a formula made from them using only the boolean operations; it is a *positive boolean combination* if only $\wedge, \vee$ are used (so $\neg$ is only used within the abbreviation $\vee$). A *quantifier-free* formula is one without

2.2. Model theory

quantifiers: it is a boolean combination of atomic formulas. An *existential (universal)* formula is one of the form $\exists \bar{x}\psi$ (or $\forall \bar{x}\psi$), where ψ is quantifier-free. A formula is in *prenex normal form,* and is said to be a *prenex formula,* if it is of the form $Q_0 x_0 \ldots Q_{n-1} x_{n-1} \psi$, where each Q_i is $\forall$ or $\exists$ and ψ is quantifier-free. Any formula can be translated to a logically equivalent one (see below for the meaning of this) in prenex normal form.

A *closed term* is one with no variables. An *L-sentence* is an *L*-formula with no free variables, and an *L-theory* is a set T of *L*-sentences. A *sentence* will mean an *L*-sentence where L is either arbitrary or given by the context, and similarly for 'theory' and other terms too. To say that T is *consistent* means that $T \not\vdash \bot$ in some standard sound and complete proof system $\vdash$ for first-order logic; but it is equivalent to say that T has a model. T is *complete* if it is consistent and for each *L*-sentence σ, either $T \vdash \sigma$ or $T \vdash \neg\sigma$. T is *decidable* if there exists an algorithm that decides whether $\sigma \in T$ or $\sigma \notin T$, for each sentence σ of the signature of T (the signature must be recursive; see section 9.4.3 for more details). T is *finitely axiomatisable* if there exists a finite theory U (in the same signature) with the same models as T. T is *universal* if it consists of universal sentences, and *equational* if it consists of sentences of the form $\forall \bar{x} \varepsilon(\bar{x})$ where ε is an equation.

2.2.2 Semantics — structures

If L is a signature, an *L-structure* is an object of the form $M = \langle D, I \rangle$, where D is a non-empty set (called the *domain* of M and written $\mathrm{dom}(M)$), and I provides an appropriate *interpretation* in D of each symbol in L. The interpretation of an n-ary relation symbol is an n-ary relation on D (i.e., a subset of D^n), the interpretation of an n-ary function symbol is a function : $D^n \to D$, and the interpretation of a constant is an element of D. If $\sigma \in L$ is a symbol, we write σ^M for the interpretation $I(\sigma)$ of σ in M. We may denote structures by $\langle D, \sigma_1, \ldots, \sigma_n \rangle$, where D is a non-empty set and $\sigma_1, \ldots, \sigma_n$ are constants, functions from D to D (of various arities), or relations on D (of various arities), as specified separately. A *structure* will mean an *L*-structure where L is arbitrary or determined by context. For many-sorted L, each element of an *L*-structure will have a unique sort, and the interpretations of the symbols of L will respect these sorts.

The first thing to say is that we will generally identify (notationally) a structure with its domain. We *do not* use the common algebraic logic convention that the domains of algebras $\mathfrak{A}, \mathfrak{B}, \ldots$ are $A, B, \ldots$. Thus, above, we often write $a \in M$ instead of the more precise $a \in D$ or $a \in \mathrm{dom}(M)$. This is common practice in model theory and algebra, though it is uncommon in algebraic logic. We occasionally extend this notation to tuples of elements, as already outlined: thus, we write $\bar{a} \in M$ to mean that each entry in the tuple $\bar{a}$ is an element of M. However, where necessary we will write $\bar{a} \in {}^n M$ and/or use $\mathrm{dom}(M)$ for the domain of the structure M. We write $|M|$ for the cardinality of $\mathrm{dom}(M)$, and say that M is finite, countable, etc., if dom(M) is.

A *relational structure* is a structure in a relational signature. The *disjoint union* of similar relational structures M_i ($i \in I$) is the structure in the same signature whose domain is the disjoint union of the domains of the M_i, and in which each relation symbol R is interpreted as the disjoint union of its interpretations R^{M_i} in the individual structures.

An *algebra* is a structure in a functional signature. It may sound obvious but is easily forgotten that the term 'algebra' is used to denote a structure that is *closed under the designated functions*. So, for example, when we write 'the algebra (A, c, f, g)', it should be checked that A contains the element c and is closed under the functions f and g. We generally write M, N, A, B etc., for structures, and $\mathcal{A}, \mathcal{B}, \mathcal{C}, \ldots$ for algebras. This is just for readability; an algebra is (a special case of) a structure.

Structures M, N are said to be *similar* if they are structures for the same signature. If L is a subsignature of L' (i.e., $L \subseteq L'$) and M is an L'-structure, we write $M \!\upharpoonright_L$ for the *L-reduct* of M obtained by discarding the interpretations of symbols of $L' \setminus L$. An *expansion* of an L-structure M is a structure N, in a signature containing L, of which M is the L-reduct. Note that M and N have the same domain.

2.2.3 Models, validity

Given an L-structure M and a set V of variables, an *assignment* of the variables in V to elements of M is simply a map $h : V \to M$. For an L-formula φ with free variables in V, and an L-structure M, we write $M, h \models \varphi$ if φ is true in M under the assignment h. This is defined in the usual Tarskian way, by induction on φ. In brief, we first extend h to terms via $h(c) = c^M$ for a constant $c \in L$, and $h(f(t_1, \ldots, t_n)) = f^M(h(t_1), \ldots, h(t_n))$ for an n-ary function symbol $f \in L$ and L-terms $t_1, \ldots, t_n$ on which we have defined h inductively. Then for L-terms $t, u, t_1, \ldots, t_n$ made with variables from V, and an n-ary relation symbol $R \in L$, we let $M, h \models t = u$ iff $h(t) = h(u)$, and $M, h \models R(t_1, \ldots, t_n)$ iff $(h(t_1), \ldots, h(t_n)) \in R^M$. If we have defined $M, h \models \varphi$ and $M, h \models \psi$ for all assignments h, we let $M, h \models \neg \varphi$ iff $M, h \not\models \varphi$, $M, h \models \varphi \wedge \psi$ iff $M, h \models \varphi$ and $M, h \models \psi$, and $M, h \models \exists x \varphi$ iff $M, g \models \psi$ for some $g : V \to M$ with $g\!\upharpoonright_{V \setminus \{x\}} = h\!\upharpoonright_{V \setminus \{x\}}$. If $\varphi(\bar{x})$ is a formula, and $\bar{a} \in M$ is a tuple of length $|\bar{x}|$, we write $M \models \varphi(\bar{a})$ to denote that φ is true in M under any assignment h defined on $\mathrm{rng}(\bar{x})$ and with $h(\bar{x}) = \bar{a}$. We often tacitly assume that $|\bar{a}| = |\bar{x}|$ in such situations. When φ is a sentence, of course, we just write $M \models \varphi$, and say that M is a model of φ.

For a theory T, $M \models T$ ('M is a model of T') means that $M \models \sigma$ for all sentences $\sigma \in T$. We write $\mathrm{Mod}(T)$ (or $\mathrm{Mod}(\sigma)$) for the class of all models of T (or σ). For any class K of L-structures, we write $\mathrm{Th}(\mathsf{K})$, the *theory of* K, for the set of L-sentences true in every structure in K, and for a structure M we write $\mathrm{Th}(M)$ for

2.2. Model theory

Th($\{M\}$). The *universal theory of* K is the set of universal sentences true in every structure in K.

L-structures M and N are said to be *elementarily equivalent,* written $M \equiv N$, if $M \models \sigma$ iff $N \models \sigma$ for every L-sentence σ. The *elementary closure* of a class K of similar structures is the class $\{M : \exists N \in \mathsf{K}(N \equiv M)\}$.

An *elementary class* is a class of similar structures of the form $\mathrm{Mod}(T)$ for some first-order theory T. We say that the class is *axiomatised* by T and that T is an *axiomatisation* of it. Any elementary class is closed under elementary equivalence (i.e., if $M \equiv N \models T$ then $M \models T$), but not always conversely. A class is said to be *finitely axiomatisable* if it has the form $\mathrm{Mod}(T)$ for some finite theory T (or equivalently, if it has the form $\mathrm{Mod}(\sigma)$), and *universal* or *universally axiomatisable* if it has the form $\mathrm{Mod}(T)$ for some universal theory T; such a T is a *universal axiomatisation* of it. The notions of *equational class (axiomatisation)* are defined analogously. A universally axiomatisable class is closed under substructures (and if the class is elementary, the converse holds: this is the Łoś-Tarski theorem [ChaKei90, 3.2.2], [Hod93, 6.5.4]). For classes $\mathsf{K} \subseteq \mathsf{K}'$ of similar structures, K is said to be *finitely axiomatisable over* K' if $\mathsf{K} = \mathsf{K}' \cap \mathrm{Mod}(\sigma)$ for some sentence σ.

The notion of validity of a *formula* (as opposed to a sentence) is common in algebra (at least for equations) but rare in model theory. An L-formula $\varphi(\bar{x})$ is said to be *valid* in an L-structure M if $M \models \forall \bar{x} \varphi(\bar{x})$. φ is said to be valid in a class K of L-structures if it is valid in each structure in K. The *equational theory* of K is the set of all L-equations that are valid in K. We write $M \models \varphi$ to denote that φ is valid in M, and for a set Φ of formulas we write $M \models \Phi$ to denote that $M \models \varphi$ for every $\varphi \in \Phi$. We write $\mathsf{K} \models \Phi$ if $M \models \Phi$ for every $M \in \mathsf{K}$.

L-formulas φ, ψ are said to be *equivalent* in M if $M \models \varphi \leftrightarrow \psi$; equivalence in K is defined similarly. φ, ψ are said to be *logically equivalent* if they are equivalent in every L-structure. This is so iff $\varphi \leftrightarrow \psi$ is provable in any complete calculus: $\vdash \varphi \leftrightarrow \psi$.

2.2.4 Homomorphisms, embeddings, substructures

We will consider many kinds of map between structures. Let M, N be similar structures — say, L-structures, for some signature L.

1. A *homomorphism* from M to N is a map $h : M \to N$ such that for every atomic formula $\varphi(\bar{x})$ and $\bar{a} \in M$,

    $$M \models \varphi(\bar{a}) \Rightarrow N \models \varphi(h(\bar{a})).$$

    Equivalently, if $c \in L$ is a constant then $h(c^M) = c^N$, if $f \in L$ is an n-ary function symbol and $\bar{a} \in M$ an n-tuple then $h(f^M(\bar{a})) = f^N(h(\bar{a}))$, and if

$R \in L$ is an n-ary relation symbol, $\bar{a} \in M$ an n-tuple, and $M \models R(\bar{a})$, then $N \models R(h(\bar{a}))$. That is, 'h preserves all L-symbols forwards'.

2. N is said to be a *homomorphic image* of M if there exists a surjective (onto) homomorphism $h : M \to N$.

3. A *partial isomorphism* from M to N is a possibly partial map $p : M \to N$ that preserves all atomic formulas in both directions: for every atomic formula $\varphi(\bar{x})$ and $|\bar{x}|$-tuple $\bar{a} \in \mathrm{dom}(p)$,

$$M \models \varphi(\bar{a}) \iff N \models \varphi(p(\bar{a})).$$

Note that because the formula $x = y$ is preserved, p must be one-one.

4. A partial isomorphism is called an *embedding* if it is total. If the signature is functional, an embedding is just a one-one homomorphism.

5. We say that M is a *substructure* (or, for algebras, *subalgebra*) of N, and write $M \subseteq N$, if $\mathrm{dom}(M) \subseteq \mathrm{dom}(N)$ and the inclusion map from $\mathrm{dom}(M)$ to $\mathrm{dom}(N)$ is an embedding from M to N. In this case, we also say that N is an *extension* of M.

6. An *isomorphism* from M to N is a bijective partial isomorphism (or bijective embedding) from M onto N. If L is functional, an isomorphism is just a bijective homomorphism. We write $M \cong N$, and say that M, N are *isomorphic*, if there exists an isomorphism from M to N. If $X \subseteq M \cap N$, we say that M, N are *isomorphic over* X if there is an isomorphism $f : M \to N$ fixing every element of X (i.e., $f(x) = x$ for all $x \in X$).

7. An *automorphism* of M is an isomorphism $g : M \to M$. The set of all automorphisms of M, with the identity map on M, composition and inverse of automorphisms, forms a group, the *automorphism group of M*, denoted $\mathrm{Aut}(M)$. $\mathrm{Aut}(M)$ is evidently a group of permutations acting on M.

8. An *elementary map* from M to N is a possibly partial map $e : M \to N$ such that for every L-formula $\varphi(\bar{x})$ and $|\bar{x}|$-tuple $\bar{a} \in \mathrm{dom}(e)$, we have $M \models \varphi(\bar{a})$ iff $N \models \varphi(e(\bar{a}))$. Such a map is said to be an *elementary embedding* if it is total.

9. M is an *elementary substructure* of N, written $M \preceq N$, if $\mathrm{dom}(M) \subseteq \mathrm{dom}(N)$ and the inclusion map from $\mathrm{dom}(M)$ to $\mathrm{dom}(N)$ is an elementary embedding from M to N. In this case, we also say that N is an *elementary extension* of M. Clearly, $M \preceq N$ implies $M \subseteq N$ and $M \equiv N$.

2.2.5 Generating sets

If M is an L-structure and $A \subseteq \mathrm{dom}(M)$, we say that A *generates* M if for every element b of M, there is an L-term $t(\bar{x})$ and a tuple $\bar{a} \in A$ with $M \models b = t(\bar{a})$. Equivalently, if $N \subseteq M$ (substructure) and $A \subseteq \mathrm{dom}(N)$ (subset) then $N = M$. The *substructure of M generated by A* is the smallest substructure N of M containing A — that is, the intersection of all such substructures (this is clearly a substructure itself). Clearly, A generates N. M is said to be *finitely generated* if some finite set generates it, and *n-generated* (for some whole number n) if some set of size at most n generates it. Of course, these notions apply to algebras too, as a special case.

2.2.6 Compactness, Löwenheim–Skolem–Tarski theorems

The *compactness theorem* for first-order logic states that a first-order theory has a model iff every finite subset of it does. It is one of the most fundamental properties of first-order logic. The *downward Löwenheim–Skolem–Tarski theorem* states that if L is a first-order language of cardinality κ and M is an L-structure, then there is $N \preceq M$ with $|N| \leq \kappa$. The *upward Löwenheim–Skolem–Tarski theorem*, a consequence of the downward one and compactness, states that if M is infinite then for any cardinal $\lambda > |M| + |L|$, there is $N \succeq M$ with $|N| = \lambda$. See [ChaKei90, Hod93].

2.2.7 Relativisation, interpretations, second-order logic

Definable sets A set $S \subseteq {}^n M$ is said to be *definable* if $S = \{\bar{a} \in {}^n M : M \models \varphi(\bar{a})\}$ for some formula $\varphi(\bar{x})$. S is *definable with parameters* $\bar{b} \in M$ if $S = \{\bar{a} \in {}^n M : M \models \varphi(\bar{a}, \bar{b})\}$ for some formula $\varphi(\bar{x}, \bar{y})$.

Relativisation It is often useful to relativise quantifiers to a definable set. Given a fixed L-formula $\chi(x, \bar{y})$, for any L-formula $\varphi(\bar{z})$ not involving the variables $x\bar{y}$, we may relativise the quantifiers in φ to χ, by requiring that they range only over elements satisfying χ. Formally, we define the *relativisation* $\varphi^\chi(\bar{y}, \bar{z})$ by induction on φ, as follows. For atomic φ we let $\varphi^\chi = \varphi$. If φ^χ, ψ^χ are defined, we let $(\neg\varphi)^\chi = \neg(\varphi^\chi)$ and $(\varphi \wedge \psi)^\chi = \varphi^\chi \wedge \psi^\chi$. Finally, we let $(\exists z \varphi)^\chi = \exists z(\chi(z, \bar{y}) \wedge \varphi^\chi)$, where $\chi(z, \bar{y})$ denotes the result of freely substituting z for x in $\chi(x, \bar{y})$. (Alternatively, let $(\exists z \varphi)^\chi = \exists z(\exists x(x = z \wedge \chi) \wedge \varphi^\chi)$.) If L is relational, it is easily checked that for any L-structure M, formulas $\chi(x, \bar{y}), \varphi(\bar{z})$ as above, and $\bar{b} \in M$, if $M \models \exists x \chi(x, \bar{b})$ and N is the substructure of M with domain $\{a \in M : M \models \chi(a, \bar{b})\}$, then for all $\bar{c} \in N$ we have $N \models \varphi(\bar{c}) \iff M \models \varphi^\chi(\bar{b}, \bar{c})$.

Interpretations These are a very useful model-theoretic tool. We shall use simple interpretations from time to time. For signatures L, L', an L-structure N is (one-dimensionally) *interpretable* in an L'-structure M if there is an L'-formula $\delta(x)$ (defining the domain of N within M) such that $M \models \exists x \delta(x)$, and for each atomic L-formula $\alpha(\bar{x})$ there is an associated L'-formula $\alpha'(\bar{x})$, such that the L-structure Δ with domain δ^M got by letting $\Delta \models \alpha(\bar{a})$ iff $M \models \alpha'(\bar{a})$, for $\bar{a} \in \Delta$, is well-defined and isomorphic to N. We may extend $'$ to arbitrary L-formulas by induction: $(\neg \varphi)' = \neg(\varphi')$, $(\varphi \wedge \psi)' = \varphi' \wedge \psi'$, and $(\exists x \varphi)' = \exists x (\delta(x) \wedge \varphi')$. So for $\bar{a} \in \Delta$, we have $\Delta \models \varphi(\bar{a})$ iff $M \models \varphi'(\bar{a})$. In particular, for any L-sentence σ there is an L'-sentence σ' such that $N \models \sigma \iff M \models \sigma'$.

For more details, and more general kinds of interpretation, see [Hod93].

Second-order logic We will occasionally come across simple second-order formulas. An *existential* (respectively, *universal*) *second-order L-sentence* is a sentence of the form $\exists P_1, \ldots, P_n \psi$ (respectively, $\forall P_1, \ldots, P_n \psi$), where $P_1, \ldots, P_n$ are relation symbols not in L, and ψ is a first-order sentence of the signature $L^+ = L \cup \{P_1, \ldots, P_n\}$. For an L-structure M, $M \models \exists P_1, \ldots, P_n \psi$ iff there exists an expansion of M to an L^+-structure in which ψ is true; $M \models \forall P_1, \ldots, P_n \psi$ iff ψ is true in every L^+-structure expanding M. Full second-order logic allows arbitrary quantification over relations; for this, and higher-order logic, see [Hod93, section 2.8].

2.3 Boolean algebras

In 1847 George Boole defined a calculus of propositions which later developed into the subject we now call boolean algebra, a subject which found successful application throughout mathematics and computing [Boo51]. Relation algebras are motivationally and technically based on boolean algebras, and in this section we outline the parts of the theory of boolean algebras that will be needed later. For general reference, see, e.g., [Sik64].

2.3.1 Definition and examples

Perhaps the best way of thinking of boolean algebra is to start with the notion of a *field of sets*.

DEFINITION 2.1 We let L_{BA} be the functional signature with constants 0 and 1, a binary function symbol $+$, and a unary function symbol $-$.

DEFINITION 2.2 Let X be any set (the 'base set'). A *field of sets* over the base X is an L_{BA}-algebra $\mathcal{F} = \langle F, \emptyset, X, \cup, \setminus \rangle$, where F is a non-empty set of subsets of X (so $\emptyset \subset F \subseteq \wp(X)$) such that if $S, T \in F$ then $(S \cup T) \in F$ and $(X \setminus S) \in F$. 0 and 1 are interpreted in $\mathcal{F}$ as $\emptyset$ and X, $+$ is interpreted as $\cup$, and $-$ as the unary function $S \mapsto X \setminus S$.

2.3. Boolean algebras

Thus a field of sets is closed under finite unions and under complementation relative to X. It follows that $\mathcal{F}$ contains $\emptyset, X$, so it is an L_{BA}-algebra. Note that it is also closed under finite intersections. (The word 'field' is used to indicate this closedness.)

Now, a field of sets is a rather important example of a boolean algebra. The idea with boolean algebras is that we treat the set-theoretic operations of union and relative complement as abstract operations $+, -$. That is to say, a boolean algebra does not have to consist of sets with the operations of union and complement: the elements of the domain of a boolean algebra are arbitrary, and the operations $-$ and $+$ become formal operations that satisfy certain axioms. So we are dealing with models of a certain L_{BA}-theory. In such models $\mathcal{B}$, where no confusion is likely, we will write the interpretations $0^{\mathcal{B}}, 1^{\mathcal{B}}, +^{\mathcal{B}}, -^{\mathcal{B}}$ of the boolean symbols simply as $0, 1, +, -$.

DEFINITION 2.3 Let $\mathcal{B} = \langle B, 0, 1, +, - \rangle$ be an L_{BA}-structure.

1. $\mathcal{B}$ is a *boolean algebra* if it satisfies the following equations, for all $a, b, c \in B$:

    - $+$ is associative, commutative and idempotent
    $$
    \begin{aligned}
    (a+b)+c &= a+(b+c) \\
    a+b &= b+a \\
    a+a &= a
    \end{aligned}
    $$

    - complement
    $$
    \begin{aligned}
    -(-b) &= b \\
    b+(-b) &= 1 \\
    -1 &= 0
    \end{aligned}
    $$

    - connections of $\cdot$ and $+$
    $$a \cdot (b+c) = a \cdot b + a \cdot c,$$
    where $a \cdot b$ abbreviates $-(-a + -b)$

    - zero
    $$0 + a = a$$

2. We will use some other standard abbreviations. We let $a \leq b$ abbreviate the equation $a + b = b$. We let $a < b$ abbreviate the formula $a \leq b \wedge a \neq b$.

3. We sometimes use $-$ as a binary operator, letting $a - b = a \cdot (-b)$.

4. A boolean algebra is said to be *degenerate* if it has exactly one element. Up to isomorphism, there is a unique degenerate boolean algebra.

5. A *boolean homomorphism* is a homomorphism from a boolean algebra $\mathcal{A}$ to another, $\mathcal{B}$.

In a boolean algebra, $\leq$ defines a partial order (reflexive, transitive, and antisymmetric) on the domain, and $a \leq b$ iff $a \cdot b = a$ iff $a \cdot -b = 0$ (exercise 2 below). In a field of sets $\mathcal{F}$, we have $a \leq b$ iff $a + b = a \cup b = b$ iff $a \subseteq b$ (for all $a, b \in \mathcal{F}$), so $\leq$ is set inclusion. Because we are dealing with algebras, a boolean homomorphism is an isomorphism iff it is a bijection.

EXAMPLE 2.4 Let L be any logic including the propositional connectives $\vee, \neg$ (e.g., propositional logic, predicate logic, modal logic, etc.), with a standard notion of proof $\vdash$. Let form(L) be the set of all L-formulas. The following are boolean algebras:

- Let Γ be any consistent set of L-formulas. Γ defines an equivalence relation $\sim$ on form(L) where $\varphi \sim \psi$ if and only if $\Gamma \vdash (\varphi \leftrightarrow \psi)$. For $\varphi, \psi \in$ form(L) let $\varphi/\sim$ denote the $\sim$-equivalence class of φ, and define $\varphi/\sim + \psi/\sim = (\varphi \vee \psi)/\sim$ and $-(\varphi/\sim) = \neg\varphi/\sim$. For typical $\vdash$, these operations are well-defined and $(\text{form}(L)/\sim, \bot/\sim, \top/\sim, +, -)$ is a boolean algebra.

- Let K be a class of L-structures. K defines a semantic equivalence relation $\approx$ where $\varphi \approx \psi$ if and only if $\mathsf{K} \models (\varphi \leftrightarrow \psi)$. Then
$$(\text{form}(L)/\approx, \bot/\approx, \top/\approx, +/\approx, -/\approx),$$
defined in the obvious way, is a boolean algebra.

2.3.2 Atoms

Many properties of boolean algebras and related algebras reduce to properties of their atoms. The notion of an atom will therefore be very important for us.

DEFINITION 2.5 Let $\mathcal{B}$ be a boolean algebra.

1. An *atom* of $\mathcal{B}$ is a minimal non-zero element $b \in \mathcal{B}$: i.e., for all $b' \in \mathcal{B}$ we have $b' < b \iff b' = 0$. We write $\text{At}(\mathcal{B})$, or $\text{At}\,\mathcal{B}$, for the set of all atoms of $\mathcal{B}$.

2. $\mathcal{B}$ is said to be *atomic* if for every non-zero $b \in \mathcal{B}$ there is an atom a of $\mathcal{B}$ with $a \leq b$.

2.3. Boolean algebras

3. $\mathcal{B}$ is said to be *atomless* if it is non-degenerate and has no atoms.

A field of sets of the form $\mathcal{F} = \langle \wp(X), \emptyset, X, \cup, \setminus \rangle$ is atomic: the atoms are precisely the singletons $\{x\} \in \mathcal{F}$, for all $x \in X$. Any finite boolean algebra is atomic (a degenerate one vacuously so). There exist infinite atomic boolean algebras and infinite atomless boolean algebras.

2.3.3 Dense sets

The atoms of an atomic boolean algebra form an example of a dense set.

DEFINITION 2.6 Let $\mathcal{B}$ be a boolean algebra.

1. A *dense subset* D of $\mathcal{B}$ is one such that if $b \in \mathcal{B} \setminus \{0\}$ then there exists $d \in D \setminus \{0\}$ with $d \leq b$.

2. A subalgebra $C \subseteq \mathcal{B}$ is said to be a *dense subalgebra* of $\mathcal{B}$ if $\text{dom}(C)$ is dense in $\mathcal{B}$.

2.3.4 Ideals, filters, ultrafilters

DEFINITION 2.7 Let $\mathcal{B}$ be a boolean algebra.

1. An *ideal* over $\mathcal{B}$ is a non-empty subset I of the domain of $\mathcal{B}$ (written $I \subseteq \mathcal{B}$) such that

    (a) I is 'closed downwards': $s \in I$ and $t \leq s$ imply $t \in I$,

    (b) S is closed under $+$: if $s, t \in I$ then $s + t \in I$.

2. Dually, we define a *filter* F to be any non-empty subset of the domain of $\mathcal{B}$ such that F is 'closed upwards' (if $t \geq s \in F$ then $t \in F$) and closed under $\cdot$ (if $s, t \in F$ then $s \cdot t \in F$).

3. Clearly, $\text{dom}(\mathcal{B})$ forms a filter on $\mathcal{B}$. Any other filter is said to be *proper*. On the other side, $\{1\}$ is a (trivial) filter of $\mathcal{B}$; any other filter is said to be *non-trivial*. Similar definitions are made for ideals.

4. For any $b \in \mathcal{B}$, let $I(b) = \{c \in \mathcal{B} : b \geq c\}$ and $F(b) = \{c \in \mathcal{B} : b \leq c\}$. $I(b)$ and $F(b)$ are called, respectively, the *principal* ideal and filter *generated by* b. Any filter or ideal not of this form is said to be *non-principal*.

5. A *maximal ideal* I is a proper ideal that is not strictly contained in any other proper ideal.

6. An *ultrafilter* F is a proper filter not strictly contained in any other proper filter.

42   Chapter 2. Preliminaries

   7. We write Uf($\mathcal{B}$) for the set of all ultrafilters of $\mathcal{B}$.

A filter is proper iff it does not contain 0. An equivalent definition of an ultrafilter is a filter F such that for each $b \in \mathcal{B}$, exactly one of $b, -b$ belongs to F (see exercise 11). Exercise 15 shows that any ultrafilter of a boolean algebra is principal iff it contains an atom; necessarily, it will be generated by this atom. An ultrafilter is in a way an idealised atom.

   There is a duality between filters and ideals. Mathematicians can be divided into two camps according to whether they prefer to work with ideals or filters. In the following we mostly use filters, and those who don't like this will have to translate.

FACT 2.8 (Boolean prime ideal theorem, 'BPI') *Let $\mathcal{B}$ be a boolean algebra, and let $S \subseteq \mathcal{B}$ be a subset such that $s_0 \cdot s_1 \cdot \ldots \cdot s_{n-1} \neq 0$ for any $n < \omega$ and any $s_0, \ldots, s_{n-1} \in S$. (We say that S has the finite intersection property — for example, any proper filter has this property.) Then there exists an ultrafilter of $\mathcal{B}$ containing S.*

For a proof, see [ChaKei90, 4.1.3] or [Hod93, 6.2.1]. A degenerate boolean algebra has no ultrafilters, but it has no subset with the finite intersection property, so BPI holds vacuously in this case.

   Note on set theory: the boolean prime ideal theorem cannot be proved without some form of the axiom of choice, such as Zorn's lemma, but BPI is actually weaker than the full axiom of choice. Modulo ZF, BPI is equivalent to the compactness theorem for first-order logic, and to the statement that any set can be linearly ordered (the axiom of choice is equivalent to the statement that any set can be well-ordered). See [Hod93].

2.3.5 Representations of boolean algebras

It is quite straightforward (see exercise 1) to check that a field of sets obeys the axioms for boolean algebra. Of course, no theory will be able to distinguish between isomorphic algebras, so we should not hope that only fields of sets form boolean algebras. But, rather surprisingly, the finitely many axioms for boolean algebra achieve the next best possible result: they determine exactly when an algebra is isomorphic to a field of sets — when it is representable.

DEFINITION 2.9 A boolean algebra $\mathcal{B}$ is said to be *representable* if it is isomorphic to a field of sets. An isomorphism from a boolean algebra $\mathcal{B}$ to a field of sets $\mathcal{F}$ is called a *representation* of $\mathcal{B}$; its *base* is defined to be the base of $\mathcal{F}$.

THEOREM 2.10 (Stone, [Sto36]) *Every boolean algebra is isomorphic to a field of sets*

2.3. Boolean algebras

Proof. Let $\mathcal{B}$ be any boolean algebra. The base of our field of sets is the set $\mathrm{Uf}(\mathcal{B})$ of ultrafilters of $\mathcal{B}$. The embedding $h : \mathcal{B} \to \mathcal{F} = \langle \wp(\mathrm{Uf}(\mathcal{B})), \emptyset, \mathrm{Uf}(\mathcal{B}), \cup, \setminus \rangle$ is defined by

$$h(b) = \{\gamma \in \mathrm{Uf}(\mathcal{B}) : b \in \gamma\}.$$

Checking that this really is an embedding — here, a one-one homomorphism — is a routine exercise. Here, we only check that $+$ is preserved.

$$\begin{aligned} h(a+b) &= \{\gamma \in \mathrm{Uf}(\mathcal{B}) : a+b \in \gamma\} \\ &= \{\gamma : a \in \gamma\} \cup \{\gamma : b \in \gamma\} \\ &= h(a) \cup h(b). \end{aligned}$$

The second line follows from the first because $(a+b) \in \gamma$ if and only if $a \in \gamma$ or $b \in \gamma$ (cf. exercise 12). To see that this map h is one-one, note that we can use BPI (fact 2.8) to show that for any non-zero $a \in \mathcal{B}$ there is an ultrafilter γ with $a \in \gamma$. Therefore, if $a \neq 0$ then $h(a) \neq \emptyset$. Hence, if $h(a) = h(b)$ then $h((a-b)+(b-a)) = (h(a) - h(b)) + (h(b) - h(a)) = \emptyset$, so $(a-b)+(b-a) = 0$, whence $a = b$. (Cf. exercise 3.)

Thus, $\mathcal{B}$ is isomorphic to the subalgebra of $\mathcal{F}$ with domain $\mathrm{rng}(h)$, and any subalgebra of a field of sets is also a field of sets. $\square$

If $\mathcal{B}$ is atomic, there is an easier construction of a representation using the atoms of $\mathcal{B}$ (see corollary 2.22). For the general case, we cannot use the set of atoms as the base of a representation — after all, there may not be any atoms.

Stone's theorem can also be thought of as a completeness theorem — any first-order formula in the language of boolean algebras that is valid over all fields of sets can be deduced from the axioms of boolean algebra, using the axioms and inference rules of first-order logic.

2.3.6 Canonical extensions

Looking back at the proof of theorem 2.10, observe that the construction suggests a natural topology on the space of all ultrafilters from $\mathcal{B}$. This allows us to pick out the range of the representation h constructed in Stone's theorem.

DEFINITION 2.11 Let $\mathcal{B}$ be a boolean algebra and let $X = \mathrm{Uf}(\mathcal{B})$. For each element $b \in \mathcal{B}$, let $h(b) = \{\gamma \in X : b \in \gamma\}$. The set $C = \{h(b) : b \in \mathcal{B}\}$ forms a basis of clopen sets for the *Stone space topology* $\mathcal{T}$ over X. $\mathcal{T}$ is obtained from C by closing under arbitrary unions.

THEOREM 2.12 *For any boolean algebra $\mathcal{B}$, the Stone space topology is compact and totally disconnected.*

Proof. For compactness, let Ω be an open cover of $X = \text{Uf}(\mathcal{B})$. Since the open sets are defined to be arbitrary unions of the basis of clopen sets, we can assume that each set in Ω is of the form $h(s)$ for some $s \in \mathcal{B}$. Thus $X = \bigcup \{h(s) : s \in S\}$ for some set $S \subseteq \mathcal{B}$.

Now let

$$\begin{aligned} U(S) &= \{b \in \mathcal{B} : b \geq (-s_0) \cdot (-s_1) \cdot \ldots \cdot (-s_{k-1}) \\ &\qquad \text{for some } s_i \in S,\ i < k < \omega\} \\ &= \{b \in \mathcal{B} : b + s_0 + \ldots + s_{k-1} = 1,\ \text{some } s_i\}. \end{aligned}$$

Then $U(S)$ is a filter containing $\{-s : s \in S\}$. If $U(S)$ is not a proper filter then $0 \in U(S)$ so there are finitely many elements $s_0, \ldots s_{k-1} \in S$ with $\sum_{i<k} s_i = 1$, hence $\{h(s_i) : i < k\}$ is a finite open cover of X, as required.

Otherwise $U(S)$ is a proper filter. By BPI (fact 2.8), $U(S)$ extends to an ultrafilter γ. But for any $s \in S$, we know $-s \in U(S) \subseteq \gamma$, so $s \notin \gamma$ and $\gamma \notin h(s)$. But this means that $\{h(s) : s \in S\}$ is not an open cover of X, contrary to assumption. This is a contradiction and proves compactness.

To show that the topology is totally disconnected we have to show, for any two distinct ultrafilters γ_1, γ_2 of $\mathcal{B}$, that there is a clopen set that contains one but not the other. This is easy: if $\gamma_1 \neq \gamma_2$ then without loss of generality there exists $b \in \gamma_1 \setminus \gamma_2$. Thus, $\gamma_1 \in h(b)$ and $\gamma_2 \notin h(b)$. $\square$

Observe that the range of $\mathcal{B}$ under the representation given in the proof of theorem 2.10 consists of the clopen sets in the Stone space topology. These clopen sets form a field of sets isomorphic to $\mathcal{B}$, but they form a subalgebra of another field of sets — the power set over X.

DEFINITION 2.13 Let $\mathcal{B}$ be a boolean algebra. $\langle \wp(\text{Uf}(\mathcal{B})), \emptyset, \text{Uf}(\mathcal{B}), \cup, \cap, \setminus \rangle$ is called the *canonical extension* (or sometimes the *canonical embedding algebra*) of $\mathcal{B}$, denoted $\mathcal{B}^+$.

$\mathcal{B}^+$ is complete (arbitrary sums and products exist; see definition 2.23 below) and atomic. Up to isomorphism it is an extension of $\mathcal{B}$, since $\mathcal{B}$ embeds in $\mathcal{B}^+$ via h. We'll see shortly that any complete, atomic extension of $\mathcal{B}$ whose set of atoms is compact and totally disconnected under the natural topology must be isomorphic to $\mathcal{B}^+$ (see theorem 2.73).

These canonical extensions become more significant when we consider generalisations of boolean algebras, such as the BAOs to be seen later, and the relation algebras of the title.

2.3.7 Infinite sums and products

Although in a boolean algebra, $+$ and $\cdot$ are finitary operations, we can sometimes define infinite sums and products.

2.3. Boolean algebras

DEFINITION 2.14 Let $\mathcal{B}$ be a boolean algebra and let S be some subset of its elements. The *supremum* or *join* or *sum* of S, written $\sum S$ or $\sum^{\mathcal{B}} S$ if necessary, is the unique, least upper bound of S in $\mathcal{B}$ if the least upper bound exists, and is undefined otherwise (you have to prove uniqueness in exercise 30). Formally, $\sum S$, if defined, is the unique element of $\mathcal{B}$ satisfying:

1. if $s \in S$, then $s \leq \sum S$,

2. for any $b \in \mathcal{B}$, if $b \geq s$ for all $s \in S$ then $b \geq \sum S$.

Similarly, we let $\prod S$ or $\prod^{\mathcal{B}} S$, the *infimum* or *meet* or *product* of S in $\mathcal{B}$, be the greatest lower bound in $\mathcal{B}$ of S if it exists, and let it be undefined otherwise.

If S is finite, say $S = \{s_0, \ldots, s_{n-1}\}$, then $\sum S$ always exists and is $s_0 + \cdots + s_{n-1}$ (see exercise 28). For infinite $S \subseteq \mathcal{B}$, $\sum S$ need not exist; even if $\mathcal{B}$ is a field of sets over some set X, so that $\bigcup S$ exists, is a subset of X, and seems a good candidate for being $\sum S$, we may have $\bigcup S \notin \mathcal{B}$. Similar facts hold for products. We will pursue this in the next section.

Note that it may happen that for boolean algebras $\mathcal{B} \subseteq C$, there is (necessarily infinite) $S \subseteq \mathcal{B}$ such that $\sum S$ taken in $\mathcal{B}$ and in C are not equal, or even that one exists and the other does not. (But if both exist, we must have $\sum^{C} S \leq \sum^{\mathcal{B}} S$.) We now consider the case where this doesn't happen.

DEFINITION 2.15 Let $\mathcal{A}$ and $\mathcal{B}$ be boolean algebras. We say that a map $\iota: \mathcal{A} \to \mathcal{B}$ is a *complete embedding* of $\mathcal{A}$ into $\mathcal{B}$ if ι is a boolean algebra embedding and for any $a \in \mathcal{A}$ and any subset S of the domain of $\mathcal{A}$, if $a = \sum^{\mathcal{A}} S$ then $\iota(a) = \sum^{\mathcal{B}} \iota[S]$ (recall that $\iota[S] = \{\iota(s) : s \in S\}$).

If $\mathcal{A} \subseteq \mathcal{B}$, we will say that $\mathcal{A}$ *completely embeds into* $\mathcal{B}$, and write $\mathcal{A} \subseteq^c \mathcal{B}$, if the inclusion map $\iota: \mathcal{A} \to \mathcal{B}$ is a complete embedding. (It would be nicer but, given definition 2.23 below, could be confusing to say that $\mathcal{A}$ is a complete subalgebra of $\mathcal{B}$.)

By exercise 33, $\iota: \mathcal{A} \to \mathcal{B}$ is a complete embedding iff for any $a \in \mathcal{A}$ and $S \subseteq \mathcal{A}$, if $a = \prod^{\mathcal{A}} S$ then $\iota(a) = \prod^{\mathcal{B}} \iota[S]$.

LEMMA 2.16 *If $\mathcal{A} \subseteq^c \mathcal{B}$ and $\mathcal{B}$ is atomic then so is $\mathcal{A}$.*

Proof. Let $a \in \mathcal{A} \setminus \{0\}$, and take an atom $b \in \mathcal{B}$ with $b \leq a$. Then $f = \{a' \in \mathcal{A} : a' \geq b\}$ is an ultrafilter of $\mathcal{A}$. If $\prod^{\mathcal{A}} f = 0$ then since $\mathcal{A} \subseteq^c \mathcal{B}$, we would have $\prod^{\mathcal{B}} f = 0$, contradicting $0 < b \leq a'$ for all $a' \in f$. Hence there is non-zero $a_0 \in \mathcal{A}$ with $a_0 \leq a'$ for all $a' \in f$. Since $a_0 > 0$, $a_0 \not\leq -a_0$, so $-a_0 \notin f$. Hence, $a_0 \in f$. This means that f is principal, and a_0 is an atom of $\mathcal{A}$ beneath a. □

LEMMA 2.17 *Let $\mathcal{A} \subseteq \mathcal{B}$ and assume that $\mathcal{A}$ is atomic. Then $\mathcal{A} \subseteq^c \mathcal{B}$ if and only if $(*)$ for every non-zero $b \in \mathcal{B}$ there is an atom $a \in \mathcal{A}$ such that $b \cdot a \neq 0$.*

Proof. Suppose $\mathcal{A} \subseteq^c \mathcal{B}$ and let $0 \neq b \in \mathcal{B}$. Since $\mathcal{A}$ is atomic, $\sum^{\mathcal{A}} \text{At}(\mathcal{A}) = 1$ so by definition of complete embedding, $\sum^{\mathcal{B}} \text{At}(\mathcal{A}) = 1$. Therefore $b \cdot a \neq 0$ for some atom $a \in \text{At}(\mathcal{A})$, as required (if not, $-b$ would be an upper bound of $\text{At}(\mathcal{A})$ in $\mathcal{B}$).

Conversely, assume (∗) and suppose that $S \subseteq \mathcal{A}$ and $\sum^{\mathcal{A}} S = \sigma \in \mathcal{A}$. We show σ is the least upper bound for S in $\mathcal{B}$. Assume for contradiction that there is $b \in \mathcal{B}$ with $s \leq b < \sigma$ for all $s \in S$. Then $\sigma - b \neq 0$, so by (∗) there is an atom a of $\mathcal{A}$ with $a \cdot (\sigma - b) \neq 0$. If $a \leq s$ for some $s \in S$, then $a \leq b$, impossible. So $a \cdot s = 0$ for all $s \in S$. Hence, $a \cdot \sigma = 0$, a contradiction. □

One important case where $\mathcal{A} \subseteq^c \mathcal{B}$ is when $\mathcal{A}$ is a dense subalgebra of $\mathcal{B}$: see exercise 37 below. The other exercises give more information on arbitrary sums.

2.3.8 Complete representations

We have defined a representation of a boolean algebra $\mathcal{B}$ to be an isomorphism from $\mathcal{B}$ to a field of sets. Such a representation preserves all the boolean operators. But set theory allows us to define other infinitary operators, $\bigcap S$ and $\bigcup S$, for any set of unary relations S. On the algebra side, we just saw that we can sometimes define infima and suprema which we hope will correspond correctly to these intersections and unions. When this happens, the representation is called a *complete representation*.

DEFINITION 2.18 A *complete representation* $h : \mathcal{B} \to \mathcal{F}$, where $\mathcal{F}$ is a field of sets, is a representation such that whenever $S \subseteq \mathcal{B}$ and $\sum S$ exists then

$$h(\sum S) = \bigcup \{h(s) : s \in S\}.$$

A boolean algebra is said to be *completely representable* if it has a complete representation.

Note that whether h above is a complete representation depends only on the field of sets $\text{rng}(h) \subseteq \mathcal{F}$.

LEMMA 2.19 *A representation h is complete if and only if all infima that exist are preserved — that is, if $\prod S$ exists then $h(\prod S) = \bigcap \{h(s) : s \in S\}$.*

The proof is left as an exercise (exercise 43).

For boolean algebras, it turns out to be very easy to identify the completely representable ones: a boolean algebra has a complete representation if and only if it is atomic. To show this, first we introduce the notion of an *atomic representation,* and then show that this is equivalent to being a complete representation.

2.3. Boolean algebras

Let $h: \mathcal{B} \to \mathcal{P}$ be any representation of the boolean algebra $\mathcal{B}$ to the field of sets $\mathcal{F}$ over the base X (i.e., each element of $\mathcal{F}$ is a subset of X). If $S \subseteq \mathcal{B}$, we write $h[S]$ for the set $\{h(s) : s \in S\}$.

Observe that for each $x \in X$, the set
$$h^{-1}[x] = \{b \in \mathcal{B} : x \in h(b)\}$$
is an ultrafilter of $\mathcal{B}$. In certain cases, $h^{-1}[x]$ will be a principal ultrafilter, generated by an atom of $\mathcal{B}$. If this is so for all $x \in X$, we will say that the representation is atomic.

DEFINITION 2.20 Let $h : \mathcal{B} \to \mathcal{F}$ be a representation of the boolean algebra $\mathcal{B}$ to the field of sets $\mathcal{F}$ over the base set X. If, for all $x \in X$, there is an atom $a \in \mathrm{At}(\mathcal{B})$ such that $x \in h(a)$, then h is called an *atomic representation* of $\mathcal{B}$.

THEOREM 2.21 *Let $\mathcal{B}$ be a boolean algebra. A representation h of $\mathcal{B}$ is atomic if and only if it is complete.*[1]

Proof. Let $h : \mathcal{B} \to \mathcal{F}$ be a representation of $\mathcal{B}$ over the base X. First, suppose that h is an atomic representation. Then for any $b \in \mathcal{B}$, we have
$$h(b) = \bigcup \{h(a) : \text{atoms } a \leq b\}.$$
To show that h is a complete representation, let S be a subset of $\mathcal{B}$, and suppose that the supremum $\sum S$ exists in $\mathcal{B}$. Let $S\!\downarrow$ be the set of atoms a of $\mathcal{B}$ with $a \leq \sum S$. Note that for an atom a, $a \in S\!\downarrow$ iff $a \leq s$ for some $s \in S$ (cf. exercise 30). Then

$$\begin{aligned}
x \in h(\textstyle\sum S) &\Leftrightarrow x \in \bigcup h[S\!\downarrow] \\
&\Leftrightarrow x \in h(a) && \text{some } a \in S\!\downarrow \\
&\Leftrightarrow x \in h(a) && \text{some atom } a \text{ and } s \in S \text{ with } a \leq s \\
&\Leftrightarrow x \in h(s) && \text{some } s \in S \\
&\Leftrightarrow x \in \bigcup h[S].
\end{aligned}$$

Conversely, suppose that h is a complete representation; we show that h is atomic by an argument similar to that of lemma 2.16. Let $x \in X$; it is enough to show that the ultrafilter $h^{-1}[x] = \{b \in \mathcal{B} : x \in h(b)\}$ is principal, since it will then be generated by an atom (cf. exercise 15). Now 0 is a lower bound of $h^{-1}[x]$. If 0 is the greatest lower bound $(0 = \prod h^{-1}[x])$, then since h is complete,
$$\emptyset = h(0) = h(\textstyle\prod h^{-1}[x]) = \bigcap h[h^{-1}[x]].$$
But $x \in \bigcap h[h^{-1}[x]]$, which gives a contradiction. Therefore, there is a non-zero lower bound b of $h^{-1}[x]$. Since $b \not\leq -b$, we have $-b \notin h^{-1}[x]$. As $h^{-1}[x]$ is an ultrafilter, $b \in h^{-1}[x]$. It follows that $h^{-1}[x]$ is principal. Thus, h is an atomic representation. □

---

[1] A version of theorem 2.21, the note following it, and corollary 2.22 appeared in [HirHod97a, §2]; we thank the copyright owners Association for Symbolic Logic for granting permission for this republication. The text has been changed slightly.

Note If a boolean algebra $\mathcal{B}$ admits an atomic representation, then it is atomic. Moreover, a representation $h : \mathcal{B} \to \mathcal{P}$, over the base X, is atomic if and only if $\langle h(a) : a \in \text{At}(\mathcal{B}) \rangle$ is a partition of X. So every non-atomic representation $h : \mathcal{B} \to \mathcal{P}(X)$ of an atomic boolean algebra can be turned into an atomic representation by restricting X to the interpretations of atoms, as follows. Let $Y = \bigcup h[\text{At}(\mathcal{B})]$, and let $f : \wp(X) \to \wp(Y)$ be the map defined by $f(Z) = Z \cap Y$. The composition $f \circ h$ is an atomic representation of $\mathcal{B}$. By theorems 2.10 and 2.21, this gives us the following, which we will prove explicitly.

COROLLARY 2.22 *A boolean algebra $\mathcal{B}$ has a complete representation if and only if it is an atomic boolean algebra.*[2]

Proof. If $\mathcal{B}$ is atomic then let $X = \text{At}(\mathcal{B})$. The representation $h : \mathcal{B} \to \wp(X)$ defined by

$$h(b) = \{a \in \text{At}(\mathcal{B}) : a \leq b\}$$

is an atomic representation, and hence, by theorem 2.21, a complete representation.

Conversely, suppose that $\mathcal{B}$ has a complete representation, h. By theorem 2.21, h must be an atomic representation. For any non-zero $b \in \mathcal{B}$, pick any point $x \in h(b)$. So there is an atom a of $\mathcal{B}$ with $x \in h(a)$. Hence $a \cdot b \neq 0$, so, as a is an atom, $a \leq b$. It follows that $\mathcal{B}$ is atomic. □

Note that Stone's theorem 2.10 follows from this: any boolean algebra $\mathcal{B}$ embeds in its canonical extension $\mathcal{B}^+$, which is atomic and therefore has a (complete) representation; such a representation induces by restriction a representation of $\mathcal{B}$.

2.3.9 Completions of boolean algebras

The canonical extension of a boolean algebra (section 2.3.6) is useful in many ways. Its chief features are that it is complete and atomic and (up to isomorphism) extends the original boolean algebra. However, it does not preserve any infinitary meets and joins that exist in the original algebra, unless they are also finite joins — see exercise 26. For instance, any non-principal ultrafilter of the original algebra has zero greatest lower bound in it, but not in the extension. A different kind of extension of a boolean algebra, called the *completion,* avoids these snags. It is a complete boolean algebra extending the original one, and in which the original is dense. The completion will only be atomic if the original algebra is, but on the other hand, all existing meets and joins in the original algebra are preserved

---

[2] For complete boolean algebras, this was known to Lindenbaum and Tarski, who proved it using the fact that a complete, completely distributive boolean algebra is atomic ([Sik64, p 105]; cf. exercise 34c below). See [Tar35].

2.3. Boolean algebras

(the original algebra embeds completely into its completion). Every boolean algebra has such a completion, which is unique up to isomorphism over the original algebra.

DEFINITION 2.23 A boolean algebra $\mathcal{B}$ is said to be *complete* if $\sum S$ and $\prod S$ exist for every $S \subseteq \mathcal{B}$.

Example: a countably infinite atomic boolean algebra cannot be complete. On the other hand, any finite boolean algebra is complete, as is a field of sets of the form $\langle \wp(X), \emptyset, X, \cup, \backslash \rangle$.

Every boolean algebra is a dense subalgebra of a complete boolean algebra. The proof generalises both the construction of the canonical extension in Stone's theorem, and the simpler construction of a representation of an atomic boolean algebra in corollary 2.22: see parts 6 and 8 in the proof of theorem 2.24, below. We now sketch the idea; the details are left as exercise 41.

THEOREM 2.24 Let $\mathcal{B}$ be a boolean algebra. There exists a complete boolean algebra $C \supseteq \mathcal{B}$ such that $\mathcal{B}$ is dense in C. Consequently, $\mathcal{B} \subseteq^c C$.

Proof (sketch). By Stone's theorem 2.10, we can assume without loss of generality that $\mathcal{B}$ is a field of sets with base U, say, so $\mathcal{B} \subseteq \langle \wp(U), \emptyset, U, \cup, \backslash \rangle$. Let $\mathcal{T}$ be any topology on U such that each $b \in \mathcal{B}$ is an open set: i.e., $\mathcal{B} \subseteq \mathcal{T}$. For example:

1. the discrete topology on U, i.e., $\mathcal{T} = \wp(U)$,

2. the topology generated by the sets in $\mathcal{B}$ — as $\mathcal{B}$ is closed under $\cap$, we have $\mathcal{T} = \{\bigcup S : S \subseteq \mathcal{B}\}$.

For $X \subseteq U$, the interior $\operatorname{int}(X)$ of X is $\bigcup\{S \in \mathcal{T} : S \subseteq X\}$, and the closure $\operatorname{cl}(X)$ of X is $U \setminus \operatorname{int}(U \setminus X)$. Write X^* for $\operatorname{int}(\operatorname{cl}(X))$, the interior of the closure of X. A *regular open set* with respect to $\mathcal{T}$ is a subset $X \subseteq U$ such that $X = X^*$. Let RO be the set of regular open sets, and define operations $+, -, \cdot$ on RO by

$$\begin{aligned} X + Y &= (X \cup Y)^*, \\ -X &= U \setminus \operatorname{cl}(X), \\ X \cdot Y &= -((-X) + (-Y)). \end{aligned}$$

Then it can be checked that:

1. $\emptyset, U$ are regular open sets.

2. $-X = (U \setminus X)^*$, for $X \in RO$.

3. $X \cdot Y = X \cap Y = (X \cap Y)^*$, for $X, Y \in RO$.

4. $C \stackrel{\text{def}}{=} (RO, \emptyset, U, +, -)$ is a complete boolean algebra.

5. $\mathcal{B}$ is a subalgebra of C.

6. In the case where $U = \text{Uf}(\mathcal{B})$ and $\mathcal{T}$ is the discrete topology, we have $X^* = X$, $X + Y = X \cup Y$, and $-X = U \setminus X$ for all $X, Y \subseteq U$, and $C = \mathcal{B}^+$.

7. In the case where $\mathcal{T}$ is the topology generated by $\mathcal{B}$, $\mathcal{B}$ is a dense subalgebra of C.

8. In the case where $\mathcal{B}$ is atomic, let $U = \text{At}(\mathcal{B})$ and regard $\mathcal{B}$ as a field of sets with base U, by identifying each $b \in \mathcal{B}$ with $\{a \in \text{At}\,\mathcal{B} : a \leq b\}$. Cf. corollary 2.22. Then $\mathcal{B}$ generates the discrete topology on U, so $C = (\wp(\text{At}\,\mathcal{B}), \emptyset, \text{At}\,\mathcal{B}, \cup, \setminus)$.

The theorem follows from nos. 4 and 7; the last part is immediate from exercise 37. For more details, see [Sik64]. □

This theorem justifies the following definition.

DEFINITION 2.25 The *completion* of a boolean algebra $\mathcal{B}$ is defined to be the complete boolean algebra $C \supseteq \mathcal{B}$ constructed in theorem 2.24.

So the completion of $\mathcal{B}$ is a complete boolean algebra extending $\mathcal{B}$ and in which $\mathcal{B}$ is dense. In fact, up to isomorphism it is characterised by these properties, as we now show.

LEMMA 2.26 *If $C, C' \supseteq \mathcal{B}$ are complete boolean algebras in which $\mathcal{B}$ is dense, then C is isomorphic to C' over $\mathcal{B}$ (i.e., there is an isomorphism from C to C' fixing every element of $\mathcal{B}$ — see section 2.2.4).*

Proof. The isomorphism is $c \mapsto \sum^{C'} \{b \in \mathcal{B} : b \leq c\}$. Checking that this works is exercise 41. □

We also note that by theorem 2.24(8), if $\mathcal{B}$ is atomic then its completion is isomorphic to $(\wp(\text{At}\,\mathcal{B}), \emptyset, \text{At}\,\mathcal{B}, \cup, \setminus)$.

Summary To summarise the key results on boolean algebras: they are defined by a finite number of axioms; the theory of their representations is straightforward — every boolean algebra is representable; every boolean algebra has a completion and a canonical extension; and a boolean algebra has a complete representation iff it is an atomic boolean algebra.

2.3. Boolean algebras

Exercises

1. Show that any field of sets is a boolean algebra. Show that if $\mathcal{P}(X) = \langle \wp(X), \emptyset, X, \cup, \setminus \rangle$ for some set X, then $\mathcal{P}(X)$ is a complete atomic boolean algebra. If F is the set of subsets $X \subseteq \omega$ such that X is finite or $\omega \setminus X$ is finite, show that $\mathcal{F} = \langle F, \emptyset, X, \cup, \setminus \rangle$ is an atomic boolean algebra and $\mathcal{F} \subset \mathcal{P}(\omega)$.

2. Prove from the axioms for a boolean algebra (definition 2.3) that the binary relation $\leq$ is a reflexive ($a \leq a$), transitive ($a \leq b \wedge b \leq c \Rightarrow a \leq c$), antisymmetric ($a \leq b \wedge b \leq a \Rightarrow a = b$) lattice ordering ($\forall a, b \, \exists c, a \leq c \wedge b \leq c$). Also show that $a \leq b$ iff $a \cdot b = a$, iff $a \cdot -b = 0$.

3. Show that for elements a, b of a boolean algebra, $(a - b) + (b - a) = 0$ iff $a = b$. Deduce that for boolean algebras $\mathcal{B}, \mathcal{C}$, a homomorphism $h : \mathcal{B} \to \mathcal{C}$ is one-one iff $\{b \in \mathcal{B} : h(b) = 0\} = \{0\}$.

4. Let e be an equation valid in boolean algebras and let e^* be obtained from it by (i) replacing all 0s by 1s and all 1s by 0s and (ii) replacing all occurrences of $+$ by $\cdot$, and vice versa. Prove that e^* is also valid over boolean algebras. Conclude that $+$ is distributive over $\cdot$ in boolean algebras.

Atoms

5. Show that the class of atomic boolean algebras is elementary but not universally axiomatisable.

6. Show that any finite boolean algebra is atomic. Find or construct an example of an atomless boolean algebra.

7. Show that there is, up to isomorphism, at most one boolean algebra of any given finite cardinality κ. What are the possible κ? Is this true for infinite κ?

8. Show that, up to isomorphism, there is exactly one countable atomless boolean algebra.

9. Check that theorem 2.21 and corollary 2.22 hold for degenerate boolean algebras.

10. Show that if $\mathcal{B}$ is atomic then At $\mathcal{B}$ is dense in $\mathcal{B}$. Deduce that if $\mathcal{B} \subseteq \mathcal{C}$ and both algebras are atomic with the same atoms, then $\mathcal{B}$ is dense in $\mathcal{C}$. Show that if $\mathcal{B} \subseteq \mathcal{C}$, $\mathcal{B}$ is atomic, and every atom of $\mathcal{B}$ is an atom of $\mathcal{C}$, then $\mathcal{B}$ is dense in $\mathcal{C}$ iff $\sum^{\mathcal{C}}$ At $\mathcal{B} = 1$.

Filters

11. [ChaKei90, proposition 4.1.2] Let $\mathcal{B}$ be a boolean algebra and let $\gamma \subseteq \mathcal{B}$. Prove that the following are equivalent.

(a) γ is an ultrafilter,

(b) γ is a filter of $\mathcal{B}$, and for all $b \in \mathcal{B}$, exactly one of $b, -b$ belongs to γ.

Show that in 11b, 'filter' cannot be replaced by 'closed-upwards subset' (i.e., $b \in \gamma, b \leq c \Rightarrow c \in \gamma$).

12. Let γ be an ultrafilter of a boolean algebra $\mathcal{B}$, and let $b_1, \ldots, b_n \in \mathcal{B}$ (for some finite $n \geq 1$). Show that $b_1 + \cdots + b_n \in \gamma$ iff $b_i \in \gamma$ for some i, $1 \leq i \leq n$.

13. Let $h : \mathcal{B} \to C$ be a homomorphism of boolean algebras. Show that $\{b \in \mathcal{B} : h(b) = 1^C\}$ is a filter of $\mathcal{B}$. Under what circumstances is it an ultrafilter? Show also that for any (ultra)filter f of C, $h^{-1}[f]$ is an (ultra)filter of $\mathcal{B}$.

14. Let f be a filter on a boolean algebra $\mathcal{B}$. Define a binary relation $\sim$ on $\mathcal{B}$ by $b \sim c$ iff $(b \cdot c) + (-b \cdot -c) \in f$. Show that $\sim$ is an equivalence relation, and that there is a natural well-defined boolean algebra with domain $\mathcal{B}/\sim$ such that the map $b \mapsto b/\sim$ is a surjective homomorphism.

15. Let γ be an ultrafilter over the boolean algebra $\mathcal{B}$. Prove that γ is a principal ultrafilter iff there is an atom $b \in \mathrm{At}(\mathcal{B})$ with $b \in \gamma$, iff γ is generated by an atom.

16. Let γ be an ultrafilter of the boolean algebra $\mathcal{P}(\omega)$ as in exercise 1. Show that γ is principal iff it contains a finite subset of ω, and non-principal iff it contains no finite sets, iff it contains every cofinite subset of ω.

17. Show that every ultrafilter of a finite boolean algebra is principal. Use BPI to prove that any infinite boolean algebra has a non-principal ultrafilter. [Try $S = \{\text{complements of finite sums of atoms}\}$.]

18. Let $\mathcal{F}$ be as in exercise 1. Show that $\mathcal{F}$ has a unique non-principal ultrafilter (sometimes called the Fréchet filter).

19. Find an infinite boolean algebra $\mathcal{B}$ with $2^{|\mathcal{B}|}$ non-principal ultrafilters. [Hint: $^{<\omega}2$ or $(\mathbb{Q}, <)$.]

20. Show that any infinite boolean algebra has infinitely many ultrafilters, contains an infinite chain C ($\leq$ linearly orders C), and an infinite set S of pairwise disjoint elements ($x \cdot y = 0$ for distinct $x, y \in S$).

21. Use Zorn's lemma to prove that any non-zero element of a boolean algebra is contained in an ultrafilter. Extend the argument to prove the boolean prime ideal theorem.

2.3. Boolean algebras

Representations

22. Let $\mathcal{B} \subseteq C$ be boolean algebras and let h be a representation of C mapping it to a field of sets $\mathcal{F}$ with base X. Show that $\mathcal{B}$ also has a representation mapping it to a field of sets with base X.

23. Complete the proof that the map h in Stone's theorem (2.10) is really an isomorphism.

24. (Amalgamation for boolean algebras) Let $\mathcal{B}, C_1, C_2$ be boolean algebras and let $e_i : \mathcal{B} \to C_i$ ($i = 1, 2$) be embeddings. Show that there exist a boolean algebra $\mathcal{D}$ and embeddings $f_i : C_i \to \mathcal{D}$ ($i = 1, 2$) with $f_1 \circ e_1 = f_2 \circ e_2$. [Find representations of C_1, C_2 on the same base.]

Canonical extensions

25. Check that any boolean algebra embeds in its canonical extension.

26. Let $\mathcal{B}^+$ be the canonical extension of a boolean algebra $\mathcal{B}$, let $S \subseteq \mathcal{B}$, and suppose that $\sum^{\mathcal{B}} S$ exists in $\mathcal{B}$. Show that $\sum^{\mathcal{B}} S = \sum^{\mathcal{B}^+} S$ iff there exists finite $S_0 \subseteq S$ such that $\sum^{\mathcal{B}} S = \sum^{\mathcal{B}} S_0$. Do the same for $\prod$. Deduce that *no* 'strictly infinite' sums or products are preserved when passing to the canonical extension.

27. If $\mathcal{B} \subseteq C$ are boolean algebras, show that there is a natural complete embedding $\iota : \mathcal{B}^+ \to C^+$. [Cf. theorem 2.71.]

Infinite sums and products

28. If $\mathcal{B}$ is a boolean algebra and $S = \{s_0, \ldots, s_{n-1}\} \subseteq \mathcal{B}$ for some finite n, show that $\sum S$ exists and is $s_0 + \cdots + s_{n-1}$, and that $\prod S$ exists and is $s_0 \cdot \ldots \cdot s_{n-1}$.

29. Let $\mathcal{B}$ be an atomic boolean algebra, and $b \in \mathcal{B}$. Show that $b = \sum\{a \in \text{At}(\mathcal{B}) : a \le b\}$.

30. Prove that if S is a subset of a boolean algebra $\mathcal{B}$ and there is a least upper bound of S, then the least upper bound is unique. If b is an atom of $\mathcal{B}$, show that $b \le \sum S$ iff $b \le s$ for some $s \in S$.

31. Let $\mathcal{B}$ be a boolean algebra and let $S_i \subseteq \mathcal{B}$ with $\sum S_i = s_i \in \mathcal{B}$ for each $i \in I$. Show that $\sum\{s_i : i \in I\}$ exists iff $\sum \left(\bigcup_{i \in I} S_i\right)$ exists, and that if they do exist, they are equal. [Just show that any upper bound of $\{s_i : i \in I\}$ is an upper bound of $\bigcup_{i \in I} S_i$ and vice versa.]

32. Show that in a boolean algebra, if $\sum_{i \in I} b_i$ exists then $a \cdot \sum_{i \in I} b_i = \sum_{i \in I} (a \cdot b_i)$.

54   Chapter 2. Preliminaries

33. If $\mathcal{B}$ is a boolean algebra and $b_i \in \mathcal{B}$ for $i \in I$, show that in the following equations, if one side exists then so does the other, and they are equal:
$$-\sum_i b_i = \prod_i -b_i,$$
$$-\prod_i b_i = \sum_i -b_i.$$
When $|I| = 2$, these are called the De Morgan laws.

34. Let $\mathcal{B}$ be a boolean algebra and $b_{ij} \in \mathcal{B}$ for $i \in I, j \in J$.

    (a) If I, J are finite sets, show that
    $$\prod_i \sum_j b_{ij} = \sum_{\sigma: I \to J} \prod_i b_{i,\sigma(i)},$$
    $$\sum_i \prod_j b_{ij} = \prod_{\sigma: I \to J} \sum_i b_{i,\sigma(i)},$$
    where σ ranges over all maps from I to J.

    (b) State and prove a generalisation to arbitrary I, J when $\mathcal{B}$ is atomic.

    (c) Show that if $\mathcal{B}$ validates all equations of the form in part (a), in the sense that whenever one side exists then so does the other and they are equal, then $\mathcal{B}$ is atomic. [Show that $1 = \sum \{\prod f : f$ an ultrafilter of $\mathcal{B}\}$.]

    (d) Find $\mathcal{B}, I, J$, and $b_{ij} \in \mathcal{B}$ with $\prod_i \sum_j b_{ij} = 1$ and $\sum_{\sigma: I \to J} \prod_i b_{i,\sigma(i)} = 0$. [$S \subseteq \mathcal{B}$ partitions $b \in \mathcal{B}$ if $\sum S = b$ and $s \cdot s' = 0$ for distinct $s, s' \in S$. Find elements b_t ($t \in {}^{<\omega}2$) in the countable atomless boolean algebra such that $\{b_t : t \in {}^i 2\}$ for $i = 0, 1, 2, \ldots$ form increasingly fine partitions of 1.]

    (e) More generally, show that for every infinite cardinal κ, there exist a boolean algebra $\mathcal{B}$ and elements $a_i, b_i \in \mathcal{B}$ with $a_i + b_i = 1$ (for $i < \kappa$), such that $\prod_{i \in I} c_i = 0$ for all infinite $I \subseteq \kappa$ and $c_i \in \{a_i, b_i\}$ (all $i \in I$).

35. Find boolean algebras $\mathcal{B} \subseteq C$ and subsets $S \subseteq \mathcal{B}$ such that (i) $\sum^{\mathcal{B}} S$ exists and $\sum^C S$ does not, (ii) $\sum^C S$ exists and $\sum^{\mathcal{B}} S$ does not, (iii) $\sum^{\mathcal{B}} S, \sum^C S$ both exist but $\sum^C S < \sum^{\mathcal{B}} S$. Show that if $\sum^C S \in \mathcal{B}$ then $\sum^C S = \sum^{\mathcal{B}} S$.

36. If $\mathcal{B}, C$ are boolean algebras, $\mathcal{B} \subseteq C$, and $\mathcal{B}$ is finite, prove that $\mathcal{B} \subseteq^c C$.

37. Show that if $\mathcal{B}, C$ are boolean algebras and $\mathcal{B}$ is a dense subalgebra of C then $\mathcal{B} \subseteq^c C$. Show that the converse fails in general.

38. Let $\mathcal{B} \subseteq C$ be boolean algebras. Prove that the following are equivalent:

    (a) $\mathcal{B} \subseteq^c C$.

2.3. Boolean algebras

(b) For all subsets $S \subseteq \mathcal{B}$, if $\prod^{\mathcal{B}} S$ exists then $\prod^C S$ exists and they are equal.

(c) For all subsets $S \subseteq \mathcal{B}$, if $\sum^{\mathcal{B}} S = 1$ then $\sum^C S = 1$ (i.e. if the former exists and is 1 then the latter exists and is 1).

39. Let $\mathcal{B} \subseteq C$ be boolean algebras, and suppose that for all $c \in C$, either

    least upper bound: $\{b \in \mathcal{B} : b \geq c\}$ has a minimal element, or

    no greatest lower bound: $\{b \in \mathcal{B} : b \leq c\}$ has no maximal element.

    Show that $\mathcal{B} \subseteq^c C$. Find an example to show that the converse fails in general.

Complete boolean algebras and completions

40. Show that no countably infinite boolean algebra is complete. [Exercise 20 may help.]

41. Complete the proofs of lemma 2.26 and theorem 2.24. (Part was already done in exercise 37.)

42. Show that the completion of a boolean algebra $\mathcal{B}$ is atomic iff $\mathcal{B}$ is atomic.

Complete representations

43. Let h be a complete representation of $\mathcal{B}$. Show that h respects infima — i.e., if $\prod(S)$ exists then $h(\prod(S)) = \bigcap\{h(s) : s \in S\}$.

44. Let $\mathcal{B}$ be a finite boolean algebra. Show that any representation of $\mathcal{B}$ is a complete representation.

45. Let $\mathcal{B}$ be an infinite boolean algebra. Prove that $\mathcal{B}$ has a representation that fails to be complete.

46. Show that there exist non-complete boolean algebras with complete representations. [Exercise 40 may help.]

47. Is there a complete boolean algebra with no complete representation? [See exercise 42.]

48. Let $\mathcal{B}, C$ be boolean algebras with $\mathcal{B}$ a dense subalgebra of C. Without using corollary 2.22, show that given a complete representation of $\mathcal{B}$, we can obtain from it 'in a natural way' a complete representation of C, and vice versa. Deduce that $\mathcal{B}$ has a complete representation iff its completion does. (This remains true for other classes of algebra where a simple characterisation of the completely representable ones is lacking.)

2.4 Products and ultraproducts

Having studied boolean algebras, we now introduce two general model-theoretic constructions heavily used in universal algebra and much needed later in the book. The first one, products, does not use boolean algebras; the second, ultraproducts, does.

2.4.1 Products

Let L be a signature, let Λ be an index set, and let M_λ ($\lambda \in \Lambda$) be a set of L-structures. The L-structure

$$M = \prod_{\lambda \in \Lambda} M_\lambda,$$

the *product* of the M_λ, is defined as follows. Its domain is the Cartesian product of the domains of the M_λ — so if $a \in M$ then a is a function from Λ to the union of the domains of the M_λ such that $a(\lambda) \in M_\lambda$ for all $\lambda \in \Lambda$. If $a_\lambda \in M_\lambda$ for each $\lambda \in \Lambda$, we write the function mapping λ to a_λ as $\langle a_\lambda : \lambda \in \Lambda \rangle$. We define the interpretations of the L-symbols by:

- If c is a constant of L, then $c^M = \langle c^{M_\lambda} : \lambda \in \Lambda \rangle$.

- If $f \in L$ is an n-ary function symbol, then f^M is defined by:
  $f^M(a_1, \ldots, a_n) = \langle f^{M_\lambda}(a_1(\lambda), \ldots, a_n(\lambda)) : \lambda \in \Lambda \rangle$, for any $a_1, \ldots, a_n \in M$.

- If $R \in L$ is an n-ary relation symbol, then R^M is defined by:
  $M \models R(a_1, \ldots, a_n)$ iff $M_\lambda \models R(a_1(\lambda), \ldots, a_n(\lambda))$ for all $\lambda \in \Lambda$ and $a_1, \ldots, a_n \in M$.

For each $\lambda \in \Lambda$, we can define the λ*th projection* $p_\lambda : M \to M_\lambda$ to be the homomorphism given by

$$p_\lambda(a) = a(\lambda), \text{ for } a \in M.$$

If $\Lambda = \emptyset$ then, $\emptyset$ being the only function from $\emptyset$ to $\emptyset$, it can be checked that M is a one-element structure with domain $\{\emptyset\}$, with $M \models R(\emptyset, \ldots, \emptyset)$ for all relation symbols $R \in L$.

2.4.2 Ultraproducts, ultrapowers

DEFINITION 2.27 Let Λ be any non-empty index set and let D be any ultrafilter of the boolean algebra $\langle \wp(\Lambda), \emptyset, \Lambda, \cup, \setminus \rangle$. As a shorthand, we call D an ultrafilter *over* Λ. Intuitively, a subset of Λ is *large* iff it is in D.

2.4. Products and ultraproducts

Let L be a signature and let M_λ ($\lambda \in \Lambda$) be L-structures. To define the *ultraproduct* over D of the M_λ, consider the product $M = \prod_{\lambda \in \Lambda} M_\lambda$. Define an equivalence relation $\sim$ on $\text{dom}(M)$ by

$$a \sim b \iff \{\lambda \in \Lambda : a(\lambda) = b(\lambda)\} \in D.$$

That is, a and b agree on a large set of indices. The fact that D is an ultrafilter guarantees that this is an equivalence relation. Let U be the set of $\sim$-equivalence classes in M. This will be the domain of the ultraproduct over D. For $a \in M$, we write a/D for the $\sim$-class of a.

We can now define interpretations of the L-symbols over this domain, so obtaining an L-structure U, the *ultraproduct* of the M_λ over D. We write this ultraproduct as $\prod_{\lambda \in \Lambda} M_\lambda/D$, or for short, $\prod_D M_\lambda$. ($\Lambda = \bigcup D$ is recoverable from D.)

- If $c \in L$ is a constant, let $c^U = c^M/D$.

- If $f \in L$ is an n-ary function symbol, and $a_1, \ldots, a_n \in M$, we let
  $f^U(a_1/D, \ldots, a_n/D) = f^M(a_1, \ldots, a_n)/D$.

- If $R \in L$ is an n-ary relation symbol, and $a_1, \ldots, a_n \in M$, we let
  $U \models R(a_1/D, \ldots, a_n/D)$ iff $\{\lambda \in \Lambda : M_\lambda \models R(a_1(\lambda), \ldots, a_n(\lambda))\} \in D$.

The ultraproduct is said to be *principal* if D is a principal ultrafilter, and *non-principal* otherwise.

You have to prove that these operations are well-defined in exercise 4.

DEFINITION 2.28 If all the M_λ are isomorphic to a single structure, M say, then their ultraproduct over D is called the *ultrapower* of M over D, written $\prod_D M$ or M^Λ/D.

If M is any structure such that $\prod_D M$ is isomorphic to N (for some ultrafilter D), we call M an *ultraroot* of N.

The value of this construction comes from the next few theorems and corollaries.

THEOREM 2.29 (Łoś) *Let $\Lambda \neq \emptyset$, let M_λ ($\lambda \in \Lambda$) be L-structures, and let $U = \prod_D M_\lambda$ be an ultraproduct of the M_λ, for some ultrafilter D over Λ. Then for any first-order L-formula $\varphi(x_1, \ldots, x_n)$ and elements $a_1/D, \ldots, a_n/D \in U$, we have:*

$$U \models \varphi(a_1/D, \ldots, a_n/D) \iff \{\lambda \in \Lambda : M_\lambda \models \varphi(a_1(\lambda), \ldots, a_n(\lambda))\} \in D,$$

independently of the choice of representatives $a_1, \ldots, a_n$ of the equivalence classes $a_1/D, \ldots, a_n/D$.

In particular, if σ is an L-sentence then $U \models \sigma$ iff $\{\lambda \in \Lambda : M_\lambda \models \sigma\} \in D$.

Proof. See [ChaKei90, theorem 4.1.9] or [Hod93, theorem 9.5.1]. □

COROLLARY 2.30 (Ultrapower embedding) *Let $\prod_D M$ be any ultrapower of M. For $a \in M$, let $\bar{a}$ be the constant function in $\prod_{\lambda \in \Lambda} M$ given by $\bar{a}(\lambda) = a$ (all $\lambda \in \Lambda$). The map $a \mapsto \bar{a}/D$ is an elementary embedding of M into $\prod_D M$.*

Proof. Immediate; see [ChaKei90, corollary 4.1.13] or [Hod93, corollary 9.5.2]. □

The following is a deep theorem of Keisler and Shelah; see [ChaKei90, She71] for details.

FACT 2.31 Any two elementarily equivalent structures have isomorphic ultrapowers.

As a corollary, we have:

THEOREM 2.32 *Let K be a class of similar structures.*

1. *K is elementary if and only if K is closed under taking isomorphic copies, ultraproducts, and ultraroots.*

2. *K is closed under elementary equivalence (i.e., $M \equiv N \in K \Rightarrow M \in K$) if and only if K is closed under taking isomorphic copies, ultrapowers, and ultraroots.*

3. *$K = mod(\sigma)$ for some single sentence σ if and only if both K and its complement are closed under ultraproducts and isomorphic copies.*

Proof. See [Hod93, corollary 9.5.10] or [ChaKei90, theorem 4.1.12 and corollary 4.3.13]. □

Exercises

1. Let $\mathcal{B}_i$ ($i \in I$) be boolean algebras and let $h_i : \mathcal{B}_i \to \mathcal{F}_i$ be a representation of $\mathcal{B}_i$, where $\mathcal{F}_i$ is a field of sets with base X_i, say, for each i. Show that $\prod_{i \in I} \mathcal{B}_i$ has a representation mapping it to a field of sets whose base is the disjoint union of the X_i.

2. Let $\mathcal{B}$ be a boolean algebra and $h : \mathcal{B} \to \mathcal{F}$ a representation, where $\mathcal{F}$ is a field of sets with base X. Let C be a homomorphic image of $\mathcal{B}$. Under what circumstances does C have a representation mapping it in a natural way to a field of sets with base $\subseteq X$? [As an example, consider the boolean algebra $\mathcal{F}$ of exercise 2.3(1). Take the representation h to be the identity map, C to be the two-element boolean algebra, and let the homomorphism $\theta : \mathcal{F} \to C$ be given by $\theta(Y) = 1$ iff Y is infinite and $\theta(Y) = 0$ if Y is finite, for $Y \subseteq \omega$. What if we took h to be the representation of $\mathcal{F}$ constructed in theorem 2.10?]

2.4. Products and ultraproducts

3. Let $n < \omega$ and let $\mathcal{A}_i$ ($i < n$) be boolean algebras. Show that $\prod_{i<n}(\mathcal{A}_i^+) \cong (\prod_{i<n}\mathcal{A}_i)^+$.

   Find boolean algebras $\mathcal{A}_i$ ($i < \omega$) such that $\prod_{i<\omega}(\mathcal{A}_i^+)$ is not isomorphic to $(\prod_{i<\omega}\mathcal{A}_i)^+$. (See [AndGiv+95, §3].)

4. Prove that the ultraproduct operations of definition 2.27 are well-defined. So, for example, in the case of boolean algebras $\mathcal{B}_\lambda$ ($\lambda \in \Lambda$), to show that $+$ is well-defined, suppose that $b, b_1, b', b'_1 \in \prod_\lambda \mathcal{B}_\lambda$ and that $b \sim b_1$ and $b' \sim b'_1$. You have to show that $(b+b') \sim (b_1+b'_1)$. Then prove that negation is also well-defined.

5. Show that any principal ultraproduct of structures M_λ is isomorphic to one of the M_λ.

6. Let $M = \prod_D M_i$ be a non-principal ultraproduct of L-structures M_i ($i < \omega$) over ω, and let σ be an L-sentence. Show that if $M \models \sigma$ then $M_i \models \sigma$ for infinitely many $i < \omega$. Deduce that if there is $n < \omega$ such that $M_i \models \sigma$ for all $i \geq n$, then $M \models \sigma$.

7. Prove theorem 2.32(3) using the Keisler–Shelah theorem.

8. (a) Prove that any existential second-order sentence (see section 2.2.7) is preserved under ultraproducts: that is, if M_i ($i \in I$) are L-structures, σ is an existential second-order sentence, $M_i \models \sigma$ for all $i \in I$ and D is an ultrafilter over I, then $\prod_D M_i \models \sigma$.

   (b) Find a universal second-order sentence that is not preserved under ultraproducts.

   (c) Deduce that not every universal second-order sentence is equivalent to an existential second-order sentence. (Remark: the statement that every universal second-order sentence is equivalent to an existential second-order sentence *over finite structures* is equivalent to the open complexity-theoretic problem $NP = \text{co-}NP$. See [Fag74].)

9. Show that an ultraproduct of ultraproducts is (isomorphic to) an ultraproduct. Formally, if I is a non-empty set, J_i ($i \in I$) are non-empty sets, M_{ij} ($i \in I$, $j \in J_i$) are L-structures for some L, E_i ($i \in I$) is an ultrafilter over J_i, D is an ultrafilter over I, and $M_i = \prod_{E_i} M_{ij}$, show that $\prod_D M_i$ is isomorphic to an ultraproduct of the structures M_{ij} ($i \in I$, $j \in J_i$).

10. Let G be an undirected graph, and let $n < \omega$. G is said to be n-colourable if each node of G can be coloured with a unique colour, using at most n colours altogether, such that there is no edge between any two nodes of the same colour. Using ultraproducts, show that for any finite n, if every finite

subgraph of G is n-colourable then so is G. [Embed G in an ultraproduct of finite subgraphs of G.] Repeat using compactness. Which method is easier? Honestly?

11. Use Łoś' theorem to prove the compactness theorem for first-order logic. (Cf. [Hod93, theorem 9.5.9].)

2.5 Boolean algebras with operators

We now generalise the notion of boolean algebras by adding extra functions (or operators). What results is called a *boolean algebra with operators,* or BAO. This more general setting was proposed by Jónsson and Tarski in [JónTar51], and many of the definitions and results below are due to them. BAOs will be helpful when we come to consider algebras of binary and higher-order relations later in the book. Many of the results from universal algebra apply to arbitrary algebras, not just BAOs. Since the scope of this book is restricted to BAOs, we do not give these results in their full generality.

In this section we introduce some basic aspects of BAOs and operators. In the next, we consider varieties of BAOs, and after that we briefly consider some aspects of duality theory — relationships between atomic BAOs and certain structures induced on their atoms.

2.5.1 Definitions

DEFINITION 2.33 Let $\mathcal{B} = \langle B, 0, 1, +, - \rangle$ be a boolean algebra.

- An *operator* Ω with *arity* or *rank* $\mathrm{rk}(\Omega) = n < \omega$ on $\mathcal{B}$ is a function $\Omega : B^n \to B$ such that:

  1. Ω is *normal:* for all $b_0, \ldots, b_{n-1} \in \mathcal{B}$ and $i < n$, if $b_i = 0$ then $\Omega(b_0, \ldots, b_{n-1}) = 0$,

  2. Ω is *additive:* for any $b_0, \ldots, b_{n-1}, b, b' \in \mathcal{B}$ and $i < n$, we have
  $$\begin{aligned}&\Omega(b_0, \ldots, b_{i-1}, (b+b'), \ldots, b_{n-1}) \\&= \Omega(b_0, \ldots, b_{i-1}, b, b_{i+1}, \ldots, b_{n-1}) \\&\quad + \Omega(b_0, \ldots, b_{i-1}, b', b_{i+1}, \ldots, b_{n-1}).\end{aligned}$$

  We may call Ω an *n-ary operator.*

- A *boolean algebra with operators* (BAO) is an algebra $\langle B, 0, 1, +, -, \Omega_\lambda : \lambda \in \Lambda \rangle$, where $\mathcal{B} = \langle B, 0, 1, +, - \rangle$ is a boolean algebra, Λ is a set (perhaps uncountable), and Ω_λ ($\lambda \in \Lambda$) are operators on $\mathcal{B}$.

2.5. Boolean algebras with operators

This notion was introduced in [JónTar51, definition 2.13] under the name of 'normal boolean algebra with operators'.

- If $\mathcal{B} = \langle B, 0, 1, +, -, \Omega_\lambda : \lambda \in \Lambda \rangle$ is a BAO then $\mathrm{bool}(\mathcal{B}) \stackrel{\mathrm{def}}{=} \langle B, 0, 1, +, - \rangle$ is called the *boolean part* or *boolean reduct* of $\mathcal{B}$.

$n = 0$ is quite acceptable above: when $n = 0$, there is no i with $i < n$, so normality and additivity hold vacuously, and hence a nullary operator is just an element of $\mathcal{B}$.

Any operator is monotonic (order-preserving) in each argument:

LEMMA 2.34 *Let $\mathcal{B}$ be a BAO and Ω an n-ary operator of $\mathcal{B}$, where $n > 0$. Then for all $b_0, \ldots, b_{n-1}, b, b' \in \mathcal{B}$ and $i < n$, we have*

$$b \leq b' \Rightarrow \Omega(b_0, \ldots, b_{i-1}, b, b_{i+1}, \ldots, b_{n-1}) \leq \Omega(b_0, \ldots, b_{i-1}, b', b_{i+1}, \ldots, b_{n-1}).$$

Proof. The lemma holds vacuously when $n = 0$. Assume that $n > 0$; for brevity, write $\Omega(x)$ for $\Omega(b_0, \ldots, b_{i-1}, x, b_{i+1}, \ldots, b_{n-1})$. If $b \leq b'$ then $b + b' = b'$, so $\Omega(b') = \Omega(b + b') = \Omega(b) + \Omega(b')$, giving $\Omega(b) \leq \Omega(b')$. □

We will apply the terms and definitions of boolean algebra to BAOs — so, for example, an atom of a BAO is an atom of its boolean part, and a BAO is said to be atomic, complete, or degenerate if its boolean part is. A filter (ultrafilter) of a BAO is a filter (ultrafilter) of its boolean reduct (but it is different for ideals; see definition 2.36). If $\mathcal{A}, \mathcal{B}$ are BAOs we say that $\mathcal{A}$ is a *dense* subalgebra of $\mathcal{B}$ if it is a subalgebra and the boolean part of $\mathcal{A}$ is dense in the boolean part of $\mathcal{B}$. A map $\iota : \mathcal{A} \to \mathcal{B}$ is called a *complete embedding* if ι is a complete embedding from the boolean part of $\mathcal{A}$ into the boolean part of $\mathcal{B}$. We write $\mathcal{A} \subseteq^c \mathcal{B}$ if $\mathcal{A} \subseteq \mathcal{B}$ and the inclusion map is a complete embedding of $\mathcal{A}$ into $\mathcal{B}$.

The signature L of a BAO contains the signature L_{BA} of boolean algebras; we call the function symbols in $L \setminus L_{BA}$ *operator symbols*. To economise on notation, we will often use Ω as an operator symbol. In general, we will have a fixed L and any BAO under consideration will have this similarity type. Thus, we implicitly assume that BAOs are similar. Exceptions to this include chapters 5, 13, and 15, where we consider the connections between relation algebras and cylindric algebras of different dimensions. In those chapters, we will have to be extra careful about the signature.

2.5.2 Homomorphisms and ideals

These two are closely connected in BAOs, as in boolean algebras — recall exercise 2.3(13,14).

DEFINITION 2.35 *Given similar BAOs $\mathcal{B}, C$ and a homomorphism $h : \mathcal{B} \to C$ (see section 2.2.4), the kernel of h is the subset $\ker(h) = \{b \in \mathcal{B} : h(b) = 0\}$ of $\mathcal{B}$.*

The 'abstract analogue' of a kernel is an ideal.

DEFINITION 2.36 (Cf. [Sai82, proposition 7.4]) An *ideal* of a BAO $\mathcal{B}$ is a subset $I \subseteq \mathcal{B}$ such that

- I is an ideal of the boolean reduct bool($\mathcal{B}$) (see definition 2.7),

- for any n-ary operator Ω of $\mathcal{B}$, and any $b_0, \ldots, b_{n-1} \in \mathcal{B}$, if $b_i \in I$ for some $i < n$ then $\Omega(b_0, \ldots, b_{n-1}) \in I$.

The second condition holds vacuously when $n = 0$, since there is no i with $i < n$. So the values of constant (nullary) operators need not be in I.

Homomorphism from ideal Given any ideal I of a BAO $\mathcal{B}$, define a binary relation $\sim$ on $\mathcal{B}$ by $b \sim c$ iff the symmetric difference $b \oplus c \stackrel{\text{def}}{=} (b - c) + (c - b)$ of b, c is in I. Then $\sim$ is an equivalence relation on $\mathcal{B}$. We claim that it is in fact a *congruence*: if f is an n-ary boolean function or operator of $\mathcal{B}$, $b_0, \ldots, b_{n-1}, c_0, \ldots, c_{n-1} \in \mathcal{B}$, and $b_i \sim c_i$ for all $i < n$, then $f(b_0, \ldots, b_{n-1}) \sim f(c_0, \ldots, c_{n-1})$. For the boolean functions, see exercise 2.3(14). For the operators, consider for example a unary operator Ω on $\mathcal{B}$. If $b - c \in I$, then in $\mathcal{B}$ we have

$$\begin{aligned} \Omega(b) - \Omega(c) &\leq \Omega(b) - \Omega(b \cdot c) \\ &= \Omega((b-c) + b \cdot c) - \Omega(b \cdot c) \\ &= (\Omega(b-c) + \Omega(b \cdot c)) - \Omega(b \cdot c) \\ &\leq \Omega(b-c) \in I. \end{aligned}$$

So $\Omega(b) - \Omega(c) \in I$. By symmetry, if $c - b \in I$ then $\Omega(c) - \Omega(b) \in I$ too. So if $b \oplus c \in I$, then $b - c, c - b \in I$, so $\Omega(b) \oplus \Omega(c) = (\Omega(b) - \Omega(c)) + (\Omega(c) - \Omega(b)) \in I$. If Ω is binary and $b \oplus b', c \oplus c' \in I$, then by applying the calculation above to each argument, $\Omega(b, c) - \Omega(b', c') \leq (\Omega(b, c) - \Omega(b', c)) + (\Omega(b', c) - \Omega(b', c')) \in I$. Similarly, $\Omega(b', c') - \Omega(b, c) \in I$, and so $\Omega(b, c) \oplus \Omega(b', c') \in I$. The proof for n-ary operators ($n > 2$) is by iterating this, and for $n = 0$ there is nothing to prove. This proves the claim.

Thus, if $\mathcal{B}$ has signature L, say, we may define an L-algebra $\mathcal{B}/I$ on the set $\mathcal{B}/\sim$ of $\sim$-equivalence classes, by

$$f^{\mathcal{B}/I}(b_0/\sim, \ldots, b_{n-1}/\sim) \stackrel{\text{def}}{=} f^{\mathcal{B}}(b_0, \ldots, b_{n-1})/\sim,$$

for all n-ary $f \in L$ (including the boolean functions) and $b_0, \ldots, b_{n-1} \in \mathcal{B}$. The above shows that this is well-defined. It can be checked that $\mathcal{B}/I$ is an L-BAO and that the map $b \mapsto b/\sim$ is a surjective homomorphism : $\mathcal{B} \to \mathcal{B}/I$ with kernel I. So any ideal of $\mathcal{B}$ gives rise to a homomorphic image of $\mathcal{B}$.

2.5. Boolean algebras with operators

Ideal from homomorphism Conversely, if $h : \mathcal{B} \to C$ is a surjective homomorphism, then $\ker(h)$ is an ideal of $\mathcal{B}$, and $\mathcal{B}/\ker(h)$ is naturally isomorphic to C, via $b/\sim \mapsto h(b)$, where $\sim$ is defined relative to $\ker(h)$.

2.5.3 Completely additive and conjugated algebras

Completely additive BAOs are fortunately the most common in this book: all the BAOs we will see later are of this kind. Their study originates in [JónTar51].

DEFINITION 2.37

- Let $\mathcal{B}$ be a boolean algebra and f an n-ary function on $\mathcal{B}$. f is said to be *completely additive* if it satisfies, for any $b_0, \ldots, b_{n-1} \in \mathcal{B}$, any $i < n$, and any non-empty $X \subseteq \mathcal{B}$ such that $\sum X$ exists in $\mathcal{B}$,

$$f(b_0, \ldots, \sum X, \ldots, b_{n-1}) = \sum_{x \in X} f(b_0, \ldots, x, \ldots, b_{n-1}).$$

Note that this equation implies that the right-hand sum exists in $\mathcal{B}$. The restriction to non-empty X is to ensure that the boolean $+$ is completely additive. A nullary function is vacuously completely additive.

- A BAO is said to be *completely additive* if each of its operators is completely additive.

- A class K of BAOs is *completely additive* if each BAO in K is completely additive.

LEMMA 2.38 *Any normal completely additive n-ary function f on a boolean algebra $\mathcal{B}$ is an operator on $\mathcal{B}$.*

Proof. We can suppose that $n > 0$; for simplicity of notation, assume that $n = 1$. Then we have

$$f(a+b) \;=\; f(\sum\{a,b\}) \;=\; \sum_{x \in \{a,b\}} f(x) \;=\; f(a)+f(b),$$

for all $a, b \in \mathcal{B}$. So f is normal and additive, and hence an operator on $\mathcal{B}$. □

The existence of a conjugate for a function (defined next) gives an easy way to prove that the function is a completely additive operator. We'll use this to establish (e.g., in lemma 3.13) that the non-boolean functions in relation algebras and cylindric algebras are completely additive operators, so that these algebras are BAOs. The notion of conjugate of a function is due to Tarski and the term 'conjugate' was introduced in [JónTar51, definition 1.11].

DEFINITION 2.39 Let $\mathcal{B}$ be a boolean algebra, and let $f : \mathcal{B}^n \to \mathcal{B}$ be an n-ary function on $\mathcal{B}$.

- Let $i < n$. An *ith conjugate* of f over $\mathcal{B}$ is an n-ary function $c_{f,i} : \mathcal{B}^n \to \mathcal{B}$ such that for any $a, b_0, \ldots, b_{n-1} \in \mathcal{B}$, we have
  $$a \cdot f(b_0, \ldots, b_{n-1}) = 0 \iff b_i \cdot c_{f,i}(b_0, \ldots, b_{i-1}, a, b_{i+1}, \ldots, b_{n-1}) = 0.$$

- f is said to be *conjugated* over $\mathcal{B}$ if for every $i < n$ there exists an ith conjugate of f over $\mathcal{B}$. Again, any nullary function is vacuously conjugated.

Now let $L \supseteq L_{BA}$ be a functional signature, and let $\mathcal{B}$ be an L-algebra whose L_{BA}-reduct bool($\mathcal{B}$) is a boolean algebra.

- $\mathcal{B}$ said to be *conjugated* if for every function symbol $f \in L \setminus L_{BA}$, $f^\mathcal{B}$ is conjugated over bool($\mathcal{B}$).

- A class K of L-algebras whose L_{BA}-reducts are boolean algebras (e.g., a variety of L-BAOs) is said to be *conjugated* if every algebra in K is conjugated.

The following important result forms part of [JónTar51, theorem 1.14].

THEOREM 2.40 *Any conjugated function on a boolean algebra is normal and completely additive. For any functional signature $L \supseteq L_{BA}$, any conjugated class of L-algebras whose L_{BA}-reducts are boolean algebras is a completely additive class of L-BAOs.*

Proof. Let f be a conjugated n-ary function over the boolean algebra $\mathcal{B}$. To show that f is normal and completely additive, we can assume that $n > 0$, and we must show that

$$f(a_0, \ldots, a_{i-1}, \sum_{j \in J} b_j, a_{i+1}, \ldots, a_{n-1}) = \sum_{j \in J} f(a_0, \ldots, a_{i-1}, b_j, a_{i+1}, \ldots, a_{n-1})$$

for any $i < n$, any set J, and $a_0, \ldots, a_{n-1}, b_j \in \mathcal{B}$ (all $j \in J$) such that $\sum_{j \in J} b_j$ exists. (The case $J = \emptyset$ establishes normality.)

Well, let $c_{f,i}$ be an ith conjugate of f over $\mathcal{B}$. Assuming that $\sum_j b_j$ exists, for any $z \in \mathcal{B}$ we have

$$z \geq f(a_0, \ldots, \sum_j b_j, \ldots, a_{n-1})$$
$$\Leftrightarrow \quad (-z) \cdot f(a_0, \ldots, \sum_j b_j, \ldots, a_{n-1}) = 0$$
$$\Leftrightarrow \quad (\sum_j b_j) \cdot c_{f,i}(a_0, \ldots, -z, \ldots, a_{n-1}) = 0$$
$$\Leftrightarrow \quad b_j \cdot c_{f,i}(a_0, \ldots, -z, \ldots, a_{n-1}) = 0 \text{ for all } j$$
$$\Leftrightarrow \quad (-z) \cdot f(a_0, \ldots, b_j, \ldots, a_{n-1}) = 0 \text{ for all } j$$
$$\Leftrightarrow \quad z \geq f(a_0, \ldots, b_j, \ldots, a_{n-1}) \text{ for all } j.$$

2.5. Boolean algebras with operators

This holds for all $z \in \mathcal{B}$, so $\sum_j f(a_0,\ldots,a_{i-1},b_j,a_{i+1},\ldots,a_{n-1})$ exists and is equal to $f(a_0,\ldots,\sum_j b_j,\ldots,a_{n-1})$, as required.

Hence, if K is a conjugated class of L-algebras whose L_{BA}-reducts are boolean algebras, $\mathcal{B} \in$ K, and $\Omega \in L \setminus L_{BA}$, then $\Omega^{\mathcal{B}}$ is normal and completely additive. By lemma 2.38, $\Omega^{\mathcal{B}}$ is an operator on $\mathcal{B}$. So K is a completely additive class of L-BAOs. □

2.5.4 Completions of BAOs

In section 2.3.9 we saw how to construct the completion of a boolean algebra. The completion respects all existing infima and suprema, though it is atomic iff the original boolean algebra is. This construction was extended to *completely additive* BAOs by Monk [Mon70]. Thus, for completely additive BAOs, completions provide an alternative to the 'canonical extensions' to be defined in section 2.7.3; both have their advantages and disadvantages.

THEOREM 2.41 (Monk) *Let $\mathcal{B}$ be any completely additive BAO. Then there exists a similar complete, completely additive BAO $C \supseteq \mathcal{B}$ in which $\mathcal{B}$ is dense. Any two such BAOs are isomorphic over $\mathcal{B}$.*

Proof. Let L be the signature of $\mathcal{B}$. Let C^- be the completion of $\text{bool}(\mathcal{B})$, as constructed in theorem 2.24. For $c \in C^-$ let $c{\downarrow} = \{b \in \mathcal{B} : b \leq c\}$. As $\text{bool}(\mathcal{B})$ is dense in C^-, we have $c = \sum^{C^-} c{\downarrow}$. For each n-ary operator Ω on $\mathcal{B}$, define an n-ary function Ω^* on C^- by

$$\Omega^*(c_0,\ldots,c_{n-1}) = \sum\nolimits^{C^-} \{\Omega(b_0,\ldots,b_{n-1}) : b_i \in c_i{\downarrow} \text{ for each } i < n\},$$

for every $c_0,\ldots,c_{n-1} \in C^-$. Because C^- is complete, Ω^* is a total function.

It is clear that Ω^* is normal. We check that it is completely additive. For simplicity of notation, we take Ω to be unary, and write b, b' for elements of $\mathcal{B}$; $\sum$ denotes supremum in C^- (note that this always exists), and $\sum^{\mathcal{B}}$ denotes supremum in $\mathcal{B}$.

Let $S \subseteq C^-$; we require $\Omega^*(\sum S) = \sum\{\Omega^*(s) : s \in S\}$. By definition of Ω^*, this equation is $\sum\{\Omega(b) : b \leq \sum S\} = \sum\{\sum\{\Omega(b') : b' \leq s\} : s \in S\}$, so by exercise 2.3(31) we require

$$\sum_{b \leq \sum S} \Omega(b) = \sum_{b' \leq s \in S} \Omega(b'). \tag{2.1}$$

Here and below, '$b' \leq s \in S$' is shorthand for '$b' \leq s$ for some $s \in S$'. Since if $b \leq s \in S$ then $b \leq \sum S$, '$\geq$' in (2.1) is clear. For the converse, note first the related fact that

$$\sum S = \sum\{b' : b' \leq s \in S\}. \tag{2.2}$$

(Again, '$\geq$' is clear. In the other direction, by density of bool($\mathcal{B}$) in C^- we see that if $s \in S$ then $s = \sum\{b' : b' \leq s\} \leq \sum\{b' : b' \leq s' \in S\}$. So the right-hand side in (2.2) is an upper bound for S, giving '$\leq$'.)

Now fix $b \leq \sum S$. (2.2) and distributivity (exercise 2.3(32)) yield $b = b \cdot \sum S = b \cdot \sum\{b' : b' \leq s \in S\} = \sum\{b \cdot b' : b' \leq s \in S\}$. This last is equal to $\sum^{\mathcal{B}}\{b \cdot b' : b' \leq s \in S\}$, since its value in C^- is in $\mathcal{B}$ (namely, b). So we have

$$b = \sum\nolimits^{\mathcal{B}}\{b \cdot b' : b' \leq s \in S\}. \tag{2.3}$$

By complete additivity of Ω in $\mathcal{B}$, $\Omega(b) = \sum^{\mathcal{B}}\{\Omega(b \cdot b') : b' \leq s \in S\}$. As suprema are preserved in C^- (exercise 2.3(37)), we have

$$\Omega(b) = \sum\{\Omega(b \cdot b') : b' \leq s \in S\}. \tag{2.4}$$

This holds for all $b \leq \sum S$. Now we obtain '$\leq$' in (2.1) by

$$\begin{aligned}
\sum_{b \leq \sum S} \Omega(b) &= \sum_{b \leq \sum S} \left(\sum\{\Omega(b \cdot b') : b' \leq s \in S\} \right) \\
&\leq \sum_{b \leq \sum S} \left(\sum\{\Omega(b') : b' \leq s \in S\} \right) \\
&= \sum\{\Omega(b') : b' \leq s \in S\}.
\end{aligned}$$

Hence, Ω^* is a (completely additive) operator on C^-. Moreover, it is clear that for all $b_0, \ldots, b_{n-1} \in \mathcal{B}$,

$$\Omega^*(b_0, \ldots, b_{n-1}) = \Omega(b_0, \ldots, b_{n-1}). \tag{2.5}$$

Let C be the expansion of C^- to an L-structure obtained by interpreting each operator symbol $\Omega \in L \setminus L_{BA}$ as $(\Omega^{\mathcal{B}})^*$. So C is an L-BAO, and by (2.5), $\mathcal{B}$ is a subalgebra of C. Then C has the required properties.

The proof of uniqueness is similar to that for boolean algebras. $\square$

DEFINITION 2.42 The BAO C constructed in the theorem is called the *completion* of $\mathcal{B}$. It is defined up to isomorphism over $\mathcal{B}$ by being a complete and completely additive BAO in which $\mathcal{B}$ is dense.

Exercises

1. Give an example of a non-completely additive BAO with complete boolean reduct.

2. Let $\mathcal{A}$ be a completely additive BAO. Show that any dense subalgebra of $\mathcal{A}$ is also completely additive.

3. Check theorem 2.41 for n-ary operators ($n > 1$).

2.6 Varieties and quasi-varieties of BAOs

Varieties are the most important kind of classes of BAOs. In this section we introduce the basic facts about them that we will need later, including Birkhoff's theorem, subdirect products, and discriminator varieties.

2.6.1 Basic concepts

Given a functional signature L, we know that L-formulas are built from atomic formulas $t = s$ (where s, t are terms) using boolean connectives and quantifiers. Recall from section 2.2.3 that an L-formula $\varphi(\bar{x})$ is said to be valid in an L-algebra $\mathcal{B}$ if $\mathcal{B} \models \varphi(\bar{b})$ for all $|\bar{x}|$-tuples $\bar{b}$ of elements of $\mathcal{B}$. We say that $\mathcal{B}$ validates φ in this case.

DEFINITION 2.43

1. Recall that an *equation* is an atomic formula of the form $s = t$, for terms s, t.

2. A *quasi-equation* is a quantifier-free formula of the form
$$((s_0 = t_0) \wedge \ldots \wedge (s_{n-1} = t_{n-1})) \to (s_n = t_n)$$
for some $n \geq 0$ and some terms s_i, t_i $(i \leq n)$.

3. A *variety* (or an equational variety) is a class V of similar BAOs that can be defined by a set E of equations in the signature of V. That is, V consists of precisely those BAOs of the given similarity type that validate all the equations in E.

4. Similarly, a *quasi-variety* consists of all the BAOs in a given signature that validate Q, where Q is some set of quasi-equations of the signature.

5. Given a class K of similar BAOs, the *variety generated by* K is the smallest variety of BAOs of the signature of K containing K. It may equivalently be defined as the class of all BAOs of the signature of K that satisfy all equations of this signature that are valid in K.

6. The *quasi-variety generated by* K is defined similarly.

Note that when we consider the validity of a quantifier-free formula $\varphi(\bar{x})$ in some structure $\mathcal{A}$, we have
$$\mathcal{A} \models \varphi \iff \mathcal{A} \models \forall \bar{x} \varphi(\bar{x}).$$

So varieties and quasi-varieties are universally axiomatisable and are therefore closed under subalgebras. The class of all BAOs of a given signature is a variety, since normality and additivity of the operators are expressible by equations.

THEOREM 2.44 *If $\mathcal{A}$ is an algebra of the signature of a variety* V, *then* $\mathcal{A} \in \mathsf{V}$ *iff every finitely generated subalgebra of $\mathcal{A}$ is in* V.

If V,W are equational varieties of similar algebras, containing exactly the same finitely generated algebras, then $\mathsf{V} = \mathsf{W}$.

Proof. Being defined by universal formulas, V is closed under subalgebras. So if $\mathcal{A} \in \mathsf{V}$ then every finitely generated subalgebra of $\mathcal{A}$ is in V. For the converse, let E be a set of equations defining V. If $\mathcal{A}$ is an algebra of the signature of V, and $\mathcal{A} \notin \mathsf{V}$, then there is an equation $e \in E$ such that $\mathcal{A} \not\models e$. Now $e = e(\bar{x})$ is an equation with only finitely many variables, so there is a finite tuple $\bar{a} \in \mathcal{A}$ such that $\mathcal{A} \not\models e(\bar{a})$. But then, the subalgebra $\mathcal{A}\langle\bar{a}\rangle$ of $\mathcal{A}$ generated by $\bar{a}$ also fails to satisfy the equation e, so $\mathcal{A}\langle\bar{a}\rangle \notin \mathsf{V}$.

The second half follows immediately. □

In particular, for countable signatures, if V,W contain the same countable algebras then they are equal (as a finitely generated algebra in a countable signature is certainly countable).

2.6.2 HSP notation and Birkhoff's theorem

Let K be a class of algebras in the same signature. We write $\mathbf{I}(\mathsf{K}), \mathbf{H}(\mathsf{K}), \mathbf{S}(\mathsf{K})$ and $\mathbf{P}(\mathsf{K})$ to stand for the classes of isomorphic copies of algebras in K, homomorphic images of algebras in K, subalgebras of algebras in K, and products of algebras in K, respectively. It is often conventional to let $\mathbf{S}(\mathsf{K})$ actually stand for all isomorphic copies of subalgebras of algebras in K, and we adopt that convention here. So $\mathbf{I}\mathsf{K} \subseteq \mathbf{S}\mathsf{K}$. $\mathbf{Up}(\mathsf{K})$ denotes the class of all algebras isomorphic to an ultraproduct of algebras in K.

These definitions make sense for arbitrary structures but we will only use them for algebras.

Birkhoff's theorem ((1) below) is striking because it links two quite different attributes of classes of BAOs. Tarski later proved (2).

THEOREM 2.45 (Birkhoff, Tarski) *Let K be a non-empty class of similar BAOs.*

1. *[Bir35] The following are equivalent*

   - K *is a variety*
   - K *is closed under the taking of subalgebras, homomorphic images, and products: i.e.,* $\mathbf{H}(\mathsf{K}), \mathbf{S}(\mathsf{K}), \mathbf{P}(\mathsf{K}) \subseteq \mathsf{K}$.

2. *[Tar46] $\mathbf{HSP}(\mathsf{K}) = \mathbf{H}(\mathbf{S}(\mathbf{P}(\mathsf{K})))$ is the smallest variety containing K — the variety generated by K.*

2.6. Varieties and quasi-varieties of BAOs

Quasi-varieties can also be characterised algebraically. McKinsey showed in [McKi43] that any universal class closed under products is a quasi-variety. We will use the following variant:

THEOREM 2.46 (Mal'cev, see [Mal71a, theorem 3]) *Let* K *be any non-empty class of similar BAOs.* K *is a quasi-variety if and only if it is closed under the taking of subalgebras, products, and ultraproducts, iff* K = **SPUp**K.

2.6.3 Subdirect products

An important result from universal algebra that we will need to refer to is the subdirect decomposition theorem.

DEFINITION 2.47

1. Let $\mathcal{A}_i$ ($i \in I$) be similar BAOs. A *subdirect product* of $\langle A_i : i \in I \rangle$ is a subalgebra $\mathcal{B}$ of $\prod_{i \in I} A_i$ such that for all $i \in I$, the ith projection p_i maps $\mathcal{B}$ onto $\mathcal{A}_i$.

2. A *subdirect representation* of the BAO $\mathcal{A}$ is an embedding $f : \mathcal{A} \to \prod_{i \in I} \mathcal{A}_i$ (for some index set I and some algebras $\mathcal{A}_i$) such that $p_i \circ f$ maps $\mathcal{A}$ onto $\mathcal{A}_i$, for each $i \in I$.

3. A BAO $\mathcal{A}$ is said to be *subdirectly irreducible* if for every subdirect representation $f : \mathcal{A} \to \prod_{i \in I} \mathcal{A}_i$, at least one of the homomorphisms $p_i \circ f : \mathcal{A} \to \mathcal{A}_i$ is an isomorphism.

4. For any class K of similar BAOs, Sir(K) denotes the subclass of subdirectly irreducible members of K.

5. A *subdirect decomposition* of the BAO $\mathcal{A}$ is a subdirect representation $f : \mathcal{A} \to \prod_{i \in I} \mathcal{A}_i$ (some I, some $\mathcal{A}_i$) such that each $\mathcal{A}_i$ is subdirectly irreducible. The algebras $\mathcal{A}_i$ are called *subdirectly irreducible components* of $\mathcal{A}$.

Note that according to this definition, and as in, e.g., [BurSan81], degenerate algebras are subdirectly irreducible. (Some authors do not allow degenerate algebras to be subdirectly irreducible.)

THEOREM 2.48 (Birkhoff, [Bir44]) *Every BAO has a subdirect decomposition.*

COROLLARY 2.49 *If* V *is a variety then every member of* V *is a subdirect product of subdirectly irreducible members of* V.

Proof. Let $\mathcal{A} \in \mathsf{V}$ and let $\langle A_\lambda : \lambda \in \Lambda \rangle$ be the subdirectly irreducible components of $\mathcal{A}$ in a subdirect decomposition. Since $\mathcal{A}_\lambda$ is a homomorphic image of $\mathcal{A}$, we see by theorem 2.45 that $\mathcal{A}_\lambda \in \mathsf{V}$, for each $\lambda \in \Lambda$. □

COROLLARY 2.50 *Let V be a variety and let ε be an equation. Then*

$$\mathsf{Sir}(V) \models \varepsilon \iff V \models \varepsilon.$$

Proof. Suppose ε is valid over the subdirectly irreducible members of V. Let $\mathcal{A} \in V$ and let $\langle \mathcal{A}_\lambda : \lambda \in \Lambda \rangle$ be the subdirectly irreducible components of $\mathcal{A}$ in a subdirect decomposition. Then $\mathcal{A}_\lambda \models \varepsilon$ for each λ. Since equations are preserved under products and subalgebras, $\mathcal{A} \models \varepsilon$ too. Since $\mathcal{A}$ was an arbitrary member of V, we see that $V \models \varepsilon$. The converse is trivial. □

COROLLARY 2.51 *Let W be a quasi-variety, V be a variety and let* $\mathsf{Sir}(V) \subseteq W \subseteq V$. *Then* $W = V$ *and W is a variety.*

The proof of this is left for exercise 5.

2.6.4 Discriminator varieties

These are special varieties where some term 'discriminates' the zero from the non-zero elements of their subdirectly irreducible algebras. Good references for them include [Wer78, Jip93]; a good many results about them were proved for the special cases of relation algebras and cylindric algebras by McKinsey and Tarski [Tar41, JónTar52]. Many (but not all) varieties in algebraic logic are discriminator varieties, and we will use the general properties of discriminator varieties explained below to identify them and provide equational axiomatisations for them. Discriminators also help in showing that certain classes are varieties.

DEFINITION 2.52

1. Let K be a class of similar BAOs. A *discriminator term* for K is a term $d(x)$ of the signature of K satisfying

$$d(x) = \begin{cases} 0 & \text{if } x = 0, \\ 1 & \text{otherwise} \end{cases}$$

   in each algebra in K.

2. A discriminator term for a BAO $\mathcal{A}$ is a discriminator term for the class $\{\mathcal{A}\}$.

3. A variety V is said to be a *discriminator variety* if there exists a discriminator term for $\mathsf{Sir}(V)$.

Note that, in a discriminator variety V, we do not expect to find a discriminator term for the whole of V, but only for $\mathsf{Sir}(V)$. Any variety contained in V is clearly also a discriminator variety.

Discriminator varieties are common in algebraic logic, and they have nice properties. One is that in any such variety, the 'simple' algebras are precisely the subdirectly irreducible algebras.

2.6. Varieties and quasi-varieties of BAOs

DEFINITION 2.53 A BAO $\mathcal{A}$ is said to be *simple* if any homomorphism whose domain is $\mathcal{A}$ is either an embedding, or maps $\mathcal{A}$ to a degenerate BAO (with only a single element in its domain).

Note that according to this definition (and as in, e.g., [BurSan81]), degenerate algebras are simple. (As with 'subdirectly irreducible', some writers do not call the degenerate algebra simple.) Simple algebras are related to subdirectly irreducible ones:

PROPOSITION 2.54 *Any simple BAO is subdirectly irreducible.*

Proof. See exercise 2.6(3a). □

In a discriminator variety, the converse also holds:

THEOREM 2.55 (McKinsey, Tarski, see [JónTar52, theorem 4.14]) *Any BAO with a discriminator term is simple. Hence, if* V *is a discriminator variety of BAOs, the class of subdirectly irreducible members of* V *is identical to the class of simple algebras in* V.

Proof. See exercise 2.6(3b,c). □

So we obtain from corollary 2.49:

COROLLARY 2.56 *If* V *is a discriminator variety, then every algebra in* V *is a subdirect product of simple algebras from* V.

Varieties and discriminators We saw in Birkhoff's theorem that a class K of BAOs is a variety iff it is closed under subalgebras, direct products, and homomorphic images. To check that K is a variety, it is usually easy to show that it is closed under subalgebras and products but much harder to check that it is closed under homomorphic images. In the context of discriminators, this check is not required.

THEOREM 2.57 (Givant, [Giv99, theorem 2.3]) *Let* V *be a discriminator variety and* K *a class of simple algebras in* V. *If* $\mathsf{Up}\mathsf{K} \subseteq \mathsf{S}\mathsf{K}$, *then* $\mathsf{SP}\mathsf{K}$ *is a variety and* $\mathsf{S}\mathsf{K}$ *is the universal class of non-degenerate simple algebras in* $\mathsf{SP}\mathsf{K}$.

Exercise 9 below gives approximately an alternative formulation. This theorem can be used to give a very quick proof that **RRA** is a variety (see exercise 3.4(6)).

Characterising discriminator varieties The property of a variety being a discriminator variety can be characterised by equations as follows. Such a characterisation is given in [Jip93, theorem 3] (see exercise 11 below), and adapts to the BAO case a result of McKenzie [McKe75]. The version we use comes from [AndGiv+98, lemma 2.1]. The equations are related to the axioms defining an S5-modality: see exercise 2.7(9). We will use this result in chapter 3 to show that **RA** is a discriminator variety.

PROPOSITION 2.58 *Let $L \supseteq L_{BA}$ be a functional signature, and V a variety of L-BAOs. Let $d(x)$ be a unary L-term. Then the following are equivalent:*

1. *d is a discriminator term for $\mathsf{Sir}\,V$ (so that V is a discriminator variety),*

2. *all equations of the following form are valid in V:*

   - $x \leq d(x)$
   - $d(d(x)) \leq d(x)$
   - $d(-d(x)) \leq -d(x)$
   - $\Omega(x_0,\ldots,x_{n-1}) \leq d(x_i)$ for all $n > 0$, all n-ary $\Omega \in L \setminus L_{BA}$, and all $i < n$.

Proof. $(1 \Rightarrow 2)$: if d is a discriminator term for $\mathsf{Sir}\,V$, then because $d(x)$ is always 0 or 1, it is easily checked that all equations in (2) are valid in $\mathsf{Sir}\,V$. For example, $\mathsf{Sir}\,V \models \Omega(x_0,\ldots,x_{i-1},0,x_{i+1},\ldots,x_{n-1}) = 0 \leq d(0)$ since operators are normal, while if $\mathcal{A} \in \mathsf{Sir}\,V$, $a_0,\ldots,a_{n-1} \in \mathcal{A}$, and $a_i > 0$, then trivially $\mathcal{A} \models \Omega(a_0,\ldots,a_{n-1}) \leq 1 = d(a_i)$. By corollary 2.50, the equations in (2) are valid in V, too.

$(2 \Rightarrow 1)$: First, let $\mathcal{A}$ be any L-BAO, and suppose that $\alpha \in \mathcal{A}$ is such that

$$\begin{aligned} \Omega(1,\ldots,1,\alpha,1,\ldots,1) &\leq \alpha \\ \Omega(1,\ldots,1,-\alpha,1,\ldots,1) &\leq -\alpha \end{aligned} \quad (2.6)$$

for every operator Ω of $\mathcal{A}$ of non-zero arity. For any $n < \omega$ and any function $f : \mathcal{A}^n \to \mathcal{A}$, we may define its 'restriction $f\!\restriction_\alpha$ to α' by:

$$f\!\restriction_\alpha : \{a \in \mathcal{A} : a \leq \alpha\}^n \to \{a \in \mathcal{A} : a \leq \alpha\},$$
$$f\!\restriction_\alpha(a_0,\ldots,a_{n-1}) = \alpha \cdot f(a_0,\ldots,a_{n-1}), \text{ for } a_0,\ldots,a_{n-1} \leq \alpha.$$

Define the 'relativised' L-algebra $\mathcal{A}\!\restriction_\alpha$ with domain $\{a \in \mathcal{A} : a \leq \alpha\}$ by interpreting each $f \in L$ as $f^{\mathcal{A}}\!\restriction_\alpha$. It is easily checked that $\mathcal{A}\!\restriction_\alpha$ is an L-BAO.

Now define a map $\pi_\alpha : \mathcal{A} \to \mathcal{A}\!\restriction_\alpha$ by $\pi_\alpha(a) = \alpha \cdot a$, for $a \in \mathcal{A}$. We claim that π_α is a surjective homomorphism. Surjectivity is clear, since $b = \alpha \cdot b$ for any $b \in \mathcal{A}\!\restriction_\alpha$. To check that π_α is a homomorphism, we require the following:

2.6. Varieties and quasi-varieties of BAOs

- For any constant $c \in L$, $\pi_\alpha(c^{\mathcal{A}}) = c^{\mathcal{A}\restriction\alpha}$ — for example, $\pi_\alpha(0^{\mathcal{A}}) = 0^{\mathcal{A}}$, and $\pi_\alpha(1^{\mathcal{A}}) = \alpha$. This is clear by the definition of $\mathcal{A}\restriction_\alpha$.

- $\pi_\alpha(a+b) = \pi_\alpha(a) +\restriction_\alpha \pi_\alpha(b)$, for all $a,b \in \mathcal{A}$. That is, $\alpha \cdot (a+b) = \alpha \cdot ((\alpha \cdot a) + (\alpha \cdot b))$. This is clear by distribution of $\cdot$ over $+$ in boolean algebras.

- $\pi_\alpha(-a) = -\restriction_\alpha(\pi_\alpha(a))$, for all $a \in \mathcal{A}$. That is, $\alpha \cdot (-a) = \alpha - (\alpha \cdot a)$. Again, this is a property of boolean algebras.

- $\pi_\alpha(\Omega(a_0,\ldots,a_{n-1})) = \Omega\restriction_\alpha(\pi_\alpha(a_0),\ldots,\pi_\alpha(a_{n-1}))$ for all n-ary operators Ω of $\mathcal{A}$ and all $a_0,\ldots,a_{n-1} \in \mathcal{A}$. That is,

$$\alpha \cdot \Omega(a_0,\ldots,a_{n-1}) = \alpha \cdot \Omega(\alpha \cdot a_0,\ldots,\alpha \cdot a_{n-1}).$$

If $n=0$, this is already proved. Assume $n>0$. For the proof, it is convenient to write α^0 for α and α^1 for $-\alpha$. Then $\alpha^0 + \alpha^1 = 1$, so by additivity of Ω,

$$\begin{aligned}
&\alpha \cdot \Omega(a_0,\ldots,a_{n-1}) \\
=\;& \alpha \cdot \Omega\big((\alpha^0 + \alpha^1) \cdot a_0,\ldots,(\alpha^0 + \alpha^1) \cdot a_{n-1}\big) \\
=\;& \alpha \cdot \Omega\big((\alpha^0 \cdot a_0 + \alpha^1 \cdot a_0),\ldots,(\alpha^0 \cdot a_{n-1} + \alpha^1 \cdot a_{n-1})\big) \\
=\;& \alpha \cdot \sum_{\eta \in {}^n 2} \Omega(\alpha^{\eta(0)} \cdot a_0,\ldots,\alpha^{\eta(n-1)} \cdot a_{n-1}) \\
=\;& \sum_{\eta \in {}^n 2} \big(\alpha \cdot \Omega(\alpha^{\eta(0)} \cdot a_0,\ldots,\alpha^{\eta(n-1)} \cdot a_{n-1})\big).
\end{aligned}$$

We reduce this expression to the desired $\alpha \cdot \Omega(\alpha^0 \cdot a_0,\ldots,\alpha^0 \cdot a_{n-1})$ by showing that all other terms in the sum vanish. Notice that for any $\eta \in {}^n 2$ and any $i < n$ such that $\eta(i) = 1$, we have

$$\begin{aligned}
& \alpha \cdot \Omega(\alpha^{\eta(0)} \cdot a_0,\ldots,\alpha^{\eta(n-1)} \cdot a_{n-1}) \\
\leq\;& \alpha \cdot \Omega(1,\ldots,1,\alpha^{\eta(i)},1,\ldots,1) \\
\leq\;& \alpha \cdot \alpha^{\eta(i)} && \text{(by (2.6))} \\
=\;& \alpha \cdot -\alpha \\
=\;& 0.
\end{aligned}$$

Hence indeed,

$$\begin{aligned}
& \alpha \cdot \Omega(a_1,\ldots,a_n) \\
=\;& \sum_{\eta \in {}^n 2} \alpha \cdot \Omega(\alpha^{\eta(0)} \cdot a_0,\ldots,\alpha^{\eta(n-1)} \cdot a_{n-1}) \\
=\;& \alpha \cdot \Omega(\alpha^0 \cdot a_0,\ldots,\alpha^0 \cdot a_{n-1}) \\
=\;& \alpha \cdot \Omega(\alpha \cdot a_0,\ldots,\alpha \cdot a_{n-1}),
\end{aligned}$$

as required.

So $\pi_\alpha : \mathcal{A} \to \mathcal{A}{\upharpoonright}_\alpha$ is a surjective homomorphism, as claimed.

We are now in a position to prove $(2 \Rightarrow 1)$ of the proposition. Assume (2), and let $\mathcal{A} \in \text{Sir}\,\mathsf{V}$. We require that d is a discriminator term in $\mathcal{A}$.

By the first and third equations in (2), in $\mathcal{A}$ we have $1 \leq d(1)$, so $d(1) = 1$, $-d(1) = 0$, and $d(0) = d(-d(1)) \leq -d(1) = 0$. So $d(0) = 0$.

Now take any $a \in \mathcal{A}$, $a \neq 0$, and let $\alpha = d^\mathcal{A}(a)$. We must prove that $\alpha = 1$.

Assume not. The first equation in (2) tells us that $0 < a \leq \alpha$, so we have $0 < \alpha < 1$ and $0 < -\alpha < 1$. By the rest of (2), for any operator Ω of $\mathcal{A}$ of non-zero arity, we have

$$\Omega(1,\ldots,1,\alpha,1,\ldots,1) \leq d(\alpha) = d(d(a)) \leq d(a) = \alpha,$$
$$\Omega(1,\ldots,1,-\alpha,1,\ldots,1) \leq d(-\alpha) = d(-d(a)) \leq -d(a) = -\alpha.$$

Thus, all equations of the form (2.6) for α and for $-\alpha$ hold in $\mathcal{A}$. So by the foregoing, $\pi_\alpha : \mathcal{A} \to \mathcal{A}_\alpha$ and $\pi_{-\alpha} : \mathcal{A} \to \mathcal{A}_{-\alpha}$ are surjective homomorphisms. We will define a subdirect representation $h : \mathcal{A} \to \mathcal{A}{\upharpoonright}_\alpha \times \mathcal{A}{\upharpoonright}_{-\alpha}$ such that neither of the projections $p_0 : \mathcal{A}{\upharpoonright}_\alpha \times \mathcal{A}{\upharpoonright}_{-\alpha} \to \mathcal{A}{\upharpoonright}_\alpha$, $p_1 : \mathcal{A}{\upharpoonright}_\alpha \times \mathcal{A}{\upharpoonright}_{-\alpha} \to \mathcal{A}{\upharpoonright}_{-\alpha}$ yields an isomorphism $p_i \circ h$. This will prove that $\mathcal{A}$ has a non-trivial subdirect representation, contradicting the subdirect irreducibility of $\mathcal{A}$.

So, let

$$h(a) = (\pi_\alpha(a), \pi_{-\alpha}(a)) = (\alpha \cdot a, -\alpha \cdot a), \quad \text{for } a \in \mathcal{A}.$$

Since in a product algebra the operators work 'coordinatewise' (see section 2.4), h is a homomorphism. Next we must show that h is an embedding, i.e., one-one. This is straightforward: $h(a) = h(b) \Rightarrow (\alpha \cdot a, -\alpha \cdot a) = (\alpha \cdot b, -\alpha \cdot b) \Rightarrow a = \alpha \cdot a + -\alpha \cdot a = \alpha \cdot b + -\alpha \cdot b = b$. That $p_0 \circ h$ and $p_1 \circ h$ are surjective is immediate, since $p_0 \circ h = \pi_\alpha$ and $p_1 \circ h = \pi_{-\alpha}$.

But finally, since $\alpha, -\alpha \neq 1$, neither $p_0 \circ h$ nor $p_1 \circ h$ is one-one: we have $h(\alpha) = (\alpha, 0)$ and $h(1) = (\alpha, -\alpha)$, so $(p_0 \circ h)(\alpha) = (p_0 \circ h)(1) = \alpha$. Similarly, $(p_1 \circ h)(-\alpha) = (p_1 \circ h)(1) = -\alpha$. So $\mathcal{A}$ is not subdirectly irreducible. This contradiction completes the proof. □

Translating universal sentences to equations One of the chief benefits of discriminators for us is that they allow us to replace universal sentences by equations. We will use this when axiomatising classes of algebras by games in part II of the book.

LEMMA 2.59 *Let d be a discriminator term for the class K of L-BAOs. Let $\varphi(\bar{x})$ be any quantifier-free L-formula. Then there is an L-equation, of the form $s = 0$ for some term $s(\bar{x})$, that is equivalent in K to φ. That is, $\varphi(\bar{x}) \leftrightarrow (s(\bar{x}) = 0)$ is valid in K.*

2.6. Varieties and quasi-varieties of BAOs

Proof. By induction on φ. If φ is the equation $t = u$, let $s = (t-u)+(u-t)$. Then φ is equivalent to $s = 0$. Assume inductively that φ is equivalent to $t = 0$, and ψ to $u = 0$. Then:

- $\neg\varphi$ is equivalent to $\neg(t = 0)$ and so (in K) to $d(t) = 1$, and so to $-d(t) = 0$.
- $\varphi \wedge \psi$ is clearly equivalent to $t + u = 0$.

$\square$

THEOREM 2.60 *Let* V *be a variety,* D *a discriminator variety, and suppose that* $V \subseteq D$. *Let* Σ *be a set of universal sentences that defines* V *over* D — *so for any BAO* $\mathcal{A} \in D$, *we have* $\mathcal{A} \in V$ *iff* $\mathcal{A} \models \Sigma$. *There is an effective procedure which produces from* Σ *a set of equations that, when combined with equations defining* D, *axiomatises* V.

Proof. For each universal sentence $\sigma \in \Sigma$, first find a logically equivalent prenex sentence $\forall \bar{x} \varphi_\sigma(\bar{x})$, where $\varphi_\sigma(\bar{x})$ is quantifier-free. Then, as in lemma 2.59, find an equation $\varepsilon_\sigma(\bar{x})$ equivalent to $\varphi_\sigma(\bar{x})$ in Sir D. Let $E = \{\varepsilon_\sigma(\bar{x}) : \sigma \in \Sigma\}$. By lemma 2.59, we know that if $\mathcal{A} \in \text{Sir D}$ then $\mathcal{A} \models E$ iff $\mathcal{A} \in V$. We have to show more generally that E defines V over D: that is, for all $\mathcal{A} \in D$, we have $\mathcal{A} \in V$ iff $\mathcal{A} \models E$.

Let $\mathcal{A} \in D$. By corollary 2.49, $\mathcal{A}$ has a subdirect decomposition $f : \mathcal{A} \to \prod_{i \in I} \mathcal{A}_i$, where $\mathcal{A}_i \in \text{Sir D}$ for all $i \in I$. Since V is a variety and so closed under subalgebras, products, and homomorphic images, we obtain

$$\mathcal{A} \in V \iff \mathcal{A}_i \in V \text{ for all } i \in I$$
$$\iff \mathcal{A}_i \models E \text{ for all } i \in I$$
$$\iff \mathcal{A} \models E,$$

the last equivalence holding because equations are preserved under homomorphisms, subalgebras, and products. $\square$

See exercise 12 for a somewhat stronger result. We remark that if Σ in the theorem is recursive then so is E; we leave the proof as an exercise. See chapter 9, around definition 9.13, for a discussion of recursive sets of formulas.

Exercises

All varieties in these exercises are varieties of BAOs.

1. Let K be a variety and suppose that K is finitely axiomatisable (there is a first-order sentence σ such that $K = \text{Mod}(\sigma)$; an alternative phrase is 'finitely based'). Show that K is axiomatised by a finite set of equations. [Compactness.]

2. Let $\mathcal{A}_\lambda$ ($\lambda \in \Lambda$) be BAOs, and suppose that $\mathcal{A}_\lambda \models q$, for some quasi-equation q (all λ). Prove that $\prod_\lambda \mathcal{A}_\lambda \models q$.

3. (a) Prove that a simple BAO is always subdirectly irreducible.
   (b) Prove that any BAO that has a discriminator term is simple.
   (c) Deduce theorem 2.55: the subdirectly irreducible members of a discriminator variety of BAOs are identical with the simple members of that variety.

4. Does theorem 2.44 hold for quasi-varieties? Does it hold for an arbitrary class defined by universal sentences?

5. Prove corollary 2.51: for any quasi-variety W containing all the subdirectly irreducible members of a variety V, and such that $W \subseteq V$, we have $W = V$.

6. For a countable signature L of BAOs, show that two varieties of L-BAOs are equal if they contain the same countable subdirectly irreducible algebras.

7. Show that the variety of boolean algebras is a discriminator variety.

8. Let V be a discriminator variety of BAOs, and let $\mathcal{A}, \mathcal{B} \in V$. If $\mathcal{B}$ is simple, and $\mathcal{A} \subseteq \mathcal{B}$, prove that $\mathcal{A}$ is also simple. Is this true without a discriminator?

9. Let V be a variety of BAOs generated by a class K (that is, $V = \mathbf{HSP}K$), and suppose that there exists a discriminator term for K. Show that V is a discriminator variety, and that $V = \mathbf{SPUp}K$. In words, the quasi-variety generated by a class with a discriminator term is a discriminator variety.

   Deduce theorem 2.57 from this, and vice versa.

10. Show that the equation $d(d(x)) \leq d(x)$ in proposition 2.58(2) can be replaced by $d(x) \leq d(x+y)$.

11. [Jip93, theorem 3] Show that the equations in proposition 2.58(2) can be replaced by:
    - $d(0) = 0$
    - $x \leq d(x)$
    - $\Omega(1, \ldots, 1, d(x), 1, \ldots, 1) \leq d(x)$, where $\Omega \in L \setminus L_{BA}$ has arity > 0
    - $\Omega(1, \ldots, 1, -d(x), 1, \ldots, 1) \leq -d(x)$, where $\Omega \in L \setminus L_{BA}$ has arity > 0.

12. [Jip93] Let V be a discriminator variety axiomatised by a set Σ of universal sentences, with discriminator term $d(x)$ for Sir V. Show that V is axiomatised by the set E of equations obtained from Σ as in theorem 2.60, together with the equations for d in proposition 2.58(2). Deduce theorem 2.60.

2.7 Aspects of duality for BAOs

For a completely additive operator Ω over an atomic BAO $\mathcal{B}$, we can calculate Ω if we know how it is defined on the atoms of $\mathcal{B}$. The behaviour of the operators on the atoms can be defined by specifying the *atom structure* of $\mathcal{B}$ (see definition 2.62). This is one reason why atoms are so important. The following definitions and propositions are just a (long-winded) way of saying that you can work equally well with an atomic, completely additive BAO or, if you know its boolean part, with its atom structure. (But it's easier to work with the atom structure — this is what modal logicians call 'working at the frame level'.) They also allow us to construct various new BAOs from old ones, by building atom structures.

In sections 2.7.1 and 2.7.2 we show how to extract the atom structure of an atomic BAO, and conversely to construct a BAO from an atom structure. A key example of this is the canonical extension construction, which we will meet in section 2.7.3. In the sections following, we spend a little time studying the connections between BAOs and atom structures. Complications arise because the map from an atomic BAO to its atom structure is not one-one. The question of to what extent we can get away with working with atom structures instead of BAOs has provoked extensive work, with substantial open questions remaining. We mostly confine ourselves here to outlining those results used later in the book, and giving some references for further reading.

2.7.1 Atom structures of BAOs

First, we observe that an operator really is determined by its values on atoms, in completely additive atomic BAOs.

LEMMA 2.61 [JónTar51, theorem 1.7] *Let Ω be a completely additive n-ary operator on the atomic boolean algebra $\mathcal{B}$, and let $b_0, \ldots, b_{n-1} \in \mathcal{B}$. Then*

$$\Omega(b_0, \ldots, b_{n-1}) = \sum \{\Omega(a_0, \ldots, a_{n-1}) : a_0, \ldots, a_{n-1} \text{ atoms of } \mathcal{B},\ a_i \leq b_i \text{ for each } i < n\}.$$

In particular, the sum on the right exists in $\mathcal{B}$.

Proof. We write $a_0, a_1, \ldots$ for atoms of $\mathcal{B}$. Recall from exercise 2.3(29) that as $\mathcal{B}$ is atomic, we have $b_0 = \sum \{a_0 : a_0 \leq b_0\}$. So by complete additivity of Ω, we get

$$\begin{aligned}\Omega(b_0, \ldots, b_{n-1}) &= \Omega(\sum \{a_0 : a_0 \leq b_0\}, b_1, \ldots, b_{n-1}) \\ &= \sum \{\Omega(a_0, b_1, \ldots, b_{n-1}) : a_0 \leq b_0\},\end{aligned}$$

and the right-hand side exists. Similarly, for each $a_0 \leq b_0$, we have

$$\Omega(a_0, b_1, \ldots, b_{n-1}) = \sum \{\Omega(a_0, a_1, b_2, \ldots, b_{n-1}) : a_1 \leq b_1\}.$$

So by exercise 2.3(31),

$$\begin{aligned}&\Omega(b_0,\ldots,b_{n-1})\\ &= \Sigma\{\Sigma\{\Omega(a_0,a_1,b_2,\ldots) : a_1 \leq b_1\} : a_0 \leq b_0\}\\ &= \Sigma\{\Omega(a_0,a_1,b_2,\ldots,b_{n-1}) : a_0 \leq b_0, a_1 \leq b_1\}.\end{aligned}$$

Continuing through the arguments $b_2,\ldots,b_{n-1}$ in this way, we obtain the result. □

Since by exercise 2.3(29), an element of an atomic boolean algebra is determined by the atoms beneath it, this motivates the following important definition. Cf. [JónTar51] and [HenMon⁺71, definition 2.7.32].

DEFINITION 2.62 Let L be a functional signature containing the signature L_{BA} of boolean algebras. We regard constants of L as nullary function symbols.

1. For each n-ary operator symbol $\Omega \in L \setminus L_{BA}$, introduce an $(n+1)$-ary relation symbol R_Ω. Let L^a be the relational signature $\{R_\Omega : \Omega \in L \setminus L_{BA}\}$.

2. Let $\mathcal{B}$ be an atomic BAO of signature L. The *atom structure* of $\mathcal{B}$, — in symbols, $\text{At}(\mathcal{B})$ or $\text{At}\,\mathcal{B}$ — is defined to be the L^a-structure with domain the set of atoms of $\mathcal{B}$ and with relations defined by:

$$\text{At}(\mathcal{B}) \models R_\Omega(a_0,\ldots,a_{n-1},b) \iff \mathcal{B} \models b \leq \Omega(a_0,\ldots,a_{n-1}),$$

for each n-ary $\Omega \in L \setminus L_{BA}$ and atoms $a_0,\ldots,a_{n-1},b$ of $\mathcal{B}$.[3]

REMARK 2.63 Previously, we defined $\text{At}(\mathcal{B})$ for a *boolean algebra* $\mathcal{B}$ to be the set of atoms of $\mathcal{B}$. But since the signature $(L_{BA})^a$ is empty, the atom structure of a boolean algebra, as just defined, is simply a set in the empty signature. (Also, we use the same notation for a structure as for its domain.) Therefore, the two definitions (2.5 and 2.62) are consistent with each other.

Thus, At maps an atomic BAO $\mathcal{B}$ to an atom structure, and if $\mathcal{B}$ is completely additive, its BAO structure is determined by its boolean structure and by $\text{At}(\mathcal{B})$:

PROPOSITION 2.64 *Let $\mathcal{B}$ and C be completely additive, atomic, similar BAOs. Then for any boolean isomorphism $I : \text{bool}(\mathcal{B}) \to \text{bool}(C)$ (where $\text{bool}(\mathcal{B})$ denotes the boolean part of $\mathcal{B}$), if the restriction of I to $\text{At}(\mathcal{B})$ is an isomorphism from $\text{At}(\mathcal{B})$ to $\text{At}(C)$ then I is an isomorphism from $\mathcal{B}$ to C.*

The proof of this proposition is left for exercise 1.

---

[3] As you might expect, the alternative notation $R_\Omega(b,a_0,\ldots,a_{n-1})$ also occurs in the literature, especially in modal logic.

2.7.2 Complex algebras

We can also work in the other direction, building a BAO from an atom structure. See [JónTar51, definition 3.8].

DEFINITION 2.65 Let L be as in definition 2.62, and let A be any L^a-structure. The *complex algebra* $\mathfrak{Cm}(A)$ (or $\mathfrak{Cm}A$) over A is defined to be the following BAO of signature L:

- The boolean part of $\mathfrak{Cm}(A)$ is $\langle \wp(\mathrm{dom}(A)), \emptyset, \mathrm{dom}(A), \cup, \setminus \rangle$,

- each operator symbol $\Omega \in L \setminus L_{BA}$ is interpreted as follows: if $s_0, \ldots, s_{n-1} \subseteq \mathrm{dom}(A)$, then

$$\Omega^{\mathfrak{Cm}(A)}(s_0, \ldots, s_{n-1})$$
$$= \{x \in A : A \models R_\Omega(x_0, \ldots, x_{n-1}, x) \text{ for some } x_i \in s_i \ (i < n)\}.$$

For a class C of L^a-structures, we write $\mathfrak{Cm}\,\mathsf{C}$ for $\{\mathfrak{Cm}A : A \in \mathsf{C}\}$.

Some authors also use 'complex algebra' for subalgebras of $\mathfrak{Cm}(A)$ — a 'complex' is just a subset of A. They call $\mathfrak{Cm}(A)$ the *full complex algebra* over A.

PROPOSITION 2.66 *Let L be as in definition 2.62.*

1. *For any L^a-structure A, $\mathfrak{Cm}(A)$ is a complete, atomic, completely additive L-BAO [JónTar51, theorem 3.9]. Furthermore, $A \cong \mathrm{At}\,\mathfrak{Cm}A$.*

2. *Every completely additive atomic BAO $\mathcal{B}$ (of signature L) is isomorphic to a subalgebra of $\mathfrak{Cm}\,\mathrm{At}\,\mathcal{B}$.*

3. *If $\mathcal{B}$ is a finite BAO (of signature L), then $\mathfrak{Cm}\,\mathrm{At}\,\mathcal{B} \cong \mathcal{B}$.*

Proof.

1. Easy. For the last part, the isomorphism is $a \mapsto \{a\}$, for $a \in A$.

2. The map $b \mapsto b{\downarrow} \stackrel{\mathrm{def}}{=} \{a \in \mathrm{At}\,\mathcal{B} : a \leq b\}$ embeds $\mathcal{B}$ in $\mathfrak{Cm}\,\mathrm{At}\,\mathcal{B}$. We will only check that the non-boolean operators are preserved. Let $\Omega \in L \setminus L_{BA}$ be an n-ary operator symbol, and suppose that $\mathcal{B} \models \Omega(b_0, \ldots, b_{n-1}) = b$. We must prove that

$$\mathfrak{Cm}\,\mathrm{At}\,\mathcal{B} \models \Omega(b_0{\downarrow}, \ldots, b_{n-1}{\downarrow}) = b{\downarrow}. \tag{2.7}$$

Let $a \in \mathrm{At}\,\mathcal{B}$. Then by definition of $\mathfrak{Cm}\,\mathrm{At}\,\mathcal{B}$, we have $\mathfrak{Cm}\,\mathrm{At}\,\mathcal{B} \models \{a\} \leq \Omega(b_0{\downarrow}, \ldots, b_{n-1}{\downarrow})$ iff $\mathrm{At}\,\mathcal{B} \models R_\Omega(a_0, \ldots, a_{n-1}, a)$ for some atoms $a_i \in b_i{\downarrow}$

($i < n$). By definition of At $\mathcal{B}$, this is iff $\mathcal{B} \models a \leq \Omega(a_0, \ldots, a_{n-1})$ for some atoms $a_i \in \mathcal{B}$ with $a_i \leq b_i$ (all $i < n$). Now $\mathcal{B}$ is completely additive, so by lemma 2.61 and exercise 2.3(29), this is iff $\mathcal{B} \models a \leq \Omega(b_0, \ldots, b_{n-1})$, iff $a \in b\!\downarrow$, iff $\mathfrak{Cm}\,\text{At}\,\mathcal{B} \models \{a\} \leq b\!\downarrow$. Since $\{a\}$ is an arbitrary atom of the atomic algebra $\mathfrak{Cm}\,\text{At}\,\mathcal{B}$, (2.7) follows.

3. The above embedding is an isomorphism.

□

In view of this, we sometimes identify an atom a of $\mathcal{B}$ with the atom $\{a\}$ of $\mathfrak{Cm}\,\text{At}\,\mathcal{B}$.

$\mathfrak{Cm}$ and At are therefore in some way dual to each other. If we apply At $\mathfrak{Cm}$ to an atom structure, we get back the original atom structure. As for $\mathfrak{Cm}$ At applied to a completely additive atomic algebra, we may not get back the original algebra but perhaps a bigger one. In the infinite case, it is possible for two non-isomorphic BAOs to have the same (or isomorphic) atom structures. The difference between them lies in the presence or absence of certain infima and suprema. These differences can be important: see sections 2.7.5, 14.3, and 17.7. Still, for completely additive atomic BAOs, proposition 2.64 shows that the atom structure plus the boolean part of a BAO uniquely determines the BAO.

REMARK 2.67 If $\mathcal{B}$ is a completely additive atomic BAO and C is the completion of $\mathcal{B}$ (see definition 2.42), then $C \cong \mathfrak{Cm}\,\text{At}\,\mathcal{B}$ (because $\mathfrak{Cm}\,\text{At}\,\mathcal{B}$ is complete and $\mathcal{B}$ is dense in it).

2.7.3 Canonical (perfect) extensions of BAOs

Not all BAOs are complete, completely additive, or atomic, but it is always possible to find an extension with these properties. If we also impose certain topological restrictions on the extension (compactness and total disconnectedness) it is still possible to find such an extension, and furthermore, all extensions with these properties are isomorphic. There is a natural way to construct such an extension, by generalising from the boolean case (theorem 2.10). This was done by Jónsson and Tarski in [JónTar51, theorem 2.15]. The resulting extension is called the *canonical extension* (or *canonical embedding algebra*).

The canonical extension of a BAO is useful in the same way as for boolean algebras. It is complete and atomic and (up to isomorphism) extends the original BAO. However, as with boolean algebras, it does not in general preserve all infima and suprema that exist in the original algebra (see exercise 2.3(26)). The completion of a BAO (section 2.5.4) does preserve all existing infima and suprema, but it is only atomic when the original algebra is. Both constructions have their uses.

The canonical extension is obtained from a BAO $\mathcal{B}$ by taking the power set of the set of ultrafilters over $\mathcal{B}$ (i.e., over bool($\mathcal{B}$)), as in Stone's theorem (2.10). The

2.7. Aspects of duality for BAOs

boolean operations are just the natural set-theoretic operations, as before, but now we also have to define the non-boolean operators on sets of ultrafilters. Note that the power set algebra is complete and atomic (the atoms are just singleton sets of ultrafilters), and thus we can define the operators in a completely additive way by dealing with the atom structure. Indeed, this is one of the most important uses of atom structures.

DEFINITION 2.68 Let $L \supseteq L_{BA}$ be a functional signature and $\mathcal{B}$ an L-BAO. We can define an L^a-structure $\mathrm{Uf}(\mathcal{B})$ with domain the set of ultrafilters of $\mathcal{B}$, by interpreting the $(n+1)$-ary relation R_Ω as follows, for each n-ary operator symbol $\Omega \in L$ and ultrafilters $\beta_0, \ldots, \beta_{n-1}, \gamma$ of $\mathcal{B}$:

$$\mathrm{Uf}(\mathcal{B}) \models R_\Omega(\beta_0, \ldots, \beta_{n-1}, \gamma)$$
$$\iff \{\Omega^{\mathcal{B}}(b_0, \ldots, b_{n-1}) : b_0 \in \beta_0, \ldots, b_{n-1} \in \beta_{n-1}\} \subseteq \gamma.$$

As in remark 2.63, there is no real clash involved in using the notation $\mathrm{Uf}(\mathcal{B})$ for the *set* of ultrafilters of a *boolean algebra* $\mathcal{B}$, and for the L^a-*structure* defined on the ultrafilters of a *BAO* $\mathcal{B}$, since for boolean algebras they amount to the same thing.

We can now define the notions of canonical extension and canonical variety. They are some of the most important ideas in the theory of BAOs.

DEFINITION 2.69 Let $L \supseteq L_{BA}$ be a functional signature, and let $\mathcal{B}$ be any L-BAO.

1. The *canonical extension* or *canonical embedding algebra* $\mathcal{B}^+$ of $\mathcal{B}$ is defined to be $\mathfrak{Cm}(\mathrm{Uf}(\mathcal{B}))$ — the unique completely additive BAO whose atom structure is $\mathrm{Uf}(\mathcal{B})$ and whose domain is the power set of the set of ultrafilters of $\mathcal{B}$.

2. A class K of BAOs is said to be *canonical* if it is closed under canonical extensions: i.e., if $\mathcal{B} \in \mathsf{K}$ then $\mathcal{B}^+ \in \mathsf{K}$.

3. A variety V of BAOs is said to be a *canonical variety* if it is a canonical class.

Again, if $\mathcal{B}$ is a boolean algebra then definition 2.69 and the one in section 2.3.6 agree about $\mathcal{B}^+$, so there is no clash in notation.

THEOREM 2.70 *There is an embedding from any BAO $\mathcal{B}$ into its canonical extension $\mathcal{B}^+$. If $\mathcal{B}$ is finite then $\mathcal{B} \cong \mathcal{B}^+$.*

Proof. This embedding is given by

$$I : b \mapsto \{\gamma \in \mathrm{Uf}(\mathcal{B}) : b \in \gamma\}.$$

For finite $\mathcal{B}$, it is a surjective isomorphism. □

Because of this theorem, we often identify a BAO $\mathcal{B}$ with its image under the given embedding, and so regard $\mathcal{B}$ as a subalgebra of $\mathcal{B}^+$. This justifies the use of the term 'extension' in definition 2.69.

The following theorem shows that canonical extensions behave well with respect to subalgebras. A slightly weaker version, for cylindric algebras, can be found in [HenMon⁺71, theorem 2.7.23].

THEOREM 2.71 *Let $\mathcal{A}, \mathcal{B}$ be similar BAOs. Any embedding $\iota : \mathcal{A} \to \mathcal{B}$ has a 'canonical' extension to a complete embedding $\iota^+ : \mathcal{A}^+ \to \mathcal{B}^+$ (see definition 2.15 for complete embeddings).*

Proof. For simplicity of notation, we assume that $\mathcal{A} \subseteq \mathcal{B}$ and ι is the inclusion map. We also identify ultrafilters of $\mathcal{A}$ with atoms of $\mathcal{A}^+$, and similarly for $\mathcal{B}$. Note that the intersection with $\mathcal{A}$ of an ultrafilter of $\mathcal{B}$ is an ultrafilter of $\mathcal{A}$. For each $S \in \mathcal{A}^+$ (so $S \subseteq \mathrm{Uf}(\mathcal{A})$), we let

$$\iota^+(S) = \{\beta \in \mathrm{Uf}(\mathcal{B}) : \beta \cap \mathcal{A} \in S\} \in \mathcal{B}^+.$$

We check that $\iota^+ : \mathcal{A}^+ \to \mathcal{B}^+$ is a complete embedding. First, it must preserve the boolean operations. We have $\iota^+(0^{\mathcal{A}^+}) = \iota^+(\emptyset) = \emptyset = 0^{\mathcal{B}^+}$. Because the intersection with $\mathcal{A}$ of an ultrafilter of $\mathcal{B}$ is an ultrafilter of $\mathcal{A}$, we have $\iota^+(\mathrm{At}\,\mathcal{A}^+) = \mathrm{At}\,\mathcal{B}^+$: i.e., $\iota^+(1^{\mathcal{A}^+}) = 1^{\mathcal{B}^+}$. Clearly, for $S, S' \in \mathcal{A}^+$, we have $\iota^+(S \cup S') = \iota^+(S) \cup \iota^+(S')$, and $\iota^+(-S) = \{\mu : \mu \cap \mathcal{A} \in -S\} = \{\mu : \mu \cap \mathcal{A} \notin S\} = -\iota^+(S)$. If $S \neq \emptyset$ then by BPI (fact 2.8), $\iota^+(S) \neq \emptyset$ too, so by exercise 2.3(3), ι^+ is one-one.

Now we check that ι^+ is a complete embedding. By lemma 2.17, it suffices to show that for any atom β of $\mathcal{B}^+$, there is an atom α of $\mathcal{A}^+$ with $\beta \leq \iota^+(\alpha)$. We may simply take $\alpha = \beta \cap \mathcal{A}$.

Now let Ω be an n-ary operator symbol of the signature of $\mathcal{A}$ and $\mathcal{B}$, and let $S_1, \ldots, S_n \in \mathcal{A}^+$. We show that

$$\iota^+(\Omega^{\mathcal{A}^+}(S_1, \ldots, S_n)) = \Omega^{\mathcal{B}^+}(\iota^+(S_1), \ldots, \iota^+(S_n)).$$

First assume that $\beta \in \Omega^{\mathcal{B}^+}(\iota^+(S_1), \ldots, \iota^+(S_n))$. By complete additivity of $\Omega^{\mathcal{B}^+}$, there are ultrafilters $\beta_1 \in \iota^+(S_1), \ldots, \beta_n \in \iota^+(S_n)$ such that $\Omega^{\mathcal{B}}(b_1, \ldots, b_n) \in \beta$ for all $b_i \in \beta_i$ ($i \leq n$). Then $\beta_i \cap \mathcal{A} \in S_i$ for all i, and if $a_1 \in S_1, \ldots, a_n \in S_n$ then $\Omega^{\mathcal{A}}(a_1, \ldots, a_n) = \Omega^{\mathcal{B}}(a_1, \ldots, a_n) \in \beta \cap \mathcal{A}$. Hence, $\beta \cap \mathcal{A} \in \Omega^{\mathcal{A}^+}(S_1, \ldots, S_n)$ and $\beta \in \iota^+(\Omega^{\mathcal{A}^+}(S_1, \ldots, S_n))$.

Conversely, assume that $\beta \in \iota^+(\Omega^{\mathcal{A}^+}(S_1, \ldots, S_n))$. Then we have $\beta \cap \mathcal{A} \in \Omega^{\mathcal{A}^+}(S_1, \ldots, S_n)$, so there are ultrafilters $\alpha_1, \ldots, \alpha_n$ of $\mathcal{A}$ with $\alpha_1 \in S_1, \ldots, \alpha_n \in S_n$, and $\Omega^{\mathcal{A}}(a_1, \ldots, a_n) \in \beta \cap \mathcal{A}$ for all $a_1 \in \alpha_1, \ldots, a_n \in \alpha_n$. We want to extend the α_i to similar ultrafilters of $\mathcal{B}$.

We do this one at a time. Let $\varphi_1, \ldots, \varphi_n$ be non-empty subsets of $\mathcal{B} \setminus \{0\}$ that are closed under $\cdot$ and are such that $\Omega^{\mathcal{B}}(b_1, \ldots, b_n) \in \beta$ for all $b_1 \in \varphi_1, \ldots,$

2.7. Aspects of duality for BAOs

$b_n \in \varphi_n$. E.g., $\varphi_i = \alpha_i$; or φ_i might be an ultrafilter of $\mathcal{B}$ extending α_i that has already been constructed. Fix $m \leq n$. An element $b \in \mathcal{B}$ is said to be *good* if $\Omega^{\mathcal{B}}(b_1, \ldots, b_{m-1}, b, b_{m+1}, \ldots, b_n) \in \beta$ for all $b_1 \in \varphi_1, \ldots, b_n \in \varphi_n$. b is *bad* if it is not good. Elements of φ_m are good, and 0 is bad. Since $\Omega^{\mathcal{B}}$ is additive and the φ_i are closed under $\cdot$, it is easily seen that the bad elements form an ideal of $\mathcal{B}$. So if b is good and b' bad, then $b \not\leq b'$, so $b - b' > 0$. It now follows that $\{b - b' : b \in \varphi_m, b' \in \mathcal{B} \text{ bad}\}$ has the finite intersection property (see fact 2.8). By BPI, we may choose an ultrafilter β_m of $\mathcal{B}$ containing it. Then $\varphi_m \subseteq \beta_m$ and every element of β_m is good.

Repeating this argument for successive $m \leq n$ gives ultrafilters β_i of $\mathcal{B}$ containing α_i and such that $\Omega^{\mathcal{B}}(b_1, \ldots, b_n) \in \beta$ for all $b_i \in \beta_i$ ($i \leq n$). Therefore, $\beta_i \in \iota^+(S_i)$ for each i, and $\beta \in \Omega^{\mathcal{B}^+}(\beta_1, \ldots, \beta_n) \subseteq \Omega^{\mathcal{B}^+}(\iota^+(S_1), \ldots, \iota^+(S_n))$, as desired. □

For a generalisation, see exercise 7 below.

Topological view Now we give a topological characterisation of canonical extensions. First we define this for boolean algebras without operators.

DEFINITION 2.72 Cf. [JónTar51, definitions 1.19, 2.14]

- $\mathcal{P}$ is a *perfect extension* of a boolean algebra $\mathcal{B}$ if

    - $\mathcal{P}$ is a complete, atomic boolean algebra,
    - $\mathcal{B} \subseteq \mathcal{P}$,
    - for all $p, q \in \text{At}(\mathcal{P})$, if $p \neq q$ then there exists $b \in \mathcal{B}$ with $p \leq b$ and $q \cdot b = 0$ [totally disconnected], and
    - for any subset $S \subseteq \mathcal{B}$, if $\sum^{\mathcal{P}} S = 1$ (note that the supremum is the least upper bound in $\mathcal{P}$) then there is a finite subset F of S such that $\sum^{\mathcal{P}} F = 1$ [compact].

- Let $\mathcal{B}$ and $\mathcal{P}$ be similar BAOs with $\mathcal{B} \subseteq \mathcal{P}$ (subalgebra). We say that $\mathcal{P}$ is a *perfect extension* of $\mathcal{B}$ if, as boolean algebras, the boolean reduct $\text{bool}(\mathcal{P})$ is a perfect extension of $\text{bool}(\mathcal{B})$, and $\mathcal{P}$ is completely additive.

THEOREM 2.73 *Let $\mathcal{B}$ and $\mathcal{P}$ be similar BAOs. If $\mathcal{P}$ is a perfect extension of $\mathcal{B}$ then $\mathcal{P}$ is isomorphic to the canonical extension $\mathcal{B}^+$.*

Proving this is the task of exercise 3. Cf. [JónTar51, theorems 2.16, 2.17].

2.7.4 Axiomatising the atom structures of a variety

We have seen how to pass from an atomic BAO to its atom structure, and how to get back at least one atomic BAO, the complex algebra, from such an atom structure. As it is usually easier to work with atom structures than with algebras, this is an attractive technique. However, we will usually be interested in a particular *variety* V of BAOs. If we are going to work with atom structures instead, we need to know to what extent V is determined by its associated atom structures (see definition 2.74 below), and how to characterise them.

This section is concerned with obtaining atom structures from a variety. (In section 2.7.5 we will consider how to go in the other direction.) The question we try to answer here is 'which atom structures do we get when we apply At to atomic algebras in V?' Formally,

DEFINITION 2.74 Let V be a variety of BAOs. We let $\mathrm{At}V = \{\mathrm{At}\,\mathcal{B} : \mathcal{B} \in V,\ \mathcal{B}\ \text{atomic}\}$.

At V is the class of atom structures of atomic algebras in V. Knowing axioms for V, can we then axiomatise At V?

Because atom structures are much the same as modal frames, these kinds of problem have been extensively studied by modal logicians, and as a result our question has a complete answer for completely additive varieties of BAOs. Two different methods apply.

1. We can use 'Sahlqvist correspondence'. This says the following (see section 2.7.6 for the definitions): if V is a conjugated Sahlqvist variety of BAOs, then At V is axiomatised by the 'first-order correspondents' of the Sahlqvist equations defining V. This produces a finite, 'natural' axiomatisation of At V from one for V, and often the requirement that V is conjugated can be replaced by requiring only complete additivity.

2. In full generality, for any completely additive variety V of BAOs, At V is elementary and axioms defining it can be synthesised from equations defining V. This approach is due to Venema [Ven97a]. While this method is more general than the first, the axioms obtained are 'theoretical' and are infinite in number even if V is finitely axiomatised.

In this section we outline these two methods. Sahlqvist correspondence is very general and useful, but we will not need its full power in this book, so we will only prove a 'cut-down' version adequate for our later applications. In section 2.7.6 we will describe the general theory, stating its results and referring the reader to the literature for details of the proofs. Venema's method ((2) above) will be needed later to axiomatise the 'weakly representable atom structures' of relation algebras, so we give full details of it.

2.7. Aspects of duality for BAOs

Fix a functional signature $L \supseteq L_{BA}$, and let L^a be the signature of atom structures of L-algebras, as in definition 2.62.

Standard translation

We begin by defining a 'standard translation' of L-terms into L^a-formulas, commonly used in modal logic (cf. [Ben85, BlaRij$^+$01]). This will be needed in both methods.

DEFINITION 2.75 For each L-term $t = t(\bar{x})$ written with variables from $\bar{x}$, and any variable y not in $\bar{x}$, we define an L^a-formula $\delta_t(\bar{x}, y)$ by induction on t as follows:

- $\delta_0 \stackrel{\text{def}}{=} y \neq y$
- $\delta_1 \stackrel{\text{def}}{=} y = y$
- For a variable x, we let δ_x be $y = x$
- $\delta_{t+u} \stackrel{\text{def}}{=} \delta_t \vee \delta_u$
- $\delta_{-t} \stackrel{\text{def}}{=} \neg \delta_t$
- $\delta_{\Omega(t_0,\ldots,t_{n-1})}(\bar{x}, y) \stackrel{\text{def}}{=} \exists z_0 \ldots z_{n-1} \left(R_\Omega(z_0, \ldots, z_{n-1}, y) \wedge \bigwedge_{i<n} \delta_{t_i}(\bar{x}, z_i) \right)$,

  where $\Omega \in L \setminus L_{BA}$ is an n-ary operator symbol and $z_0, \ldots, z_{n-1}$ are new variables.

Then we have:

LEMMA 2.76 *For any L-term $t(\bar{x})$, if $\mathcal{B}$ is any completely additive atomic BAO of signature L and $\bar{a} \in \text{At}\,\mathcal{B}$, then for all atoms $b \in \mathcal{B}$,*

$$\mathcal{B} \models b \leq t(\bar{a}) \quad \textit{iff} \quad \text{At}\,\mathcal{B} \models \delta_t(\bar{a}, b).$$

This says that in a completely additive atomic L-BAO, given any L-term with atoms as parameters, the set of atoms beneath the value of the term is definable in the atom structure with the same parameters. The defining formula is independent of the BAO.

Proof. By induction on t. The base cases and boolean cases are straightforward and left to the reader. We check the case of operator symbols. Let $t_0, \ldots, t_{n-1}$ be terms with variables among $\bar{x}$, and assume inductively that the $\delta_{t_i}(\bar{x}, y)$ (for $i < n$) meet the conditions. Let $\Omega \in L \setminus L_{BA}$ be an n-ary operator symbol, let $\mathcal{B}$ be a completely additive atomic L-BAO, and let $\bar{a}, b \in \text{At}\,\mathcal{B}$. Then by lemma 2.61 and exercise 2.3(29), $\mathcal{B} \models b \leq \Omega(t_0(\bar{a}), \ldots, t_{n-1}(\bar{a}))$ iff there exist atoms $c_0, \ldots, c_{n-1}$ with $\mathcal{B} \models b \leq \Omega(c_0, \ldots, c_{n-1})$ and $\mathcal{B} \models c_i \leq t_i(\bar{a})$ for each $i < n$. We use complete additivity here. By definition of At $\mathcal{B}$ and the inductive hypothesis, this is iff there exist $c_0, \ldots, c_{n-1} \in$ At $\mathcal{B}$ such that At $\mathcal{B} \models R_\Omega(c_0, \ldots, c_{n-1}, b)$ and At $\mathcal{B} \models \delta_{t_i}(\bar{a}, c_i)$ for each $i < n$. By definition of δ, this is iff At $\mathcal{B} \models \delta_{\Omega(t_0, \ldots, t_{n-1})}(\bar{a}, b)$. □

Simple Sahlqvist method of axiomatising At V

DEFINITION 2.77 We define the following kinds of L-term.

1. A *positive (negative) term* is one in which every variable lies under an even (respectively, odd) number of occurrences of $-$.

2. A *simple term* is one built from variables and constants using only $\cdot$ and the operator symbols of $L \setminus L_{BA}$ (i.e., $+, -$ only occur in the abbreviation '$\cdot$', we defined $x \cdot y = -(-x + -y)$).

3. A *simple equation* is an equation of the form

$$s(\bar{x}, v_0(\bar{x}, \bar{z}), \ldots, v_{n-1}(\bar{x}, \bar{z})) = 0,$$

where $s(\bar{x}, u_0, \ldots, u_{n-1})$ is a simple term, $\bar{x}$ is a tuple of distinct variables, each of which occurs exactly once in $s(\bar{x}, u_0, \ldots, u_{n-1})$, $u_0, \ldots, u_{n-1}$ are variables not occurring in $\bar{x}$, and $v_0, \ldots, v_{n-1}$ are negative terms whose variables are from $\bar{x}\bar{z}$.

Simple equations are very special kinds of Sahlqvist equation (we will discuss these in section 2.7.6). In spite of their very restrictive definition, many varieties do turn out to be axiomatised by simple equations — in particular, the variety of relation algebras (chapter 3).

Part 1 of the following lemma says that 'positive terms are monotonic'.

LEMMA 2.78 *Let $\mathcal{B}$ be any L-BAO. In the following, all terms are L-terms.*

1. *Let $\pi(x, \bar{y})$ be a positive term, $v(x, \bar{y})$ a negative one. If $a, b, \bar{c} \in \mathcal{B}$ with $a \leq b$, then $\mathcal{B} \models \pi(a, \bar{c}) \leq \pi(b, \bar{c})$ and $\mathcal{B} \models v(a, \bar{c}) \geq v(b, \bar{c})$.*

2. *Let $s(x, \bar{y})$ be a simple term in which x occurs exactly once. Assume that $\mathcal{B}$ is atomic and completely additive. If $a \in$ At $\mathcal{B}$, $b, \bar{c} \in \mathcal{B}$, and $\mathcal{B} \models a \leq s(b, \bar{c})$, then $b > 0$ and there is an atom $b' \leq b$ with $\mathcal{B} \models a \leq s(b', \bar{c})$.*

Proof. A straightforward induction on terms. In the inductive step in (2), use lemma 2.61 and exercise 2.3(29). □

2.7. Aspects of duality for BAOs

THEOREM 2.79 *Let ε be a simple equation of L. Then there is a first-order L^a-sentence χ_ε, called the correspondent of ε, such that for any completely additive atomic L-BAO $\mathcal{B}$, we have $\mathcal{B} \models \varepsilon$ iff At $\mathcal{B} \models \chi_\varepsilon$.*

Proof. We may take ε to be $s(\bar{x}, \nu_0(\bar{x},\bar{z}), \ldots, \nu_{n-1}(\bar{x},\bar{z})) = 0$, where $s(\bar{x},\bar{u})$ is built from variables with only $\cdot$ and operator symbols, the tuple $\bar{x} = (x_0, \ldots, x_{m-1})$ consists of distinct variables not occurring in $\bar{z}$, and each variable in $\bar{x}$ occurs exactly once in s. If ε is not valid in $\mathcal{B}$, there are $a \in \text{At}\,\mathcal{B}$ and $\bar{b}, \bar{c} \in \mathcal{B}$ with $\mathcal{B} \models a \leq s(\bar{b}, \nu_0(\bar{b},\bar{c}), \ldots, \nu_{n-1}(\bar{b},\bar{c}))$. Here, $\bar{b} = (b_0, \ldots, b_{m-1})$. By m applications of lemma 2.78(2), we see that for each $i < m$ there is an atom $b'_i \leq b_i$ such that if $\bar{b}' = (b'_0, \ldots, b'_{m-1})$ then $\mathcal{B} \models a \leq s(\bar{b}', \nu_0(\bar{b},\bar{c}), \ldots, \nu_{n-1}(\bar{b},\bar{c}))$.

By lemma 2.78(1) applied to s and the ν_j, if $\bar{0}$ is a sequence of $|\bar{c}|$ 0s then $\mathcal{B} \models s(\bar{b}', \nu_0(\bar{b},\bar{c}), \ldots, \nu_{n-1}(\bar{b},\bar{c})) \leq s(\bar{b}', \nu_0(\bar{b}',\bar{0}), \ldots, \nu_{n-1}(\bar{b}',\bar{0}))$, so we have

$$\mathcal{B} \models a \leq s(\bar{b}', \nu_0(\bar{b}',\bar{0}), \ldots, \nu_{n-1}(\bar{b}',\bar{0})).$$

Since clearly this condition implies that ε is not valid in $\mathcal{B}$, we see that ε is valid in $\mathcal{B}$ iff there are do not exist atoms a, b_i for $i < m$ such that if $\bar{b} = (b_0, \ldots, b_{m-1})$ then $\mathcal{B} \models a \leq s(\bar{b}, \nu_0(\bar{b},\bar{0}), \ldots, \nu_{n-1}(\bar{b},\bar{0}))$.

We now simply use the standard translation (definition 2.75) to express this in terms of atom structures. Let

$$\sigma(\bar{x}) = s(\bar{x}, \nu_0(\bar{x},\bar{0}), \ldots, \nu_{n-1}(\bar{x},\bar{0})).$$

Then by lemma 2.76, ε is valid in $\mathcal{B}$ iff At $\mathcal{B} \models \forall \bar{x}y \neg \delta_\sigma(\bar{x}, y)$. We may take the correspondent to be $\chi_\varepsilon = \forall \bar{x}y \neg \delta_\sigma(\bar{x}, y)$. □

COROLLARY 2.80 *If V is a completely additive variety of BAOs axiomatised by a set Σ of simple equations, then AtV is elementary and is axiomatised by the set $\{\chi_\varepsilon : \varepsilon \in \Sigma\}$ of their first-order correspondents.*

Proof. If $\mathcal{B} \in V$ is atomic then $\mathcal{B} \models \Sigma$ and $\mathcal{B}$ is completely additive; by theorem 2.79, At $\mathcal{B} \models \{\chi_\varepsilon : \varepsilon \in \Sigma\}$. Conversely, if A is an L^a-structure and $A \models \{\chi_\varepsilon : \varepsilon \in \Sigma\}$, then the complex algebra $\mathfrak{Cm}A$ is a completely additive BAO with $\mathfrak{Cm}A \models \Sigma$. Hence, $\mathfrak{Cm}A \in V$, so by proposition 2.66, $A \cong \text{At}\,\mathfrak{Cm}A \in \text{At}\,V$. □

EXAMPLE 2.81 Consider the simple equation $\Omega(x) \leq -\Omega(-\Omega(x))$, or equivalently, $\Omega(x) \cdot \Omega(-\Omega(x)) = 0$. It is similar to the modal S5 axiom $\Diamond p \to \Box \Diamond p$. Here, we can take $s(x, u)$ to be $\Omega(x) \cdot \Omega(u)$ and $\nu(x)$ to be $-\Omega(x)$. We see that x occurs only once in s. The correspondent is obtained as follows. The standard translation $\delta(x, y)$ of $\Omega(x) \cdot \Omega(-\Omega(x))$ is $\exists u(R_\Omega(u, y) \wedge u = x) \wedge \exists v(R_\Omega(v, y) \wedge \neg \exists w(R_\Omega(w, v) \wedge w = x))$. This is logically equivalent to $R_\Omega(x, y) \wedge \exists v(R_\Omega(v, y) \wedge \neg R_\Omega(x, v))$. So the correspondent is

$$\begin{aligned}\chi &\equiv \forall xy \neg (R_\Omega(x, y) \wedge \exists v(R_\Omega(v, y) \wedge \neg R_\Omega(x, v))) \\ &= \forall xyv(R_\Omega(x, y) \wedge R_\Omega(v, y) \to R_\Omega(x, v)).\end{aligned}$$

REMARK 2.82 A common case is equations of the form $s(\bar{x}) = t(\bar{x})$, where s, t are built from variables $\bar{x}$ using only $\cdot$ and operators and each variable in $\bar{x}$ occurs exactly once in s and in t. Such an equation is equivalent in BAOs to the conjunction of the two simple equations $s \cdot -t = 0$ and $t \cdot -s = 0$. By theorem 2.79, their correspondents are $\forall \bar{x} y \neg \delta_{s \cdot -t}(\bar{x}, y)$ and $\forall \bar{x} y \neg \delta_{t \cdot -s}(\bar{x}, y)$. By definition of the standard translation, these are equivalent to $\forall \bar{x} y (\delta_s(\bar{x}, y) \to \delta_t(\bar{x}, y))$ and $\forall \bar{x} y (\delta_t(\bar{x}, y) \to \delta_s(\bar{x}, y))$. So we see that the original equation $s = t$ has a correspondent: for any atomic completely additive BAO $\mathcal{B}$, $s = t$ is valid in $\mathcal{B}$ iff At $\mathcal{B} \models \forall \bar{x} y (\delta_s(\bar{x}, y) \leftrightarrow \delta_t(\bar{x}, y))$.

General method of axiomatising At V

This is Venema's method [Ven97a], and works for any completely additive variety, even one that is not Sahlqvist-axiomatised. We begin with the following easy consequence of the 'standard translation' definition.

COROLLARY 2.83 *If ε is an L-equation, then there is a first-order L^a-theory Σ_ε such that for any completely additive atomic BAO $\mathcal{B}$ generated by its atoms,*

$$\mathcal{B} \models \varepsilon \iff \text{At } \mathcal{B} \models \Sigma_\varepsilon.$$

Proof. Let ε be $t = t'$, where $t(\bar{z}), t'(\bar{z})$ are L-terms with variables from the tuple $\bar{z} = (z_1, \ldots, z_n)$, say. Enumerate all n-tuples of L-terms, as $\tau_i = (u_i^1, \ldots, u_i^n)$, say, for $i < \kappa = |L|$, and choose a tuple $\bar{x}_i$ of variables containing the variables of $u_i^1, \ldots, u_i^n$. For each i, let t_i and t_i' denote $t(u_i^1, \ldots, u_i^n)$ and $t'(u_i^1, \ldots, u_i^n)$, respectively. Then $\bar{x}_i$ contains the variables of t_i, t_i'. Now let

$$\Sigma_\varepsilon = \{\forall \bar{x}_i y \big(\delta_{t_i}(\bar{x}_i, y) \leftrightarrow \delta_{t_i'}(\bar{x}_i, y) \big) : i < \kappa \},$$

where the L^a-formulas $\delta_{t_i}, \delta_{t_i'}$ are as in lemma 2.76. Then for $\mathcal{B}$ as in the corollary, we have $\mathcal{B} \models \varepsilon$ iff $\mathcal{B} \models t(\bar{a}) = t'(\bar{a})$ for all n-tuples $\bar{a} = (a_1, \ldots, a_n) \in \mathcal{B}$. Since $\mathcal{B}$ is generated by atoms, for all $a_1, \ldots, a_n \in \mathcal{B}$ there are $i < \kappa$ and an $|\bar{x}_i|$-tuple $\bar{b}_i \in \text{At } \mathcal{B}$ such that $\mathcal{B} \models a_j = u_i^j(\bar{b}_i)$ for each $1 \leq j \leq n$. So $\mathcal{B} \models \varepsilon$ iff $\mathcal{B} \models t_i(\bar{b}_i) = t_i'(\bar{b}_i)$ for all $i < \kappa$ and $\bar{b}_i \in \text{At } \mathcal{B}$. Using complete additivity, by lemma 2.76 we have

$$\mathcal{B} \models t_i(\bar{b}_i) = t_i'(\bar{b}_i) \text{ iff } \text{At } \mathcal{B} \models \forall y \big(\delta_{t_i}(\bar{b}_i, y) \leftrightarrow \delta_{t_i'}(\bar{b}_i, y) \big).$$

It follows that $\mathcal{B} \models \varepsilon$ iff At $\mathcal{B} \models \Sigma_\varepsilon$. □

THEOREM 2.84 (Venema, [Ven97a]) *Let Φ be an equational axiomatisation of a completely additive variety V of L-BAOs. Then the class At V of atom structures of atomic algebras in V is elementary and is axiomatised by the L^a-theory $\Sigma_\Phi \stackrel{\text{def}}{=} \bigcup_{\varepsilon \in \Phi} \Sigma_\varepsilon$.*

2.7. Aspects of duality for BAOs

Proof. Take atomic $\mathcal{B} \in \mathsf{V}$; we show $\mathsf{At}\,\mathcal{B} \models \Sigma_\Phi$. Let $\mathcal{A}$ be the subalgebra of $\mathcal{B}$ generated by the atoms of $\mathcal{B}$. Then $\mathcal{A} \in \mathsf{V}$ as V is a variety, so $\mathcal{A} \models \Phi$. By corollary 2.83, $\mathsf{At}\,\mathcal{B} = \mathsf{At}\,\mathcal{A} \models \Sigma_\Phi$.

Conversely, let $A \models \Sigma_\Phi$; we show that $A \in \mathsf{At}\,\mathsf{V}$. Let $\mathcal{B}$ be the subalgebra of $\mathfrak{Cm}\,A$ generated by the atoms of $\mathfrak{Cm}\,A$. (Later, we will call $\mathcal{B}$ the *term algebra* over A.) Clearly, $\mathcal{B}$ and $\mathfrak{Cm}\,A$ have the same atoms. From this we deduce:

1. $\mathcal{B}$ is a dense subalgebra of $\mathfrak{Cm}\,\mathcal{A}$, so by exercise 2.5(2), $\mathcal{B}$ is completely additive,

2. $\mathcal{B}$ is atomic,

3. $\mathcal{B}$ is generated by its own atoms,

4. $\mathsf{At}\,\mathcal{B} \cong A$. For clearly, $\mathsf{At}\,\mathcal{B} = \mathsf{At}\,\mathfrak{Cm}\,A$; by proposition 2.66, $\mathsf{At}\,\mathfrak{Cm}\,A \cong A$.

By (4) and the initial assumption, $\mathsf{At}\,\mathcal{B} \models \Sigma_\Phi$. Corollary 2.83 now yields $\mathcal{B} \models \Phi$, whence $\mathcal{B} \in \mathsf{V}$ and $A \cong \mathsf{At}\,\mathcal{B} \in \mathsf{At}\,\mathsf{V}$, as required. $\square$

See proposition 2.88 and exercises 12 and 20 below for related results.

2.7.5 Recovering a variety from its atom structures?

The results of the preceding section show how to pass from a variety V to its class $\mathsf{At}\,\mathsf{V}$ of atom structures. The converse direction, however, is not so straightforward. In this section, we briefly survey the situation.

The canonical atom-canonical case: $\mathsf{At}\,\mathsf{V}$

Starting with a variety V, we would like to recover V from $\mathsf{At}\,\mathsf{V}$. Together with the results of the preceding section, this would confirm the place of $\mathsf{At}\,\mathsf{V}$ as a useful counterpart to V on the atom structure level.

How can we obtain BAOs from atom structures at all? We saw in definition 2.65 how to form a completely additive BAO $\mathfrak{Cm}\,A$, the complex algebra of A, from an atom structure A. So the obvious approach is to consider the smallest variety containing $\mathfrak{Cm}\,\mathsf{At}\,\mathsf{V}$, the class of complex algebras of atom structures in $\mathsf{At}\,\mathsf{V}$. That is, we consider **HSP** $\mathfrak{Cm}\,\mathsf{At}\,\mathsf{V}$, the *variety generated by* $\mathsf{At}\,\mathsf{V}$, and try to show that it is V.

An obvious necessary condition for this is *atom-canonicity* (from [Ven97b]):

DEFINITION 2.85 A variety V of BAOs is *atom-canonical* if $\mathfrak{Cm}\,\mathsf{At}\,\mathsf{V} \subseteq \mathsf{V}$.

Less obviously, canonicity of V is also necessary, and these two conditions together are sufficient. The proof will use the following important result in modal logic due to Fine [Fin75] and van Benthem [Ben80], reformulated by Goldblatt in terms of ultraproducts and BAOs ([Gol89, theorem 3.6.7]; see also [Gol91, Gol95]). This reformulation states the following:

FACT 2.86 *Let* V *be a variety of BAOs. If* V *is generated by a class of atom structures that is closed under ultraproducts, then* V *is a canonical variety.*

For a proof, see exercise 3.4(14). The hypothesis cannot be weakened to closure under ultrapowers [Gol91]. In related work, Givant developed theorem 2.57 to obtain the following theorem. It gives a very neat proof that the class **SP(GRA)** of subalgebras of products of group relation algebras is a canonical variety (see exercise 4.2(2)).

THEOREM 2.87 (Givant, [Giv99, theorem 2.6]) *Let* D *be a discriminator variety, and suppose that* K *is a class of simple algebras in* D. *If there is a class* L *of relational structures that is closed under ultraproducts and such that*

$$\mathfrak{Cm}\,L \subseteq K \subseteq S\,\mathfrak{Cm}\,L$$

then **SP**K *is a canonical variety and* **S**K *is the universal class of non-degenerate simple algebras in* **SP**K.

We can now prove what we wanted. We note that [Gol95, theorem 5.8] proved that if V is any canonical atom-canonical variety then At V is elementary.

PROPOSITION 2.88 *Let* V *be a variety of BAOs.*

1. *If* V *is atom-canonical, then* At V *is elementary and can be effectively axiomatised from an axiomatisation of* V.

2. *The following are equivalent:*

    (a) V *is canonical and atom-canonical,*

    (b) V *is generated by* At V *in the strong sense that* V = **S** $\mathfrak{Cm}$ At V,

    (c) V *is generated by* At V.

Proof.

1. This is proved similarly to theorem 2.84. Assume that V is atom-canonical; we show that At V is axiomatised by the theory Σ_Φ given in theorem 2.84, where Φ is an equational axiomatisation of V. If $\mathcal{B} \in$ V is atomic, we require At $\mathcal{B} \models \Sigma_\Phi$. By atom-canonicity, $\mathfrak{Cm}$ At $\mathcal{B} \in$ V. So the subalgebra $\mathcal{A}$ of $\mathfrak{Cm}$ At $\mathcal{B}$ generated by At $\mathcal{B}$ is in V, so $\mathcal{A} \models \Phi$; also, $\mathcal{A}$ is a dense subalgebra of $\mathfrak{Cm}$ At $\mathcal{B}$, so is completely additive. By corollary 2.83, At $\mathcal{B}$ = At $\mathcal{A} \models \Sigma_\Phi$. The proof that if $A \models \Sigma_\Phi$ then $A \in$ At V is as in theorem 2.84.

2.7. Aspects of duality for BAOs 91

2. $(a) \Rightarrow (b)$ Assume that V is canonical and atom-canonical. If $\mathcal{A} \in V$ then $\mathcal{A}^+ \in V$, so $\mathrm{At}\,\mathcal{A}^+ \in \mathrm{At}\,V$, and as $\mathcal{A}^+ = \mathfrak{Cm}\,\mathrm{At}\,\mathcal{A}^+$, $\mathcal{A}^+ \in \mathfrak{Cm}\,\mathrm{At}\,V$. Since $\mathcal{A} \subseteq \mathcal{A}^+$, we have $\mathcal{A} \in \mathbf{S}\,\mathfrak{Cm}\,\mathrm{At}\,V$, and as $\mathcal{A}$ was arbitrary, we have $V \subseteq \mathbf{S}\,\mathfrak{Cm}\,\mathrm{At}\,V$. Conversely, atom-canonicity of V yields $\mathfrak{Cm}\,\mathrm{At}\,V \subseteq V$ and so $\mathbf{S}\,\mathfrak{Cm}\,\mathrm{At}\,V \subseteq \mathbf{S}V = V$.

$(b) \Rightarrow (c)$ is trivial.

$(c) \Rightarrow (a)$ Assume that V is generated by $\mathrm{At}\,V$. Then $\mathfrak{Cm}\,\mathrm{At}\,V \subseteq V$, so V is atom-canonical. By (1), $\mathrm{At}\,V$ is elementary, so by theorem 2.32 it is closed under ultraproducts. By fact 2.86, V is canonical. $\square$

The canonical case: $\mathrm{Str}\,V$

Proposition 2.88 has quite wide application. Nonetheless, there exist important canonical varieties that are not atom-canonical. We will see in chapter 14 that the variety **RRA** of representable relation algebras is one such. On the other side, [Gol01, theorem 4.2] gives an example of an atom-canonical variety that is not canonical. For such V, the variety generated by $\mathrm{At}\,V$ is not V, and so we *cannot* in general recover V as the variety generated by $\mathrm{At}\,V$.

In the general case, then, we try to find a more appropriate class K than $\mathrm{At}\,V$ for the purpose of generating V. Our first problem is therefore: 'given a variety V, which atom structures do have the property that their complex algebras are in V?' Formally, we make the following definition.

DEFINITION 2.89 For a variety V of L-BAOs, we write $\mathrm{Str}\,V$ for the class of L^a-structures A such that $\mathfrak{Cm}\,A \in V$.

It is clear that $\mathrm{Str}\,V \subseteq \mathrm{At}\,V$, with equality iff V is atom-canonical. We also note the following:

PROPOSITION 2.90 (Goldblatt) *Let V be a variety of L-BAOs. Then* $\mathrm{Str}\,V$ *is closed under ultraroots. Further,* $\mathrm{Str}\,V$ *is closed under ultrapowers iff it is closed under ultraproducts, iff it is closed under elementary equivalence, iff it is elementary.*

Proof. See exercise 16 below, or [Gol89, theorems 3.8.1, 3.8.4]. $\square$

For a *canonical* variety V, $\mathrm{Str}\,V$ generates V — and indeed in the strong sense that $V = \mathbf{S}\,\mathfrak{Cm}\,\mathrm{Str}\,V$ (see exercise 14 below). So in this case, while not all questions about V can necessarily be answered by analysis of $\mathrm{Str}\,V$, it can nonetheless shed considerable light on V. To begin, we would like an intrinsic characterisation of $\mathrm{Str}\,V$, for example by an axiomatisation perhaps similar to the one obtained for $\mathrm{At}\,V$ in theorem 2.84. So we ask:

Question: knowing axioms for V, can we axiomatise Str V?

This question is harder than the corresponding one for At V. In the case where V is atom-canonical, we have Str V = At V, and hence, by proposition 2.88, Str V is elementary; but for canonical atom-canonical V we already had a satisfactory answer through the proposition. *Conjugated Sahlqvist varieties* (see section 2.7.6, next) are canonical (theorem 2.95) and atom-canonical ([Ven97b, corollary 1]; see exercise 19 below), so again are covered by proposition 2.88. For Sahlqvist varieties in general, the answer remains 'yes':

PROPOSITION 2.91 *If V is a Sahlqvist variety of L-BAOs, then Str V is elementary and V = S𝔈m Str V.*

Proof. By theorem 2.94 below, any Sahlqvist equation ε has a first-order correspondent: an L^a-sentence η such that for any L^a-structure A, $A \models \eta$ iff $\mathfrak{Cm} A \models \varepsilon$. (We proved this for simple equations in theorem 2.79.) Hence, if V is defined by Sahlqvist equations, their correspondents axiomatise Str V. Moreover, V is canonical (theorem 2.95 below); so by exercise 14 below, V = S𝔈m Str V. □

It is therefore viable to study such varieties V by proxy, working with Str V (or At V) instead. Modal logicians do this all the time. However, the answer to the question posed above is in general negative. In chapter 14, we'll show that Str **RRA** is not an elementary class.

Elementary alternatives to Str V?

For any variety V of BAOs, Str V is plainly the largest possible class K of atom structures that generates V. Any such K is contained in Str V. It would be of interest to find an elementary class K, possibly smaller than Str V but still generating V. Such a K could then be used to study V as an alternative to Str V, which as we mentioned is not in general elementary.

It is noteworthy that existence of such a K implies by fact 2.86 that V is canonical. It is a long-open and intensely studied question whether the converse to this fact holds. Goldblatt showed it to be true for canonical atom-canonical varieties [Gol95, theorem 5.8]: we already saw this above. He also shows in [Gol95, theorem 4.13] that if a variety V is generated by *any* class K of atom structures closed under ultraproducts, then it is also generated — in the strong sense that V = S𝔈m K — by an elementary class K satisfying

$$\mathrm{Cst}\, V \subseteq K \subseteq \mathrm{Str}\, V,$$

where $\mathrm{Cst}\, V = \{\mathrm{At}\,\mathcal{A}^+ : \mathcal{A} \in V\}$. So at least we know where to look for a suitable K. For **RRA** there are several candidate classes, and we will see some later — e.g., in chapters 3 and 11. This means that to a large degree we can study **RRA**

2.7. Aspects of duality for BAOs

using atom structures. But for arbitrary canonical varieties our search remains incomplete given the current state of knowledge, and for non-canonical varieties we are asking the impossible. Readers interested in pursuing these matters could well begin by studying [Gol89, Gol91, Gol95, Gol01].

2.7.6 Sahlqvist varieties

These are varieties axiomatised by 'Sahlqvist equations', of a special syntactic form. The simple equations of definition 2.77 are Sahlqvist equations. Sahlqvist equations have first-order correspondents and are preserved in canonical extensions (canonical embedding algebras) and completions of BAOs. We cannot describe this work in detail here — we will only present those results that will be needed later, and suggest some further reading. We thank Yde Venema for help in preparing this section.

We begin by defining Sahlqvist equations. There are many definitions of these in the literature, though most of them are equivalent. We give one that is simple but adequate for our purposes.

DEFINITION 2.92 Consider terms in a functional signature $L \supseteq L_{BA}$.

1. A *boxed*[4] *variable* is a term of the form $-\Omega_1 \Omega_2 \ldots \Omega_n(-x)$, for some $n \geq 0$ and unary operator symbols $\Omega_1, \ldots, \Omega_n \in L \setminus L_{BA}$.

2. A *strictly positive term* is one built from variables and constants using only $+, \cdot,$ and the operator symbols of $L \setminus L_{BA}$ (i.e., the only negations occur in the abbreviation '$\cdot$' — we defined $x \cdot y = -(-x + -y)$).

3. A *Sahlqvist equation* is an equation of the form
$$s(\beta_0, \ldots, \beta_{m-1}, \nu_0, \ldots, \nu_{n-1}) = 0,$$
where $s(x_0, \ldots, x_{m-1}, u_0, \ldots, u_{n-1})$ is a strictly positive term, $\beta_0, \ldots, \beta_{m-1}$ are boxed variables, and $\nu_0, \ldots, \nu_{n-1}$ are negative terms (definition 2.77).

4. A *Sahlqvist variety* is one axiomatised by Sahlqvist equations.

EXAMPLE 2.93 If t, u are strictly positive terms, the equation $t = u$ is equivalent in BAOs to the Sahlqvist equation $t \cdot -u + u \cdot -t = 0$.

Any simple equation (definition 2.77) is a Sahlqvist equation.

If $s = 0$ is a Sahlqvist equation and t is a positive term (see definition 2.77), then $s \leq t$ is equivalent in BAOs to the Sahlqvist equation $s \cdot -t = 0$.

---
[4] The term $-\Omega(-x)$ is called the *dual* of $\Omega(x)$, and is sometimes written $\Omega_\delta(x)$, $\Omega^d(x)$, etc. It corresponds to $\Box x$ in modal logic, whereas $\Omega(x)$ corresponds to $\Diamond x$.

Some Sahlqvist equations can be seen as such in more than one way: in $-\Omega(z) \cdot -\Omega(z') = 0$, for example, we can take s to be a variable u and v to be $-\Omega(z) \cdot -\Omega(z')$, or we can take s to be $u \cdot u'$, v to be $-\Omega(z)$, and v' to be $-\Omega(z')$.

The first important property of Sahlqvist equations is 'correspondence'. Theorem 2.79 is a special case.

THEOREM 2.94 *Any Sahlqvist L-equation has a first-order 'correspondent' — a first-order L^a-sentence that holds in an atom structure of a conjugated atomic BAO iff the original equation holds in the BAO. For BAOs of the form $\mathfrak{Cm}A$ (i.e., complex algebras), the 'conjugated' condition can be dropped. For Sahlqvist equations where the boxed variables are just variables, the 'conjugated' condition can be replaced by the weaker requirement of being completely additive.*

The correspondent can be easily (and effectively, by a simple algorithm) obtained from the equation.

Proof. See, e.g., [Ben85, SamVac89, RijVen95, Kra99, BlaRij$^+$01]. □

The second aspect of Sahlqvist theory is 'persistence'.

THEOREM 2.95 *Let $\mathcal{B}$ be a BAO and ε a Sahlqvist equation. If $\mathcal{B} \models \varepsilon$ then $\mathcal{B}^+ \models \varepsilon$ (that is, Sahlqvist equations are canonical). Hence, any Sahlqvist variety is canonical.*

Proof. [JónTar51] proved that strictly positive equations, and some implications, are preserved in canonical extensions. [Sah75] extended this result to equations similar to the Sahlqvist equations defined above, using modal logical techniques. [RijVen95, theorem 3.5] proves the result for BAOs from this 'modal' viewpoint. [Jón95] gives an algebraic treatment; [GivVen99] point out that this can potentially be applied to other kinds of extension. □

An analogue of this theorem covers completions. Later, we will use it to show that the variety of representable relation algebras is not a Sahlqvist variety.

THEOREM 2.96 *Suppose that $\mathcal{B}$ is a conjugated BAO with completion C, and let ε be a Sahlqvist equation. If $\mathcal{B} \models \varepsilon$ then $C \models \varepsilon$. Hence, a conjugated Sahlqvist variety is closed under completions.*

Proof. For the case of strictly positive equations (example 2.93), and without the 'conjugated' restriction, see [Mon70]. For the case of atomic $\mathcal{B}$, see [Ven97b, theorem 2], using [Ven93, theorem 3.5]; or use theorem 2.94 and remark 2.67. See [GivVen99, corollary 34] for the general case. This paper also shows that the 'conjugated' condition is necessary, and proves that in any (not necessarily conjugated) BAO, all Sahlqvist equations in which the 'boxed variables' of definition 2.92(3) are actually variables (for example, strictly positive equations) are preserved in completions; this generalises Monk's result. □

2.7. Aspects of duality for BAOs

Sahlqvist theory is now quite a large field, involving work of many authors. Two important pioneering papers are [JónTar51, Sah75]. See [SamVac89] for a topological proof of Sahlqvist's theorem in modal logic, and [RijVen95, Jón95] for algebraic proofs. [Kra99] gives a characterisation of the first-order formulas that are equivalent to Sahlqvist formulas, which, among other things, helps to make sense of the plethora of definitions of Sahlqvist equation in the literature. [GivVen99] discusses the results and extends them to completions. General references for the area include [ChagZak97, Kra99, BlaRij$^+$01].

Exercises

All varieties in the exercises are of BAOs.

1. Prove proposition 2.64: if $\mathcal{B}, C$ are completely additive, atomic, similar BAOs, then $\mathcal{B} \cong C$ iff there is a boolean isomorphism from $\mathcal{B}$ onto C whose restriction to the atoms of $\mathcal{B}$ is an isomorphism between At $\mathcal{B}$ and At C.

2. Let $L \supseteq L_{BA}$ be a functional signature, let $n < \omega$, let $\Omega \in L$ be an n-ary operator symbol, let $\mathcal{B}$ be any L-BAO, and let $\beta_0, \ldots, \beta_{n-1}, \gamma \in \mathrm{Uf}(\mathcal{B})$. Show that the following are equivalent:

   - $\mathrm{Uf}(\mathcal{B}) \models R_\Omega(\beta_0, \ldots, \beta_{n-1}, \gamma)$,
   - $\mathcal{B} \models \Omega(b_0, \ldots, b_{n-1}) \cdot c \neq 0$ for all $b_0 \in \beta_0, \ldots, b_{n-1} \in \beta_{n-1}, c \in \gamma$.

3. Prove theorem 2.73: if $\mathcal{P}$ is a perfect extension of the BAO $\mathcal{A}$ then $\mathcal{P}$ is isomorphic to the canonical extension $\mathcal{A}^+$. For the proof, show first that $\mathcal{P}$ and $\mathcal{A}^+$ have isomorphic atom structures. For this isomorphism, consider the functions $f : \mathrm{Uf}(\mathcal{A}) \to \mathrm{At}\,\mathcal{P}$ and $g : \mathrm{At}\,\mathcal{P} \to \mathrm{Uf}(\mathcal{A})$ defined by

$$f(\gamma) = \prod \gamma,$$
$$g(b) = \{a \in \mathcal{A} : a \geq b\}.$$

   Show that f and g are inverses and hence bijections. Then check that they are isomorphisms. Finally use proposition 2.64 to show that $\mathcal{A}^+$ and $\mathcal{P}$ are isomorphic.

4. [JónTar51, definition 2.14, theorem 2.15] Let $\mathcal{A}$ be a BAO, regarded as a subalgebra of $\mathcal{A}^+$ as usual. Show that if Ω is an n-ary operator of $\mathcal{A}^+$, and $a_0, \ldots, a_{n-1} \in \mathrm{At}\,\mathcal{A}^+$, then

$$\Omega(a_0, \ldots, a_{n-1}) = \prod \{\Omega(b_0, \ldots, b_{n-1}) : b_i \in \mathcal{A},\, b_i \geq a_i,\, i < n\}.$$

5. Let $\mathcal{A} \subseteq \mathcal{B}$ be BAOs. Identifying $\mathcal{A}$ with its image in $\mathcal{B}^+$ under the embedding given in theorem 2.70, prove that $\mathcal{A}^+$ is naturally isomorphic to the complete subalgebra of $\mathcal{B}^+$ generated by $\mathcal{A}$ — i.e., the closure of $\mathcal{A}$ in $\mathcal{B}^+$ under arbitrary sums and products. (See [AndGiv$^+$95, corollary 1.7], or for the special case of cylindric algebras, see [HenMon$^+$71, theorem 2.7.23].)

6. We can split theorem 2.71 in two (parts b and c below). Let $L \supseteq L_{BA}$ be a functional signature and let A, B be L^a-structures (see definition 2.62). A map $f : A \to B$ is called a *bounded morphism* if it is a homomorphism (see section 2.2.4) and for every n-ary operator symbol $\Omega \in L \setminus L_{BA}$, $a \in A$, and $b_0, \ldots, b_{n-1} \in B$, if $B \models R_\Omega(b_0, \ldots, b_{n-1}, f(a))$ then there are $a_0, \ldots, a_{n-1} \in A$ with $A \models R_\Omega(a_0, \ldots, a_{n-1}, a)$ and $f(a_i) = b_i$ for each $i < n$. See [Gol89, BlaRij$^+$01] for much more on this topic.

   (a) Let $\mathcal{A}, \mathcal{B}$ be completely additive atomic L-BAOs and let $f : \mathcal{A} \to \mathcal{B}$ be a complete homomorphism — i.e., one that respects arbitrary suprema and infima. Find a bounded morphism $f' : \text{At}\,\mathcal{B} \to \text{At}\,\mathcal{A}$.

   (b) Let $\mathcal{A}, \mathcal{B}$ be arbitrary L-BAOs and let $f : \mathcal{A} \to \mathcal{B}$ be any embedding. Show that $f^- : \text{Uf}(\mathcal{B}) \to \text{Uf}(\mathcal{A})$ given by $f^-(\gamma) = \{a \in \mathcal{A} : f(a) \in \gamma\}$ (for any ultrafilter γ of $\mathcal{B}$) is a surjective bounded morphism.

   (c) Let A, B be L^a-structures, and let $f : A \to B$ be a surjective bounded morphism. Show that $f^* : \mathfrak{Cm}\,B \to \mathfrak{Cm}\,A$ given by $f^*(S) = \{a \in A : f(a) \in S\}$ (for $S \subseteq B$) is a complete embedding of BAOs.

7. Let $\mathcal{A}, \mathcal{B}$ be BAOs and let $h : \mathcal{A} \to \mathcal{B}$ be a homomorphism. Define $h^+ : \mathcal{A}^+ \to \mathcal{B}^+$ by letting
$$h^+(S) = \{\rho \in \text{Uf}(\mathcal{B}) : \exists \gamma \in S(h[\gamma] \subseteq \rho)\}$$
for any set of ultrafilters $S \in \mathcal{A}^+$. Prove that h^+ is a complete homomorphism (defined in exercise 6a) from $\mathcal{A}^+$ to $\mathcal{B}^+$. (This extends theorem 2.71. It was known to Jónsson and Tarski and is proved in [AndGiv$^+$95, corollary 1.6].) Show that if h is surjective then so is h^+. Deduce that if a class K of BAOs is canonical then so is H K.

8. Let $\{\mathcal{A}_i : i < n\}$ be a finite set of similar BAOs. Prove that $\prod_{i<n}(\mathcal{A}_i^+) \cong (\prod_{i<n}\mathcal{A})^+$. (Cf. exercise 2.4(3). See [AndGiv$^+$95, corollary 1.8], and for the case of cylindric algebras see [HenMon$^+$71, theorem 2.7.29].)

9. [We heard this from Venema] Consider the equations for a discriminator variety in proposition 2.58(2). Regarding the term $d(x)$ as a unary operator symbol, show that the equations are simple (see definition 2.77) and calculate their first-order correspondents. Show that if Ω is a unary operator

2.7. Aspects of duality for BAOs

symbol, the correspondents taken together imply that $R_d(x,y)$ contains the equivalence relation generated by $R_\Omega(x,y)$.

10. Let $s(x, \bar y)$ be an L-term built from variables $x, \bar y$ using only $\cdot$ and operator symbols, where x does not occur in $\bar y$. Let $\mathcal{B}$ be a completely additive atomic L-BAO and let $a \in \text{At}\,\mathcal{B}$ and $b, \bar c \in \mathcal{B}$ be such that $b > 0$ and $\mathcal{B} \models a \leq s(b, \bar c)$. Let k be the number of distinct occurrences of x in s. Show that there are atoms $b'_0, \ldots, b'_{k-1} \leq b$ in $\mathcal{B}$, not necessarily distinct, satisfying $\mathcal{B} \models a \leq s(\sum_{i<k} b'_i, \bar c)$.

11. Consider an equation $s(\bar x, v_0(\bar x, \bar z), \ldots, v_{n-1}(\bar x, \bar z)) = 0$, where $s(\bar x, \bar u)$ is a term built from variables and constants with only $\cdot$ and operator symbols, $\bar x = (x_0, \ldots, x_{m-1})$ is a tuple of distinct variables not occurring in $\bar u$ or $\bar z$, all variables in $\bar x$ occur in s, and the v_i are negative terms. For $i < m$, let k_i denote the number of distinct occurrences of x_i in $s(\bar x, \bar u)$. Introduce new distinct variables x_i^j for $i < m$, $j < k_i$, and define the following (tuples of) terms:

$$\begin{aligned}
\tilde x &= (x_0^0, \ldots, x_{m-1}^{k_{m-1}-1}), \\
t_i &= x_i^0 + \cdots + x_i^{k_i-1} \quad \text{for each } i < m, \\
\bar t &= (t_0, \ldots, t_{m-1}), \\
\sigma(\bar x) &= s(\bar t, v_0(\bar t, \bar 0), \ldots, v_{n-1}(\bar t, \bar 0)).
\end{aligned}$$

Now let $\mathcal{B}$ be a completely additive atomic BAO. Show that ε is valid in $\mathcal{B}$ iff $\text{At}\,\mathcal{B} \models \forall \tilde x y \neg \delta_\sigma(\tilde x, y)$.

12. [Gol95, Gol01] Let V be any variety of BAOs. Show that At V is closed under isomorphism and ultraproducts. Assuming further that V is completely additive, show that At V is closed under ultraroots. Deduce in this case that At V is an elementary class (theorem 2.84).

13. Prove theorem 2.87 [use theorem 2.57 and fact 2.86].

14. Show that for a canonical variety V, we have $V = \mathbf{S}\,\mathfrak{Cm}\,\text{Str}\,V$.

15. Let V be a variety of BAOs. Show that Str V is closed under disjoint unions. [This is the dual of V being closed under products.]

16. Prove proposition 2.90. [See [Gol89]. Hint for the second part: show that the complex algebra $\mathfrak{Cm}(\prod_D A_i)$ of an ultraproduct of atom structures A_i ($i \in I$) is a homomorphic image of $\mathfrak{Cm}\,B$, where B is an ultrapower of the disjoint union of the A_i.]

17. [Gol01] Let $L \supseteq L_{BA}$ be a functional signature. Let Φ be an operation assigning a subalgebra of $\mathfrak{Cm}\,A$ to any L^a-structure A, and satisfying

- $A \cong B \Rightarrow \Phi(A) \cong \Phi(B)$
- For any L^a-structures A_λ ($\lambda \in \Lambda$) and any ultrafilter D on Λ, $\Phi(\prod_D A_\lambda)$ embeds into the ultraproduct $\prod_D(\Phi(A_\lambda))$
- For any L^a-structure A and any ultrafilter D on a set Λ, $\Phi(A)$ embeds into $\Phi(\prod_D A)$.

Show that for any universal class V of L-BAOs, $\{A : \Phi(A) \in V\}$ is elementary. Deduce that the class $\mathsf{Wst}\,V$ of L^a-structures whose term algebras are in V is elementary, and satisfies $\mathsf{Str}\,V \subseteq \mathsf{Wst}\,V \subseteq \mathsf{At}\,V$. [Using work of Venema, [Gol01, theorem 4.1] shows that the variety V defined by the simple equation $f(x) \leq g(f(x))$ satisfies $\mathsf{Str}\,V = \mathsf{Wst}\,V \subset \mathsf{At}\,V$.]

18. Let V be a variety of BAOs such that whenever $A_\lambda \in \mathsf{Cst}\,V$ ($\lambda \in \Lambda$) and D is an ultrafilter on Λ, then $\prod_D A_\lambda \in \mathsf{Str}\,V$. Show that V is canonical. [Use fact 2.86.]

19. Show that any completely additive variety is atom-canonical if it is
    (a) closed under completions (cf. [Ven97b, corollary 1]), or
    (b) axiomatised by simple equations (definition 2.77), or
    (c) a conjugated Sahlqvist variety [use theorem 2.96].

20. Let $L \supseteq L_{BA}$ be a functional signature and let $2 \leq r \leq \omega$ be such that all function symbols of $L \setminus L_{BA}$ have arity $< r$. Let $r \leq k \leq \omega$. For an L^a-structure A, let $\Delta^k(A)$ denote the subalgebra of $\mathfrak{Cm}\,A$ whose domain consists of all subsets of A of the form $\{a \in A : A \models \varphi(a, \bar{b})\}$, where $\varphi(x, \bar{y})$ is a first-order L^a-formula written with variables $\{x_i : i < k\}$ only, and $\bar{b}$ is a tuple of elements of A. In short, $\Delta^k(A)$ is the algebra of k-variable first-order definable (with parameters) subsets of A. For $k = \omega$, it is often called the *first-order algebra* over A.

    (a) Show that $\Delta^k(A)$ is indeed a subalgebra of $\mathfrak{Cm}\,A$ (i.e., it is closed under the boolean and L-operators).
    (b) If $a \in A$, show that $\{a\} \in \Delta^k(A)$. Deduce that $\Delta^k(A)$ is atomic with atom structure isomorphic to A.

    Now let V be a variety of L-BAOs, and let
    $$\mathsf{FO}^k V = \{L^a\text{-structures } A : \Delta^k(A) \in V\}.$$

    (c) Show that $\mathsf{Str}\,V \subseteq \mathsf{FO}^\omega V \subseteq \cdots \subseteq \mathsf{FO}^{r+1} V \subseteq \mathsf{FO}^r V \subseteq \mathsf{At}\,V$.
    (d) For $r \leq k \leq \omega$, show that $\mathsf{FO}^k V$ is elementary and axioms for it can be obtained effectively from equations defining V.

Chapter 3

Binary relations and relation algebra

3.1 Algebraic logic

We could motivate algebraic logic by surveying its numerous applications in temporal reasoning [AllKau83, All84, AllHay85a, Lig90, Mei91, DecMei+91] and [KauLad91, Lig94, NebBür94, Hir96, Hir95], in planning [AllKoo83, Pel88], in databases [DeaMcD87], in modal logic, and elsewhere. Instead, we prefer to recommend the part of algebraic logic most relevant to our investigations here as a very well established mathematical procedure — *algebraisation* — applied to a fundamental entity, viz. a *relation*.

Typically in algebra, we take some well known, concrete structure or class of structures S; we focus our attention on some of the associated operations while ignoring all others; we then try to handle these selected operations abstractly — usually this means providing some axioms A for the operations from which we can derive many of the important properties of S. So, for example, if we start with the integers $\mathbb{Z}$ and consider the addition operation $+$ and the constant 0, we can ignore all properties of the integers which refer to its ordering $<$, multiplication or division, etc. Considering $+$ as an abstract operation, we might put forward the axioms for abelian groups. These are certainly valid in $(\mathbb{Z}, +, 0)$.

It would be unusual to choose axioms A that were not valid in S. Immediately we face the following question: are the axioms we have chosen *complete* for S — that is, if ϕ is any formula, using only symbols for the chosen operations, that is valid over S, can we logically derive ϕ from A? In the example just mentioned, the axioms for abelian groups are not complete over $(\mathbb{Z}, +, 0)$: for example any formula $\forall x(x + x + \cdots + x = 0 \rightarrow x = 0)$ is valid over the integers but not in abelian groups. And if A is not complete over S, is it possible to choose an alternative set of axioms A' that is complete over S?

This is precisely the process that is undertaken with the algebraisation of relations. We start with a very basic mathematical object, a *relation* of a certain rank (arity). In order that we have natural operations, we actually look at a *field of relations:* a set of relations, over some fixed domain, which is closed under certain concrete operations. These operations will generally include the boolean operations: so given any two relations in a field of relations, we can form the union and intersection of the two relations; we can also form the complement of a relation, though this is generally taken relative to some biggest relation over the domain. So we have a field of sets as defined in definition 2.2. But there will also be other operations pertaining to the relational properties. With binary relations, for example, we are most interested in the identity relation, the converse of a relation, and the composition of two binary relations. These operations will be defined in a moment. A field of binary relations should then include the identity and be closed under conversion and composition.

Having chosen the rank or arity of our relations, and the operations on them, we then write out some axioms for these operations. These should include the axioms for boolean algebra but will include other axioms involving the relational operators. The axioms will almost invariably be valid over fields of relations of the appropriate type, but the question of completeness is more tricky. Let us momentarily call an abstract structure that obeys these axioms an *algebra*. The question of completeness can now be thought of in another way: is an arbitrary algebra isomorphic to one of the original, concrete structures — in this case, is it isomorphic to a field of relations of the appropriate type? An isomorphism from an algebra to a field of relations is called a *representation*.

What are the advantages of working with the algebras rather than the concrete structures? There are a number of possible answers to this rather deep question. Firstly, the technique of focusing our whole attention on a small number of properties to the exclusion of all others is quite general in scientific analysis. In the newtonian theory of gravitation, for example, it is very helpful to restrict our attention to problems involving only two bodies, at least initially. Secondly, and precisely because we are ignoring many of the concrete features of our structures, we may find that our approach is more general, and that the complexity of certain problems may be lower. Finally, algebra is a very thoroughly studied discipline and there is a range of methods and results that we can apply to the study of relations. Universal algebra and model theory, in particular, have techniques and theorems that can be exploited fruitfully here.

Outline of chapter In this chapter, we concentrate solely on the case of binary relations. We introduce concrete fields of binary relations, called 'proper relation algebras', and their algebraic counterparts, called 'relation algebras'. The question of completeness of the axioms for relation algebras will not be resolved until later, but we can make a start by establishing the basic properties of the algebras that will be needed. Both universal algebra and model theory will be helpful in doing this.

3.2 Binary relations

In the previous chapter, we considered fields of unary relations (or fields of sets), and saw that the corresponding algebra was boolean algebra. Let us start our investigation of binary relations by defining a field of binary relations. Recall that a binary relation on a set B is by definition a set of pairs of elements of B: that is, a subset of $B \times B$.

3.2.1 Proper relation algebras

DEFINITION 3.1 Let B be a set. A *proper relation algebra* (**PRA**) *with base set* B is an algebra of the form $S = (S, \emptyset, U, \cup, \setminus, \mathrm{Id}_B, ^{-1}, |)$, where S is a non-empty set of binary relations over B (so $S \subseteq \wp(B \times B)$), and the following hold:

- $(S, \emptyset, U, \cup, \setminus)$ forms a field of sets: i.e., if $r, s \in S$ then $r \cup s \in S$ and $U \setminus r \in S$.

  It follows that S contains $\emptyset, U$ and is closed under $\cap$. Also, $U \subseteq B \times B$ is the biggest binary relation in S: we have $U = \bigcup S$. U is called the *unit* of S, or the 'top element'.

- $\mathrm{Id}_B \stackrel{\mathrm{def}}{=} \{(b,b) : b \in B\} \in S$. Id_B is the identity over B.

- S is closed under taking converses: $s \in S$ implies $s^{-1} \in S$, where $s^{-1} = \{(c,b) : (b,c) \in s\}$.

- S is closed under composition of binary relations: $r, s \in S$ implies $r|s \in S$, where
  $r|s = \{(b,c) : \exists d((b,d) \in r \wedge (d,c) \in s)\}$.

Remember that $\setminus$ is a unary operation, its value on s being $U \setminus s$, but we sometimes write $r \setminus s$ as an abbreviation for $r \cap (U \setminus s)$ — see definition 2.3.

EXAMPLE 3.2 *Let $B = \{0, 1, 2\}$ and let S be the set consisting of the following subsets of $B \times B$:*

$\emptyset$ $\{(0,0),(1,1)\}$ $\{(2,2)\}$
$\{(0,1),(1,0)\}$ $\{(0,0),(1,1),(2,2)\}$ $\{(0,0),(1,1),(0,1),(1,0)\}$
$\{(2,2),(0,1),(1,0)\}$ $\{(0,0),(1,1),(2,2),(0,1),(1,0)\}$.

Here, the identity and unit are

$$\text{Id}_B = \{(0,0),(1,1),(2,2)\},$$
$$U = \{(0,0),(1,1),(2,2),(0,1),(1,0)\}.$$

All elements are self-converse, and it is easy to check that S is closed under composition. So $\langle S, \emptyset, U, \cup, \setminus, \text{Id}_B, -^{-1}, | \rangle$ *is a proper relation algebra with base B and unit U.*

We will see more examples in chapter 4.

3.2.2 Square proper relation algebras

It is important to remember that the complement $\setminus$ in a proper relation algebra is always taken relative to the unit, U. As we saw in example 3.2, the unit of a proper relation algebra with base B need not be $B \times B$. In terms of definition 3.1, $\langle S, \emptyset, U, \cup, \setminus \rangle$ is not necessarily a subalgebra of the field of sets $\langle \wp(B \times B), \emptyset, B \times B, \cup, \setminus \rangle$. It is instead a subalgebra of the field of sets $\langle \wp(U), \emptyset, U, \cup, \setminus \rangle$. Strictly, complementation depends on U, so we should perhaps write $\setminus_U$; but we mostly use the context to determine the relative complement. See exercise 8.

There are several reasons for not insisting that the unit has the form $B \times B$. The most obvious is to ensure that the class of algebras isomorphic to proper relation algebras is a variety (theorem 3.37 below), so that the results of the preceding chapter about good behaviour of varieties can be applied. Another, which we mentioned in the introduction and which we will see in chapter 5, is that by relativising not just complement but all the operations to the unit, we can obtain other classes of algebras which are in many ways even better-behaved.

Nonetheless, those proper relation algebras where the unit does have the form $B \times B$ are very important, and most natural examples are like this. We call these proper relation algebras *square*. We will see that the simple and subdirectly irreducible proper relation algebras are those isomorphic to square ones.

DEFINITION 3.3 A proper relation algebra S with base B is said to be *square* if its unit U is equal to $B \times B$.

An arbitrary proper relation algebra S need not be square, but its unit will always be an equivalence relation, and we can obtain a square proper relation algebra by restricting to a single equivalence class of the unit.

LEMMA 3.4 *Let* $S = (S, \emptyset, U, \cup, \setminus, \text{Id}_B, ^{-1}, |)$ *be a proper relation algebra with base B. Then the unit U is an equivalence relation over B.*

Proof. **Reflexive** U is the biggest element of S, so $\text{Id}_B = \{(b,b) : b \in B\} \subseteq U$.

3.2. Binary relations

Symmetric Since S is closed under conversion, $U^{-1} \in S$, and since U is the top element, $U^{-1} \subseteq U$. So if $(b,c) \in U$ then $(c,b) \in U^{-1} \subseteq U$.

Transitive Since S is closed under composition, $U \mid U \in S$, and hence $U \mid U \subseteq U$. So if $(b,c), (c,d) \in U$, then by definition of composition, $(b,d) \in U \mid U \subseteq U$.

$\square$

DEFINITION 3.5 Let $S = (S, \emptyset, U, \cup, \setminus, \mathrm{Id}_B, ^{-1}, \mid)$ be a proper relation algebra with base set B. Let $E \subseteq B$ be an equivalence class of U. Define the *restricted proper relation algebra* $S{\upharpoonright}_E \stackrel{\mathrm{def}}{=} (S{\upharpoonright}_E, \emptyset, E \times E, \cup, \setminus_E, \mathrm{Id}_E, ^{-1}, \mid)$, where

- $S{\upharpoonright}_E = \{s \cap (E \times E) : s \in S\}$
- $\setminus_E(s) = (E \times E) \setminus s$.

In exercise 2, you have to check that $S{\upharpoonright}_E$ is a square proper relation algebra.

So, starting from a proper relation algebra S, it is possible to obtain a square one by restricting to any of the equivalence classes of the unit. The way to reverse this process, recovering S from all these 'square components', is, roughly, to take disjoint unions (or, viewed algebraically, products). However, there are subtleties arising from the fact that $E \times E$ may not belong to S and, further, S may not be complete as a boolean algebra.

DEFINITION 3.6 Let $S_i = (S_i, \emptyset, U_i, \cup, \setminus_i, \mathrm{Id}_{B_i}, ^{-1}, \mid)$ be a proper relation algebra with base B_i for each $i \in I$ (some index set I), and suppose that if $i \neq j$ in I then $B_i \cap B_j = \emptyset$. We define

$$\bigotimes_{i \in I} S_i = (S, \emptyset, U, \cup, \setminus, \mathrm{Id}_B, ^{-1}, \mid)$$

to be the proper relation algebra with base $B = \bigcup_{i \in I} B_i$, where:

- $S = \{\bigcup_{i \in I} s_i : s_i \in S_i \text{ for each } i \in I\}$. More formally, $S = \{\bigcup \mathrm{rng}(s) : s \in {}^I(\bigcup_{i \in I} S_i), s(i) \in S_i \text{ for each } i \in I\}$. (Note: here we are including arbitrary unions of the binary relations.)

- $U = \bigcup_{i \in I} U_i$.

LEMMA 3.7 *Let $S = (S, \emptyset, U, \cup, \setminus, \mathrm{Id}_B, ^{-1}, \mid)$ be a proper relation algebra with base B, and write $\mathcal{E}$ for the set B/U of equivalence classes of the unit U. Then*

1. *S is a subalgebra of $\bigotimes_{E \in \mathcal{E}}(S{\upharpoonright}_E)$, and*

2. *for each $E \in \mathcal{E}$, there is a homomorphism mapping S onto $S{\upharpoonright}_E$.*

Proof.

1. Each element of S is an element of $\bigotimes_{E \in \mathcal{E}} S{\restriction}_E$, S contains the constants $\emptyset, U, \mathrm{Id}_B$, and since $\bigcup_{E \in \mathcal{E}} U_E = \bigcup_{E \in \mathcal{E}} (E \times E) = U$, S contains the constants of $\bigotimes_{E \in \mathcal{E}} S{\restriction}_E$. Its operations are obtained by restricting the operations of $\bigotimes_{E \in \mathcal{E}} S{\restriction}_E$ to the elements in S. Thus, the inclusion map is an embedding from S into $\bigotimes_{E \in \mathcal{E}} S{\restriction}_E$.

2. The required homomorphism is given by $s \mapsto s \cap (E \times E)$, for $s \in S$.

$\square$

In exercises below, you are asked to check that proper relation algebras are boolean algebras with operators, square proper relation algebras are subdirectly irreducible, and that in the notation of definition 3.6, $\bigotimes_{i \in I} S_i$ is isomorphic to a direct product of the S_i. So lemma 3.7 is a special instance of the subdirect decomposition theorem (theorem 2.48). This will permit us to focus our attention on square proper relation algebras.

Exercises

1. How many equivalence classes does U have in example 3.2? Write down the square proper relation algebras obtained by restricting to each of these equivalence classes.

2. Let S be a proper relation algebra and let E be an equivalence class of the unit. Show that the restriction $S{\restriction}_E$ is a square proper relation algebra, i.e., check that it is closed under all the operations and show that it is square.

3. Find a proper relation algebra S where the unit has equivalence classes E_i : $i \in I$ such that $S \neq \bigotimes_{i \in I} S{\restriction}_{E_i}$. Under what conditions can we be sure that $S = \bigotimes_{i \in I} S_i$?

4. Show that in the circumstances of definition 3.6, $\bigotimes_{i \in I} S_i \cong \prod_{i \in I} S_i$.

5. Show that any proper relation algebra is a boolean algebra with operators.

6. Show that any square proper relation algebra is simple and subdirectly irreducible. [Is there a discriminator term?] Is there a converse to this?

7. Define the *full proper relation algebra* $\mathcal{F}(B)$ over the base B to be $(\wp(B \times B), \emptyset, B \times B, \cup, \backslash, \mathrm{Id}_B, ^{-1}, |)$. In terms of $|B|$, how many binary relations are there in $\mathcal{F}(B)$? Is $\mathcal{F}(B)$ necessarily square? Is a square proper relation algebra necessarily the full proper relation algebra over some base set?

8. Let $\sim$ be the binary set operation $a \sim b = \{x \in a : x \notin b\}$. Show that the unary operation $\backslash$ can defined by $\sim$ and that $\sim$ can be defined by $\backslash$ and $\cup$.

3.3. Relation algebras

The next three exercises illustrate examples of generalised subreducts of proper relation algebras. See section 6.4 for more information.

9. Define the *relative sum* $r \dagger s = \{(x, y) : \forall z \in B, \text{ either } (x, z) \in r \text{ or } (z, y) = s\}$ of two binary relations r, s over the base B. Show how to define † using \ and | only. Show that composition can be defined using relative sum and the other operations.

10. Define the binary operators $\triangleright, \triangleleft$ of 'proper sequential algebras' by

    $r \triangleright s = \{(x, y) : \exists z \in B \ (z, x) \in r \text{ and } (z, y) \in s\}$
    $r \triangleleft s = \{(x, y) : \exists z \in B \ (x, z) \in r \text{ and } (y, z) \in s\}.$

    Show how to define $\triangleright$ and $\triangleleft$ using composition and conversion only.

11. Repeat the preceding exercise for the *residual operators* (see figure 3.1):

    $r \setminus s = \{(x, y) : \forall z \in B((z, x) \in r \rightarrow (z, y) \in s)\}$
    $r / s = \{(x, y) : \forall z \in B((y, z) \in s \rightarrow (x, z) \in r)\}.$

    (The notation \ clashes with that for boolean complement, but we use residuals very rarely so this does not cause problems in practice.)

Figure 3.1: Residuals

3.3 Relation algebras

Having defined a proper relation algebra, we now seek an algebraic counterpart. We will write down certain axioms, defining (abstract) relation algebras. We will then show, among other things, that relation algebras form a conjugated discriminator variety of boolean algebras with operators. These results are mostly due to Tarski: see [ChiTar51, JónTar52].

3.3.1 Definition of relation algebras

The following definition, due to Tarski, is now more or less standard.

DEFINITION 3.8 A *relation algebra* is an algebra $\mathcal{A} = \langle A, 0, 1, +, -, 1', \breve{\ }, ; \rangle$ satisfying the following axioms, for all $a, b, c \in A$:

R0. The equations defining a boolean algebra (see definition 2.3)

R1. ; is an associative binary operation on A: that is, $a;(b;c) = (a;b);c$

R2. $(a+b);c = a;c + b;c$

R3. $a;1' = a$

R4. $\breve{\breve{a}} = a$

R5. $(a+b)\breve{\ } = \breve{a} + \breve{b}$

R6. $(a;b)\breve{\ } = \breve{b};\breve{a}$

R7. $\breve{a};(-(a;b)) \leq -b$.

RA denotes the class of all relation algebras.

Note that all these axioms are equations — even R7, since it abbreviates $-b + \breve{a};(-(a;b)) = -b$. (We will have more to say about R7 in section 3.3.2.)

Notation The operations $1', \breve{\ }$, and ; are called 'identity', 'conversion', and 'composition', respectively. We write $a\breve{\ }$ or $\breve{a}$, as convenient. Binding conventions: $\breve{\ }$ is tighter than $-$, so $-\breve{a}$ should be read as $-(\breve{a})$, all unary operations are tighter than binary ones, and among them, ; is the tightest, then $\cdot$ (so $a;b \cdot c$ abbreviates $(a;b) \cdot c$), then $+$. So the operations in descending order of priority are $\breve{\ }, -, ;, \cdot, +$. However, to avoid confusion we will often use parentheses.

What is the language that we use to describe relation algebras?

DEFINITION 3.9 L_{RA} is the functional signature consisting of constants $0, 1, 1'$, unary function symbols $-, \breve{\ }$, and the binary function symbols $+, ;$.

L_{RA}-structures are often called *relation-type algebras*. By our conventions in section 2.2, the first-order language in this signature is also denoted L_{RA}, and the interpretations of the symbols $0, 1, 1'$, etc., in an L_{RA}-structure $\mathcal{A}$ are denoted by $0^{\mathcal{A}}, 1^{\mathcal{A}}, 1'^{\mathcal{A}}$, etc. However, as is common practice in algebra, we will often simplify (i.e., abuse) notation by dropping the superfix '$\mathcal{A}$' and leaving the reader to infer it from the context. An L_{RA}-*term* (or 'relation algebra term') is either a constant ($1, 0,$ or $1'$), a variable symbol, or is built up from these using the functions symbols $+, -, \breve{\ }$, and ;. The atomic L_{RA}-*formulas* have the form $t = u$, for terms t, u (i.e., equations). Arbitrary formulas are built from atomic ones using the boolean connectives $\wedge, \neg$, and quantifiers.

3.3. Relation algebras

In view of Birkhoff's theorem (theorem 2.45) we are particularly interested in L_{RA}-equations. The axioms for relation algebras are L_{RA}-equations. Later (chapter 8) we will provide an equational axiomatisation of the isomorphism-closure of the class of proper relation algebras. This axiomatisation will also be in the language L_{RA}.

3.3.2 Peircean law

It will help to rewrite the last relation algebra axiom, R7, in a more intuitive and useful form, known as the Peircean law[1] or De Morgan's Theorem K.

DEFINITION 3.10 The *Peircean law* is the L_{RA}-sentence

$$(PL) \quad \forall xyz\big((x;y) \cdot \breve{z} = 0 \leftrightarrow (y;z) \cdot \breve{x} = 0\big).$$

Of course, because the variables are universally quantified, we can continue and add a third equivalence, $(z;x) \cdot \breve{y} = 0$. In relation algebras, by taking converses and applying axioms R4 and R6 and lemma 3.11 below, we can add another three equivalences, $(\breve{y};\breve{x}) \cdot z = 0$, $(\breve{z};\breve{y}) \cdot x = 0$, and $(\breve{x};\breve{z}) \cdot y = 0$. The whole says that any 'triangle' of three elements of a relation algebra can be equivalently looked at in any of the six ways resulting from applying symmetries to it. The property that the composition of the first and second elements intersects the third is invariant under the symmetries. See figure 3.2.

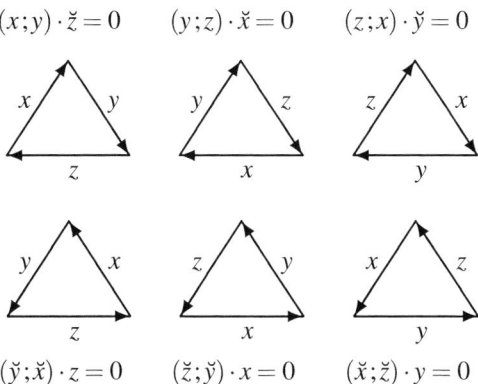

Figure 3.2: Peircean law

We begin the proof that (PL) is equivalent to (R7) by establishing that conversion is well behaved in relation algebras.

---
[1] Maddux uses the term slightly differently in [Madd82].

LEMMA 3.11 *In any relation algebra $\mathcal{A}$, $\smile$ is an automorphism of the boolean reduct* $\mathrm{bool}(\mathcal{A})$.

Proof. We must show that $\smile$ is bijective and that $\breve{0} = 0$, $\breve{1} = 1$, and $(a+b)\breve{\ } = \breve{a} + \breve{b}$ and $(-a)\breve{\ } = -\breve{a}$ for all $a, b \in \mathcal{A}$. Note that by the relation algebra axiom R0, $\mathrm{bool}(\mathcal{A})$ is a boolean algebra.

Bijectivity follows trivially from axiom R4, and $(a+b)\breve{\ } = \breve{a} + \breve{b}$ is axiom R5. Now we have $a \leq b$ iff $a + b = b$ iff $\breve{a} + \breve{b} = \breve{b}$ iff $\breve{a} \leq \breve{b}$, so we see that $\smile$ is order-preserving. Hence, since clearly $\breve{1} \leq 1$, we have $1 = (\breve{1})\breve{\ } \leq \breve{1}$ by axiom R4, so that $\breve{1} = 1$. Similarly, $0 \leq \breve{0}$, and so $\breve{0} \leq (\breve{0})\breve{\ } = 0$, so $\breve{0} = 0$.

For complement, we have $a + (-a) = 1$. So by R5, $\breve{a} + (-a)\breve{\ } = \breve{1} = 1$, giving $-\breve{a} \leq (-a)\breve{\ }$. This holds for all a, so taking the case of $\breve{a}$ we get $-\breve{\breve{a}} \leq (-\breve{a})\breve{\ }$. By R4, this is $-a \leq (-\breve{a})\breve{\ }$. We know $\smile$ preserves $\leq$, so by R4 again, $(-a)\breve{\ } \leq ((-\breve{a})\breve{\ })\breve{\ } = -\breve{a}$. Thus, $-\breve{a} \leq (-a)\breve{\ } \leq -\breve{a}$, giving $-\breve{a} = (-a)\breve{\ }$ as required. Note that only axioms R0, R4, and R5 were used. □

LEMMA 3.12 *Axioms R0, R2, R4, R5, R6* $\vdash$ *(axiom R7* $\leftrightarrow$ *PL)*.

Proof. We prove the implication right-to-left first. So let $\mathcal{A}$ be an L_{RA}-algebra satisfying axioms R0–R6. If $\mathcal{A}$ also satisfies (PL), then for all $a, b \in \mathcal{A}$, we know

$$(a;b)\breve{\ } \cdot -((a;b)\breve{\ }) = 0$$
$$\Rightarrow (\breve{b};\breve{a}) \cdot (-(a;b))\breve{\ } = 0 \quad \text{(axiom R6 and lemma 3.11)}$$
$$\Rightarrow (\breve{a};(-(a;b))) \cdot b = 0 \quad \text{(PL and axiom R4)}$$
$$\Rightarrow \breve{a};(-(a;b)) \leq -b \quad \text{(axiom R0 and def. of } \leq\text{)}.$$

So axiom R7 is valid in $\mathcal{A}$. For the converse, assume that

$$\breve{x};(-(x;y)) \leq -y \quad \text{for all } x, y \in \mathcal{A}. \tag{3.1}$$

Suppose for some $a, b, c \in \mathcal{A}$ that $a;b \cdot \breve{c} = 0$ — or equivalently, $-(a;b) \geq \breve{c}$. Then by axioms R4, R6, and lemma 3.11,

$$-(\breve{b};\breve{a}) \geq c. \tag{3.2}$$

By equation 3.1 with $x = \breve{b}$, $y = \breve{a}$, and using axiom R4, we get

$$b;(-(\breve{b};\breve{a})) \cdot \breve{a} = 0. \tag{3.3}$$

By right monotonicity of composition (using lemma 2.34 and the axioms R2, R4, R5 and R6), equations 3.2 and 3.3 yield $b;c \cdot \breve{a} = 0$. Repeating this, we get

$$a;b \cdot \breve{c} = 0 \Rightarrow b;c \cdot \breve{a} = 0$$
$$\Rightarrow c;a \cdot \breve{b} = 0$$
$$\Rightarrow a;b \cdot \breve{c} = 0,$$

which proves (PL). We did not use axioms R1 or R3. □

We conclude that (PL) holds in **RA**.

3.3.3 RA is a completely additive variety of BAOs

RA is evidently a variety, being defined by equations. We can now show easily that it is conjugated (definition 2.39), so that the non-boolean functions $\breve{\ }$ and ; are completely additive operators.

LEMMA 3.13 *In any relation algebra:*

- *The conversion operator $\breve{\ }$ is self-conjugate.*
- *The 0th conjugate of $x_0;x_1$ is $x_0;\breve{x}_1$ and the 1st conjugate of $x_0;x_1$ is $\breve{x}_0;x_1$.*

Hence, **RA** *is conjugated.*

Proof. The following are true for any elements of a relation algebra:

- For converse, we have $a \cdot \breve{b} = 0 \Leftrightarrow b \cdot \breve{a} = 0$, by axioms R0, R4, and the fact that $\breve{\ }$ preserves · (lemma 3.11).
- For composition, we have $a \cdot (b_0;b_1) = 0 \Leftrightarrow b_0 \cdot (a;\breve{b}_1) = 0 \Leftrightarrow b_1 \cdot (\breve{b}_0;a) = 0$ by axioms R0, R4, R6, and the Peircean law (PL).

Cf. the operators $\triangleright, \triangleleft$ of exercise 3.2(10). □

We now apply theorem 2.40 to obtain:

THEOREM 3.14 **RA** *is a completely additive variety of boolean algebras with operators.*

This allows us to use the relevant definitions and results from the preceding chapter. As we remarked after definition 2.33, we apply the terms and definitions for boolean algebra to the boolean part of a relation algebra. So an *atom* of $\mathcal{A}$ is a minimal, non-zero element with respect to the ordering $\leq$ and we say that $\mathcal{A}$ is *atomic* if every non-zero element is above an atom. A relation algebra is *complete* if its boolean part is a complete boolean algebra. We can define a homomorphism from one relation algebra to another, a subalgebra of a relation algebra, and a product or subdirect product of relation algebras. Other notions from the theory of boolean algebras with operators, such as a simple or subdirectly irreducible relation algebra and the canonical extension or completion of a relation algebra, are defined similarly.

The following definition will also occasionally be useful.

DEFINITION 3.15 Let $\mathcal{A}$ be an atomic relation algebra and a an atom of $\mathcal{A}$. We say that a is an *identity atom* if $a \leq 1'$, and a *diversity atom* if $a \leq -1'$. 0' may sometimes be used as an abbreviation for $-1'$.

3.3.4 RA is a canonical variety

We have

THEOREM 3.16 **RA** *is a conjugated Sahlqvist variety, and hence is canonical, closed under completions, and atom-canonical.*

Proof. By its definition and lemma 3.13, **RA** is a conjugated variety. To prove canonicity, we will use Sahlqvist theory (see [JónTar52, theorem 4.21] for the original proof). Axioms R1–R6 are strictly positive equations (hence equivalent to Sahlqvist equations), and axiom R7 is equivalent to the Sahlqvist equation $\breve{a}\,;(-(a\,;b))\cdot b = 0$. Therefore, by theorem 2.95, **RA** is a canonical variety and by theorem 2.96 it is closed under completions. It follows by remark 2.67 (cf. exercise 2.7(19)) that **RA** is atom-canonical. $\square$

3.3.5 RA is a discriminator variety

Next, we apply proposition 2.58 to show that **RA** is a discriminator variety with discriminator term $1\,;x\,;1$. We need to do some calculations first; note that the associativity axiom R1 is not used in these.

LEMMA 3.17 *Axioms R3, R4, R6* $\vdash (\breve{1\text{'}} = 1\text{'}) \wedge \forall x (1\text{'}\,;x = x)$.

Proof. $\breve{1\text{'}} =_{R3} \breve{1\text{'}}\,;1\text{'} =_{R4} \breve{1\text{'}}\,;\breve{\breve{1\text{'}}} =_{R6} (\breve{1\text{'}}\,;1\text{'})^{\breve{}} =_{R3} \breve{\breve{1\text{'}}} =_{R4} 1\text{'}$. So for any x, we have $1\text{'}\,;x =_{R4,\,R6} (\breve{x}\,;\breve{1\text{'}})^{\breve{}} = (\breve{x}\,;1\text{'})^{\breve{}} =_{R3} (\breve{x})^{\breve{}} =_{R4} x$. $\square$

LEMMA 3.18 *Let $\mathcal{A}$ be a relation algebra. For all $a \in \mathcal{A}$ we have $(a\,;\breve{a})\,;a \geq a$.*

Proof. First observe that for any non-zero $x \in \mathcal{A}$ we have $x\,;\breve{x} \neq 0$. For, if it were zero, then $x\,;\breve{x} \cdot 1\text{'} = 0$, and by (PL) and $\breve{1\text{'}} = 1\text{'}$ (from lemma 3.17) we would have $x = 1\text{'}\,;x \cdot x = 0$ which is false.

Now let $0 < x \leq a$ in $\mathcal{A}$. We know $x\,;\breve{x} \neq 0$. By monotonicity of composition (theorem 3.14 and 2.34), $x\,;\breve{x} \leq a\,;\breve{x}$ and $x\,;\breve{x} \leq a\,;\breve{a} = (a\,;\breve{a})^{\breve{}}$ (the last equality uses axioms R6 and R4). So $(a\,;\breve{x}) \cdot (a\,;\breve{a})^{\breve{}} \neq 0$. By (PL), we obtain $((a\,;\breve{a})\,;a) \cdot x \neq 0$, for all non-zero $x \leq a$. This suffices to prove $(a\,;\breve{a})\,;a \geq a$: if it fails, take $x = a - (a\,;\breve{a})\,;a$. $\square$

Now, using associativity, we obtain our desired result.

THEOREM 3.19 (McKinsey and Tarski) *The variety **RA** of relation algebras is a discriminator variety with discriminator term $1\,;x\,;1$.*

Proof. We check that **RA** validates the equations of proposition 2.58 for the term $d(x) = 1\,;x\,;1$:

3.3. Relation algebras   111

1. $x \leq 1;x;1$

2. $1;1;x;1;1 \leq 1;x;1$

3. $1;-(1;x;1);1 \leq -(1;x;1)$

4. $\breve{x} \leq 1;x;1$, $x;y \leq 1;x;1$, and $x;y \leq 1;y;1$.

By theorem 3.14, ; is additive, so by lemma 2.34 it is monotonic in each argument. We obtain $x = 1';x;1' \leq 1;x;1$, proving (1). (2) is clear, using associativity and the fact that $1;1 = 1$.

For (3), write χ for $1;x;1$. First, we show $1;-\chi \leq -\chi$. If this fails, then $(1;-\chi) \cdot \chi \neq 0$. By (PL), $(\breve{1};\chi) \cdot -\chi \neq 0$. By lemma 3.11 and using associativity and $1;1 = 1$, we have $\breve{1};\chi = 1;\chi = \chi$, so we get $\chi \cdot -\chi \neq 0$, which is false. That $-\chi;1 \leq -\chi$ is proved similarly. So $1;-\chi;1 \leq -\chi;1 \leq -\chi$, as required for (2).

For (4), by lemma 3.18 and R4 we have $\breve{x} \leq \breve{x};\breve{x};\breve{x} \leq 1;x;1$. For the composition equations, we have $x;y = 1';x;y \leq 1;x;1$, and $x;y \leq 1;y;1$ is proved similarly.

We conclude by proposition 2.58 that **RA** is a discriminator variety with discriminator term $1;x;1$. □

COROLLARY 3.20

1. A relation algebra is simple if and only if it is subdirectly irreducible.

2. Every relation algebra is a subdirect product of simple relation algebras.

3. A relation algebra $\mathcal{A}$ is simple if and only if for all non-zero $x \in \mathcal{A}$ we have $1;x;1 = 1$.

Proof. (1) is by theorems 3.19 and 2.55. (2) follows from (1) and theorem 2.48. For (3), if $\mathcal{A}$ is a relation algebra satisfying the condition then $1;x;1$ is a discriminator term for $\mathcal{A}$, so by theorem 2.55, $\mathcal{A}$ is simple. Conversely, any simple $\mathcal{A}$ is subdirectly irreducible, by (1), so theorem 3.19 tells us that $1;x;1$ is a discriminator term in it. □

3.3.6 Atom structures of relation algebras

Since relation algebras are (completely additive) BAOs, we can apply our definitions and results about atom structures (section 2.7.1). The most convenient way to specify an atomic relation algebra is by giving the atom structure. This means listing the atoms underneath the identity, listing the pairs of atoms (a,b) such that $b \leq \breve{a}$, and listing the triples of atoms (a,b,c) such that $c \leq a;b$. We end up with a structure $(X, R_{1'}, R_{\smile}, R_;)$ as in definition 2.62. This structure determines the relation algebra structure of a finite relation algebra, or more generally (see proposition 2.64) any atomic relation algebra whose boolean structure is known.

Because we are dealing with relation algebras, we can simplify the format of atom structures slightly. By lemma 3.11, if a is an atom of a relation algebra then so is $\breve{a}$. Therefore, the interpretation of $R_\smile$ in the atom structure of an atomic relation algebra will always be a function — the restriction of $\smile$ to the atoms. So we will replace the binary relation symbol $R_\smile$ by the unary function symbol $\smile$, and read $R_\smile(x,y)$ as $\breve{x} = y$.

DEFINITION 3.21 The *atom structure*, written At($\mathcal{A}$) or At$\mathcal{A}$, *of an atomic relation algebra* $\mathcal{A}$ is the structure

$$\langle S,\ \{a \in S : a \leq 1'\},\ (a \mapsto \breve{a})_{a \in S},\ \{(a,b,c) \in S : c \leq a;b\}\rangle,$$

where S is the set of atoms of $\mathcal{A}$. The signature of this structure consists of a unary relation symbol $R_{1'}$, a unary function symbol $\smile$, and a ternary relation symbol $R_;$.

Working in the other direction, given any structure A in the signature $\{R_{1'},\smile,R_;\}$, we can define a conventional L_{RA}^a-structure A' (as in definition 2.62) by letting $(R_\smile)^{A'} = \{(a,\breve{a}) : a \in A\}$. So from A', we can define an L_{RA}-BAO, $\mathfrak{Cm}A'$, as in definition 2.65. For short, we will write this BAO as $\mathfrak{Cm}A$.

$\mathfrak{Cm}A$ is of the signature of relation algebras, but it may fail to be a relation algebra. For example, $\breve{\breve{x}} = x$ may not be valid in it, ';' may fail to be associative in it, and axioms R4 and R6 have implications for the interpretation of $R_;$ which for an arbitrary interpretation may well fail. So we want to find axioms for those structures in the signature $\{R_{1'},\smile,R_;\}$ that are the atom structures of relation algebras. (Lyndon first expressed some of the relation algebra axioms in terms of atoms [Lyn50, §4], though very similar ideas can be found in [JónTar52, section 5]. [Jón59, pp. 460–461] gives a characterisation of atom structures of symmetric atomic relation algebras.)

DEFINITION 3.22 A structure S for the signature $\{R_{1'},\smile,R_;\}$ is called a *relation algebra atom structure* if it satisfies:

1. identity: for all $x,y \in S$, $x = y$ iff there is $e \in S$ such that $R_{1'}(e) \wedge R_;(x,e,y)$.

2. for all $x,y,z \in S$, if $R_;(x,y,z)$ then $R_;(\breve{x},z,y)$ and $R_;(\breve{y},\breve{x},\breve{z})$.

3. Associativity: for all $x,y,z,t \in S$,

$$\exists u\bigl(R_;(x,y,u) \wedge R_;(u,z,t)\bigr) \leftrightarrow \exists v\bigl(R_;(y,z,v) \wedge R_;(x,v,t)\bigr).$$

LEMMA 3.23 *If S is a relation algebra atom structure then $\breve{\breve{a}} = a$ for all $a \in S$.*

3.3. Relation algebras

Proof. Let $a \in S$. By (1), there is $e \in S$ with $R_{1'}(e)$ and $R_;(a,e,a)$. Successive applications of (2) give $R_;(\breve{a},a,e)$ and $R_;(\breve{a},e,a)$. By (1), $\breve{\breve{a}} = a$. □

LEMMA 3.24 *Let S be a structure for the signature $\{R_{1'}, \breve{\ }, R_;\}$ of definition 3.21. Then S is a relation algebra atom structure (definition 3.22) iff it is isomorphic to $\text{At}(\mathcal{A})$ for some atomic relation algebra $\mathcal{A}$ (definition 3.21).*

Proof. The relation algebra axioms R1, R3, R4, R6, and R7 can be written as simple equations (see definition 2.77). We begin by calculating their first-order correspondents.

R1. The correspondent of the associativity axiom $(x;y);z = x;(y;z)$ can be calculated using remark 2.82 as

$$\forall xyzt \Big(\exists u z'\big(R_;(u,z',t) \wedge \exists x'y'(R_;(x',y',u) \wedge x' = x \wedge y' = y) \wedge z' = z\big) \\ \leftrightarrow \exists x'v\big(R_;(x',v,t) \wedge x' = x \wedge \exists y'z'(R_;(y',z',v) \wedge y' = y \wedge z' = z)\big) \Big).$$

The definition of the standard translation (definition 2.75) tends to produce correspondents with a lot of equalities. We can simplify them by replacing equals by equals. (Below, we will do this automatically.) In the instance above, we obtain

$$\forall xyzt \big(\exists u(R_;(u,z,t) \wedge R_;(x,y,u)) \leftrightarrow \exists v(R_;(x,v,t) \wedge R_;(y,z,v)) \big).$$

See figure 3.3.

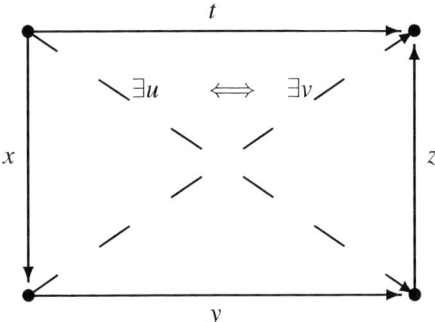

Figure 3.3: The associativity condition

R3. Applying remark 2.82 to $x;1' = x$, we obtain the correspondent

$$\forall xy(\exists z(R_;(x,z,y) \wedge R_{1'}(z)) \leftrightarrow x = y).$$

R4. Applying remark 2.82 to $\breve{\breve{x}} = x$ gives the correspondent

$$\forall xy(\exists z(R_{\smile}(z,y) \wedge R_{\smile}(x,z)) \leftrightarrow x = y).$$

It is easy to check that this is equivalent to the statement that the binary relation $R_{\smile}$ is a self-inverse bijection. We are already reading $R_{\smile}(x,y)$ as the function $\breve{x} = y$, so for our atom structures, the above correspondent is equivalent to $\forall x(\breve{\breve{x}} = x)$.

R6. Applying the remark to $(x;y)^{\smile} = \breve{y};\breve{x}$ gives the correspondent

$$\forall xyz\bigl(\exists u(R_{\smile}(u,z) \wedge R_;(x,y,u)) \\ \leftrightarrow \exists vw(R_;(v,w,z) \wedge R_{\smile}(y,v) \wedge R_{\smile}(x,w))\bigr).$$

In our notation for atom structures, this reduces to

$$\forall xyz\bigl(\exists u(\breve{u} = z \wedge R_;(x,y,u)) \leftrightarrow R_;(\breve{y},\breve{x},z)\bigr).$$

R7. In the presence of R0, $\breve{x};(-(x;y)) \leq -y$ is equivalent to the simple equation $\breve{x};(-(x;y)) \cdot y = 0$. By theorem 2.79, we see that its correspondent is $\forall xyz \neg\bigl(\exists uv(R_;(u,v,z) \wedge R_{\smile}(x,u) \wedge \neg R_;(x,y,v)) \wedge z = y\bigr)$. This reduces to $\forall xyuv\bigl(R_;(u,v,y) \wedge R_{\smile}(x,u) \to R_;(x,y,v)\bigr)$, and then to

$$\forall xyv\bigl(R_;(\breve{x},v,y) \to R_;(x,y,v)\bigr).$$

Claim. The correspondents of axioms R1, R3, R4, R6, and R7 are equivalent to the conditions given in definition 3.22.

Proof of claim. Conditions 1 and 3 of definition 3.22 are plainly the correspondents of axioms R3 and R1. By lemma 3.23 and the correspondent of axiom R4, any relation algebra atom structure and any model of the above correspondents satisfies $\forall x(\breve{\breve{x}} = x)$. So it suffices to show that a model of $\forall x(\breve{\breve{x}} = x)$ satisfies condition 2 of definition 3.22 iff the correspondents of R6 and R7 are true in it. Now $\forall x(\breve{\breve{x}} = x)$ implies that the map $x \mapsto \breve{x}$ is bijective, so we can equivalently replace any universally quantified variable x in the correspondents by $\breve{x}$, and $\breve{x}$ by x. Similarly, we can replace '$\leftrightarrow$' in the R6 correspondent by '$\to$'. Hence, modulo $\forall x(\breve{\breve{x}} = x)$, the correspondents of R6 and R7 are equivalent to

$$\forall xyu\bigl(R_;(x,y,u) \to R_;(\breve{y},\breve{x},\breve{u})\bigr),\\ \forall xyv\bigl(R_;(x,v,y) \to R_;(\breve{x},y,v)\bigr),$$

and hence to condition 2 of definition 3.22. This proves the claim.

By the claim, it only remains to check that a structure S of signature $\{R_1, \breve{}, R_;\}$ satisfies the above correspondents iff it is of the form $\text{At}\,\mathcal{A}$ for some atomic $\mathcal{A} \in \mathbf{RA}$,

3.3. Relation algebras

as in definition 3.21. First, if $S = \text{At}\,\mathcal{A}$ for atomic $\mathcal{A} \in \mathbf{RA}$, then as $\mathbf{RA}$ is a completely additive variety of BAOs (theorem 3.14), by corollary 2.80 the correspondents hold in S. Conversely, if the correspondents hold in S then by corollary 2.80 again, $\mathfrak{Cm}\,S$ satisfies the relation algebra axioms R1, R3, R4, R6, and R7. $\mathfrak{Cm}\,S$ is a BAO, so it satisfies axioms R0, R2, and R5 too. So $\mathfrak{Cm}\,S \in \mathbf{RA}$, and $\mathfrak{Cm}\,S$ is atomic with atom structure isomorphic to S. □

3.3.7 Consistent and forbidden triples of atoms

In practice, we will often specify the composition relation $R_;$ of a relation algebra atom structure by listing its *consistent triples* ([Lyn50] calls them 'cycles'). Alternatively, we can list the *inconsistent* or *forbidden triples* instead.

DEFINITION 3.25 Let $\mathcal{A}$ be an atomic relation algebra, and S a relation algebra atom structure.

1. A triple (a,b,c) of atoms of $\mathcal{A}$ is said to be *consistent* if $\breve{c} \leq a;b$.

2. A triple (a,b,c) of elements of S is said to be *consistent* if $R_;(a,b,\breve{c})$.

3. A triple (a,b,c) of atoms of $\mathcal{A}$ or elements of S is said to be *inconsistent, or forbidden,* if it is not consistent — i.e., if $(a;b) \cdot \breve{c} = 0$, or $\neg R_;(a,b,\breve{c})$, respectively.

4. The six *Peircean transforms* of a triple (a,b,c) of atoms are:

$$(a,b,c), (b,c,a), (c,a,b), (\breve{a},\breve{c},\breve{b}), (\breve{b},\breve{a},\breve{c}), (\breve{c},\breve{b},\breve{a}).$$

Consistent and forbidden triples have the advantage over $R_;$ of simpler symmetry. As we are dealing with atoms, the Peircean law (PL) for relation algebras, and lemma 3.24, yield that (a,b,c) is consistent iff (b,c,a) is consistent, iff $(\breve{c},\breve{b},\breve{a})$ is consistent. So the set of consistent triples is closed under Peircean transforms, and so is the set of forbidden triples. Also, using consistent triples saves having to remember whether $R_;(a,b,c)$ means $a \leq b;c$ or $c \leq a;b$.[2]

Clearly, in the presence of $\breve{}$, the consistent triples determine the relation $R_;$ on $\text{At}\,\mathcal{A}$, so we can and will specify a relation algebra atom structure by stating which atoms are under the identity (so specifying $R_{1'}$), which atoms are the converse of which, and by listing the consistent (or forbidden) triples of atoms.

---

[2] For an atomic BAO with an operator Ω and atoms $a, b_0, \ldots, b_{n-1}$ with $a \leq \Omega(b_0, \ldots, b_{n-1})$, some workers form the atom structure by letting $R_\Omega(a, b_0, \ldots, b_{n-1})$ in this case, while others write $R_\Omega(b_0, \ldots, b_{n-1}, a)$.

Atom structure of canonical extension As an illustration, the next lemma specifies the atom structure of the canonical extension $\mathcal{A}^+$ of a relation algebra $\mathcal{A}$. Recall that $\mathcal{A}^+$ is atomic: it is the complex algebra over the atom structure $\mathrm{Uf}(\mathcal{A})$ (see definitions 2.68, 2.69).

LEMMA 3.26 *Let $\mathcal{A}$ be any relation algebra, and let $\varphi, \psi, \chi \in \mathrm{At}(\mathcal{A}^+)$ (they are ultrafilters over $\mathcal{A}$). Then:*

1. *$\mathrm{At}\,\mathcal{A}^+ \models R_{1'}(\varphi)$ iff $1'^{\mathcal{A}} \in \varphi$.*

2. *$\breve{\varphi} = \{\breve{a} : a \in \varphi\}$.*

3. *The following are equivalent:*
   - *(a) The triple (φ, ψ, χ) is a consistent triple of atoms of $\mathcal{A}^+$.*
   - *(b) $\{r;s : r \in \varphi,\ s \in \psi\} \subseteq \chi$.*
   - *(c) $\mathcal{A} \models (r;s) \cdot \breve{t} \neq 0$ for all $r \in \varphi$, $s \in \psi$, and $t \in \chi$.*

Proof. Part 1 is by definition of $\mathrm{Uf}(\mathcal{A})$ (definition 2.68). This definition also tells us that $R_{\smile}(\varphi, \psi)$ iff $\{\breve{a} : a \in \varphi\} \subseteq \psi$. Since $a \mapsto \breve{a}$ is an automorphism of $\mathrm{bool}(\mathcal{A})$ (lemma 3.11), $\{\breve{a} : a \in \varphi\}$ is actually an ultrafilter of $\mathcal{A}$. Part 2 now follows.

For part 3, (φ, ψ, χ) is consistent iff $\mathcal{A}^+ \models \varphi;\psi \geq \chi$. By definition of composition in $\mathcal{A}^+$, $\varphi;\psi = \{\xi \in \mathrm{At}(\mathcal{A}^+) : \xi \supseteq \{r;s : r \in \varphi,\ s \in \psi\}\}$. So (φ, ψ, χ) is consistent iff χ is in this set, iff $\{r;s : r \in \varphi,\ s \in \psi\} \subseteq \chi$. This proves $a \Leftrightarrow b$.

Now we prove $b \Rightarrow c$. If $\{r;s : r \in \varphi,\ s \in \psi\} \subseteq \chi$, let $r \in \varphi$, $s \in \psi$. As $r;s \in \chi$ and χ is a filter, $(r;s) \cdot \breve{t} \neq 0$ in $\mathcal{A}$ for any $t \in \chi$.

For $c \Rightarrow b$, assume (c) and let $r \in \varphi$, $s \in \psi$; we require $r;s \in \chi$. If not, then as χ is an ultrafilter, $-(r;s) \in \chi = \{\breve{t} : t \in \chi\}$, so $t = (-(r;s))^{\smile} \in \chi$. Then $r;s \cdot \breve{t} = r;s \cdot (-(r;s)) = 0$, contradicting (c). □

Exercises

1. Show that any proper relation algebra is a relation algebra.

2. Consider the axiomatisation of relation algebras of definition 3.8. Show, in the presence of axioms R0 and R5, that validity of the equation $(a - b)^{\smile} = \breve{a} - \breve{b}$ is equivalent to the conjunction of the following three axioms

   (a) $\breve{0} = 0$
   (b) $\breve{1} = 1$
   (c) $(a \cdot b)^{\smile} = \breve{a} \cdot \breve{b}$

   for all $a, b \in \mathcal{A}$.

3.3. Relation algebras

3. Modulo the other relation algebra axioms, show that axiom R1 is equivalent to $x;(y;z) \leq (x;y);z$. Hence, show that a structure S for the signature $\{R_1, \breve{}, R_;\}$ is a relation algebra atom structure iff it satisfies conditions 1–2 of definition 3.22 and
$$\forall xyzt \left(\exists u (R_;(u,z,t) \wedge R_;(x,y,u)) \rightarrow \exists v (R_;(x,v,t) \wedge R_;(y,z,v)) \right).$$

4. Show that in a relation algebra atom structure, R_1, and $\breve{}$ are determined by (are definable in terms of) $R_;$.

5. Let a and b be atoms of a relation algebra $\mathcal{A}$. Using only the axioms for relation algebra (definition 3.8) prove formally that if $a;b \cdot 1' \neq 0$ then $a = \breve{b}$. Which axioms do you need?

6. Show that if a, b are elements of a relation algebra with $a, b \leq 1'$, then $a = \breve{a}$ and $a;b = a \cdot b$.

7. Show that a symmetric relation algebra (satisfying the equation $x = \breve{x}$) must also be commutative (satisfy $x;y = y;x$).

8. [JónTar52, theorem 4.17] A relation algebra is *integral* if it satisfies the condition $\forall xy(x;y = 0 \rightarrow x = 0 \vee y = 0)$.³ Show that a non-degenerate relation algebra is integral iff 1' is an atom. [Cf. lemma 5.13 later.]

9. Show that any simple commutative relation algebra is integral [AndGiv⁺97, p6].

10. [JónTar51, theorem 1.15] Show that in the presence of axioms R0 and R4–R6, the Peircean law (PL) is equivalent to the conjunction of the following equations.

    (a) $a;0 = 0;a = 0$
    (b) $a;b \cdot c \leq a;(b \cdot \breve{a};c)$
    (c) $b;a \cdot c \leq (b \cdot c;\breve{a});a$
    (d) $\breve{a};b \cdot c \leq \breve{a};(b \cdot a;c)$ and
    (e) $b;\breve{a} \cdot c \leq (b \cdot c;a);\breve{a}$.

11. [JónTar52, definition 4.1] and [ChiTar51]. Let $\mathcal{A} = (A, +, -, 0, 1, 1', \breve{}, ;)$ be a relation-type algebra. Show that $\mathcal{A}$ is a relation algebra iff:

    - $(A, +, -, 0, 1)$ is a boolean algebra,

---
³Some authors additionally require that an integral relation algebra be non-degenerate.

- ; is associative and $x;1' = 1';x = x$, for all $x \in A$ (i.e., $(A, ;, 1')$ is a monoid),
- $(x;y) \cdot z = 0$ iff $(\breve{x};z) \cdot y = 0$ iff $x \cdot (z;\breve{y}) = 0$, for all $x, y, z \in A$.

PROBLEM 3.27 [Jónsson] *Find all simple relation algebras with no subalgebras other than the whole algebra and the degenerate relation algebra with just one element in its domain.*

This is [Madd94a, problem P2]. Maddux stated there that 22 simple relation algebras with no non-trivial proper subalgebras had been found (this does not include the degenerate relation algebra itself).

3.4 Representations of relation algebras

Proper relation algebras are the concrete, set-theoretic objects we wish to study and relation algebras form the algebraic version. To compare these two, we define the classical notion of representation (due to Tarski).

3.4.1 The class RRA

DEFINITION 3.28 Let $\mathcal{A}$ be a relation algebra.

- A *representation of* $\mathcal{A}$ is an isomorphism from $\mathcal{A}$ to a proper relation algebra. $\mathcal{A}$ is said to be *representable* if it has a representation.

- A representation $h : \mathcal{A} \to \mathcal{P}$ is said to be *square* if $\mathcal{P}$ is a square proper relation algebra (see definition 3.3). (Do not confuse square representations with the n-square relativised representations to be seen later.)

- If $h : \mathcal{A} \to \mathcal{P}$ is a representation of $\mathcal{A}$, where $\mathcal{P}$ is a proper relation algebra with base B, we call B the *base,* and $h(1)$ the *unit,* of the representation.

- The *size* or *cardinality* of a representation is the cardinality of its base. A representation is therefore said to be *finite* (respectively, *infinite*) if it has finite (infinite) base.

- **RRA** is the class of all representable relation algebras: i.e., it is the closure of **PRA** under isomorphism.

When dealing with representations, there is another kind of isomorphism, often called *base-isomorphism*.[4]

---
[4] It is called 'equivalence' in [Lyn61], where the notion probably originates. Some writers use 'base-isomorphism' to denote an isomorphism between two relation algebras that is induced by a bijection between the base sets of representations of them.

3.4. Representations of relation algebras

DEFINITION 3.29 Two representations

$$g : \mathcal{A} \to \langle P, \emptyset, U, \cup, \setminus, \mathrm{Id}_B, ^{-1}, | \rangle$$
$$g' : \mathcal{A} \to \langle P', \emptyset, U', \cup, \setminus, \mathrm{Id}_{B'}, ^{-1}, | \rangle$$

with bases B, B', respectively, are said to be *isomorphic* (or *base-isomorphic*) if there is a bijection ι from B to B' (a *base-isomorphism*) such for any $x, y \in B$ and $a \in \mathcal{A}$ we have $(x, y) \in g(a)$ if and only if $(\iota(x), \iota(y)) \in g'(a)$.

Recall from chapter 2 that every boolean algebra is representable: i.e., isomorphic to a field of sets. Unfortunately, it turns out that Tarski's axioms for relation algebras (definition 3.8) are not complete over **RRA**. Roger Lyndon constructed the first non-representable relation algebra [Lyn50], which was a little complicated with 56 atoms. Ralph McKenzie constructed the smallest non-representable relation algebra with just four atoms [McKe70], which we will see in section 4.4.

Lyndon's example raised the problem of finding additional axioms so as to exactly define the isomorphism types of proper relation algebras — that is, to axiomatise **RRA**. In particular, it was asked if it is possible to define this class with only finitely many axioms. But Monk showed [Mon64] that **RRA** is not finitely axiomatisable. This is the chief negative result in the subject; proofs will be given in exercise 11.5(3), remark 15.13, and section 17.1.

3.4.2 Model-theoretic view of representations

Fix a relation algebra $\mathcal{A}$. We have presented the classical definition of a representation of $\mathcal{A}$ as an isomorphism to a proper relation algebra. But model theory suggests an alternative approach in which a representation of $\mathcal{A}$ is defined to be a model of a certain first-order theory (see [Tar41, McKe66]). We find this approach very natural, and it allows us to use some results from model theory.

DEFINITION 3.30 Let $\mathcal{A}$ be a relation algebra.

1. The first-order relational language $L(\mathcal{A})$ has equality plus one binary predicate symbol for each element of $\mathcal{A}$. We use the same symbol for an element of $\mathcal{A}$ as for the corresponding binary predicate in $L(\mathcal{A})$. This will not lead to ambiguity: for $R \in \mathcal{A}$, if we write $R(x, y)$, we are thinking of R as a relation symbol, but if we write simply R, we are thinking of R as an element of $\mathcal{A}$.

2. The $L(\mathcal{A})$-theory $T_{\mathcal{A}}$ consists of the following sentences:

$$\begin{array}{rcll}
\sigma_{1'} & = & \forall x, y\, [1'^{\mathcal{A}}(x, y) \leftrightarrow (x = y)] & \\
\sigma_{+}(R, S, T) & = & \forall x, y\, [R(x, y) \leftrightarrow S(x, y) \vee T(x, y)] & : R = S + T \\
\sigma_{\neg}(R, S) & = & \forall x, y\, [1^{\mathcal{A}}(x, y) \to (R(x, y) \leftrightarrow \neg S(x, y))] & : R = -S \\
\sigma_{\smile}(R, S) & = & \forall x, y\, [R(x, y) \leftrightarrow S(y, x)] & : R = \breve{S} \\
\sigma_{;}(R, S, T) & = & \forall x, y\, [R(x, y) \leftrightarrow \exists z (S(x, z) \wedge T(z, y))] & : R = S; T \\
\sigma_{\neq 0}(R) & = & \exists x, y\, R(x, y) & : R \neq 0^{\mathcal{A}},
\end{array}$$

for each $R, S, T \in \mathcal{A}$ satisfying the conditions on each line.

In line with our comments after definition 3.9, we shall generally write the relation symbols $1^{\mathcal{A}}(x,y)$, $1'^{\mathcal{A}}(x,y)$ of $L(\mathcal{A})$ simply as $1(x,y)$, $1'(x,y)$, respectively. This minimises the use of unpleasant notations such as $(1'^{\mathcal{A}})^M$ for the interpretation in a $L(\mathcal{A})$-structure M of the relation symbol $1'^{\mathcal{A}}$ of $L(\mathcal{A})$. The fact that $1'$ (for example) now has three possible meanings — a constant in L_{RA} (syntactic), the algebra element $1'^{\mathcal{A}}$ (semantic), and the relation symbol $1'^{\mathcal{A}} \in L(\mathcal{A})$ (again syntactic) — causes remarkably little confusion in practice. The most dangerous places are when we are working with the signatures L_{RA}, $L(\mathcal{A})$ at the same time (so that dropping the superfix might confuse), or when we are dealing with more than one algebra at once (so, e.g., $1'^{\mathcal{A}}$ and $1'^{\mathcal{B}}$ may be different). In these places we will be careful with the notation. Also, we will sometimes write $1'^{\mathcal{A}}$ simply to emphasise the precise meaning.

It is clear that an $L(\mathcal{A})$-structure is essentially a representation of $\mathcal{A}$ if and only if it is a model of $T_{\mathcal{A}}$. If $h: \mathcal{A} \to \mathcal{B}$ is a representation of $\mathcal{A}$, where $\mathcal{B}$ is a proper relation algebra with base X, then making X into an $L(\mathcal{A})$-structure M by defining $r^M = h(r)$, for $r \in \mathcal{A}$, we see that $M \models T_{\mathcal{A}}$. Conversely, if $M \models T_{\mathcal{A}}$ then we can obtain a representation h of $\mathcal{A}$ by defining $h(r) = r^M$, for $r \in \mathcal{A}$: the axioms of $T_{\mathcal{A}}$ ensure that h respects the algebra operations and the axioms σ_+, σ_-, and $\sigma_{\neq 0}$ ensure that it is one-one. Under this correspondence of h and M, h is a square representation iff $M \models \forall xy 1(x,y)$. In this book, we will frequently consider a representation of a relation algebra $\mathcal{A}$ to be a model of $T_{\mathcal{A}}$ rather than an isomorphism from $\mathcal{A}$ onto a proper relation algebra. (Another fruitful model-theoretic approach is to form a two-sorted structure consisting of a relation algebra and a representation of it; see chapter 9.) Thinking of representations g, h as two models of $T_{\mathcal{A}}$, we see that a base-isomorphism between the representations is the same as an isomorphism between the models.

This model-theoretic framework yields several interesting results, beginning with the following easy one from [Jón82]:

THEOREM 3.31 *If $\mathcal{A}$ is a relation algebra with an infinite square representation, then it has square representations of size κ for each infinite cardinal $\kappa \geq |\mathcal{A}|$.*

Proof. Recall from definition 3.28 that in a square representation h over the base X we have $h(1) = X \times X$. Evidently, the theory $T^+ = T_{\mathcal{A}} \cup \{\forall x \forall y 1(x,y)\}$ defines the class of square representations of $\mathcal{A}$. The upward Löwenheim–Skolem–Tarski theorem (see section 2.2.4) tells us that if T^+ has an infinite model then it has a model of cardinality κ. □

The following quantifier elimination result for square representations is also interesting (cf. [TarGiv87, §3.9]).

3.4. Representations of relation algebras

THEOREM 3.32 *Let $\mathcal{A}$ be a relation algebra. Then for any $L(\mathcal{A})$-formula φ written with at most three distinct variables, possibly re-used, there is a quantifier-free formula that is equivalent to φ in any square representation M of $\mathcal{A}$, viewed as a model of $T_\mathcal{A}$.*

Proof. In fact we show that any $L(\mathcal{A})$-formula φ written with variables x_0, x_1, x_2 is M-equivalent to a positive boolean combination of atomic formulas not involving equality, and not depending on M. The proof is by induction on φ. If φ is atomic, it is $r(x,y)$ or $x = y$, for $x, y \in \{x_0, x_1, x_2\}$. The latter is equivalent in M to $1'(x,y)$, and we are done. Assume the result for φ, ψ. The result for $\varphi \wedge \psi$ follows trivially from the inductive hypothesis. For $\neg \varphi$, we put φ in disjunctive normal form, which inductively can be taken to be $\bigvee_i \bigwedge_j r_{ij}(x_{ij}, y_{ij})$ for various $r_{ij} \in \mathcal{A}$ and $x_{ij}, y_{ij} \in \{x_0, x_1, x_2\}$. Then because M is square, $\neg \varphi$ is equivalent to $\bigwedge_i \bigvee_j -r_{ij}(x_{ij}, y_{ij})$, which has the required form.

Finally, $\exists z \bigvee_i \bigwedge_j r_{ij}(x_{ij}, y_{ij})$ is equivalent to $\bigvee_i \exists z \bigwedge_j r_{ij}(x_{ij}, y_{ij})$, so it suffices to handle each disjunct. Such a disjunct is of the form $\exists z \bigwedge_j r_j(x_j, y_j)$, where $r_j \in \mathcal{A}$ and $x_j, y_j \in \{x_0, x_1, x_2\}$. Since for any x, y, $r(x,y) \wedge s(x,y)$ is equivalent in M to $(r \cdot s)(x,y)$, and $r(y,x)$ is equivalent to $\breve{r}(x,y)$, we can collect up all atomic subformulas of $\exists z \bigwedge_j r_j(x_j, y_j)$ with the same free variables, to obtain a formula of the form

$$\exists z \big(r_x(x,x) \wedge r_y(y,y) \wedge r_z(z,z) \wedge r_{xy}(x,y) \wedge r_{xz}(x,z) \wedge r_{zy}(z,y)\big),$$

where $\{x, y, z\} = \{x_0, x_1, x_2\}$. (We can assume that all six conjuncts are present, by adding conjuncts $1(x,y)$, etc., if necessary.) This is equivalent in M to

$$r_x(x,x) \wedge r_y(y,y) \wedge [r_{xy} \cdot r_{xz}; (r_z \cdot 1'); r_{zy}](x,y),$$

as required. □

Of course, this result is trivial if $\mathcal{A}$ is not representable. However, there are similar versions of quantifier elimination for 'relativised representations' that avoid this triviality. See exercise 13.1(4). We note that if x_2 is not free in φ, then φ is equivalent to an atomic formula, and if φ is a sentence, it is equivalent to $0(x,y)$ or $1(x,y)$.

3.4.3 Saturation

Our next application of this model-theoretic view of representations is to show that **RRA** is a canonical variety. (We will use similar canonicity proofs for other classes of algebra later.) The essential idea of the proof of canonicity is to try to turn a representation of a relation algebra $\mathcal{A}$ (a model of $T_\mathcal{A}$) into a representation of its canonical extension $\mathcal{A}^+$. The problem in doing so is that the original representation may have 'gaps', which may prevent it being a faithful representation of $\mathcal{A}^+$ or stop composition working correctly. To fill up the gaps, we will use the model-theoretic notion of *saturation*, and we start by recalling the definition of this.

DEFINITION 3.33 Let L be a first-order signature and M an L-structure. Let S be a subset of the domain of M.

1. The language $L(S)$ is obtained by augmenting L with one constant for each element of S, and of course all these new constants must be distinct from each other and from all the symbols of L. We will identify each constant with the corresponding element of S. We regard M canonically as an $L(S)$-structure by interpreting each new constant as itself.

2. An *n-type over S,* or *with parameters in S,* is a set p of $L(S)$-formulas $\varphi(\bar{x})$, where $\bar{x}$ is a fixed n-tuple of variables. We may indicate the variables by writing $p(\bar{x})$. (Though we will hardly use it, the notion extends to α-types for any ordinal α.) A *type* is an n-type for some n.

3. A type $p(\bar{x})$ is said to be *realised in M* if there is $\bar{a} \in M$ with $M \models \varphi(\bar{a})$ for every $\varphi \in p$. If p is not realised in M, we say that M *omits* p.

4. A type p is said to be *finitely satisfiable* in M, or *M-consistent*, or if the context is clear, *consistent*, if every finite subtype $p_0 \subseteq p$ is realised in M. (That is, $M \models \exists \bar{x} \bigwedge p_0$.)

    By compactness, p is M-consistent iff some elementary extension $N \succeq M$ realises p.

5. For a theory T, a type p over $\emptyset$ is said to be *T-consistent* if there exists $M \models T$ such that p is M-consistent.

6. Let κ be a cardinal. M is said to be κ-*saturated* if for any $S \subseteq \text{dom}(M)$ with $|S| < \kappa$, every consistent 1-type over S is realised in M.

Examples The structure $M = (\{0, 1, \ldots\}, <)$, the ordered natural numbers, omits the type $p(x) = \{x > n : n < \omega\}$, but p is finitely satisfiable in M. The order type of the rationals (countable, dense, without endpoints) is ω-saturated.

FACT 3.34 Let κ be a cardinal.

1. A structure M is κ-saturated iff every M-consistent α-type with $< \kappa$ parameters is realised in M, for any $\alpha \leq \kappa$.

2. A κ-saturated structure is λ-saturated for any cardinal $\lambda < \kappa$.

3. A finite structure is κ-saturated for all κ.

3.4. Representations of relation algebras

4. Any consistent theory has a κ-saturated model, for any κ (for a proof, see [ChaKei90, lemma 5.1.4]). Any structure has a κ-saturated elementary extension [Hod93, theorems 10.2.1, 10.1.2].

5. If D is a non-principal ultrafilter over ω, then for any similar structures A_i ($i < \omega$) in a countable signature, the ultraproduct $\prod_D A_i$ is ω_1-saturated [ChaKei90, theorem 6.1.1].

   See [ChaKei90, 6.1.4, 6.1.8] for a (much harder) generalisation to arbitrary signatures and higher degrees of saturation. A consequence that we will occasionally use is that for any structure A and cardinal κ, there exists a κ-saturated ultrapower of A.

We are generally only interested in the case $\kappa = \omega$ (i.e., in finite sets of parameters).

3.4.4 RRA is a canonical variety

We are going to prove that like **RA**, the class **RRA** of representable relation algebras is a canonical variety. However, unlike for **RA** (see theorem 3.16), canonicity is not immediate, since we do not have to hand an explicit positive or Sahlqvist axiomatisation of **RRA**. (In section 17.2, we will see that **RRA** is not closed under completions, so that there is no such axiomatisation.) Luckily, we can use saturation to show that **RRA** is canonical (closed under the map $\mathcal{A} \mapsto \mathcal{A}^+$); this result will help us to prove that **RRA** is actually a variety (theorem 3.37). A similar proof of canonicity, using compactness, may be found in [HirHod97b, theorem 22]. The method is well known in modal logic: cf. [Fin75, lemma 9], and fact 2.86 in the preceding chapter (we cannot use this fact directly as we do not yet know that **RRA** is a variety).

Though the first published proof that **RRA** is canonical is in [Madd83], the result itself was first proved by J. D. Monk. It was reported by McKenzie in [McKe66, theorem 2.12], where it is stated (roughly) that Monk's proof used the neat embedding theorem.[5] The theorem below proves something slightly stronger. To state it, we need the following definition.

DEFINITION 3.35 A representation h of a relation algebra $\mathcal{A}$ is, inter alia, a representation of the boolean reduct bool($\mathcal{A}$) of $\mathcal{A}$. The representation h is said to be *complete* if it is complete as a representation of bool($\mathcal{A}$) (see definition 2.18). A relation algebra is said to be *completely representable* if it has a complete representation. We write **CRA** for the class of all relation algebras that have a complete representation.

---
[5] In the notation of chapter 5, McKenzie says that Monk combined **RRA** = $S\mathfrak{Ra}CA_\omega$ (proposition 13.48) with a version of theorem 5.43 having '$\subseteq$' in place of '$\subseteq^c$'.

By theorem 2.21, any complete representation h of a relation algebra $\mathcal{A}$ is atomic: if $(x,y) \in h(1)$ then $(x,y) \in h(a)$ for some atom $a \in \mathcal{A}$. It follows that any completely representable relation algebra must be atomic. As with boolean algebras, any representation of a finite relation algebra is necessarily complete.

THEOREM 3.36 (Monk) *A relation algebra $\mathcal{A}$ is representable if and only if its canonical extension $\mathcal{A}^+$ has a complete representation.*

Proof. Since $\mathcal{A}$ can be identified with a subalgebra of its canonical extension, any representation of the extension automatically induces a representation of $\mathcal{A}$.

For the converse, recall from definition 3.30 that $T_{\mathcal{A}}$ is a theory of the first-order language $L(\mathcal{A})$ that exactly characterises the representations of $\mathcal{A}$. Since $\mathcal{A}$ is assumed representable, $T_{\mathcal{A}}$ is consistent. Let M be an ω-saturated model of $T_{\mathcal{A}}$ (see fact 3.34). We will show that M induces in a natural way a (complete) representation of $\mathcal{A}^+$.

Define a map $h : \mathcal{A}^+ \to \wp(M \times M)$ as follows. First observe that by definition of $T_{\mathcal{A}}$, the set

$$f_{(x,y)} \stackrel{\text{def}}{=} \{a \in \mathcal{A} : M \models a(x,y)\}$$

is an ultrafilter of $\mathcal{A}$, whenever $x, y \in M$ and $M \models 1(x,y)$. This already gives a natural representation of the atoms of $\mathcal{A}^+$. We extend it to arbitrary elements in the obvious way. Recalling that the elements of $\mathcal{A}^+$ are the sets of ultrafilters of $\mathcal{A}$, define

$$h(S) = \{(x,y) \in 1^M : f_{(x,y)} \in S\},$$

for any $S \in \mathcal{A}^+$.

We will show that h is indeed a complete representation of $\mathcal{A}^+$. Certainly, $h(0^{\mathcal{A}^+}) = h(\emptyset) = \emptyset$. Also, from the definition of h we obtain $h(1^{\mathcal{A}^+}) = 1^M$, and it follows that h respects complementation, simply because for any $(x,y) \in 1^M$ and $S \in \mathcal{A}^+$, $(x,y) \notin h(S)$ iff $f_{(x,y)} \notin S$ iff $f_{(x,y)} \in -S$ iff $(x,y) \in h(-S)$. The definition of h, in particular the way h is defined on sets of ultrafilters, guarantees that h respects arbitrary unions. More formally, if $S_i \in \mathcal{A}^+$ ($i \in I$) then for all $(x,y) \in 1^M$ we have $(x,y) \in h(\bigcup_i S_i)$ iff $f_{(x,y)} \in \bigcup_i S_i$, iff $f_{(x,y)} \in S_i$ for some $i \in I$, iff $(x,y) \in h(S_i)$ for some $i \in I$. Thus, $h(\bigcup_i S_i) = \bigcup_i h(S_i)$.

We now check that h is injective. Let γ be any ultrafilter of $\mathcal{A}$; we show that $h(\{\gamma\}) \neq \emptyset$. It is clear that the 2-type

$$p(x,y) = \{a(x,y) : a \in \gamma\}$$

is finitely satisfiable in M. For, if $\{a_0(x,y), \ldots, a_{n-1}(x,y)\}$ is an arbitrary finite subset of $p(x,y)$ then $a = a_0 \cdot a_1 \cdots a_{n-1} \in \gamma$, so $a > 0$. By axiom $\sigma_{\neq 0}(a)$ of $T_{\mathcal{A}}$, we

3.4. Representations of relation algebras

have $M \models \exists xy\, a(x,y)$. Since $a \leq a_i$ for each $i < n$, using the $\sigma_+(a_i, a, a_i)$-axiom of $T_{\mathcal{A}}$ we obtain $M \models \exists xy \bigwedge_{i<n} a_i(x,y)$, showing that $p(x,y)$ is finitely satisfiable in M as required. Hence, by ω-saturation, p is realised in M by some pair (x,y), say. Clearly, $M \models 1(x,y)$ and $f_{(x,y)} \supseteq \gamma$; since both these sets are ultrafilters, $f_{(x,y)} = \gamma$. So $(x,y) \in h(\{\gamma\})$ and $h(\{\gamma\}) \neq \emptyset$. It follows that $h(S) \neq \emptyset$ for any non-zero $S \in \mathcal{A}^+$, and since h is a boolean homomorphism, this proves that h is one-one.

It remains to check that the non-boolean operations — identity, conversion and composition — are preserved by h. For identity, let $x, y \in M$. Then $(x,y) \in h(1'^{\mathcal{A}^+})$ if and only if $(x,y) \in 1^M$ and $f_{(x,y)} \in 1'^{\mathcal{A}^+}$. Now the identity $1'^{\mathcal{A}^+}$ of $\mathcal{A}^+$ is the set of all ultrafilters containing the identity $1'^{\mathcal{A}}$ of $\mathcal{A}$. So the above holds iff $(x,y) \in 1^M$ and $1'^{\mathcal{A}} \in f_{(x,y)}$, which is iff $M \models 1'^{\mathcal{A}}(x,y)$. By definition of $T_{\mathcal{A}}$, this is equivalent to $x = y$, as required.

For conversion, let $(x,y) \in 1^M$. Then $f_{(y,x)} = f_{(x,y)}^{\smile}$. (For, if $a \in \mathcal{A}$, we have $a \in f_{(y,x)}$ iff $M \models a(y,x)$ iff $M \models \breve{a}(x,y)$ iff $\breve{a} \in f_{(x,y)}$, iff $a = \breve{b}$ for some $b \in f_{(x,y)}$.) Let S be a set of ultrafilters. Then $(x,y) \in h(S)$ iff $f_{(x,y)} \in S$, iff $f_{(y,x)} = f_{(x,y)}^{\smile} \in \breve{S}$, iff $(y,x) \in h(\breve{S})$, as required.

Finally, for composition, let $x, y \in M$ and let S, T be sets of ultrafilters. Suppose first that $(x,y) \in h(S;T)$. We have to show that there is a $z \in M$ with $(x,z) \in h(S)$ and $(z,y) \in h(T)$. Well, the assumption yields $M \models 1(x,y)$ and $f_{(x,y)} \in S;T$. So by definition of $\mathcal{A}^+$ as a complex algebra, there must be ultrafilters $\sigma \in S$, $\tau \in T$ with $f_{(x,y)} \in \{\sigma\};\{\tau\}$. Recall that a product of singleton sets of ultrafilters is defined by

$$\{\sigma\};\{\tau\} = \{\text{ultrafilters } \gamma \text{ of } \mathcal{A} : \gamma \supseteq \{s;t : s \in \sigma, t \in \tau\}\}.$$

So for all $s \in \sigma$ and $t \in \tau$, we have $s;t \in f_{(x,y)}$. Hence, $M \models [s;t](x,y)$. By the axiom $\sigma_;(s;t,s,t) \in T_{\mathcal{A}}$, $M \models \exists z(s(x,z) \wedge t(z,y))$. This holds for all $s \in \sigma, t \in \tau$. It now follows that the 1-type

$$q(z) = \{s(x,z) \wedge t(z,y) : s \in \sigma, t \in \tau\},$$

with free variable z and parameters $x, y \in M$, is finitely satisfiable in M. For, let $\{s_i(x,z) \wedge t_i(z,y) : i < n\}$ be a finite subset of q. As before, setting $s = s_0 \cdot s_1 \cdots s_{n-1}$ and $t = t_0 \cdot t_1 \cdots t_{n-1}$, we have $s \in \sigma$ and $t \in \tau$, since σ, τ are filters. Hence, $M \models \exists z(s(x,z) \wedge t(z,y))$. The boolean axioms of $T_{\mathcal{A}}$ now yield $M \models \exists z \bigwedge \{s_i(x,z) \wedge t_i(z,y) : i < n\}$, as required. We now use saturation again to show that q is realised in M, at a point z, say. The definition of q yields $M \models 1(x,z) \wedge 1(z,y)$, $f_{(x,z)} \supseteq \sigma$, and $f_{(z,y)} \supseteq \tau$. Since $f_{(x,z)}, f_{(z,y)}, \sigma, \tau$ are ultrafilters, we obtain $f_{(x,z)} = \sigma$ and $f_{(z,y)} = \tau$. So $(x,z) \in h(S)$ and $(z,y) \in h(T)$, which gives our result.

To prove that composition is preserved we have to show the converse too: if there is a z in M with $(x,z) \in h(S)$ and $(z,y) \in h(T)$, then $(x,y) \in h(S;T)$. Assume there is such a z. Write σ for $f_{(x,z)}$ and τ for $f_{(z,y)}$. Since $\sigma \in S$ and $\tau \in T$, it suffices to prove that $(x,y) \in h(\{\sigma\};\{\tau\})$, i.e., that $M \models 1(x,y)$ and $f_{(x,y)} \in \{\sigma\};\{\tau\}$. Let

126   Chapter 3. Binary relations and relation algebra

$s \in \sigma, t \in \tau$. Now $M \models s(x,z) \wedge t(z,y)$, so by definition of $T_{\mathcal{A}}$, $M \models [s;t](x,y)$. This implies that $M \models 1(x,y)$ and $s;t \in f_{(x,y)}$. This holding for all $s \in \sigma$, $t \in \tau$ yields $f_{(x,y)} \in \{\sigma\};\{\tau\}$, as required.

Since all the operations are preserved by h, h must be a representation of $\mathcal{A}^+$, and since h preserves arbitrary unions it must be a complete representation. (Alternatively, use theorem 2.21 to show that h is complete because it is atomic.) □

This allows us to prove the following important result.

THEOREM 3.37 (Tarski, [Tar55]) *The class* **RRA** *forms an equational variety.*

Proof. By Birkhoff's theorem (theorem 2.45), it is enough to show that **RRA** is closed under direct products, subalgebras, and homomorphic images. For direct products, let $\mathcal{A}_\lambda \in$ **RRA** ($\lambda \in \Lambda$). Then, for each $\lambda \in \Lambda$, there is an isomorphism $h_\lambda : \mathcal{A}_\lambda \to \mathcal{B}_\lambda$ for some $\mathcal{B}_\lambda \in$ **PRA**. Let the base set of $\mathcal{B}_\lambda$ be B_λ (each λ). By taking copies, if necessary, we can assume that the base sets are pairwise disjoint from each other. Let B be the disjoint union of the base sets B_λ. Define a map $h : \prod_\lambda \mathcal{B}_\lambda \to \wp(B \times B)$ by

$$h(\langle b_\lambda : \lambda \in \Lambda \rangle) \mapsto \bigcup_{\lambda \in \Lambda} h_\lambda(b_\lambda).$$

Clearly, h respects all the operators and is injective. So let $\mathcal{B} = \mathrm{rng}(h)$. Then $\prod_\lambda \mathcal{A}_\lambda \cong \mathcal{B} \in$ **PRA**, hence $\prod_\lambda \mathcal{A}_\lambda \in$ **RRA**.

Showing that **RRA** is closed under subalgebras is even easier — after all, a subalgebra of a proper relation algebra $\mathcal{P}$ is already a proper relation algebra.

It remains to show that **RRA** is closed under homomorphic images. So let $\mathcal{A} \in$ **RRA** and let $\mathcal{B}$ be a homomorphic image of $\mathcal{A}$. We must find a proper relation algebra isomorphic to $\mathcal{B}$.

Since $\mathcal{A}$ is representable, by theorem 3.36 $\mathcal{A}^+$ is representable too; so there is an isomorphism χ from $\mathcal{A}^+$ into some proper relation algebra with base set X, say. If f is any ultrafilter of $\mathcal{A}$, then $\{f\}$ is an atom of $\mathcal{A}^+$. Since χ is a representation of $\mathcal{A}^+$, there must be $(x,y) \in \chi(\{f\})$ (this is the only consequence of theorem 3.36 that is needed here). By definition of the natural embedding of $\mathcal{A}$ into $\mathcal{A}^+$, we have $\mathcal{A}^+ \models f \leq a$ for all $a \in f$, and so $(x,y) \in \chi(a)$ for all $a \in f$.

For notational simplicity, from now on we will identify $\mathcal{A}$ with its image under χ. Thus, we will take $\mathcal{A}$ to be a proper relation algebra with base X. By the above,

$$\bigcap f \neq \emptyset \quad \text{for any ultrafilter } f \text{ of } \mathcal{A}. \tag{3.4}$$

By assumption, there is a surjective relation algebra homomorphism $h : \mathcal{A} \to \mathcal{B}$. First, define the *kernel* of h, and corresponding subsets of X:

$$\begin{aligned} \ker(h) &= \{a \in \mathcal{A} : h(a) = 0\}, \\ K &= \{x \in X : (x,x) \in k \text{ for some } k \in \ker(h)\}, \\ E &= X \setminus K. \end{aligned}$$

3.4. Representations of relation algebras

Claim. If $x \in E$ and $(x,y) \in a$ for some $a \in \mathcal{A}$, then $y \in E$.

Proof of claim. Assuming the hypotheses, suppose for contradiction that $y \in K$. So there is some $k \in \ker(h)$ with $(y,y) \in k$. Now $(x,y) \in a$, so $(y,x) \in \breve{a}$, and evidently, $(x,x) \in a;k;\breve{a}$. But $h(a;k;\breve{a}) = h(a);h(k);h(\breve{a}) = h(a);0;h(\breve{a}) = 0$. Thus, $a;k;\breve{a} \in \ker(h)$, so $x \in K$, contradicting our assumption that $x \in E$. This proves the claim.

We will show that we can represent $\mathcal{B}$ by the restriction of elements of $\mathcal{A}$ to $E \times E$. To do this, we will 'factor' the restriction map through h:

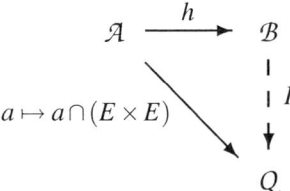

I, Q will be defined below. The claim tells us that any $a \in \mathcal{A}$ is the disjoint union of $a \cap (E \times E)$ and of $a \cap (K \times K)$, and it will follow that this restriction map is a homomorphism. The faithfulness of the representation will follow from (3.4).

Let Q be the proper relation algebra

$$\langle \wp(1^{\mathcal{A}} \cap (E \times E)), \emptyset, 1^{\mathcal{A}} \cap (E \times E), \cup, \setminus, 1'^{\mathcal{A}} \cap (E \times E), {}^{-1}, | \rangle.$$

Define a mapping $I: \mathcal{B} \to Q$ by

$$I(h(a)) = a \cap (E \times E), \quad \text{for } a \in \mathcal{A}.$$

This map is well-defined for the following reason. Let $a, b \in \mathcal{A}$ and suppose that $h(a) = h(b)$; we require $a \cap (E \times E) = b \cap (E \times E)$. Suppose that $x, y \in X$ and $(x,y) \in (a-b)$. Then $(y,x) \in (a-b)\breve{\,}$, so $(x,x) \in (a-b);(a-b)\breve{\,}$. But $h(a) = h(b)$ implies $h(a-b) = 0$, and so $h((a-b);(a-b)\breve{\,}) = h(a-b);h((a-b)\breve{\,}) = 0;h((a-b)\breve{\,}) = 0$. So $x \in K$, $x \notin E$. By the claim, $y \notin E$. Hence, there are no points $x, y \in E$ with $(x,y) \in (a-b)$. Similarly there are no pairs $(x,y) \in (E \times E) \cap (b-a)$. Thus, $a \cap (E \times E) = b \cap (E \times E)$. This shows that I is well-defined on its domain. Since $h: \mathcal{A} \to \mathcal{B}$ is surjective, the domain of I is $\mathcal{B}$.

We now show that I is a homomorphism. We omit the rather easy proofs that I respects the constants, $+$, $-$, and conversion. For composition, we must show that $I(h(a);h(b)) = I(h(a)) | I(h(b))$ for all $a, b \in \mathcal{A}$. Since h is a homomorphism, we have $I(h(a);h(b)) = I(h(a;b))$. Let $x,y \in E$.

$$\begin{aligned}
(x,y) \in I(h(a;b)) &\Leftrightarrow (x,y) \in a;b \text{ (since } x,y \in E\text{)} \\
&\Leftrightarrow \exists z \in X, (x,z) \in a, (z,y) \in b \\
&\Leftrightarrow \exists z \in E, (x,z) \in a, (z,y) \in b \text{ (by claim)} \\
&\Leftrightarrow \exists z \big((x,z) \in I(h(a)) \wedge (z,y) \in I(h(b))\big) \\
&\Leftrightarrow (x,y) \in I(h(a)) \,|\, I(h(b)).
\end{aligned}$$

Hence, I is a homomorphism from $\mathcal{B}$ into Q. We'll show that I is in fact one-one, so that the subalgebra of Q with domain rng(I) is a proper relation algebra isomorphic to $\mathcal{B}$.

The proof that I is one-one is the reverse of the argument that it is well-defined, and uses the fact (3.4) that the intersection of any ultrafilter of $\mathcal{A}$ is a non-empty subset of $X \times X$. Because I is a homomorphism, it suffices to prove that if $a \in \mathcal{A}$ and $I(h(a)) = 0$ then $h(a) = 0$: i.e., $a \in \ker(h)$. Assume that $a \notin \ker(h)$. It is clear that the boolean reduct bool($\mathcal{A}$) is a boolean algebra and $\ker(h)$ is an ideal of it, so $a \not\leq k_0 + \cdots + k_{n-1}$ for any $k_0, \ldots, k_{n-1} \in \ker(h)$. Hence, the boolean product of any finite number of elements of $\{a\} \cup \{-k : k \in \ker(h)\}$ is non-zero. By (BPI) (fact 2.8), there is an ultrafilter f of $\mathcal{A}$ containing $\{a\} \cup \{-k : k \in \ker(h)\}$. By (3.4), there are $x,y \in X$ with $(x,y) \in b$ for all $b \in f$.

Since $a \in f$, we have $(x,y) \in a$. If $x \in K$, then $(x,x) \in k$ for some $k \in \ker(h)$. Then $(x,y) \in k;a$, and $h(k;a) = h(k);h(a) = 0;h(a) = 0$ so $k;a \in \ker(h)$. But this means that $-(k;a) \in f$ and $(x,y) \in -(k;a)$, a contradiction. So $x \in E$. Similarly, $y \in E$. So $(x,y) \in a \cap (E \times E) = I(h(a))$, whence $I(h(a)) \neq 0$, as required. □

This tells us that **RRA** can be axiomatised by equations, even though Monk's theorem (theorem 17.3; also see exercises 11.5(3) and 14.2(9) and remark 15.13) tells us that no finite set of formulas can define the class. We'll see an infinite axiom scheme in chapter 8.

COROLLARY 3.38 **RRA** *is a canonical variety.*

Proof. By theorems 3.36 and 3.37. □

COROLLARY 3.39 *Let $\mathcal{A}$ be an arbitrary relation algebra. $\mathcal{A}$ is representable if and only if every countable subalgebra of $\mathcal{A}$ is representable, if and only if every finitely generated subalgebra of $\mathcal{A}$ is representable.*

Proof. From theorem 2.44. □

3.4. Representations of relation algebras

Exercises

1. Show that any simple, representable relation algebra has a representation onto a square proper relation algebra.

2. Show that any representable relation algebra has a representation with base of cardinality at least κ, for any cardinal κ. Show that any relation algebra with a finite representation has a representation with arbitrarily large finite base. [The representation need not be square!]

3. Let $\mathcal{A}$ be a full proper relation algebra (i.e., the domain of $\mathcal{A}$ consists of all binary relations over some base set). Assuming that $\mathcal{A}$ is finite, prove that all square representations of $\mathcal{A}$ are base-isomorphic to each other. Is this still true if $\mathcal{A}$ is infinite? For the infinite case, are all complete square representations base-isomorphic?

4. Let $\mathcal{A}, \mathcal{B}$ be atomic relation algebras with $\mathrm{At}\,\mathcal{A} \cong \mathrm{At}\,\mathcal{B}$. Show that if $\mathcal{A}$ is completely representable then so is $\mathcal{B}$.

The following few exercises give alternative proofs that **RRA** is a variety.

5. In this exercise, which is closely related to exercise 6, you have to prove that **RRA** is a variety by proving closure under ultraproducts instead of canonical extensions and homomorphic images. It also gives additional information about **RRA**. See [Jón91] for a similar proof.

    (a) Show that **RRA** = **SPUp Sir RRA**.
    [Hint: **RRA** ⊆ **SP Sir RRA** ⊆ **SPUp Sir RRA** ⊆ **SPUp(RRA)** ⊆ **RRA**.]

    (b) Show that Sir **RRA** is a discriminator class (i.e., there exists a discriminator term for it).

    (c) Show that **SPUp Sir RRA** is a variety (cf. exercise 2.6(9)).

    (d) Deduce that **RRA** is a (discriminator) variety and is **SP Sir RRA**.

6. Cf. [Giv99, p. 579]. Use theorem 2.87 to prove that **RRA** is a canonical variety. [As in exercise 3.2(7), let $\mathcal{F}(B)$ be the full proper relation algebra over the base set B; take L in the theorem to be the class of all atom structures of relation algebras of this form.]

7. Let $\mathcal{A}$ be an infinite, representable relation algebra. Consider representations of $\mathcal{A}$ as models of $T_{\mathcal{A}}$. By compactness or otherwise, prove that $\mathcal{A}$ has a representation that fails to be complete. [Use theorem 2.21.]

8. Prove corollary 3.39 using compactness. [See definition 3.30.]

9. Show that the set of isomorphism types of finite non-representable relation algebras is recursively enumerable.

10. [TarGiv87, p137] Show that the 'pairing axiom'
$$\forall xy \exists z \forall t (R(t,z) \leftrightarrow t = x \lor t = y)$$
is logically equivalent to a sentence written with only three variables.

11. Prove the claim in fact 3.34 that if D is a non-principal ultrafilter over ω and A_i ($i < \omega$) are structures in a countable signature L, then $\prod_D A_i$ is ω_1-saturated [See [ChaKei90, theorem 6.1.1]].

12. A representation $M \models T_{\mathcal{A}}$ of a relation algebra $\mathcal{A}$ is said to be *permutational* if for any $x, y \in M$, there is an automorphism of M taking x to y. (See section 2.2.4 for automorphisms, and example 9.2 later for more on permutational representations.) That is, the automorphism group of M acts transitively on M.

    Write P for the class of relation algebras with a permutational representation. Clearly, P ⊆ **RRA**.

    (a) Show that if $\mathcal{A} \in $ P then $\mathcal{A}$ is integral.

    (b) Show that P is closed under subalgebras but not products.

    FACT: Let T be a consistent first-order theory. There exists an ω-saturated model M of T such that any finite partial elementary map from M to M extends to an automorphism of M. (Take M to be ω-big; see theorem 10.1.2, exercise 10.1.4, and corollary 10.2.3 of [Hod93].)

    (c) Let $\mathcal{A} \in $ P. Show that there is a unique maximal $T_{\mathcal{A}}$-consistent 1-type. Deduce that the canonical extension $\mathcal{A}^+$ of $\mathcal{A}$ is also in P. Cf. [AndMon+91, problem 30, p. 736].

13. Cf. [Fin75, Ben80, Gol89]. Let L be a functional signature containing the signature of boolean algebras, and let L^a be formed as in definition 2.62. Let S be an L^a-structure, or perhaps an expansion of one, and let S' be an ω-saturated elementary extension of S. Define the first-order algebra $\Delta^\omega(S)$ over S as in exercise 2.7(20). Show that there is an embedding of $\Delta^\omega(S)^+$ into $\mathfrak{Cm}\,S'$. [Consider the map π that maps an element $a \in S'$ to its type over S — that is, $\pi(a)$ is the set of all first-order formulas $\varphi(x)$ with parameters in S such that $S' \models \varphi(a)$. Identify types with ultrafilters.]

    Deduce that if K is a class of BAO atom structures that is closed under ultrapowers then **S**$\mathfrak{Cm}$K is canonical. [Use fact 3.34.]

3.4. Representations of relation algebras

14. [Gol91] Let K be a class of BAO atom structures that is closed under ultraproducts. Write $\mathbf{U_d}$ K for the class of disjoint unions of structures in K.

    (a) Show that $\mathbf{P}\, \mathfrak{Cm}\, \mathsf{K} = \mathfrak{Cm}\, \mathbf{U_d}\, \mathsf{K}$.
    (b) Show that $\mathfrak{Cm}\, \mathbf{Up}\, \mathbf{U_d}\, \mathsf{K} \subseteq \mathbf{S}\, \mathfrak{Cm}\, \mathbf{U_d}\, \mathsf{K}$.
    (c) Deduce that $\mathbf{SP}\, \mathfrak{Cm}\, \mathsf{K} = \mathbf{S}\, \mathfrak{Cm}\, \mathbf{Up}\, \mathbf{U_d}\, \mathsf{K}$.
    (d) Use exercise 13 and exercise 2.7(7) to obtain fact 2.86 — the variety $\mathbf{HSP}\, \mathfrak{Cm}\, \mathsf{K}$ generated by K is canonical.

Chapter 4

Examples of relation algebras

Before we go any further, let's get our hands on some relation algebras. In this chapter we'll show some methods of constructing them and give some examples of these algebras.

Recall, from chapter 3, that **RA** and **RRA** are varieties and so are closed under homomorphic images, subalgebras and direct products. Consequently, we can build bigger relation algebras from the ones given by taking products. Furthermore, we saw that every relation algebra is a subdirect product of simple relation algebras. We therefore concentrate most of our attention on simple algebras.

4.1 Set algebras

The easiest way of constructing a relation algebra is to take a proper relation algebra. So first pick some base set B. Then choose a set of binary relations over B and check that it is closed under all the operations for relation algebra. One possible construction here is to take all possible binary relations over B.

DEFINITION 4.1 The *full proper relation algebra over B* is the proper relation algebra

$$\mathcal{F}(B) = \langle \wp(B \times B), \emptyset, (B \times B), \cup, \setminus, 1', {}^{-1}, | \, \rangle.$$

Any relation algebra of this form is called a *full proper relation algebra*.

Another common notation for $\mathcal{F}(B)$ is $\mathfrak{Re}\, B$. If $|B| = n$ then $|B \times B| = n^2$, so the full proper relation algebra over B has n^2 atoms and 2^{n^2} elements. Smaller proper relation algebras can also be defined on the base B. In fact, an arbitrary set of binary relations S over B generates a unique proper relation algebra.

DEFINITION 4.2 Let $S \subseteq \wp(B \times B)$. The proper relation algebra *generated* by S is the smallest subalgebra of the full proper relation algebra over B containing all the binary relations in S. If $t(\bar{x})$ is any term in the language L_{RA} and if $\bar{s}$ is any finite sequence from S of the same length as the sequence of variables $\bar{x}$ then $t(\bar{s})$ belongs to the algebra generated by S, and all elements of this algebra are of this form.

A further generalisation of full proper relation algebras is obtained by removing the requirement that the unit 1 should be $B \times B$. The unit will always be an equivalence relation (lemma 3.4). So given any equivalence relation E and any set of relations $S \subseteq \wp(E)$, we can define the *proper relation algebra generated by S in E*. For the special case where $E = B \times B$, we call the proper relation algebra *square* (see definition 3.3).

4.2 Group relation algebras

A second source of representable relation algebras is groups.

DEFINITION 4.3

- Let $\mathcal{G} = (G, \circ, ^{-1}, e)$ be any group. Define the *complex algebra* $\mathfrak{Cm}(\mathcal{G})$ to be the tuple $\langle \wp(G), \emptyset, G, \cup, \setminus, 1', ^{-1}, \star \rangle$, where
    - $1' = \{e\}$
    - $S^{-1} = \{s^{-1} : s \in S\}$ for $S \subseteq G$
    - $S \star T = \{s \circ t : s \in S, t \in T\}$ for $S, T \subseteq G$.

- Any algebra of the form $\mathfrak{Cm}(\mathcal{G})$ for some group $\mathcal{G}$ is called a *complex group*.

- Let $\mathcal{A}$ be a relation algebra. If there exist a group $\mathcal{G}$ and an embedding from $\mathcal{A}$ into $\mathfrak{Cm}(\mathcal{G})$ then we call $\mathcal{A}$ a *group relation algebra*.

- **GRA** denotes the class of all group relation algebras.

McKinsey showed that group relation algebras are relation algebras; they are representable [JónTar48]:

LEMMA 4.4 (Cayley representation) *For any group $\mathcal{G}$, the complex algebra $\mathfrak{Cm}\,\mathcal{G}$ has a complete representation. Hence, every group relation algebra is representable.*

Proof. We show that $\mathfrak{Cm}\,\mathcal{G}$ is isomorphic to a proper relation algebra whose base is G. Define a map $h : \mathfrak{Cm}\,\mathcal{G} \to \wp(G \times G)$ by

$$h(S) = \{(f, g) : f^{-1} \circ g \in S\}.$$

4.2. Group relation algebras

Clearly h preserves the boolean operations and is one-one, $h(\{e\}) = \{(g,g) : g \in G\}$, and $h(S^{-1}) = h(S)^{-1}$. To check that $h(S \star T) = h(S) | h(T)$, let $f, g \in G$. Then $(f,g) \in h(S \star T)$ iff $f^{-1} \circ g \in S \star T$, iff $f^{-1} \circ g = s \circ t$ for some $s \in S$, $t \in T$, iff $f^{-1} \circ (g \circ t^{-1}) \in S$ and $(g \circ t^{-1})^{-1} \circ g = t \in T$ for some $t \in G$, iff $(f, g \circ t^{-1}) \in h(S)$ and $(g \circ t^{-1}, g) \in h(T)$ for some $t \in G$, iff $(f,g) \in h(S) | h(T)$.

So h is an embedding and its image is a proper relation algebra. Therefore, $\mathfrak{Cm}\, G$ is a representable relation algebra. Observe that for any two points f, g in the base of the representation, there is a unique atom $\{f^{-1} \circ g\}$ such that $(f,g) \in h(\{f^{-1} \circ g\})$. Thus, h is an atomic representation. Restricting to just the boolean operators, we may consider h as a boolean representation of the boolean part of $\mathfrak{Cm}\, G$. So we apply theorem 2.21 to deduce that h is a complete boolean representation. But that means that h is a complete relation algebra representation of $\mathfrak{Cm}\, G$. □

If G is finite then this is the only square representation (up to isomorphism) that $\mathfrak{Cm}\, G$ can have. For the infinite case there is a problem to do with incomplete representations, which always exist (see exercise 3.4(7)), and are not isomorphic to the Cayley representation. The following lemma handles complete representations (it is known by Andréka–Givant and probably others).

LEMMA 4.5 *Let G be a group and let k be any complete, square representation of $\mathfrak{Cm}\, G$ with base set B. Then k is base-isomorphic to the Cayley representation h.*

Proof. By squareness and completeness of k, for any two points $x, y \in B$ there is an atom $\{g\} \in \mathfrak{Cm}\, G$ such that $(x,y) \in k(\{g\})$. Pick an arbitrary point $x_0 \in B$. Define a map $j : B \to G$ by letting $j(x)$ be the unique element of G such that $(x_0, x) \in k(\{j(x)\})$, for each $x \in B$. We must check that j is a base-isomorphism. Let $x, y \in B$. Since $(x_0, x) \in k(\{j(x)\})$ and $(x_0, y) \in k(\{j(y)\})$ it follows, since k is a representation, that $(x,y) \in k(\{j(x)\}^{-1}) | k(\{j(y)\}) = k(\{j(x^{-1}) \circ j(y)\})$. Hence, for any $g \in G$ we have

$$(x,y) \in k(\{g\}) \Leftrightarrow g = j(x)^{-1} \circ j(y)$$
$$\Leftrightarrow (j(x), j(y)) \in h(\{g\}).$$

j is clearly one-one, since $j(x) = j(y) \Leftrightarrow (j(x), j(y)) \in h(\{e\}) \Leftrightarrow (x,y) \in k(\{e\}) \Leftrightarrow x = y$. j is also onto, since if $g \in G$ then $(x_0, x_0) \in k(\{e\}) = k(\{g\} \star \{g^{-1}\}) = k(\{g\}) | k(\{g^{-1}\})$, so that there is $x \in B$ with $(x_0, x) \in k(\{g\})$. Clearly, $j(x) = g$. So j is a base-isomorphism. □

COROLLARY 4.6 *If G is finite then any square representation of $\mathfrak{Cm}\, G$ is isomorphic to the Cayley representation.*

Proof. Immediate, since any representation of a finite relation algebra is complete.
□

Every group relation algebra is an integral representable relation algebra (see exercise 3.3(8) for integral relation algebras). Whether the converse held was a long-open question. (For example, [JónTar48] asked whether every integral relation algebra was a group relation algebra; at that time, it was not known that non-representable relation algebras existed.) Eventually, McKenzie showed in [McKe66, McKe70] that **GRA** is not finitely axiomatisable even over the class of relation algebras with a permutational representation (exercise 3.4(12) showed that these algebras are integral). **GRA** was already known not to be finitely axiomatisable, by Monk's original proof [Mon64] that **RRA** is not finitely axiomatisable.

GRA was axiomatised in [HodMik+01], and we will see how to do it in chapter 9. See, e.g., [AndGiv+97] for more information on **GRA**.

Exercises

1. Show that any group relation algebra has a permutational representation (see exercise 3.4(12)).

2. (See [Tar55] and [Giv99, example 2.8].) Use theorem 2.87 to prove that **SP(GRA)**, the class of subalgebras of products of relation algebras that embed into a complex group $\mathfrak{Cm}\,\mathcal{G}$ for some group $\mathcal{G}$, is a canonical variety and that **GRA** is the (universal) class of simple algebras of **SP(GRA)**. Hence or otherwise, show that if $\mathcal{A} \in$ **GRA** then $\mathcal{A}^+ \in$ **GRA**.

4.3 n-variable logic

A third method of finding algebras is to build one out of formulas of a logic. Relation algebra will correspond naturally to formulas with two free variables.

Notation Take any first-order language with equality L. Let $L(x,y)$ denote the set of all L-formulas whose free variables must be taken from $\{x,y\}$ (they may have any number of bound variables though). Let L^n denote the set of all L-formulas using only the variables (whether bound or free) $x_0 = x, x_1 = y, x_2 = z, x_3, \ldots, x_{n-1}$ ($n \geq 3$). $L^n(x,y)$ denotes $L(x,y) \cap L^n$.

4.4. Examples

Now take any non-empty set S of L-structures (possibly S consists of just a single structure). Define an equivalence relation $\sim$ on $L(x,y)$ by $\phi \sim \psi$ if and only if $S \models \phi \leftrightarrow \psi$. Then we form a relation algebra form$(x,y)/S$ from the equivalence classes:

$$\langle L(x,y)/\sim,\ \bot/\sim,\ \top/\sim,\ +,\ -,\ (x=y)/\sim,\ \breve{\ },\ *\rangle,$$

where the operations are (well-)defined as follows:

$$
\begin{array}{rrcl}
+: & (\phi/\sim, \psi/\sim) & \mapsto & (\phi \vee \psi)/\sim \\
-: & \phi/\sim & \mapsto & (\neg\phi)/\sim \\
\breve{\ }: & \phi(x,y)/\sim & \mapsto & \phi(y,x)/\sim \\
*: & (\phi(x,y)/\sim, \psi(x,y)/\sim) & \mapsto & [\exists z(\phi(x,z) \wedge \psi(z,y))]/\sim.
\end{array}
$$

Above, $\phi(y,x), \phi(x,z)$ are obtained from $\phi(x,y)$ by swapping all occurrences of x and y (respectively, y and z), and $\psi(z,y)$ is obtained from $\psi(x,y)$ by swapping x and z. If L is a relational signature, then form$(x,y)/S$ is a representable relation algebra. To obtain a representation assume, without loss, that the domains of the structures in S are pairwise disjoint. Define a natural representation

$$\phi/\sim\ \mapsto\ \bigcup_{M \in S} \{(x,y) \in {}^2M : M \models \phi(x,y)\}.$$

Also, form$(x,y)/S$ has a subalgebra form$^n(x,y)/S$ (any $n \geq 3$) which can be obtained by restricting to formulas in $L^n(x,y)$. We need at least three variables because composition requires an extra bound variable (z) for its definition.

4.4 Examples

Now we are going to list a few examples of relation algebras, mostly finite, all atomic. We will be brief, since the topic is well-covered in, e.g., [AndMad94, Madd].

Recall that an atomic relation algebra is determined by its boolean structure together with its atom structure (proposition 2.64). For finite relation algebras the atom structure alone determines the relation algebra, up to isomorphism. In any case, for an atomic relation algebra, if we know the atom structure we can determine composition and conversion on any elements of the algebra. To define the atom structure we must state the set of atoms and then list the atoms under the identity, the converse of each atom, and the set of all consistent triples (a,b,c) of atoms — those such that $\breve{a} \leq b;c$. In fact, the conversion operator is already determined by composition and identity (exercise 3.3(4)). In most cases the identity will itself be an atom. So the important thing is to specify the composition of any two atoms. We can do this by providing a composition table, as in the first few examples below. The format of such tables will be

;	s
r	$r;s$

Secondly, we can list all the *consistent* triples of atoms: those triples (a,b,c) of atoms such that $\breve{a} \leq b;c$. Or we can list all the *inconsistent* or *forbidden* triples (a,b,c) of atoms such that $\breve{a} \cdot (b;c) = 0$. This is sometimes more convenient, and we use forbidden triples a number of times in this and in later chapters.

The first few examples are proper relation algebras, so certainly representable. The 'point' algebra and the Allen interval algebras are introduced — both very widely used in temporal reasoning and planning. We include an example (due to McKenzie) that is not isomorphic to any proper relation algebra: i.e., not representable.

The smallest relation algebra This is of course the one-element (degenerate) relation algebra. It is unique up to isomorphism, and a member of any variety of BAOs of the signature of relation algebras, so a representable relation algebra. A representation with base $\emptyset$ is obtained by mapping its unique element to $\emptyset$. The algebra is isomorphic to $\mathcal{F}(\emptyset)$.

The smallest non-degenerate relation algebra The smallest non-degenerate relation algebra has one atom (1) and two elements, 0, 1. It is based on the two-element boolean algebra with the following operators: 1' = 1, both elements are self-converse, and composition is defined by $1;1 = 1$, $0;a = a;0 = 0$ (all $a \in \{0,1\}$). For a representation, take a set X with exactly one point x. Let $h(0) = \emptyset$ and $h(1) = \{(x,x)\}$. The algebra is isomorphic to $\mathcal{F}(\{x\})$.

There are two non-isomorphic relation algebras with two atoms.

Two-atom algebra (a) This relation algebra, $\mathcal{T}$, has two atoms, 1', #, and therefore has four elements. Both atoms are self-converse and composition is defined by the table below.

;	1'	#
1'	1'	#
#	#	1

Alternatively, the forbidden triples of atoms are just

$$(1', 1', \#), (1', \#, 1'), (\#, 1', 1').$$

As often, this is the easiest way to specify the algebra. A square representation X of $\mathcal{T}$ can be obtained by taking for its base any set B with $|B| > 2$

4.4. Examples

and setting $X(1') = \{(b,b) : b \in B\}$, $X(\#) = \{(b_1, b_2) \in B \times B : b_1 \neq b_2\}$. Indeed, any square representation of $\mathcal{T}$ arises in this way.

Two-atom algebra (b) This relation algebra $\mathcal{T}'$ is very similar to the previous one, but the composition table is

;	1'	#
1'	1'	#
#	#	1'

This time, there is one more forbidden triple: $(\#, \#, \#)$ together with the permutations of $(1', 1', \#)$ (as before). $\mathcal{T}'$ has a representation X consisting of just two points a, b. Here $X(\#) = \{(a,b), (b,a)\}$.

The point algebra (Tarski) $\mathcal{P}$ has three atoms, 1', <, and >, so that $1 = 1' + < + >$. The identity is 1' (self-converse of course), and the converse of < is >. Composition is defined by the table below.

;	1'	<	>
1'	1'	<	>
<	<	<	1
>	>	1	>

The forbidden triples are all Peircean transforms of $(1', 1', <)$, $(1', <, <)$, and $(<, <, <)$. The rational numbers form the base of a representation of $\mathcal{P}$ in which < is interpreted as the usual ordering on the rationals. Though finite and representable, $\mathcal{P}$ has no finite representation (it has a unique countable square representation up to isomorphism). See exercise 3 below; also see [AndGiv$^+$94b] which deals with this topic.

The pentagonal algebra $\mathcal{PA}$ [Madd91a] has three atoms $1', e, d$, all of which are self-converse; composition is defined by

;	e	d
e	1'+d	e+d
d	e+d	1'+e

(we omit the entries for 1').

The consistent triples of atoms are those associated with 1' (here, the permutations of $(1', x, x)$ for all x, since atoms are self-converse), and Peircean transforms of (e, e, d) and (d, d, e).

It is not hard to check that $\mathcal{PA}$ has exactly one square representation X, up to base-isomorphism. The base of X has five points $0, \ldots, 4$. $X(e) = \{(i, j) : |i - j| = 1 \pmod 5\}$ and $X(d) = \{(i, j) : |i - j| = 2 \pmod 5\}$.

Relation algebra from graph This has three self-converse atoms, $1', e, d$ (e is intended to mean there *exists* a graph edge and d means that there *doesn't* exist an edge). Composition is defined by the table

;	e	d
e	1	$e+d$
d	$e+d$	1

The forbidden triples of atoms are just those arising from the identity (permutations of $(1', x, y)$ for $x \neq y$).

A graph G defines a representation of this algebra if $|G| \geq 2$ and for every pair of nodes $g_1, g_2 \in G$ there is a node connected to g_1 and g_2; there is a node other than g_1, g_2 connected to neither g_1 nor g_2; and if g_1 and g_2 are distinct, there is a node connected to g_1 but not g_2.

For more about relation algebras with at most three atoms, see [AndMad94], where the interested reader can find the smallest representations.

The smallest non-representable relation algebra The following relation algebra $\mathcal{K}$ was constructed by McKenzie [McKe70, page 286]. It has four atoms: $1', <, >, \#$. Conversion is defined by $\breve{<} = >, \breve{>} = <$, and $\breve{\#} = \#$. The composition table is below.

;	$<$	$>$	$\#$
$<$	$<$	1	$<+\#$
$>$	1	$>$	$>+\#$
$\#$	$<+\#$	$>+\#$	$1'+<+>$

The forbidden triples of atoms are all Peircean transforms of:

$$(x, 1', y) \text{ where } x \neq \breve{y}, (<,<,<), (<,<,\#), \text{ and } (\#,\#,\#).$$

The reader might want to verify that $\mathcal{K}$ is a relation algebra (i.e., a model of the Tarski axioms of definition 3.8). To see that it is not representable, suppose for contradiction that h is an embedding of $\mathcal{K}$ into a proper relation algebra on base B, say. Since $\# \neq 0$, there are $b_0, b_1 \in B$ with $(b_0, b_1) \in h(\#)$. As $(<,>,\#), (>,<,\#)$ are consistent, there are $c, d \in B$ with $(c, b_0), (c, b_1), (b_0, d), (b_1, d) \in h(<)$. Then we must have $(c, d) \in h(<)$. Now, $(<,\#,\#)$ is consistent, so there is $e \in B$ with $(c, e), (e, d) \in h(\#)$. Now $(b_0, e) \in h(x)$ and $(b_1, e) \in h(y)$ for some atoms $x, y \in \mathcal{K}$. Considering the triangles (c, b_0, e) and (d, b_0, e), we see that $(<, x, \#)$ and $(>, x, \#)$ must be consistent; so $x = \#$ is the only choice. Similarly, $y = \#$. But triangle (b_0, b_1, e) shows that

4.4. Examples

$(x, \breve{y}, \#)$ must also be consistent, which is not the case if $x = y = \#$. See figure 4.1.

All smaller relation algebras are representable, but there are in all 30 non-representable relation algebras with four atoms.

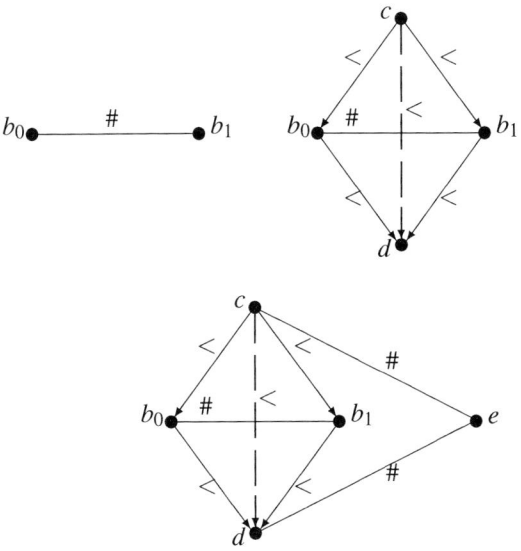

Figure 4.1: McKenzie's algebra is not representable

Algebras with flexible atoms An atom of a relation algebra is said to be *flexible* if it is consistent in any triple not involving 1'. Many relation algebras have flexible atoms. We give one example, with four atoms: 1', r, b, g. They are all self-converse. The forbidden triples of atoms not involving 1' are the Peircean transforms of (b, b, b) (b, b, g), and (b, g, g). The atom r is flexible.

This algebra is representable, as an easy argument using the game-theoretic techniques of chapter 11 will show — see exercise 11.4(1). We believe it is not known whether it has a finite representation. For a related example, see exercise 11.4(2). ([Com84] proved that any relation algebra with a flexible atom is representable; this also follows from [Madd82, theorem 5.19].)

'Monk algebras' The following collection of algebras might be called Monk algebras or Maddux algebras — both authors have used them. They have an important role to play in chapters 14 and 15. For $2 \leq n < \omega$, the Monk algebra $\mathcal{M}_n$ has n atoms: $\{1', a(p) : p < n-1\}$, all self-converse, and the composition table is summarised by

;	$a(p)$	$a(q)$
$a(p)$	$-a(p)$	$-1'$
$a(q)$	$-1'$	$-a(q)$

for any distinct $p, q < n-1$. The forbidden triples in $\mathcal{M}_n$ not involving $1'$ are $\{(a(p), a(p), a(p)) : p < n-1\}$. Thus, if we think of p as the 'colour' of the atom $a(p)$, there are no 'monochromatic triangles' permitted in representations of Monk algebras. So Ramsey's theorem (section 2.1) tells us that if $\mathcal{M}_n$ is representable, then there is a finite upper bound on the size of any representation.

A more general kind of Monk algebra has the set of atoms $\{1', a^k(p) : p < n-1, k \in I_p\}$, where each of the sets I_p ($p < n-1$) is non-empty. All atoms are symmetric, and the forbidden triples not involving $1'$ are $(a^k(p), a^l(p), a^m(p))$ for any $p < n-1$ and $k, l, m \in I_p$. Again, there is a finite upper bound on the size of a representation of this algebra.

The left linear point algebra The left linear point algebra was first presented and proved representable in [Com83], where it is referred to as $\mathcal{N}_1$. A concrete representation of it appeared in [Dün91]; see also [AndGiv$^+$94a, page 642].

A *left-linear* structure $(L, <)$ is a partial order such that if $s, t, u \in L$ and $s, t < u$ then either $s < t, t < s$ or $s = t$. The algebraic counterpart to this type of structure is the relation algebra $\mathcal{L}$ which has four atoms: $1', <, >, \sharp$ and the composition table is:

;	$<$	$>$	$\sharp$
$<$	$<$	$-\sharp$	$< + \sharp$
$>$	1	$>$	$\sharp$
$\sharp$	$\sharp$	$> + \sharp$	1

A representation of this is more difficult to define but it is determined by a dense partial order, left linear, densely branching and without any endpoints.

The Allen interval algebra This algebra (I) has thirteen atoms: $1'$, precedes, meets, overlaps, starts, during, ends, together with the converses of the last six. The composition table can be found in [All83] — e.g., overlaps ; overlaps = overlaps + meets + precedes. A natural representation of I is obtained by taking as base all intervals of rational numbers of the form $(p, q) = \{r \in \mathbb{Q} : p < r < q\}$, where $p, q \in \mathbb{Q}$ and $p < q$. Each of the thirteen atoms is then interpreted in the obvious way: for example, (p, q) meets

4.4. Examples

(r,s) if and only if $q = r$. It was proved in [LadMad94] that all countable, square representations of I are base-isomorphic to the one just outlined. In model-theoretic terms, the theory $T_I \cup \{\forall xy\ 1(x,y)\}$ (see definition 3.30) is ω-categorical. The Allen interval algebra can be built out of the point algebra (above) by 'moving from one to two dimensions' by a generalised method of constructing interval algebras out of relation algebras, given in [Hir96].

The containment algebra (Ladkin–Hayes) The Allen interval algebra I has a subalgebra C with five atoms:

$$1'$$

$$\begin{aligned}
\text{'contained-in'} &= \text{starts} + \text{during} + \text{ends} \\
\text{'contains'} &= \text{starts}^{\smile} + \text{during}^{\smile} + \text{ends}^{\smile} \\
\cap = \text{'intersects'} &= \text{meets} + \text{overlaps} + \text{meets}^{\smile} + \text{overlaps}^{\smile} \\
\# = \text{'disjoint'} &= \text{precedes} + \text{precedes}^{\smile}.
\end{aligned}$$

One way of finding a representation of C is to take any representation of I and then take the restriction to C. Thus, two intervals are related by the atom 'contained-in' if and only if they are related by 'starts', 'during' or 'ends'. But note that the containment algebra has representations that cannot be obtained from representations of the Allen interval algebra in this way. See exercise 4.

The metric point algebra This relation algebra — say, M — was first defined in [DecMei+91]. The elements of M are all finite unions of intervals of real numbers with rational endpoints together with unbounded intervals e.g. $[2,3) \cup (7, 9\frac{1}{2}) \cup [11, \infty)$. Open, closed and semi-open intervals are included. The identity element is $[0,0]$; the top element is $(-\infty, \infty)$; the converse of $[p,q]$ is $[-q,-p]$ (with similar definitions for open and semi-open intervals); negation is defined by $-[p,q] = (-\infty, p) \cup (q, \infty)$ and composition is defined by

$$[p,q];[r,s] = [p+r, q+s].$$

Exercise 5 gives a representation of M with base set $\mathbb{R}$ and asks for a proof that this representation is not complete. However, M has other complete representations (exercise 6). That still leaves the question: are there any representable, atomic relation algebras that possess *no* complete representation? In fact, there are: we'll say more about this in sections 14.3.2 and 17.2.

There are 18 relation algebras with at most three atoms, and all of them are representable. (Hence, McKenzie's example (example 4.4) of a non-representable relation algebra is smallest possible.) All integral relation algebras with at most four atoms have been classified (by work of many people). See [AndMad94] for more details. Peter Jipsen has available some powerful computer programs for relation algebra work. See [Madd85, Madd], and (as of April 2002) http://www.math.vanderbilt.edu/~pjipsen/gap/ramaddux.html for some of the programs.

We have two more important classes of relation algebras to introduce: the Lyndon algebras, given next, and the rainbow construction. Both types of algebras form useful sources of counter-examples to various conjectures — for example, they give rich sources of non-representable relation algebras and can be used to prove the non-finite axiomatisability of the class of representable relation algebras. The rainbow construction is of central importance in this book and is the topic of chapter 16.

Exercises

1. Prove that *any* square representation of the algebra $\mathcal{T}'$ has exactly two points in its base set. Conclude that all its square representations are isomorphic to each other.

2. Show that $\mathcal{M}_2 \cong \mathcal{T}'$ and that $\mathcal{M}_3 \cong \mathcal{PA}$.

3. Prove that if h is a square representation of the point algebra $\mathcal{P}$, then $h(<)$ is a dense linear order without endpoints. Deduce that all the countable, square representations of $\mathcal{P}$ are base-isomorphic.

4. Find a representation of the containment algebra that is not the reduct of any representation of the Allen interval algebra. [If stuck, see [LadMad94, §4.3].]

5. Consider the representation h of the metric point algebra $\mathcal{M}$ over the base $\mathbb{R}$, defined by $h([p,q]) = \{(r,s) \in \mathbb{R} : s - r \in [p,q]\}$ and with similar definitions for the representation of open and semi-open intervals. Use theorem 2.21 to prove that h is not a complete representation of $\mathcal{M}$.

6. Construct another representation g of $\mathcal{M}$ over a different base set and show that g is a complete representation. [Again, theorem 2.21 should be useful.]

7. Show that McKenzie's 4-atom relation algebra is generated by a single element. [See [TarGiv87, p. 54] for an application.]

4.5 The Lyndon algebras

These are relation algebras based on *projective geometries,* and they have always been important in the representation theory of relation algebras. In [Jón59],

4.5. The Lyndon algebras

Jónsson obtained relation algebras from projective planes, using a step-by-step construction to arrange that composition was associative. He used the algebras to show that not every relation algebra is 'weakly representable' (see section 5.2 for more on this). In [Lyn61], Lyndon modified Jónsson's construction in two ways. First, he allowed not just planes but projective spaces of arbitrary dimension, and second, by a small alteration to Jónsson's definition of composition he was able to avoid the need for a step-by-step construction and could build an associative relation algebra whose diversity atoms corresponded directly to the points of a projective geometry. We will call such an algebra a *Lyndon algebra*.

Lyndon proved that such an algebra is completely representable iff the geometry embeds as a hyperspace into a geometry of one higher dimension. In [Mon64], Monk used Lyndon algebras based on projective lines, together with the Bruck–Ryser theorem (see below), to prove that the class **RRA** of representable relation algebras is not finitely axiomatisable; see exercise 11.5(3). Jónsson also used them in [Jón91] to show that **RRA** is not axiomatisable by equations using only finitely many variables; see exercise 11.5(4).

Let us write L_n, $L_{\leq n}$, $L_{\geq n}$, L_∞ for the classes of Lyndon algebras of dimension n, at most n, at least n, and infinity, respectively. In [AndGiv+97], an analysis of which subclasses of L_1 (those Lyndon algebras based on projective lines) have decidable equational theories is given. The universal class $\mathbf{SL}_1$ is axiomatised there (pp. 57–59). In [Giv99], Givant showed using theorem 2.87 that for any finite $n > 0$, $\mathbf{SL}_n$, $\mathbf{SL}_{\leq n}$, $\mathbf{SL}_{\geq n}$ are universal classes, and that **SP** of them are varieties; the same is shown for L_∞. Givant also axiomatises the atom structures of Lyndon algebras. In [Ste00], Stebletsova showed among other things that given any class $\mathcal{G}$ of projective geometries that contains an infinite geometry of dimension at least three, the equational theory of the class of Lyndon algebras over geometries in $\mathcal{G}$ is undecidable. She also used a Gabbay-style inference rule to axiomatise the equational theory of the class of Lyndon algebras of dimension d, for any finite $d \geq 2$. In section 6.4.2 we will discuss Sági's application of Lyndon algebras to the finitisation problem.

We will only discuss Lyndon algebras based on projective lines. Since their representability is determined by whether the line embeds in a projective plane, we list some facts about these planes first.

Projective planes A projective plane is an incidence system of points and lines, such that any two distinct points lie on a unique line, and (dually) any two distinct lines meet in a unique point. Further, the plane must contain four points, no three of which lie on a single line. Figure 4.2 illustrates the smallest projective plane. It has seven points and seven lines (one of the lines being drawn as a circle).

In a projective plane, all lines must contain the same number of points. The *order* of a projective plane is defined to be one less than the number of points

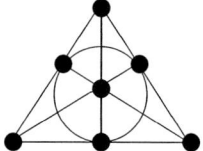

Figure 4.2: The projective plane of order 2

on a line. The plane shown in figure 4.2 has order 2. For any prime p, and any $m \geq 1$, there is a projective plane of order p^m. One such plane may be obtained from the three-dimensional vector space over the finite field of order p^m, its points and lines being the one-dimensional and two-dimensional subspaces, respectively, and the point x lying on the line l iff $x \subseteq l$. These planes are the only ones that embed into projective geometries of dimension 3. There are others ('non-desarguesian' ones) not obtainable like this; but all known finite projective planes do have prime power order. By the Bruck–Ryser theorem [BruRys49], if there is a projective plane of order n, and $n \equiv 1$ or $2 \pmod{4}$, then n must be the sum of the squares of two non-negative whole numbers. (One of them could be zero if n is a square.) So, for example, there is no projective plane of order 6 or 14, and the same holds for infinitely many more numbers (see exercise 1 below). The cases $n = 10, 12, 15, 18, \ldots$ are not covered by this result. While it is known that for some of these n there is no projective plane of order n, most cases remain open. See, e.g., [Cam91, HugPip73, Stevens72] for information on projective planes.

Lyndon algebras based on projective lines We take n to be a whole number, at least 2. The Lyndon algebra $\mathcal{A}_n$ is a finite relation algebra with $n+2$ atoms, say $1', a_0, \ldots, a_n$. All elements are self-converse and composition is defined by:

- $a_i ; a_i = a_i + 1'$ if $n \geq 3$, and $a_i ; a_i = 1'$ if $n = 2$.[1]

- $a_i ; a_j = \sum\limits_{k \neq i,j} a_k$ if $i \neq j$,

where $i, j, k \leq n$. Alternatively, we can list the forbidden triples to determine composition. The forbidden triples are all the permutations of $(x, 1', y)$ for $x \neq y$ and of (a_i, a_i, a_j) for $i \neq j$.

If there is a projective plane $\mathcal{P}_n$ of order n, choose a projective line l (called 'the line at infinity'), and let $\mathcal{S}_n$ be the 'affine' plane obtained from $\mathcal{P}_n$ by deleting l and all points on l. Then the set of points of $\mathcal{S}_n$ can be made into a representation

---

[1] Jónsson [Jón59] took $a_i ; a_i = 1'$ in all cases.

4.5. The Lyndon algebras

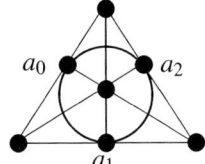 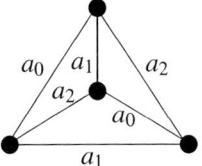

Figure 4.3: Lyndon algebra from projective plane of order 2

of $\mathcal{A}_n$ by identifying the diversity (non-'1') atoms of $\mathcal{A}_n$ with the points of l (in any fashion), and defining, for any atom a_i, and any distinct points x, y of $\mathcal{S}_n$,

$$a_i(x,y) \text{ holds iff } \overline{xy} \cap l = \{a_i\}.$$

Here, $\overline{xy}$ is the unique projective line of $\mathcal{P}_n$ through x and y. (We can recover the interpretation of any $r \in \mathcal{A}_n$ from this.)

As an example, we obtain a representation of $\mathcal{A}_2$ from the projective plane pictured in figure 4.2, as follows. We choose l to be the 'circular' line, and identify its points with the atoms a_0, a_1, a_2 of $\mathcal{A}_2$, as shown on the left of figure 4.3. The affine plane on the right of figure 4.3, obtained by deleting l and its points, gives a representation of $\mathcal{A}_2$, the atomic relations holding between its points being as illustrated. For example, the top two points are related by a_1, because, on the left, the (vertical) line joining them cuts the circular line in the point identified with the atom a_1 of $\mathcal{A}_2$.

Lyndon proved in [Lyn61, theorem 1] that any representation of any $\mathcal{A}_n$ must arise as an affine plane in the way described. So:

PROPOSITION 4.7 (Lyndon) *Any given Lyndon algebra $\mathcal{A}_n$ is representable iff there is a projective plane of order n.*

In particular, if there is no projective plane of order n (for example, $n = 6$) then $\mathcal{A}_n$ is not representable (Lyndon, [Lyn61]). This is how some of the early non-representable relation algebras were found: at the time of [Lyn61], the 8-atom algebra $\mathcal{A}_6$ was the smallest-known non-representable relation algebra. Infinitely many Lyndon algebras are representable (e.g., $\mathcal{A}_p$ for prime p) and infinitely many are not (see exercise 1 below). Their superficial similarity illustrates the subtlety of the representation problem.

Non-isomorphic representations of Lyndon algebras [Lyn61, corollary 1.3] observed that the base-isomorphism type of the representation of a Lyndon algebra obtained above depends in general on the chosen enumeration of the line l at infinity. If any two enumerations gave base-isomorphic representations, it would mean that any permutation of l is induced by a collineation (line-preserving permutation) of the projective plane. But this is not always the case. For large n, e.g., $n=29$, the group of collineations of $\mathcal{P}_n$ fixing l does not induce the full symmetric group on the points of l, so the isomorphism type of S_n varies with the choice of association of atoms with these points. This was used in [HirHod97b, proposition 18] to prove that the Lyndon algebra $\mathcal{A}_{29}$ has at least two non-base-isomorphic, square representations. This somewhat understated the position; Givant showed in [Giv01] that for $n=p^k$ where p is prime, $\mathcal{A}_n$ has at least $(n-2)!/k$ non-base-isomorphic square representations (in fact he gives an exact formula for the number of representations, using notions from projective geometry). Thus, $\mathcal{A}_{29}$ has at least $27!$ (about 10^{28}) such representations.

Exercises

1. [Mon64] Show that any number of the form $2 \cdot 3^{2n+1}$ (for $n < \omega$) is equal to 2 (mod 4) and is not the sum of two squares. Deduce using the Bruck–Ryser theorem that infinitely many Lyndon algebras $\mathcal{A}_n$ are not representable.

2. This exercise is to show that the classes At**RRA** and Str**RRA** of relation algebra atom structures (definitions 2.74, 2.89) are not finitely axiomatisable. (See exercise 11.5(3) and chapter 14 for further results.) We will use first-order interpretations: see section 2.1 and [Hod93] for information. For $2 \leq n < \omega$, let $\mathcal{A}_n$ be the Lyndon algebra described above.

   (a) Show that the atom structure $S_n = \text{At}\,\mathcal{A}_n$ is interpretable in the structure $(\text{dom}\,S_n, 1')$, in the signature L consisting of a single constant $1'$.

   (b) Assume that σ is a first-order sentence axiomatising At**RRA**. Let σ' be an L-sentence such that $(\text{dom}\,S_n, 1') \models \sigma'$ iff $S_n \models \sigma$. Show that the L-theories

   $$\begin{aligned} R &= \{\sigma'\} \cup \{\exists_{\geq n} x(x=x) : n < \omega\}, \\ N &= \{\neg\sigma'\} \cup \{\exists_{\geq n} x(x=x) : n < \omega\} \end{aligned}$$

   are consistent. Here, $\exists_{\geq n} x(x=x)$ abbreviates the sentence $\exists_{i<n} x_i (\bigwedge_{i<j<n} x_i \neq x_j)$.

   (c) Show that any two countably infinite L-structures are isomorphic. Deduce that there is no first-order sentence axiomatising At**RRA**.

   (d) Repeat for Str**RRA**. (Actually, this class is non-elementary: see theorem 14.3.)

3. Consider the equational theories of the following two classes of relation algebras: the group relation algebras (**GRA**) and the class of all Lyndon algebras (see section 4.5). Does either equational theory include the other?

4.5. The Lyndon algebras

4. Let $\mathcal{A}_n$ be the Lyndon algebra with $n+2$ atoms (where $n \geq 2$). Is it true that $\mathcal{A}_n \in \mathbf{RRA} \Rightarrow \mathcal{A}_n \in \mathbf{GRA}$?

5. Let $\mathcal{L}$ be the lattice of subvarieties of **RRA** ordered by inclusion. Show that there are exactly three atoms (minimal non-empty members) of this lattice. For each of these three atoms, find a relation algebra $\mathcal{A}$ from the examples listed in this chapter such that the variety generated by $\mathcal{A}$ is the chosen atom of the lattice. ([AndGiv$^+$94b] shows how very complicated the lattice $\mathcal{L}$ is further up.)

Chapter 5

Relativisation and cylindric algebras

Outline of chapter In chapter 3 we investigated the basic properties of proper relation algebras, relation algebras, and representations. Here we continue that investigation as follows:

- We generalise from proper relation algebras so that the unit or top element is not required to be an equivalence relation. The definition of a representation is generalised to that of a *relativised representation*. Corresponding to this, we generalise relation algebras by weakening or dropping the associativity axiom, obtaining the classes **NA**, **WA**, and **SA** of non-associative, weakly associative, and semi-associative algebras. We examine their basic algebraic properties ('arithmetic'). This investigation will continue in chapters 7 and 13.

- We briefly consider an alternative kind of representation of relation algebras, where the boolean operations $+, -$ need not be respected. These representations are called 'weak' representations. We will discuss the 'weakly representable relation algebras' in greater depth in chapters 17 and 18.

- We briefly introduce cylindric algebras. Since the focus of the book is on relation algebras, we only introduce those parts of the theory of cylindric algebras that will be needed later on. We give references to the extensive literature on them as we go. We also discuss some connections between relation algebras and cylindric algebras, a topic that will be developed further in chapter 13.

5.1 Relativisation

Several times already we mentioned Monk's theorem that the class of representable relation algebras cannot be defined with only finitely many axioms. The *finitisation problem* seeks ways around this and other similar negative results. There has been a great deal of research into this problem. One approach has been to consider alterations to the similarity type: can we achieve finite axiomatisability if we add additional operators; what happens if we delete some of the operators? We will discuss some of this work in section 6.4. An important alternative approach is to alter the definition of a representation, by *relativisation*. This is our topic in this section.

5.1.1 Relativised representations

A representation is an isomorphism from a relation algebra to a proper relation algebra, and in this way, proper relation algebras provide 'classical' semantics for relation algebras. Recall the definition of a proper relation algebra $\mathcal{P}$ over the base B: it is a field of binary relations on B containing the identity relation on B and closed under the boolean operations, conversion, and composition — defined by $r \mid s = \{(x,y) \in B \times B : \exists z \in B((x,z) \in r \wedge (z,y) \in s)\}$. There is a unit 1, the largest relation in the algebra. The unit is not necessarily $B \times B$, but lemma 3.4 showed using closure under the operations that it is an equivalence relation over B.

In the relativised approach to representations, the semantics are generalised, or weakened: the unit is some fixed binary relation W (not necessarily an equivalence relation) over the base, and the operations are relativised to W. The hope is that with such non-classical representations we may make inroads into the finitisation problem, or at any rate deepen our understanding of relation algebras.

We will present these relativised representations here as models of a first-order theory. Recall from definition 3.30 that a (classical) representation of an relation algebra $\mathcal{A}$ can be viewed as a model of the first-order theory $T_{\mathcal{A}}$. Now we generalise that definition.

DEFINITION 5.1 Let $\mathcal{A}$ be a BAO of the signature L_{RA} of relation algebras (see definition 3.9; we call such an $\mathcal{A}$ a *relation-type BAO*).

1. Write $\mathcal{L}(\mathcal{A})$ for the first-order signature consisting of a binary relation symbol for each element of $\mathcal{A}$.

2. $R_{\mathcal{A}}$ is the $\mathcal{L}(\mathcal{A})$-theory consisting of the following axioms:

5.1. Relativisation

$$\forall xy[1\text{'}^{\mathcal{A}}(x,y) \leftrightarrow x = y] \quad \text{for each } r,s,t \in \mathcal{A} \text{ with:}$$
$$\forall xy[r(x,y) \leftrightarrow s(x,y) \vee t(x,y)] \quad r = s+t$$
$$\forall xy[1^{\mathcal{A}}(x,y) \rightarrow (r(x,y) \leftrightarrow \neg s(x,y))] \quad r = -s$$
$$\forall xy[r(x,y) \leftrightarrow s(y,x)] \quad r = \breve{s}$$
$$\forall xy[1^{\mathcal{A}}(x,y) \rightarrow (r(x,y) \leftrightarrow \exists z(s(x,z) \wedge t(z,y)))] \quad r = s;t$$
$$\exists xy\, r(x,y) \quad r \neq 0.$$

3. A *relativised representation* of $\mathcal{A}$ is a model of $R_{\mathcal{A}}$.

4. The *base* of a relativised representation M of $\mathcal{A}$ is $\text{dom}(M)$. The *unit* of M is the binary relation $(1^{\mathcal{A}})^M = \{(x,y) : M \models 1^{\mathcal{A}}(x,y)\}$ on $\text{dom}(M)$.

5. A *complete relativised representation* of $\mathcal{A}$ is a model M of $R_{\mathcal{A}}$ such that for all $x,y \in M$ and each set S of elements of $\mathcal{A}$ such that the supremum $\sum S$ exists in $\mathcal{A}$, we have $M \models (\sum S)(x,y)$ iff $M \models s(x,y)$ for some $s \in S$.

6. Where no confusion is likely, we will write the elements $0^{\mathcal{A}}, 1^{\mathcal{A}}, 1\text{'}^{\mathcal{A}}$ of $\mathcal{A}$ and $\mathcal{L}(\mathcal{A})$ simply as $0, 1, 1\text{'}$.

Relativised representations generalise classical representations. Any model of $T_{\mathcal{A}}$ (definition 3.30) is a model of $R_{\mathcal{A}}$. A classical representation M of a BAO $\mathcal{A}$ of the type of relation algebras is a special case of a relativised representation where 1^M is an equivalence relation. But not all relativised representations are classical representations. In the $R_{\mathcal{A}}$ axiom for composition, for example, we cannot conclude from $s(x,z) \wedge t(z,y)$ that $(s;t)(x,y)$. The reason is that (x,y) might not be an 'edge' in the relativised representation: we may not have $M \models 1(x,y)$. We will see examples later.

General relativised representations

In the most general kind of relativised representation, all operations would be relativised to 1. We will temporarily call such representations 'general relativised representations'. A general relativised representation of a relation-type BAO $\mathcal{A}$ would be a model of the theory

$$\forall xy(1(x,y) \rightarrow [1\text{'}(x,y) \leftrightarrow x = y])$$
$$\forall xy(1(x,y) \rightarrow [(r+s)(x,y) \leftrightarrow r(x,y) \vee s(x,y)])$$
$$\forall xy(1(x,y) \rightarrow [(-r)(x,y) \leftrightarrow \neg r(x,y)])$$
$$\forall xy(1(x,y) \rightarrow [\breve{r}(x,y) \leftrightarrow 1(y,x) \wedge r(y,x)])$$
$$\forall xy(1(x,y) \rightarrow [(r;s)(x,y) \leftrightarrow \exists z(1(x,z) \wedge r(x,z) \wedge 1(z,y) \wedge s(z,y))])$$
$$\exists xy[1(x,y) \wedge t(x,y)]$$

for all $r,s,t \in \mathcal{A}$ with $t \neq 0$. It can be checked that every model of $R_{\mathcal{A}}$ is a model of this more complicated theory, but the converse is false. We will take a few moments to explain why we did not define $R_{\mathcal{A}}$ like this.

First, observe that because all quantifiers in the above theory are relativised to the unit, without loss of generality we can restrict the domain and relations to the unit. Formally, if M is a general relativised representation of $\mathcal{A}$ then so is the $\mathcal{L}(\mathcal{A})$-structure M' with domain $\{x \in M : \exists y \in M (M \models 1(x,y) \vee 1(y,x))\}$ and defined by $r^{M'} = r^M \cap 1^M$ for all $r \in \mathcal{A}$. Then $M' \models \forall xy(r(x,y) \rightarrow 1(x,y))$ for all $r \in \mathcal{A}$. So the $R_{\mathcal{A}}$-axioms $\forall xy[(r+s)(x,y) \leftrightarrow r(x,y) \vee s(x,y)])$, $\forall xy(1(x,y) \rightarrow [(r;s)(x,y) \leftrightarrow \exists z(r(x,z) \wedge s(z,y))])$, and $\exists xy\, t(x,y)$ are valid in M', for all $r,s,t \in \mathcal{A}$ with $t \neq 0$. So these axioms are (without loss of generality) valid in general relativised representations, and there was no need to (fully) relativise them in $R_{\mathcal{A}}$.

The remaining difference from $R_{\mathcal{A}}$ is in the axioms for identity and conversion, which in $R_{\mathcal{A}}$ are not relativised to 1. M' may not validate these. It can be checked that M' validates the identity axiom of $R_{\mathcal{A}}$ iff the unit $1^{M'}$ is reflexive on $\operatorname{dom}(M')$, while M' validates the conversion axioms of $R_{\mathcal{A}}$ iff $1^{M'}$ is symmetric.

Let us consider the merits of general relativised representations as semantics for relation algebras. It turns out that any relation algebra has a general relativised representation (completeness), but not every relation-type BAO with such a representation is a relation algebra (no soundness). The relation algebra axioms R0, R2, R5, and R7 of definition 3.8 do hold over relation-type BAOs with general relativised representations, but the other relation algebra axioms may fail. For a general relativised representation M of a relation-type BAO $\mathcal{A}$, if 1^M is not reflexive over its range then axiom R3, $a;1' = a$, is not valid in $\mathcal{A}$, and if 1^M is not symmetric then axiom R4, $\breve{\breve{a}} = a$, is not valid. Axiom R1, associativity of composition, and axiom R6 may also fail. So general relativised representations are too general for our purposes here. However, they are important in other contexts. A finite axiomatisation for the isomorphism class of relation-type BAOs with general relativised representations has been defined in [Kram91]. In this paper and in, e.g., [And88, Mar99b], the reader will find a thorough investigation into various kinds of relativisation.

General relativised representations with reflexive symmetric unit

Let us now, bit by bit, impose restrictions on the unit of a general relativised representation of a relation-type BAO $\mathcal{A}$, and observe that each restriction causes $\mathcal{A}$ to validate a larger set of axioms. If we insist that the unit is reflexive over its range, then the identity axiom (axiom R3) becomes valid in $\mathcal{A}$. If we also insist that 1^M is symmetric, then axioms R4 and R6 become true. Thus, if $\mathcal{A}$ has a general relativised representation whose unit is reflexive and symmetric then $\mathcal{A}$ validates all the axioms for relation algebra except perhaps the associativity axiom (axiom R1).

5.1. Relativisation

There is a converse: if axioms R3 and R4 hold in $\mathcal{A}$ then the unit of any general relativised representation of $\mathcal{A}$, if there is one, is reflexive on its range and symmetric.

So, we have informally established that for any relation-type BAO $\mathcal{A}$ that has a general relativised representation, the following are equivalent:

- $\mathcal{A}$ has a relativised representation in the sense of definition 5.1 (i.e., a model of $R_{\mathcal{A}}$),

- $\mathcal{A}$ has a general relativised representation M whose unit is reflexive on $\text{dom}(M)$ and symmetric,

- axioms R0 and R2–R7 of definition 3.8 are valid in $\mathcal{A}$.

Thus, the effect of our failing to relativise the axioms for 1' and ˘ in $R_{\mathcal{A}}$ is to limit ourselves to relativised representations of relation-type BAOs satisfying all the relation algebra axioms except perhaps associativity. Such algebras are called 'non-associative algebras'. This book is about relation algebras and is not concerned with algebras failing axioms R3, R4, or R6. In many cases, however, we are dealing with algebras of the type of relation algebras that we know satisfy all the relation algebra axioms, except that we do not know that they are associative. Non-associative algebras provide a useful catch-all generalisation. Unless otherwise stated, we will use 'relativised representation' as in definition 5.1 from now on, as we have seen that this is the appropriate definition for non-associative algebras.

5.1.2 Non-associative algebras

We now introduce these formally.

DEFINITION 5.2 [Madd82, definition 1.2]   A *non-associative algebra* is a relation-type algebra (i.e., an algebra of the same signature as relation algebras) that obeys all the relation algebra axioms (see definition 3.8) except perhaps the associativity axiom R1. **NA** denotes the class of all non-associative algebras.

Of course, the term 'non-associative algebra' does not entail that the algebra is not associative, merely that it need not be.

We next prove that the unit of any relativised representation of a relation-type BAO $\mathcal{A}$ is reflexive and symmetric. This follows from the fact, important in its own right, that a relativised representation of $\mathcal{A}$ is a boolean representation of the boolean reduct $\text{bool}(\mathcal{A})$. We also show a little more formally than before that $\mathcal{A}$ must be a non-associative algebra.

LEMMA 5.3 *Let $\mathcal{A}$ be a relation-type BAO (or more generally, a relation-type algebra whose boolean reduct is a boolean algebra). Suppose that $\mathcal{A}$ has a relativised representation $M \models R_{\mathcal{A}}$. Then*

1. *As boolean algebras, we have*

$$(\mathcal{A}, 0, 1, +, -) \cong (\{r^M : r \in \mathcal{A}\}, \emptyset, 1^M, \cup, \setminus),$$

   where, for $r \in \mathcal{A}$, r^M denotes the interpretation of the binary relation symbol r as a binary relation on M (so $r^M \subseteq {}^2 M$).

2. *$\mathcal{A}$ is a non-associative algebra.*

3. *The unit 1^M is a reflexive and symmetric relation on M.*

Proof.

1. The isomorphism is evidently $r \mapsto r^M$. By the axioms for $+$ and $-$ in $R_{\mathcal{A}}$, $(r+s)^M = r^M \cup s^M$ and $(-r)^M = 1^M \setminus r^M$. Obviously, $1 \mapsto 1^M$. We check that 0 maps to $\emptyset$. If $x,y \in M$ and $M \models 0(x,y)$, then by the $+$-axiom, $M \models 1(x,y)$, so by the axiom for $-$, $M \models \neg 0(x,y)$, a contradiction. So $0^M = \emptyset$. Finally, we check that the map $r \mapsto r^M$ is one-to-one. We have just seen that this map is a boolean homomorphism, so it is enough to show that $r^M \neq \emptyset$ if $r \neq 0$ in $\mathcal{A}$. But there is an axiom of $R_{\mathcal{A}}$ saying exactly this.

2. The axioms R2–R7 are easily verified. As an example, consider the axiom R7: $\breve{a};(-(a;b)) \leq -b$ for all $a,b \in \mathcal{A}$. By part 1, it suffices to check that

$$M \models \forall xy (\breve{a};(-(a;b)))(x,y) \to -b(x,y)).$$

   Assume for contradiction that $M \models \breve{a};(-(a;b))(x,y) \wedge \neg(-b)(x,y)$ for some $x,y \in M$. Since $\breve{a};(-(a;b)) \leq 1$, part 1 yields $M \models 1(x,y)$. The negation axiom of $R_{\mathcal{A}}$ now yields $M \models b(x,y)$. Also, by the composition axiom of $R_{\mathcal{A}}$, we may choose $z \in M$ with $M \models \breve{a}(x,z) \wedge (-(a;b))(z,y)$. So as above, $M \models 1(z,y)$. Also, the 'converse' axiom of $R_{\mathcal{A}}$ yields $M \models a(z,x)$. So $M \models 1(z,y) \wedge a(z,x) \wedge b(x,y)$, whence $M \models (a;b)(z,y)$ by the composition axiom of $R_{\mathcal{A}}$. Since $M \models 1(z,y) \wedge (-(a;b))(z,y)$, the negation axiom yields $M \models \neg(a;b)(z,y)$, a contradiction.

3. If $x \in M$ then $M \models 1'(x,x)$ by the identity axiom; and as $1' \leq 1$, we have $M \models 1(x,x)$ by part 1. So 1^M is reflexive. For symmetry, if $M \models 1(x,y)$ then by the axiom for conversion in $R_{\mathcal{A}}$, $M \models \breve{1}(y,x)$, so as $\breve{1} \leq 1$, the axiom for $+$ shows that $M \models 1(y,x)$, too, proving symmetry.

□

5.1.3 Weakly associative algebras

However, even using relativised representations in which the unit is symmetric and reflexive, we find that the axioms for non-associative algebras are sound but not complete: there are non-associative algebras that have no relativised representations of this type. This is because just as relativised representations enforce conditions on 1' and ˘, they also enforce a little associativity of ';'. This amount of associativity is encapsulated in the following definition.

DEFINITION 5.4 [Madd82, definition 1.2] A *weakly associative algebra* $\mathcal{A}$ is an algebra of the type of relation algebras that obeys all the Tarski axioms for relation algebra (see definition 3.8) except perhaps the associativity axiom R1 (i.e., it is a non-associative algebra), but satisfying instead the *weak associativity law* (WL)

$$((1' \cdot x); 1); 1 = (1' \cdot x); (1; 1)$$

for all $x \in \mathcal{A}$. Of course, $1; 1 = 1$, so the right-hand side may be simplified.

WA denotes the class of all weakly associative algebras.

See exercise 5.1(6) for an example of a weakly associative algebra that is not a relation algebra, and (e.g.) exercise 12.6(5) for a non-associative algebra that is not weakly associative.

PROPOSITION 5.5 *If a relation-type BAO $\mathcal{A}$ has a relativised representation then $\mathcal{A} \in$ **WA**.*

Proof. By lemma 5.3, an algebra with a relativised representation must belong to **NA**, so the proposition is proved by verifying that the weak associativity law holds for such an algebra. This is the task of exercise 5.1(1). □

The converse of proposition 5.5 also holds: this was proved by Maddux in [Madd82, theorem 5.20], and we will prove it by games in theorem 7.5. Thus, **WA** is precisely the class of relation-type BAOs with relativised representations (with reflexive and symmetric unit).

5.1.4 Semi-associative algebras

A stronger associativity axiom is the semi-associative law:

DEFINITION 5.6 [Madd82, definition 1.2] A *semi-associative algebra* $\mathcal{A}$ is a non-associative algebra that satisfies

$$(x; 1); 1 = x; (1; 1),$$

and, again, the right-hand side may be simplified to $x; 1$. **SA** denotes the class of all semi-associative algebras.

It was shown by Maddux in [Madd91b, theorem 25] that given any term built from variables and 1 using only composition ';', and in which at least one 1 occurs, its brackets can be rearranged arbitrarily without changing its value in a semi-associative algebra under any given assignment to variables. So, for example, **SA** $\models (x;1);(y;(z;v)) = ((x;(1;y));z);v$.

SA characterises a slightly stronger[1] class of relativised representations called 3-square relativised representations, defined below. Do not confuse these with the square classical representations of definition 3.3.

DEFINITION 5.7 Let $\mathcal{A}$ be a non-associative algebra and let M be a model of $R_\mathcal{A}$ — i.e., a relativised representation of $\mathcal{A}$.

- A *clique* C of M is a subset of the domain of M such that for all $x, y \in C$ we have $M \models 1(x,y)$.

- Let n be a cardinal. M is said to be *n-square* if for all cliques C of M with $|C| < n$, all $x, y \in C$, and all $a, b \in \mathcal{A}$ with $M \models (a;b)(x,y)$, there exists $z \in M$ such that $C \cup \{z\}$ is a clique and $M \models a(x,z) \wedge b(z,y)$.

Any classical representation is an n-square relativised representation, for any n. The class of relation-type BAOs with a 3-square (respectively, 4-square) relativised representation turns out to be precisely **SA** (respectively, **RA**). We'll see how to prove these results in chapters 13, where n-square and other 'locally classical' relativised representations will be used to capture classes of algebras nearer and nearer to **RRA**.

Finally, what happens when we insist that the unit of a relativised representation of a non-associative algebra is transitive? Then the unit is an equivalence relation and any such relativised representation is a disjoint union of classical representations. Thus, the class of non-associative algebras with such relativised representations is just **RRA**, and this class cannot be defined using finitely many axioms. (In fact, the class of relation-type BAOs with a general relativised representation with transitive unit (even if not reflexive or symmetric) is not finitely axiomatisable [Mar99b].)

5.1.5 Basic facts about NA, WA, SA

These will be needed later. Many of their analogues for relation algebras were proved in [ChiTar51, JónTar52]. Clearly, we have **NA** $\supseteq$ **WA** $\supseteq$ **SA** $\supseteq$ **RA**. Recall that **RA** is a canonical variety (theorem 3.16). So are the others:

THEOREM 5.8 [Madd82, theorem 4.2] *The classes* **NA**, **WA**, *and* **SA** *are canonical varieties.*

---

[1] See exercise 5.1(6).

5.1. Relativisation

Proof. By the proof of theorem 3.16, **NA** can be defined by Sahlqvist equations. Since **NA** can be defined by equations, it is a variety. Furthermore, for any $\mathcal{A} \in \mathbf{NA}$, the canonical extension $\mathcal{A}^+$ must satisfy all the Sahlqvist equations that are true in $\mathcal{A}$ (by theorem 2.95), so $\mathcal{A}^+$ satisfies all the equations defining **NA** and therefore must belong to **NA**. Hence, **NA** is a canonical variety.

Similarly, each of the other classes can be defined by the equations defining **NA** plus, in each case, a single equation that does not involve negation. By the same argument, each of these classes is therefore a canonical variety. □

Many facts about relation algebras hold for non-associative algebras too. The following extended version of the Peircean law (PL), saying in some sense that consistency of a triple is independent of its orientation, is a very useful example.

LEMMA 5.9 (PL) *Let* $\mathcal{A} \in \mathbf{NA}$. *If* $a \,;b \cdot \check{c} \neq 0$, *then* $b\,;c \cdot \check{a} \neq 0$ *and* $\check{b}\,;\check{a} \cdot c \neq 0$.

Proof. The first part is proved by lemma 3.12, and the second by the relation algebra axioms R4, R6, and lemma 3.11. The lemmas did not require associativity. □

Note also that in any non-associative algebra, $\check{1} = 1$ and $\check{1'} = 1'$, by lemmas 3.11 and 3.17.

The following lemma contains some handy 'arithmetic facts' about weakly associative algebras, and will be useful on several occasions later. See exercise 10 below for the converse of item 6 when we are dealing with a semi-associative algebra.

LEMMA 5.10 (Maddux 1982) *Let* $\mathcal{A}$ *be an atomic weakly associative algebra and let* $a \in \mathrm{At}\,\mathcal{A}$. *Define* $\mathrm{st}(a) = a\,;\check{a} \cdot 1'$ *and* $\mathrm{end}(a) = \check{a}\,;a \cdot 1'$ *(the start and end of a, respectively). Then for all* $a,b \in \mathrm{At}\,\mathcal{A}$,

1. $\mathrm{st}(a), \mathrm{end}(a) \in \mathrm{At}\,\mathcal{A}$,

2. $\mathrm{st}(a)\check{{}} = \mathrm{st}(a)$, $\mathrm{st}(a)\,;\mathrm{st}(a) = \mathrm{st}(a)$, *and similar equations for* $\mathrm{end}(a)$,

3. $\mathrm{st}(a)\,;a = a$, $\mathrm{end}(a)\,;\check{a} = \check{a}$,

4. *if* $a \leq 1'$ *then* $\mathrm{st}(a) = \mathrm{end}(a) = a$,

5. $\mathrm{st}(a) = \mathrm{end}(\check{a})$ *and* $\mathrm{end}(a) = \mathrm{st}(\check{a})$,

6. *if* $a\,;b \neq 0$ *then* $\mathrm{end}(a) = \mathrm{st}(b)$.

Proof. The following was originally proved in [Madd82, theorems 2.2, 3.5]. We use the Peircean law (PL, lemma 5.9) repeatedly in this lemma.

1. First, $\mathrm{st}(a) = a;\breve{a}\cdot 1' \neq 0$, else by the Peircean law we would obtain $1';a\cdot a = 0$ which is false (we use $\breve{1'} = 1'$ here, for example). Let $x,x' \in \mathrm{At}\,\mathcal{A}$ with $x,x' \leq \mathrm{st}(a)$. We must show that $x = x'$. Since $x \leq \mathrm{st}(a)$, we have $x\cdot a;\breve{a} \neq 0$, so by (PL), $a\cdot \breve{x};a \neq 0$. Since a is an atom, $a \leq \breve{x};a \leq 1';a = a$. Thus,
$$\breve{x};a = a, \tag{5.1}$$
and similarly, $\breve{x'};a = a$. Hence, $\breve{x'};(\breve{x};a) = a$, so by (PL) we obtain $(\breve{x};a);\breve{a}\cdot x' \neq 0$. As x' is an atom, $(\breve{x};a);\breve{a} \geq x'$. Thus, $x' \leq (\breve{x};1);1 \leq \breve{x};1$, by the weak associativity law (WL). By complete additivity of composition, there is an atom y such that $x' \leq \breve{x};y$. But then, using (PL) and because y is an atom, we get $y \leq x;x'$. Finally, using $x,x' \leq 1'$, we see that $y \leq x;1'\cdot 1';x' = x\cdot x'$. Since y is non-zero and x,x' are atoms, we deduce that $x = x'$. Thus, $\mathrm{st}(a)$ is an atom; and similarly, $\mathrm{end}(a)$ is an atom too.

2. Easy exercise.

3. Follows from (2) and equation 5.1 above.

4. By (3), $a = \mathrm{st}(a);a \leq \mathrm{st}(a);1' = \mathrm{st}(a)$, and as $a,\mathrm{st}(a)$ are atoms we have $a = \mathrm{st}(a)$. The proof that $a = \mathrm{end}(a)$ is similar.

5. Immediate from axiom R4 and the definitions of $\mathrm{st}(a),\mathrm{end}(\breve{a})$.

6. Suppose $a;b \neq 0$, and let z be an atom with $z \leq a;b$. Then

$$\begin{aligned}
z \leq a;b &\Rightarrow \breve{a} \leq b;\breve{z} && \text{by (PL)} \\
&\Rightarrow \breve{a} \leq (\mathrm{st}(b);b);\breve{z} && \text{by part (3)} \\
&\Rightarrow \breve{a} \leq \mathrm{st}(b);w \leq w && \text{some atom } w, \\
&&& \text{by (WL) and } \mathrm{st}(b) \leq 1' \\
&\Rightarrow \breve{a} = \mathrm{st}(b);\breve{a} && \text{since } \breve{a},w \text{ are atoms.}
\end{aligned}$$

Proceeding in the same way,

$$\begin{aligned}
&\mathrm{end}(a) \leq \breve{a};a \\
\Rightarrow\ &\mathrm{end}(a) \leq (\mathrm{st}(b);\breve{a});a && \text{by the above} \\
\Rightarrow\ &\mathrm{end}(a) \leq \mathrm{st}(b);v \leq v && \text{some atom } v, \text{ as above} \\
\Rightarrow\ &\mathrm{end}(a) = \mathrm{st}(b);\mathrm{end}(a) && \text{since } \mathrm{end}(a),\mathrm{st}(b) \text{ are atoms} \\
\Rightarrow\ &\mathrm{end}(a) \leq \mathrm{st}(b) && \text{since } \mathrm{end}(a) \leq 1' \\
\Rightarrow\ &\mathrm{end}(a) = \mathrm{st}(b) && \text{since they are atoms.}
\end{aligned}$$

□

The following lemma will be very useful when we come on to chapter 7, because it shows that there is a consistent way of labelling the edges of 3-graphs with atoms in such a way that a given product of atoms is realised.

5.1. Relativisation

LEMMA 5.11 [Madd91b, lemma 49] *Let $\mathcal{A}$ be an atomic weakly associative algebra, and let $a, b, c \in \text{At}\,\mathcal{A}$ with $a \leq b\,;c$.*

1. *Let x, y be objects such that $x = y$ iff $a \leq 1$'. There is a function $f : \{x, y\} \times \{x, y\} \to \text{At}\,\mathcal{A}$ such that $f(x, y) = a$ and for all $i, j, k \in \{x, y\}$ we have $f(i, i) \leq 1$' and $f(i, k) \leq f(i, j)\,;f(j, k)$.*

2. *Let x, y, z be objects such that $x = y$ iff $a \leq 1$', $x = z$ iff $b \leq 1$' and $z = y$ iff $c \leq 1$'. There is a function $g : \{x, y, z\} \times \{x, y, z\} \to \text{At}\,\mathcal{A}$ such that $g(x, y) = a$, $g(x, z) = b$, $g(z, y) = c$, and for all $i, j, k \in \{x, y, z\}$ we have $g(i, i) \leq 1$' and $g(i, k) \leq g(i, j)\,;g(j, k)$.*

Proof. We consider just the second part; the first part is a special case. We define g on the remaining pairs from $\{x, y, z\}$ as follows. $g(x, x) = \text{st}(a)$, $g(z, z) = \text{st}(c)$, $g(y, y) = \text{end}(a)$, $g(y, x) = \breve{a}$, $g(z, x) = \breve{b}$, and $g(y, z) = \breve{c}$. These are atoms, by lemmas 3.17 and 5.10. Now use lemmas 5.9 (the Peircean law) and 5.10 to check that g is well-defined and satisfies the consistency conditions for all $i, j, k \in \{x, y, z\}$. □

DEFINITION 5.12 A non-associative algebra is said to be *integral* if it satisfies[2]

$$x\,;y = 0 \iff (x = 0) \vee (y = 0).$$

LEMMA 5.13 [Madd90b, theorem 4] *Let $\mathcal{A} \in \mathbf{SA}$ be non-degenerate. Then $\mathcal{A}$ is integral if and only if the identity 1' is an atom.*

Proof. Since $\mathcal{A}$ is non-degenerate, we have $0 \neq 1 = 1\,;1$'. Since $1\,;0 = 0$, we have 1' > 0. If 1' is not an atom, there is $x \in \mathcal{A}$ with $0 < x < 1$'. Then 1' $- x \neq 0$, and $x\,;(1' - x) \leq 1'\,;(1' - x) = 1' - x$, but also $x\,;(1' - x) \leq x\,;1' = x$. Therefore, $x\,;(1' - x) \leq x \cdot (1' - x) = 0$, so $\mathcal{A}$ is not integral. This holds for any non-associative algebra.

Conversely, suppose 1' is an atom. Let $x, y \in \mathcal{A}$, $x, y \neq 0$. We must show that $x\,;y \neq 0$. From $\breve{x} = 1'\,;\breve{x} \cdot \breve{x} \neq 0$, the Peircean law (PL, lemma 5.9) gives $\breve{x}\,;x \cdot 1' \neq 0$. Since 1' is an atom, we see that $1' \leq \breve{x}\,;x$. Hence $y = 1'\,;y \leq (\breve{x}\,;x)\,;y \leq (\breve{x}\,;1)\,;1 = \breve{x}\,;1$, the last equality using the semi-associative law. Therefore, $\breve{x}\,;1 \cdot y \neq 0$, so by (PL), $x\,;y \cdot 1 \neq 0$, as required. □

---

[2] Some authors additionally require that an integral non-associative algebra be non-degenerate.

Exercises

1. Let $\mathcal{A}$ be a relation-type BAO with a relativised representation. Prove that $\mathcal{A}$ satisfies the weak associativity law (see definition 5.4), and so is a weakly associative algebra.

2. Check that any relation-type BAO with a general relativised representation satisfies axioms R0, R2, R5, and R7 of definition 3.8. Find a relation-type BAO with a general relativised representation that does not satisfy axioms R3, R4, and R6. [Start with a general relativised representation and construct the algebra from it.]

3. [Madd85] Prove that a non-associative algebra is integral if and only if $x; 1 = 1$, for all non-zero x.

4. [JónTar52] Prove that an integral non-associative algebra is simple.

5. Find an example of a simple representable relation algebra that is not integral.

6. Let $\mathcal{A}$ be the relation-type BAO with atoms $1', a, b$, defined by the following relativised representation M on the set $\{0, 1, 2, 3\}$ (see figure 5.1).

$$a^M = \{(0,1),(1,0),(2,3),(3,2)\},$$
$$b^M = \{(0,2),(2,0),(1,3),(3,1)\}.$$

Show that $\mathcal{A} \in \mathbf{WA} \setminus \mathbf{SA}$. Show directly that M is not 3-square. Show that

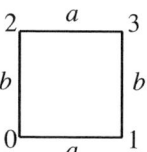

Figure 5.1: The weakly associative algebra $\mathcal{A}$

lemma 5.13 fails over weakly associative algebras. Show that $\mathcal{A}$ is simple and subdirectly irreducible. [$\mathcal{A}$ was discovered by McKinsey; see, e.g., [Madd85, Madd90b].]

7. Prove that **NA** is a conjugated variety.

8. Prove that **SA** is a discriminator variety.
[It may help to look up [Madd91b, theorems 8, 13].]

5.2. Weakly representable relation algebras

9. Prove that **WA** is not a discriminator variety.
   [It may help to look up [Madd91b, theorem 29].]

10. Let $\mathcal{A}$ be an atomic semi-associative algebra and let a,b be atoms of $\mathcal{A}$. Prove that if $\text{end}(a) = \text{st}(b)$ then $a;b \neq 0$. (See lemma 5.10 for definitions of $\text{st}(a), \text{end}(a)$).

11. Let M be a relativised representation of the relation algebra $\mathcal{A}$: i.e., a model of $R_{\mathcal{A}}$. Prove that M is a complete relativised representation iff for every $x,y \in M$, if $M \models 1(x,y)$ then $M \models a(x,y)$ for some atom a of $\mathcal{A}$.

12. [Madd82, definition 5.10, lemma 5.12.1]. An *identity atom* e in a weakly associative algebra is an atom such that $e \leq 1$'. Let h be a complete, relativised representation of the weakly associative algebra $\mathcal{W}$. Prove that the domain of the relativised representation can be partitioned into $\{S_e : e \text{ is an identity atom of } \mathcal{W}\}$ in such a way that for any atom $a \in \text{At}(\mathcal{W})$ we have $(x,y) \in h(a) \Rightarrow x \in S_{\text{st}(a)} \wedge y \in S_{\text{end}(a)}$.

13. Show that the weak associative and semi-associative laws are simple equations in the sense of definition 2.77, find their first-order correspondents, and compare with the correspondent of the associativity axiom given in lemma 3.24. Discover why the semi-associative law is so-called.

14. [Madd82, theorem 2.2(3)]. Let A be a set, $E \subseteq A$, and let $\breve{} : A \to A$ be any function. Let C be any ternary relation on A. Consider the complex algebra with domain $\wp(A)$, identity E, conversion defined by $\breve{S} = \{\breve{s} : s \in S\}$, and composition defined by $S;T = \{a \in A : \exists s \in S, \, t \in T((s,t,a) \in C)\}$, for any $S, T \subseteq A$ (see definition 2.65). Prove that this algebra is a weakly associative algebra iff the following conditions are met, for all $a,b,c \in A$:

    (a) If $(a,b,c) \in C$ then $(\breve{a},c,b), (c,\breve{b},a) \in C$.

    (b) $a = b$ iff there is $e \in E$ such that $(a,e,b) \in C$.

    (c) If $e \in E$ and $(e,a,a), (a,b,c) \in C$ then $(e,c,c) \in C$.

5.2 Weakly representable relation algebras

Jónsson [Jón59] considered reducts of relation algebras obtained by dropping the operations $+, -$. The notion of representation for such algebras is weaker than for relation algebras, since representations need not respect the missing operations. Jónsson called the ones representable in this way 'weakly representable relation algebras'.

164   Chapter 5. Relativisation and cylindric algebras

Algebras of relations   Jónsson's paper starts off by considering algebras closed only under the operations $\cdot, 1', \smile, ;$. Roughly, he gave the name 'algebras of relations'[3] to algebras of the form $\mathcal{A} = (A, |, \cap, ^{-1}, I)$ where A is a set of binary relations on some set X, the relational operations $\cap, ^{-1}$, and $|$ have their usual meanings ($^{-1}$ is conversion and $|$ is relational composition), and $I \in A$ satisfies $a | I = a$ for all $a \in A$. Such algebras need not be BAOs and need not even have interpretations of $+, -, 0, 1$ at all.

Although I is not required to be the identity relation (equality) on X, Jónsson nonetheless proves [Jón59, p. 450] that any such $\mathcal{A}$ is isomorphic to an algebra $\mathcal{B} = (B, |, \cap, ^{-1}, \mathrm{Id}_X)$ of binary relations on some set X; here, Id_X denotes the identity relation $\{(x, x) : x \in X\}$ on X. Moreover, the proof shows that if $\emptyset \in A$ then the constructed isomorphism from $\mathcal{A}$ to $\mathcal{B}$ takes $\emptyset$ to $\emptyset$. (See exercise 2 below.)

Jónsson [Jón59, theorem 1] gave an axiomatisation of the class of all algebras isomorphic to some algebra of relations, using a method that we view as rather game-theoretic: see exercise 7.6(7). His axioms consist of nine basic equations together with an infinite set Σ of quasi-equations. He also gave a similar characterisation of the lattices of commuting equivalence relations.

Weakly representable relation algebras   Moving to relation algebras now, Jónsson [Jón59, p. 459] defines a relation algebra to be weakly representable if it is isomorphic to a relation algebra $\mathcal{B}$ whose elements are binary relations on some set, and in which the operations $0, \cdot, ;$, and $\smile$ have their usual relation-operational meanings. Again, 1' need not be the identity relation on the base of $\mathcal{B}$. However, since $x; 1' = x$ is an axiom (R3) of relation algebras, the argument mentioned above shows that any relation algebra that is weakly representable in this sense is isomorphic to a relation algebra consisting of binary relations on some set X in which not only do $0, \cdot, ;,$ and $\smile$ have their usual meanings but also in which 1' is Id_X. This means that the following definition of weak representability of a relation algebra is equivalent to Jónsson's; it has been used in later papers such as [And94], and it is the definition we adopt in this book.

DEFINITION 5.14   A relation algebra $\mathcal{A}$ is said to be *weakly representable* if there exists an isomorphism $h : \mathcal{A} \to \mathcal{B}$, where $\mathcal{B}$ is a relation algebra, the domain of $\mathcal{B}$ consists of binary relations on some set X, and the operations $0, \cdot, 1', \smile$, and $;$ have their usual set-theoretic meanings: i.e., $0^{\mathcal{B}} = \emptyset$, $\cdot^{\mathcal{B}}$ is intersection of relations ($\cap$), $1'^{\mathcal{B}} = \mathrm{Id}_X$, $\smile^{\mathcal{B}}$ is relational conversion $^{-1}$, and $;^{\mathcal{B}}$ is relational composition $|$. Such a map h is called a *weak representation* of $\mathcal{A}$. We let **wRRA** denote the class of weakly representable relation algebras.

---

[3] We will only use the term 'algebra of relations' in this section, as in the context of this book it could well be confused with other notions.

5.2. Weakly representable relation algebras

We could define weak representations for non-associative algebras too, but it is easily checked that any weakly representable non-associative algebra is a relation algebra, so we do not need to do so.

As with ordinary and relativised representability, we can view the notion of weak representability in terms of a first-order theory. Let $\mathcal{A}$ be a relation algebra. As in definition 3.30, the signature $L(\mathcal{A})$ will consist of a binary relation symbol for each element of $\mathcal{A}$.

DEFINITION 5.15 Given a relation algebra $\mathcal{A}$, the $L(\mathcal{A})$-theory $W_\mathcal{A}$ is defined as follows:

$$\begin{array}{rll}
\sigma_{1'}: & \forall x,y \left[1'^{\mathcal{A}}(x,y) \leftrightarrow (x=y)\right] & \text{for all } R,S,T \in \mathcal{A} \text{ with} \\
\sigma_{\cdot}(R,S,T): & \forall x,y \left[R(x,y) \leftrightarrow S(x,y) \wedge T(x,y)\right] & R = S \cdot T \\
\sigma_{\smile}(R,S): & \forall x,y \left[R(x,y) \leftrightarrow S(y,x)\right] & R = \breve{S} \\
\sigma_{;}(R,S,T): & \forall x,y \left[R(x,y) \leftrightarrow \exists z(S(x,z) \wedge T(z,y))\right] & R = S;T \\
\sigma_0: & \forall x,y \, \neg 0^{\mathcal{A}}(x,y) & \\
\sigma_{1-1}(R,S): & \exists x,y \left[R(x,y) \wedge \neg S(x,y)\right] & R \not\leq S
\end{array}$$

As usual, we will write simply 1' for $1'^{\mathcal{A}}$, etc.

It is easily seen that a relation algebra $\mathcal{A}$ is weakly representable if and only if $W_\mathcal{A}$ has a model; indeed, any weak representation of $\mathcal{A}$ gives rise to a model of $W_\mathcal{A}$, and vice versa, just as for representable relation algebras. Note that for $M \models W_\mathcal{A}$ and $(x,y) \in 1^M$, the set $\{a \in \mathcal{A} : M \models a(x,y)\}$ is a filter of $\mathcal{A}$, but not necessarily an ultrafilter.

Results of Jónsson and Andréka Jónsson remarked in [Jón59] that the non-representable relation algebra constructed by Lyndon [Lyn50, p. 715] is not weakly representable. He then gave an example of a symmetric, integral relation algebra that is not weakly representable. Strikingly, his construction involved non-desarguesian projective planes. As we said in section 4.5, the projective plane construction of relation algebras was later modified by Lyndon [Lyn61]; Lyndon algebras were used by Monk in his proof of non-finite axiomatisability of **RRA** [Mon64].

The set Σ of quasi-equations mentioned above, combined with the equations defining **RA**, was shown to axiomatise **wRRA** in [Jón59, theorem 3]. Hence, **wRRA** is a quasi-variety. [Jón59, problem 1] asks whether Σ can be replaced [in these results] by a finite set of axioms, or by a set of equations.

Clearly, **RRA** $\subseteq$ **wRRA**. Jónsson asked [Jón59, problem 3] whether **wRRA** = **RRA**. Andréka answered this negatively in [And94]; indeed, she proved that **RRA** is not finitely axiomatisable over **wRRA**. **wRRA** was shown to be non-finitely axiomatisable implicitly in [Hai87, Hai91] and explicitly in [HodMik00]; the latter paper also proved analogous results for other classes of algebras. This provided a

166   Chapter 5. Relativisation and cylindric algebras

negative answer to the first part of Jónsson's problem 1 for **wRRA** and (since **RA** is finitely axiomatised) for the class of 'algebras of relations'.

A result of Jónsson and Tarski A contrasting result was obtained in [JónTar52]. Jónsson and Tarski consider a different kind of 'weak representation' — this time, it is required that the following operations are respected: $0, 1, +, 1', \smile$, and $;$. That is, all operations except perhaps the boolean complement and product $(-, \cdot)$ must be interpreted in the natural, set-theoretic way. In [JónTar52, theorem 4.22] they prove that every relation algebra has a representation of this type.

Looking ahead We will return to **wRRA** in exercise 7.6(7), where we describe Jónsson's axiomatisation method in terms of games, and also in chapter 17, where we prove non-finite axiomatisability of **wRRA** by approximately the method of [HodMik00]. In chapter 18, we will prove that it is undecidable whether a finite relation algebra is weakly representable, and indeed that there is no class K with **RRA** $\subseteq$ K $\subseteq$ **wRRA** such that it is decidable whether a finite relation-type algebra is in K. However, **wRRA** is not so well understood as **RRA**, and some basic questions about it remain open as far as we know. For example, is it a variety? Is it canonical?

Exercises

1. Show that McKenzie's algebra $\mathcal{K}$ (see section 4.4) is not weakly representable.

2. [Jón59] Show that if a relation algebra $\mathcal{A}$ has a representation respecting $\cdot, 1', \smile$, and $;$, then $\mathcal{A}$ is weakly representable (i.e., it has a representation respecting 0 as well).

   Show that if $\mathcal{A}$ has a representation respecting $\cdot, \smile$, and $;$, then it is weakly representable.

3. Show that **wRRA** is a quasi-variety.

5.3 Cylindric algebras

So far, we have dealt with algebras corresponding to binary relations. A class of algebras corresponding to α-ary relations, for any ordinal α, is α-*dimensional cylindric algebras*. This subject, which was launched by Tarski and his students Louise Chin and Frederick Thompson as an algebraic counterpart to the semantics of first-order logic, is comprehensively dealt with in [HenMon+71, HenMon+85].

5.3. Cylindric algebras

An alternative approach, using *polyadic algebras* with substitution and permutation operators, was initiated by Halmos in [Hal62], and will be discussed a little in section 6.2.

This book is focused on relation algebras and we do not propose to repeat the results of those volumes here, but we do give some of the elementary definitions and facts (often referring the reader to [HenMon+71] for proofs). This is for two reasons. First, it is generally acknowledged that results for relation algebras can in the main be replicated for cylindric algebras, though this is not always straightforward to do. Much of the work of parts II–V later can be carried through for cylindric algebras (we will refer the reader to published papers for the proofs). Second, the relationship between relation algebras and algebras of higher-dimensional relations forms an important subject later in this book (in chapters 13 and 15 especially), and we will need to know the basic properties of cylindric algebras in order to develop this material.

We must distinguish two cases, roughly according to whether equality is in the signature or not: with equality we get cylindric algebra, without equality we get diagonal-free algebra (to be discussed later, in section 6.1). In first-order logic, a relation always has a finite arity. However in the algebraic version it turns out that it is unnecessary to impose this restriction.

DEFINITION 5.16 [HenMon+71] Let α be an ordinal.

1. Recall that if U is a non-empty set, $^\alpha U$ denotes the set of functions from α to U. We write such functions as x, y, and for $i < \alpha$ we write $x(i)$ more concisely as x_i. A subset of $^\alpha U$ is called an α-*ary relation* on U. For $i, j < \alpha$, the i, jth diagonal D_{ij} denotes the set of all elements y of $^\alpha U$ such that $y_i = y_j$. Given $i < \alpha$ and an α-ary relation X on U, we define the *ith cylindrification* $C_i X$ to be the set of all elements of $^\alpha U$ that agree with some element of X on each coordinate except, perhaps, on the ith coordinate. That is, $C_i X = \{y \in {}^\alpha U : \exists x \in X \, \forall j < \alpha (j \neq i \rightarrow y_j = x_j)\}$.

2. A *cylindric set algebra* of dimension α is an algebra consisting of a set S of α-ary relations on some base set U, equipped with the operations $0, 1 \, (= {}^\alpha U), \cup, \setminus$ (complement in $^\alpha U$), the diagonal elements D_{ij} $(i, j < \alpha)$, and the cylindrifications C_i $(i < \alpha)$. S must of course be closed under all these operations.

   The signature of α-dimensional cylindric set algebras consists of the boolean symbols $0, 1, +, -$, constants d_{ij} $(i, j < \alpha)$, and unary functions c_i $(i < \alpha)$. A structure for this signature is called a *cylindric-type algebra,* and a BAO of this signature is called a *cylindric-type BAO*.

   The class of all cylindric set algebras of dimension α is denoted $\mathbf{Cs}_\alpha$.

3. A *generalised cylindric set algebra of dimension* α is a subdirect product of cylindric set algebras of dimension α.

4. A *cylindric algebra* of dimension α is defined to be an algebra

$$C = (C, 0, 1, +, -, c_i, d_{ij})_{i,j<\alpha}$$

obeying the following axioms for every $x, y \in C$, $i, j, k < \alpha$:

C0. $(C, 0, 1, +, -)$ is a boolean algebra
C1. $c_i 0 = 0$
C2. $x \leq c_i x$
C3. $c_i(x \cdot c_i y) = c_i x \cdot c_i y$
C4. $c_i c_j x = c_j c_i x$
C5. $d_{ii} = 1$
C6. if $k \neq i, j$, then $d_{ij} = c_k(d_{ik} \cdot d_{kj})$
C7. if $i \neq j$, then $c_i(d_{ij} \cdot x) \cdot c_i(d_{ij} \cdot -x) = 0$.

These axioms are valid over cylindric set algebras.

We write $\mathbf{CA}_\alpha$ for the class of all cylindric algebras of dimension α.

5. An α-dimensional cylindric algebra C is said to be *representable* if it is isomorphic to a generalised cylindric set algebra of dimension α; such an isomorphism is called a *representation* of C. $\mathbf{RCA}_\alpha$ denotes the class of all representable cylindric algebras of dimension α.

Cylindric set algebras correspond, more or less, to square proper relation algebras. The cylindric algebraic counterpart of proper relation algebras is a subdirect product of cylindric set algebras. Every cylindric set algebra is a cylindric algebra (of the appropriate dimension), so every subdirect product of cylindric set algebras is a cylindric algebra, but in dimensions higher than 1 there are non-representable cylindric algebras. Most of the universal-algebraic results for relation algebras carry over to cylindric algebras. So $\mathbf{CA}_\alpha$ is a conjugated variety of BAOs (fact 5.17 below), and hence is completely additive. The axioms are simple equations, hence Sahlqvist equations, so $\mathbf{CA}_\alpha$ is a Sahlqvist variety; by theorems 2.95 and 2.96 it is canonical and closed under completions. For finite (but not infinite) α, $\mathbf{CA}_\alpha$ is a discriminator variety: a discriminator term is $c_0 c_1 \ldots c_{\alpha-1} x$. See, e.g., [Madd91a] and exercise 3.

It is known that $\mathbf{RCA}_\alpha$ is a variety [HenMon+85, corollary 3.1.108]. For $\alpha \leq 2$ it is finitely axiomatisable [HenMon+85, 3.2.56, 3.2.65], since $\mathbf{RCA}_\alpha = \mathbf{CA}_\alpha$ for $\alpha \leq 1$, and

$$\mathbf{RCA}_2 = \mathbf{CA}_2 \cap \mathrm{Mod}\{c_i(x \cdot y \cdot c_j(x \cdot -y)) \leq c_j(c_i x \cdot -d_{01}) : \{i, j\} = 2\}.$$

5.3. Cylindric algebras

(The two formulas here are known as *Henkin's two equations*.) Monk proved that $\mathbf{RCA}_\alpha$ is not finitely axiomatisable for finite $\alpha \geq 3$ (see [Mon69], or [HenMon+85, theorem 4.1.3]). Many stronger negative results concerning axiomatisations of $\mathbf{RCA}_\alpha$ are known: see, e.g., [And97a, Ven97b].

We will be using several simple consequences of the cylindric algebra axioms, and we list the main ones below. The proofs can be done as exercises or can be found in [HenMon+71, chapter 1].

FACT 5.17 The following statements are valid in $\mathbf{CA}_\alpha$ for any ordinal α and any $i, j, k < \alpha$.

1. $c_i c_i x = c_i x$ [HenMon+71, 1.2.3].

2. $x \cdot c_i y = 0$ iff $y \cdot c_i x = 0$ [HenMon+71, 1.2.5]. Hence, c_i is conjugated and so (by theorem 2.40) completely additive.

3. $d_{ij} = d_{ji}$ [HenMon+71, 1.3.1].

4. $c_i d_{ij} = 1$ [HenMon+71, 1.3.2]. If $k \neq i, j$ then $c_k d_{ij} = d_{ij}$ [HenMon+71, 1.3.3].

5. $c_i(-c_i x) = -c_i x$ [HenMon+71, 1.2.11].

6. If x, y are atoms, then $x \leq c_i y$ iff $c_i x = c_i y$ (by items 1 and 2 above).

7. If x, y are atoms and $x \leq c_i c_j y$, then $c_i c_j x = c_i c_j y$, and there is an atom z with $x \leq c_i z$ and $z \leq c_j y$ (by items 1, 2 above, axiom C4 and complete additivity of cylindrification).

Exercises

1. Calculate the Sahlqvist correspondents of the cylindric algebra axioms and of Henkin's two equations. (Cf. exercise 2.7(11).)

2. For $n \leq 2$, show that $\mathbf{RCA}_n$ is a conjugated Sahlqvist variety. Deduce that

    (a) $\mathbf{RCA}_n$ is canonical,
    (b) $\mathbf{RCA}_n$ is atom-canonical and closed under completions,
    (c) the class $\mathrm{Str}\,\mathbf{RCA}_n$ of $\mathbf{CA}_n$-atom structures whose complex algebras are in $\mathbf{RCA}_n$ is elementary.

    ((2b) fails for finite $n \geq 3$ [Hodk97]. It seems likely that (2c) does too, but we do not currently have a proof.)

3. For infinite α show that $\mathbf{CA}_\alpha$ is not a discriminator variety. Consider and investigate α-dimensional cylindric algebras with an additional 'discriminator' operator: $c_\alpha(x) = 0$ if $x = 0$ and $c_\alpha(x) = 1$, otherwise.

PROBLEM 5.18 *If $C \in \mathbf{RCA}_\omega$ does it follow that C^+ has a complete representation? For information on complete representations of ω-dimensional cylindric algebras, see [Mon00, section 15].*

5.4 Substitutions in cylindric algebras

We will shortly investigate how to relativise a cylindric algebra and to form a relation algebra from a cylindric-type algebra. We will be using *substitutions* in cylindric algebras here and later in this book, so we begin with a short primer on them. For a fuller study, see [Jón62, HenMon$^+$71, Res75, HenMon$^+$85, Tho93], for example.

Fix some dimension α for our cylindric algebras. α-dimensional cylindric algebras arise naturally from first-order logic restricted to α variables. Diagonals d_{ij} correspond to formulas $x_i = x_j$, and cylindrifications c_i correspond to quantifiers $\exists x_i$. Now in first-order logic, we are all familiar with the process of substituting one free variable by another. Substitution can be achieved in cylindric algebra using the *substitution operator* s^i_j, defined as follows.

DEFINITION 5.19 For $i, j < \alpha$, and x any term of the signature of α-dimensional cylindric algebras, we define

$$s^i_j x = \begin{cases} x, & \text{if } i = j; \\ c_i(d_{ij} \cdot x), & \text{otherwise.} \end{cases}$$

The corresponding formula, if $i \neq j$, is $\exists x_i(x_i = x_j \wedge \chi)$.

5.4.1 Basic facts about substitutions

We will use the following facts about substitutions, taken from [HenMon$^+$71].

FACT 5.20 Let α be an ordinal and $i, j, k, l < \alpha$. The following are valid equations in $\mathbf{CA}_\alpha$.

1. $s^i_j x \leq c_i x$ (by definition of s^i_j and additivity of c_i).

2. $s^i_j(x \cdot y) = s^i_j x \cdot s^i_j y$, $s^i_j(-x) = -s^i_j x$, $s^i_j(x+y) = s^i_j x + s^i_j y$. Indeed, whenever $C \in \mathbf{CA}_\alpha$, $X \subseteq C$, and $\sum X$ exists, then $C \models s^i_j \sum X = \sum_{x \in X} s^i_j x$. Also, in consequence, $s^i_j 0 = 0$ and $s^i_j 1 = 1$. That is, s^i_j is completely additive and acts as a complete boolean endomorphism. [HenMon$^+$71, 1.5.3].

5.4. Substitutions in cylindric algebras

3. If $i \neq k$, then $\mathsf{s}^i_j \mathsf{d}_{ik} = \mathsf{d}_{jk}$. If $i \neq k, l$ then $\mathsf{s}^i_j \mathsf{d}_{kl} = \mathsf{d}_{kl}$. [HenMon+71, 1.5.4].

4. $\mathsf{d}_{jk} \cdot \mathsf{s}^i_j x = \mathsf{d}_{jk} \cdot \mathsf{s}^i_k x$ [HenMon+71, 1.5.6].

5. $\mathsf{s}^i_j \mathsf{c}_i x = \mathsf{c}_i x$ [HenMon+71, 1.5.8(i)].

6. $\mathsf{s}^i_j \mathsf{c}_k x = \mathsf{c}_k \mathsf{s}^i_j x$ if $k \neq i, j$ [HenMon+71, 1.5.8(ii)].

7. $\mathsf{c}_j \mathsf{s}^i_j x = \mathsf{c}_i \mathsf{s}^j_i x$ [HenMon+71, 1.5.9(i)].

8. $\mathsf{c}_i \mathsf{s}^i_j x = \mathsf{s}^i_j x$ if $i \neq j$ [HenMon+71, 1.5.9(ii)].

9. $\mathsf{s}^i_j \mathsf{s}^i_k x = \mathsf{s}^i_k x$ if $i \neq k$ [HenMon+71, 1.5.10(i)].

10. $\mathsf{s}^i_j \mathsf{s}^k_l x = \mathsf{s}^k_l \mathsf{s}^i_j x$ if either $i \notin \{k, l\}$ and $k \notin \{i, j\}$, or $j = l$. [HenMon+71, 1.5.10(iii,iv)].

11. $\mathsf{s}^i_j \mathsf{s}^j_i x = \mathsf{s}^i_j x$ [HenMon+71, 1.5.10(v)].

12. $\mathsf{s}^i_k \mathsf{s}^j_i x = \mathsf{s}^i_k \mathsf{s}^j_k x = \mathsf{s}^j_k \mathsf{s}^i_j x$ [HenMon+71, 1.5.10(ii,vi)].

These facts suffice for most elementary work with substitutions — one uses them again and again. Thinking of $\mathsf{s}^i_j x$ as $\exists x_i (x_i = x_j \wedge \chi)$ and d_{ij} as $x_i = x_j$, the statements corresponding to these identities are easily seen to be valid. Or just think of the indices $i < \alpha$ as variables x_i, the operation s^i_j as setting x_i to (the contents of) x_j — that is, '$x_i := x_j$' — and the operation c_i as 'randomising' x_i. In this light, the facts above seem intuitively natural. Considering fact 5.20(12), for example, we see that

- $\exists x_i (x_i = x_k \wedge \exists x_j (x_j = x_i \wedge \chi))$ is logically equivalent to
  $\exists x_j (x_j = x_k \wedge \exists x_i (x_i = x_j \wedge \chi))$,

- setting x_i to x_k and then x_j to x_i has the same effect as setting x_j to x_k and then x_i to x_j.

Similarly, randomising x_i and then setting it to x_j is the same as setting it to x_j straight off (if $j \neq i$; cf. no. 8). However, the reader should beware: for finite α, the $\mathbf{CA}_\alpha$ axioms are not strong enough to prove all logical equivalences between α-variable formulas, and not all 'natural facts' about substitutions are valid in cylindric algebras. For example, $\mathbf{CA}_\alpha \not\models \mathsf{s}^2_0 \mathsf{s}^0_1 \mathsf{s}^1_2 \mathsf{c}_2 x = \mathsf{s}^2_1 \mathsf{s}^1_0 \mathsf{s}^0_2 \mathsf{c}_2 x$. Below, we will give additional conditions that are sufficient to ensure validity.

5.4.2 More valid substitution-cylindrification identities

The rest of this section is devoted to obtaining a more complete picture of which equations involving substitutions and cylindrifications are valid in $\mathbf{CA}_\alpha$. This will be needed in chapter 13; we will also use it in theorem 5.44, though for that proof we could get by with just fact 5.20.

For the rest of the section, fix an ordinal $\alpha \geq 2$. Unless otherwise stated, i, j, k, l, p, q denote arbitrary ordinals $< \alpha$.

s-c-words

DEFINITION 5.21

1. An *s-word* is a finite string of substitutions (s^i_j), a *c-word* is a finite string of cylindrifications (c_k), and an *s-c-word* is a finite string of substitutions and cylindrifications, all of the signature of $\mathbf{CA}_\alpha$.

2. We write ε for the empty s-c-word. If u, w are s-c-words, we write simply uw for their concatenation. The length of w (number of symbols $\mathsf{c}_k, \mathsf{s}^i_j$ in w) is written $|w|$. For example, $|\mathsf{s}^2_0 \mathsf{s}^0_1 \mathsf{s}^1_2 \mathsf{c}_2| = 4$.

Clearly, for any s-c-word w and term t of the signature of α-dimensional cylindric algebras, wt is also a term of this signature. And if $x \in C \in \mathbf{CA}_\alpha$, then $wx \in C$.

Program view of s-c-words

Viewing s^i_j and c_k as 'instructions' `set x_i to x_j` (i.e., `x_i := x_j`) and `randomise x_k`, respectively, we obtain a 'computational' view of any s-c-word: the x_i ($i < \alpha$) are thought of as computer storage variables, and the word is thought of as a 'program' by reading it in this way as a sequence of instructions to be executed from left to right, one after the other. We mentioned this above as intuition for fact 5.20. As an example, the 'program' corresponding to $\mathsf{s}^2_0 \mathsf{s}^0_1 \mathsf{s}^1_2 \mathsf{c}_2$ is:

```
begin
  x_2 := x_0
  x_0 := x_1
  x_1 := x_2
  randomise x_2
end
```

After such a program has been run, starting with arbitrary values of the variables, the following possibilities arise for each x_i:

- The final value of x_i is the initial value of some x_j, this being forced by the program in the sense that it will always hold regardless of the initial values of the variables. For example, if $i \neq j$ then $\mathsf{s}^i_j \mathsf{c}_j \mathsf{s}^k_i$ (or strictly, the program associated with it) ensures that the final value of x_k is the initial value of x_j.

5.4. Substitutions in cylindric algebras

- The final value of x_i is necessarily equal to the final value of some other variable x_j. For example, $c_j s_j^i$ ensures that the final values of x_i and x_j are equal.

- The initial value of x_i is destroyed during the execution of w (or equivalently of w'), by being overwritten or randomised. So there is no variable whose final value is forced to be the initial value of x_i. For example, if $i \neq j$ then both s_j^i and $s_i^j c_j c_i$ destroy the initial value of x_i.

Informally, we aim to prove, using fact 5.20 and a theorem of Thompson [Tho93], that whenever two s-c-words w, w' satisfy, for all $i, j < \alpha$:

1. if w forces that the final value of x_i is the initial value of x_j, then so does w', and vice versa,

2. if w forces that the final value of x_i is equal to the final value of x_j, then so does w', and vice versa,

3. the initial values of at least two variables x_i, x_j (some $i < j < \alpha$) are destroyed during the execution of w (or equivalently of w'),

then $wx = w'x$ is a valid equation in $\mathbf{CA}_\alpha$. The first two conditions express that the non-deterministic programs associated with w, w' are 'equivalent'. The reader may check that they hold for each side of all equations of s-c-words in fact 5.20. For $\alpha \geq 4$, the words $s_0^2 s_1^0 s_2^1 c_2 c_3$ and $s_1^2 s_0^1 s_2^0 c_2 c_3$ satisfy all three conditions, so that $\mathbf{CA}_\alpha \models \forall x(s_0^2 s_1^0 s_2^1 c_2 c_3 x = s_1^2 s_0^1 s_2^0 c_2 c_3 x)$.

Formalisation

We now proceed to formalise this 'program' view of s-c-words, using the approach of Resek–Thompson [ResTho91]. We will need some standard notation.

DEFINITION 5.22 Define the substitution map $[i/j] : \alpha \to \alpha$ by:

$$[i/j](k) = \begin{cases} j, & \text{if } k = i, \\ k, & \text{otherwise,} \end{cases}$$

where $i, j < \alpha$.

Of course, this definition depends implicitly on α. We write maps on the left, and $\circ$ denotes map composition, so that for example, $([1/2] \circ [2/3])(1) = [1/2]([2/3](1)) = 2$. The identity map on a set X is as usual denoted by Id_X.

A possibly partial map $g : \alpha \to \alpha$ is said to be *finitary* if $\{i < \alpha : i \notin \mathrm{dom}(g)$ or $g(i) \neq i\}$ is finite. $\ker(g)$, the *kernel* of g, denotes the binary relation on α (and equivalence relation on $\mathrm{dom}(g)$) given by $(i, j) \in \ker(g)$ iff $i, j \in \mathrm{dom}(g)$ and $g(i) = g(j)$. For $\Gamma \subseteq \alpha$, we write $g^{-1}[\Gamma]$ for the set $\{i \in \mathrm{dom}(g) : g(i) \in \Gamma\}$. When $\Gamma = \{i\}$ we write $g^{-1}[i]$ instead of $g^{-1}[\{i\}]$.

DEFINITION 5.23

1. With each s-c-word w, we associate a partial finitary map $\widehat{w}: \alpha \to \alpha$ by induction on $|w|$ as follows:

    - $\widehat{\varepsilon} = \mathrm{Id}_\alpha$.
    - $\widehat{ws_j^i} = \widehat{w} \circ [i/j]$.
    - $\widehat{wc_i} = \widehat{w} \circ \mathrm{Id}_{\alpha \setminus \{i\}} = \widehat{w} \restriction_{\alpha \setminus \{i\}}$.

2. We define $w^* = \bigcup_{w=uv} \ker(\widehat{v})$. Equivalently, $w^* = \{(i,j) : \exists u, v (w = uv \wedge i, j \in \mathrm{dom}(\widehat{v}) \wedge \widehat{v}(i) = \widehat{v}(j))\}$.

3. For s-c-words w, w', we write

    - $w \sim w'$ if $\widehat{w} = \widehat{w'}$ and $w^* = w'^*$,
    - $w \simeq w'$ if $w \sim w'$ and $\mathbf{CA}_\alpha \models \forall x (wx = w'x)$.

Exercise 2 below shows that w^* is an equivalence relation on α. Of course, $\sim, \simeq$ depend implicitly on α. The reader may wish to verify that w, w' satisfy the three 'program' conditions given above iff $w \sim w'$ and $|\alpha \setminus \mathrm{rng}(\widehat{w})| \geq 2$: see exercise 7 below.

EXAMPLE 5.24 Let $w = uv$, where u is an s-word and $v = c_{i_0} \ldots c_{i_{n-1}}$ a c-word. Let $K = \{i_0, \ldots, i_{n-1}\}$. Then $w^* = \ker(\widehat{w} \cup \mathrm{Id}_K)$. Hence, for two words w, w' of this form, we have $w \sim w'$ iff $\widehat{w} = \widehat{w'}$.

Equal s-c-words modulo $\mathbf{CA}_\alpha$

We will prove the following theorem:

THEOREM 5.25 Let w, w' be s-c-words with $w \sim w'$ and $|\alpha \setminus \mathrm{rng}(\widehat{w})| \geq 2$. Then $\mathbf{CA}_\alpha \models \forall x (wx = w'x)$. Hence, $w \simeq w'$.

This extends to s-c-words the following theorem of Thompson:

FACT 5.26 [Tho93, theorem 3.6] Let $\alpha \geq 2$ be an ordinal and let $q, r < \omega$. Assume that $i_1, j_1, \ldots, i_q, j_q, k_1, m_1, \ldots, k_r, m_r < \alpha$ are such that $[i_1/j_1] \circ \cdots \circ [i_q/j_q] = [k_1/m_1] \circ \cdots \circ [k_r/m_r] = f \in {}^\alpha\alpha$ and $|\alpha \setminus \mathrm{rng}(f)| \geq 2$. Then

$$\mathbf{CA}_\alpha \models \forall x (\mathsf{s}_{j_1}^{i_1} \ldots \mathsf{s}_{j_q}^{i_q}(x) = \mathsf{s}_{m_1}^{k_1} \ldots \mathsf{s}_{m_r}^{k_r}(x)).$$

Clearly, this is the restriction of theorem 5.25 to s-words. ([HenMon+85, 3.2.52] is a similar result, based on a semigroup theorem of Jónsson.) We will rely on it heavily in the proof.

5.4. Substitutions in cylindric algebras

Basic properties

We begin with the following easy lemma which will be useful later.

LEMMA 5.27 *Let u, w be s-c-words.*

1. $\widehat{uw} = \widehat{u} \circ \widehat{w}$.

2. $\mathbf{CA}_n \models w\mathsf{d}_{ij} = \mathsf{d}_{\widehat{w}(i)\widehat{w}(j)}$, *for all $i, j \in \mathrm{dom}(\widehat{w})$.*

Proof.

1. We prove $\widehat{uw} = \widehat{u} \circ \widehat{w}$ for all u, by a trivial induction on the length of w. If this is zero, then $\widehat{w} = \mathrm{Id}_\alpha$ and we are done. Assume the result for w, and let $i, j < \alpha$. Then $\widehat{uw\mathsf{s}^i_j} = \widehat{uw} \circ [i/j] = (\widehat{u} \circ \widehat{w}) \circ [i/j]$ (by the inductive hypothesis) $= \widehat{u} \circ (\widehat{w} \circ [i/j]) = \widehat{u} \circ \widehat{w\mathsf{s}^i_j}$, as required. The proof for wc_i is similar.

2. The proof is by induction on the length of w. If this is zero, there is nothing to prove. Assume the result for u. We first prove it for $w = u\mathsf{s}^k_l$. Let $i, j \in \mathrm{dom}(\widehat{w})$. If $i = j$, then in any α-dimensional cylindric algebra, we have $\mathsf{d}_{ij} = 1 = \mathsf{d}_{\widehat{w}(i),\widehat{w}(j)}$. So suppose that $i \neq j$. Now, by fact 5.20(3), $u\mathsf{s}^k_l \mathsf{d}_{ij} = u\mathsf{d}_{[k/l](i),[k/l](j)}$; by assumption, $[k/l](i), [k/l](j) \in \mathrm{dom}(\widehat{u})$, so by the induction hypothesis this is $\mathsf{d}_{\widehat{u}([k/l](i)),\widehat{u}([k/l](j))} = \mathsf{d}_{\widehat{w}(i),\widehat{w}(j)}$.

   Next let $w = uc_k$. Assume that $i, j \in \mathrm{dom}(\widehat{w}) = \mathrm{dom}(\widehat{u}) \setminus \{k\}$. Then $\mathbf{CA}_\alpha \models c_k \mathsf{d}_{ij} = \mathsf{d}_{ij}$, so $w\mathsf{d}_{ij} = u\mathsf{d}_{ij}$, which by the inductive hypothesis is $\mathsf{d}_{\widehat{u}(i),\widehat{u}(j)}$; and this is clearly equal to $\mathsf{d}_{\widehat{w}(i),\widehat{w}(j)}$, as required.

□

PROPOSITION 5.28 $\simeq$ *is a congruence on s-c-words: an equivalence relation such that if u, u', w, w' are s-c-words and $u \simeq u'$, $w \simeq w'$, then $uw \simeq u'w'$.*

Proof. $\simeq$ is clearly an equivalence relation. Assume that $u \simeq u'$ and $w \simeq w'$. Then $\mathbf{CA}_\alpha \models \forall x(wx = w'x)$, so $\mathbf{CA}_\alpha \models \forall x(uwx = u'w'x)$. By lemma 5.27(1), $\widehat{uw} = \widehat{u} \circ \widehat{w} = \widehat{u'} \circ \widehat{w'} = \widehat{u'w'}$. So it remains to check that $(uw)^* = (u'w')^*$.

Let $(i, j) \in (uw)^* = \bigcup_{uw=xy} \ker(\widehat{y})$. Pick s-c-words x, y such that $uw = xy$ and $(i, j) \in \ker(\widehat{y})$. If $|y| \leq |w|$, then $(i, j) \in w^* = w'^* \subseteq (u'w')^*$. Otherwise, $y = zw$ for some z. By lemma 5.27(1), $\widehat{y} = \widehat{z} \circ \widehat{w}$. So $(i, j) \in \ker(\widehat{y})$ implies $(\widehat{w}(i), \widehat{w}(j)) \in \ker(\widehat{z})$. But z is a final segment of u, so $\ker(\widehat{z}) \subseteq u^*$ and $(\widehat{w}(i), \widehat{w}(j)) \in u^*$. Since $\widehat{w} = \widehat{w'}$ and $u^* = u'^*$, we obtain $(\widehat{w'}(i), \widehat{w'}(j)) \in u'^*$, so clearly, $(i, j) \in (u'w')^*$. Thus, $(uw)^* \subseteq (u'w')^*$; the converse is similar. □

The following is easily checked, using the definition of $\sim$ and fact 5.20.

LEMMA 5.29

1. $c_i c_j \simeq c_j c_i$ and $c_i c_i \simeq c_i$.

2. $s_j^i c_i \simeq c_i$

3. $s_j^i c_k \simeq c_k s_j^i$ if $k \neq i,j$

4. $c_j s_j^i \simeq c_i s_i^j$

5. $c_i s_j^i \simeq s_j^i$ if $i \neq j$

6. $s_j^i s_k^i \simeq s_k^i$ if $i \neq k$

7. $s_j^i s_l^k \simeq s_l^k s_j^i$ if either $i \notin \{k,l\}$ and $k \notin \{i,j\}$, or $j = l$.

8. $s_j^i s_i^j \simeq s_j^i$

9. $s_k^i s_i^j \simeq s_k^i s_k^j \simeq s_k^j s_j^i$

DEFINITION 5.30 For any $n < \omega$ and $i_0, \ldots, i_{n-1} < \alpha$, we define the following s-c-word:

$$\rho_{(i_0,\ldots,i_{n-1})} = \begin{cases} \varepsilon, & \text{if } n = 0, \\ c_{i_0} s_{i_0}^{i_0} \ldots s_{i_0}^{i_{n-1}}, & \text{otherwise.} \end{cases}$$

LEMMA 5.31 *Assume that* $\Gamma = \{i_0, \ldots, i_{n-1}\} = \{j_0, \ldots, j_{m-1}\} \subseteq \alpha$, *for* $n, m < \omega$. *Then:*

1. $\rho_{(i_0,\ldots,i_{n-1})} \simeq \rho_{(j_0,\ldots,j_{m-1})}$.

2. *If* $k \in \Gamma$, $l < \alpha$, *then* $c_k \rho_{(i_0,\ldots,i_{n-1})} \simeq s_l^k \rho_{(i_0,\ldots,i_{n-1})} \simeq \rho_{(i_0,\ldots,i_{n-1})}$.

Proof. By the definition and lemma 5.29(6,7), we have $s_i^i \simeq \varepsilon$, $s_j^i s_j^i \simeq s_j^i$, and $s_{i_0}^{i_j} s_{i_0}^{i_k} \simeq s_{i_0}^{i_k} s_{i_0}^{i_j}$. So by proposition 5.28, we may assume that $i_0, \ldots, i_{n-1}$ are pairwise distinct, similarly for $j_0, \ldots, j_{m-1}$, and $n = m$. If $n \leq 1$, the result is clear, since $\rho_{(i_0)} = \rho_{(j_0)}$. Assume that $n > 1$; let $j_0 = i_k$, say. Note that by lemma 5.29(9,7),

$$s_{i_k}^{i_0} s_{i_0}^{i_s} \simeq s_{i_k}^{i_s} s_{i_k}^{i_0}, \text{ for all } s < n. \tag{$*$}$$

5.4. Substitutions in cylindric algebras

So we have

$$
\begin{aligned}
& \rho_{(i_0,\ldots,i_{n-1})} \\
= \; & c_{i_0} s_{i_0}^{i_0} \ldots s_{i_0}^{i_{n-1}} \\
\simeq \; & c_{i_0} s_{i_0}^{i_k} s_{i_0}^{i_0} \ldots s_{i_0}^{i_{n-1}} && \text{by lemma 5.29(7)} \\
\simeq \; & c_{i_k} s_{i_k}^{i_0} s_{i_0}^{i_1} \ldots s_{i_0}^{i_{n-1}} && \text{by lemma 5.29(4) and } s_{i_0}^{i_0} \simeq \varepsilon \\
\simeq \; & c_{i_k} s_{i_k}^{i_1} s_{i_k}^{i_0} s_{i_0}^{i_2} \ldots s_{i_0}^{i_{n-1}} && \text{by } (*) \\
\simeq \; & c_{i_k} s_{i_k}^{i_1} s_{i_k}^{i_2} s_{i_k}^{i_0} s_{i_0}^{i_3} \ldots s_{i_0}^{i_{n-1}} && \text{by } (*) \text{ again} \\
\ldots \simeq \; & c_{i_k} s_{i_k}^{i_1} \ldots s_{i_k}^{i_{n-1}} s_{i_k}^{i_0} && \text{continuing} \ldots \\
= \; & c_{j_0} s_{j_0}^{i_1} \ldots s_{j_0}^{i_{n-1}} s_{j_0}^{i_0} \\
\simeq \; & c_{j_0} s_{j_0}^{j_0} s_{j_0}^{j_1} \ldots s_{j_0}^{j_{m-1}} && \text{by lemma 5.29(7) and } s_{j_0}^{j_0} \simeq \varepsilon \\
= \; & \rho_{(j_0,\ldots,j_{m-1})}.
\end{aligned}
$$

For the second part, if $k \in \Gamma$ then using the above and lemma 5.29(2),

$$
\begin{aligned}
c_k \rho_{(i_0,\ldots,i_{n-1})} &\simeq c_k c_k s_k^{i_0} \ldots s_k^{i_{n-1}} \simeq c_k s_k^{i_0} \ldots s_k^{i_{n-1}} = \rho_{(i_0,\ldots,i_{n-1})}, \\
s_l^k \rho_{(i_0,\ldots,i_{n-1})} &\simeq s_l^k c_k s_k^{i_0} \ldots s_k^{i_{n-1}} \simeq c_k s_k^{i_0} \ldots s_k^{i_{n-1}} \simeq \rho_{(i_0,\ldots,i_{n-1})}.
\end{aligned}
$$

□

DEFINITION 5.32 We write ρ_Γ for any s-c-word of the form $c_{i_0} s_{i_0}^{i_0} \ldots s_{i_0}^{i_{n-1}}$ where $\Gamma = \{i_0, \ldots, i_{n-1}\} \subseteq \alpha$ for some $n < \omega$. By lemma 5.31, ρ_Γ is well-defined up to $\simeq$-equivalence.

We want to move cylindrifications rightwards within s-c-words, preserving $\simeq$-equivalence. The next lemma shows how to do this. Cf. [HenMon+85, theorem 3.2.51(vi,vii)].

LEMMA 5.33 Let w be an s-c-word. Then $c_i w \simeq w \rho_{(\widehat{w}^{-1}[i])}$.

Proof. The proof is by induction on $|w|$. We use proposition 5.28 and lemma 5.29 freely in the proof. If $|w| = 0$, there is nothing to prove. Assume the result for u, and let $\Gamma = \widehat{u}^{-1}[i]$. First we prove it for $w = u c_j$. Clearly, $\widehat{w}^{-1}[i] = \Gamma \setminus \{j\}$. Inductively, $c_i u c_j \simeq u \rho_\Gamma c_j$. So we need only check that $\rho_\Gamma c_j \simeq c_j \rho_{\Gamma \setminus \{j\}}$. If $j \notin \Gamma$, then c_j commutes modulo $\simeq$ with every item in ρ_Γ, giving the result. If $\Gamma = \{j\}$, the result is trivial as $c_j s_j^j c_j \simeq c_j$. If $\{j\} \subset \Gamma$, then take $k \in \Gamma \setminus \{j\}$ and observe that $c_j \rho_{\Gamma \setminus \{j\}} \simeq \rho_{\Gamma \setminus \{j\}} c_j \simeq \rho_{\Gamma \setminus \{j\}} s_k^j c_j \simeq \rho_\Gamma c_j$.

Now we prove it for $w = u s_k^j$. Observe here that

$$
\Gamma' \stackrel{\text{def}}{=} \widehat{w}^{-1}[i] = [j/k]^{-1}[\Gamma] = \begin{cases} \Gamma \cup \{j\}, & \text{if } k \in \Gamma, \\ \Gamma \setminus \{j\}, & \text{otherwise.} \end{cases}
$$

Inductively, $c_i u s_k^j \simeq u \rho_\Gamma s_k^j$, so we need to show that $\rho_\Gamma s_k^j \simeq s_k^j \rho_{\Gamma'}$.

If $k \in \Gamma$, then by the definition and lemma 5.31, $\rho_\Gamma s_k^j \simeq \rho_{\Gamma'} \simeq s_k^j \rho_{\Gamma'}$. So assume that $k \notin \Gamma$. If also $j \notin \Gamma$, then s_k^j commutes with every term in ρ_Γ modulo $\simeq$, so $\rho_\Gamma s_k^j \simeq s_k^j \rho_\Gamma = s_k^j \rho_{\Gamma'}$. If $\Gamma = \{j\}$, then $j \neq k$, so lemma 5.29(5) gives $\rho_\Gamma s_k^j = c_j s_j^j s_k^j \simeq s_k^j \simeq s_k^j \rho_{\Gamma'}$. If $\{j\} \subset \Gamma$, then Γ has the form $\{j, m, l_0, \ldots, l_{n-1}\}$, where $j, m, l_0, \ldots, l_{n-1}$ are distinct. Then $s_m^j s_k^j \simeq s_k^j$ and (modulo $\simeq$) s_k^j commutes with every entry in $\rho_{\Gamma'}$. So

$$\rho_\Gamma s_k^j \simeq c_m s_m^{l_0} \ldots s_m^{l_{n-1}} s_m^j s_k^j \simeq c_m s_m^{l_0} \ldots s_m^{l_{n-1}} s_k^j \simeq \rho_{\Gamma'} s_k^j \simeq s_k^j \rho_{\Gamma'},$$

as required. □

We can now prove theorem 5.25.

Proof. Let w, w' be s-c-words with $w \sim w'$ and $|\alpha \setminus \mathrm{rng}(\widehat{w})| \geq 2$. We show that $w \simeq w'$ by induction on the number n of cylindrifications c_k (any k) in ww'. If this is zero, so that w, w' are s-words, then the result follows from Thompson's theorem (fact 5.26). Let $n > 0$ and assume the result for smaller n. Without loss of generality, we have $w = uc_k v$, where v is an s-word. Let $K = \widehat{v}^{-1}[k]$. By lemma 5.33, $w = uc_k v \simeq uv\rho_K$.

If $K = \emptyset$, then lemma 5.33 gives $w' \sim w \simeq uv\rho_K = uv$. Then $w' \sim uv$ and uvw' has fewer c_k than ww', so inductively, $w' \simeq uv \simeq w$, as required. Assume then that $K \neq \emptyset$.

Claim. w' also has the form $u'c_{k'}v'$, with $K = \widehat{v'}^{-1}[k']$.

Proof of claim. Let $i, j \in K$. Then $i, j \notin \mathrm{dom}(\widehat{w}) = \mathrm{dom}(\widehat{w'})$, and since $\widehat{v}(i) = \widehat{v}(j) = k$, we have $(i, j) \in w^* = w'^*$. Therefore, we may write $w' = u'xv'$ where u', v' are s-c-words, x is an s-c-word of length 1, $(i, j) \in \ker(\widehat{v'})$, and $|v'|$ is maximal subject to this. Since $\widehat{v'}(i) = \widehat{v'}(j) = k'$, say, by maximality of v' and lemma 5.27(1) we must have $k' \notin \mathrm{dom}(\widehat{x})$, so $x = c_{k'}$.

Thus, $w' = u'c_{k'}v'$ with $\widehat{v'}(i) = \widehat{v'}(j) = k'$. Clearly, v' is the final segment of w' of maximal length such that $i \in \mathrm{dom}(\widehat{v'})$. This definition is independent of j. Hence, fixing i and letting j range over K, we see that $K \subseteq \widehat{v'}^{-1}[k']$. The converse inclusion follows by running the same argument backwards. This proves the claim.

So by lemma 5.33, $w' \simeq u'v'\rho_K$. If $K = \alpha$, then repeated use of lemma 5.31 gives $w \simeq uv\rho_K \simeq \rho_K \simeq u'v'\rho_K \simeq w'$, and we are done. Assume otherwise. Fix $l \notin K$, let $K = \{k_0, \ldots, k_{n-1}\}$, and define $t = s_l^{k_0} \ldots s_l^{k_{n-1}}$. We show that $uvt \sim u'v't$.

5.4. Substitutions in cylindric algebras

1. $\widehat{uvt} = \widehat{u'v't}$. For, it is clear that $\widehat{\rho_K} = \mathrm{Id}_{\alpha\setminus K}$. Lemma 5.27(1) now gives $\widehat{uv}\restriction_{\alpha\setminus K} = \widehat{uv} \circ \widehat{\rho_K} = \widehat{uv\rho_K} = \widehat{w} = \widehat{w'} = \widehat{u'v'\rho_K} = \widehat{u'v'} \circ \widehat{\rho_K} = \widehat{u'v'}\restriction_{\alpha\setminus K}$. $\widehat{uvt} = \widehat{u'v't}$ now follows from lemma 5.27(1), since $\mathrm{rng}(\hat{t}) \subseteq \alpha\setminus K$.

2. We check $(uvt)^* = (u'v't)^*$ as follows. Let $i,j < \alpha$. If $(i,j) \in (uvt)^*$, then since t is an s-word, $(\hat{t}(i),\hat{t}(j)) \in (uv)^*$. Since $\hat{t}(i) \notin K$, we see that $\hat{t}(i) = \widehat{\rho_K}(\hat{t}(i))$, and similarly for j. So $(\hat{t}(i),\hat{t}(j)) \in (uv\rho_K)^* = w^* = w'^* = (u'v'\rho_K)^*$. As before, this implies that $(\hat{t}(i),\hat{t}(j)) \in (u'v')^*$, and it follows that $(i,j) \in (u'v't)^*$. The converse is similar.

We have established that $uvt \sim u'v't$. Also, $uvtu'v't$ has two fewer c_m than ww'. Inductively, we obtain $uvt \simeq u'v't$. Now repeated use of lemma 5.31 (as when $K = \alpha$ above) gives $t\rho_K \simeq \rho_K$. We now obtain

$$w \simeq uv\rho_K \simeq uvt\rho_K \simeq u'v't\rho_K \simeq u'v'\rho_K \simeq w',$$

completing the proof. □

Exercises

In the exercises, α is an ordinal and s-c-words are of the signature of $\mathbf{CA}_\alpha$.

1. Check that the facts in 5.20 are valid in $\mathbf{RCA}_\alpha$. Prove that some of the more interesting ones are valid in $\mathbf{CA}_\alpha$.

2. Let w be any s-c-word. Show that w^* is an equivalence relation on α.

3. Let w be an s-c-word such that $\widehat{w}$ is one-one on its domain. Let $\alpha \setminus \mathrm{dom}(\widehat{w}) \subseteq \{k_0,\ldots,k_{n-1}\}$, and $v = wc_{k_0}\ldots c_{k_{n-1}}$. Show that $v^* = \ker(\mathrm{Id}_\alpha)$.

4. Prove that for any s-c-word w, any $i < \alpha$ with $\widehat{w}^{-1}[i] = \{i_0,\ldots,i_{k-1}\}$, and $j \in \mathrm{dom}(\widehat{w})$, we have $\mathsf{s}^i_{\widehat{w}(j)}w \sim w\mathsf{s}^{i_0}_j\ldots\mathsf{s}^{i_{k-1}}_j$.

   Can you show $\mathsf{s}^i_{\widehat{w}(j)}w \simeq w\mathsf{s}^{i_0}_j\ldots\mathsf{s}^{i_{k-1}}_j$ too?

5. Show that the condition $w^* = w'^*$ in theorem 5.25 is necessary: that is, show that $\widehat{w} = \widehat{w'}$ and $|\alpha \setminus \mathrm{rng}(\widehat{w})| \geq 2$ do not imply that $w \simeq w'$. [Try $c_0\mathsf{s}^1_0$ and c_0c_1.]

6. There are special circumstances in which the condition $w^* = w'^*$ in theorem 5.25 is not necessary. An s-c-word w is said to be *modest* if for all s-c-words u,v and $i < \alpha$, if $w = uc_iv$ then $|\widehat{v}^{-1}[i]| \leq 1$. For example, the concatenation of an s-word and a c-word is modest. Show that for modest w we have $w^* = \ker(\widehat{w} \cup \mathrm{Id}_{\alpha\setminus\mathrm{dom}(\widehat{w})})$. Deduce that if w,w' are modest s-c-words and $\widehat{w} = \widehat{w'}$ then $w \sim w'$.

7. Show that after running the 'program' associated with an s-c-word w, variable x_i always ends up with the initial value of x_j iff $i \in \mathrm{dom}(\widehat{w})$ and $\widehat{w}(i) = j$, and that variables x_i, x_j always end up equal iff $(i,j) \in w^*$. Deduce that s-c-words w, w' satisfy the three 'program' conditions of section 5.4.2 iff $w \sim w'$ and $|\alpha \setminus \mathrm{rng}(\widehat{w})| \geq 2$.

5.5 Relativised cylindric algebras

Just as with relation algebras, we can generalise the notion of a cylindric set algebra by relativisation. The following definition is due to Henkin (cf., e.g., [Hen68]).

DEFINITION 5.34 Let α be an ordinal.

1. Let x, y be two α-ary sequences of elements of some set, and let $i < \alpha$. We write $x \equiv_i y$ if for each $j < \alpha$, if $j \neq i$ then $x_j = y_j$.

2. A *cylindric relativised set algebra of dimension* α consists of a set S of α-ary relations over some base set U, equipped with constants 0 and 1 (here, the unit 1 is any α-ary relation over U) and for $i, j < \alpha$, the operations $\cup, \setminus$ (complement relative to 1), diagonal elements $D_{ij} = \{x \in 1 : x_i = x_j\}$, and the cylindrifications C_i where $C_i(X) = \{u \in 1 : \exists x \in X, u \equiv_i x\}$. S must contain the constants and be closed under the operations.

3. The class of all α-dimensional cylindric relativised set algebras is denoted by $\mathbf{Crs}_\alpha$.

With relation algebras, we started with general relativisation and then imposed restrictions (symmetry, reflexivity) on the unit. We can do the same here with cylindric set algebras to obtain interesting classes of algebras. Fix an ordinal α. Recall that the substitution map $[i/j] : \alpha \to \alpha$ is given by:

$$[i/j](k) = \begin{cases} j, & \text{if } k = i; \\ k, & \text{otherwise.} \end{cases}$$

DEFINITION 5.35 For any ordinal α,

- $\mathbf{D}_\alpha$ is the class of all algebras in $\mathbf{Crs}_\alpha$ whose unit is closed under substitutions $[i/j]$: if x is in the unit and $i, j < \alpha$, then

$$x \circ [i/j] = (x_0, \ldots, x_{i-1}, x_j, x_{i+1}, \ldots)$$

is in the unit, too.

5.5. Relativised cylindric algebras

- G_α is the class of all algebras in Crs_α (and of D_α) with unit closed under all maps: if $\pi : \alpha \to \alpha$ is any map, and x is in the unit, then so is $x \circ \pi = (x_{\pi(0)}, x_{\pi(1)}, \ldots)$. Such units, or relativised representations with such a unit, are called 'locally square', or 'locally cubic'.

For infinite α, an intermediate class between D_α and G_α can be obtained by requiring that if x is in the unit and $\pi : \alpha \to \alpha$ is finitary (i.e., it moves finitely many points), then $x \circ \pi$ is in the unit. See, e.g., [Sai95].

LEMMA 5.36 *Cylindrifications and substitutions are normal and completely additive in any algebra in Crs_α (any α).*

Proof. Let $C \in Crs_\alpha$. We show that cylindrifications are conjugated in C — just as in cylindric algebras. Let $c, d \in C$ and $i < \alpha$. Then $c \cdot c_i d \neq 0$ iff there is $\bar{a} \in c \cdot c_i d$, iff there are $\bar{a} \in c$ and $\bar{b} \in d$ with $\bar{a} \equiv_i \bar{b}$. This condition is symmetrical, so $c \cdot c_i d = 0$ iff $c_i c \cdot d = 0$. By theorem 2.40, c_i is normal and completely additive.

Similarly, for distinct $i, j < \alpha$ we have $c \cdot s^i_j d = 0$ iff $c \cdot c_i(d_{ij} \cdot d) = 0$, iff $c_i c \cdot (d_{ij} \cdot d) = 0$, iff $(c_i c \cdot d_{ij}) \cdot d = 0$. So s^i_j is also conjugated, and so is normal and completely additive. The result for s^i_i is trivial. □

Of course, $\sum S$ in the lemma will be different from $\bigcup S$ if the latter is not an element of C. The question of when they coincide is related to complete representations (cf. section 2.3.8 for boolean algebras and definition 3.35 for relation algebras). We know that the class of completely representable boolean algebras coincides with the class of atomic boolean algebras (corollary 2.22). But we'll see later that the class of completely representable relation algebras is not elementary (theorem 17.6). Similarly, it can be shown that the class of completely representable α-dimensional cylindric algebras is not an elementary class [HirHod97a, theorem 34], for $\alpha \geq 3$. So this might surprise you:

PROPOSITION 5.37 *Let α be given and let $C \in D_\alpha$ be atomic. There is $D \in D_\alpha$ such that $C \cong D$ and such that for any $\bar{a} \in 1^D$, there is an atom $d \in D$ with $\bar{a} \in d$. Hence, $\sum^D S = \bigcup S$ for all $S \subseteq D$ such that $\sum^D S$ exists.*

That is, any atomic $C \in D_\alpha$ has a *complete* relativised representation in the D_α sense. By corollary 2.22, any algebra with such a representation must be atomic, so this result is best possible. A similar result can be proved for Crs_α, but we will not use it.

Proof. If U is the base of C, let $I = \bigcup At\, C \subseteq {}^\alpha U$, and define a new Crs_α D with unit I and domain $\{c \cap I : c \in C\}$. We check that D is indeed in Crs_α (i.e., it is closed under the natural operations), that I is closed under substitutions so that $D \in D_\alpha$, and that $c \mapsto c \cap I$ is an isomorphism from C to D.

By distributivity of operations over $\cap$ in boolean algebras, it follows that $\mathcal{D}$ is closed under the boolean operations and that $c \mapsto c \cap I$ preserves them. Clearly, $\{\bar{a} \in I : a_i = a_j\} = \mathsf{d}_{ij}^C \cap I$, so $\mathcal{D}$ contains the required diagonals and $c \mapsto c \cap I$ preserves these.

For cylindrifications, we let $i < \alpha$, $x \in C$, and check that

$$I \cap \mathsf{c}_i^C x = \{\bar{a} \in I : \bar{a} \equiv_i \bar{b} \text{ for some } \bar{b} \in I \cap x\}.$$

The inclusion '$\supseteq$' is clear. Conversely, assume that $\bar{a} \in I \cap \mathsf{c}_i^C x$. Take an atom γ of C with $\bar{a} \in \gamma$. Then $\bar{a} \in \gamma \cdot \mathsf{c}_i^C x$, so $\gamma \cdot \mathsf{c}_i^C x \neq 0$ and $\gamma \leq \mathsf{c}_i^C x$. By lemma 5.36 and atomicity of C, there is $\delta \in \mathrm{At}\, C$ with $\delta \leq x$ and $\gamma \leq \mathsf{c}_i^C \delta$. So $\bar{a} \equiv_i \bar{b}$ for some $\bar{b} \in \delta$. Since $\bar{b} \in I \cap x$ and $\bar{a} \in I$, we are done.

The foregoing shows that $\mathcal{D}$ is closed under the **Crs** operations and that $c \mapsto c \cap I$ is a homomorphism from C onto $\mathcal{D}$. This map is clearly one-one since it leaves atoms invariant.

To show that $\mathcal{D} \in \mathbf{D}_\alpha$, it suffices to show that $\mathcal{D} \models \mathsf{s}_j^i 1 = 1$ for all $i, j < \alpha$. (Cf. proposition 5.39(1) below.) This is clear, since $C \in \mathbf{D}_\alpha$ so $C \models \mathsf{s}_j^i 1 = 1$, and $C \cong \mathcal{D}$.

Since every $\bar{a} \in I$ is contained in some atom of $\mathcal{D}$, theorem 2.21 shows that whenever $S \subseteq \mathcal{D}$ and $\sum^\mathcal{D} S$ exists, $\sum^\mathcal{D} S = \bigcup S$. □

The classes **Crs, D, G** have been very intensively studied (some references will be given at the end of the section). Fact 5.38 below records some information about them. The facts we quote will not be used heavily later on, but we feel it is worth including them for their intrinsic interest and because they foreshadowed developments in the theory of relation algebras to some degree — for example, the theorem of Maddux ([Madd82], and theorem 7.5) that **WA** is the class of relation-type BAOs with relativised representations is perhaps the relation algebra analogue of fact 5.38. They also have wider significance. Consideration of **Crs** led to the decidable *guarded fragment* of first-order logic, which was introduced in [AndBen[+]98] and has generated an important field of research: see, e.g., [Ben96, Ben97, AreMon[+]99, Grä99a, Grä99b, HodOtt01] for more information. We will use the 'loosely guarded fragment' in chapter 19. This fragment has the finite model property, and this yields 'finite base property' results for **Crs, D, G**. Such results are not included in fact 5.38 since chapter 19 will cover this topic.

Recall that for a class K of algebras, IK denotes the closure of K under isomorphism.

FACT 5.38 *Let $\alpha \geq 3$ be an ordinal.*

1. $\mathbf{CA}_\alpha \cap \mathbf{ICrs}_\alpha$ *is a variety, axiomatised by the equations defining* $\mathbf{CA}_\alpha$, *plus the two 'merry-go-round' axioms:*

5.5. Relativised cylindric algebras

MGR1: $s_\kappa^\lambda s_\mu^\kappa s_\lambda^\mu c_\lambda x = s_\mu^\lambda s_\kappa^\mu s_\lambda^\kappa c_\lambda x$ *for all distinct* $\kappa, \lambda, \mu < \alpha$,

MGR2: $s_\kappa^\lambda s_\mu^\kappa s_\nu^\mu s_\lambda^\nu c_\lambda x = s_\nu^\lambda s_\kappa^\nu s_\mu^\kappa s_\lambda^\mu c_\lambda x$ *for all distinct* $\kappa, \lambda, \mu, \nu < \alpha$.

Since $\mathbf{CA}_\alpha \cap \mathbf{ICrs}_\alpha$ *is axiomatised by simple equations, theorem 2.95 shows that it is a canonical variety.*

2. $\mathbf{ICrs}_\alpha$ *is a non-finitely axiomatisable conjugated canonical variety.*

   Its equational (and indeed, universal) theory is decidable.

3. $\mathbf{ID}_\alpha$ *is a canonical variety, axiomatised by the axioms defining* $\mathbf{CA}_\alpha$, *with C4* ($c_i c_j x = c_j c_i x$) *replaced by the weaker* $c_i c_j x \geq c_j c_i x \cdot d_{jk}$ *for* $k \neq i, j$, *plus MGR2. For finite* α, *it is finitely axiomatisable and has decidable equational and universal theory.*

4. $\mathbf{IG}_2 = \mathbf{ID}_2 \cap \mathrm{Mod}\{x - d_{01} \leq c_0 c_1(-d_{01} \cdot s_1^0 c_1 x \cdot s_0^1 c_0 x)\}$, *and for finite* $\alpha \geq 3$,

$$\mathbf{IG}_\alpha = \mathbf{ID}_\alpha \cap \mathrm{Mod}\left\{ x \leq c_i c_j \left(s_j^i c_j x \cdot s_i^j c_i x \cdot \prod_{k < \alpha, k \neq i,j} s_i^k s_j^i s_k^j c_k x \right) : \begin{array}{c} i, j < \alpha, \\ i \neq j \end{array} \right\}.$$

Hence, $\mathbf{IG}_\alpha$ *is a finitely axiomatisable canonical variety. Its universal theory is decidable. For arbitrary* α, *its equational theory is decidable.*

We now make some remarks on these facts.

1. The line of research of items 1–4 began with Resek, who showed in [Res75], *inter alia*, that $\mathbf{CA}_\alpha \cap \mathbf{ICrs}_\alpha$ is axiomatised by the cylindric algebra axioms plus the so-called 'merry-go-round' axioms:

   $MGR \quad s_{\kappa_0}^\lambda s_{\kappa_1}^{\kappa_0} \ldots s_{\kappa_{n-1}}^{\kappa_{n-2}} s_\lambda^{\kappa_{n-1}} c_\lambda x = s_{\kappa_{n-1}}^\lambda s_{\kappa_0}^{\kappa_{n-1}} s_{\kappa_1}^{\kappa_0} \ldots s_{\kappa_{n-2}}^{\kappa_{n-3}} s_\lambda^{\kappa_{n-2}} c_\lambda x,$

   for any $2 \leq n < \omega$, $\kappa \in {}^n\alpha$ a one-one sequence, and $\lambda < \alpha$, $\lambda \notin \mathrm{rng}(\kappa)$.

   Thompson simplified Resek's axioms by showing that all MGR axioms are implied by the instances MGR1 and MGR2; this showed that $\mathbf{CA}_\alpha \cap \mathbf{ICrs}_\alpha$ is finitely axiomatisable for finite α. See [AndTho88] and [HenMon+85, 3.2.88] for more information. We note that MGR1 holds in a cylindric algebra with $\kappa = 0, \mu = 1, \lambda = 2$ iff its relation algebra reduct (see section 5.6 below) satisfies $\breve{\breve{x}} = x$.

2. Németi proved that $\mathbf{ICrs}_\alpha$ for $\alpha \geq 2$ is a variety [Ném81]; a proof is also given in [HenMon+85, 5.5.10]. A construction of Andréka–Németi shows that $\mathbf{ICrs}_\alpha$ is not finitely axiomatisable for $\alpha \geq 3$ [HenMon+85, 5.5.12]. Thompson adapted Resek's equations to obtain an explicit equational axiomatisation of $\mathbf{ICrs}_\alpha$ (see [ResTho91, p. 520]); a different proof using a

step-by-step technique of Andréka can be found in [Mon93] and the updated [Mon00]. The equations required are a variant of MGR together with some cylindric algebra axioms (excluding C4) and simple consequences of them.

Canonicity of $\mathbf{ICrs}_\alpha$ was proved by Goldblatt in [Gol95, theorem 4.6]. For conjugation see lemma 5.36.

The decidability of the equational theory of $\mathbf{Crs}_\alpha$ for $\alpha \leq \omega$ was proved by Németi in [Ném86] and also in [Ném95, theorem 4.3]. Decidability of the universal theory of $\mathbf{Crs}_\alpha$ for all α, plus other results, are established in [Ném96, theorem 4.1, remark 4.14].

See [AndGol⁺98] and [Ném95, §4], [Ném96] for more information.

3. The axiomatisation of $\mathbf{ID}_\alpha$ was established by Thompson, adapting Resek's axiomatisation of $\mathbf{CA}_\alpha \cap \mathbf{ICrs}_\alpha$. A short proof, by the technique of Andréka mentioned above, can be found in [AndTho88]. The axioms are Sahlqvist equations, and so $\mathbf{ID}_\alpha$ is canonical. Decidability of the equational and universal theories of $\mathbf{D}_\alpha$ for finite α was proved by Németi in [Ném86, theorem 10(ii)] and [Ném95, theorem 4.3], and in [Ném96, theorem 4.1], respectively.

[AndTho88, p. 681], [Ném95, remark 4.15], and [Ném96, remark 4.14] state that decidability of the equational theory of $\mathbf{D}_\alpha$ for infinite α is an open problem.

4. The axiomatisation of $\mathbf{IG}_\alpha$ was established in [And01], using the axiomatisation of $\mathbf{ID}_\alpha$. The axioms are Sahlqvist equations, so $\mathbf{IG}_\alpha$ is canonical. Decidability of the equational theory of $\mathbf{G}_\alpha$ for $\alpha \leq \omega$ was proved by Németi [Ném95, theorem 4.3]. Among other things, decidability of the universal theory of $\mathbf{G}_\alpha$ for finite α was proved by Németi in [Ném96, theorem 4.1]; remark 4.14 of this paper conjectures that the universal theory of $\mathbf{G}_\alpha$ for infinite α is also decidable.

For infinite α, it is not known whether $\mathbf{IG}_\alpha$ is a variety, or even closed under ultraproducts. (This was asked in [Ném96, AndGol⁺98, And01].)

We also lay out some simple properties of the classes, which will be occasionally useful.

PROPOSITION 5.39 *Let $\alpha \geq 3$ be an ordinal.*

1. $\mathbf{D}_\alpha = \mathbf{Crs}_\alpha \cap \mathrm{Mod}\{c_i d_{ij} = 1 : i,j < \alpha\}$.
2. $\mathbf{IG}_\alpha \subset \mathbf{ID}_\alpha \subset \mathbf{ICrs}_\alpha$.
3. $\mathbf{G}_\alpha \not\subseteq \mathbf{CA}_\alpha \not\subseteq \mathbf{ICrs}_\alpha$.

5.5. Relativised cylindric algebras

4. $\mathbf{CA}_\alpha \cap \mathbf{Crs}_\alpha = \mathbf{CA}_\alpha \cap \mathbf{D}_\alpha$, and at least for $\alpha = 3$, $\mathbf{CA}_\alpha \cap \mathbf{D}_\alpha \supset \mathbf{CA}_\alpha \cap \mathbf{G}_\alpha$.

5. $\mathbf{G}_\alpha \cap \mathrm{Mod}(C4) \subseteq \mathbf{D}_\alpha \cap \mathrm{Mod}(C4) \subseteq \mathbf{Crs}_\alpha \cap \mathrm{Mod}(C4, c_i d_{ij} = 1) \subseteq \mathbf{CA}_\alpha$, where C4 denotes the $\mathbf{CA}_\alpha$ axiom $c_i c_j x = c_j c_i x$ for all $i, j < \alpha$.

Proof (sketch).

1. It is clear that $\mathbf{D}_\alpha \models c_i d_{ij} = 1$ for all $i, j < \alpha$. Conversely, if $\mathcal{A} \in \mathbf{Crs}_\alpha$ satisfies these equations, then for any x in the unit 1 of $\mathcal{A}$ and $i, j < \alpha$, $x \in c_i d_{ij}$, so there is $y \equiv_i x$ with $y \in d_{ij}$. Thus, $y \in 1$ and $y_i = y_j$, so $x \circ [i/j] = y \in 1$.

2. Clearly, $\mathbf{G}_\alpha \subseteq \mathbf{D}_\alpha \subseteq \mathbf{Crs}_\alpha$, so the same holds for the closures of these classes under isomorphism. That $\mathbf{G}_\alpha \neq \mathbf{D}_\alpha$ can be shown by constructing an algebra in $\mathbf{D}_\alpha$ as follows. We take $\alpha = 3$, let $x = (0, 1, 2) \in {}^3 3$, and let X be the smallest subset of ${}^3 3$ containing x and closed under substitutions $[i/j]$ for $i, j < 3$. Let $\mathcal{A}$ have domain $\wp(X)$. It can be checked that $\mathcal{A} \in \mathbf{D}_3$. But $\mathcal{A}$ does not satisfy the axioms of (4), because $c_0 c_1 (s_1^0 c_1 \{x\} \cdot s_0^1 c_0 \{x\} \cdot s_0^2 s_1^0 s_2^1 c_k \{x\}) = \emptyset \not\supseteq \{x\}$. So $\mathcal{A} \notin \mathbf{IG}_3$.

   To see that $\mathbf{D}_\alpha \neq \mathbf{Crs}_\alpha$, choose any $\mathcal{A} \in \mathbf{Crs}_\alpha$ whose unit is not closed under substitutions. By the argument of (1), $\mathcal{A}$ does not satisfy the axioms $c_i d_{ij} = 1$, so certainly $\mathcal{A} \notin \mathbf{ID}_\alpha$.

3. Consider the algebra $\mathcal{A} \in \mathbf{G}_2$ with domain

$$\wp({}^2\{0,1\}) \cup \wp({}^2\{1,2\}) \cup \wp({}^2\{2,3\}).$$

   Then since $(0,1) \equiv_0 (2,1) \equiv_1 (2,3)$, we have $\mathcal{A} \models \{(0,1)\} \leq c_0 c_1 \{(2,3)\}$. But $(0,1) \equiv_1 (x,y) \equiv_0 (2,3)$ implies $(x,y) = (0,3) \notin 1$. So $\mathcal{A} \not\models \{(0,1)\} \leq c_1 c_0 \{(2,3)\}$, $\mathcal{A} \not\models \forall x (c_0 c_1 x = c_1 c_0 x)$, and $\mathcal{A} \notin \mathbf{CA}_2$. A similar argument works for larger α. We are assuming $\alpha \geq 3$, and in this case, $\mathbf{CA}_\alpha \not\subseteq \mathbf{ICrs}_\alpha$: see [HenMon⁺85, 5.5.9]. Note however that $\mathbf{CA}_2 \subseteq \mathbf{ICrs}_2$ [HenMon⁺85, 3.2.61, 5.5.6].

4. By [HenMon⁺71, 1.3.2], $\mathbf{CA}_\alpha \models c_i d_{ij} = 1$. So any algebra in $\mathbf{CA}_\alpha \cap \mathbf{Crs}_\alpha$ satisfies this axiom and hence is in $\mathbf{D}_\alpha$ by (1). We will see that $\mathbf{CA}_3 \cap \mathbf{D}_3 \supset \mathbf{CA}_3 \cap \mathbf{G}_3$ in fact 5.46 below.

5. The first inclusion is trivial, the second follows from (1), and the third is simply a matter of checking the $\mathbf{CA}_\alpha$ axioms. $c_i d_{ij} = 1$ is needed to prove the $d_{jk} \leq c_i (d_{ji} \cdot d_{ik})$ part of C6.

□

There are finite-schema analogues of the finite axiomatisation results, for infinite α. For further information on these classes (there is much), see [HenMon⁺81, HenMon⁺85, Ném85, Ném86, AndTho88, Ném91, ResTho91, Ném95, Mar95, Ném96, MarVen97, And97a, AndGol⁺98, Mon00, And01], for example.

Exercises

1. Which parts of fact 5.20 are valid in $\mathbf{D}_\alpha$ or $\mathbf{G}_\alpha$?

2. Is the analogue of proposition 5.37 true for $\mathbf{Crs}, \mathbf{G}$?

3. Let L be a finite signature and let K be a class of L-structures with K = SK and with decidable universal theory. Show that there is an algorithm to decide whether a finite L-structure is in K. [Cf. corollary 18.27.] Deduce that for $n < \omega$, it is decidable whether a finite $\mathbf{CA}_n$-type algebra is in $\mathbf{ICrs}_n$, $\mathbf{ID}_n$, or $\mathbf{IG}_n$ [Ném96, theorem 4.1(ii, v)].

5.6 Relation algebra reducts of cylindric algebras

Cylindric algebras are an attempt to algebraise relations of arbitrary arity, just as relation algebras are for binary relations. We now examine the connections between them; this will be followed up at greater length in chapter 13.

5.6.1 Neat reducts and relation algebra reducts

There is a well known method of obtaining a relation-type algebra $\mathfrak{Ra}(C)$ from an α-dimensional cylindric algebra C (for any $\alpha \geq 3$): $\mathfrak{Ra}(C)$ is constructed by taking the two-dimensional elements of C and using the spare dimensions to define conversion and composition. $\mathfrak{Ra}(C)$ is called the relation algebra reduct of C. More formally, this is done as follows.

DEFINITION 5.40 [HenMon$^+$85, 5.3.7] Let C be any α-dimensional cylindric algebra, where $\alpha \geq 3$.

1. For $\beta \leq \alpha$, the *neat β-reduct of C* (in symbols, $\mathfrak{Nr}_\beta C$) is the β-dimensional cylindric algebra with domain $\{a \in C : c_j a = a \text{ for all } \beta \leq i < \alpha\}$ and with operations $+, -, 0, 1, c_j, d_{jk}$ for $j, k < \beta$ induced from C.

2. The *relation algebra reduct of C* — in symbols, $\mathfrak{Ra}(C)$ — is the algebra

$$\langle \mathrm{dom}(\mathfrak{Nr}_2 C), 0, 1, +, -, 1', \breve{\ }, ; \rangle,$$

    where

    - $+, -, 0, 1$ are as in C
    - $1' = d_{01} \,(\in \mathfrak{Nr}_2 C)$
    - conversion is defined by $\breve{r} = s_0^2 s_1^0 s_2^1 r$, for $r \in \mathfrak{Nr}_2 C$

5.6. Relation algebra reducts of cylindric algebras

- composition is defined by $r;s = c_2(s_2^1 r \cdot s_2^0 s)$, for $r, s \in \mathfrak{Nr}_2 C$.

We generally identify notationally the algebras $\mathfrak{Nr}_\beta C$, $\mathfrak{Ra} C$ with their domains.

LEMMA 5.41 $\mathfrak{Nr}_\beta C$ and $\mathfrak{Ra}(C)$ are closed under these operations.

Proof. First we consider $\mathfrak{Nr}_\beta C$. Clearly, $0, 1 \in \mathfrak{Nr}_\beta C$. In cylindric algebras we have $c_k d_{ij} = d_{ij}$ whenever $k \neq i, j$. Hence $d_{ij} \in \mathfrak{Nr}_\beta C$ for all $i, j < \beta$. Let $r, s \in \mathfrak{Nr}_\beta C$, and $\beta \leq i < \alpha$. Then using the axioms defining $\mathbf{CA}_\alpha$ (definition 5.16), we have $c_i(r+s) = c_i r + c_i s = r + s$, and $c_i(-r) = -c_i r = -r$ (see fact 5.17 or [HenMon$^+$71, 1.2.11]). So $\mathfrak{Nr}_\beta C$ is closed under the boolean operations. Also, for $j < \beta \leq i$, we have $c_i c_j r = c_j c_i r = c_j r$, using commutativity of c_i, c_j in $\mathbf{CA}_\alpha$ (axiom C4): so $\mathfrak{Nr}_\beta C$ is closed under appropriate cylindrifications, too.

Now consider $\mathfrak{Ra} C$. By the above, it is closed under the boolean operations and contains 0, 1, and $1' = d_{01}$. Let $r, s \in \mathfrak{Ra} C$ and $i \geq 2$. If $i \geq 3$, then because c_i commutes with all three substitutions (fact 5.20(6)), $c_i \breve{r} = c_i s_0^2 s_1^0 s_2^1 r = s_0^2 s_1^0 s_2^1 c_i r = s_0^2 s_1^0 s_2^1 r = \breve{r}$. If $i = 2$, then $c_2 s_0^2 x = s_0^2 x$ for any $x \in C$ (fact 5.20(8)), so again, $c_2 \breve{r} = \breve{r}$.

Finally, consider $c_i(r;s) = c_i c_2(s_2^1 r \cdot s_2^0 s)$. If $i = 2$, then clearly this is $c_2(s_2^1 r \cdot s_2^0 s) = r;s$ by idempotence of cylindrification (an easy consequence of axioms C2, C3; see fact 5.17). If $i \geq 3$, then c_i commutes with c_2 (axiom C4), so the above is $c_2 c_i(s_2^1 r \cdot s_2^0 s) = c_2 c_i(s_2^1 c_i r \cdot s_2^0 c_i s)$. As c_i commutes with the two substitutions here, this is equal to $c_2 c_i(c_i s_2^1 r \cdot c_i s_2^0 s)$. By axiom C3, this is $c_2(c_i s_2^1 r \cdot c_i s_2^0 s)$. Wrapping up again, this is $c_2(s_2^1 c_i r \cdot s_2^0 c_i s) = r;s$, as required. □

The following result will be needed (and generalised) in chapter 13. Recall that bool(C) denotes the boolean reduct of the cylindric algebra C.

LEMMA 5.42 If $m < n < \omega$ and $C \in \mathbf{CA}_n$, then bool($\mathfrak{Nr}_m C$) $\subseteq^c$ bool(C). Hence, if C is atomic then so are $\mathfrak{Nr}_m C$ and (if $n \geq 3$) $\mathfrak{Ra} C$.

Proof. Let $S \subseteq \mathfrak{Nr}_m C$ and suppose $\sum^{\mathfrak{Nr}_m C} S = \sigma$ exists. Assume for contradiction that $d \in C$ and $s \leq d < \sigma$ for all $s \in S$. Let

$$\tau \stackrel{\text{def}}{=} \sigma \cdot -c_m c_{m+1} \ldots c_{n-1}(-d).$$

We claim that $\tau \in \mathfrak{Nr}_m C$ and $s \leq \tau < \sigma$ for all $s \in S$; this will contradict $\sigma = \sum^{\mathfrak{Nr}_m C} S$.

We use $\sigma \in \mathfrak{Nr}_m C$ throughout. First, we show that $\tau \in \mathfrak{Nr}_m C$. If $m \leq i < n$, we have:

$$\begin{aligned}
c_i \tau &= c_i(c_i \sigma \cdot -c_m \ldots c_{n-1} -d) && \text{as } \sigma = c_i \sigma \\
&= c_i \sigma \cdot c_i -c_m \ldots c_{n-1} -d && \text{by } \mathbf{CA}_n\text{-axiom C3} \\
&= c_i \sigma \cdot c_i -c_i c_m \ldots c_{n-1} -d && \text{by fact 5.17(1), axiom C4} \\
&= c_i \sigma \cdot -c_i c_m \ldots c_{n-1} -d && \text{by fact 5.17(5)} \\
&= \sigma \cdot -c_m \ldots c_{n-1} -d && \text{as before} \\
&= \tau.
\end{aligned}$$

This holds for all i, so $\tau \in \mathfrak{Nr}_m C$.

Second, if $s \in S$, we show $s \leq \tau$. We know $s \leq \sigma$. Also, $s \leq d$, so $s \cdot -d = 0$. Hence $0 = c_m \ldots c_{n-1}(s \cdot -d) = s \cdot c_m \ldots c_{n-1} -d$. We obtain $s \leq -c_m \ldots c_{n-1} -d$. So $s \leq \tau$ as required.

Finally we check that $\tau < \sigma$. If not, then we have $\tau = \sigma$, so $\sigma \leq -c_m \ldots c_{n-1} -d$, so $\sigma \cdot c_m \ldots c_{n-1} -d = 0$. Moving the cylindrifications over, using fact 5.17(2), shows that $\sigma \cdot -d = 0$. This contradicts $d < \sigma$.

The last part of the lemma follows immediately from lemma 2.16. □

5.6.2 Relation algebra reducts and canonical extensions

The following result is due to Henkin and Monk (cf. [HenMon+71, 2.7.23] for cylindric algebras) and we will need it in chapter 13.

THEOREM 5.43 *Let $\alpha \geq 3$, and let $\mathcal{A}$ be a non-associative algebra. If $\mathcal{A} \subseteq \mathfrak{Ra}\mathcal{B}$ for some $\mathcal{B} \in \mathbf{CA}_\alpha$, then $\mathcal{A}^+ \subseteq^c \mathfrak{Ra}(\mathcal{B}^+)$ up to isomorphism.*

Proof. Note that since $\mathbf{CA}_\alpha$ is canonical, $\mathcal{B}^+ \in \mathbf{CA}_\alpha$, so that $\mathfrak{Ra}(\mathcal{B}^+)$ is defined.

For $C \in \mathbf{CA}_\alpha$ let $R(C)$ be the expansion of C by a constant 1' and function symbols $\check{\ }, ;$ defined by $1' = \mathsf{d}_{01}$, $\check{c} = \mathsf{s}_0^2 \mathsf{s}_1^0 \mathsf{s}_2^1 c$, and $c;d = \mathsf{c}_2(\mathsf{s}_2^1 c \cdot \mathsf{s}_2^0 d)$, for $c,d \in C$. By facts 5.17 and 5.20, $R(C)$ is a BAO.

We claim that $R(C)^+ = R(C^+)$. Since $R(C)$ is just an expansion of C, $R(C)^+$ and $R(C^+)$ have the same reduct to the signature of C. Now

$$\left(x;y \cdot -\mathsf{c}_2(\mathsf{s}_2^1 x \cdot \mathsf{s}_2^0 y)\right) + \left(\mathsf{c}_2(\mathsf{s}_2^1 x \cdot \mathsf{s}_2^0 y) \cdot -(x;y)\right) = 0$$

is a Sahlqvist equation and is valid in $R(C)$, so by theorem 2.95, it is valid in $R(C)^+$ too. So for any $a,b \in R(C)^+$, $a;^{R(C)^+} b = \mathsf{c}_2^{C^+}(\mathsf{s}_2^{1\,C^+} a \cdot \mathsf{s}_2^{0\,C^+} b)$. This says that the identity map from $R(C)^+$ to $R(C^+)$ preserves ';'. It can be shown similarly that it preserves 1' and $\check{\ }$. Hence, $R(C)^+ = R(C^+)$ as claimed.

Let $R(C)\!\upharpoonright_{L_{RA}}$ be the reduct of $R(C)$ to the signature of relation algebras. Then obviously $(R(C)\!\upharpoonright_{L_{RA}})^+ = (R(C)^+)\!\upharpoonright_{L_{RA}} = R(C^+)\!\upharpoonright_{L_{RA}}$.

Now $\mathcal{A}$ is a subalgebra of $R(\mathcal{B})\!\upharpoonright_{L_{RA}}$. By the above and theorem 2.71, the map $\iota^+ : \mathcal{A}^+ \to R(\mathcal{B}^+)\!\upharpoonright_{L_{RA}}$ given by $\iota^+(S) = \{\beta \in \mathrm{Uf}(\mathcal{B}) : \beta \cap \mathcal{A} \in S\}$ is a complete embedding. To check that it embeds $\mathcal{A}^+$ into $\mathfrak{Ra}(\mathcal{B}^+)$, it only remains to check that if $2 \leq i < \alpha$ and $S \in \mathcal{A}^+$, then in $R(\mathcal{B}^+)$ we have $c_i(\iota^+(S)) = \iota^+(S)$. For '⊇', we recall (section 5.3) that $\mathbf{CA}_\alpha$ is canonical; hence, $\mathcal{B}^+ \in \mathbf{CA}_\alpha$, and $\mathcal{B}^+ \models c_i x \geq x$. We prove the converse inclusion. Given $\beta \in c_i(\iota^+(S))$, there must be some $\gamma \in \mathrm{Uf}(\mathcal{B})$ with $\gamma \cap \mathcal{A} \in S$ and $\beta \leq c_i^{\mathcal{B}^+} \gamma$. By fact 5.17, $\gamma \leq c_i^{\mathcal{B}^+} \beta$. So $\{c_i b : b \in \beta\} \subseteq \gamma$. So if $b \in \beta \cap \mathcal{A}$ then $b = c_i b \in \gamma$, whence $\beta \cap \mathcal{A} = \gamma \cap \mathcal{A} \in S$ and $\beta \in \iota^+(S)$.

So $\iota^+ : \mathcal{A}^+ \to \mathfrak{Ra}(\mathcal{B}^+)$ is a complete embedding. Hence, up to isomorphism, $\mathcal{A}^+ \subseteq^c \mathfrak{Ra}(\mathcal{B}^+)$. □

5.6. Relation algebra reducts of cylindric algebras

5.6.3 Relation algebra reducts are relation algebras

Clearly, if $C \in \mathbf{CA}_\alpha$ and $\beta \leq \alpha$ then $\mathfrak{Nr}_\beta C \in \mathbf{CA}_\beta$. We will now prove a similar result for relation algebra reducts, using the work on substitutions of the preceding section.

THEOREM 5.44 (Henkin–Tarski, [HenMon$^+$85, 5.3.8]) *For any ordinal $\alpha \geq 4$, if $C \in \mathbf{CA}_\alpha$ then $\mathfrak{Ra}(C)$ is a relation algebra.*

Proof. Let $\alpha \geq 4$ and $C \in \mathbf{CA}_\alpha$. We check that $\mathfrak{Ra}C$ satisfies the relation algebra axioms of definition 3.8. Let $x, y, z \in \mathfrak{Nr}_2 C$. We frequently use that $x = c_2 c_3 x$, and similarly for $y, \check{x}, \check{y}, x; y$. We use the following standard notation:

$$_k \mathsf{s}(i,j) \stackrel{\text{def}}{=} \mathsf{s}_i^k \mathsf{s}_j^i \mathsf{s}_k^j, \text{ for distinct } i, j, k < \alpha.$$

By definition, we have $\check{x} = {}_2\mathsf{s}(0,1) x$.

R0. Clearly, $\mathrm{bool}(\mathfrak{Ra}C) \subseteq \mathrm{bool}(C)$, so the boolean axioms are satisfied in $\mathfrak{Ra}C$.

R2. $(x+y); z = c_2(\mathsf{s}_2^1(x+y) \cdot \mathsf{s}_2^0 z) = c_2(\mathsf{s}_2^1 x \cdot \mathsf{s}_2^0 z) + c_2(\mathsf{s}_2^1 y \cdot \mathsf{s}_2^0 z) = x; z + y; z$.

R3. $x; 1' = c_2(\mathsf{s}_2^1 x \cdot \mathsf{s}_2^0 \mathsf{d}_{01}) = c_2(\mathsf{s}_2^1 x \cdot \mathsf{d}_{12}) = \mathsf{s}_1^2 \mathsf{s}_2^1 x = \mathsf{s}_1^2 x = x$.

R4. Let w be the s-c-word ${}_2\mathsf{s}(0,1)c_2 c_3$. Note that $\check{x} = wx \in \mathfrak{Ra}C$. Calculation shows that $\widehat{ww} = \widehat{c_2 c_3} = \mathrm{Id}_{\alpha \setminus \{2,3\}}$ and that $(i,j) \in (ww)^*$ iff $i = j$ iff $(i,j) \in (c_2 c_3)^*$. Thus, $ww \sim c_2 c_3$, and $|\alpha \setminus \mathrm{rng}(\widehat{ww})| \geq 2$. By theorem 5.25, $ww \simeq c_2 c_3$, so $\check{\check{x}} = wwx = c_2 c_3 x = x$.

R5. Using fact 5.20(2), we have
$(x+y)\check{} = {}_2\mathsf{s}(0,1)(x+y) = {}_2\mathsf{s}(0,1)x + {}_2\mathsf{s}(0,1)y = \check{x} + \check{y}$.

R6. First, note that $\mathsf{s}_2^1 {}_2\mathsf{s}(0,1)c_2 c_3 \sim {}_3\mathsf{s}(0,1)\mathsf{s}_2^0 c_2 c_3$, and if v is either of these words, $|\alpha \setminus \mathrm{rng}(\widehat{v})| \geq 2$. So by theorem 5.25, we have $\mathsf{s}_2^1 {}_2\mathsf{s}(0,1)c_2 c_3 \simeq {}_3\mathsf{s}(0,1)\mathsf{s}_2^0 c_2 c_3$. In the same way, we obtain $\mathsf{s}_2^0 {}_2\mathsf{s}(0,1)c_2 c_3 \simeq {}_3\mathsf{s}(0,1)\mathsf{s}_2^1 c_2 c_3$ and ${}_2\mathsf{s}(0,1)c_2 c_3 \simeq {}_3\mathsf{s}(0,1)c_2 c_3$. Hence,

$$\begin{aligned}
\check{y}; \check{x} &= c_2\left(\mathsf{s}_2^1({}_2\mathsf{s}(0,1)y) \cdot \mathsf{s}_2^0({}_2\mathsf{s}(0,1)x)\right) & \text{by definition} \\
&= c_2\left({}_3\mathsf{s}(0,1)\mathsf{s}_2^0 y \cdot {}_3\mathsf{s}(0,1)\mathsf{s}_2^1 x\right) & \text{as } \mathsf{s}_2^i {}_2\mathsf{s}(0,1)c_2 c_3 \\
& & \simeq {}_3\mathsf{s}(0,1)\mathsf{s}_2^{1-i} c_2 c_3 \\
&= c_2 {}_3\mathsf{s}(0,1)(\mathsf{s}_2^1 x \cdot \mathsf{s}_2^0 y) & \text{by fact 5.20(2)} \\
&= {}_3\mathsf{s}(0,1)c_2(\mathsf{s}_2^1 x \cdot \mathsf{s}_2^0 y) & \text{by fact 5.20(6)} \\
&= {}_3\mathsf{s}(0,1)(x;y) & \text{by definition} \\
&= {}_2\mathsf{s}(0,1)(x;y) & \text{as } {}_3\mathsf{s}(0,1)c_2 c_3 \\
& & \simeq {}_2\mathsf{s}(0,1)c_2 c_3 \\
&= (x;y)\check{} & \text{by definition.}
\end{aligned}$$

R7. We prove that $\breve{x};(-(x;y))\cdot y = 0$. By theorem 5.25, we have $s_2^1 2s(0,1)c_2c_3 \simeq {}_3s(0,2)s_2^1 c_2 c_3$, and ${}_3s(0,2)s_2^0 c_2 c_3 \simeq c_2 c_3$. So

$$\begin{aligned} s_2^1 \breve{x} \cdot y &= s_2^1 2s(0,1) x \cdot y && \text{by definition} \\ &= {}_3s(0,2)s_2^1 x \cdot {}_3s(0,2)s_2^0 y && \text{by the above} \\ &= {}_3s(0,2)(s_2^1 x \cdot s_2^0 y) && \text{by fact 5.20(2)} \\ &\leq c_3 s_2^0 c_2 (s_2^1 x \cdot s_2^0 y) && \text{by definition of } s_0^3, s_3^2 \\ &= s_2^0(x;y) && \text{by fact 5.20(6)} \\ & && \text{and } c_3(x;y) = x;y. \end{aligned}$$

So by the cylindric algebra axioms C1, C3, and $c_2 y = y$, we obtain the required

$$\begin{aligned} \breve{x};(-(x;y)) \cdot y &= c_2(s_2^1 \breve{x} \cdot s_2^0 - (x;y)) \cdot y \\ &= c_2(s_2^1 \breve{x} \cdot -s_2^0(x;y) \cdot y) = c_2 0 = 0. \end{aligned}$$

R1. Finally, we prove associativity. As with conversion, we redefine composition equivalently using the third dimension. It is easily checked using fact 5.20 or theorem 5.25 that $c_3 s_3^2 c_3 \simeq c_3 c_2$, $s_3^2 s_2^1 c_2 c_3 \simeq s_3^1 c_2 c_3$, and $s_3^2 s_2^0 c_2 c_3 \simeq s_3^0 c_2 c_3$. So by fact 5.20(2,6), $x;y = c_3(x;y) = c_3 c_2(s_2^1 x \cdot s_2^0 y) = c_3 s_3^2 c_3(s_2^1 x \cdot s_2^0 y) = c_3(s_3^2 s_2^1 x \cdot s_3^2 s_2^0 y) = c_3(s_3^1 x \cdot s_3^0 y)$. (Cf. lemma 13.31 later.)

Now, we get $x;(y;z) = (x;y);z$, because

$x;(y;z) = c_2(s_2^1 x \cdot s_2^0 c_3(s_3^1 y \cdot s_3^0 z)) = c_2(c_3 s_2^1 x \cdot c_3 s_2^0(s_3^1 y \cdot s_3^0 z)) = c_2 c_3(c_3 s_2^1 x \cdot s_2^0(s_3^1 y \cdot s_3^0 z)) = c_2 c_3(s_2^1 x \cdot s_2^0 s_3^1 y \cdot s_3^0 z)$, and

$(x;y);z = c_3(s_3^1 c_2(s_2^1 x \cdot s_2^0 y) \cdot s_3^0 z) = c_3(c_2 s_3^1(s_2^1 x \cdot s_2^0 y) \cdot c_2 s_3^0 z) = c_3 c_2(s_3^1(s_2^1 x \cdot s_2^0 y) \cdot c_2 s_3^0 z) = c_3 c_2(s_2^1 x \cdot s_3^1 s_2^0 y \cdot s_3^0 z)$,

and by fact 5.20(10), $s_2^0 s_3^1 y = s_3^1 s_2^0 y$. $\square$

5.6.4 The classes $\mathbf{S\mathfrak{Nr}_\beta CA_\alpha}$ and $\mathbf{S\mathfrak{Ra}CA_n}$

DEFINITION 5.45 Let α be an ordinal and let $\mathsf{K} \subseteq \mathbf{CA}_\alpha$.

1. For $\beta \leq \alpha$, $\mathfrak{Nr}_\beta \mathsf{K}$ denotes the class $\{\mathfrak{Nr}_\beta \mathcal{C} : \mathcal{C} \in \mathsf{K}\}$, and $\mathbf{S}\mathfrak{Nr}_\beta \mathsf{K}$ the closure of $\mathfrak{Nr}_\beta \mathsf{K}$ under subalgebras, as usual. In particular, $\mathbf{S}\mathfrak{Nr}_\beta \mathbf{CA}_\alpha = \{\mathcal{B} : \mathcal{B} \subseteq \mathfrak{Nr}_\beta \mathcal{C} \text{ for some } \mathcal{C} \in \mathbf{CA}_\alpha\}$, the class of subalgebras of neat β-reducts of α-dimensional cylindric algebras.

2. For $\alpha \geq 3$, we define $\mathfrak{Ra}\mathsf{K}$ to be the class $\{\mathfrak{Ra}\mathcal{C} : \mathcal{C} \in \mathsf{K}\}$. The class $\mathbf{S}\mathfrak{Ra}\mathsf{K}$ is the closure of $\mathfrak{Ra}\mathsf{K}$ under subalgebras. In particular, $\mathbf{S}\mathfrak{Ra}\mathbf{CA}_\alpha = \{\mathcal{A} : \mathcal{A} \subseteq \mathfrak{Ra}\mathcal{C} \text{ for some } \mathcal{C} \in \mathbf{CA}_\alpha\}$, the class of subalgebras of relation algebra reducts of α-dimensional cylindric algebras.

5.6. Relation algebra reducts of cylindric algebras

Some authors use the notation $\mathfrak{Ra}^*\mathbf{CA}_\alpha$ instead, to denote that the range of the (proper class) map $\mathfrak{Ra}$ is intended (and similarly for neat reducts). We see no ambiguity in $\mathfrak{Ra}\mathsf{K}$ when, as will usually be the case, K is a proper class and so not an algebra.

The classes $\mathbf{S}\mathfrak{Ra}\mathbf{CA}_n$ for finite $n \geq 3$, and to a lesser extent the corresponding neat reduct classes, are a major object of study in chapters 13 and 15. For now, we list some results about them.

FACT 5.46

1. By theorem 5.44, $\mathbf{S}\mathfrak{Ra}\mathbf{CA}_4 \subseteq \mathbf{RA}$; in fact, $\mathbf{RA} = \mathbf{S}\mathfrak{Ra}\mathbf{CA}_4$.

    See [HenMon+85, 5.3.17] for a stronger result.

2. One might guess that $\mathbf{S}\mathfrak{Ra}\mathbf{CA}_3 = \mathbf{SA}$, or perhaps $\mathbf{WA}$. But the relation algebra axiom R4 ($\breve{\breve{x}} = x$), among others, fails in $\mathbf{S}\mathfrak{Ra}\mathbf{CA}_3$. Hence, $\mathbf{S}\mathfrak{Ra}\mathbf{CA}_3 \not\subseteq \mathbf{NA}$, so by exercise 1 below, $\mathbf{S}\mathfrak{Ra}\mathbf{CA}_3 \supset \mathbf{S}\mathfrak{Ra}\mathbf{CA}_4$. See [Sim97] for more information about $\mathbf{S}\mathfrak{Ra}\mathbf{CA}_3$; we quote some of the results in theorem 6.7 below.

    It has long been known that the condition $\alpha \geq 4$ in theorem 5.44 is necessary for $\mathfrak{Ra}C$ to have associative composition. But R1 is not the only relation algebra axiom that fails in $\mathbf{S}\mathfrak{Ra}\mathbf{CA}_3$. For example, in [Sim97, theorem 4.10], an example is given of a 3-dimensional cylindric algebra satisfying the merry-go-round axioms (see fact 5.38) but whose relation algebra reduct does not satisfy R7. (This example further shows that R6 $\not\vdash$ R7.) Monk and Fuhrken proved [Mon61b, theorem 9.10] that $\mathbf{RA} = \mathbf{S}\mathfrak{Ra}\mathbf{CA}_3 \cap \mathrm{Mod}\{\mathrm{R1}, \mathrm{R4}, \mathrm{R6}\}$. Németi and Simon [NémSim97, Sim97] improved this by showing that a subset of the $\mathbf{RA}$-axioms defines $\mathbf{RA}$ within $\mathbf{S}\mathfrak{Ra}\mathbf{CA}_3$ (or within $\mathfrak{Ra}\mathbf{CA}_3$) iff it includes $\{\mathrm{R1}, \mathrm{R6}\}$ or $\{\mathrm{R1}, \mathrm{R7}\}$, and that a subset of the $\mathbf{SA}$ axioms defines $\mathbf{SA}$ within $\mathbf{S}\mathfrak{Ra}\mathbf{CA}_3$ (or within $\mathfrak{Ra}\mathbf{CA}_3$) iff it contains R6 and R7. See also theorems 6.7 and 6.8 later.

    We use these results in theorem 13.49 to show that $\mathbf{SA} = \mathbf{S}\mathfrak{Ra}(\mathbf{CA}_3 \cap \mathbf{G}_3)$.

    Since it is easily seen that if $C \in \mathbf{G}_3$ then $\mathfrak{Ra}C \models \mathrm{R7}$, this establishes that at least for $\alpha = 3$, $\mathbf{CA}_\alpha \cap \mathbf{D}_\alpha \neq \mathbf{CA}_\alpha \cap \mathbf{G}_\alpha$. Cf. proposition 5.39.

3. If $1 < \alpha < \beta$ then $\mathfrak{Nr}_\alpha\mathbf{CA}_\beta$ is not closed under forming subalgebras. This was proved by Németi in [Ném83], solving problem 2.11 of [HenMon+71]. Furthermore, this class is not even closed under elementary subalgebras, and hence is not an elementary class [Say01, Theorem 1]. [SayNém01] shows that its elementary closure is strictly contained in $\mathbf{S}\mathfrak{Nr}_\alpha\mathbf{CA}_\beta$. This paper also extends Németi's result to Pinter's substitution algebras and quasi-polyadic algebras (see chapter 6 for some details of these), and shows that in contrast, the neat reducts of infinite-dimensional polyadic algebras form a variety.

192   Chapter 5. Relativisation and cylindric algebras

4. Analogous results are known for relation algebra reducts. Maddux and Németi independently proved that $\mathfrak{Ra}\mathbf{CA}_n \subset \mathbf{S}\mathfrak{Ra}\mathbf{CA}_n$ for $n \geq 4$ [Madd90a, Ném86]; Simon proved the same for $n = 3$ in [Sim97].

5. For ordinals $\alpha \geq \beta \geq 3$, $\mathbf{S}\mathfrak{Ra}\mathbf{CA}_\alpha \subseteq \mathbf{S}\mathfrak{Ra}\mathbf{CA}_\beta$. For finite $n \geq 4$, $\mathbf{S}\mathfrak{Ra}\mathbf{CA}_{n+1}$ is not finitely axiomatisable over $\mathbf{S}\mathfrak{Ra}\mathbf{CA}_n$. See exercise 1 below, and chapter 15.

6. $\bigcap_{3 \leq n < \omega} \mathbf{S}\mathfrak{Ra}\mathbf{CA}_n = \mathbf{RRA}$ (see proposition 13.48 and [HenMon$^+$85, 5.3.13, 5.3.16]). It follows by exercises 1–2 below that for any infinite α, we have $\mathbf{S}\mathfrak{Ra}\mathbf{CA}_\alpha = \mathbf{RRA}$. (Cf. [Mon61b, theorem 4.1] for a corresponding result, due to Henkin and Tarski, for cylindric algebras.) For this reason, we are mainly interested in $\mathbf{S}\mathfrak{Ra}\mathbf{CA}_n$ for finite n.

7. Analogous results for neat reducts hold, more or less. For any $m < \omega$, $\mathbf{CA}_m = \mathbf{S}\mathfrak{Nr}_m\mathbf{CA}_m \supset \mathbf{S}\mathfrak{Nr}_m\mathbf{CA}_{m+1} \supset \mathbf{S}\mathfrak{Nr}_m\mathbf{CA}_{m+2} \supset \cdots$. Andréka showed that $\mathbf{S}\mathfrak{Nr}_m\mathbf{CA}_{m+1}$ is finitely axiomatisable [And90a], but (see chapter 15) for finite $n > m$, $\mathbf{S}\mathfrak{Nr}_m\mathbf{CA}_{n+1}$ is not finitely axiomatisable over $\mathbf{S}\mathfrak{Nr}_m\mathbf{CA}_n$. We will see that the classes $\mathbf{S}\mathfrak{Nr}_m\mathbf{CA}_n$ are associated with n-variable proof theory and that these non-finite axiomatisability results have consequences in that field.

By the neat embedding theorem of Henkin–Tarski (see, e.g., [Mon61b, theorem 4.1]),

$$\mathbf{S}\mathfrak{Nr}_\beta\mathbf{CA}_{\beta+\alpha} = \bigcap_{n<\omega} \mathbf{S}\mathfrak{Nr}_\beta\mathbf{CA}_{\beta+n} = \mathbf{RCA}_\beta$$

for any infinite α.

PROBLEM 5.47 (Németi–Sayed Ahmed) *Is $\mathfrak{Ra}\mathbf{CA}_\alpha$ elementary for $\alpha \geq 3$?*

We now show that for all ordinals $\alpha \geq 3$, $\mathbf{S}\mathfrak{Ra}\mathbf{CA}_\alpha$ is a canonical variety. The analogue of this for $\mathbf{S}\mathfrak{Nr}_\beta\mathbf{CA}_\alpha$ was proved by Monk [Mon61a]; a proof is also given in [HenMon$^+$71, 2.6.32(ii)] and we modified it for the proof here.

PROPOSITION 5.48 *For $\alpha \geq 3$, $\mathbf{S}\mathfrak{Ra}\mathbf{CA}_\alpha$ is a canonical variety.*

Proof. We show that $\mathbf{HSP}\mathfrak{Ra}\mathbf{CA}_\alpha \subseteq \mathbf{S}\mathfrak{Ra}\mathbf{CA}_\alpha$. Evidently, if $\mathcal{B}_i \in \mathbf{CA}_\alpha$, $i \in I$, then $\prod_{i \in I} \mathfrak{Ra}\mathcal{B}_i = \mathfrak{Ra}\prod_{i \in I} \mathcal{B}_i \in \mathfrak{Ra}\mathbf{CA}_\alpha$. From this we see that $\mathbf{P}\mathfrak{Ra}\mathbf{CA}_\alpha \subseteq \mathfrak{Ra}\mathbf{CA}_\alpha$, and hence, $\mathbf{SP}\mathfrak{Ra}\mathbf{CA}_\alpha \subseteq \mathbf{S}\mathfrak{Ra}\mathbf{CA}_\alpha$. So it suffices to check that $\mathbf{S}\mathfrak{Ra}\mathbf{CA}_\alpha$ is closed under homomorphic images. By the results of section 2.5.2, we can work with ideals instead of homomorphisms. Let $\mathcal{A} \subseteq \mathfrak{Ra}\mathcal{B}$ for $\mathcal{B} \in \mathbf{CA}_\alpha$, and let I be an ideal of $\mathcal{A}$ (definition 2.36). Plainly, I is a subset of $\mathcal{B}$. Let J be the ideal of $\mathcal{B}$ generated by I (i.e., the intersection of all ideals of $\mathcal{B}$ containing I). By [HenMon$^+$71,

5.6. Relation algebra reducts of cylindric algebras

theorem 2.3.8], $J = \{b \in \mathcal{B} : b \leq c_{i_0} \ldots c_{i_l}(x_0 + \cdots + x_{k-1})$ for some $i_0, \ldots, i_l < \alpha$ and $x_0, \ldots, x_{k-1} \in I\}$. Now since I is an ideal of $\mathcal{A}$, it is closed under $+$, and if $x \in I$ then $1; x \in I$ and $x; 1 \in I$. By fact 5.20(2, 7, 5) and the fact that $c_2 x = x$,

$$1; x = c_2(s_2^1 1 \cdot s_2^0 x) = c_2(1 \cdot s_2^0 c_2 x)$$
$$= c_2 s_2^0 c_2 x = c_0 s_0^2 c_2 x = c_0 c_2 x = c_0 x.$$

So $c_0 x \in I$, and similarly, $c_1 x \in I$. So the above expression simplifies to $J = \{b \in \mathcal{B} : b \leq x$ for some $x \in I\}$. It follows that $J \cap \mathcal{A} = I$.

Now define a homomorphism from $\mathcal{A}/I$ into $\mathfrak{Ra}(\mathcal{B}/J)$ by $a/I \mapsto a/J$ (for $a \in \mathcal{A}$). As $J \cap \mathcal{A} = I$, this map is one-one. Since $\mathcal{B}/J$ is a homomorphic image of $\mathcal{B}$, we have $\mathcal{B}/J \in \mathbf{CA}_\alpha$ and $\mathcal{A}/I \in \mathbf{S\mathfrak{Ra}CA}_\alpha$.

So $\mathbf{S\mathfrak{Ra}CA}_\alpha$ is closed under **H**, **S**, and **P**. By Birkhoff's theorem (2.45), this shows that $\mathbf{S\mathfrak{Ra}CA}_\alpha$ is a variety and can be equationally axiomatised.

If $\mathcal{A} \in \mathbf{S\mathfrak{Ra}CA}_\alpha$, let $C \in \mathbf{CA}_\alpha$ with $\mathcal{A} \subseteq \mathfrak{Ra}C$. By theorem 5.43, $\mathcal{A}^+ \subseteq \mathfrak{Ra}C^+$. As $\mathbf{CA}_\alpha$ is a Sahlqvist variety, it is canonical, so $C^+ \in \mathbf{CA}_\alpha$ too. Hence $\mathcal{A}^+ \in \mathbf{S\mathfrak{Ra}CA}_\alpha$ and $\mathbf{S\mathfrak{Ra}CA}_\alpha$ is canonical. □

Exercises

1. Show that if $3 \leq \beta \leq \alpha$ then $\mathbf{S\mathfrak{Ra}CA}_\alpha \subseteq \mathbf{S\mathfrak{Ra}CA}_\beta$.

2. Show that $\mathbf{RRA} = \mathbf{S\mathfrak{Ra}RCA}_\alpha \subseteq \mathbf{S\mathfrak{Ra}CA}_\alpha$ for all ordinals $\alpha \geq 3$.

3. For a class K of BAOs, let $S^d K$ denote the closure under isomorphism of the class of dense subalgebras of algebras in K. Show that
   (a) $\mathbf{S\mathfrak{Ra}CA}_\omega = \bigcap_{3 \leq n < \omega} \mathbf{S\mathfrak{Ra}CA}_n$,
   (b) $S^d \mathfrak{Ra} \mathbf{CA}_\omega = \bigcap_{3 \leq n < \omega} S^d \mathfrak{Ra} \mathbf{CA}_n$.

   [If $\mathcal{A} \subseteq \mathfrak{Ra}C_n$ for all finite $n \geq 3$, where $C_n \in \mathbf{CA}_n$, embed $\mathcal{A}$ in an ultraproduct of arbitrary expansions of the C_n to the signature of $\mathbf{CA}_\omega$. For (3b), consider the subalgebra of the ultraproduct generated by the image of $\mathcal{A}$ under this embedding.]

4. Check the congruences $_2 s(0,1) c_2 c_3 \simeq c_2 c_3$, etc. in the proof of theorem 5.44.

5. Prove theorem 5.44 using only fact 5.20. (See [HenMon$^+$85, 5.3.8–9] for a solution.)

6. For finite n, show that if $C \in \mathbf{CA}_n$ is atomic then the set of atoms of $\mathfrak{Nr}_m C$ is $\{c_m c_{m+1} \ldots c_{n-1} x : x$ an atom of $C\}$. (By lemma 5.42, $\mathfrak{Nr}_m C$ is atomic.)

7. Show that for $\beta \leq \alpha$, $\mathbf{S\mathfrak{Nr}_\beta CA}_\alpha$ is a variety.

5.7 Relation algebra reducts of other cylindric-type algebras

For $\mathcal{D} \in \mathbf{D}_\alpha$ and $\beta \leq \alpha$, we might try to define $\mathfrak{Nr}_\beta \mathcal{D}$ to be a $\mathbf{CA}_\beta$-type algebra with domain $\{x \in \mathcal{D} : c_i x = x \text{ for } \beta \leq i < \alpha\}$, and operations induced from $\mathcal{D}$ as before. However, this may not be closed under the c_i for $i < \beta$. Similarly, the problem we meet in trying to define $\mathfrak{Ra}C$ for $\mathcal{D} \in \mathbf{D}_\alpha$ or $\mathcal{D} \in \mathbf{G}_\alpha$ is that $\{x \in \mathcal{D} : c_i x = x \text{ for } i \geq 2\}$ may not be closed under composition as defined by $x;y = c_2(s_2^1 x \cdot s_2^0 y)$. There are alternative definitions known (see the exercises), but they do not necessarily yield associative composition. So we abuse notation somewhat, by the following definition.

DEFINITION 5.49 Let $\alpha \geq 3$.

1. Let $\mathcal{A}$ be a relation-type algebra, and let $\mathcal{D} \in \mathbf{D}_\alpha$. A *relation algebra reduct embedding* from $\mathcal{A}$ into $\mathcal{D}$ is a boolean algebra embedding $\iota : \text{bool}(\mathcal{A}) \to \text{bool}(\mathcal{D})$ satisfying, for all $r, s \in \mathcal{A}$:

    (a) $c_i \iota(r) = \iota(r)$, for all i with $2 \leq i < \alpha$.
    (b) $\iota(1') = d_{01}$.
    (c) $\iota(\breve{r}) = s_0^2 s_1^0 s_2^1 \iota(r)$.
    (d) $\iota(r;s) = c_2(s_2^1 \iota(r) \cdot s_2^0 \iota(s))$.

2. We write $\mathbf{S}\mathfrak{Ra}\mathbf{D}_\alpha$ for the class of all relation-type algebras $\mathcal{A}$ such that there is a relation algebra reduct embedding from $\mathcal{A}$ into some $\mathcal{D} \in \mathbf{D}_\alpha$.

3. We define $\mathbf{S}\mathfrak{Ra}\mathbf{G}_\alpha$ similarly.

REMARK 5.50

1. Of the many notation abuses in the book, $\mathbf{S}\mathfrak{Ra}\mathbf{D}_\alpha$ and $\mathbf{S}\mathfrak{Ra}\mathbf{G}_\alpha$ are probably the worst. They behave similarly to $\mathbf{S}\mathfrak{Ra}\mathbf{CA}_\alpha$ and are nice to use, once used to them; and they are very important classes. But the reader should always be aware that they do *not* denote the result of successively applying $\mathfrak{Ra}$ and $\mathbf{S}$ to algebras in $\mathbf{D}_\alpha, \mathbf{G}_\alpha$, and indeed that for C in $\mathbf{D}_\alpha$ or $\mathbf{G}_\alpha$, $\mathfrak{Ra}C$ is in general undefined. To limit confusion, we will not go on to write $\mathbf{S}\mathfrak{Ra}\mathsf{K}$ for subclasses K of $\mathbf{D}_\alpha, \mathbf{G}_\alpha$ — except of course when $\mathsf{K} \subseteq \mathbf{CA}_\alpha$, in which case definition 5.45 applies.

2. Of course, $\mathbf{S}\mathfrak{Ra}\mathbf{D}_\alpha$ and $\mathbf{S}\mathfrak{Ra}\mathbf{G}_\alpha$ are closed under taking subalgebras. So although we have not defined the classes $\mathfrak{Ra}\mathbf{D}_\alpha, \mathfrak{Ra}\mathbf{D}_\alpha$, in a way the definition behaves as though we had.

5.7. Relation algebra reducts of other cylindric-type algebras

3. Unlike for cylindric algebras, we have not been able to define $\mathfrak{Ra}C$ for an algebra $C \in \mathbf{D}_\alpha$ (or $\mathbf{G}_\alpha$). (The nearest we might get is to define the class $\mathbf{S\mathfrak{Ra}}\{C\}$.) Because of this, we do not find definition 5.49 entirely satisfactory, and we set a problem (5.55 below) to improve it.

4. Suppose that we were to extend definition 5.49 in the natural way to subclasses K of $\mathbf{D}_\alpha, \mathbf{G}_\alpha$. Suppose we have such a K, and $K \subseteq \mathbf{CA}_\alpha$ too. If $\mathcal{A} \in \mathbf{S\mathfrak{Ra}}K$, then there would be some $\mathcal{D} \in K$ and a relation algebra reduct embedding $\iota: \mathcal{A} \to \mathcal{D}$. Then as $\mathcal{D} \in \mathbf{CA}_\alpha$, $\mathfrak{Ra}\mathcal{D}$ is defined according to definition 5.40, $\mathrm{rng}(\iota) \subseteq \mathfrak{Ra}\mathcal{D}$, and (if $\alpha \geq 4$) $\iota: \mathcal{A} \to \mathfrak{Ra}\mathcal{D}$ is a relation algebra embedding; so $\mathcal{A} \in \mathbf{S\mathfrak{Ra}}K$ by definition 5.45. Conversely, if $\mathcal{A} \in \mathbf{S\mathfrak{Ra}}K$ as in definition 5.45, then there is $C \in K$ with $\mathcal{A} \subseteq \mathfrak{Ra}C$. Clearly, the inclusion map from $\mathcal{A}$ to C is a relation algebra reduct embedding, so that $\mathcal{A} \in \mathbf{S\mathfrak{Ra}}K$ via the extended definition 5.49.

The conclusion is that there is no potential clash of notation between definition 5.45 and the natural extension of definition 5.49 to subclasses of $\mathbf{D}_\alpha \cap \mathbf{CA}_\alpha$.

THEOREM 5.51 *Let $\alpha \geq 3$, and let $\mathcal{A}$ be a non-associative algebra. If $\iota: \mathcal{A} \to \mathcal{B}$ is a relation algebra reduct embedding, for some $\mathcal{B} \in \mathbf{D}_\alpha$, then $\iota^+: \mathcal{A}^+ \to \mathcal{B}^+$ (as defined in theorem 2.71) is a complete relation algebra reduct embedding.*

Proof. The proof is similar to that of theorem 5.43; we leave it as an exercise. □

We will see in chapter 13 that for finite $\alpha \geq 4$, these classes $\mathbf{S\mathfrak{Ra}D}_\alpha$ and $\mathbf{S\mathfrak{Ra}G}_\alpha$ are well behaved and indeed well known by another name: the canonical varieties $\mathbf{RA}_\alpha$, as defined in [Madd83]. In particular, $\mathbf{S\mathfrak{Ra}D}_4 = \mathbf{S\mathfrak{Ra}G}_4 = \mathbf{RA}$. Also, $\mathbf{S\mathfrak{Ra}G}_3 = \mathbf{WA}$. (We will show in theorem 13.49 that $\mathbf{S\mathfrak{Ra}}(\mathbf{CA}_3 \cap \mathbf{G}_3) = \mathbf{SA}$.) Here we begin by proving the analogue of theorem 5.44.

NOTATION 5.52 *For an α-tuple $\bar{x}$ of elements of some set, and $i, j < \alpha$, we write x_i for the ith entry of $\bar{x}$, and $\bar{x}_j^i$ for the α-tuple $(x_0, \ldots, x_{i-1}, x_j, x_{i+1}, \ldots)$. For example, $\bar{x}_{012}^{201} = ((\bar{x}_0^2)_1^0)_2^1 = (x_1, x_0, x_0, x_3, \ldots)$.*

LEMMA 5.53 *Let $\alpha \geq 4$. Then $\mathbf{S\mathfrak{Ra}D}_\alpha \subseteq \mathbf{RA}$.*

Proof. Let $\mathcal{D} \in \mathbf{D}_\alpha$. We let $\mathcal{D}$ be a cylindric relativised set algebra of α-ary relations on the set M. Let $\mathcal{A}$ be a relation-type algebra that embeds in $\mathcal{D}$ by a relation algebra reduct embedding. We may assume without loss of generality that $\mathrm{dom}(\mathcal{A}) \subseteq \mathrm{dom}(\mathcal{D})$.

We check that $\mathcal{A} \in \mathbf{RA}$ by running through the relation algebra axioms. Let $a, b, c \in \mathcal{A}$, and let $\bar{x}$ be an arbitrary element of the unit 1 of $\mathcal{D}$. Note that for $i, j < \alpha$, $\bar{x} \in \mathsf{s}_j^i a$ iff $\bar{x}_j^i \in a$ (because $\bar{x} \in 1$); see notation 5.52.

196   *Chapter 5. Relativisation and cylindric algebras*

R0. As $\mathrm{bool}(\mathcal{A}) \subseteq \mathrm{bool}(\mathcal{D})$, $\mathrm{bool}(\mathcal{A})$ is a boolean algebra.

R2. $\bar{x} \in (a+b); c = c_2(s_2^1(a+b) \cdot s_2^0 c)$ iff $\bar{y} \in s_2^1(a+b) \cdot s_2^0 c$ for some $\bar{y} \in 1$, $\bar{y} \equiv_2 \bar{x}$, iff $\bar{y}_2^1 \in a+b$ and $\bar{y}_2^0 \in c$ for some $\bar{y} \equiv_2 \bar{x}$ with $\bar{y} \in 1$, iff ($\bar{y}_2^1 \in a$ or $\bar{y}_2^1 \in b$) and $\bar{y}_2^0 \in c$, for some $\bar{y} \equiv_2 \bar{x}$ with $\bar{y} \in 1$, iff $\bar{x} \in a; c + b; c$. Hence, $(a+b); c = a; c + b; c$.

R3. $\bar{x} \in a; 1' = c_2(s_2^1 a \cdot s_2^0 d_{01})$ iff there is $\bar{y} \equiv_2 \bar{x}$ with $\bar{y} \in 1$, $\bar{y}_2^1 \in a$, and $(\bar{y}_2^0)_0 = (\bar{y}_2^0)_1$. Such a $\bar{y}$ must satisfy $y_2 = x_1$ — i.e., we must have $\bar{y} = \bar{x}_1^2$. Since indeed $\bar{x}_1^2 \in 1$, we see that $\bar{x} \in a; 1'$ iff $\bar{y}_2^1 = (\bar{x}_1^2)_2^1 = \bar{x}_1^2 \in a$, iff $\bar{x} \in a$ as $a = c_2 a$. So $a; 1' = a$.

R4. $\bar{x} \in \breve{\breve{a}} = s_0^2 s_1^0 s_2^2 s_0^0 s_1^2 s_2^1 a$ iff $\bar{x}_{012012}^{201201} \in a$. Evaluating, this is iff $\bar{x}_1^2 \in a$, iff $\bar{x} \in a$ as a is 2-dimensional. We conclude that $\breve{\breve{a}} = a$.

R5. $\bar{x} \in (a+b)^{\breve{\ }} = s_0^2 s_1^0 s_2^1 (a+b)$ iff $\bar{x}_{012}^{201} \in a+b$, iff $\bar{x}_{012}^{201} \in a$ or $\bar{x}_{012}^{201} \in b$, iff $\bar{x} \in \breve{a} + \breve{b}$.

R6. Using that $\alpha \geq 4$ and $(a;b)^{\breve{\ }}$ is 2-dimensional, we have $\bar{x} \in (a;b)^{\breve{\ }}$ iff $\bar{x}_0^3 \in (a;b)^{\breve{\ }} = s_0^2 s_1^0 s_2^1 c_2(s_2^1 a \cdot s_2^0 b)$, iff for some $\bar{y} \in 1$ we have $\bar{x}_{0012}^{3201} \equiv_2 \bar{y}$, $\bar{y}_2^1 \in a$, and $\bar{y}_2^0 \in b$.

Since $\bar{y} = (x_1, x_0, y_2, x_0, x_4, \ldots)$, this is iff

(∗) for some $m \in M$ we have $\bar{y} = (x_1, x_0, m, x_0, x_4, \ldots) \in 1$, $(x_1, m, m, x_0, x_4, \ldots) \in a$, and $(m, x_0, m, x_0, x_4, \ldots) \in b$.

Similarly, as $\breve{b}; \breve{a}$ is 2-dimensional, $\bar{x} \in \breve{b}; \breve{a}$ iff $\bar{x}_0^3 \in \breve{b}; \breve{a}$, iff

(∗∗) for some $n \in M$ we have $(x_0, x_1, n, x_0, x_4, \ldots) \in 1$, $(n, x_0, x_0, x_0, x_4, \ldots) \in b$, and $(x_1, n, n, x_0, x_4, \ldots) \in a$.

Since 1 is closed under substitutions and b is 2-dimensional, we have

$(m, x_0, m, x_0, x_4, \ldots) \in b \quad \Rightarrow \quad (m, x_0, x_0, x_0, x_4, \ldots) \in b,$
$(n, x_0, x_0, x_0, x_4, \ldots) \in b \quad \Rightarrow \quad (n, x_0, n, x_0, x_4, \ldots) \in b.$

Now it is clear that (∗∗) ⇔ (∗), as we can let $m = n$.

R7. We prove that $\breve{a}; (-(a;b)) \cdot b = 0$. Assume for contradiction that there is some $\bar{x} \in \breve{a}; (-(a;b)) \cdot b$. By 2-dimensionality, we may suppose that $x_0 = x_3$. Then $(x_0, x_1, x_2, x_3, \ldots) \in b$, and for some $m \in M$, we have $\bar{y} = (x_0, x_1, m, x_3, \ldots) \in 1$, $(m, x_0, x_0, x_3, \ldots) \in a$, and $(m, x_1, b, x_3, \ldots) \in -(a;b)$. So there is no $n \in M$ with $(m, x_1, n, x_3, \ldots) \in 1$, $(m, n, n, x_3, \ldots) \in a$, $(n, x_1, n, x_3, \ldots) \in b$.

5.7. Relation algebra reducts of other cylindric-type algebras

Now take $n = x_0 = x_3$. By the above, we have $(m,x_1,n,x_3,\ldots) = \bar{y}_{23}^{02} \in 1$, $(m,n,n,x_3,\ldots) = (m,x_0,x_0,x_3,\ldots) \in a$, and $(n,x_1,x_2,x_3,\ldots) = \bar{x} \in b$. As b is 2-dimensional, we obtain $(n,x_1,n,x_3,\ldots) \in b$, a contradiction.

R1. It remains to prove associativity. Let $\bar{x} \in (a;b);c$; we show $\bar{x} \in a;(b;c)$. By 2-dimensionality of $(a;b);c$ and $a;(b;c)$, we can assume that $x_3 = x_1$. There are $m,n \in M$ with both $\bar{y} = (x_0,x_1,m,x_3,\ldots)$ and $\bar{z} = (x_0,m,n,x_3,\ldots)$ in the unit 1, and $(x_0,n,n,x_3,\ldots) \in a$, $(n,m,n,x_3,\ldots) \in b$, $(m,x_1,m,x_3,\ldots) \in c$.

Now $\bar{x} \equiv_2 \bar{z}' \stackrel{\text{def}}{=} (x_0,x_1,n,x_3,\ldots) = \bar{z}_3^1 \in 1$ and $(\bar{z}')_2^1 = (x_0,n,n,x_3,\ldots) \in a$. But also, $(\bar{z}')_2^0 = (n,x_1,n,x_3,\ldots) \equiv_2 (n,x_1,m,x_3,\ldots) = \bar{z}_{213}^{021} \in 1$. Plainly, $(n,m,m,x_3,\ldots) \in b$ and $(m,x_1,m,x_3,\ldots) \in c$, so we have $(\bar{z}')_2^0 \in b;c$. Thus, $\bar{z}' \in \mathsf{s}_2^1 a \cdot \mathsf{s}_2^0 (b;c)$, and $\bar{x} \in a;(b;c)$, as required. The converse is similar. □

In chapter 13 we will see that $\mathbf{RA} = \mathbf{S\Re aD}_4 = \mathbf{S\Re aG}_4$ and $\mathbf{SA} = \mathbf{S\Re a}(\mathbf{CA}_3 \cap \mathbf{G}_3)$.

We can prove similarly that $\mathbf{S\Re aG}_3 \subseteq \mathbf{WA}$. Exercise 4 below shows that $\mathbf{WA} \subseteq \mathbf{S\Re aG}_3$, so $\mathbf{WA} = \mathbf{S\Re aG}_3$ in fact.

LEMMA 5.54 $\mathbf{S\Re aG}_3 = \mathbf{WA}$.

Proof. For '$\subseteq$', let $\mathcal{G} \in \mathbf{G}_3$ be a cylindric relativised set algebra of 3-ary relations on the set M, say, and let $\mathcal{A}$ be a relation-type algebra that embeds in $\mathcal{G}$ by a relation algebra reduct embedding. As in lemma 5.53, we assume that $\text{dom}(\mathcal{A}) \subseteq \text{dom}(\mathcal{G})$, and check that $\mathcal{A} \in \mathbf{WA}$. As before, $\text{bool}(\mathcal{A}) \subseteq \text{bool}(\mathcal{G})$ so $\text{bool}(\mathcal{A})$ is a boolean algebra. R2–R5 are proved as in the lemma, since only dimensions 0, 1, 2 were used there. To prove R6, R7, and WL, let $a,b \in \mathcal{A}$, and let $\bar{x} = (x_0,x_1,x_2)$ be an arbitrary element of the unit 1 of $\mathcal{G}$.

R6. We have $\bar{x} \in (a;b)^\smile$ iff $(x_1,x_0,x_0) \in a;b$, iff there is $m \in M$ with $(x_1,x_0,m) \in 1$, $(x_1,m,m) \in a$, and $(m,x_0,m) \in b$. Since 1 is closed under substitutions and permutations and b is 2-dimensional, this is iff there is $m \in M$ such that $(x_0,x_1,m) \in 1$, $(x_1,m,m) \in a$, and $(m,x_0,x_0) \in b$. Equivalently, there is $m \in M$ with $(x_0,x_1,m) \in 1$, $(x_0,m,m) \in \breve{b}$, and $(m,x_1,m) \in \breve{a}$. This is equivalent to $\bar{x} \in \breve{b};\breve{a}$, as required.

R7. Assume for contradiction that $\bar{x} \in \breve{a};(-(a;b)) \cdot b$. So there is $m \in M$ with $(x_0,x_1,m) \in 1$, $(x_0,m,m) \in \breve{a}$, and $(m,x_1,m) \notin a;b$. So $(m,x_0,x_0) \in a$. But also, $(x_0,x_1,x_2) \in b$, b is 2-dimensional, and 1 is closed under substitutions, so $(x_0,x_1,x_0) \in b$. The closure of 1 under permutations yields $(m,x_1,x_0) \in 1$. But now, $(m,x_1,x_0) \in 1$, $(m,x_0,x_0) \in a$, and $(x_0,x_1,x_0) \in b$ witness that $(m,x_1,m) \in a;b$. This is a contradiction, so $\breve{a};(-(a;b)) \cdot b = 0$ as required.

WL. It remains to check weak associativity, $((a \cdot 1');1);1 = (a \cdot 1');1$. Certainly, $(a \cdot 1');1 = ((a \cdot 1');1);1' \subseteq ((a \cdot 1');1);1$. For the converse, first note that because 1 is closed under substitutions, for any $b \in \mathcal{A}$ and $\bar{y} = (y_0, y_1, y_2) \in 1$ we have $\bar{y} \in b; 1$ iff there is $m \in M$ with $(y_0, y_1, m) \in 1$ and $(y_0, m, m) \in b$. Now assume that $\bar{x} \in ((a \cdot 1'); 1); 1$. So there is $m \in M$ with $(x_0, x_1, m) \in 1$ and $(x_0, m, m) \in (a \cdot 1'); 1$. Hence, there is $n \in M$ with $(x_0, m, n) \in 1$ and $(x_0, n, n) \in a \cdot 1'$. Since $1' = d_{01}$, $n = x_0$, so $(x_0, x_0, x_0) \in a \cdot 1'$. Certainly, $(x_0, x_1, x_0) \in 1$. Hence, $\bar{x} = (x_0, x_1, x_2) \in (a \cdot 1'); 1$.

The converse inclusion is left to exercise 4. $\square$

PROBLEM 5.55 Is there a good definition of $\mathfrak{Ra}\mathcal{D}$ for $\mathcal{D} \in \mathbf{D}_\alpha$? Composition should be defined by $x; y = c_2(s_2^1 x \cdot s_2^0 y)$, as usual, so it would be more helpful to find a 'nice' $\mathsf{K} \subseteq \mathbf{D}_\alpha$ such that $\mathfrak{Nr}_2 \mathcal{D}$ is closed under the relation algebra operations, for $\mathcal{D} \in \mathsf{K}$. This may be difficult, since for finite $n \geq 4$, the equational theories of $\mathbf{D}_n, \mathbf{G}_n$ are decidable (fact 5.38), while the equational theories of $\mathbf{S}\mathfrak{Ra}\mathbf{D}_n$ and $\mathbf{S}\mathfrak{Ra}\mathbf{G}_n$ (as in definition 5.49) are not (theorem 18.28 and corollary 13.47).

PROBLEM 5.56 How are $\mathbf{S}\mathfrak{Ra}\mathbf{D}_3$ and $\mathbf{S}\mathfrak{Ra}(\mathbf{CA}_3 \cap \mathbf{D}_3)$ related to **NA**, **WA**, **SA**, and **RA**?

Exercises

1. Suppose that $\mathcal{A} \in \mathbf{RA}$, $C \in \mathbf{CA}_n$ for some $n \geq 4$, and $\mathrm{dom}(\mathcal{A}) \subseteq \mathrm{dom}(C)$. Show that $\mathcal{A} \subseteq \mathfrak{Ra}C$ iff the inclusion map $\iota : \mathcal{A} \to C$ is a relation algebra reduct embedding. Show further that $\mathcal{A} \subseteq^c \mathfrak{Ra}C$ iff ι is a complete relation algebra reduct embedding.

2. Prove theorem 5.51.

3. [Andréka] For finite $n \geq 4$, let ζx abbreviate $s_1^2 s_1^3 \ldots s_1^{n-1} x$. Let $\mathcal{D} \in \mathbf{D}_n$. Show that $\zeta \zeta x = \zeta x$ for all $x \in \mathcal{D}$.
   Writing $\zeta(\mathcal{D})$ for $\{\zeta x : x \in \mathcal{D}\}$, define $1' = \zeta d_{01}$, $\check{z} = s_0^2 s_1^0 s_2^1 z$ and $z; w = \zeta c_2(s_2^1 z \cdot s_2^0 w)$, for $z, w \in \zeta(\mathcal{D})$. Show that $\zeta(\mathcal{D})$ is closed under these and the boolean operations, so that $\mathcal{A} = (\zeta(\mathcal{D}), +, -, 0, 1, 1', \check{}, ;)$ is a relation-type algebra. Show that $\mathcal{A}$ is a non-associative algebra. Is it in **WA**? **SA**? **RA**?

4. Assuming that every weakly associative algebra has a relativised representation (it does — see [Madd82] or theorem 7.5), prove that $\mathbf{WA} \subseteq \mathbf{S}\mathfrak{Ra}\mathbf{G}_3$. [Given a relativised representation M of $\mathcal{A} \in \mathbf{WA}$, consider the algebra $\mathcal{G} \in \mathbf{G}_3$ with unit $1^\mathcal{G} = \{(x_0, x_1, x_2) \in {}^3M : M \models \bigwedge_{i,j<3} 1(x_i, x_j)\}$ and domain $\wp(1^\mathcal{G})$, and the map $\iota : \mathcal{A} \to \mathcal{G}$ given by $\iota(a) = \{(x_0, x_1, x_2) \in 1^\mathcal{G} : M \models a(x_0, x_1)\}$, for $a \in \mathcal{A}$. See also exercise 13.2(1).]

5. Show that $\mathbf{S}\mathfrak{Ra}(\mathbf{CA}_3 \cap \mathbf{G}_3) \subseteq \mathbf{SA}$. (In fact we have equality: see theorem 13.49.)

Chapter 6

Other approaches to algebras of relations

We end this introductory part of the book by briefly mentioning some important related approaches to algebraic logic. They will not be heavily used later, so we will confine ourselves to a short survey, referring the reader to the literature for more details. We are very grateful indeed to Hajnal Andréka, Steven Givant, Ágnes Kurucz, Roger Maddux, Istvan Németi, and Tarek Sayed Ahmed for providing much of the information; responsibility for errors is of course solely our own.

6.1 Diagonal-free algebras

We mentioned when introducing cylindric algebras in section 5.3 that there is an option to exclude equality (the diagonals d_{ij}). Doing this leads us to *diagonal-free algebras,* which we briefly discuss now.

DEFINITION 6.1 Let α be an ordinal.

1. A *diagonal-free set algebra* of dimension α is an algebra consisting of a set S of α-ary relations on some domain U, equipped with the operations $0, 1 \, (= {}^{\alpha}U), \cup, \setminus$ (complement in ${}^{\alpha}U$), and the cylindrifications C_i ($i < \alpha$). These operations are defined as for cylindric algebras (see definition 5.16). S must of course be closed under all these operations.

    The signature of α-dimensional diagonal-free set algebras consists of the boolean symbols 0, 1, $+$, $-$, and unary functions c_i ($i < \alpha$).

2. A *generalised diagonal-free set algebra of dimension* α is a subdirect product of diagonal-free set algebras of dimension α.

3. A *diagonal-free algebra* of dimension α is an algebra of signature $\{+, -, 1, 0, c_i : i < \alpha\}$ satisfying axioms C0 to C4 of definition 5.16: i.e., all the cylindric algebra axioms except those which involve diagonals. **Df**$_\alpha$ denotes the class of all α-dimensional diagonal-free algebras.

4. An α-dimensional diagonal-free algebra is said to be *representable* if it is isomorphic to a generalised diagonal-free set algebra of dimension α. **Rdf**$_\alpha$ denotes the class of all representable diagonal-free algebras of dimension α.

See [HenMon⁺85, §5.1] for information on these algebras. We mention the following important results.

THEOREM 6.2 (Johnson) *For $\alpha \geq 3$, **Rdf**$_\alpha$ is not finitely axiomatisable, nor axiomatisable by a finite schema of equations.*

See [Joh69] and [HenMon⁺85, 5.1.63]. For information on finite schemas see [HenMon⁺85, p. 260ff].

THEOREM 6.3 *For $3 \leq \alpha \leq \omega$, the equational theory of any class K with **Rdf**$_\alpha \subseteq$ K $\subseteq$ **Df**$_\alpha$ is undecidable.*

See [Madd80] and [HenMon⁺85, 5.1.66]. Stebletsova applied this in [Ste00, theorem 5.1.1] to prove the undecidability of the equational theory of Lyndon algebras over any class of projective geometries containing an infinite geometry of dimension at least three.

In [Ven98], Venema proved that any rectangularly dense diagonal-free algebra is representable.

THEOREM 6.4 (Halmos, Johnson) *Let $3 \leq \alpha < \omega$ and let $C \in \mathbf{CA}_\alpha$ be generated by its $< \alpha$-dimensional elements (i.e., by the set $\{c \in C : c_i c = c$ for some $i < \alpha\}$). Then $C \in \mathbf{RCA}_\alpha$ iff the diagonal-free reduct of C is in **Rdf**$_\alpha$.*

See [Hal57, Joh69], and [HenMon⁺85, 5.1.51]. A variant of this for reducts without $0, 1, +$ was proved by Mikulás:

THEOREM 6.5 (Mikulás) *Let $3 \leq \alpha < \omega$ and $\mathcal{A} \in \mathbf{CA}_\alpha$. Let $\mathcal{B}$ be a subalgebra of the $\{\cdot, c_i, d_{ij} : i, j < \alpha\}$-reduct of $\mathcal{A}$, and assume that $\mathcal{B}$ is generated by $\{b \in \mathcal{B} : c_i b = b$ for some $i < \alpha\}$. Let C be the diagonal-free reduct of $\mathcal{B}$ and suppose that C is representable as a set algebra. Then $\mathcal{B}$ is representable as well.*

See [HodMik00, theorem 4.9]. The proof shows that if C has a finite representation then so does $\mathcal{B}$, and similarly for theorem 6.4. These results have applications in many-dimensional modal logic: see, e.g., [HirHod⁺02b], where they were used with results of chapter 18 to show that any α-dimensional modal logic ($3 \leq \alpha < \omega$) between K^α and $S5^\alpha$ is undecidable and non-finitely axiomatisable.

6.2 Polyadic algebra

An alternative algebraisation of the semantics of first-order logic is Halmos' *polyadic algebra*. Two main references are [Hal62] and [HenMon+85, §5.4] We will summarise some interesting results about polyadic algebras and their connections to other algebras, but our main interest in this book is in relation algebras so we will not go further than that.

DEFINITION 6.6 Let α be an ordinal.

- An algebra $\mathcal{A} = \langle A, 0, 1, +, -, c_\Gamma, s_\tau \rangle_{\Gamma \subseteq \alpha, \, \tau \in {}^\alpha\alpha}$ is an α-*dimensional polyadic algebra* if for all $x, y \in A$, $\Gamma \subseteq \alpha$, and $\sigma, \tau \in {}^\alpha\alpha$,

  1. $\langle A, 0, 1, +, - \rangle$ is a boolean algebra
  2. $c_\Gamma 0 = 0$
  3. $x \leq c_\Gamma x$
  4. $c_\Gamma(x \cdot c_\Gamma y) = c_\Gamma x \cdot c_\Gamma y$
  5. $c_\emptyset x = x$
  6. $c_\Gamma c_\Delta x = c_{\Gamma \cup \Delta} x$
  7. $s_{\mathrm{Id}} x = x$ (Id denotes the identity map on α)
  8. $s_{\sigma \circ \tau} x = s_\sigma s_\tau x$
  9. $s_\sigma(x + y) = s_\sigma x + s_\sigma y$
  10. $s_\sigma(-x) = -s_\sigma x$
  11. If $\sigma(i) = \tau(i)$ for all $i \in \alpha \setminus \Gamma$ then $s_\sigma c_\Gamma x = s_\tau c_\Gamma x$
  12. If τ is one-one over $\tau^{-1}[\Gamma] = \{i < \alpha : \tau(i) \in \Gamma\}$, then $c_\Gamma s_\tau x = s_\tau c_{\tau^{-1}[\Gamma]} x$.

  PA$_\alpha$ denotes the class of all α-dimensional polyadic algebras.

- Let $\langle A, 0, 1, +, -, c_\Gamma, s_\tau \rangle_{\Gamma \subseteq \alpha, \, \tau \in {}^\alpha\alpha} \in \mathbf{PA}_\alpha$.
  Then $\langle A, 0, 1, +, -, c_\Gamma, s_\tau, d_{ij} \rangle_{\Gamma \subseteq \alpha, \, \tau \in {}^\alpha\alpha, \, i,j < \alpha}$ is an α-*dimensional polyadic equality algebra* if

  13. $d_{ii} = 1$
  14. $x \cdot d_{ij} \leq s_{[i/j]} x$ (recall that $[i/j](k) = k$ for $k < \alpha$, $k \neq i$, and $[i/j](i) = j$)
  15. $s_\tau d_{ij} = d_{\tau(i)\tau(j)}$

  for all $x \in A$, $i, j < \alpha$, $\tau \in {}^\alpha\alpha$. **PEA**$_\alpha$ denotes the class of all α-dimensional polyadic equality algebras.

- Let U be a non-empty set and let $\Gamma \subseteq \alpha$. For $S \subseteq {}^{\alpha}U$ we let

$$\begin{aligned} c_\Gamma S &= \{x \in {}^{\alpha}U : \exists s \in S \, \forall i \in \alpha \setminus \Gamma \, (x_i = s_i)\}, \\ s_\sigma(S) &= \{x \in {}^{\alpha}U : x \circ \sigma \in S\}, \\ d_{\kappa\lambda} &= \{x \in {}^{\alpha}U : x_\kappa = x_\lambda\}. \end{aligned}$$

An α-*dimensional polyadic set algebra* has the form

$$\langle A, \cup, \setminus, {}^{\alpha}U, \emptyset, c_\Gamma, s_\tau \rangle_{\Gamma \subseteq \alpha, \, \tau \in {}^{\alpha}\alpha},$$

for some set $A \subseteq \wp({}^{\alpha}U)$ closed under the operations. An α-*dimensional polyadic equality set algebra* has the form

$$\langle A, \cup, \setminus, {}^{\alpha}U, \emptyset, c_\Gamma, s_\tau, d_{ij} \rangle_{\Gamma \subseteq \alpha, \, \tau \in {}^{\alpha}\alpha, \, i,j < \alpha}.$$

- A **PA**$_\alpha$ (**PEA**$_\alpha$) is said to be *representable* if it is isomorphic to a subdirect product of α-dimensional polyadic (equality) set algebras. We write **RPA**$_\alpha$ (**RPEA**$_\alpha$) for the class of representable algebras in **PA**$_\alpha$ (**PEA**$_\alpha$).

There are some obvious connections between polyadic and cylindric algebras. From an arbitrary $\mathcal{P} \in \mathbf{PEA}_\alpha$ it is possible to form its *cylindric reduct* $\mathfrak{Cr}\mathcal{P}$ by ignoring all the non-cylindric operators and cylindrifications c_Γ for $|\Gamma| \neq 1$, and letting $c_i = c_{\{i\}}$. It is easy to check that $\mathfrak{Cr}\mathcal{P}$ is indeed an α-dimensional cylindric algebra. Similarly, the *diagonal-free reduct* of a polyadic algebra is a diagonal-free algebra of the same dimension.

Working in the opposite direction (from cylindric or diagonal-free algebras to polyadic equality algebras or polyadic algebras) is more tricky. The first problem is that in a cylindric algebra we have cylindric operators c_i which work along a single dimension i, whereas polyadic algebras have operators c_Γ that cylindrify along all the dimensions in Γ. If α is infinite this difference will be significant. So **PEA**$_\alpha$ is always a discriminator variety (the discriminator is c_α) but **CA**$_\alpha$ has no discriminator, for infinite α. Halmos defines an α-dimensional *quasi-polyadic algebra* (**QPA**$_\alpha$) to be an algebra obtained from a **PA**$_\alpha$ by restricting the cylindrifiers c_Γ to finite sets Γ of dimensions, and the transformations s_τ to those 'finitary' τ which move only finitely many dimensions. A *quasi-polyadic equality algebra* (**QPEA**$_\alpha$) is defined similarly. For finite α, the **Q** has no effect. Clearly, **CA**$_\alpha$ is more closely related to **QPEA**$_\alpha$ than **PEA**$_\alpha$.

The other problem is to recover the polyadic operators s_τ from the cylindric operations. We have seen that the substitutions s^i_j are definable in a cylindric algebra (though they are not in a diagonal-free algebra) and if α is finite it is possible to define s_τ from these substitutions, provided τ is not a permutation of α. But if τ is a permutation then s_τ may not be definable by cylindric operations.

Nonetheless, suitable interpretations of the s_τ can sometimes be imposed, if not in the original algebra then perhaps in an extension of it:

6.2. Polyadic algebra

THEOREM 6.7 (Simon, [Sim97]) *Let $C \in \mathbf{CA}_3$. The following are equivalent:*

1. $\mathfrak{Ra}C \in \mathbf{SA}$,

2. *The conditions $x; y \cdot z = 0$, $z; \breve{y} \cdot x = 0$, and $\breve{x}; z \cdot y = 0$ are equivalent in $\mathfrak{Ra}C$,*

3. *C satisfies the axiom MGR^+:*

$$\bigwedge_{\{k,l,m\}=3} \mathsf{s}_l^k \mathsf{s}_m^l \mathsf{s}_k^m \mathsf{c}_k(\mathsf{c}_l x \cdot \mathsf{c}_m y) = \mathsf{c}_k(\mathsf{s}_l^m \mathsf{c}_l x \cdot \mathsf{s}_m^l \mathsf{c}_m y) \quad (MGR^+),$$

4. $C \in \mathbf{S}\mathfrak{Cr}\mathbf{PEA}_3$.

If C is generated by $\{c \in C : c_i c = c$ for some $i < 3\}$, its set of 2-dimensional elements, then the four conditions are further equivalent to $C = \mathfrak{Cr}\mathcal{P}$ for some $\mathcal{P} \in \mathbf{PEA}_3$.

THEOREM 6.8 *Let $\alpha < \omega$ and let $C \in \mathbf{CA}_\alpha$. The following are equivalent:*

1. $C \in \mathbf{S}\mathfrak{Cr}\mathbf{PEA}_\alpha$,

2. $C \in \mathbf{IG}_\alpha$.

Proof (sketch). It can be checked that if $C \in \mathbf{CA}_\alpha$ and $C \subseteq \mathfrak{Cr}\mathcal{P}$ for some $\mathcal{P} \in \mathbf{PEA}_\alpha$ then C satisfies the axioms for $\mathbf{IG}_\alpha$ given in fact 5.38. Hence, $C \in \mathbf{IG}_\alpha$. For the converse, check that if $C \in \mathbf{CA}_\alpha \cap \mathbf{G}_\alpha$ has unit 1, then

$$\mathcal{P} = \langle \wp(1), \cup, \setminus, \wp(1), \emptyset, \mathsf{c}_\Gamma, \mathsf{s}_\tau, \mathsf{d}_{ij} \rangle_{\Gamma \subseteq \alpha,\ \tau \in {}^\alpha\alpha,\ i,j < \alpha}$$

satisfies the $\mathbf{PEA}_\alpha$ axioms (because 1 is closed under substitutions and permutations), and $C \subseteq \mathfrak{Cr}\mathcal{P}$. We need $C \in \mathbf{CA}_\alpha$ to establish $\mathbf{PEA}$ axiom 12 in the case where τ is not a permutation; cf. lemma 5.33. □

For related work, see [AndNém84b]. The situation for diagonals is similar, but *any* polyadic algebra can be extended to one with suitable (perhaps non-definable) interpretations of the diagonals:

THEOREM 6.9 ([Hal62, theorem 14]) *Every polyadic algebra embeds in a polyadic equality algebra of the same dimension.*

Now we discuss the relation of polyadic algebras to representable ones.

THEOREM 6.10 (Daigneault, Keisler, Monk) *For $\alpha \geq \omega$, $\mathbf{PA}_\alpha = \mathbf{RPA}_\alpha$.*

See [DaiMon63]. It follows that not every α-dimensional diagonal-free algebra embeds in the diagonal-free reduct of an α-dimensional polyadic algebra, for $\alpha \geq \omega$, as not all diagonal-free algebras are representable. A contrasting result for **PEA** is proved in [Ság99, chapter 4], [NémSág00].

THEOREM 6.11 (Johnson, [Joh69]) *Let* $3 \leq \alpha < \omega$. *Neither* **RPA**$_\alpha$ *nor* **RPEA**$_\alpha$ *is finitely axiomatisable.*

Things improve a little if we restrict attention to the polyadic algebras with representable cylindric reduct. Let $\alpha < \omega$. [HenMon$^+$85, theorem 5.4.26] proves that for $\mathcal{P} \in$ **PEA**$_\alpha$, if every element of $\mathfrak{Cr}(\mathcal{P})$ is a boolean combination of elements x satisfying $\{i : \mathsf{c}_i x = x\} \neq \emptyset$, then $\mathfrak{Cr}(\mathcal{P}) \in$ **RCA**$_\alpha$ if and only if $\mathcal{P} \in$ **RPEA**$_\alpha$ (cf. theorem 6.7 above). But what happens in the general case, where not every element of $\mathfrak{Cr}(\mathcal{P})$ is a boolean combination of such elements? Then for each finite $\alpha \geq 4$, Andréka and Németi found a non-representable **PEA**$_\alpha$ whose cylindric reduct is representable [AndNém84a], [HenMon$^+$85, remarks 5.4.40]. Andréka extended this to $\alpha = 3$ later ([And87]; cf. [AndMon$^+$91, p 725]), solving [HenMon$^+$85, problem 5.7]. Johnson proved that the same holds for infinite α [HenMon$^+$85, remarks 5.4.41]. There is worse:

THEOREM 6.12 *For finite* $\alpha \geq 3$, *there is no set of prenex universal sentences using finitely many variables that axiomatises* **RPEA**$_\alpha$ *over* **RPA**$_\alpha$ *[And97a, theorem 6], or over* **RCA**$_\alpha$ *(Andréka for* $\alpha \geq 4$, *and Andréka–Tuza for* $\alpha \geq 3$; *cf.* *[AndTuz88], [AndMon$^+$91, p 725]).*

This solved [Joh69, problem 2] and [HenMon$^+$85, problem 5.8], and strengthened the result mentioned above that there is a non-representable **PEA**$_\alpha$ with representable cylindric reduct. For further reading on polyadic algebras, see, e.g., [HenMon$^+$85, Ném91, SaiTho91].

Exercises

1. Prove that $\mathsf{s}_{[i/j]} x = \mathsf{c}_{\{i\}}(\mathsf{d}_{ij} \cdot x)$ is valid in polyadic equality algebras.

2. For $\mathcal{P}$ in **PEA**$_\alpha$ or **QPEA**$_\alpha$, check that $\mathfrak{Cr}\mathcal{P} \in$ **CA**$_\alpha$ and that $\mathfrak{Cr}\mathcal{P}$ satisfies MGR. Hence, by fact 5.38, $\mathfrak{Cr}\mathcal{P} \in$ **ICrs**$_\alpha$.

3. Show that MGR^+ implies MGR1, MGR2 in dimension 3.

4. Show that MGR^+ implies the axiom for **G**$_\alpha$ in fact 5.38(4).

6.3 Pinter's substitution algebras

There are many other kinds of algebras of relations beyond the ones we have mentioned (see [HenMon$^+$85, p. 263–271] for a survey). For example, Pinter's substitution algebras have as primitive connectives the substitutions s^i_j rather than the cylindrifications. They are strictly weaker than both cylindric algebras and quasi-polyadic algebras, but they are still an interesting algebraisation of first-order logic.

The main representation result due to Pinter in this connection is that every such locally finite algebra is representable. [Pin73, Ném91] have more information.

6.4 Finitisation problem

The non-finite axiomatisability results of Monk in the 1960s for relation algebras and cylindric algebras have been hugely influential in algebraic logic. They were strengthened by Jónsson [Jón91], who showed that **RRA** is not axiomatisable by equations using finitely many variables (see exercise 11.5(4)). This result was known by Tarski in 1974 — he mentions it at the end of a video of a lecture given in Campinas, Brazil. Andréka [And97a] proved similar and more detailed results for cylindric algebras. Venema [Ven97b] showed using results of chapter 17 that **RRA** and **RCA**$_n$ for finite $n \geq 3$ are not axiomatisable by Sahlqvist equations. The undecidability of representability of finite relation algebras (chapter 18, theorem 18.13) implies that **RRA** is not finitely axiomatisable in second-order, or for that matter, 112th-order logic.

We said in chapter 1 that Monk's results led to several other reactions. A very important one, though not central to this book, can be crudely summarised as investigating whether finite axiomatisability can be obtained by changing the signature. If so, it will help in pinning down the 'cause' of non-finite axiomatisability (but note that neither finite nor non-finite axiomatisability need be preserved by taking a smaller signature). We now outline some of the work in this field. The reader may refer to [Sim93] or [Ném91] for an introduction.

6.4.1 Reducts, subreducts, generalised subreducts

First, we consider *reducts* obtained by dropping some of the operations. Once this is done, more subalgebras of an algebra appear, since the closure requirements on being a subalgebra have been relaxed. Therefore we are in the area of *subreducts* — subalgebras of reducts. The question now is which subreducts of **RRA** and related classes are finitely axiomatisable.

A great deal of work has been done on this kind of question. See [Sch91] for a survey of the area. We already discussed Jónsson's 'algebras of relations' in section 5.2, but many other reducts have also been considered. Andréka showed that any subreduct of **RRA** whose operations include union, intersection and composition is not finitely axiomatisable [And91], and that the {union, composition}-subreduct is a non-finitely axiomatisable quasi-variety [And89]. It was shown by Bredikhin in [Bre77] that the {composition, conversion}-subreduct is also not finitely axiomatisable.

Various generalised subreducts, where the operations of the new signature are merely term-definable in the old one, have been studied too. One example is *sequential algebras*. The variety of representable sequential algebras is not finitely axiomatisable: see [Kar94, JipMad97]. It was shown in [HodMik00] that the union-free subreducts of representable sequential algebras are also non-finitely axiomatisable.

On the other hand, some subreducts of **RRA** are finitely axiomatisable. (Obviously, boolean algebras are, but this is going too far.) Bredikhin and Schein [BreSch78] showed that the {intersection, composition}-subreduct of **RRA** coincides with the class of semilattice-ordered semigroups. Another example is the generalised subreduct with the similarity type of intersection, composition and its two residuals: see [AndMad94]. Andréka [And90b] has shown that the equational theory of many positive reducts of representable algebras is decidable. See [Bre93] for more information on axiomatisability of the equational theories of reducts of **RRA**, and, e.g., [Mik93b, AndMik94] for 'game-style' results on the Lambek calculus, another generalised subreduct of **RRA**.

Finite axiomatisability of subreducts of **RCA**$_\alpha$ has also been investigated — cf. [Com91, Han95, And97b]. See also [Dün93] for lattice reducts of cylindric algebras and their connections to databases.

6.4.2 Expansions

Next, what happens if we expand the signature? For example, can we add operations to relation algebras and write a finite set Σ of equations in the new language such that the representable relation algebras are precisely the relation algebra reducts of models of Σ? Since it is undecidable whether a finite relation algebra is representable (theorem 18.13), the answer is 'no'.

However, the problem can be finessed a little. First, what if we only require that the relation algebra is a *subalgebra* of the relation algebra reduct of a model of Σ (a subreduct)? This line of investigation is associated with Craig, and there have been some positive finitisation results for infinite-dimensional diagonal-free algebras. E.g., [Sai95, SaiGyu97, Sai00] selects a 'nice' semigroup of substitutions, leading to a finitely axiomatisable Sahlqvist variety of reducts of polyadic algebras. [Say] surveys this approach and proves that such algebras have strong amalgamation, giving the Craig and Beth properties for the corresponding logics.

Nonetheless, there are strong negative results about **RRA**, **RCA**$_n$ in [Birò92, MadaNém+97, Madaa, Madab]; see also [And94, Ném91].

[Birò92, Madaa, Madab] consider whether permutationally invariant operations can be added to **RRA** so that the expanded class, say **RRA**$^+$, is finitely axiomatisable. [Birò92] uses relation algebras given in [Madd89b] to prove that the finitisation problem is not solvable by adding finitely many first-order definable operations. In [Madab], it is shown that if each new operation is either first-order definable or is completely additive (on at least one full proper relation algebra

6.4. Finitisation problem

with uncountable base set), then **RRA**$^+$ is not finitely axiomatisable, even with the new operations. In [Madaa] it is proved that if the extra operations are binary and additive, then **RRA**$^+$ cannot be axiomatised by any set Σ of universally quantified formulas if Σ involves only finitely many variables. Madarász also proves here the stronger version of this theorem where we allow some of the new operations to be binary and additive while others to be $L^3_{\infty\omega}$-definable, where $L^3_{\infty\omega}$ is the 3-variable fragment of $L_{\infty\omega}$. (As an example, we note that transitive closure is $L^3_{\infty\omega}$-definable.) [Madaa, Madab] extend these negative results from **RRA** to the class **DS** of distributive lattice-ordered semigroups. **DS** consists of subalgebras of reducts of **RRA** with operations $+, \cdot, ;$ — i.e., the lattice-operations and relation composition. She proves that if you are in **DS** then you have already lost finite axiomatisability and you cannot regain it by adding to **DS** new operations of the kind described above.

In [Ság99], Sági reduces the permutation-invariant problem to working inside relation algebras. The problem is reduced to finding a certain kind of sequence of finite relation algebras. If such a sequence can be found then there does not exist any permutation-invariant extension of **RRA** that is axiomatisable by finitely many universal formulas. Sági also finds a sequence of relation algebras based on Lyndon algebras that give a negative solution to a weaker form of the problem: he shows that there is no permutation-invariant extension of **RRA** axiomatisable by finitely many strongly balanced universal formulas. A universal formula φ in a signature extending that of **RRA** is strongly balanced if for every subterm $f(t_1, \ldots, t_n)$ of φ, where f is not an **RRA** operation symbol, every variable occurring in φ *is* (not just *occurs in*) one of the t_i. This notion (and some variants of it) has been investigated by Jónsson, McNulty, and others. Intuitively, the balanced formulas are the 'simple' ones, because the new operation symbols can only be used in a special way.

These are strong results suggesting that probably

$$\text{\textbf{RRA} + permutation-invariant new operations} \atop \text{cannot be finitely axiomatisable.} \tag{6.1}$$

However, in its full power (6.1) remains an open problem because the results do not handle all possible kinds of permutationally invariant new operations. But they handle many. So, they (together with [MadaNém$^+$97]) make it very likely that (6.1) above is true. Put another way, if someone does produce a finite axiomatisation of **RRA** using extra permutationally invariant operators, then these new operators will be very strange.

Second, can we obtain 'nicer' axioms by adding operations? There are positive results here. In [SteVen98, Ste00], Stebletsova and Venema expanded **RRA** with slight variants of the Q_n-operators (all $n < \omega$) of Jónsson (see [Jón91]; related operators are discussed in chapters 8, 9, and 19). They axiomatised the resulting expansion; the axioms so obtained arise from games in the style of chapter 8, and have a clear intuitive meaning in terms of them. Furthermore, they allow good n-dimensional approximations to **RRA**. The expansion by a single Q_n ($5 \leq n < \omega$) allows (loosely speaking) a finite axiomatisation of the variety **RA**$_n$ of 'n-dimensional' relation algebras in terms of subreducts: a relation-type algebra is in **RA**$_n$ iff it is a subalgebra of the relation algebra reduct of a Q_n-type algebra satisfying a certain finite set of equations. (For the definition of **RA**$_n$, see definition 12.30; by results of section 17.4, this variety is not finitely axiomatisable in its natural signature.) For further work on Q-operators, see [NémAnd91, Ven91].

More still can be achieved if we allow relativisation. We already mentioned in section 5.1 that every weakly associative algebra has a relativised representation. Weakly associative algebras correspond closely to (and can be studied via) arrow logic. [Kur00a] considered arrow logic augmented with various kinds of infinite counting modalities, such as 'much more', 'of good quantity', 'many times'. It was shown that the addition of these modal operators to weakly associative arrow logic results in finitely axiomatisable and decidable logics — the first such extensions of weakly associative arrow logic that do not have the finite model property. Arrow logic with *projections* is extremely expressive. Still, it is quite surprising that adding projections can even spoil the robust decidability of weakly associative arrow logic: [Kur00b] proved that some of these systems are even not recursively enumerable, using a reduction of unsolvable Diophantine equations. This negative property was shown in the radical departure [KurNém00] to be an artifact of the underlying set theory — certain non-well-founded set theories interpret the meaning of projections so as to allow finite axiomatisability even of full arrow logic with projections, and hence of relation algebras with distinguished projection elements. See also [Kur97]. The so-called directed cylindric algebras form a related topic: [Ság99, chapter 5] and [Ság00] prove representability of these algebras in the absence of the axiom of foundation.

Similar approaches for other algebras such as **Crs** have been fruitfully taken by various authors: see, e.g., [Mar95, Mik95, MarVen97, VenMar98, Ben96]. This leads into the field of dynamic logic.

6.4.3 Special conditions for representability

An important related approach is to seek special conditions under which a relation algebra is representable. We do not try to finitely axiomatise all of **RRA**, but only part of it. Early results in this direction include:

1. Any atomic relation algebra whose atoms are functional is representable. (This is a result of Jónsson and Tarski [JónTar48, JónTar52]. An element a of a relation algebra is said to be *functional* if $\breve{a}; a \leq 1'$.) Note that any representable relation algebra is a subalgebra of an atomic relation algebra

6.4. Finitisation problem

with functional atoms.

2. Any relation algebra in which the unit 1 is the sum of finitely many functional elements is representable. ([JónTar52, theorem 4.32]; a corollary to (1).)

3. Any atomic relation algebra in which every atom is a point is representable [JónTar52, theorem 4.30]. An element a of a relation algebra is called a *point* if $a > 0$ and $a;1;a \leq 1$'.

4. Any relation algebra containing a pair p,q of functional elements such that $\breve{p};q = 1$ is representable (Tarski, 1942, announced in [Tar53], published in [TarGiv87, theorem 8.4(iii)].) Such p,q are called *quasi-projections*.

Maddux extended several of these results.

5. Any relation algebra in which the unit is the sum of (any number of) functional elements is representable [MaddTar76, Madd78a].

6. Any relation algebra in which the unit is the sum of elements of the form $\breve{p};q$, where p and q are functional, is representable [Madd78a]. Such algebras are called *tabular*.

(3) above essentially requires the unit to be the sum of points. Maddux strengthened this result by proving:

7. [Madd91b, theorems 54, 52] If, in a semi-associative algebra, the identity element 1' (and not necessarily the whole unit) is the sum of points (i.e., the algebra is 'point-dense'), then the semi-associative algebra is representable.[1] If in addition the algebra is simple, it is completely representable.

8. [Madd91b, theorems 54, 51] If, in a relation algebra, the identity element is the sum of points and pairs (i.e., the algebra is 'pair-dense'), then the algebra is representable. (An element a of a relation algebra is called a *pair* if $a > 0$ and $a;0';a;0';a \leq 1$'.) If in addition the algebra is simple, it is completely representable.

9. Let $\mathcal{A}$ be a relation algebra, let $x \in \mathcal{A}$ with $0 < x \leq 1$', and let $n < \omega$. Say that x is *n-max* if $x;1;x$ is the sum of at most n non-zero functional elements. $\mathcal{A}$ is said to be *n-dense* if 1' is the sum of (any number of) n-max elements. It is not difficult to show that 1- and 2-dense relation algebras are the same as point-dense and pair-dense relation algebras, respectively; by (7) and (8) above, they are representable. Givant and Andréka recently showed in [GivAnd02] that for $n \leq 7$, any n-dense relation algebra is representable, but that there is a finite 8-dense relation algebra which is not representable.

---

[1] The word 'completely' in the statement of [Madd91b, theorem 54] should be deleted.

Related work includes that of Andréka et al. [AndGiv+98], Maddux and Tarski [MaddTar76], and Venema [Ven98], looking at dense subsets of the algebra, and Maddux with the notions of quasi-products [Madd78b]. See [Madd96] for recent applications to semantical issues in computer science. Fork algebras are a related approach: see, e.g., [FriBau+95, FriBau+96, FriBau+97, BauFri+96, HaeFri+97, Fri02]. Strong generalisations of several of the results above — (9) above is one example — were recently established by Givant and Andréka; their statement requires additional group-theoretic terminology which would take us some way from our path, so we refer the reader to [GivAnd02, theorems 5, 6] for details. The ideas referred to here have helped to provide finite non-orthodox ('Gabbay-style') axiomatisations of classes such as **RRA** and the related arrow logic [Ven92, Ven93, Madd93, VenMar98, Mik93a, Mik96b, Ven98, MarPól+96, BlaRij+01, MarVen97, Ste00]. The material has a strong game-theoretic flavour, but we have not the space to examine it here.

6.5 Decidability

We refer the reader to [AndGiv+97] for a comprehensive treatment of decidability of equational theories of varieties in algebraic logic. [Kur97] has many relevant results. Marx's [Mar99a] shows that for proving the undecidability of the equational theory of relation algebras, interpreting tiling problems is an alternative to interpreting semigroups (the approach initiated in CA_3 by Maddux [Madd80] and enhanced by the Budapest group). We will use tiling in chapter 18. [Ném87b] proves undecidability results for varieties of CA_ω by interpreting the set of satisfiable Diophantine equations into the equational theory of the class of minimal cylindric algebras of dimension ω.

6.6 Amalgamation

Amalgamation is the algebraic analogue of logical interpolation. An excellent introduction covering Beth definability as well can be found in [Hoo01].

RA and **RRA** do not have the interpolation or Beth properties [Com69, Sai90]. However, in [Mar01a], Marx added to **RRA** the apparatus of hybrid logic, which allows 'naming' of points of the base of a representation. The process is rather like quantification in first-order logic, but only those points arising during the evaluation of a relation algebra term in the representation may be named. Marx showed that the resulting formalism is exactly as expressive as first-order logic, and hence has the interpolation and Beth properties. He asked if there exists a variety of hybrid relation algebras from which RA_n can be obtained by subreducts.

Which subclasses of the representable algebras have (strong) amalgamation?

Pigozzi [Pig71] gives a comprehensive picture of the amalgamation property in subclasses of cylindric algebras. Problems raised there concerning strong amalgamation were solved in [MadaSay01], which contains a great deal of information about this kind of problem for both relation algebras and cylindric algebras. It also unifies two techniques used in algebraic logic for proving amalgamation. One is due to Németi [Ném85], addressing strong amalgamation in BAOs similar to cylindric algebras and relativised versions of representable cylindric algebras like $\mathbf{Crs}_\alpha$ and $\mathbf{D}_\alpha$. This technique was generalised by Marx [Mar98], stressing the modal aspect. The other technique is due to Pigozzi and addresses classes of cylindric algebras. The unification consists of presenting both techniques as transforming a diagram of the algebras to be (strongly) amalgamated into certain saturated representations of these algebras that can be (strongly) amalgamated, and then returning to the original diagram using an inverse operator. Both can be described functorially by an adjoint situation making the notion of inverse precise. In the case of Pigozzi it is the neat reduct functor (an inverse to a neat embedding functor taking an algebra to one in ω extra dimensions, i.e., to a classical representation), while in Németi's case it is basically the operation of forming atom structures that is an inverse of taking an algebra to its canonical extension (which can be seen as a modal representation). So this takes the representation problem expressed by a two-sorted defining theory (chapter 9) a step further, asking that the second sort be a saturated representation.

6.7 Technical innovations

Here we briefly mention three techniques from the theory of cylindric algebras: relativisation, splitting, and twisting. *Relativisation* is the abstract algebraic analogue of the concrete relativisation of section 5.1. It can be applied to any relation algebra (or cylindric algebra) $\mathcal{A}$, representable or not, and involves defining a new algebra $\mathcal{A}_a$ whose domain consists of the elements below some given element $a \in \mathcal{A}$, and defining the algebraic operations in $\mathcal{A}_a$ by restriction to a: for example, $x;^{\mathcal{A}_a} y = a \cdot (x;^{\mathcal{A}} y)$. Cf. the proof of proposition 2.58. References here include [HenMon+85, Kram91, Mar99b]. Marx has shown that relativisation to non-transitive units give positive results concerning amalgamation and complexity.

Splitting, or *dilation*, originates with Henkin and involves replacing (typically) an atom of an algebra by a whole set of atoms, and defining the algebraic operations on these new atoms by and large in terms of the old one. This can have the effect of destroying representability of the algebra, and is useful in constructing examples. 'Monk algebras' ([Mon69] and section 4.4) can be viewed this way, and we will use what is in effect splitting in chapters 14 and 17. See [HenMon+85, 3.2.68] and [Sim97] for splitting in cylindric algebras, and [AndMad+91] for splitting in relation algebras.

Twisting, a more sophisticated technique also due to Henkin, involves altering the definition of cylindrification in a cylindric algebra in a controlled way. It has numerous uses in the theory of cylindric algebras. See [HenMon+85, 3.2.71] for a detailed description of it, and [Sim97] for an excellent introduction to both splitting and twisting, and an important application (indeed, generalisation) of twisting, showing that every 3-dimensional cylindric algebra can be obtained from a representable one by relativisation and twisting. Formally, for every $\mathcal{A} \in \mathbf{CA}_3$ there are $\mathcal{A}_1, \mathcal{A}_2, \mathcal{A}_3 \in \mathbf{CA}_3$ and a set algebra $\mathcal{A}_4 \in \mathbf{Cs}_3 \subseteq \mathbf{RCA}_3$ such that $\mathcal{A}_3$ is a subalgebra of a relativisation of $\mathcal{A}_4$, $\mathcal{A}_2$ is a twisted version of $\mathcal{A}_3$, $\mathcal{A}_1$ a relativisation of $\mathcal{A}_2$, and $\mathcal{A} \subseteq \mathcal{A}_1$.

6.8 Applications

The connection of algebraic logic to modal and other logics is well known. This can be very direct: arrow logic [MarPól+96], for example, is a modal version of relation algebra. By algebraically reformulating problems of (say) modal logic, one may apply known results in algebraic logic to resolve them. For example, in chapter 15 we will use algebra to show that $(n + 1)$-variable classical first-order proof theory is not finitely axiomatisable over n-variable proof theory, for all $n \geq 4$. Conversely, modal and classical techniques can be fruitfully applied in algebraic logic: Sahlqvist's theorem provides an example, and we hope that much of this book does too. For further information the reader is referred to [Ném91, CsiGab+95, MarPól+96, Ben96, AndBen+98].

Blok and Pigozzi's work on abstract algebraic logic and algebraisable logics [BloPig89] is another example of this connection. One aim of research in this field is to find an adequate criterion for algebraisability. (There is a duality between algebraisable logics and quasi-varieties — analogous to that existing between BAOs and modal logics, or rather extending it.) Another is to give a general 'universal' definition of a logic (L). One may then translate metalogical properties like completeness, compactness, definability, omitting types, etc., to purely algebraic properties of the corresponding quasi-variety Alg(L), like finite axiomatisability, (strong) amalgamation, and complete representability (in case we have a boolean reduct).

Algebraic logic is central to first-order proof theory, especially finite-variable proof theory. See theorem 15.17 for one application, and [TarGiv87] for a compendium of information; this book develops set theory via relation algebras, showing that relation algebras may serve as a foundation for all mathematics.

Relation algebras have also been directly applied in theoretical computer science and databases: see, e.g., [Madd96, Dün93, BriKah+97].

Part II

Games

Introduction to games

In the first part of this book, we met relation algebras, and other algebraic formalisms such as cylindric algebras. We saw that one of the prime problems associated with these algebras is to characterise which of them are representable. All boolean algebras are representable, but for relation algebras (and cylindric algebras) this is not the case, and indeed, telling which relation algebras are representable is a hard problem. We saw examples, due to Lyndon, of relation algebras made out of projective planes. Though each of the algebras in this infinite collection is superficially similar to the next, it is extremely difficult to decide which of them are representable, since the representation problem for them is equivalent to the existence of a projective plane of a given order; and this problem is in the general case unsolved. In fact, it was these algebras that Monk [Mon64] used in showing that the variety **RRA** of representable relation algebras is not finitely axiomatisable in the first-order language of relation algebras. (See exercise 11.5(3). We will give a different proof of this result in section 17.1.)

Monk's theorem has been refined by various authors to show that **RRA** has no 'nice' axiomatisation. Jónsson showed in [Jón91] that there is no finite bound on the number of variables required in an equational axiomatisation of **RRA**; Andréka [And97a] showed the same for the variety of representable n-dimensional cylindric algebras, **RCA**$_n$ ($3 \leq n < \omega$). Venema showed [Ven97b] that these varieties cannot be axiomatised by Sahlqvist equations. We will see in theorem 18.13 that it is undecidable whether a finite relation algebra is representable. All this suggests that the notion of representability for relation algebras is unruly, subtle, and difficult to capture and to work with. Any such impression may be reinforced by a look at existing axiomatisations of **RRA** and related classes, for example in [Lyn56, Mon69, HenMon+71, HenMon+85]: they are not very easy to understand. Indeed, [HenMon+71, p.461] identifies one of the two outstanding problems of the representation theory as ' ... the problem of providing a simple intrinsic characterisation for all representable cylindric algebras ... '.

We will approach the problem of representability by stealth: we set up some finite *approximations* to representability, and study how they approximate it and what happens as the degree of approximation improves. The approximations will be called *networks,* and the ways in which the degree of approximation can be improved will be studied by combinatorial *games.*

We will discuss some of the history of this method in chapter 9; it has all kinds of antecedents. It can be seen as a variation on the argument of Lyndon [Lyn50]. In so far as the game construction is a 'step-by-step' argument, it has affinities with the work of many authors, such as Andréka, Henkin, Maddux (notably [Madd82], for example), Németi, Venema, to name only a few within algebraic logic. The method of games is well known in model theory: see [Hod85, Hod93], for example. For the topological connections, via the Banach–Mazur theorem, see [Oxt71].

The advantage of using this network–game approach to characterise the representable relation algebras is that, in our view, the proofs are simpler than those used in previous characterisations. We believe that the explicit use of games concentrates attention on the essential concepts rather than notational details, and so permits more lucid proofs. Further, the method of games can be used to solve many other problems related to representability. We hope the reader will be convinced of its utility after having read the remainder of the book.

Outline

Chapters 7 and 8 introduce the game technique by focusing on the representability problem for relation algebras; generalisations are saved for later. In chapter 7, we define networks and how to refine them, and give a first substantial example of the game method in action, by proving Maddux's result of [Madd82] that every weakly associative algebra has a relativised representation. We develop a similar game for testing the representability of relation algebras (section 7.5), and another game for testing the representability of finite-dimensional cylindric algebras (section 7.8). In section 7.9 we give an application of games to constraint handling in temporal reasoning. In chapter 8 we use the games to obtain an equational axiomatisation for **RRA** and equational axiomatisations for the representable cylindric algebras of any given dimension greater than two.

The remaining chapters in this part of the book deal with generalisations of this game-theoretic method. Chapter 9 shows how to obtain an explicit axiomatisation of the elementary closure of any pseudo-elementary class from its defining theory. In chapter 10 we define 'game trees', based on Hintikka games, and directly generalising the games we used to test representability of relation algebras. Later on, we will use several variants of the **RRA** games: we investigate complete representations by using networks with edges labelled by atoms in chapter 11; and we will use a number of 'n-pebble games' for testing membership of $\mathbf{RA}_n$ and $S\mathfrak{R}a\mathbf{CA}_n$ from chapter 12 onwards. All these games can be represented as game trees.

Chapter 7

Games and networks

In this chapter, we introduce games. Section 7.1 discusses networks, the 'pieces' used in the games. Section 7.2 shows how networks may be changed during a game. In section 7.3 we use games to prove Maddux's important result that every weakly associative algebra has a relativised representation. Section 7.4 describes games for classical representations of relation algebras. Section 7.5 briefly discusses strategies in games, and the key section 7.6 relates them to representations of relation algebras. Sections 7.7 and 7.8 do the same for cylindric algebras. Finally, section 7.9 describes an application to temporal constraints in computer science.

7.1 Networks

Our notion of approximation to a representation, whether of a relation algebra, a cylindric algebra, a weakly associative algebra (where we approximate a relativised representation), or whatever, will be the *network*. Notions related to networks have been used by Lyndon [Lyn50] and Maddux [Madd78b, Madd82], for example; Németi [Ném86] uses a similar notion, called 'mosaic', in the context of other algebras. The notion of a network is an absolutely central one for us, and it merits some discussion. In our description below we concentrate on relation algebras, leaving the other cases aside for a while. For ease of description, we will confine our attention for now to *simple* relation algebras $\mathcal{A}$, as these, if representable, have square representations of the form $h : \mathcal{A} \to (\wp(B \times B), 0, B \times B, \cup, \setminus, \mathrm{Id}_B, -^{-1}, \mid)$; see exercise 3.4(1).

We know from chapter 3 that a representation of a relation algebra $\mathcal{A}$ provides a base set B and an interpretation $h(a)$ of each element $a \in \mathcal{A}$ as a binary relation on B. The interpretation $h(1)$ of the top element 1 of $\mathcal{A}$ is called the *unit*, which we are assuming is of the form $B \times B$. The interpretations respect the algebraic operations: for example, the interpretation $h(a;b)$ of the composition $a;b$ of elements $a,b \in \mathcal{A}$ is the relative product of $h(a)$ and $h(b)$. It is helpful to note that, viewing pairs of elements of B (i.e., elements of the unit) as *edges* or *arcs,* a representation of the algebra amounts to stating which elements of the algebra, when interpreted as binary relations, hold on each edge.

A network *approximates* a representation in the sense that it contains only a *finite amount* of information about it. First, a network is only concerned with a finite part of the base set of the representation. If the representation has infinite base, then clearly the network will be missing something; but even if the base is finite, the network may not describe all of the base. Second, even on the part of the representation's base that it is concerned with, the network only describes *crudely* which relations from the algebra hold or fail on edges. For each pair of elements from the part of the base that it 'knows about', the network will specify some single element of the algebra that holds on that edge in the representation. In the full representation, of course, *all* of the elements of the algebra that hold on the edge can be determined; with the network, we only get one, plus of course by inference all larger algebra elements, which must automatically hold as well. Again, in certain circumstances this one element may determine all the rest — for example, if it is an atom — but not every representation will have atoms holding on arcs.

On the second point, it might seem more natural if a network specified finitely many algebra elements that hold on an arc, together with finitely many elements that do not. But the same information is conveyed by networks that only specify a single algebra element to hold on an arc. For, as relation algebras are closed under the boolean operations, and these are respected in any representation, to say that relations $a_1, \ldots, a_n$ hold on an arc and that relations $b_1, \ldots, b_m$ fail is equivalent to saying that the single relation $c = \prod_i a_i \cdot \prod_j -b_j$ holds on the arc. It is easier to associate a single algebra element with each arc, and this is how we define networks here. For algebras without all the boolean operations, this definition might need to be revised.

Although it is only an approximation of a representation, a network will inherit some properties from the representation. For example, no network could ever deem the bottom element 0 to hold on an arc — this never happens in a representation, and a network that said that it did would not be a true approximation. This restriction on what a network may say comes about simply because a representation respects the algebra operations. Here, 0 must be interpreted in the representation as the empty relation.

We may expect the preservation of each of the other algebra operations to be reflected in some way in properties of networks. This will indeed be so. Consider the conversion operation. In any representation, we know that if an algebra relation

7.1. Networks

holds on an arc then its converse holds on the reversed arc. Thus, a network should never say anything to contradict this. However, this does not mean that if the network says that a holds on an arc e, and b on the reversed arc e', then $\breve{a} = b$. This is because the network's 'knowledge' of what holds on e may be more or less refined than its corresponding knowledge for e'. There is no reason that $\breve{a}$ should be exactly b. But at least, $\breve{a}$ and b should overlap: we should always have that $a \cdot \breve{b} \neq 0$, because this is true in a representation whenever a holds on an arc and b on the reverse arc.

Similarly, the fact that representations preserve relational composition is reflected in networks: if the network says that the relations a,b,c hold on the arcs $(x,y),(x,z),(z,y)$, respectively, then we should have $a \cdot (b\,;c) \neq 0$.

Preservation of the boolean operations in representations is embodied in the definition of networks, in the sense that they only need name a single relation that holds on an arc, and not a finite number.

There is nothing much to say about the top element 1. Preservation by representations of identity, 1', should perhaps emerge as the requirement that if the network deems that a holds on a reflexive arc (one of the form (x,x) for some x in the base of the representation) then $a \cdot 1' \neq 0$. But in the light of the boolean discussion, we may as well require that $a \leq 1'$. For 1' will definitely hold on the arc in a representation, so a slightly more refined network would certainly admit this and restrict a to the part beneath 1'.

The requirement that elements beneath 1', rather than just not disjoint from 1', hold on reflexive arcs in networks is at variance with our treatment of conversion and composition, above. This is one of several points where we have a choice: we could just require non-zero intersection with 1', and occasionally this has advantages; but on balance the stronger option seems to make things simpler, e.g., in definition 7.1 and lemma 7.2 below. In any case, if we wanted to be consistent we would have to relax our restriction on 1', not tighten that on the other operations. For, unlike 1', conversion and (especially) composition are not local operations: their effects distribute themselves around the points of a representation. If we were to insist that whenever the arc e is the composition of the arcs e',e'' on which the network deems b,c to hold, respectively, then it should say that $b\,;c$ holds on e, the network would 'be' the solution of a certain finite set of simultaneous equations in the algebra. It is not clear that such a solution would always exist. But this is quite immaterial to the issue of approximating a representation. If we are content to use approximations at all, we can agree not to mind if different parts of a representation are approximated to different degrees. After all, approximations can always be improved. This is the issue we discuss in the next section.

But first, it is crucial to note that a network is not tied to a representation and can be regarded abstractly, as a set of nodes and edges labelled by elements of $\mathcal{A}$ and satisfying the formal conditions we have described. This makes sense even if $\mathcal{A}$ has no representation. We now write down these conditions; this will formalise the discussion so far. We continue to take relation algebras as our main examples; we will generalise to other algebras later.

DEFINITION 7.1 Let $\mathcal{A}$ be a relation algebra.

1. An $\mathcal{A}$-*network*, or simply a *network*, is a non-empty complete directed graph with edges labelled by elements of $\mathcal{A}$. Formally, it is a pair $N = (N_1, N_2)$, where N_1 is a set of nodes, and $N_2 : N_1 \times N_1 \to \mathcal{A}$ is a map assigning an element of $\mathcal{A}$ to each pair of nodes. We require that:

    (a) $N_2(x,x) \leq 1'$ for all $x \in N_1$;

    (b) $N_2(x,y) \cdot (N_2(x,z); N_2(z,y)) \neq 0$ for all $x,y,z \in N_1$ (this condition is called 'triangle-consistency').

    We should strictly write $1'^{\mathcal{A}}$ and $0^{\mathcal{A}}$ here, but as we said after definition 3.9, we often drop the superfix '$\mathcal{A}$'.

2. To free up some suffixes, and in line with our identifying structures with their domains, we will abuse notation and write simply N for any of N, N_1, N_2 above, distinguishing them by context. Where necessary, we will write nodes(N) for the set N_1 of nodes of the network N. We will also write $|N|$ for $|N_1|$, the cardinality of this set of nodes.

3. We will refer to ordered pairs of nodes of N_1 as *edges* or *arcs*, in keeping with our definition of N as a labelled graph. If e is such an edge, we will of course write $N(e)$ for the label on it.

The properties (a) and (b) in part 1 of the definition are as we discussed them before. They are *motivated* by what would hold if the network were approximating an actual representation; but they are *defined* without reference to a representation. So even non-representable relation algebras have associated networks.

There is one minor point to note: we do not require that networks N satisfy:

if $x, y \in N$ and $N(x,y) \leq 1'$ then $x = y$.

Networks with this additional property are said to be *strict*. In non-strict networks, several different nodes of a network can stand for the same point of a representation, which is often convenient. But sometimes, as with weakly associative algebras in section 7.3, strict networks are easier to use.

The foregoing discussion would lead us to expect that a network N should have two additional properties:

1. $N(x,y) \neq 0$ for all $x, y \in N$,

2. $N(x,y)^{\smile} \cdot N(y,x) \neq 0$, for all $x, y \in N$.

7.1. Networks

Why did we not include these in the definition of 'network'? The answer is that they follow anyway.

LEMMA 7.2 *For all nodes x,y in any network N:*

1. $N(x,y) \neq 0$,
2. $N(x,y)^{\smile} \cdot N(y,x) \neq 0$.

Proof. Assume for contradiction that $N(x,y) = 0$ for some $x,y \in N$. Then clearly, $N(x,y) \cdot (N(x,x) ; N(x,y)) = 0$, contradicting property 2 of definition 7.1.

Let $x,y \in N$. By property 2 of definition 7.1, $N(x,x) \cdot (N(x,y); N(y,x)) \neq 0$. By property 1 and lemma 3.17, $N(x,x) \leq 1' = \breve{1'}$, so $(N(x,y); N(y,x)) \cdot \breve{1'} \neq 0$. By the Peircean law in relation algebras (lemma 3.12), $(N(y,x); 1') \cdot N(x,y)^{\smile} \neq 0$. By the law for 1' in relation algebras (definition 3.8, axiom R3), $N(y,x); 1' = N(y,x)$, so we obtain $N(y,x) \cdot N(x,y)^{\smile} \neq 0$, as required. $\square$

Exercises

1. Let $\mathcal{P}$ be the point algebra of section 4.4. How many $\mathcal{P}$-networks with two given nodes (say 0, 1) are there? How many of them are such that all labels on arcs are atoms? (Such networks will be called *atomic;* we will have a lot more to say about them later.)

2. Let $\mathcal{A}$ be a relation algebra and let $0 < a \leq 1'$ in $\mathcal{A}$. Let X be any set. Define N, with domain X, by $N(x,y) = a$ for all $x,y \in X$. Show that N is an $\mathcal{A}$-network.

3. Let h be a representation of the relation algebra $\mathcal{A}$, with base B, say. Let X be a set, let $\iota : X \to B$ be a map, and let $\nu : X \times X \to \mathcal{A}$ be a map satisfying $\nu(x,x) \leq 1'$ and $(\iota(x), \iota(y)) \in h(\nu(x,y))$, for all $x,y \in X$. Show that (X, ν) is an $\mathcal{A}$-network. (This formalises the notion of a network approximating a representation.) Name circumstances under which (X,ν) determines h.

4. Suppose that $\mathcal{A}$ is a relation algebra and $N = (N_1, N_2)$ where N_1 is a set and $N_2 : N_1 \times N_1 \to \mathcal{A}$ satisfies:

    - $N_2(x,x) = 1'$ for all $x \in N_1$,
    - for some enumeration $(\{x_i, y_i\} : i < \alpha)$ of the subsets of N_1 of size 2, we have $N_2(x_i, y_i) = N_2(y_i, x_i)^{\smile} \neq 0$, for all $i < \alpha$,
    - for some enumeration $(\{x_i, y_i, z_i\} : i < \beta)$ of the subsets of N_1 of size 3, we have $N_2(x_i, y_i) \cdot (N_2(x_i, z_i); N_2(z_i, y_i)) \neq 0$ for all $i < \beta$.

    Show that N is an $\mathcal{A}$-network.

5. Suppose that N is an atomic network over some atomic relation algebra (see exercise 1). Show that:

   (a) $N(y,x) = N(x,y)^{\smile}$ and $N(x,y) \leq N(x,z)\,;N(z,y)$ for all $x,y,z \in N$.
   (b) $N(x_1,x_n) \leq N(x_1,x_2)\,;N(x_2,x_3)\,;\cdots;N(x_{n-1},x_n)$, for all $x_1,\ldots,x_n \in N$. (For arbitrary networks, this property is called path-consistency in the AI literature.)

6. Assume N to be an atomic $\mathcal{A}$-network, where $\mathcal{A}$ is an atomic relation algebra. Define a binary relation $\sim$ on the nodes of N by $x \sim y$ iff $N(x,y) \leq 1'$. Prove that $\sim$ is an equivalence relation on the nodes of N. Show that it is a *congruence* — i.e., if $x \sim x'$ and $y \sim y'$ then $N(x,y) = N(x',y')$. Deduce that the quotient $N/\!\sim$ is well-defined, and a strict $\mathcal{A}$-network.

7.2 Refining networks

There is a natural notion of one network refining another — i.e., being a better approximation to a representation. It could 'know' about more of the base of the representation than the first network, or it could know more about the relations that hold on the arcs, or both. This leads to a view of a set of networks as a direct system, whose direct limit presumably yields a representation. But at this juncture we must remember that *there may not be a representation of the given algebra* — the whole point of our study is to determine whether there is! So we look at the notion of refinement more abstractly. We continue to let $\mathcal{A}$ denote a relation algebra.

DEFINITION 7.3 If N, N' are $\mathcal{A}$-networks, we say that N' *refines* N, that N is a *subnetwork* of N', and write $N \subseteq N'$, if every node of N is a node of N', and, for all nodes x,y of N, we have $N'(x,y) \leq N(x,y)$.

Evidently, $\subseteq$ is reflexive and transitive. As with networks themselves, the notion of refinement ('$N \subseteq N'$') is defined abstractly, without reference to a representation.

We hope to characterise when a relation algebra $\mathcal{A}$ has a representation (and maybe even to build one) by what kinds of refinements can be made to $\mathcal{A}$-networks. As we know, an $\mathcal{A}$-network specifies a single element of $\mathcal{A}$ that holds on each of its edges or arcs. A representation of $\mathcal{A}$ determines *for every* $a \in \mathcal{A}$ whether or not a holds on any arc (pair of points of the base). So if a network N really is an approximation to a representation of $\mathcal{A}$, then for each arc e of N and each $a \in \mathcal{A}$, we would be able to find a refinement $N' \supseteq N$ that approximates the representation well enough to decide whether or not a holds on e. That is, either $N'(e) \leq a$ or $N'(e) \leq -a$.

7.2. Refining networks

However, even if we know exactly which elements of $\mathcal{A}$ hold on the arcs in a finite subset S of the base of a representation, it does not follow that restricting the representation to S yields another representation. For example, there may be an arc $e = (x,y) \in S \times S$ on which $a\,;b$ holds, for some $a,b \in \mathcal{A}$; and yet there is no witness $z \in S$ such that a holds on (x,z) and b on (z,y). (Such tuples (x,y,a,b) are often called 'defects'.) Now there certainly will be a suitable z somewhere in the base, because the representation respects composition in $\mathcal{A}$. So if our network N describes S in sufficient detail to reveal that $a\,;b$ holds on e, we should be able to refine N to another network with a witness z as above. That is, for each arc (x,y) of N and each $a,b \in \mathcal{A}$ with $N(x,y) \leq a\,;b$, there should exist $N' \supseteq N$ with $N'(x,z) \leq a$ and $N'(z,y) \leq b$, for some $z \in N'$.

Now, given an arbitrary $\mathcal{A}$-network N, should these two kinds of refinements always exist? Our argument for their existence assumed that N was a genuine approximation to a representation. If $\mathcal{A}$ is not representable, or if it is but N does not itself approximate a representation, then we may not be able to find such refinements. But notwithstanding this, whether all required refinements of a given network N exist or not is a well-defined property of N that makes sense regardless of whether N approximates a representation of $\mathcal{A}$.

Suppose then that all required refinements (as above) exist for the network N. Does this mean that N approximates a real representation of $\mathcal{A}$? To say so would be too hasty. If a network truly approximates a representation then all the refinements exist, because the representation contains the required information. But the refinements themselves will approximate the representation in this case; so they, too, should have appropriate refinements, and so on. If all this is true, we could say that the original N is *'hereditarily refinable'*.

Any network N that approximates a representation is hereditarily refinable. And if N is hereditarily refinable, we may use a 'step-by-step' process to progressively refine it, resolving the defects of it and its refinements, one by one. In the 'limit', we may hope to obtain, essentially, a representation. There are certain technical considerations to be borne in mind here, as we will see, but our optimism will turn out to be justified.

Although deciding whether a network is hereditarily refinable may seem a daunting task (and indeed it can be — in general it is undecidable whether a given relation algebra network is hereditarily refinable: see exercise 18.10(3)), the notion has two great strengths. First, there is a natural way to approximate it, by saying that a network can be refined, and the refinements refined, and so on — but only for n steps, say, not forever. These approximations are of crucial value to us, and will allow us to solve a range of problems of representability in part IV of the book. A second, related, point is that the notion lends itself to a treatment by stages — ranks, or games.

It is well known that the approaches by ranks and games are equivalent. We prefer games for their transparency, making it easy to control the refinement process, but before we concentrate on games we briefly describe a corresponding rank treatment. It could have the following form. Every $\mathcal{A}$-network is given a rank, either an ordinal or ∞.

DEFINITION 7.4 Every network N has rank at least 0. If δ is a limit ordinal, N has rank at least δ iff it has rank at least α for every ordinal $\alpha < \delta$. And N has rank at least $\alpha + 1$ iff

1. for every $a \in \mathcal{A}$ and every edge e of N, there is a network $N' \supseteq N$ of rank at least α and such that $N'(e) \leq a$ or $N'(e) \leq -a$;

2. for every $a, b \in \mathcal{A}$ and every edge (x, y) of N with $N(x,y) \leq a;b$, there is a network $N' \supseteq N$ of rank at least α that contains a node z with $N'(x,z) \leq a$ and $N'(z,y) \leq b$.

A network N has rank α if it has rank at least α but not at least $\alpha + 1$. Finally, N is said to have rank ∞ if it has rank at least α for every ordinal α.

We write $\mathrm{rk}(N)$ for the rank of a network N.

Networks of rank ∞ are the ones that truly approximate representations. They can be equivalently characterised by games, and this is our topic in the next section. (We could use 'rank ∞' as a formal definition of hereditary refinability of a network, but we prefer the game definitions (7.7, 7.15) below.) Properties of the rank are investigated in the exercises.

Exercises

Below, N is an $\mathcal{A}$-network, for some relation algebra $\mathcal{A}$.

1. Let N' be a complete $\mathcal{A}$-labelled graph, each node of which is also a node of N, and such that the label of every reflexive edge of N' lies under the identity and for every edge e of N, $N'(e) \geq N(e)$. Show that N' is an $\mathcal{A}$-network, with $N' \subseteq N$.

2. Suppose that $\mathcal{A}$ is finite and $\mathrm{rk}(N) = \infty$. Show that N has a refinement N', say, an $\mathcal{A}$-network on the same nodes as N, such that for every edge e of N', $N'(e)$ is an atom of $\mathcal{A}$.

3. (a) If $\mathrm{rk}(N) \geq \alpha$, show that $\mathrm{rk}(N) \geq \beta$ for all $\beta < \alpha$.

   (b) Let N' be an $\mathcal{A}$-network that is isomorphic to N: i.e., there is a bijection $\theta : \mathrm{nodes}(N) \to \mathrm{nodes}(N')$ such that for all $x, y \in N$, $N(x,y) = N'(\theta(x), \theta(y))$. Show that $\mathrm{rk}(N) = \mathrm{rk}(N')$.

   (In a sense this exercise is trivial, but it is also important, and gives practice in formal proofs by induction on rank. In 3b, do not forget the case where the rank is ∞.)

7.3. All weakly associative algebras have relativised representations

4. If $\mathrm{rk}(N) = \infty$, $a \in \mathcal{A}$, and e is an edge of N, show that there is a network $N' \supseteq N$ of rank ∞ and such that $N'(e) \leq a$ or $N'(e) \leq -a$. Do the same for the second clause of definition 7.4.

5. Let N' be a network with $N \subseteq N'$, so that N' refines N. Show that $\mathrm{rk}(N') \leq \mathrm{rk}(N)$.

6. (a) For $a \in \mathcal{A}$ and e an edge of N, show that $\{\mathrm{rk}(N') : N' \supseteq N, N'(e) \leq a$ or $N'(e) \leq -a\}$ has a maximum — that is, there is $N' \supseteq N$, with $N'(e) \leq a$ or $N'(e) \leq -a$, of maximum possible rank, perhaps ∞. Do the same for clause 2 of the definition of 'rank'. [This says there is a coarsest-possible refinement. Use exercise 5.]

   (b) Show that if α is infinite and $\mathrm{rk}(N) \geq \alpha$ then $\mathrm{rk}(N) \geq \alpha + 1$. Deduce that for any network N, $\mathrm{rk}(N) < \omega$ or $\mathrm{rk}(N) = \infty$.

7.3 All weakly associative algebras have relativised representations

In order to show how to handle the notion of 'hereditarily refinable' (section 7.2) by games, we will prove the following theorem, important in its own right.

THEOREM 7.5 (Maddux) *Every weakly associative algebra has a relativised representation.*

See definitions 5.4 and 5.1 for weakly associative algebras and relativised representations. This theorem was first proved by Maddux in [Madd78b, pp 76–87], for semi-associative algebras; the same argument works for weakly associative algebras too, and [Madd82] proves it explicitly for weakly associative algebras. Maddux used a 'step-by-step' process rather similar to the game-based one we now describe.

Relativised representations present some differences from the relation algebra representations in the discussion of the preceding sections. The unit in a relativised representation is just a (reflexive symmetric) binary relation on the base, not necessarily of the form $B \times B$, and composition and complement are relativised to the unit. So we begin by modifying the notion of network to fit these relativised representations.

DEFINITION 7.6 Let $\mathcal{A}$ be a weakly associative algebra. A (strict) *relativised $\mathcal{A}$-pre-network* is a pair $N = (N_1, N_2)$, where N_1 is a finite (possibly empty) set of nodes, and $N_2 : N_1 \times N_1 \to \mathcal{A}$ is a partial map whose domain is a reflexive and symmetric binary relation on N_1. N is said to be *atomic* if $\mathrm{rng}(N_2) \subseteq \mathrm{At}\,\mathcal{A}$.

As before, we will write N for any of N, N_1, N_2; we write nodes(N) for N_1 and edges(N) for dom(N_2).

N is said to be a *relativised network* if it satisfies:

1. for all $(x,y) \in$ edges(N), we have $N(x,y) \leq 1$' iff $x = y$,

2. $N(x,y) \cdot (N(x,z); N(z,y)) \neq 0$ for all $(x,y), (x,z), (y,z) \in$ edges(N).

We write $\emptyset$ for the relativised network $(\emptyset, \emptyset)$. For relativised pre-networks N, N', we write $N \subseteq N'$ if nodes$(N) \subseteq$ nodes(N'), edges$(N) \subseteq$ edges(N'), and $N'(x,y) \leq N(x,y)$ for all $(x,y) \in$ edges(N).

A relativised network approximates a relativised representation of $\mathcal{A}$. Pairs $(x,y) \in$ edges(N) are called *edges* or *arcs* of N, and they correspond to pairs of elements of the base of a relativised representation lying in the unit. We define only *strict* relativised networks because it makes life easier when the unit is not an equivalence relation, but non-strict ones could also be used at the cost of some messiness. A strict network as in definition 7.1 is a special case of a non-empty relativised network where *all* pairs of nodes form edges: this is appropriate for (square) classical representations. Lemma 7.2 continues to hold for relativised networks, since lemmas 3.17 and 3.12 were proved without using the associativity axiom R1 of relation algebras.

The next step is to define a game to refine relativised networks. There are two players, $\forall$ and $\exists$. It will become apparent that these names (symbols) are appropriate to their roles in the game. Following Hodges [Hod85], we take $\forall$ to be male, and $\exists$, female; this gives them distinctive identities and makes games involving them easier to think about. See chapter 1 for a discussion.

DEFINITION 7.7 Let $\mathcal{A}$ be a weakly associative algebra. We define a game, denoted by $G_\omega^{\mathrm{rel}}(\mathcal{A})$, with ω rounds $(0, 1, \dots)$, in which the players $\forall, \exists$ build an infinite chain of relativised $\mathcal{A}$-pre-networks

$$\emptyset = N_0 \subseteq N_1 \subseteq \cdots$$

In round t $(t < \omega)$, assuming that N_t is the current pre-network, the players move as follows.

1. $\forall$ may choose non-zero $a \in \mathcal{A}$, and $\exists$ must respond with a relativised pre-network $N_{t+1} \supseteq N_t$ containing an edge e with $N_{t+1}(e) \leq a$.

2. Alternatively, $\forall$ may choose an edge (x,y) of N_t and an element $a \in \mathcal{A}$. $\exists$ must respond with a relativised pre-network $N_{t+1} \supseteq N_t$ such that either $N_{t+1}(x,y) \leq a$ or $N_{t+1}(x,y) \leq -a$.

7.3. All weakly associative algebras have relativised representations 227

3. Or ∀ may choose an edge (x,y) of N_t and elements $b, c \in \mathcal{A}$ with $N_t(x,y) \leq b;c$. ∃ must respond with a relativised pre-network $N_{t+1} \supseteq N_t$ such that for some $z \in N_{t+1}$ (possibly $z \in N_t$ already), $N_{t+1}(x,z) \leq b$, $N_{t+1}(z,y) \leq c$, and $N_{t+1}(x,y) \leq b;c$.

∃ wins if each relativised pre-network $N_0, N_1, \ldots$ played during the game is actually a relativised *network*. Otherwise, ∀ wins. There are no draws.

The first kind of move is to do with faithfulness of the representation (with relation algebras we handle this differently). ∀'s first move must obviously be of this kind. The second and third kinds of move were discussed before.

What we have really defined are the rules of the game. There are many different plays or matches of the game that satisfy the rules. The idea of the game is that a relativised network N is refined in each round of the game. ∀ can challenge ∃ to find a refinement of N in any of the three ways he likes, and she must either do so, or lose. Since the game then continues from there, her response must itself be refinable in any way, if she is not to lose. So we have a notion that should serve as (a game-theoretic version of) the 'hereditary refinability' of section 7.2, namely that ∃ can survive forever in this game, whatever moves ∀ makes.

We now show that every relativised network is hereditarily refinable. This is a surprising result, and the proof uses that not all pairs of nodes of a relativised network need be edges. The analogous result for networks fails.

LEMMA 7.8 *Let $\mathcal{A}$ be a weakly associative algebra. Then ∃ can win any play of $G_\omega^{\mathrm{rel}}(\mathcal{A})$ regardless of what moves ∀ makes.*

Proof. Recall from theorem 5.8 that **WA** is a canonical variety, so that $\mathcal{A}^+$ is an (atomic) weakly associative algebra of which $\mathcal{A}$ is (up to isomorphism) a subalgebra. Hence, any relativised $\mathcal{A}$-pre-network is also a relativised $\mathcal{A}^+$-pre-network. In each round t of the game $G_\omega^{\mathrm{rel}}(\mathcal{A})$, where N_t is as above, ∃ will construct an atomic relativised $\mathcal{A}^+$-network M_t satisfying

$$M_t \supseteq N_t, \quad \mathrm{nodes}(M_t) = \mathrm{nodes}(N_t), \quad \mathrm{edges}(M_t) = \mathrm{edges}(N_t).$$

Then if $(x,y), (x,z), (z,y)$ are edges of N_t, we have

$$N_t(x,y) \cdot (N_t(x,z); N_t(z,y)) \geq M_t(x,y) \cdot (M_t(x,z); M_t(z,y)) \neq 0.$$

So if ∃ also arranges that $N_t(x,x) \leq 1$' for all $x \in N_t$, N_t will be a relativised $\mathcal{A}$-network. If ∃ can only do this, she will win the game.

∃ starts by letting $M_0 = N_0 = \emptyset$. M_0 clearly meets the requirements.

Suppose that we are in round $t < \omega$ of $G_\omega^{\mathrm{rel}}(\mathcal{A})$, and assume inductively that ∃ has managed to construct $M_t \supseteq N_t$ as above. We consider the possible moves that

∀ can make in this position, and how ∃ may respond with N_{t+1} and build M_{t+1} to keep the inductive hypothesis. Notice how she uses M_t to guide her response in case 2.

1. Suppose that ∀ picks non-zero $a \in \mathcal{A}$. ∃ chooses an atom $a^- \in \operatorname{At}\mathcal{A}^+$ with $a^- \leq a$. She chooses new nodes $x, y \notin N_t$, with $x = y$ iff $a^- \leq 1$', and creates new relativised networks N_{t+1}, M_{t+1} with nodes those of N_t plus x and y, and edges those of N_t plus $(x,y), (y,x), (x,x)$, and (y,y). The labels in M_{t+1} are defined as in lemma 5.11(1): $M_{t+1}(x,y) = a^-$, $M_{t+1}(y,x) = (a^-)^\smile$, $M_{t+1}(x,x) = \operatorname{st}(a^-)$, and $M_{t+1}(y,y) = \operatorname{end}(a^-)$. It follows from the lemma that this is well-defined and that M_{t+1} is an atomic relativised $\mathcal{A}^+$-network. Labels in N_{t+1} are given by:

   - $N_{t+1}(x,y) = a \cdot 1$' if $x = y$,
   - $N_{t+1}(x,y) = a$, $N_{t+1}(y,x) = 1$, and $N_{t+1}(x,x) = N_{t+1}(y,y) = 1$' if $x \neq y$.

   ∃ responds to ∀'s move in round t with N_{t+1}; clearly, $N_t \subseteq N_{t+1} \subseteq M_{t+1}$, as required.

2. If ∀ picks an edge (x,y) of N_t and an element $a \in \mathcal{A}$ for his move, ∃ lets $M_{t+1} = M_t$, and lets N_{t+1} be the same as N_t except that

$$N_{t+1}(x,y) = \begin{cases} N_t(x,y) \cdot a, & \text{if } M_t(x,y) \leq a \\ N_t(x,y) \cdot -a, & \text{otherwise.} \end{cases}$$

   Because $M_t(x,y)$ is an atom of $\mathcal{A}^+$, it follows that if $M_t(x,y) \not\leq a$ then $M_t(x,y) \leq -a$, so this is satisfactory.

3. Alternatively, suppose that ∀ picks $x, y \in N_t$ and $b, c \in \mathcal{A}$ with $N_t(x,y) \leq b;c$. Let $M_t(x,y) = a^-$. As $M_t \supseteq N_t$, we have $a^- \leq b;c$.

   There are two cases. First suppose that there is $z \in M_t$ with $M_t(x,z) \leq b$ and $M_t(z,y) \leq c$. ∃ again lets $M_{t+1} = M_t$, and defines N_{t+1} to be the same as N_t except that $N_{t+1}(x,z) = N_t(x,z) \cdot b$, $N_{t+1}(z,y) = N_t(z,y) \cdot c$, and $N_{t+1}(x,y) = N_t(x,y) \cdot (b;c)$. Then clearly, $M_{t+1} \supseteq N_{t+1}$.

   Now assume there is no such z. Since composition is completely additive and $\mathcal{A}^+$ is atomic, $b;c = \sum\{b^-;c^- : b^-, c^- \in \operatorname{At}\mathcal{A}^+, b^- \leq b, c^- \leq c\}$. So we may choose atoms $b^-, c^- \in \operatorname{At}\mathcal{A}^+$ with $b^- \leq b, c^- \leq c$, and $a^- \leq b^-;c^-$. Our assumption implies that $b^-, c^- \leq -1$', for if (for example) $b^- \leq 1$' we could take $z = x$. Noting that $x = y$ iff $a^- \leq 1$', we see by lemma 5.11(2) that there is an atomic relativised $\mathcal{A}^+$-network G with nodes$(G) = \{x,y,z\}$ for some new node $z \notin M_t$, edges$(G) = {}^2\{x,y,z\}$, and $G(x,y) = a^-$, $G(x,z) = b^-, G(z,y) = c^-$. Now it is easily checked that $M_t(x,x) = \operatorname{st}(a^-)$, $M_t(y,x) = (a^-)^\smile$, and $M_t(y,y) = \operatorname{end}(a^-)$. So by definition of G in lemma 5.11, we see

7.3. All weakly associative algebras have relativised representations

that $G(i, j) = M_t(i, j)$ for all $i, j \in \{x, y\}$. That is, the subnetworks of M_t and G with nodes $\{x, y\}$ are isomorphic, so we may amalgamate or 'glue' M_t and G together along x, y and define M_{t+1} as the outcome (see figure 7.1). The

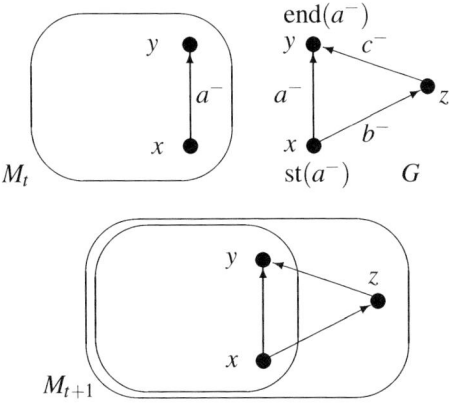

Figure 7.1: Making M_{t+1} by amalgamating M_t and G

nodes of M_{t+1} are those of M_t, plus z; its edges are those of M_t plus (x, z), (z, z), (z, y), and their converses, and the labels are given by $M_{t+1}(i, j) = M_t(i, j)$ if $i, j \in M_t$, and $G(i, j)$ otherwise. Clearly, M_{t+1} as so defined is an atomic relativised $\mathcal{A}^+$-network. N_{t+1} is now defined to have the same nodes and edges as M_{t+1}, with labelling as for N_t except that $N_{t+1}(x, z) = b$, $N_{t+1}(z, z) = 1'$, $N_{t+1}(z, y) = c$, and $N_{t+1}(x, y) = N_t(x, y) \cdot (b; c)$. This is well-defined since $z \notin N_t$. If $x \neq y$, we also need to specify $N_{t+1}(z, x)$ and $N_{t+1}(y, z)$, and we define them to be 1.

Then $N_t \subseteq N_{t+1} \subseteq M_{t+1}$, and N_{t+1} is appropriate to be played by $\exists$ in response to $\forall$'s move (and she does so).

$\square$

This lemma fails for non-relativised networks, since $\exists$ would have to label all pairs of nodes of M_{t+1}, including pairs (z, v) for all $v \in N_t$. If she is forced to do this throughout the game, she may inevitably lose at some stage by failing to create a network.

Recall from definition 5.1 that a relativised representation of a relation-type algebra $\mathcal{A}$ is a model of the following theory $R_{\mathcal{A}}$ whose signature consists of a

230   Chapter 7. Axiomatising representable relation algebras

binary relation symbol a for each $a \in \mathcal{A}$:

$$\begin{array}{rl}
\forall xy[1'(x,y) \leftrightarrow x = y] & \text{for each } a,b,c \in \mathcal{A} \text{ with:} \\
\forall xy[a(x,y) \leftrightarrow b(x,y) \vee c(x,y)] & a = b+c \\
\forall xy[1(x,y) \rightarrow (a(x,y) \leftrightarrow \neg b(x,y))] & a = -b \\
\forall xy[a(x,y) \leftrightarrow b(y,x)] & a = \breve{b} \\
\forall xy[1(x,y) \rightarrow (a(x,y) \leftrightarrow \exists z(b(x,z) \wedge c(z,y)))] & a = b\,;c \\
\exists xy\, a(x,y) & a \neq 0.
\end{array}$$

LEMMA 7.9 *Any countable weakly associative algebra $\mathcal{A}$ has a relativised representation.*

Proof. Consider a play $N_0 \subseteq N_1 \subseteq \cdots$ of $G^{\mathrm{rel}}_\omega(\mathcal{A})$ in which $\exists$ plays as in lemma 7.8 (so that all the N_t are relativised networks), and $\forall$ plays every possible legal move at some stage of play. That is, (i) each non-zero $a \in \mathcal{A}$ is played by him in some round, (ii) for each pair x,y of nodes that arise during the game and every $a \in \mathcal{A}$, $\forall$ plays x,y,a in some round, and (iii) for every $b,c \in \mathcal{A}$, every $t < \omega$, and each pair x,y of nodes of N_t with (x,y) an edge of N_t and $N_t(x,y) \leq b\,;c$, $\forall$ plays x,y,b,c in some round. He can do all this, as $\mathcal{A}$ and the networks N_t in the game are countable (cf. exercise 1 below). The outcome of play is essentially a relativised representation of $\mathcal{A}$, defined as follows. Let $N = \bigcup_{t<\omega} \mathrm{nodes}(N_t)$, and $\mathrm{edges}(N) = \bigcup_{t<\omega} \mathrm{edges}(N_t) \subseteq N \times N$. We make N into a model of $R_\mathcal{A}$ by interpreting each $a \in \mathcal{A}$ as follows:

$$a^N = \{(x,y) \in \mathrm{edges}(N) : \exists t < \omega(x,y \in N_t \text{ and } N_t(x,y) \leq a)\}.$$

Note that for any a,x,y, and $t' \geq t$, since $N_t \subseteq N_{t'}$ we have $N_t(x,y) \leq a \Rightarrow N_{t'}(x,y) \leq a$. So if $N \models a(x,y)$ then $N_t(x,y) \leq a$ for all large enough $t < \omega$. We use this below without explicit mention. Note also that $1^N = \mathrm{edges}(N)$.

Let us check that the axioms of $R_\mathcal{A}$ hold in N. The $1'$-axiom holds by definition of (strict) relativised network. For the $+$-axiom, it is clear that if $b,c \in \mathcal{A}$, $x,y \in N$, and $N \models b(x,y) \vee c(x,y)$ then $N \models (b+c)(x,y)$. For the converse, suppose $N \models (b+c)(x,y)$. Choose $t < \omega$ sufficiently large that $x,y \in N_t$, $N_t(x,y) \leq b+c$, and $\forall$ has played the moves x,y,b and x,y,c in some rounds before t. Then $N_t(x,y) \leq b$ or $N_t(x,y) \leq -b$, and similarly for c. If $N_t(x,y) \leq -b$ and $N_t(x,y) \leq -c$, then $N_t(x,y) = 0$, which is impossible by lemma 7.2(1). So $N_t(x,y) \leq b$ or $N_t(x,y) \leq c$, proving that $N \models b(x,y)$ or $N \models c(x,y)$, as required.

For complement, let $a \in \mathcal{A}$; we show $N \models (-a)(x,y) \leftrightarrow \neg a(x,y)$ for all $(x,y) \in 1^N$. Take $t < \omega$ with $(x,y) \in \mathrm{edges}(N_t)$, such that $\forall$ played the move x,y,a in some round before t. Then either $N_t(x,y) \leq a$ or $N_t(x,y) \leq -a$; and not both, by lemma 7.2(1). Hence, by definition of h, each edge (x,y) is in exactly one of $a^N, (-a)^N$.

7.3. All weakly associative algebras have relativised representations

We check conversion and composition. Let $a \in \mathcal{A}$, $x, y \in N$, and suppose $N \models \breve{a}(x,y)$. Choose $t < \omega$ with $x, y \in N_t$ and $N_t(x,y) \leq \breve{a}$, such that $\forall$ played the move y, x, a in some round before round t. So we have either $N_t(y,x) \leq a$ or $N_t(y,x) \leq -a$. But $N_t(x,y) \leq \breve{a}$, so by lemma 3.11 and axiom R4 of weakly associative algebras, $N_t(x,y)^{\smile} \leq \breve{\breve{a}} = a$, whence by lemma 7.2, $N_t(y,x) \cdot a \neq 0$. So we must have $N_t(y,x) \leq a$. Thus, $N \models a(y,x)$. The converse, that $N \models a(y,x) \to \breve{a}(x,y)$, follows from the same argument, because $\breve{\breve{a}} = a$.

Finally, composition. First let $x, y, z \in N$ with $(x,y), (x,z), (z,y) \in \text{edges}(N)$, $b, c \in \mathcal{A}$, and assume that $N \models b(x,z) \wedge c(z,y)$. We show $N \models (b;c)(x,y)$. If not, then because $N \models 1(x,y)$, we have $N \models (-(b;c))(x,y)$ by the axiom for complement verified above. So there is $t < \omega$ with $x, y, z \in N_t$ and $N_t(x,z) \leq b$, $N_t(z,y) \leq c$, $N_t(x,y) \leq -(b;c)$. This means that $N_t(x,y) \cdot (N_t(x,z) ; N_t(z,y)) = 0$, contradicting the networkhood of N_t.

Now let $x, y \in N$, $b, c \in \mathcal{A}$, and suppose $N \models (b;c)(x,y)$. Choose $t < \omega$ with $x, y \in N_t$ and such that $N_t(x,y) \leq b;c$. There is $u \geq t$ such that $\forall$ played x, y, b, c before round u. So there is $z \in N_u$ with $N_u(x,z) \leq b$, $N_u(z,y) \leq c$. Consequently, $N \models b(x,z) \wedge c(z,y)$, as required. □

We can now prove theorem 7.5 — that any weakly associative algebra has a relativised representation.

Proof. We have already proved it for countable weakly associative algebras. For the general case, let $\mathcal{A} \in \mathbf{WA}$ be arbitrary. We must show that the theory $R_\mathcal{A}$ has a model. By compactness (see section 2.2.6), it suffices to show that any finite $S \subseteq R_\mathcal{A}$ has a model. Let $\mathcal{B}$ be the subalgebra of $\mathcal{A}$ generated by the finite set of elements of $\mathcal{A}$ that occur as relation symbols in S. Then $\mathcal{B}$ is a countable weakly associative algebra, and by lemma 7.9, $R_\mathcal{B}$ has a model. But $S \subseteq R_\mathcal{B}$, so S has a model too, proving the theorem. □

REMARK 7.10 We make the following observations on the proof just given:

1. It illustrates how successive refinements of networks can be made to converge to a representation, the aim of all our discussion so far.

2. It illustrates how games allow a fairly simple intuitive proof of this, without too much complicated notation (of course, this is a matter of opinion). However, it does not conclusively establish the case for using games, since the same construction could have been effected without $\forall, \exists$ by enumerating all possible defects and eliminating them one by one. It is simply (a classic example of) a step-by-step construction.

3. It demonstrates the idea of a *winning strategy* for $\exists$ in a game (this is what we described in lemma 7.8), and our typical level of detail in describing it, in a concrete and interesting case.

4. It shows how countable algebras can be easier to handle in games, and how easy it can be to remove the countability restriction at the end.

5. It introduces some useful techniques such as atomic networks, and 'private games' — $\exists$ won $G_\omega^{\text{rel}}(\mathcal{A})$ in lemma 7.8 essentially by pretending the game was being played on the atomic relativised networks M_t instead of the N_t. This trick is very common and valuable in game arguments; cf. [Hod85]. However, private games like the one in lemma 7.8 can often be got rid of simply by taking them to be the main 'public' game: see exercises 2 and 3.

Exercises

1. Confirm that it is possible for $\forall$ to pick every pair of nodes and algebra elements during the play of $G_\omega^{\text{rel}}(\mathcal{A})$ in the proof of lemma 7.9. [This is a scheduling problem. Each round of the game potentially introduces ω tasks that $\forall$ has to undertake at some stage later in the game (check this). There are ω rounds, so he has ω opportunities to finish a task. Suggest how he might plan his work. For hints, see [Hod85].]

2. Show that any atomic weakly associative algebra has a complete relativised representation. We suggest two ways of doing this:

    (a) We may define a game like G_ω^{rel} except that all relativised networks played must be atomic. This game can be used with the ideas in the proofs of lemmas 7.8 and 7.9 to show that any countable atomic weakly associative algebra has a complete relativised representation. The required result follows by way of the downward Löwenheim–Skolem–Tarski theorem (see section 2.2.6) and compactness.

    (b) Alternatively, it may be helpful to consider a game like G_ω^{rel}, except that it has κ rounds for some suitable cardinal κ, and again all edges of the relativised networks are required to be labelled by atoms of the algebra. Then, as in the proof of lemma 7.9, we can construct a relativised representation of the algebra. But this time we have also arranged that every edge labelled in the relativised representation is labelled by an atom.

    We will have much more to say about atomic games and networks in chapter 11 and the following chapters.

3. From the preceding exercise, deduce theorem 7.5 that any weakly associative algebra has a relativised representation.

7.4 Games on relation algebra networks

We now attempt to set up analogous machinery for representations of relation algebras as in definition 3.28. This is going to be more difficult than for relativised representations of weakly associative algebras, because not every relation algebra is representable. We will use the networks of definition 7.1, and we begin by defining a suitable game to refine them. Fix a relation algebra $\mathcal{A}$. The following definition is useful.

DEFINITION 7.11 An ($\mathcal{A}$-)*pre-network* is a complete directed finite graph with edges labelled by elements of $\mathcal{A}$. Formally, it is a pair $N = (N_1, N_2)$, where N_1 is a finite non-empty set of nodes, and $N_2 : N_1 \times N_1 \to \mathcal{A}$ is a map assigning an element of $\mathcal{A}$ to each pair of nodes.

Thus, a network is simply a pre-network that satisfies conditions (a) and (b) of definition 7.1. As usual, we will use 'N' to denote any of N, N_1, N_2 above, and we define $N \subseteq N'$ for pre-networks N, N' in the same way as for networks.

DEFINITION 7.12 Let N be any pre-network, and let $n \leq \omega$. We define a game, denoted by $G_n(N, \mathcal{A})$. In it, the players $\forall$ and $\exists$ build a chain

$$N = N_0 \subseteq N_1 \subseteq \cdots \subseteq N_n$$

of pre-networks if n is finite, and an infinite chain

$$N = N_0 \subseteq N_1 \subseteq \cdots$$

if $n = \omega$. There are n rounds, numbered $0, 1, \ldots, t, \ldots$ for $t < n$. In each round, t, assuming that N_t is the current network, the players move as follows.

- $\forall$ chooses $x, y \in N_t$, and elements $r, s \in \mathcal{A}$. See figure 7.2.

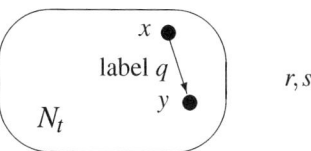

Figure 7.2: $\forall$'s choice

- $\exists$ responds with a pre-network $N_{t+1} \supseteq N_t$ such that one of the following holds (see figure 7.3):

234   *Chapter 7. Axiomatising representable relation algebras*

(**reject**) N_{t+1} is the same as N_t, except that $N_{t+1}(x,y) = N_t(x,y) \cdot -(r;s)$

(**accept**) The nodes of N_{t+1} are those of N_t, plus a new one, z. The labels on edges of N_{t+1} are given as follows:

- $N_{t+1}(x,z) = r$
- $N_{t+1}(z,z) = 1'$
- $N_{t+1}(z,y) = s$
- $N_{t+1}(x,y) = N_t(x,y) \cdot (r;s)$
- If $x', y' \in N_t$ and $(x', y') \neq (x, y)$ then $N_{t+1}(x', y') = N_t(x', y')$.
- All other labels in N_{t+1} not yet mentioned — viz. $N_{t+1}(z,x)$ and $N_{t+1}(y,z)$, if $x \neq y$, and $N_{t+1}(z,w)$ and $N_{t+1}(w,z)$ for $w \in N_t \setminus \{x,y\}$ — are 1.

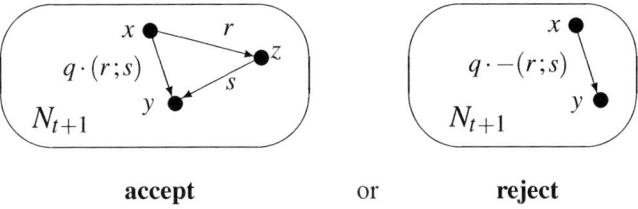

       accept   or   **reject**

Figure 7.3: $\exists$'s response

$\exists$ wins if each pre-network $N_0, N_1, \ldots$ played during the game is actually a network; otherwise, $\forall$ wins.

There are some differences from the game G_ω^{rel} for weakly associative algebras of definition 7.7. $G_n(N, \mathcal{A})$ has an arbitrary starting pre-network, and can last any pre-stated number of rounds $n \leq \omega$. (This will be used later when we axiomatise representability.) The first kind of move of $\forall$ in G_ω^{rel} has been omitted here (we will handle faithfulness of representations in another way), and his second and third kinds of move have been combined into one, for simplicity.

Most importantly, $\exists$'s possible responses are strictly limited now. In each round, she simply has to choose whether to accept or reject. The network she plays in response to $\forall$ is determined up to isomorphism by this information (and by the previous network and $\forall$'s move).[1] The fact that $\exists$ only has two choices in a round is used in proposition 7.24 below, and the fact that she does not choose elements of the algebra is used in the next chapter to provide a *universal* axiomatisation of **RRA**. The use of possibly non-strict networks, where $N(x,y) \leq 1'$ does

---

[1] This approach was suggested by Yde Venema.

7.4. Games on relation algebra networks

not imply $x = y$, allows 'z' always to be a new node and simplifies these axioms. The price we pay is having to factor the limiting network by a suitably defined congruence relation (i.e., to prove lemma 7.22) in order to obtain a genuine representation respecting identity as well as the other operators. However, in the case where $\exists$ accepts, the new node z can be any set, so that the suitable N_{t+1} form a proper class of isomorphic networks! This is rather inconvenient, so by convention we take this node z to be $\{N_t\}$ (say). Thus, N_{t+1} really is determined. This is a technical point and is only important if we wish to formalise the games in set theory. We will rarely want to do this, as it runs against the intuitiveness of games — their key attraction.

In each round of a particular play, $\forall$ challenges $\exists$ to add a certain 'triangle' to an edge of the network, and $\exists$ must either do so, or claim instead that the relation on that edge is disjoint from the relation on the proposed side of the triangle. The pre-network she responds with conveys just this information, and no more. It is the coarsest permissible refinement — cf. exercise 7.2(6). She will lose the game if, at some stage, both options lead to pre-networks violating the second, triangle-consistency condition of the definition of 'network'. She also loses $G_0(N, \mathcal{A})$ if N is not a network, but only a pre-network.

As with the game for weakly associative algebras, 'hereditary refinability' of a network N means that $\exists$ can survive forever in this game, whatever moves $\forall$ makes. Formally, this is expressed by saying that $\exists$ *has a winning strategy in the game* $G_\omega(N, \mathcal{A})$. The approximations to hereditary refinability mentioned earlier will then have the form $\exists$ *has a winning strategy in* $G_n(N, \mathcal{A})$, for various $n < \omega$.

As we said in remark 7.10, games allow a quite simple proof that every weakly associative algebra has a relativised representation, but since the argument there is just a 'step-by-step' construction, this amounts to notational convenience and does not conclusively establish the case for using games. For (non-relativised) representations of relation algebras, the situation is different. Hereditary refinability is not a tautologous property of a network as in definition 7.1 — there is no analogue of lemma 7.8, it is not always possible to construct a representation just by fixing defects one by one, and $\exists$ does not always have a winning strategy in $G_\omega(N, \mathcal{A})$. If we were confined to thinking in terms of step-by-step constructions of representations of relation algebras, we might easily say at this point that they just don't work in general, and give up. With games, however, it is natural to shift the problem from showing that $\exists$ *has* a winning strategy, to asking *when* she has a winning strategy. It will turn out that she has such a strategy precisely when the given relation algebra is representable, and we can use this to axiomatise **RRA**. But first, we had better be clear about what a winning strategy actually is. This is the subject of the next section.

Exercises

1. Suppose, in round 0 of $G_n(N, \mathcal{A})$, $\forall$ chooses nodes $x, y \in N$ and the algebra elements 1', 1'. When should $\exists$ accept (if she is not to lose)? When should she reject? In the case where N is a strict network, can you express your answer differently?

   Does she ever have a choice in this situation?

2. In the same circumstances, let $x, y, w \in N$, and suppose that $\forall$ picks the nodes x, y and the algebra elements $N(x, w), N(w, y)$ as his move in round 0. What should $\exists$ do, to avoid losing?

3. Suppose that the outcome of her decision in exercise 2 is the network $N_1 \supseteq N$. Assuming that $n > 1$, suppose that $\forall$ now picks $y, x, 1', N(x, y)^\smile$. What should $\exists$ do?

7.5 Strategies

A *strategy* for a player in a game of the form $G_n(N, \mathcal{A})$ is simply a set of rules telling the player what move to make in each situation. A strategy for $\forall$ will tell him which edge and which algebra elements to pick, and one for $\exists$ will tell her whether to accept or reject. The advice given by a strategy may depend on the entire course of the game so far: the networks played, and the moves of the players, in prior rounds. It must not depend on the future of play!

The notion of a strategy is quite intuitive, and it is not often helpful to have available a formal definition. However, there are one or two exceptions to this, and besides, readers may feel uneasy without a full definition. So we will give one. Here, $n \leq \omega$ as usual.

DEFINITION 7.13 Let PreNet($\mathcal{A}$) denote the class of all $\mathcal{A}$-pre-networks. A strategy for $\forall$ in $G_n(N, \mathcal{A})$ is a function σ defined on $^{<n}$PreNet($\mathcal{A}$). The value of σ on a sequence $(N_1, \ldots, N_t)$ consists of a pair of nodes $x, y \in N_t$ and a pair of elements $r, s \in \mathcal{A}$. Here, we let $N_0 = N$: the value of σ on the empty sequence is now obtained by taking $t = 0$.

Let us explain what it means for $\forall$ to use the strategy σ in round t of the game, for $t < n$. Suppose that the pre-networks played so far are $N = N_0, N_1, \ldots, N_t$. Then if $\forall$ uses σ in round t, he will play $(x, y, r, s) = \sigma(N_1, \ldots, N_t)$ in this round.

N_0, being fixed as part of the game definition, is not included as a parameter of σ. We do not need to include in the domain of σ information about which round the game is at — t is obviously recoverable from the sequence $(N_0, \ldots, N_t)$. Nor need we include a description of the prior course of play, as this too is recoverable from

7.5. Strategies

σ if ∀ has used it throughout. For in round 0, ∀ will have played σ(∅), where ∅ ∈ $^{<n}$PreNet($\mathcal{A}$) is the empty sequence. ∃ then responded with the pre-network N_1, and ∀ can determine from this and σ(∅) whether she accepted or rejected (she accepted iff $|N_1| > |N_0|$). He can then determine his own move in round 1: it was σ(N_1). Continuing this process inductively, he can recover the entire course of play from $(N_1, \ldots, N_t)$ and σ. This means that the form of the strategy given is fully general in that its advice may depend on anything that happened in the game so far.

DEFINITION 7.14 A strategy for ∃ in the game $G_n(N, \mathcal{A})$ is a map τ that takes arguments of the form

$$\tau = (x_0, y_0, r_0, s_0, x_1, y_1, r_1, s_1, \ldots, x_t, y_t, r_t, s_t) \quad (t < n)$$

and whose value is either accept or reject.

∃ uses τ in round t of play if, given that $N = N_0$ was the starting network and ∀ selected nodes $x_u, y_u \in N_u$ and elements $r_u, s_u \in \mathcal{A}$ in round u for each $u \leq t$, she accepts or rejects in round t according to the value $\tau(x_0, y_0, r_0, s_0, \ldots, x_t, y_t, r_t, s_t)$.

Rather as for σ, above, if ∃ uses τ throughout the game then she can recover the whole play so far from this: her own move in round u for $u < t$ was $\tau(x_0, y_0, r_0, s_0, \ldots, x_u, y_u, r_u, s_u)$; and given the network N_u existing at the start of round u, this determines N_{u+1}. So the form of strategy given is fully general.

We have defined *deterministic* strategies that always dictate a unique move for a player. This is for simplicity only; one may also consider non-deterministic ones that offer a choice of moves, but we will not do so here.

A strategy is said to be *used* by a player in a play of the game if that player uses it in every round — so his or her moves always accord with what the strategy suggests. One may note that the use of a strategy throughout play will (naturally) restrict the possible positions that may be arrived at during play. Hence, some sequences of networks never arise in plays where σ is used, and σ need not be defined on such sequences. This is of no consequence for us. One may either define a strategy to be a partial map (and face the problem of defining its domain formally!) or ignore its value on arguments on which it will never be used. We prefer the latter approach, as it is simpler and fits better with our usual informal descriptions of strategies.

DEFINITION 7.15 A strategy in the game $G_n(N, \mathcal{A})$ is said to be *winning* for its owner if the owner wins all matches in which the strategy is used, regardless of how the opposing player decides to move.

To say that an $\mathcal{A}$-network N is *hereditarily refinable* means formally that ∃ has a winning strategy in $G_\omega(N, \mathcal{A})$.

Clearly, only one player can have a winning strategy. There are (other kinds of) games in which neither player has a winning strategy. However, in our games, the condition for $\exists$ to win is just that she survives every round. In [GalSte53], such games are called *closed*, and the Gale–Stewart theorem proves that closed games are *determined* — one of the players always does have a winning strategy. Of course, who it is will depend on $\mathcal{A}, N$, and n. We will be able to see this in more concrete terms in our situation, so we do not divert into a proof of the Gale–Stewart theorem. (It is similar to exercise 6 below.) Readers interested in these matters may consult [Hod85, Kec95].

We will frequently use the following lemma, whose proof is left as an exercise.

LEMMA 7.16 *Let $\mathcal{A}$ be a relation algebra, N an $\mathcal{A}$-pre-network, and $n \leq \omega$.*

1. *If N' is an $\mathcal{A}$-network, $N \subseteq N'$, and $\exists$ has a winning strategy in $G_n(N', \mathcal{A})$, then she has one in $G_n(N, \mathcal{A})$.*

2. *$\exists$ has a winning strategy in $G_{1+n}(N, \mathcal{A})$ iff whatever move $\forall$ makes in round 0, $\exists$ has a winning strategy in $G_n(N^+, \mathcal{A})$ or in $G_n(N^-, \mathcal{A})$, where N^+, N^- are the pre-networks resulting from her accepting and rejecting $\forall$'s move. (Here, $1 + \omega = \omega$.)*

For later use, we also state and prove a rather obvious lemma.

LEMMA 7.17 *Let $n < \omega$, let $\mathcal{A}$ be a relation algebra, and let N be an $\mathcal{A}$-network. If $\forall$ has a winning strategy in the game $G_n(N, \mathcal{A})$ then there is a finite set S of pairs of elements of $\mathcal{A}$ such that his winning strategy only ever directs him to choose elements from S.*

Proof. Let σ be a fixed and deterministic winning strategy for $\forall$. In each round, $\exists$ has at most two choices for her moves. Therefore, there are at most 2^n possible plays of $G_n(N, \mathcal{A})$ in which $\forall$ uses σ. In each round, $\forall$ is directed to choose a pair of elements in his moves. So σ can direct him to choose at most $n \times 2^n$ different pairs. $\square$

Exercises

1. Define a variant $G'_n(N, \mathcal{A})$ of $G_n(N, \mathcal{A})$ in the style of the game G^{rel}_ω, with two kinds of move (corresponding to moves 2 and 3 of definition 7.7) instead of just one. Show that $\exists$ has a winning strategy in $G'_n(N, \mathcal{A})$ iff she has one in $G_n(N, \mathcal{A})$.

2. Let $n \leq \omega$, let $\mathcal{B} \subseteq \mathcal{A}$ be relation algebras, and let N be a $\mathcal{B}$-pre-network. Show that if $\exists$ has a winning strategy in $G_n(N, \mathcal{A})$ then she also has one in $G_n(N, \mathcal{B})$.

7.6. Games and representations of relation algebras

3. Let $n \leq \omega$, let $\mathcal{A}$ be a relation algebra, and let N be an $\mathcal{A}$-pre-network. If $\forall$ has a winning strategy in $G_n(N,\mathcal{A})$, show that he has one in $G_n(N,\mathcal{B})$ for some finitely generated subalgebra $\mathcal{B} \subseteq \mathcal{A}$.

4. Prove lemma 7.16. [Cf. exercise 7.2(4, 5).]

5. Let $\mathcal{A}$ be a relation algebra, N be an $\mathcal{A}$-network, and $n < \omega$. Show that the game $G_n(N,\mathcal{A})$ is determined.

6. Let N be an $\mathcal{A}$-network, for some relation algebra $\mathcal{A}$.

    (a) Suppose that $\mathrm{rk}(N) = \infty$ (in terms of definition 7.4). Show that whatever move $\forall$ makes in round 0 of $G_\omega(N,\mathcal{A})$, $\exists$ has a response N' of rank ∞.

    (b) Suppose that $\mathrm{rk}(N) < \infty$. Show that there is some move that $\forall$ can make in round 0 of $G_\omega(N,\mathcal{A})$ such that every response N' of $\exists$ has rank $< \mathrm{rk}(N)$. (Here, if N is a pre-network but not a network, define $\mathrm{rk}(N) = -1$.)

    (c) Show that $\exists$ has a winning strategy in $G_\omega(N,\mathcal{A})$ ('N is hereditarily refinable') if and only if $\mathrm{rk}(N) = \infty$.

    (d) Deduce that the game $G_\omega(N,\mathcal{A})$ is determined.

7. Propose a formalisation of a non-deterministic strategy for $\forall$ and for $\exists$ in $G_n(N,\mathcal{A})$.

8. Let $\mathcal{A}$ be any relation algebra, and $n < \omega$. Show that for any $\mathcal{A}$-pre-network N, $\exists$ has a winning strategy in $G_n(N,\mathcal{A})$ iff $\mathrm{rk}(N) \geq n$. Deduce again (cf. exercise 7.2(6)) that $\mathrm{rk}(N) < \omega$ or $\mathrm{rk}(N) = \infty$. [Proposition 7.24 below may help.]

7.6 Games and representations of relation algebras

Now comes the key proposition, in which we show that the hereditarily refinable networks are indeed exactly those that approximate representations.

DEFINITION 7.18 Let $\mathcal{A}$ be a relation algebra. For each non-zero $a \in \mathcal{A}$, let I_a denote the network consisting of two distinct nodes, 0, 1, with $I_a(0,1) = a$, $I_a(1,0) = 1$, and $I_a(0,0) = I_a(1,1) = 1'$.

THEOREM 7.19 *Let $\mathcal{A}$ be any relation algebra. Then $\mathcal{A}$ is representable if and only if $\exists$ has a winning strategy in the game $G_\omega(I_a,\mathcal{A})$ for all non-zero $a \in \mathcal{A}$.*

Proof. '⇒': Let $h: \mathcal{A} \to \mathcal{B}$ be a representation of $\mathcal{A}$, where $\mathcal{B}$ is a proper relation algebra with base B, say, and h is an isomorphism from $\mathcal{A}$ onto $\mathcal{B}$. (So the unit $h(1)$ is an equivalence relation on the base B of $\mathcal{B}$, and each $h(a)$, for $a \in \mathcal{A}$, is a binary relation on B.) ∃ can use h to help her decide whether to accept or reject in each round of $G_\omega(I_a, \mathcal{A})$ for any non-zero a. In more detail, she preserves the condition that for each $t < \omega$, there is a map $': N_t \to B$ satisfying

$$(x', y') \in h(N_t(x, y)) \text{ for all nodes } x, y \text{ of } N_t. \qquad (*)$$

This formalises the idea that N_t approximates h.

Since $a \neq 0$ and h is a representation, there are points $0', 1' \in B$ with $(0', 1') \in h(a)$. This defines $': N_0 = I_a \to B$, and it clearly satisfies $(*)$.

Assume that $(*)$ holds for the current network N_t at the beginning of round t. If ∀ plays x, y, r, s in this round, then:

- If $(x', y') \in h(r; s)$, ∃ accepts. This determines the $(t+1)$th network in the game, $N_{t+1} \supseteq N_t$. We have $N_{t+1} \setminus N_t = \{z\}$ for some new node z. ∃ extends the map $'$ by mapping z to any $z' \in B$ with $r(x', z') \in h(r)$ and $(z', y') \in h(s)$. Such a z' exists because h is a representation of $\mathcal{A}$ and preserves composition.

- If $(x', y') \notin h(r; s)$, she rejects. This leads to N_{t+1} having the same nodes as N_t; the only difference between N_{t+1} and N_t is that $N_{t+1}(x, y) = N_t(x, y) \cdot -(r; s)$. But $(x', y') \in h(N_t(x, y)) \subseteq h(1)$, and as h is a representation, we have $(x', y') \in h(1) \setminus h(r; s) = h(-(r; s))$. So

$$\begin{aligned}(x', y') &\in h(N_t(x, y)) \cdot h(-(r; s)) \\ &= h(N_t(x, y) \cdot -(r; s)) \\ &= h(N_{t+1}(x, y)).\end{aligned}$$

Thus, the conditions $(*)$ on the map $'$ are kept into the next round.

It follows from $(*)$ that each N_t ($t < \omega$) is a network. The condition $N_t(x, x) \leq 1'$ holds by definition of N_t. For triangle consistency, let $x, y, z \in N_t$ and write $a = N_t(x, y), b = N_t(x, z), c = N_t(z, y)$. We require $a \cdot (b; c) \neq 0$. First, $(*)$ gives $(x', z') \in h(b)$ and $(z', y') \in h(c)$, so $(x', y') \in h(b) | h(c) = h(b; c)$. Since also $(x', y') \in h(a)$, by $(*)$ again, we obtain $(x', y') \in h(a) \cap h(b; c) = h(a \cdot (b; c))$, so that the right-hand side is non-empty. Consequently, $a \cdot b; c \neq 0$, proving triangle consistency.

We have therefore given a winning strategy for ∃, since she clearly never loses in any round. (This will be the usual level of detail at which we describe strategies: it can be formalised but it is not helpful to the argument to do so.)

'⇐': It suffices to prove the result for countable algebras $\mathcal{A}$. For, given this special case of the theorem, assume that ∃ has a winning strategy in all games $G_\omega(I_a, \mathcal{A})$

7.6. Games and representations of relation algebras

for non-zero $a \in \mathcal{A}$. Since **RRA** is a variety (theorem 3.37), to show that $\mathcal{A}$ is representable it suffices by corollary 3.39 to show that any countable subalgebra $\mathcal{B}$ of $\mathcal{A}$ is representable. Take such a $\mathcal{B}$. Then $\exists$ has a winning strategy in $G_\omega(I_b, \mathcal{B})$ for all non-zero $b \in \mathcal{B}$, for this game can be viewed as a restricted game $G_\omega(I_b, \mathcal{A})$ in which $\forall$ is only allowed to choose algebra elements from $\mathcal{B}$, so $\exists$ may win by simply applying her winning strategy in $G_\omega(I_b, \mathcal{A})$. (See exercise 7.5(2).) By the (assumed) theorem for countable algebras, $\mathcal{B}$ is representable, as required.

So assume from now on that $\mathcal{A}$ is countable. Let a be a non-zero element of $\mathcal{A}$. Consider a play $N_0 \subseteq N_1 \subseteq \cdots$ of $G_\omega(I_a, \mathcal{A})$, in which $\exists$ uses her winning strategy (so that all the N_t are networks), and $\forall$ plays at some stage x, y, r, s, for each pair x, y of nodes that arise during the game, and for every $r, s \in \mathcal{A}$. This is possible because $\mathcal{A}$ and the N_t are countable — cf. exercise 7.3(1). The play yields a representation of $\mathcal{A}$, defined as follows. Let $N^* = \bigcup_{t<\omega} \operatorname{nodes}(N_t)$, let $\mathcal{P}$ be the full proper relation algebra with base N^*,

$$\mathcal{P} = (\wp(N^* \times N^*), \emptyset, N^* \times N^*, \cup, \setminus, \operatorname{Id}_{N^*}, {}^{-1}, \mid),$$

and define a map $h^* : \mathcal{A} \to \mathcal{P}$ by

$$h^*(r) = \{(x, y) : \exists t < \omega (x, y \in N_t \text{ and } N_t(x, y) \leq r)\}, \text{ for each } r \in \mathcal{A}.$$

Note that $N^*, \mathcal{P}, h^*$ depend on a. Write $\mathcal{A}{\upharpoonright}_{-1'}$ for the reduct of $\mathcal{A}$ to the signature $\{0, 1, +, -, \breve{}, ;\}$ without $1'$, and similarly define $\mathcal{P}{\upharpoonright}_{-1'}$.

LEMMA 7.20 h^* is a homomorphism from $\mathcal{A}{\upharpoonright}_{-1'}$ to $\mathcal{P}{\upharpoonright}_{-1'}$.

Proof. We check that h^* preserves each relation algebra operation in turn. The argument is almost identical to that of lemma 7.9. Let $x, y, z \in N^*$ and $r, s \in \mathcal{A}$.

0 If $(x, y) \in h^*(0)$, then for some $t < \omega$ we have $x, y \in N_t$ and $N_t(x, y) \leq 0$. Since N_t is a network, this contradicts lemma 7.2. So as x, y were arbitrary, $h^*(0) = \emptyset$.

1 Choose $t < \omega$ so that $x, y \in N_t$. Then $N_t(x, y) \in \mathcal{A}$, so obviously $N_t(x, y) \leq 1$. By definition of h^*, we obtain $(x, y) \in h^*(1)$. As x, y were arbitrary, this shows that $h^*(1) = N^* \times N^*$, as required.

Sum We require $h^*(r) \cup h^*(s) = h^*(r+s)$. '$\subseteq$' holds by definition of h^*. Conversely, if $(x, y) \in h^*(r+s)$, pick $t < \omega$ such that $x, y \in N_t$, $N_t(x, y) \leq r+s$, and $\forall$ played $(x, y, r, 1')$ before round t of the game. If $\exists$ accepted when he did so, then $N_t(x, y) \leq r; 1' = r$, so $(x, y) \in h^*(r)$. If she rejected, $N_t(x, y) \leq -(r; 1') = -r$, which, with $N_t(x, y) \leq r+s$, implies $N_t(x, y) \leq s$. So $(x, y) \in h^*(s)$. Either way, $(x, y) \in h^*(r) \cup h^*(s)$, completing the proof.

Complement We require $h^*(-r) = h^*(1) \setminus h^*(r)$. By the result for $+$, $h^*(r) \cup h^*(-r) = h^*(r + -r) = h^*(1)$, so we only need check that $h^*(r) \cap h^*(-r) =$

0. If $(x,y) \in h^*(r) \cap h^*(-r)$, we may choose $t < \omega$ large enough so that $x, y \in N_t$, $N_t(x,y) \leq r$, and $N_t(x,y) \leq -r$. Thus, $N_t(x,y) = 0$, contradicting lemma 7.2 (as N_t is a network).

Conversion Here we require $h^*(\check{r}) = h^*(r)^{-1}$: i.e., $(x,y) \in h^*(r)$ iff $(y,x) \in h^*(\check{r})$. Suppose $(x,y) \in h^*(r)$. Choose $t < \omega$ so large that $x, y \in N_t$, $N_t(x,y) \leq r$, and so that $\forall$ has already played $(y, x, \check{r}, 1')$ in some round before t. If $\exists$ rejected in that round, we will have $N_t(y,x) \leq -(\check{r}; 1') = -\check{r}$. But $N_t(x,y) \leq r$. Monotonicity of $\check{}$ (lemma 3.11) now yields

$$N_t(x,y)^{\check{}} \cdot N_t(y,x) \leq \check{r} \cdot (-\check{r}) = 0,$$

contradicting lemma 7.2. Hence she accepted, so that $N_t(y,x) \leq \check{r}; 1' = \check{r}$, whence $(y,x) \in h^*(\check{r})$.

By the same argument, if $(y,x) \in h^*(\check{r})$ then $(x,y) \in h^*(\check{\check{r}}) = h^*(r)$. Hence we have $(x,y) \in h^*(r)$ iff $(y,x) \in h^*(\check{r})$, as required.

Composition We require $h^*(r;s) = h^*(r) \,|\, h^*(s)$. For '$\supseteq$', suppose that $(x,z) \in h^*(r)$ and $(z,y) \in h^*(s)$. Pick $t < \omega$ so that $x, y, z \in N_t$, $N_t(x,z) \leq r$, $N_t(z,y) \leq s$, and $\forall$ has already played $(x, y, (r;s), 1')$. If $\exists$ rejected when he did this, then $N_t(x,y) \leq -((r;s); 1') = -(r;s)$. So

$$N_t(x,y) \cdot (N_t(x,z); N_t(z,y)) \leq (-(r;s)) \cdot (r;s) = 0,$$

and by definition, N_t is not a network. This is a contradiction, so she must have accepted. Hence, $N_t(x,y) \leq (r;s); 1' = r;s$. So by definition of h^*, $(x,y) \in h^*(r;s)$.

Conversely, if $(x,y) \in h^*(r;s)$, choose $t < \omega$ so large that $x, y \in N_t$, $N_t(x,y) \leq r;s$, and $\forall$ has played (x, y, r, s) already, before round t. If $\exists$ rejected, then we would have $N_t(x,y) \leq -(r;s)$, so that $N_t(x,y) = 0$, contradicting lemma 7.2. So she accepted — so there is some $z \in N_t$ with $N_t(x,z) \leq r$ and $N_t(z,y) \leq s$. By definition of h^*, we obtain $(x,z) \in h^*(r)$ and $(z,y) \in h^*(s)$, so $(x,y) \in h^*(r) \,|\, h^*(s)$ as required.

□

So we almost have a representation of $\mathcal{A}$. It remains to fix up the problems with 1' and the infidelity of h^*. For the first, define a binary relation $\sim$ on N^* by $x \sim y$ iff $(x,y) \in h^*(1')$.

LEMMA 7.21 $\sim$ *is an equivalence relation on* N^*.

Proof. For reflexivity, let $x \in N^*$; we require $x \sim x$. Let $t < \omega$ be large enough so that $x \in N_t$. Then as N_t is a network, $N_t(x,x) \leq 1'$. So by the definitions, $x \sim x$.

7.6. Games and representations of relation algebras

For symmetry, let $x, y \in N^*$. If $x \sim y$ then $(x,y) \in h^*(1')$. By lemmas 7.20 and 3.17, $(y,x) \in h^*(\breve{1'}) = h^*(1')$, and $y \sim x$.

For transitivity, if $x \sim y \sim z$ in N^* then by lemma 7.20, $(x,z) \in h^*(1') ; h^*(1') = h^*(1' ; 1') = h^*(1')$, so $x \sim z$. □

LEMMA 7.22 $\sim$ *is an h^*-congruence on N^*, in the sense that if $x \sim x'$, $y \sim y'$, then $(x,y) \in h^*(r)$ iff $(x',y') \in h^*(r)$ for all $r \in \mathcal{A}$.*

Proof. Suppose that $x \sim x'$ and $y \sim y'$. If $(x,y) \in h^*(r)$, then by lemmas 7.20 and 7.21 and the definition of $\sim$, $(x',y') \in h^*(1') ; h^*(r) ; h^*(1') = h^*(1' ; r ; 1') = h^*(r)$. The converse follows similarly. □

So we may factor N^* out by $\sim$. Formally, if $x \in N^*$, we write $x/\sim$ for the $\sim$-class $\{y \in N^* : y \sim x\}$ of x. We write B_a for the set of all $\sim$-classes (recall that a is the initial non-zero element of $\mathcal{A}$), and $\mathcal{B}_a$ for the proper relation algebra

$$(\wp(B_a \times B_a), \emptyset, B_a \times B_a, \cup, \setminus, \mathrm{Id}_{B_a}, -^{-1}, \mid).$$

Define $h_a : \mathcal{A} \to \mathcal{B}_a$ by

$$h_a(r) = \{(x/\sim, y/\sim) : x,y \in h^*(r)\},$$

LEMMA 7.23 $h_a : \mathcal{A} \to \mathcal{B}_a$ *is a homomorphism, and $h_a(a) \neq 0$.*

Proof. Let $x, y \in N^*$ and let $u = x/\sim$, $v = y/\sim$. As $\sim$ is a congruence, we have

$$(u,v) \in h_a(r) \iff (x,y) \in h^*(r), \text{ for all } r \in \mathcal{A}.$$

(For the '$\Rightarrow$'-direction, if $(u,v) \in h_a(r)$ then $(x',y') \in h^*(r)$ for some $x' \sim x$, $y' \sim y$, so that $(x,y) \in h^*(r)$ by the preceding lemma.) Armed with this information, let us check preservation of each of the operations of the algebra. We use lemma 7.20 all the time. Let $r, s \in \mathcal{A}$ and u, v, x, y be as above.

Boolean operations Clearly, $(u,v) \in h_a(1) \setminus h_a(0)$, since $(x,y) \in h^*(1) \setminus h^*(0)$. Hence $h_a(1) = B_a \times B_a$, and $h_a(0) = \emptyset$.

We have $(u,v) \in h_a(r+s)$ iff $(x,y) \in h^*(r+s)$ iff $(x,y) \in h^*(r) \cup h^*(s)$, iff $(u,v) \in h_a(r) \cup h_a(s)$. So '+' is preserved.

Similarly, we have $(u,v) \in h_a(-r)$ iff $(x,y) \in h^*(-r)$ iff $(x,y) \notin h^*(r)$ iff $(u,v) \notin h_a(r)$. So '$-$' is preserved.

Identity $(u,v) \in h_a(1')$ iff $(x,y) \in h^*(1')$ iff $x \sim y$ iff $u = v$.

Conversion We have $(u,v) \in h_a(\breve{r})$ iff $(x,y) \in h^*(\breve{r})$ iff $(y,x) \in h^*(r)$ iff $(v,u) \in h_a(r)$, as required.

Composition Suppose that $(u,v) \in h_a(r) | h_a(s)$. Let $w \in B_a$ be such that $(u,w) \in h_a(r)$ and $(w,v) \in h_a(s)$, and choose any $z \in N^*$ with $w = z/\sim$. Then $(x,z) \in h^*(r)$ and $(z,y) \in h^*(s)$. Hence $(x,y) \in h^*(r;s)$, so that $(u,v) \in h_a(r;s)$.

Conversely, if $(u,v) \in h_a(r;s)$ then $(x,y) \in h^*(r;s)$, so there is some $z \in N^*$ with $(x,z) \in h^*(r)$ and $(z,y) \in h^*(s)$. Let $w = z/\sim \in B_a$. Then $(u,w) \in h_a(r)$ and $(w,v) \in h_a(s)$, so that $(u,v) \in h_a(r)|h_a(s)$, as required.

$h_a(a)$ **is non-empty** $I_a = N_0$ and $I_a(0,1) = a$, so $0,1 \in N^*$ and $(0,1) \in h^*(a)$. Thus, $(0/\sim, 1/\sim) \in h_a(a)$. □

Thus, $h_a : \mathcal{A} \to \mathcal{B}_a$ is a relation algebra homomorphism from $\mathcal{A}$ into a proper relation algebra. Let $\mathcal{B} = \prod_{a \in \mathcal{A}\setminus\{0\}} \mathcal{B}_a$. Then as **RRA** is a variety (theorem 3.37) and so closed under products, $\mathcal{B} \in$ **RRA**. Define $h : \mathcal{A} \to \mathcal{B}$ by $h(r) = \langle h_a(r) : a \in \mathcal{A}\setminus\{0\}\rangle$. Then h is a homomorphism from $\mathcal{A}$ into $\mathcal{B}$. Moreover, it is injective. For if $r \neq 0$ in $\mathcal{A}$, then the sequence $\langle h_a(r) : a \in \mathcal{A}\setminus\{0\}\rangle$ contains the term $h_r(r)$, which by the preceding lemma is non-zero. Hence, $\mathcal{A}$ is isomorphic to a subalgebra of $\mathcal{B}$. As **RRA** is closed under isomorphism and under subalgebras, we have $\mathcal{A} \in$ **RRA**, as required. In fact, the 'product' of the h_a yields a representation of $\mathcal{A}$ with base $\bigcup_{a \in \mathcal{A}\setminus\{0\}} B_a$.

This completes the proof of theorem 7.19. □

So we have connected games to representability. If each network I_a ($a \in \mathcal{A} \setminus \{0\}$) is hereditarily refinable then $\mathcal{A}$ is representable. But we can go further, and only assume that $\exists$ has a winning strategy in all games $G_n(I_a, \mathcal{A})$ of finite length. This is proved in the next proposition, and it will be needed in the next chapter when we come to axiomatise **RRA** — the condition that $\exists$ has a winning strategy in a given finite-length game will turn out to be a universal first-order property of the algebra.

PROPOSITION 7.24 *Let $\mathcal{A}$ be any relation algebra, and N any $\mathcal{A}$-pre-network. Then $\exists$ has a winning strategy in $G_\omega(N, \mathcal{A})$ if and only if she has a winning strategy in $G_n(N, \mathcal{A})$ for all $n < \omega$.*

Proof. '⇒' is clear: $\exists$ may use her strategy for $G_\omega(N, \mathcal{A})$ in the game $G_n(N, \mathcal{A})$, so long as it lasts. Since the strategy is winning for her, she does not lose $G_\omega(N, \mathcal{A})$ in its first n rounds. So she wins $G_n(N, \mathcal{A})$.

The converse is similar to König's tree lemma: see [Hod93, exercise 5.6(5)], and exercise 10.1(2) later. We begin with a claim.

Claim. Let N be any $\mathcal{A}$-pre-network. Assume that $\exists$ has a winning strategy in $G_n(N, \mathcal{A})$ for infinitely many $n < \omega$. Then whatever move $\forall$ makes in round 0 of $G_\omega(N, \mathcal{A})$, $\exists$ has a response $N' \supseteq N$ such that she has a winning strategy in $G_n(N', \mathcal{A})$ for infinitely many $n < \omega$.

7.6. Games and representations of relation algebras

Proof of claim. $\exists$ can apply her winning strategies in the games $G_n(N, \mathcal{A})$, for infinitely many $n > 0$, to $\forall$'s move in round 0 of the game $G_\omega(N, \mathcal{A})$. These strategies say whether to accept or reject. If infinitely many of them say to accept, then she accepts in $G_\omega(N, \mathcal{A})$. If not, then infinitely many will advise rejection, and she rejects in $G_\omega(N, \mathcal{A})$. Say this course of action yields the network $N' \supseteq N$. Recall that given N and $\forall$'s move, N' is completely determined by whether $\exists$ accepts or rejects. Thus, N' is also her response to $\forall$ in infinitely many games $G_n(N, \mathcal{A})$, using a winning strategy. So she has a winning strategy in $G_{n-1}(N', \mathcal{A})$ for infinitely many $n < \omega$ (all those n for which she followed the strategy in $G_n(N, \mathcal{A})$), namely 'continue with the winning strategy in progress in $G_n(N, \mathcal{A})$'. See lemma 7.16. The claim is proved.

Assume now that N is an $\mathcal{A}$-pre-network such that $\exists$ has a winning strategy in $G_n(N, \mathcal{A})$ for all $n < \omega$. In a play $N = N_0, N_1, \ldots$ of the game $G_n(N, \mathcal{A})$, for any $t < \omega$, $\exists$ is said to be in *pole position* in round t if she has a winning strategy in $G_n(N_t, \mathcal{A})$ for infinitely many $n < \omega$. We propose the following strategy for $\exists$ in $G_\omega(N, \mathcal{A})$:

'play so as to be in pole position in each round'.

Let us show that $\exists$ can implement this strategy. By assumption, she is in pole position in round 0. Assume inductively that she is in pole position in round t. As $\forall$ makes his move in round t, $\exists$ pretends that it is his first move in $G_\omega(N_t, \mathcal{A})$. By the claim, she can respond with a pre-network $N_{t+1} \supseteq N_t$ such that she has a winning strategy in $G_n(N_{t+1}, \mathcal{A})$ for infinitely many $n < \omega$. Of course, N_{t+1} is also a legal response by $\exists$ to $\forall$ in round t of $G_\omega(N, \mathcal{A})$, so if she plays it as her response she keeps in pole position. Thus, by induction, she can stay in pole position throughout play.

Now if $\forall$ plays $G_\omega(N, \mathcal{A})$ as advised, then for each $t < \omega$ she certainly has a winning strategy in $G_0(N_t, \mathcal{A})$. Hence, N_t is a network. This holds for all t, and consequently $\exists$ wins the play. Thus, we have described a winning strategy for her in $G_\omega(N, \mathcal{A})$. □

Exercises

1. Confirm, for countable $\mathcal{A}$, that it is possible for $\forall$ to pick every pair of nodes and algebra elements during the play of $G_\omega(I_a, \mathcal{A})$ in the proof of theorem 7.19.

2. Let $\mathcal{A}$ be a simple relation algebra. Show that $\mathcal{A}$ is representable iff $\exists$ has a winning strategy in $G_\omega(I_{1^\mathcal{A}}, \mathcal{A})$. Here, $I_{1^\mathcal{A}}$ is as in definition 7.18.

3. Let $\mathcal{A}$ be an atomic relation algebra. Show that $\mathcal{A}$ is representable iff $\exists$ has a winning strategy in $G_\omega(I_a, \mathcal{A})$ for every atom $a \in \mathcal{A}$.

4. Deduce directly from lemma 7.17 and theorem 7.19 the result of corollary 3.39 that a relation algebra $\mathcal{A}$ is representable iff every finitely generated subalgebra of $\mathcal{A}$ is representable.

5. Let N be a pre-network over a simple relation algebra $\mathcal{A}$. Prove that the following are equivalent.

    (a) For each $n < \omega$, $\exists$ has a winning strategy in the game $G_n(N, \mathcal{A})$.

    (b) There is a representation M of $\mathcal{A}$ and a map $\prime : \mathrm{nodes}(N) \to M$ such that for all $i, j \in N$ we have $M \models N(i,j)(i', j')$.

6. In [Lyn56], Lyndon specified first-order axioms characterising **RRA** in a way corresponding to the following n-round game $\Lambda_n^m(\alpha, \beta, \mathcal{A})$, where $1 \leq n \leq \omega$, $2 \leq m \leq \omega$, $\alpha, \beta \in {}^n\omega$, $\alpha(t) < \beta(t) \leq t$ for all $0 < t < n$, and $\mathcal{A}$ is a relation algebra. The exercise will be to show that it works.

    Play constructs $\mathcal{A}$-pre-networks $N_1 \subseteq N_2 \subseteq \cdots \subseteq N_n$ when $n < \omega$, and $N_1 \subseteq N_2 \subseteq \cdots$ when $n = \omega$. See definition 7.11 for pre-networks. In round zero of $\Lambda_n^m(\alpha, \beta, \mathcal{A})$, $\forall$ chooses non-zero $a \in \mathcal{A}$, and $\exists$ responds with a pre-network N_1 with nodes $\{0, 1\}$ and such that $N_1(0, 1) = a$ and $N_1(1, 0) = \breve{a}$. Labels on reflexive edges are not mentioned in Lyndon's axioms; we will suppose that all reflexive labels (here, $N_1(0,0)$ and $N_1(1,1)$) are 1'.

    In each subsequent round t ($1 \leq t < n$), suppose that the currently existing pre-network at the start of the round is N_t, with nodes $\{0, 1, \ldots, t\}$. During round t, a new pre-network N_{t+1} with nodes $\{0, 1, \ldots, t+1\}$ is built as follows:

    - To open the round, $\forall$ picks the edge $(\alpha(t), \beta(t))$ of N_t and algebra elements $a, b \in \mathcal{A}$ with $N_t(\alpha(t), \beta(t)) \leq a;b$. He then defines the labels $N_{t+1}(\alpha(t), t+1) = a$ and $N_{t+1}(\beta(t), t+1) = \breve{b}$. Note that this is well-defined since $\alpha(t) \neq \beta(t)$.

    - The remainder of the round is a skirmish consisting of $t - 1$ 'sub-rounds' to decide the values $N_{t+1}(i, t+1)$ for successive nodes $i \in \{0, 1, \ldots, t\} \setminus \{\alpha(t), \beta(t)\}$. For such an i, let $\Phi_i = \{0, 1, \ldots, i-1\} \cup \{\alpha(t), \beta(t)\}$, and assume that the labels $N_{t+1}(j, t+1)$ for all $j \in \Phi_i$ are already defined.

        (a) $\forall$ chooses $X \subseteq \mathcal{A}$ with $|X| < m$ and
        $$\sum X \geq \prod_{j \in \Phi_i} N_t(i, j); N_{t+1}(j, t+1).$$

        (b) $\exists$ chooses some $x \in X$, and sets $N_{t+1}(i, t+1) = x$.
        The players then move on to deal with the next i.

7.6. Games and representations of relation algebras

After all i have been covered, round t is finished. The labelling of N_{t+1} is then completed by defining $N_{t+1}(i,j) = N_t(i,j)$ for $i,j \leq t$, $N_{t+1}(t+1,i) = N_{t+1}(i,t+1)\breve{}$ for $i \leq t$, and $N_{t+1}(t+1,t+1) = 1$'. The result is the new pre-network N_{t+1}.

This continues for n rounds. $\exists$ wins iff the label of each edge of each pre-network played during the game is non-zero.

And now for the exercise:

(a) If $\mathcal{A}$ is any representable relation algebra, prove that $\exists$ has a winning strategy in $\Lambda_n^m(\alpha,\beta,\mathcal{A})$ for all n,m,α,β as above.

Let $\mathcal{A}$ be a countable simple relation algebra, and suppose that $\exists$ has a winning strategy in $\Lambda_n^m(\alpha,\beta,\mathcal{A})$ for all $n < \omega$, all finite $m \geq 2$, and all $\alpha,\beta \in {}^n\omega$ satisfying $\alpha(t) < \beta(t) \leq t$ whenever $1 \leq t < n$.

(b) Show that $\exists$ has a winning strategy in $\Lambda_\omega^\omega(\alpha,\beta,\mathcal{A})$ for all $\alpha,\beta \in {}^\omega\omega$ such that $\alpha(t) < \beta(t) \leq t$ for all $t \geq 1$.

(c) Show that $\mathcal{A}$ is representable.

(d) Can you generalise to arbitrary relation algebras?

7. In [Jón59], Jónsson studied the class **wRRA** of *weakly representable* relation algebras. As we saw in section 5.2, **wRRA** is the class of all relation algebras $\mathcal{A}$ that have a representation respecting $0, 1, \cdot, 1', \breve{}$, and $;$, but not necessarily $+$ or $-$. Such a representation takes the form of an embedding

$$h: \mathcal{A} \to (\wp(U \times U), \cap, \emptyset, E, \mathrm{Id}_U, {}^{-1}, |)$$

for some non-empty set U and some equivalence relation E on U.

[Jón59] characterised **wRRA** by quasi-equational axioms corresponding to the following n-round game $\mathcal{J}_n(\mathcal{A})$ (for $n \leq \omega$ and $\mathcal{A}$ a relation algebra), played by $\forall, \exists$. Actually, it is a solitaire game, as $\exists$ has no real role. The game produces $\mathcal{A}$-pre-networks $N_1 \subseteq \cdots \subseteq N_t \subseteq N_{t+1} \subseteq \cdots$ ($t < n$), where N_t has nodes $\{0,1,\ldots,t\}$.

- In round 0 of the game, $\forall$ chooses non-zero $a \in \mathcal{A}$, and $\exists$ responds with a pre-network N_1 with $N_1(0,1) = a$, $N_1(1,0) = \breve{a}$, and $N_1(0,0) = N_1(1,1) = 1$'.
- At the start of round t, where $0 < t < n$ and the existing pre-network is N_t, $\forall$ chooses an edge (x,y) of N_t, and algebra elements $a,b \in \mathcal{A}$, with $N_t(x,y) \leq a;b$. $\exists$ must then play the new pre-network N_{t+1} given by
  - $N_{t+1}(u,v) = \prod_{w \in N_t} N_t(u,w);N_t(w,v)$, for all $u,v \in N_t$,

○ $N_{t+1}(u,t+1) = (N_t(u,x);a) \cdot (N_t(u,y);\check{b})$, for all $u \in N_t$,
○ $N_{t+1}(t+1,u) = N_{t+1}(u,t+1)\breve{}$,
○ $N_{t+1}(t+1,t+1) = 1'$.

$\exists$ wins iff $N_1(0,1) = N_{t+1}(0,1)$ for all $t < n$.

(a) Let $\mathcal{A} \in \mathbf{wRRA}$. Show that $\exists$ has a winning strategy in $\mathcal{J}_n(\mathcal{A})$ for all $n \leq \omega$.

Let $\mathcal{A}$ be a countable relation algebra.

(b) Show that if $\exists$ has a winning strategy in $\mathcal{J}_n(\mathcal{A})$ for all $n < \omega$, then she has a winning strategy in $\mathcal{J}_\omega(\mathcal{A})$.

(c) Show that if $\exists$ has a winning strategy in $\mathcal{J}_\omega(\mathcal{A})$ then $\mathcal{A} \in \mathbf{wRRA}$.

(d) Can you generalise to arbitrary relation algebras?

7.7 Networks for cylindric algebras

Now we do the same for cylindric algebras. We only cover the finite-dimensional case here. Let C be a cylindric algebra of finite dimension $d \geq 2$. For relation algebra, our approximations to representations were called 'networks', and we use the same term here.

Since representations of d-dimensional cylindric algebras assign d-ary relations to algebra elements (rather than binary relations as in the case of relation algebras), a network should assign an algebra element to each d-tuple of its nodes. As before, the properties that cylindric algebra networks inherit from representations come from the preservation of the cylindric algebra operations in representations. The first two (2a and 2b in definition 7.25(2) below) concern preservation of the diagonals d_{ij} and the cylindrifications c_i, respectively. The third, (2c), says that if two d-tuples of nodes are identical except at the ith place, where their values are x and y, say, then the network's idea of which algebra elements hold on them should be consistent with each other, if it considers that x and y are equal. This last means that there is a third d-tuple of nodes of the network involving x and y, say at the jth and kth places, and the network deems d_{jk} to hold on this tuple.

Notation Let $\bar{x}, \bar{y}$ be d-tuples of elements of some set. We write x_i for the ith element of $\bar{x}$, for $i < d$, so that $\bar{x} = (x_0, \ldots, x_{d-1})$. For $i < d$, we write $\bar{x} \equiv_i \bar{y}$ if $x_j = y_j$ for all $j < d$ with $j \neq i$.

DEFINITION 7.25 Let C, d be as above.

7.8. Games for cylindric algebra networks

1. A *C-pre-network* N is a complete directed d-dimensional hypergraph with each hyperedge labelled by an element of C. Formally, N consists of a finite non-empty set of nodes, and a map assigning an element of C to each d-tuple of nodes. As before, we use the same symbol N to denote the set of nodes, the mapping, and the graph itself.

2. A *C-network* (or simply a network) is a C-pre-network N satisfying:

   (a) for each $i, j < d$ and any d-tuple $\bar{x}$ from nodes(N) (written '$\bar{x} \in {}^d N$'), if $x_i = x_j$ then $N(\bar{x}) \leq d_{ij}$ in C,

   (b) for any $\bar{x}, \bar{y} \in {}^d N$ and any $i < d$, if $\bar{x} \equiv_i \bar{y}$ then $N(\bar{x}) \cdot c_i N(\bar{y}) \neq 0$,

   (c) if $\bar{x}, \bar{y}, \bar{z} \in {}^d N$, $i, j, k < d$, $\bar{x} \equiv_i \bar{y}$, $x_i = z_j$, $y_i = z_k$, and $N(\bar{z}) \leq d_{jk}$, then $N(\bar{x}) \cdot N(\bar{y}) \neq 0$.

3. As in the relation algebra case, for networks N, N', we write $N \subseteq N'$ if the nodes of N' include those of N, and $N'(\bar{x}) \leq N(\bar{x})$ for all $\bar{x} \in {}^d N$.

Exercises

1. Let N be a C-network, and let $\bar{x} \in {}^d N$. Show that $N(\bar{x}) \neq 0$.

2. Let h be a representation of C with base B. Let N be a C-pre-network satisfying condition 2a above, and suppose that $\theta : N \to B$ is a map satisfying $(\theta(x_0), \ldots, \theta(x_{d-1})) \in h(N(\bar{x}))$, for all $\bar{x} \in {}^d N$. Show that N is a C-network. (N approximates h.)

3. Give an example of a network satisfying properties 2a and 2b but not 2c of definition 7.25. Why did we not need a property corresponding to 2c for relation algebra networks?

7.8 Games for cylindric algebra networks

Here, we define an appropriate game G_n for cylindric algebras. Here, $d \geq 2$ and $C \in \mathbf{CA}_d$ remain fixed.

DEFINITION 7.26 Let N be a pre-network (over the fixed cylindric algebra C), and let $n \leq \omega$. The game $G_n(N, C)$ is of length n. We can use the same notation 'G_n' for this game because the argument C will serve to tell it apart from the relation algebra game. As before, $N_0 = N$, a play of $G_n(N, C)$ is a sequence of C-pre-networks: $N_0 \subseteq \cdots \subseteq N_n$ if n is finite, and $N_0 \subseteq N_1 \subseteq \cdots$ if $n = \omega$.

In the tth round ($t < n$), let the last pre-network played be N_t. For his move in this round, $\forall$ picks $i \leq d$, $\bar{x} \in {}^d N_t$, and $a \in C$.

1. If $i < d$, $\exists$ must respond to $\forall$'s move with a pre-network $N_{t+1} \supseteq N_t$ given by one of the following:

   reject N_{t+1} is the same as N_t, except that $N_{t+1}(\bar{x}) = N_t(\bar{x}) \cdot -c_i a$.

   accept The nodes of N_{t+1} are those of N_t plus a new node z. Let $\bar{z}$ be given by $\bar{z} \equiv_i \bar{x}$, $z_i = z$. The labels of d-tuples of nodes are given by:
   - $N_{t+1}(\bar{z}) = a \cdot \prod_{\substack{j,k<d \\ z_j = z_k}} \mathsf{d}_{jk}$,
   - $N_{t+1}(\bar{x}) = N_t(\bar{x}) \cdot c_i a$,
   - $N_{t+1}(\bar{y}) = N_t(\bar{y})$ for all $\bar{y} \in {}^d N_t \setminus \{\bar{x}\}$,
   - $N_{t+1}(\bar{y}) = \prod_{\substack{j,k<d \\ y_j = y_k}} \mathsf{d}_{jk}$ for all $\bar{y} \in {}^d N_{t+1} \setminus (\{\bar{z}\} \cup {}^d N_t)$ (i.e., all 'new' $\bar{y}$ other than $\bar{z}$).

   Here, of course, the empty product is defined to be 1^C.

2. If $i = d$, $\exists$ responds with a pre-network $N_{t+1} \supseteq N_t$ which is the same as N_t except that either

   accept $N_{t+1}(\bar{x}) = N_t(\bar{x}) \cdot a$, or

   reject $N_{t+1}(\bar{x}) = N_t(\bar{x}) \cdot -a$.

The 'i' plays no role here: it is just an indicator of the type of move $\forall$ makes.

If each N_t $(t < n)$ is a C-network, $\exists$ has won the play of $G_n(N,C)$. Otherwise, $\forall$ won.

The notion of (winning) strategy is defined as for relation algebra games. If $a \in C \setminus \{0\}$, let I_a be the C-network with set of nodes d, given by $I_a(0, 1, \ldots, d-1) = a$ and $I_a(\bar{x}) = \prod_{x_i = x_j} \mathsf{d}_{ij}$ for each $\bar{x} \in {}^d d$ with $\bar{x} \neq (0, 1, \ldots, d-1)$.

THEOREM 7.27 *Let C be a d-dimensional cylindric algebra, for finite $d \geq 2$. C is representable if and only if $\exists$ has a winning strategy in the game $G_n(I_a, C)$ for each $n < \omega$ and each non-zero $a \in C$.*

Proof. If C is representable then, as with theorem 7.19, $\exists$ can use a representation to give her a winning strategy in each of the games. Conversely, suppose that she can win each game $G_t(I_a, C)$ for every $t < \omega$ and each non-zero algebra element a. Fix such an a. Then $\exists$ also has a winning strategy for $G_\omega(I_a, C)$; the proof is the same as in proposition 7.24. Next, we want to use $\exists$'s winning strategy in $G_\omega(I_a, C)$ to construct a representation of C. **RCA**$_d$ is a variety, so as in theorem 7.19 we can suppose that C is countable. Consider a play of $G_\omega(I_a, C)$ in

7.8. Games for cylindric algebra networks

which $\exists$ uses her winning strategy and $\forall$ picks every d-tuple of nodes ever constructed, every $i \leq d$, and every $a \in C$ eventually during the game. Write M^* for the set of all nodes introduced during the game. Define a d-dimensional set algebra $\mathcal{D} = (\wp(^dM^*), \emptyset, {}^dM^*, \cup, \setminus, C_i, D_{ij} : i, j < d)$ (see definition 5.16 for C_i, D_{ij}), and a map $h^* : C \to \mathcal{D}$ as follows. For $r \in C$, let

$$h^*(r) = \{\bar{x} \in {}^dM^* : \exists t < \omega (\bar{x} \in {}^dN_t \text{ and } N_t(\bar{x}) \leq r)\}.$$

$\forall$-moves of the second kind ('$i = d$') guarantee that for any n-tuple $\bar{x}$ and any $a \in C$, for sufficiently large t we have either $N_t(\bar{x}) \leq a$ or $N_t(\bar{x}) \leq -a$. This ensures that h^* preserves the boolean operations. $\forall$-moves of the first kind ('$i < d$') ensure that the cylindrifications are respected by h^*. So h^* is a homomorphism from the diagonal-free reduct of C into the diagonal-free reduct of $\mathcal{D}$.

Now define a binary relation $\sim$ on the nodes of M^*, by

$$x \sim y \quad \Leftrightarrow \quad \exists \bar{z} \in {}^dM^* \, \exists k, l < d \, (z_k = x \wedge z_l = y \wedge \bar{z} \in h^*(\mathsf{d}_{kl})).$$

Claim. If $\bar{x}, \bar{y} \in {}^dM^*$, $i < d$, $\bar{x} \equiv_i \bar{y}$, and $x_i \sim y_i$, then $\bar{x} \in h^*(r)$ iff $\bar{y} \in h^*(r)$, for all $r \in C$.

Proof of claim. Let $r \in C$ and choose $t < \omega$ so large such that $\bar{x}, \bar{y} \in {}^dN_t$, there is $\bar{z} \in {}^dN_t$ and $k, l < d$ with $z_k = x_i$, $z_l = y_i$, and $N_t(\bar{z}) \leq \mathsf{d}_{kl}$, and $\forall$ has already played the moves $d, \bar{x}, r$ and $d, \bar{y}, r$ before round t of the game. Therefore, $N_t(\bar{x}) \leq r$ or $N_t(\bar{x}) \leq -r$, and similarly for $\bar{y}$. But by condition 2c of the definition of network and the existence of $\bar{z}$, we have $N_t(\bar{x}) \cdot N_t(\bar{y}) \neq 0$. Therefore, we must have $N_t(\bar{x}) \leq r$ iff $N_t(\bar{y}) \leq r$. Since this holds for all large enough t, we obtain $\bar{x} \in h^*(r)$ iff $\bar{y} \in h^*(r)$, as claimed.

It follows that $\sim$ is an equivalence relation on M^*. It is clearly reflexive and symmetric. For transitivity, suppose that $x \sim y \sim z$ in M^*. We may suppose that $x \neq y$. Take $\bar{w} \in {}^dM^*$ witnessing $x \sim y$: say $w_k = x$, $w_l = y$, and $\bar{w} \in h^*(\mathsf{d}_{kl})$. We must have $k \neq l$. Define $\bar{v} \in {}^dM^*$ by $\bar{v} \equiv_l \bar{w}$, $v_l = z$. Since $k \neq l$, we have $v_k = w_k = x$. By the claim, $\bar{v} \in h^*(\mathsf{d}_{kl})$, too, so $x \sim z$.

Similarly, it follows that $\sim$ is an h^*-congruence on M^*, in the sense that if $\bar{x}, \bar{y} \in {}^dM^*$ and $x_i \sim y_i$ for all $i < d$ then $\bar{x} \in h^*(r)$ iff $\bar{y} \in h^*(r)$ for all $r \in C$. To see this, for $j \leq d$ define $\bar{z}^j \in {}^dM^*$ by

$$z_i^j = \begin{cases} y_i, & \text{if } i < j \\ x_i, & \text{if } j \leq i < d. \end{cases}$$

Then

$$\bar{x} = \bar{z}^0 \equiv_0 \bar{z}^1 \equiv_1 \bar{z}^2 \equiv_2 \cdots \equiv_{d-1} \bar{z}^d = \bar{y}.$$

For all $i \leq d$, since $\bar{z}^i \equiv_i \bar{z}^{i+1}$ and $z_i^i = x_i \sim y_i = z_i^{i+1}$, the claim gives $\bar{z}^i \in h^*(r)$ iff $\bar{z}^{i+1} \in h^*(r)$ for all $r \in C$. Thus, by taking $i = 0, 1, \ldots, d$ in turn, we obtain $\bar{x} \in h^*(r)$ iff $\bar{y} \in h^*(r)$ for all $r \in C$, as required.

Now we construct $B_a = M^*/\sim$ and h_a as in theorem 7.19. It is easy to show that $h_a : C \to (\wp(^d B_a), \emptyset, {}^d B_a, \cup, \setminus, C_i, D_{ij} : i, j < d)$ is a cylindric algebra homomorphism with $h_a(a) \neq 0$, and that $\prod_{a \in C \setminus \{0\}} h_a$ defines a representation of C. □

7.9 Games for temporal constraint handling

Let us take a short excursion into one of the many applications of relation algebras. Relation algebras have their origins in nineteenth century mathematical logic, but they were rediscovered in the 1970s by computer scientists working with temporal constraints. The point algebra and the metric point algebra [DeaMcD87, DecMei+91] and the Allen interval algebra [All81, AllKau83, All84, VilKau86a, AllHay85b, KauLad91, LadRei93], were all used for temporal reasoning and turned out to form relation algebras (see chapter 4). One of the main applications for this type of interval reasoning was in planning where actions could be concurrent and were given non-zero duration [AllKoo83, Pel88].

A conjunction of binary constraints on points or intervals was represented as a pre-network — a labelled graph with edge labels taken from a given relation algebra (the term 'network' originates here, but in this book we reserve this word 'network' for a pre-network satisfying the consistency conditions of definition 7.1). The key problem is to test the consistency of a pre-network.

DEFINITION 7.28 Let $\mathcal{A}$ be a relation algebra and let N be a pre-network over $\mathcal{A}$. We say that N is *satisfiable* if there is a representation M of $\mathcal{A}$ and an embedding $\prime : \text{nodes}(N) \to M$ such that for all $i, j \in N$ we have $M \models N(i,j)(i', j')$.

The problem of determining whether a pre-network is satisfiable or not is called the *network satisfaction problem* (NSP).

In some of the research the problem of testing satisfiability in a fixed, given representation M is considered, but we do not investigate that problem here.

Allen defined a *propagation algorithm* as an approximate test of satisfiability — see figure 7.4 [All84, VilKau86b, VilKau+89]. The propagation algorithm gives a *sound* inference system in the sense that the algorithm will output a pre-network with 0s on the edges only if the input pre-network is unsatisfiable. And elementary calculations show that the algorithm is tractable: the run-time is $O(E^3)$, where E is the number of edges in the pre-network, i.e., the square of the number of nodes [VilKau86b, theorem 1]. The propagation algorithm can be seen to be a kind of shortest-path algorithm and indeed the Floyd–Warshall algorithm computes the same result with run-time $O(n^3)$. But these algorithms are not *complete* — there

7.9. Games for temporal constraint handling

```
propagate(N)
{
make empty queue (Q);
for each edge (i,j) of N add (i,j) to Q;
while Q is not empty
      { dequeue edge (i,j) from Q;
      forall l ∈ nodes(N)
          { temp := N(i,l) ∩ N(i,j)|N(j,l);
          if N(i,l) ≠ temp then enqueue (i,l) in Q;
          N(i,l) = temp;
          temp := N(l,j) ∩ N(l,i)|N(i,j);
          if N(l,j) ≠ temp then enqueue (l,j) in Q;
          N(l,j) := temp ;
          }
      }
}
```

Figure 7.4: The propagation algorithm

are pre-networks for which the propagation algorithm (or the Floyd–Warshall algorithm) will not produce a 0 on any arc but which are unsatisfiable. For an example (due to Patrick Hayes which appeared in [All84]) of an unsatisfiable (pre-)network over the Allen interval algebra whose inconsistency is not detected by the propagation algorithm, see figure 7.5.

Indeed [VilKau86b, theorem 2] showed that the network satisfiability problem (NSP) was NP-complete for the Allen interval algebra and so, assuming $P \neq NP$, there can be no tractable satisfiability-checker for pre-networks over the Allen interval algebra. An additional problem with the propagation algorithm is that for infinite relation algebras there is no guarantee that the algorithm terminates (see [LadMad94, theorem 6.4] for an example of a network where propagation does not terminate).

In this section we can apply games to provide better, tractable approximations for the NSP over finite relation algebras.[2] These algorithms improve on the propagation algorithm and converge on a complete satisfiability-checker. Before we proceed, we need two definitions.

DEFINITION 7.29 A pre-network N is said to be *3-consistent* if for all $i,j,k \in N$ we have $N(i,k) \leq N(i,j);N(j,k)$.

---

[2] A version of these approximating algorithms and the results of this section appeared in [Hir00]; we thank the copyright owners Elsevier Science for granting permission for this republication.

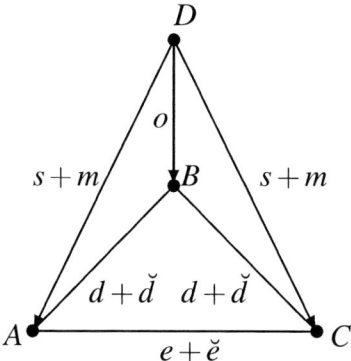

Figure 7.5: A 3-consistent but unsatisfiable network

3-consistency is sometimes referred to as *path-consistency*. Clearly, if the propagation algorithm terminates, it will output a 3-consistent pre-network refining its input.

DEFINITION 7.30 Let $\mathcal{A}$ be an atomic relation algebra. $\mathcal{A}$ is called *network-universal* if every atomic network is satisfiable.

LEMMA 7.31 *If the finite relation algebra $\mathcal{A}$ is network-universal then the NSP over $\mathcal{A}$ is in* NP.

Proof. Here is a non-deterministic algorithm that solves the NSP over $\mathcal{A}$. Given an $\mathcal{A}$-pre-network N, non-deterministically pick one atom from the label of each edge of N to obtain a pre-network labelled by atoms. If there is a choice of atoms yielding an atomic pre-network which forms an atomic network (which can be checked in cubic time) then return 'yes'; else return 'no'. If there is an atomic network L refining N then L is satisfiable (by network-universality) and hence N is too. Conversely, if N is satisfiable via the embedding $\prime$ in some representation M then, as $\mathcal{A}$ is finite, for each edge (i,j) of N there must be an atom $a \leq N(i,j)$ such that $M \models a(i',j')$. This defines an atomic network L refining N ($L(i,j) = a$) so there is a possible run of the algorithm that results in an atomic network. □

[LadMad94] shows that the Allen interval algebra is network-universal, so its NSP is in NP. However there are other relation algebras which fail to be network-universal. For these algebras the non-deterministic algorithm given in the previous

7.9. Games for temporal constraint handling

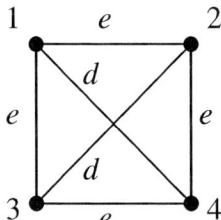

Figure 7.6: A 3-consistent but unsatisfiable atomic network over the pentagonal algebra

lemma will not be correct. An example is the pentagonal algebra with three self-converse atoms $1', e, d$ (see chapter 4 for the definition of the pentagonal algebra). For this algebra, it is possible to construct a 3-consistent atomic network where the network is not satisfiable in any representation of the algebra. See figure 7.6. To see why this network is unsatisfiable, suppose instead that $'$ is an embedding of the nodes into some representation M. Then $M \models d(1', 4')$. Since $d \leq d\,; e$, we see that $M \models (d\,; e)(1', 4')$. By definition of composition, there must be a point z in the representation such that $M \models d(1', z) \wedge e(z, 4')$. Now consider $(2', z)$. There must be some atom a of the algebra with $M \models a(2', z)$ (here we use the fact that the algebra is finite). Consideration of the triangle $(1', 2', z)$ shows that $a = 1'$ is impossible, and the triangle $(2', 4', z)$ proves that $a = e$ is also impossible. This forces $a = d$. Similarly, we see that $M \models d(3', z)$. But then the triangle $(2', 3', z)$ is inconsistent, as each side is labelled by the atom d, contrary to the fact that $d \cdot d\,; d = 0$. So we have a contradiction.

Thus, even for atomic networks, 3-consistency does not ensure satisfiability and the non-deterministic p-time algorithm given above may fail to detect an inconsistency. Indeed, it can be shown, using the tiling construction of chapter 18, that there is a finite relation algebra with an undecidable NSP (see exercise 18.10(3), or [Hir99]).

Well, the previous work of this chapter tells us what to do.

THEOREM 7.32 *Let $\mathcal{A}$ be a finite, representable[3] relation algebra and let N be a pre-network over $\mathcal{A}$.*

1. *If $\forall$ has a winning strategy in the game $G_k(N, \mathcal{A})$ (any k) then N is not satisfiable in any representation of $\mathcal{A}$.*

2. *If $\exists$ has a winning strategy in each game $G_k(N, \mathcal{A})$ (all integers $k \geq 0$) then N is satisfiable.*

---
[3] If $\mathcal{A}$ is not representable then the NSP is trivial because no network is satisfiable.

256   Chapter 7. Axiomatising representable relation algebras

3. There is an algorithm A_k to calculate if $\forall$ has a winning strategy in the game $G_k(N)$, for any pre-network N. This algorithm runs in time $O(n^{2k+3})$ where n is the number of nodes of N.

Proof. The first two parts are dealt with in exercise 7.6(5), essentially theorem 7.19.

Now, for part 3, let $|\mathcal{A}| = c$. If $k = 0$, the winner of the game is determined simply by checking if N is a network — this takes cubic time.

Assume inductively, for some $k \geq 0$, that there is an algorithm A_k that runs in time $O(n^{2k+3})$, such that $A_k(N) = \forall$ if $\forall$ has a winning strategy in $G_k(N)$ and $A_k(N) = \exists$ if $\exists$ has a winning strategy in $G_k(N)$. We now define the algorithm A_{k+1}.

```
A_{k+1}(N)
{
for each move ((i,j),a,b) that ∀ can play
    { let N⁺ be the network that results from N if ∃ accepts
    and N⁻ be the result of ∃ rejecting;
    if A_k(N⁺) = A_k(N⁻) = ∀ then return(∀) and halt;
    }
return(∃);
}
```

The number of possible $\forall$-moves is at most $|\mathrm{nodes}(N)|^2 \times c^2 = n^2.c^2$ and the networks N^+, N^- have at most $n+1$ nodes. Thus the maximum run-time for A_k is $O(n^2.c^2.(n+1)^{2(k-1)+3}) = O(n^{2k+3})$.   □

As an instructive example consider the network given in figure 7.5. If, in a play of G_5, $\forall$ plays $((A,C),1',e)$, $((A,B),1',d)$, $((B,C),1',d)$, $((D,A),1',s)$, and $((D,C),1',s)$, then however $\exists$ responds the result will fail to be an atomic network[4]. This inconsistency can be detected by the algorithm A_5.

The unsatisfiability of the network in figure 7.6 can be detected by A_4. If $\forall$ picks the edge $(1,4)$, which is labelled by d, and the pair of elements e, d then if the resulting network is fed into the propagation algorithm it will produce a network with 0 on each edge. This inconsistency is detected by A_4.

We can improve the complexity of our algorithms if we restrict to finite network-universal relation algebras. First we define a modified game $G'_k(N, \mathcal{A})$ in which $\forall$

---

[4]If $\forall$ plays $((A,C), 1', e)$ in A_1 and if the result is fed into the propagation algorithm it will produce 0s on all the edges, whether $\exists$ accepts or rejects his move. For our algorithms, it takes a few extra moves to simulate the working of the propagation algorithm. Certainly A_1 does not detect the inconsistency and A_5 does. If the reader wishes to discover the minimum k such that A_k detects this inconsistency then let it be an exercise and we wish you good luck in solving it.

7.9. Games for temporal constraint handling

does not demand witnesses but can instead refine existing edges. We also insist that he operates a 'binary search'. In $G'(N, \mathcal{A})$ $\forall$ must pick two nodes (i, j) of N and a single element a of $\mathcal{A}$ such that $|N(i, j)|$ (the number of atoms under $N(i,j)$) is as large as possible and he picks a single element $a \in \mathcal{A}$. If there is more than one edge with the maximum sized label then he picks the first, in some fixed but arbitrary enumeration of the edges of N. In response $\exists$ must either accept by playing N^+ with the same nodes and labels as N except that $N^+(i,j) = N(i,j).a$ or she can reject by playing N^-, again with the same nodes and edges as N except that $N^-(i,j) = N(i,j) - a$.

The algorithms A'_k corresponding to this modified game are defined by

$A'_{k+1}(N)$
{
find first edge (i,j) such that $|N(i,j)|$ is maximal;
for each $a \in \mathcal{A}$
    { let N^+, N^- be the same as N except that
        $N^+(i,j) = N(i,j) \cdot a$ and $N^-(i,j) = N(i,j) \cdot (-a)$;
    if $A_k(N^+) = A_k(N^-) = \forall$ then return($\forall$) and halt;
    }
return($\exists$);
}

As before, $A'_0(N)$ simply checks if N is a network.

THEOREM 7.33 *Let $\mathcal{A}$ be a finite relation algebra, say $|\mathcal{A}| = c$, and let N be a pre-network over $\mathcal{A}$ with n nodes.*

1. *If $A'_k(N) = \forall$ (any k) then N is not satisfiable.*

2. *If $\mathcal{A}$ is network-universal and if $A'_k(N) = \exists$ (for all $k \leq n^2 \cdot \lceil \log_2(\log_2(c)) \rceil$) then N is satisfiable.*

3. *$A'_k(N)$ runs in time $O(n^3)$.*

Proof.

1. Since we have restricted only $\forall$'s moves, this follows from part 1 of theorem 7.32.

2. Assume that $\exists$ has a winning strategy in $G_{n^2 \cdot \lceil \log_2(\log_2(c)) \rceil}(N)$ and that all 3-consistent atomic networks are satisfiable. Note that the number of atoms of $\mathcal{A}$ is $\log_2(c)$. In a play of the game $G_{n^2 \cdot \lceil \log_2(\log_2(c)) \rceil}(N)$, let $\forall$ always choose an element a for his move such that $|N(i,j).a| \leq |N(i,j) - a| \leq |N(i,j).a| + 1$, where (i,j) is the edge for that move. (His strategy here is to perform a binary search on the edge (i,j), so he picks an element a that divides

$N(i,j)$ into two nearly equal halves.) Then each edge (i,j) will be chosen $\lceil \log_2 \log_2(c) \rceil$ times, or until $|N(i,j)| = 1$. Each time an edge is chosen the size of the label roughly halves, or more precisely if any edge (i,j) has been chosen j times then $|N(i,j)| \leq \lceil \frac{\log_2(c)}{2^j} \rceil$. So by the end of the play we get $|N(i,j)| = 1$ for all edges (i,j). Thus at the end of play we get an atomic pre-network and since A'_0 accepts the pre-network, it must be an atomic network. Since $\mathcal{A}$ is network-universal this must be satisfiable.

3. Assume that $A'_k(N)$ runs in time λn^3 for some constant λ. Then for $A'_{k+1}(N)$ we first pick an edge (i,j) such that $|N(i,j)|$ is maximal. This takes linear time. In the 'for' loop there are c choices of a. Thus, the 'for' loop runs in time $c \times 2 \times \lambda n^3$. This is still cubic time.

□

REMARK 7.34 The algorithms can be applied to infinite atomic relation algebras with countably many atoms so long as we restrict to networks whose edges are labelled by recursive sets of atoms. The theorem also generalises to such algebras, except part 2 is replaced by: if $\exists$ has a winning strategy in each game $G_k(N) : k < \omega$ then N is satisfiable. For some infinite algebras however (e.g. the metric point algebra), only finitely many algorithms are needed to prove satisfiability.

Observe that A'_5 detects the inconsistency in the network in figure 7.5. Note that although each algorithm A'_k runs in cubic time, it takes exponential time to do a complete satisfiability check by running the algorithm A'_k for $k = n^2 \cdot \lceil \log_2 \log_2(c) \rceil$. (To see why this does not run in cubic time note that k here depends on n. In fact the time taken to run $A'_k(N)$ is bound by $\alpha^k \cdot n^3$ for some constant α, so the time taken to run $A'_{n^2 \cdot \lceil \log_2 \log_2(c) \rceil}$ is bound by $\alpha^{n^2 \cdot \lceil \log_2 \log_2(c) \rceil} \cdot n^3$, an exponential function.) Still, we have a series of cubic-time algorithms that approximate satisfiability checking for algebras like the Allen interval algebra. Adopting an incomplete inference system like the Allen propagation algorithm may well fail to detect inconsistencies. In some applications this could lead to serious errors and in these cases it may be necessary to run higher-order algorithms to detect the inconsistency.

7.10 Summary of chapter

This has been a long chapter, but (we hope) a profitable one. We have introduced games to construct relativised and non-relativised representations of weakly associative algebras, relation algebras, and cylindric algebras. The relativised case was rather simple, since the second player, $\exists$, always has a winning strategy in the relevant game. For non-relativised representations of relation algebras and cylindric algebras, this is no longer true; instead, we were able to show that her having a

7.10. Summary of chapter

winning strategy is equivalent to the representability of the algebra in question. In the next chapter, we will give first-order sentences expressing precisely that $\exists$ has a winning strategy in these games. These sentences will axiomatise the classes of representable relation algebras and cylindric algebras.

Chapter 8

Axiomatising representable relation algebras and cylindric algebras

The standard reference work [HenMon⁺71, page 461] identifies one of the two outstanding problems of the representation theory as ' … the problem of providing a simple intrinsic characterization for all representable cylindric algebras … ' Games already provide a useful characterisation. But another interpretation of this is as a request for a simple first-order axiomatisation of $\mathbf{RCA}_\alpha$ for all ordinals α. Axiomatisations do exist — for example, see [HenMon⁺85, page 112], and for relation algebras, [Lyn56] — but the axioms are certainly not simple. Here, we will find axioms for the class **RRA** of representable relation algebras, and then for the representable cylindric algebras. Our methods seem very simple: they just involve writing down axioms expressing that player $\exists$ has a winning strategy in the games $G_n(I_a, \mathcal{A})$ of the preceding chapter, for each finite n and each non-zero a. (Refer to theorem 7.19, proposition 7.24, and theorem 7.27 for the adequacy of this.) The correspondence with games ensures that the axioms have a clear intuitive meaning. By expanding the signature of relation algebras by new operators similar to the Q-connectives of Jónsson, we can obtain even simpler axioms.

There are of course many precedents for axiomatising the conditions for success in a game. We will discuss some of them in chapter 9, where we describe a more general game than $G_n(I_a, \mathcal{A})$ and show how to axiomatise the statement that $\exists$ has a winning strategy in it. None the less, in the current chapter we proceed from first principles and show how to do the axiomatisation in the concrete case of $G_n(I_a, \mathcal{A})$ and its cylindric algebra analogues. This will motivate our work on the more general games to follow.

8.1 The relation algebra case

We need a little notation.

DEFINITION 8.1

1. A *term network* is a complete directed finite graph N, each of whose edges is labelled with a term of the signature $\{0, 1, +, -, 1', \smile, ;\}$ of relation algebras.

    If x, y are nodes of N, we write $N(x, y)$ for the term labelling the edge (x, y) of N.

2. Let N be any term network, and let ι be an assignment that maps the variables occurring in the labels of N to elements of some relation algebra, say $\mathcal{A}$. Then we obtain, in the obvious way, an $\mathcal{A}$-pre-network, which we will denote by N^ι. Explicitly, N^ι has the same nodes as N, and for all pairs x, y of nodes, $N^\iota(x, y)$ is the value in $\mathcal{A}$ of the term $N(x, y)$ under the assignment ι.

3. Let N be a term network, x, y, nodes of N, and u, v, any relation algebra terms. We define two term networks, $\mathsf{Acc}(N, x, y, u, v)$ and $\mathsf{Rej}(N, x, y, u, v)$,

    - $N' = \mathsf{Acc}(N, x, y, u, v)$ has, as nodes, the nodes of N plus a new one, z. We have $N'(x, z) = u$, $N'(z, z) = 1'$, $N'(z, y) = v$, $N'(x, y) = N(x, y) \cdot (u; v)$, $N'(n, m) = N(n, m)$ for all other pairs of nodes n, m of N, and all other edges of N' are labelled by 1.
    - $\mathsf{Rej}(N, x, y, u, v) = N''$ has the same nodes as N, and the same labels, except that $N''(x, y) = N(x, y) \cdot -(u; v)$.

PROPOSITION 8.2 *For any $n < \omega$ there is a universal first-order sentence ρ_n such that for any relation algebra $\mathcal{A}$, we have $\mathcal{A} \models \rho_n$ iff $\exists$ has a winning strategy in $G_n(I_a, \mathcal{A})$ for all non-zero $a \in \mathcal{A}$. Here, the network I_a is as in theorem 7.19.*

Proof. We will first define formulas $\rho_n(N)$, where N is any term network. The free variables of $\rho_n(N)$ will be the variables occurring in the terms in the labels of N. We will require that for all assignments ι of these variables into a relation algebra $\mathcal{A}$, and for all $n < \omega$,

$$\mathcal{A}, \iota \models \rho_n(N) \text{ iff } \exists \text{ has a winning strategy for } G_n(N^\iota, \mathcal{A}). \quad (*)$$

We let $\rho_0(N)$ be a quantifier-free formula saying that N^ι is a network:

$$\bigwedge_{x,y,z \in N} (N(x,x) \leq 1') \wedge ((N(x,z); N(z,y)) \cdot N(x,y) \neq 0).$$

8.1. The relation algebra case

It is clear that $(*)$ holds for ρ_0. Inductively, we define $\rho_{n+1}(N)$ to be:

$$\bigwedge_{x,y \in N} \forall u,v \Big(\rho_n(\mathsf{Acc}(N,x,y,u,v)) \vee \rho_n(\mathsf{Rej}(N,x,y,u,v)) \Big),$$

where u,v are new variables not occurring in the labels on N. Assuming inductively that $(*)$ holds for ρ_n, we show that it holds for ρ_{n+1}. Let ι assign the variables in the labels on edges of N to elements of some relation algebra $\mathcal{A}$. By lemma 7.16, $\exists$ has a winning strategy in $G_{n+1}(N^\iota, \mathcal{A})$ iff for all $x,y \in N$ and all $r,s \in \mathcal{A}$, she has a winning strategy in $G_n(N^+, \mathcal{A})$ or in $G_n(N^-, \mathcal{A})$, where N^+, N^- are, respectively, the pre-networks created from N^ι if she accepts or rejects when $\forall$ plays (x,y,r,s). Let ι' extend ι, with $\iota'(u) = r$ and $\iota'(v) = s$. By definition of $\mathsf{Acc}(N,x,y,u,v)$ and $\mathsf{Rej}(N,x,y,u,v)$, we see that modulo the choice of new node z, $N^+ = (\mathsf{Acc}(N,x,y,u,v))^{\iota'}$ and $N^- = (\mathsf{Rej}(N,x,y,u,v))^{\iota'}$. By the inductive hypothesis, $\exists$ has a winning strategy in the game $G_n((\mathsf{Acc}(N,x,y,u,v))^{\iota'}, \mathcal{A})$ iff $\mathcal{A}, \iota' \models \rho_n(\mathsf{Acc}(N,x,y,u,v))$, and similarly for Rej. Hence, $\exists$ has a winning strategy in $G_{n+1}(N^\iota, \mathcal{A})$ iff for all nodes x,y and all extensions ι' of ι to the variables u,v,

$$\mathcal{A}, \iota' \models \rho_n(\mathsf{Acc}(N,x,y,u,v)) \vee \rho_n(\mathsf{Rej}(N,x,y,u,v)).$$

That is, $\mathcal{A}, \iota \models \rho_{n+1}(N)$. This completes the induction.

Let x be any variable, and let J_x be the term network with nodes $0,1$ and $J_x(0,1) = x$, $J_x(1,0) = \breve{1}$, and $J_x(0,0) = J_x(1,1) = 1$'. We now obtain the sentence ρ_n as $\forall x(x \neq 0 \to \rho_n(J_x))$ (for each $n < \omega$). This is clearly a universal ($\forall_1$) sentence. $\square$

We now obtain:

THEOREM 8.3 *The class of representable relation algebras is recursively axiomatised by the universal sentences $\{\rho_n : n < \omega\}$ of proposition 8.2, together with the axioms for relation algebras given in definition 3.8.*

Proof. By theorem 7.19 and propositions 7.24 and 8.2, the proposed axiomatisation is sound and complete. The axioms for **RA** are certainly recursive. We can see that $\{\rho_n : n < \omega\}$ is recursive as follows. To decide whether a given sentence σ of the signature of relation algebras is in $\{\rho_n : n < \omega\}$, recursively enumerate the sequence $\rho_0, \rho_1, \ldots$, stopping when either we find a ρ_n equal to σ (when we respond 'yes'), or when we find a ρ_n that is longer (has more symbols) than σ (we then respond 'no'). This can certainly be done, and it terminates. If we respond 'no', then $\sigma \notin \{\rho_m : m \leq n\}$ (or we would already have said 'yes'), and since each successive sentence ρ_n is clearly longer than the preceding one, $\sigma \notin \{\rho_m : m \geq n\}$. So the procedure gives the right answer. $\square$

The ρ_n of proposition 8.2 are universal sentences essentially because $\exists$ never has to choose any elements of $\mathcal{A}$. By results of chapters 2 and 3 we can replace them by equations and obtain a recursive, equational axiomatisation of **RRA**.

THEOREM 8.4 *There is a recursive set $\{\varepsilon_n : n < \omega\}$ of equations such that for any relation algebra $\mathcal{A}$, $\mathcal{A}$ is representable iff $\mathcal{A} \models \varepsilon_n$ for all n.*

Proof. Recall from theorem 3.19 that **RA** is a (finitely axiomatised) discriminator variety. Now take $\mathsf{D} = \mathbf{RA}$ and $\mathsf{V} = \mathbf{RRA}$ in theorem 2.60. □

Exercises

1. An equation ε in a signature L of BAOs is said to be *canonical* if for any L-BAO $\mathcal{A}$, if $\mathcal{A} \models \varepsilon$ then $\mathcal{A}^+ \models \varepsilon$. Show that if **RRA** has a (recursively enumerable) axiomatisation by canonical equations, then there is a (recursive) function $f : \omega \to \omega$ such that for any relation algebra $\mathcal{A}$ and $n < \omega$, if $\exists$ has a winning strategy in $G_{f(n)}(\mathcal{A})$ then she has a winning strategy in $G_n(\mathcal{A}^+)$. (It is unknown whether **RRA** has a canonical axiomatisation: see problem 17.39.)

8.2 An axiomatisation using 'Q-operators'

In our opinion, the axioms for **RRA** just obtained are reasonably simple. However, readers not convinced of this may prefer the axiomatisation we are now going to give, which is even simpler, though at a price — we axiomatise not **RRA**, but (roughly) the class of 'Q-algebras'. We will show that a relation algebra is representable iff it expands to a model of the new axioms.

The axioms will be for expansions of representable relation algebras to interpret infinitely many new function symbols. These new symbols will determine whether or not their arguments form the labels of a network, and the new axioms will simply express that each network I_a for non-zero a is hereditarily refinable (that is, $\exists$ has a winning strategy in $G_\omega(I_a, \mathcal{A})$; see definition 7.18 for I_a). The new symbols are similar to Jónsson's Q-operators [Jón91], and we will discuss the connection in section 8.2.4 below.

8.2.1 The new function symbols

Fix distinct variables x_{ij} ($i, j < \omega$). Also fix for each $n < \omega$ an order of enumeration of the variables x_{ij} for $i, j < n$ — e.g., lexicographically. When we write tuples $(x_{ij} : i, j < n)$ or $(a_{ij} : i, j < n)$ below, it will be implicit that the entries are enumerated in this chosen order.

8.2. An axiomatisation using 'Q-operators'

Recall from definition 8.1 the notion of a term network. Here we will consider only term networks N of a restricted form, satisfying the following conditions:

1. $nodes(N) = \{0, 1, \ldots, n-1\}$ for some $2 \leq n < \omega$.

2. $N(i,j) = x_{ij}$ for each distinct $i, j < n$.

3. $N(i,i) = x_{ii} \cdot 1'$ for each $i < n$.

Therefore, N is determined by n, and below we sometimes write it as N_n. For each such $N = N_n$, we add to the signature of relation algebras a new n^2-ary function symbol Q_N. (We call it Q_N and not Q_n to try to avoid confusion with the Jónsson Q-operators below.) Because we constrained the nodes and labels of term networks, this entails adding countably many new symbols. The idea is that if $\mathcal{A}$ is a relation algebra and $a_{ij} \in \mathcal{A}$ for each $i, j < n$, the value of $Q_N(a_{ij} : i, j < n)$ in $\mathcal{A}$ should be non-zero iff the $\mathcal{A}$-pre-network N', obtained from N by assigning each x_{ij} to a_{ij}, is a network.

8.2.2 Equations using these function symbols

We now write down equations, using the new function symbols, to enforce that $\exists$ has a winning strategy in $G_r(I_a, \mathcal{A})$ for all $r < \omega$ and all $a \in \mathcal{A} \setminus \{0\}$, and hence by theorem 7.19 and proposition 7.24 that $\mathcal{A}$ is representable. For each term network $N = N_n$ as above (for each finite $n \geq 2$), we include the following equations. Below, $\bar{x}$ denotes the tuple $(x_{ij} : i, j < n)$.

1. If $n = 2$ and $\bar{y}$ is the 4-tuple of relation algebra terms obtained from $\bar{x}$ by replacing x_{00}, x_{11}, and x_{10} by the constant symbol 1 of the signature of relation algebras, we add the equation $Q_N(\bar{y}) \geq x_{01}$ (this abbreviates $Q_N(\bar{y}) + x_{01} = Q_N(\bar{y})$, of course).

2. Now we add equations to express that N defines a network:

    (a) For each $i, j, k < n$, let the tuple of terms $\bar{y} = (y_{lm} : l, m < n)$ be the same as $\bar{x}$ except that
    $$y_{ij} = N(i,j) \cdot (N(i,k); N(k,j))$$
    (so $\bar{y}$ is a tuple of terms made from the variables in $\bar{x}$ and perhaps 1', and depends on i, j, and k), and add the equation $Q_N(\bar{x}) \leq Q_N(\bar{y})$.

    (b) For this to work, we also have to say that Q_N is normal (see definition 2.33): for each tuple $\bar{y}$ obtained by replacing one or more variables in $\bar{x}$ by 0, we add the equation $Q_N(\bar{y}) = 0$.

3. Finally, we add axioms expressing that whatever move ∀ makes in round 0 of the game $G_\omega(N, \mathcal{A})$, ∃ can respond to leave a network for which she still has a winning strategy. For each $i, j < n$, let $\bar{y}$ be the same as $\bar{x}$ except that x_{ij} has been replaced by $x_{ij} \cdot -(x_{in}; x_{nj})$, and let $\bar{z}$ be the same as $(x_{kl} : k, l < n+1)$ except that:

   - x_{ij} has been replaced by $x_{ij} \cdot (x_{in}; x_{nj})$,
   - x_{nn}, x_{kn}, and x_{nl}, for all $k, l < n$ with $k \neq i$ and $l \neq j$, have been replaced by 1.

   Then we add the equation $Q_N(\bar{x}) \leq Q_N(\bar{y}) + Q_{N_{n+1}}(\bar{z})$.

8.2.3 Proof that the equations characterise representability

THEOREM 8.5 *Let $\mathcal{A}$ be a relation algebra. Then $\mathcal{A}$ is representable iff there is an expansion of $\mathcal{A}$ to interpret the function symbols Q_N that satisfies the above equations.*

Proof. Assume first that $\mathcal{A}$ is representable. Let h be a representation of $\mathcal{A}$ with base D, say, as in definition 3.28. Let $2 \leq n < \omega$, $\bar{x} = (x_{kl} : k, l < n)$, and $N = N_n$. For any n^2-tuple $\bar{a} = (a_{ij} : i, j < n)$ of elements of $\mathcal{A}$, define $\bar{a}$ to be *compatible* (with h) if there are elements $d_i \in D$ ($i < n$) such that $(d_i, d_j) \in h(a_{ij})$ for all $i, j < n$. By definition of N, it is equivalent to say that if ι is an assignment such that $\iota(x_{ij}) = a_{ij}$ for all $i, j < n$, then $(d_i, d_j) \in h(N^\iota(i, j))$ for all $i, j < n$. Now let

$$Q_N(\bar{a}) = \begin{cases} 1, & \text{if } \bar{a} \text{ is compatible,} \\ 0, & \text{otherwise.} \end{cases}$$

We check that the equations hold in the expansion of $\mathcal{A}$ just defined. Below, $\bar{x} = (x_{kl} : k, l < n)$ as usual. For $n = 2$ and N as in the first equation, we have to check that any 4-tuple $\bar{a}$ consisting of a non-zero algebra element a and three 1s, with the 'a' in the place occupied by x_{01} in $\bar{x}$, satisfies $Q_N(\bar{a}) \geq a$. But clearly, any such $\bar{a}$ is compatible, since $a > 0$ implies $h(a) \neq \emptyset$, and any $(d_0, d_1) \in h(a)$ witnesses compatibility of $\bar{a}$. So $Q_N(\bar{a}) = 1 \geq a$, as required. Verifying the normality equations 2b is equally easy, since clearly no $\bar{a}$ involving 0 can be compatible.

Next consider equations 2a. Let n, i, j, k be given, let $\bar{y}$ be obtained from $\bar{x}$ as stated, let $\bar{a}$ be any n-tuple from $\mathcal{A}$ and let $\bar{b}$ be obtained from $\bar{a}$ as $\bar{y}$ is from $\bar{x}$. We have to show that $Q_N(\bar{a}) \leq Q_N(\bar{b})$ — that is, if $\bar{a}$ is compatible then so is $\bar{b}$. But clearly, if $d_0, \ldots, d_{n-1} \in D$ satisfy $(d_l, d_m) \in h(a_{lm})$ for all $l, m < n$, then also $(d_l, d_m) \in h(b_{lm})$ for all $l, m < n$ — in particular, if $\iota(\bar{x}) = \bar{a}$, then $(d_i, d_j) \in h(N_n^\iota(i, j))$, $(d_i, d_k) \in h(N_n^\iota(i, k))$, and $(d_k, d_j) \in h(N_n^\iota(k, j))$, so that $(d_i, d_j) \in h(N_n^\iota(i, j) \cdot (N_n^\iota(i, k); N_n^\iota(k, j))) = h(b_{ij})$. So $\bar{b}$ is indeed compatible.

8.2. An axiomatisation using 'Q-operators'

The type 3 equations are verified in the same way. Fix n and $i, j < n$, let $\bar{a}$ be an n-tuple from $\mathcal{A}$, and let $a, b \in \mathcal{A}$ be arbitrary. Let $\bar{b}, \bar{c}$ be obtained from $\bar{a}, a, b$ as are $\bar{y}, \bar{z}$ from $\bar{x}, x_{in}, x_{nj}$ in equation 3. We have to show that $Q_N(\bar{a}) \leq Q_N(\bar{b}) + Q_{N_{n+1}}(\bar{c})$ — i.e., if $\bar{a}$ is compatible then at least one of $\bar{b}, \bar{c}$ is. Take $d_0, \ldots, d_{n-1} \in D$ with $(d_k, d_l) \in h(a_{kl})$ for all $k, l < n$. If $(d_i, d_j) \in h(-(a;b))$, then $(d_i : i < n)$ already witnesses compatibility of $\bar{b}$. Otherwise, there is $d_n \in D$ with $(d_i, d_n) \in h(a)$ and $(d_n, d_j) \in h(b)$. Then $(d_i : i \leq n)$ witnesses compatibility of $\bar{c}$. So the equations are valid in this expansion of $\mathcal{A}$.

Conversely, assume we have an expansion of $\mathcal{A}$ that interprets the Q-operators so as to satisfy the given equations.

Claim. Let $2 \leq n < \omega$ and $N = N_n$, let $\bar{a}$ be an n^2-tuple in $\mathcal{A}$, and let ι assign $\bar{x} = (x_{kl} : k, l < n)$ to $\bar{a}$. If $Q_N(\bar{a}) > 0$, then $\exists$ has a winning strategy in $G_r(N^\iota, \mathcal{A})$ for all $r < \omega$.

Proof of claim. The proof is by induction on r. For $r = 0$, we have to show that if $Q_N(\bar{a}) > 0$ then N^ι is a network. Let $i < n$; we require $N^\iota(i, i) \leq 1'$. But this is clear since $N(i, i) = x_{ii} \cdot 1'$. Let $i, j, k < n$; we require $\beta \stackrel{\text{def}}{=} N^\iota(i, j) \cdot (N^\iota(i, k); N^\iota(k, j)) \neq 0$. Let $\bar{b}$ be the same as $\bar{a}$ except that $b_{ij} = \beta$. Then equation 2a tells us that $Q_N(\bar{a}) \leq Q_N(\bar{b})$, so $Q_N(\bar{b}) \neq 0$. So by axiom 2b, $\beta = b_{ij} \neq 0$.

Now we assume the claim for r and prove it for $r+1$. Let $\forall$ move in round 0 of $G_{r+1}(N^\iota, \mathcal{A})$ by picking nodes $i, j < n$ and elements $a, b \in \mathcal{A}$. Let ι' be any assignment extending ι to x_{in}, x_{nj} and with $\iota'(x_{in}) = a, \iota'(x_{nj}) = b$. By equation 3, under ι' we have $Q_N(\bar{x}) \leq Q_N(\bar{y}) + Q_{N_{n+1}}(\bar{z})$, where $\bar{y}, \bar{z}$ are as specified in equation 3. So as $Q_N(\bar{a}) > 0$, at least one of $Q_N(\bar{y}), Q_{N_{n+1}}(\bar{z})$ is non-zero under assignment ι'. By definition of $\bar{y}, \bar{z}$ and the inductive hypothesis, this implies that $\exists$ has a winning strategy in $G_r(N^+, \mathcal{A})$ or in $G_r(N^-, \mathcal{A})$, where N^+, N^- are the pre-networks she plays to accept and (respectively) reject his round 0 move in $G_{r+1}(N^\iota, \mathcal{A})$. By lemma 7.16, she has a winning strategy in $G_{r+1}(N^\iota, \mathcal{A})$. This completes the induction and proves the claim.

Now let $a \in \mathcal{A} \setminus \{0\}$, and let ι assign x_{01} to a, x_{00} and x_{11} to $1'$, and x_{10} to 1. By equation 1, $Q_{N_2}(\iota(\bar{x})) \geq a > 0$. By the claim, $\exists$ has a winning strategy in $G_r(N_2^\iota, \mathcal{A})$ for each $r < \omega$. As clearly, $N_2^\iota = I_a$, proposition 7.24 yields that $\exists$ has a winning strategy in $G_\omega(I_a, \mathcal{A})$. This holds for all non-zero a, so by theorem 7.19, $\mathcal{A}$ is representable. $\square$

8.2.4 The Jónsson Q-operators

In [Jón91] Jónsson defined a notion of generalised relational composition. For $1 \leq n < \omega$ and n^2 binary relations r_{ij} $(i, j < n)$ on a set D, $Q_n(r_{ij} : i, j < n)$ is the binary relation on D given by

$$Q_n(r_{ij}:i,j<n)(x,y) \iff \exists z_0,\ldots,z_{n-1} \in D\Big(z_0 = x \wedge z_1 = y \wedge \bigwedge_{i,j<n} r_{ij}(z_i,z_j)\Big).$$

For example, $\breve{r}$ is equivalent to $Q_2(r_{ij}:i,j<2)$, where $r_{10}=r$ and the other r_{ij} are 1. Similarly, $r;s$ is equivalent to $Q_3(r_{ij}:i,j<3)$, where $r_{02}=r$, $r_{21}=s$, and all other r_{ij} are 1.

In [SteVen98, Ste00], more general operators $Q_n^{ij}(r_{ij}:i,j<n)$, for $i,j<n$, were defined, with semantics

$$Q_n^{ij}(r_{ij}:i,j<n)(x,y) \iff \exists z_0,\ldots,z_{n-1} \in D\Big(z_i = x \wedge z_j = y \wedge \bigwedge_{i,j<n} r_{ij}(z_i,z_j)\Big).$$

Equational axioms were given for these operators. These axioms include a generalised Peircean Law which implies that the operators are conjugated and hence completely additive. It was shown that the axioms are sound and complete for relation algebras with representations interpreting the Q_n^{ij} as shown above.

Clearly, there is some similarity between these Q-operators and the Q_N defined above. However, there are technical differences. Suppose we have a proper relation algebra $\mathcal{A}$ consisting of binary relations on D. Given elements $r_{ij} \in \mathcal{A}$, let N be the pre-network with nodes $n = \{0,1,\ldots,n-1\}$ and edge labels given by $N(i,j)=r_{ij}$. Then our $Q_{N_n}(r_{ij}:i,j<n)$ essentially expresses only whether or not N embeds in D — i.e., whether there is $\nu:N\to D$ such that $(\nu(i),\nu(j)) \in N(i,j)$ for all $i,j \in N$. We can expand $\mathcal{A}$ to interpret Q_{N_n} as in the proof of theorem 8.5, by defining Q_{N_n} to be either 0 or 1, as appropriate, and its value carries no further information.

The Jónsson-style $Q_n(r_{ij}:i,j<n)$, on the other hand, is a full-blown *binary relation* on D; it relates points $x,y \in D$ iff N embeds into D with its first two nodes mapping to x and y. While this is a well-defined binary relation on D, there is no guarantee that $\mathcal{A}$ contains this relation. The same goes for the Q_n^{ij}. Further, unlike the Q_n^{ij}, the Q_N are not in general completely additive, because not every representable relation algebra has a complete representation. It is not in general possible to expand such a relation algebra to satisfy the axioms of [SteVen98, Ste00]: see exercise 17.2(1). However, one could show that a relation algebra is representable iff its canonical extension expands to a model of these axioms.

Exercises

1. Show that we can add axioms stating that the Q_N are additive, and thus *operators* in the sense of definition 2.33, and keep theorem 8.5.

8.3 Axiomatising $\mathbf{RCA}_d$ for $3 \leq d < \omega$

In the same way, we can obtain an equational axiomatisation of the representable finite-dimensional cylindric algebras. Though what follows works for any finite $d \geq 2$, we are only interested in $d \geq 3$, since $\mathbf{RCA}_0$, $\mathbf{RCA}_1$, and $\mathbf{RCA}_2$ are finitely axiomatisable and were dealt with in section 5.3. We will generalise to cylindric algebras of infinite dimension using algebraic techniques.[1]

As for relation algebras, the first step is to define sentences ψ_t, in the signature of d-dimensional cylindric algebras, saying that $\exists$ has a winning strategy in the game $G_t(N,C)$.

DEFINITION 8.6 We write L_c^d for the signature $\{0, 1, +, -, c_i, d_{ij} : i, j < d\}$ of d-dimensional cylindric algebras.

A (d-dimensional cylindric) *term network* is a pair consisting of a finite non-empty set N or nodes(N) of 'nodes', and a map, also written N, assigning an L_c^d-term $N(\bar{x})$ to each d-tuple $\bar{x}$ of nodes.

Given a term network N, an index $i \leq d$, a d-tuple $\bar{x} \in {}^d N$, and an L_c^d-term τ, we define two term networks, $\mathsf{Out} = \mathsf{Out}(N, i, \bar{x}, \tau)$ and $\mathsf{In} = \mathsf{In}(N, i, \bar{x}, \tau)$, corresponding to the two ways that $\exists$ can respond in the game. For $i < d$, we define In and Out as follows.

- nodes(Out) = nodes(N), and nodes(In) = nodes(N) $\cup \{z\}$ for some new node z.

- For all $\bar{y} \in {}^d N \setminus \{\bar{x}\}$, $\mathsf{In}(\bar{y}) = \mathsf{Out}(\bar{y}) = N(\bar{y})$.

- $\mathsf{Out}(\bar{x}) = N(\bar{x}) \cdot -c_i \tau$ (this completes the definition of Out for $i < d$).

- $\mathsf{In}(\bar{x}) = N(\bar{x}) \cdot c_i \tau$.

- Let $\bar{z} \equiv_i \bar{x}$ with $z_i = z$. Then $\mathsf{In}(\bar{z}) = \tau \cdot \prod\limits_{\substack{j,k<d \\ x_j = x_k}} d_{jk}$.

- For all other d-tuples $\bar{y}$ involving z, $\mathsf{In}(\bar{y}) = \prod\limits_{\substack{j,k<d \\ y_j = y_k}} d_{jk}$ (and that defines In

  for $i < d$).

For $i = d$, we let

- nodes(Out) = nodes(In) = nodes(N).

- $\mathsf{In}(\bar{x}) = N(\bar{x}) \cdot \tau$.

---

[1] These results appeared in [HirHod97b, §11.5]; we thank the copyright owners Association for Symbolic Logic for granting permission for this republication. The notation has been changed slightly.

- $\mathsf{Out}(\bar{x}) = N(\bar{x}) \cdot -\tau$.
- For all $\bar{y} \in {}^d N \setminus \{\bar{x}\}$, $\mathsf{In}(\bar{y}) = \mathsf{Out}(\bar{y}) = N(\bar{y})$.

DEFINITION 8.7 Let N be a d-dimensional cylindric term network. We define the formula $\mathrm{CNet}^d(N)$ to be the conjunction of the following three formulas:

$$\bigwedge_{\bar{x} \in {}^d N, i,j<d, x_i=x_j} N(\bar{x}) \leq \mathsf{d}_{ij}, \qquad \bigwedge_{\bar{x},\bar{y} \in {}^d N, i<d, \bar{x} \equiv_i \bar{y}} N(\bar{x}) \cdot \mathsf{c}_i N(\bar{y}) \neq 0,$$

$$\bigwedge_{\substack{\bar{x},\bar{y},\bar{z} \in {}^d N, i,j,k<n \\ \bar{x} \equiv_i \bar{y}, x_i=z_j, y_i=z_k}} (N(\bar{z}) \leq \mathsf{d}_{jk}) \to (N(\bar{x}) \cdot N(\bar{y}) \neq 0).$$

Clearly, if ι is an assignment of the variables occurring in the terms labelling the hyperedges of N to elements of a d-dimensional cylindric algebra C, then

$$C, \iota \models \mathrm{CNet}^d(N) \text{ iff } N^\iota \text{ is a } C\text{-network}.$$

Now we define

$$\begin{aligned} \psi_0^d(N) &= \mathrm{CNet}^d(N) \\ \psi_{n+1}^d(N) &= \forall y \bigwedge_{\bar{x} \in {}^d N, i \leq d} \psi_n^d(\mathsf{In}(N,i,\bar{x},y)) \vee \psi_n^d(\mathsf{Out}(N,i,\bar{x},y)), \end{aligned}$$

where y is a new variable not occurring in any of the terms of N. One can show by induction on $n < \omega$ that for all assignments ι of the variables in the terms of N into C, we have $C, \iota \models \psi_n^d(N)$ if and only if $\exists$ has a winning strategy for $G_n(N^\iota, C)$.

Now let x be any variable, and define J_x to be the graph with nodes d, with $J_x(0, 1, \ldots, d-1) = x$, and $J_x(\bar{y}) = \prod_{y_i=y_j} \mathsf{d}_{ij}$ for all $\bar{y} \in {}^d d \setminus \{(0, 1, \ldots, d-1)\}$. Then by theorem 7.27, C is representable if and only if $C, \iota \models \psi_n^d(J_x)$ for all $n < \omega$ and all assignments $\iota : \{x\} \to C \setminus \{0\}$.

Let ϕ_n^d be the sentence $\forall x (x \neq 0 \to \psi_n^d(J_x))$. Then ϕ_n^d is a universal sentence of the signature of d-dimensional cylindric algebras, and we have

PROPOSITION 8.8 *For any $3 \leq d < \omega$, a d-dimensional cylindric algebra C is representable iff $C \models \phi_n^d$ for all $n < \omega$.*

Since for finite d, $\mathbf{CA}_d$ is a discriminator variety (with discriminator term $c_0 c_1 \ldots c_{d-1} x$ — see section 5.3), we can use theorem 2.60 to convert these universal sentences ϕ_n^d into equations ε_n^d that axiomatise $\mathbf{RCA}_d$ within $\mathbf{CA}_d$. (Take $\mathsf{D} = \mathbf{CA}_d$ and $\mathsf{V} = \mathbf{RCA}_d$ in theorem 2.60.) We obtain:

THEOREM 8.9 *For finite $d \geq 3$, $\mathbf{RCA}_d$ is axiomatised by the equations $\{\varepsilon_n^d : n < \omega\}$ together with the equations of definition 5.16 defining $\mathbf{CA}_d$.*

8.4 Axiomatising $\mathbf{RCA}_\alpha$ for infinite α

The axiomatisation carries over to the α-dimensional representable cylindric algebras, for any $\alpha \geq \omega$. All we have to do is take all equations obtained by replacing the finite indices i,j of symbols $\mathsf{d}_{ij}, \mathsf{c}_i$ in the ε_n^d (for $n < \omega$, $3 \leq d < \omega$) by any indices $< \alpha$ in a one-one fashion, and throw in the axioms defining $\mathbf{CA}_\alpha$. That this works can be easily read off from the following known algebraic results.

DEFINITION 8.10 For an ordinal α, let CA_α denote the set of axioms defining α-dimensional cylindric algebras given in definition 5.16. Let Ax_α consist of CA_α together with all L_c^α-sentences of the following form:

$$\forall v_1, \ldots, v_n (\exists v_0 \varphi(v_0, \mathsf{c}_i v_1, \ldots, \mathsf{c}_i v_n) \to \exists v_0 \varphi(\mathsf{c}_i v_0, \mathsf{c}_i v_1, \ldots, \mathsf{c}_i v_n)),$$

where $\varphi(v_0, \ldots, v_n)$ is any formula of L_c^α and $i < \alpha$ is such that $\mathsf{c}_i, \mathsf{d}_{ij}, \mathsf{d}_{ji}$ do not occur in φ for any $j < \alpha$. (See [HenMon+85, definition 4.1.11].)

FACT 8.11 ([HenMon+85, corollaries 4.1.15, 4.1.16]) *Let α be an ordinal, and ε an equation of L_c^α. Then ε is valid in $\mathbf{RCA}_\alpha$ iff $Ax_{\max(\alpha,\omega)} \vdash \varepsilon$.*

To exploit this, we need some notation for renaming symbols of a signature. Write L_c^α for the signature of α-dimensional cylindric algebras. Let α, β be ordinals, and let $\mu : \alpha \to \beta$ be any 1–1 partial map. For an L_c^α-formula φ, we define φ^μ to be the L_c^β-formula obtained by replacing every c_i in φ by $\mathsf{c}_{\mu(i)}$, and replacing every d_{ij} by $\mathsf{d}_{\mu(i)\mu(j)}$. So φ^μ is only defined if μ is defined on i,j,k for every $\mathsf{c}_i, \mathsf{d}_{jk}$ occurring in φ.

For sets Φ of formulas on which μ is defined, we define $\Phi^\mu = \{\varphi^\mu : \varphi \in \Phi\}$.

LEMMA 8.12 *Let α, β be ordinals, and $\mu : \alpha \to \beta$ be a partial 1–1 map.*

1. *Let Φ be any set of L_c^α-sentences, and suppose that μ is defined on them and also on some other L_c^α-sentence σ. Then $\Phi \vdash \sigma$ iff $\Phi^\mu \vdash \sigma^\mu$.*

2. *If μ is a total map, then $(CA_\alpha)^\mu \subseteq CA_\beta$, and $(Ax_\alpha)^\mu \subseteq Ax_\beta$.*

Proof. One can prove (1) by applying μ or its inverse to every sentence in a proof of σ from Φ, for example. (2) follows from the definitions. $\square$

We use this lemma implicitly in the following corollary to theorem 8.9.

COROLLARY 8.13 *Let $\alpha \geq \omega$ be an ordinal. Then $\mathbf{RCA}_\alpha$ is axiomatised by the set*

$$\Sigma = CA_\alpha \cup \{(\varepsilon_n^d)^\mu : d, n < \omega, \ d \geq 3, \ \mu : d \to \alpha \text{ is a 1–1 map}\},$$

where the ε_n^d are as in theorem 8.9.

Proof. We begin with a claim.

Claim. For any L_c^α-equation ε, we have $Ax_\alpha \vdash \varepsilon$ iff $\Sigma \vdash \varepsilon$.

Proof of claim. For '$\Rightarrow$', fix an L_c^α-equation ε such that $Ax_\alpha \vdash \varepsilon$. There is a finite set $\Phi \subseteq Ax_\alpha$ such that $\Phi \vdash \varepsilon$. Choose a partial, surjective 1–1 map $\mu : \alpha \to \omega$ such that $\varepsilon^\mu, \varphi^\mu \, (\varphi \in \Phi)$ are all defined. Then $\Phi^\mu \vdash \varepsilon^\mu$. By lemma 8.12, $\Phi^\mu \subseteq Ax_\omega$. Thus, $Ax_\omega \vdash \varepsilon^\mu$. Choose finite $d \geq 3$ such that ε^μ is a sentence of L_c^d. By fact 8.11, ε^μ is valid in $\mathbf{RCA}_d$. By theorem 8.9, $CA_d \cup \{\varepsilon_n^d : n < \omega\} \vdash \varepsilon^\mu$. So $(CA_d)^{\mu^{-1}} \cup \{(\varepsilon_n^d)^{\mu^{-1}} : n < \omega\} \vdash \varepsilon$, whence by lemma 8.12, $\Sigma \vdash \varepsilon$, as required.

The converse is easier. It is enough to show that any $(\varepsilon_n^d)^\mu \in \Sigma$ is a consequence of Ax_α. Well, by theorem 8.9, ε_n^d is an equation valid in $\mathbf{RCA}_d$, so, by fact 8.11, $Ax_\omega \vdash \varepsilon_n^d$. Let $\nu : \omega \to \alpha$ be any 1–1 extension of μ to ω. Then $(Ax_\omega)^\nu \vdash (\varepsilon_n^d)^\mu$. By lemma 8.12, $(Ax_\omega)^\nu \subseteq Ax_\alpha$. So $Ax_\alpha \vdash (\varepsilon_n^d)^\mu$, as required. This proves the claim.

As $\mathbf{RCA}_\alpha$ is a variety, any L_c^α-structure C is in $\mathbf{RCA}_\alpha$ iff it satisfies all equations of L_c^α that are valid in $\mathbf{RCA}_\alpha$. By fact 8.11, an L_c^α-equation is valid in $\mathbf{RCA}_\alpha$ iff it is a logical consequence of Ax_α. By the claim, this is iff it is a logical consequence of the set Σ given in the corollary. So C is in $\mathbf{RCA}_\alpha$ iff it satisfies all equational consequences of Σ. Since Σ itself consists of equations, this is iff $C \models \Sigma$, as required. □

Exercises

1. Generalise theorem 8.5 to finite-dimensional cylindric algebras.

Chapter 9

Axiomatising pseudo-elementary classes

9.1 Introduction

It is an important problem in algebraic logic to find explicit axiomatisations of various classes of algebras, atom structures, etc. In the preceding chapters, we axiomatised the classes **RRA** and **RCA**$_n$ by games. But as we go on, we will see many more classes with interesting representations — for instance, **S$\Re a$CA**$_n$, **RA**$_n$, **GRA**, **Crs**$_n$, and so on. In this chapter, we generalise the game-theoretic axiomatisation method to these other classes. In fact, we show how to obtain an explicit axiomatisation of the elementary closure of any *pseudo-elementary class* of structures from its defining theory. If the class is closed under substructures, we can obtain an explicit universal axiomatisation of it. As we already know (theorem 2.60), in the context of discriminator varieties this can be refined to an equational axiomatisation.

History As discussed in chapter 1 and illustrated in chapters 7 and 8, the methodology has four ingredients:

1. Building a 'representation' of an algebra or structure by a 'step by step' construction.

2. Doing this by an (infinite) game played on the algebra.

3. Approximating the infinite game by finite ones.

4. Writing out first-order axioms expressing that the finite games can be won.

As the history of these ideas is quite interesting, we will now detail some of the many precursors for each of these four ingredients in turn. Useful general references are [AddHen+65, Hod93].

1. There is now a lot of work in algebraic logic and modal and temporal logic on building a representation or model in a 'step by step' fashion. Samples include [Gab81, Bur82, Madd82, AndTho88, And01, VenMar98]. Though such arguments can usually be expressed in terms of a game, this was not done explicitly in the original. The step-by-step construction is carried out in a setting with predetermined properties sufficient to ensure it can be concluded successfully.

2. Games to build structures are familiar in model-theoretic forcing [Hod85], and even as well known a device as Henkin's completeness theorem for first-order logic can be seen this way. As in (1), these constructions generally take place in a setting satisfying certain properties carefully chosen in advance, and the game is simply used to build the required structure; the 'properties' ensure that this can be done. Things are not usually taken further by trying to characterise precisely which properties are needed for the game construction to succeed, which is what we essentially do when writing axioms expressing a winning strategy. For example, [Ven98] used games to prove representability of any diagonal-free cylindric algebra that has the well chosen property of *rectangular density;* and Henkin's construction to prove that a *consistent* first-order theory has a model is not usually thought of as a test for consistency of the given theory, though in tableau proofs it is used that way.

3. We cannot think of any precedents for approximating infinite games by finite ones in isolation, but [Kei65] does it in combination with (4) below. Work of Svenonius and Vaught (see [Hod93]) is similar.

4. Expressing a winning strategy in a game by logical sentences has a long history, notably exemplified in the famous Ehrenfeucht–Fraïssé characterisation of equivalence in first-order logic of a given quantifier depth [Ehr61] and the related theorem of Karp [Karp65] for fragments of $L_{\infty\omega}$. These games (perhaps disguised as bisimulations or back-and-forth systems) do not usually construct anything: they simply compare two existing structures. For corresponding games on a single structure, see [Hin73], building on earlier work of Henkin (1971), and in an infinitary context, [Kei65, Sco65]. The finite model theorists have pursued these lines recently, especially using 'n-pebble games' [Bar77, ImmKoz87, DawLin+95].

As we said, combinations and blends of the ingredients (3) and (4) also occur in the literature. Van Benthem's theorem showing that two modal models satisfying

9.1. Introduction

the same modal formulas have bisimilar elementary extensions [Ben76, Ben85] seems to fit here, as does work of de Rijke [Rij95a, Rij95b]. The very striking [Kei65] should certainly be mentioned too; there was a great deal of interest in games and infinitary logic in the 1960s and 70s, leading (e.g.) to work in admissible set theory. The construction aspect of (1,2) is generally missing in these examples; it is present in Lyndon's papers [Lyn50, Lyn56], which, although they don't explicitly use games, are notable (among other things) for combining (1,3,4). Indeed, the whole method (1–4) amounts to 'Lyndon's approach plus games'. Examples of its use in the literature include [SteVen98, Ste00], axiomatising Q-operators.

Generalising the method We know from chapters 7 and 8 how games can be *used* to determine representability of algebras, but we have not described in general how to *devise* a game suitable for axiomatising a given class of algebras. In fact, the process of defining new games *ad hoc* for each kind of algebra and representation encountered, while never very difficult, is tedious, especially when we have to derive an explicit axiomatisation as well. It still treats games at the meta-level, whereas the spirit of Lyndon's approach seems to be to move them into the object level. Can we do this, and take further advantage of it by developing a more general meta-style reasoning about games?

Some more general results in this vein exist. In [Kei65], for example, a whole class of games is defined, one for each sentence of a certain infinitary logic. But these games do not construct objects. Closer to algebraic logic is [HodMik$^+$01], where the class of subalgebras of complex algebras over an arbitrary variety is axiomatised. Here, the 'representation' is a certain expansion of a structure in the given variety, and the game to construct it is defined by the variety. The method covers many classes of algebras from algebraic logic.

In this chapter, we will discuss a related, more general approach. If we want to use games to build representations, we should consider what a representation is. It turns out that being a representation of an algebra is usually easily definable by a suitable first-order theory in a 2-sorted language. The first sort of a model of this *defining theory* will be the algebra itself, and the second sort will be a representation of it. The defining theory specifies the relationship between the two, and its axioms depend on what kind of representation we are considering. Thus, the representable algebras are simply those algebras that are the first sort of some model of the defining theory.

Not all kinds of representation are so easily written, but — as a practical observation — many of them are. We will see many examples later (starting with examples 9.2), but to give the idea, we note that **RRA** is one. The parts of the defining theory expressing that representations respect boolean join and relative product of binary relations will read as follows:

$$\forall abxy(\text{holds}(a+b,x,y) \leftrightarrow \text{holds}(a,x,y) \vee \text{holds}(b,x,y))$$
$$\forall abxy(\text{holds}(a;b,x,y) \leftrightarrow \exists z(\text{holds}(a,x,z) \wedge \text{holds}(b,z,y))),$$

where holds is a ternary predicate with first argument of the 'algebra' sort and the others of the 'representation' sort, and $\text{holds}(a,x,y)$ is intended to mean that the element a of the algebra is a binary relation that holds on (x,y), for elements x,y of the representation. Compare with definition 3.30.

The class of all structures that arise as the first sort of a model of a fixed two-sorted first-order theory is of course a venerable old notion in model theory, introduced by Mal'cev in the 1940s and studied by Makkai, Svenonius, Tarski, and others. It is often used in algebraic logic when showing that various classes are closed under ultraproducts (see the recent survey [AndNém+01]). It is known as a *pseudo-elementary class*. Many classes of representable algebras from algebraic logic can be expressed as pseudo-elementary classes. The defining theory is usually finite and very natural and simple, and essentially always recursively enumerable, because we certainly expect that a Turing machine should be able to write down what we mean by a representation. In the light of the examples below, we would go so far as to say that a (fairly but not completely general) *definition* of the notion of representation of an algebra is just the second sort of a model of some two-sorted (perhaps recursively enumerable) first-order theory, the first sort of the model being the algebra.

Now model-theoretic forcing, as seen for example in [Hod85] and indeed in the classical construction of Henkin to show completeness of first-order logic, typically involves constructing a model of some first-order theory by a game. The game builds the model in just the same way as the games of chapter 7 build a representation of an algebra — step-by-step, elements of the model being introduced by the second player $\exists$ in response to criticisms by $\forall$. (In forcing, during the game itself the elements of the model are treated syntactically, e.g. as Henkin witnesses; only at the end of play do they become semantic as elements of the actual model. Here, this is only of notational significance and we don't worry about it further.)

This suggests that we combine forcing-games with the pseudo-elementary view of representations, by *using a game to build the second sort of a model of the defining theory whose first sort is already fixed to be the algebra whose representability is at issue*. We take the defining theory of the pseudo-elementary class to be given — this defines the notion of representation to be axiomatised. The game is just as in many forcing arguments. We apply the process seen earlier, approximating the infinite-length game by finite ones and writing down axioms expressing that $\exists$ has a winning strategy in them. These axioms can be obtained recursively from the defining theory of the class.

We will carry forward this plan in the rest of the chapter, but we stress that we can hardly claim much originality for it. As we have seen, the ideas behind it are

9.2. Pseudo-elementary classes

some years old — Keisler in 1965 would not have been surprised by it. Indeed, axiomatising pseudo-elementary classes is not new: theorem 9.14 below gives a similar result. (When we come to this theorem, we will remark on why we think the game-theoretic approach is preferable.)

Outline of chapter In section 9.2, we define pseudo-elementary and 'pseudo-universal' classes, and in section 9.3 we give examples of these classes familiar from algebraic logic. Section 9.4 covers their basic model-theoretic properties. The real work starts in section 9.5, where we universally axiomatise any pseudo-universal class. In section 9.6, we axiomatise the elementary closure of any pseudo-elementary class. In the final section, 9.7, we generalise the 'Q-operators' of section 8.2 to pseudo-universal and pseudo-elementary classes.

9.2 Pseudo-elementary classes

We will define pseudo-elementary-type classes using two-sorted first-order logic. It is not always done this way, but we choose to do so for three reasons. First, it keeps the syntax simple in the examples. Second, physically restricting quantifiers 'in the hardware' gives more concise game-theoretic axiomatisations. Third, it makes the definition of a 'pseudo-universal class' (definition 9.1(3) below) simple and intuitive. The notation in the definition will be explained afterwards.

DEFINITION 9.1 Let L be a first-order signature, and let K be a class of L-structures.

1. K is said to be a *pseudo-elementary class* if there are

    (a) a sorted language L^s with two disjoint sorts **a** and **r** (standing for 'algebra' and 'representation'), containing **a**-sorted copies of all symbols of L,

    (b) an L^s-theory U, called the *defining theory*,

    such that $\mathsf{K} = \{M^{\mathbf{a}}{\upharpoonright}_L : M \models U\}$.

2. K is said to be a PC^s *class* if the above holds for some finite U.

3. K is said to be a *pseudo-universal class* if (1) above holds with the following restrictions:

    (a) Each function symbol and constant in L^s, other than the copies of the ones in L, takes values in sort **r**.

    (b) Each sentence in U is built from atomic and negated-atomic formulas using $\wedge, \vee, \forall x^{\mathbf{a}}, \forall x^{\mathbf{r}}, \exists x^{\mathbf{r}}$ (no negation or $\exists x^{\mathbf{a}}$).

Here, we write $x^{\mathbf{a}}$ to indicate that the variable x has sort $\mathbf{a}$, and similarly for $\mathbf{r}$. Though the point is not important, we can allow the $\mathbf{r}$-sort of L^s-structures to be empty if L^s has no $\mathbf{r}$-valued function symbols; but the $\mathbf{a}$-sort must be non-empty. We identify the symbols of L with their $\mathbf{a}$-sorted copies in L^s, so that $L \subseteq L^s$. Let $L^{\mathbf{a}}$ denote the set of symbols of L^s of sole sort $\mathbf{a}$. Then for an L^s-structure M, we write $M^{\mathbf{a}}$ for the $L^{\mathbf{a}}$-structure whose domain is the set of elements of M of sort $\mathbf{a}$. Since $L \subseteq L^{\mathbf{a}}$, $M^{\mathbf{a}}$ has an L-reduct, written $M^{\mathbf{a}}\!\restriction_L$ as usual. We define $M^{\mathbf{r}}$ similarly, when it is non-empty.

We tend to write M for an L^s-structure and $\mathcal{A}$ for an L-structure, since in applications the latter will usually be an algebra while the former is more general.

It may be worth listing some facts about these classes in advance. They will be established in the theorems and exercises in later sections. Any elementary class is (obviously) pseudo-elementary, but not vice versa — see example 9.2(2) and exercise 9.4.(9) below. Any pseudo-elementary class is closed under ultraproducts. Any pseudo-universal class is elementary and universally axiomatisable (hence closed under substructures).

9.3 Examples

Many familiar classes in algebraic logic are pseudo-elementary. In example 9.2 below, we list several of them. As they are often even pseudo-universal, the work in section 9.5 below provides a natural universal recursive axiomatisation of them, and we can often go on to obtain a recursive equational axiomatisation, using discriminators. Others are at least elementary, and for these the techniques of section 9.6 provide an explicit first-order axiomatisation.

In the examples of pseudo-universal classes, we will present the defining theory in a readable form, and leave it to the reader to check that equivalent axioms satisfying the conditions of definition 9.1(3) can be obtained by replacing $\varphi \leftrightarrow \psi$ by $(\varphi \to \psi) \wedge (\psi \to \varphi)$, $\varphi \to \psi$ by $\neg \varphi \vee \psi$, and then pushing negations down next to atomic formulas. It will be clear that no new $\mathbf{a}$-sorted function symbols or constants have been introduced, so the only thing to watch for is that no $\mathbf{a}$-sorted existential quantifiers are implicitly present in the original axioms.

Of course, many of the classes in the examples are known to be elementary and even varieties, so are necessarily pseudo-elementary. The point of the example is to practise stating the defining theory explicitly, so that we can obtain explicit axioms from it later on. Also, showing a class to be pseudo-universal is a fast way of showing it to be elementary and universally axiomatisable (see corollary 9.15 below).

EXAMPLE 9.2 Let L be the signature L_{RA} of relation algebras.

9.3. Examples

1. The class **RRA** of representable relation algebras is a pseudo-universal class. We let $L^s = L \cup \{\texttt{holds}(\mathbf{a}, \mathbf{r}, \mathbf{r})\}$ (so L-symbols have sole sort $\mathbf{a}$, and holds is a ternary relation symbol with first argument of sort $\mathbf{a}$ and the others of sort $\mathbf{r}$). We let U be the L^s-theory consisting of the equations of definition 3.8 for relation algebras, plus the following axioms:

$$\begin{aligned}
\forall xx'yz(\texttt{holds}(x+x',y,z) &\leftrightarrow \texttt{holds}(x,y,z) \vee \texttt{holds}(x',y,z)) \\
\forall xyz(\texttt{holds}(-x,y,z) &\leftrightarrow \texttt{holds}(1,y,z) \wedge \neg\texttt{holds}(x,y,z)) \\
\forall yz(\texttt{holds}(1',y,z) &\leftrightarrow y = z) \\
\forall xyz(\texttt{holds}(\check{x},y,z) &\leftrightarrow \texttt{holds}(x,z,y)) \\
\forall xx'yz(\texttt{holds}(x;x',y,z) &\leftrightarrow \exists t(\texttt{holds}(x,y,t) \wedge \texttt{holds}(x',t,z))) \\
\forall x(x \neq 0 &\rightarrow \exists yz\,\texttt{holds}(x,y,z)).
\end{aligned}$$

The variables x, x' are of sort $\mathbf{a}$ and the other variables are of sort $\mathbf{r}$. As remarked above, these axioms are easily seen to be equivalent to ones in the form of definition 9.1(3). For example, the penultimate one is equivalent to the axioms

$$\forall xx'yz(\neg\texttt{holds}(x;x',y,z) \vee \exists t(\texttt{holds}(x,y,t) \wedge \texttt{holds}(x',t,z))),$$
$$\forall xx'yz(\texttt{holds}(x;x',y,z) \vee \forall t(\neg\texttt{holds}(x,y,t) \vee \neg\texttt{holds}(x',t,z))).$$

The reader may compare this theory U with the characterisation of relation algebra representations by the single-sorted first-order theory $T_{\mathcal{A}}$ of definition 3.30.

We claim that $\{M^{\mathbf{a}}{\restriction}_L : M \models U\} = \mathbf{RRA}$. For, let $\mathcal{A} \in \mathbf{RRA}$. A representation of $\mathcal{A}$ on a base set X provides, for each $a \in \mathcal{A}$, a binary relation $h(a)$ on X. We require that the map h takes the algebra operations to the corresponding operations on binary relations: so, e.g., $h(a;b)$ is the relative product of the binary relations $h(a)$ and $h(b)$, for all $a, b \in \mathcal{A}$. Let M be the disjoint union of $\mathcal{A}$ (taken to be of sort $\mathbf{a}$) and X (sort $\mathbf{r}$), and make M an L^s-structure by letting $\texttt{holds}(a, b, c)$ hold iff $(b, c) \in h(a)$. Evidently, $M \models U$ and $M^{\mathbf{a}}{\restriction}_L = \mathcal{A}$. Conversely, if $\mathcal{A} = M^{\mathbf{a}}{\restriction}_L$ for some $M \models U$, then $\mathcal{A}$ is a relation algebra, and $M^{\mathbf{r}}$ yields a representation h of $\mathcal{A}$ via $h(a) = \{(b, c) : M \models \texttt{holds}(a, b, c)\}$, for $a \in \mathcal{A}$. So $\mathcal{A} \in \mathbf{RRA}$. See (8) below for another way to the same result.

2. Recall from definition 3.35 that **CRA** denotes the class of completely representable relation algebras — those relation algebras with a representation respecting arbitrary meets and joins. Now any representation of a relation algebra yields a boolean representation of the boolean reduct of the relation algebra, simply by ignoring all the non-boolean operators. Theorem 2.21 then states that this boolean representation is complete if and only if it is atomic. Thus, the completely representable relation algebras are precisely the relation algebras with an 'atomic' representation — i.e., a representation

280   Chapter 9. Axiomatising pseudo-elementary classes

h such that $h(1) = \bigcup\{h(a) : a$ an atom of $\mathcal{A}\}$. This gives us a pseudo-elementary characterisation of **CRA**: to the theory U for **RRA** above, we add a sentence saying that the representation is atomic:

$$\forall yz(\texttt{holds}(1,y,z) \to \exists x(\texttt{holds}(x,y,z) \wedge \forall x'(x' < x \to x' = 0))).$$

Note that we no longer formally have a pseudo-universal class, because of the quantification $\exists x^{\mathbf{a}}$. In fact, **CRA** is not pseudo-universal, as it is not even elementary (section 17.2), while any pseudo-universal class is elementary (theorem 9.14 below). The elementary closure of **CRA** can be axiomatised by the techniques of section 9.6: the resulting (non-universal) axioms are essentially the 'Lyndon conditions' of [Lyn50], to be discussed in chapter 11.

3. The class of relation algebras that have a square representation with infinite base is pseudo-universal. To define it, we can take the theory U for **RRA** above and add the axioms $\forall yz\,\texttt{holds}(1,y,z)$ and $\exists y_0,\ldots y_{n-1} \bigwedge_{i<j<n} y_i \neq y_j$ for $n < \omega$. Alternatively, for a finite theory, we can add a new binary relation symbol $\mathbf{r} \sqsubset \mathbf{r}$ to L^s and add to U an axiom stating that $\sqsubset$ is a non-trivial dense linear order on the **r**-sort.

4. The class **GRA** of group relation algebras (see definition 4.3) is a pseudo-universal class. To see this, we may take

$$L^s = L \cup \{\texttt{holds}(\mathbf{a},\mathbf{r}),\ (-)^{-1}: \mathbf{r} \to \mathbf{r},\ *: \mathbf{r} \times \mathbf{r} \to \mathbf{r},\ e^{\mathbf{r}}\},$$

and write an L^s-theory expressing that:

(a) the **a**-sort (of its models) is a boolean algebra with operations $0, 1, +, -$ from L.

(b) the **r**-sort is a group with group product operation '$*$', inverse $-^{-1}$, and identity e.

(c) $\forall xx'y(\texttt{holds}(x+x',y) \leftrightarrow \texttt{holds}(x,y) \vee \texttt{holds}(x',y))$

(d) $\forall xy(\texttt{holds}(-x,y) \leftrightarrow \neg\texttt{holds}(x,y))$

(e) $\forall y(\texttt{holds}(1',y) \leftrightarrow y = e)$

(f) $\forall xy(\texttt{holds}(\check{x},u) \leftrightarrow \texttt{holds}(x,u^{-1}))$

(g) $\forall xx'y(\texttt{holds}(x;x',y) \leftrightarrow \exists zz'(y = z * z' \wedge \texttt{holds}(x,z) \wedge \texttt{holds}(x',z')))$

(h) $\forall x(x \neq 0 \to \exists y\,\texttt{holds}(x,y)).$

As before, x, x' here are of sort **a** and the other variables are of sort **r**.

The variety **SP(GRA)** of exercise 4.2(2) can be checked to be pseudo-universal as well.

9.3. Examples

5. The class **wRRA** of weakly representable relation algebras (definition 5.14) is pseudo-universal. This can be shown in the same way as for **RRA**.

6. A representation of a relation algebra $\mathcal{A}$ on the set X is said to be *permutational* if for all $y, z \in X$, there is a permutation of X leaving all $\mathcal{A}$-relations invariant (an 'automorphism') and taking y to z. (See exercise 3.4(12).) Regarding X as a model of the theory $T_{\mathcal{A}}$ of definition 3.30, an equivalent definition of permutational representation is to say that the automorphism group of X acts transitively on X. Every relation algebra with a permutational representation is integral, but not every integral representable relation algebra has a permutational representation [AndDün$^+$92]. Every group relation algebra has a permutational representation, but the class **GRA** of group relation algebras is not finitely axiomatisable over the class of relation algebras with a permutational representation [McKe70]. See the discussion at the end of section 4.2.

   The class of relation algebras with a permutational representation can be seen to be pseudo-universal as follows. The **r**-sort will consist of two disjoint components: a representation X of the algebra occupying the **a**-sort, and a set Q of automorphisms of this representation. We include in L^s an **r**-sorted ternary relation symbol $\texttt{eval}(q,x,x')$ to record that the automorphism q takes x to x'. The defining L^s-theory will express that X is a representation of $\mathcal{A}$, as in (1) above, that Q is a set of automorphisms of it, via the axiom

   $$\forall (qxx'yy')^{\mathbf{r}}(\texttt{eval}(q,x,x') \wedge \texttt{eval}(q,y,y') \\ \to \forall a^{\mathbf{a}}(\texttt{holds}(a,x,y) \leftrightarrow \texttt{holds}(a,x',y'))),$$

   and that for any two **r**-elements, there is an automorphism in Q that takes one to the other. The form of this statement is

   $$\forall x^{\mathbf{r}} y^{\mathbf{r}} \big(X(x) \wedge X(y) \to \exists q^{\mathbf{r}} (Q(q) \wedge \texttt{eval}(q,x,y)) \big).$$

   Note that we could just as easily (and more neatly) have used *three* sorts, for **a**, X, and Q. This is harmless to the theory of pseudo-elementary classes, and we will sometimes do it later on.

7. A *homogenous* representation of a relation algebra $\mathcal{A}$ is rather similar to a permutational representation. We require that any two n-tuples $(y_1, \ldots, y_n)$, $(z_1, \ldots, z_n)$ of elements of X (for any finite n) such that (y_i, y_j) and (z_i, z_j) satisfy the same $\mathcal{A}$-relations, for all $i, j < n$, must be taken one to the other by an automorphism of the representation.

   The class '**HRA**' of relation algebras with a homogeneous representation (not to be confused with **H**(**RA**)) can be shown to be pseudo-elementary in a similar way to the class of relation algebras with a permutational representation. We leave the details to exercise 3 below. But **HRA** is not closed under substructures (exercise 4 below) and so is not pseudo-universal.

The method of section 9.6 will explicitly axiomatise the elementary closure of **HRA**; the axioms hold in a finite relation algebra $\mathcal{A}$ iff $\mathcal{A} \in$ **HRA**, since $\mathcal{A} \equiv \mathcal{B} \Rightarrow \mathcal{A} \cong \mathcal{B}$. It may be interesting to compare these axioms with those in [HirHod97b, theorem 14].

PROBLEM 9.3 Is **HRA** an elementary class?

8. The class $\mathbf{S\mathfrak{R}aCA}_\alpha$ (see definition 5.45) is pseudo-universal (for any ordinal $\alpha \geq 3$). Here, and below, e.g., in proposition 9.7, we will embed the **a**-sort into the **r**-sort. Let L^s consist of L (the signature of **RA**), a unary function symbol $f : \mathbf{a} \to \mathbf{r}$, and **r**-sorted copies of the function symbols of the signature of α-dimensional cylindric algebras. We do not distinguish notationally between the boolean symbols $+, -, 0, 1$ of sort **a** and those of sort **r**. Our theory U' will express that the **r**-sort is a cylindric algebra and that the **a**-sort embeds via f into the relation algebra reduct of the **r**-sort. It consists of the axioms of $\mathbf{CA}_\alpha$, written with the **r**-sorted symbols, and the following sentences:

$$\forall x(\mathsf{c}_i(f(x)) = f(x)), \text{ for each } i \text{ with } 2 \leq i < \alpha$$
$$\forall xy(f(x+y) = f(x) + f(y))$$
$$\forall x(f(-x) = -f(x))$$
$$f(1') = \mathsf{d}_{01}$$
$$\forall x(f(\breve{x}) = \mathsf{s}_0^2 \mathsf{s}_1^0 \mathsf{s}_2^1 f(x))$$
$$\forall xy(f(x;y) = \mathsf{c}_2(\mathsf{s}_2^1 f(x) \cdot \mathsf{s}_2^0 f(y)))$$
$$\forall xy(f(x) = f(y) \to x = y).$$

Here, x, y have sort **a**. Recall from definition 5.19 that $\mathsf{s}_j^i x$ is the standard abbreviation for $\mathsf{c}_i(\mathsf{d}_{ij} \cdot x)$, when $i \neq j$. It is easy to see that $\mathbf{S\mathfrak{R}aCA}_\alpha = \{M^\mathbf{a}\!\upharpoonright_L : M \models U'\}$.

Note that by proposition 13.48 (or by [HenMon$^+$85, 3.2.10, 5.3.13, 5.3.16]), $\mathbf{S\mathfrak{R}aCA}_\omega = \mathbf{RRA}$, so this is another way of seeing that **RRA** is pseudo-universal. However, the defining theory U' is now infinite.

In chapter 13, alternative characterisations of $\mathbf{S\mathfrak{R}aCA}_n$ by relativised representations and 'hyperbases' will be given; these also yield pseudo-universal definitions of $\mathbf{S\mathfrak{R}aCA}_n$.

For finite α, we can also obtain $\mathfrak{R}\mathbf{aCA}_\alpha$ as a pseudo-elementary class, by adding the axiom

$$\forall y^\mathbf{r} \left(\left(\bigwedge_{2 \leq i < \alpha} \mathsf{c}_i y = y \right) \to \exists x^\mathbf{a}(y = f(x)) \right).$$

9.3. Examples

9. The class of relation algebras that have a dense representable subalgebra (see definition 2.6) is pseudo-elementary (exercise).

10. Let us now change L to be the signature of relation algebra atom structures. It is easily seen that the class of *weakly representable atom structures* — that is, the class At**RRA** (definition 2.74) of atom structures of atomic, representable relation algebras — is pseudo-elementary. We may write a theory that divides the **r**-sort into two parts, namely, Q, and the rest (where Q is a unary relation symbol), and says that the **a**-sort is the atom structure of a relation algebra $\mathcal{A}$ given by the Q-part of the **r**-sort, and that the 'rest' of the **r**-sort is a representation of $\mathcal{A}$. Or we can use three sorts, of course.

11. More generally, if V is any variety of boolean algebras with operators, the class AtV (definition 2.74) of atom structures of atomic algebras in V is pseudo-elementary. The defining theory simply says that the **a**-sort is the atom structure of an atomic BAO occupying the **r**-sort and satisfying the defining equations of V. (Of course, by theorem 2.84 and proposition 2.88, AtV is frequently already known to be elementary with an explicit axiomatisation.)

12. Similarly, **RCA**$_n$ (for any fixed finite n) is a pseudo-universal class, as are **ICrs**$_n$, **ID**$_n$, **IG**$_n$, $\mathbf{S\mathfrak{Nr}}_m\mathbf{CA}_n$, and **Rdf**$_n$, for finite $m \leq n$. Polyadic algebras, Jónsson Q-algebras, sequential algebras, and others can also be handled.

In all these cases, the theory $(U, U'$, etc.) in the extended language L^s is explicit and almost always recursive. In many cases it is finite, but this does not seem to help when we axiomatise the class.

Difficult cases include **RCA**$_\alpha$ and $\mathfrak{RaCA}_\alpha$ for $\alpha \geq \omega$, and algebras with the Kleene star operator from PDL, whose definition is not first-order. Another is the class of atom structures of representable relation algebras that are full complex algebras (the 'strongly representable atom structures', or Str**RRA** in terms of section 2.7.4). We will see in theorem 14.3 that this class is not closed under ultraproducts and hence is not pseudo-elementary.

PROBLEM 9.4 (Németi) *Is* **IG**$_\omega$ *a pseudo-elementary class? Is it a variety, or even closed under ultraproducts? (This was asked by Németi; see [Ném96, AndGol*$^+$*98, And01].)*

Exercises

1. Which of the equations for relation algebras are really necessary to put in 'U' in example 9.2(1)?

2. Check that **SP**(**GRA**) is pseudo-universal.

3. Complete the argument to show that the class of all relation algebras with a homogeneous representation is pseudo-elementary.

4. Show that the class of relation algebras with a homogeneous representation is not closed under subalgebras. [Take a representation h of some relation algebra $\mathcal{A}$ that has no homogeneous representation, and embed $\mathcal{A}$ in the relation algebra with domain $\wp(h(1))$.]

5. Let $n < \omega$. Show that the class of relation algebras with an 'n-homogeneous' representation — any partial isomorphism of size $< n$ extends to one of size n — is pseudo-elementary.

6. Which of the examples in 9.2 are PC^s classes?

9.4 Model theory of pseudo-elementary classes

In this section, we recall some model-theoretic results about pseudo-elementary classes, mostly taken from [Hod93]. Our main interest is in axiomatisations, so we will not give proofs unless they are directly related to this.

9.4.1 Alternative single-sorted view

Because pseudo-elementary classes are usually defined using single-sorted logic, in order to quote results from the literature we must first show that our two-sorted approach is equivalent. Actually, four kinds of pseudo-elementary class have been studied: their technical names are PC'_Δ, PC_Δ, PC', and PC. Roughly, a PC'_Δ class consists of every L-structure (for some first-order signature L) that arises in a uniform way as a definable part of a model of some first-order theory in a language extending L. Dropping the $'$ denotes that this model has no elements except those in the original structure, and dropping the Δ indicates that the defining theory is finite. [ChaKei90, exercise 4.1.17] and [Hod93, §5.2] define a pseudo-elementary class to be a PC_Δ one. See remark 9.12 for the relations between the notions.

DEFINITION 9.5 Let L, L^+ be first-order signatures with $L \subseteq L^+$. Recall that for an L^+-structure M, we write $M\!\restriction_L$ for the *L-reduct* of M, obtained by discarding from M interpretations of all symbols of $L^+ \setminus L$. Let $P \in L^+ \setminus L$ be a distinguished unary relation symbol.

1. For any L-formula φ, we define φ^P, its *relativisation* to P, by induction as in section 2.2.7:

    - For atomic φ, we let $\varphi^P = \varphi$.
    - $(\neg\varphi)^P = \neg\varphi^P$ and $(\varphi \wedge \psi)^P = \varphi^P \wedge \psi^P$.

9.4. Model theory of pseudo-elementary classes

- $(\exists x \varphi)^P = \exists x (P(x) \wedge \varphi^P)$.

Thus, all quantification is restricted or 'relativised' to the elements that satisfy P.

2. Let M be an L^+-structure, and suppose that $\{a \in M : M \models P(a)\}$ is a substructure of $M\!\upharpoonright_L$. That is, it is non-empty, for every constant $c \in L$ we have $M \models P(c)$, and for every n-ary function symbol $f \in L$ we have $M \models \forall x_1, \ldots, x_n (P(x_1) \wedge \ldots \wedge P(x_n) \rightarrow P(f(x_1, \ldots, x_n)))$. In that case, we write this L-substructure as M_P and call it the *P-part* of M.

Evidently (an easy exercise), for any L^+-structure M whose P-part exists, any L-formula $\varphi(\bar{x})$, and any tuple $\bar{a} \in M_P$, we have $M_P \models \varphi(\bar{a})$ iff $M \models \varphi^P(\bar{a})$.

DEFINITION 9.6 Let L be a first-order signature and K a class of L-structures.

1. K is said to be a *PC_Δ class* if there are a first-order signature $L^+ \supseteq L$ and a first-order L^+-theory T such that $\mathsf{K} = \{M\!\upharpoonright_L : M \models T\}$.

2. K is said to be a *PC'_Δ class* if there are a first-order signature $L^+ \supseteq L$, a distinguished unary relation symbol $P \in L^+ \setminus L$, and a first-order L^+-theory T, such that M_P exists for all $M \models T$, and $\mathsf{K} = \{M_P : M \models T\}$.

3. As a special case, K is said to be a *PC class* if it is PC_Δ and the defining theory T, as above, is finite; the notion of *PC' class* is defined similarly.

For L, P, L^+ as above, any theory T such that $T \vdash \exists x P(x)$, $T \vdash P(c)$ for every constant $c \in L$, and $T \vdash \forall x_1, \ldots, x_n (P(x_1) \wedge \ldots \wedge P(x_n) \rightarrow P(f(x_1, \ldots, x_n)))$ for every n-ary function symbol $f \in L$, gives rise to the PC'_Δ class $\{M_P : M \models T\}$.

9.4.2 Equivalence of sorted and unsorted approaches

PROPOSITION 9.7 *A class of L-structures is pseudo-universal iff it is PC'_Δ and is closed under substructures.*

Proof. Any pseudo-universal class $\mathsf{K} = \{M^{\mathbf{a}}\!\upharpoonright_L : M \models U\}$ can be viewed as a PC'_Δ class as follows. We remove the sorts from the language L^s, and relativise quantifiers over all variables that were of sort $\mathbf{a}$ to P and all that were of sort $\mathbf{r}$ to $\neg P$. Doing this to the sentences in U, and adding axioms stating that L^s-function symbols and constants that took values in sort $\mathbf{a}$ have values satisfying P and that those that had values in sort $\mathbf{r}$ have values satisfying $\neg P$, yields a theory T that defines K as a PC'_Δ class. For if $M \models U$, then for each sorted function symbol f^s with value of sort $\mathbf{a}$, say, interpret the corresponding unsorted function symbol f by $f(\bar{a}) = f^s(\bar{a})$, if $f^s(\bar{a})$ is defined, and $f(\bar{a})$ is an arbitrary element of sort $\mathbf{a}$, otherwise. This yields a model of T with P-part $M^{\mathbf{a}}\!\upharpoonright_L$. Conversely, if $N \models T$ then interpreting the sorted function symbols by restricting the interpretations of the corresponding unsorted ones to the appropriate sorts yields a model M of U such that $M^{\mathbf{a}}\upharpoonright_L = N_P$. Hence, $\mathsf{K} = \{N_P : N \models T\}$.

To see that K is closed under substructures, let $\mathcal{A} = M^{\mathbf{a}}\upharpoonright_L \in \mathsf{K}$ for $M \models U$ and let $\mathcal{A}'$ be a substructure of $\mathcal{A}$. Define M' to be the substructure of M whose domain consists of the elements of $\mathcal{A}'$ together with all elements of $M^{\mathbf{r}}$. This is evidently a substructure of M. As all quantifiers in U-sentences of sort $\mathbf{a}$ are universal, we easily see that $M' \models U$. Thus, $\mathcal{A}' = (M')^{\mathbf{a}}\upharpoonright_L \in \mathsf{K}$.

For the converse, assume that $\mathsf{K} = \{N_P : N \models T\}$ (for some L^+-theory T as in definition 9.6) is a PC'_Δ class closed under substructures. Define the language L^s by including $\mathbf{a}$-sorted copies of all L-symbols, $\mathbf{r}$-sorted copies of all L^+-symbols, and a unary function symbol $f : \mathbf{a} \to \mathbf{r}$. Define the theory U to contain an $\mathbf{r}$-sorted copy of T, plus universal axioms stating that f is an L-embedding in the obvious sense — e.g., $\forall \bar{x}^{\mathbf{a}}(R^{\mathbf{a}}(\bar{x}^{\mathbf{a}}) \leftrightarrow R^{\mathbf{r}}(f(\bar{x}^{\mathbf{a}})))$, where $R \in L$ is a relation symbol — and with range contained in the P-part of the $\mathbf{r}$-sort — $\forall x^{\mathbf{a}} P^{\mathbf{r}}(f(x^{\mathbf{a}}))$. By pushing negations down next to atomic subformulas, we can assume that every sentence in U is built from atomic and negated-atomic formulas with $\wedge, \vee, \forall, \exists$. Evidently, all existential quantifiers in U, and all values of function symbols other than the $\mathbf{a}$-sorted copies of L-symbols, have sort $\mathbf{r}$.

We show that $\mathsf{K} = \{M^{\mathbf{a}}\upharpoonright_L : M \models U\}$. Let $\mathcal{A} \in \mathsf{K}$, so that $\mathcal{A} = N_P$ for some $N \models T$. Define an L^s-structure M whose $\mathbf{a}$-sorted part is $\mathcal{A}$ and whose $\mathbf{r}$-sorted part is a copy of N disjoint from $\mathcal{A}$. Let f be interpreted in M as an isomorphism $: \mathcal{A} \to N_P$. See figure 9.1. Clearly, $M \models U$ and $M^{\mathbf{a}}\upharpoonright_L = \mathcal{A}$.

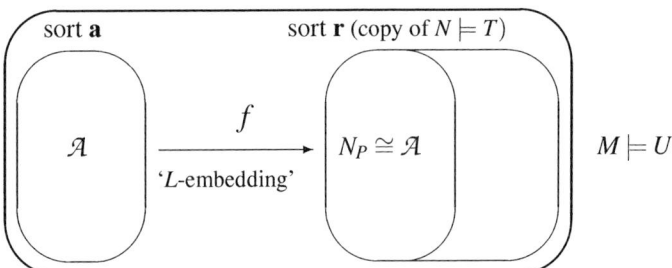

Figure 9.1: A model of U from a model of T

Conversely, if $\mathcal{A} = M^{\mathbf{a}}\upharpoonright_L$ for some $M \models U$, then by construction of U, up to renaming of symbols we have that $M^{\mathbf{r}} \models T$ (so that $M_P^{\mathbf{r}} \in \mathsf{K}$) and $f : \mathcal{A} \to M_P^{\mathbf{r}}$ is an L-embedding. Since K is assumed closed under substructures, $\mathcal{A} \in \mathsf{K}$. □

REMARK 9.8 In the proof, T is obtained explicitly from U, and vice versa. Hence, T is a recursive theory in a recursive language (see definition 9.13 below)

9.4. Model theory of pseudo-elementary classes

iff U is so. Also, if the underlying signatures have finitely many symbols then T is finite iff U is finite. Analogous considerations apply to proposition 9.9 below.

PROPOSITION 9.9 *A class of L-structures is pseudo-elementary (respectively, PC^s) iff it is PC'_Δ (respectively, PC').*

Proof. This can be proved similarly to proposition 9.7. □

9.4.3 Survey of known results

We are now in a position to gather some known results on pseudo-elementary classes. First, though we will not use it very much here, it is worth noting the following:

FACT 9.10 Let L be a finite signature. A class of finite L-structures is NP (recognisable in non-deterministic polynomial time) iff it is existential second-order definable (i.e., it is the class of all models of some existential second-order L-sentence) [Fag74], and hence (evidently) iff it is the class of all finite structures in some PC class.

The next, at first sight rather surprising result is that PC_Δ and PC'_Δ classes coincide.

FACT 9.11 (Makkai, [Mak64]) Any PC'_Δ class is also a PC_Δ class. Any PC' class consisting only of infinite structures is PC.

There is a proof in [Hod93, 5.2.1]. Since 'pseudo-elementary' traditionally means PC_Δ, we see by proposition 9.9, fact 9.11, and exercise 2 below that our definition (9.1) of 'pseudo-elementary' is equivalent to the usual one in the literature.

REMARK 9.12 Any PC_Δ class is PC'_Δ (by exercise 2 below). PC'_Δ classes needing an infinite signature, such as $\mathbf{RCA}_\omega$, cannot be PC'. See also exercises 8 and 10. **RRA** is an example of a PC' class that is not PC (exercise 11). So we see that informally, the 'classes of classes' defined so far satisfy the relations:

$$PC \subset PC' = PC^s \subset PC'_\Delta = PC_\Delta = \text{pseudo-elementary} \supset \text{pseudo-universal}.$$

The inclusions shown are strict.

[Hod93, theorem 6.6.7] proves that any pseudo-elementary class that is closed under substructures is universal. This theorem is implicit in [Mal71b] and explicit in [Tar54]; it gives an easy way of showing that pseudo-universal classes are elementary. Our third result, theorem 9.14 below, adapts it to obtain recursive axiomatisations.

A technical consideration now arises. Thus far, we have been considering particular first-order languages in usually finite signatures. We have implicitly regarded their formulas as being comprehensible by a Turing machine. For example, in chapter 8, we gave a recursive axiomatisation of **RRA** without feeling the need to define formally what 'recursive' meant in this context, since the axioms were written down explicitly. Here, our results will be for arbitrary first-order languages. So we need to define the notion of a recursive language.

Essentially, a first-order language L is recursive if a Turing machine can determine whether a given symbol of L is a variable, logical symbol, or constant, or a function or relation symbol and its arity. For this to make sense, the symbols of L have to be inputtable to a Turing machine. So we will have to assume (e.g.) that the variables are $v_0, v_1, \ldots, v_{2048}, \ldots$ (the latter being a string of five symbols), and similarly for the non-logical symbols (relation symbols etc.) of L.

DEFINITION 9.13 Consider a first-order language L using the set V of variables (e.g., $V = \{v_0, v_1, \ldots\}$). Let Symb(L) be the disjoint union of V, the set $\{=, \top, \bot, \wedge, \vee, \neg, \rightarrow, \leftrightarrow, \forall, \exists, (,)\}$ of logical symbols (or whatever subset of it we are using), and of the signature (set of non-logical symbols) of L.

For a function or relation symbol σ of L, write $\alpha(\sigma)$ for its arity (or rank), written in decimal (say) as a string of digits.

L is said to be *recursive* if each element of Symb(L) is a finite string of letters from some finite alphabet, and the sets V, $\{c : c$ a constant of $L\}$, $\{f : f$ a function symbol of $L\}$, $\{(f, \alpha(f)) : f$ a function symbol of $L\}$, $\{R : R$ a relation symbol of $L\}$ and $\{(R, \alpha(R)) : R$ a relation symbol of $L\}$, are pairwise disjoint and recursive.

See [ChaKei90] for a helpful discussion of recursive languages. If L is recursive, (distinct) L-formulas are (distinct) strings of symbols. It then makes sense to say that a set of formulas or an axiomatisation Σ is *recursive,* meaning that there exists a Turing machine that accepts precisely the formulas in Σ; or *recursively enumerable,* meaning that there exists a Turing machine that lists precisely the formulas in Σ on its output tape during its (perhaps infinite) run.

Of course, any recursive language must be countable. But for countable languages, being recursive is not really much of a restriction and in practice it is rare to see countable non-recursive languages.

THEOREM 9.14 (Mal'cev, Tarski) *Suppose that* $\mathsf{K} = \{M_P : M \models T\}$ *is a* PC'_Δ *class of L-structures, for some L^+-theory T and some signatures $L^+ \supseteq L$.*

1. Assume that K is closed under ultraroots. Then K is elementary and is axiomatised by the L-theory $\{L$-sentences $\sigma : T \vdash \sigma^P\}$.

9.4. Model theory of pseudo-elementary classes

2. Assume that K is closed under substructures. Then K is elementary and indeed universal, and is axiomatised by the universal L-theory {universal L-sentences $\sigma : T \vdash \sigma^P$}.

These theories are recursively enumerable if T is.

Proof. That K is elementary in each case follows from exercise 5 below. The remainder of the proof is by compactness. We only prove part 1; the proof of part 2 is much the same and is essentially proved in [Hod93, theorem 6.6.7]. Let

$$S = \{L\text{-sentences } \sigma : T \vdash \sigma^P\}.$$

If T is recursively enumerable then so is S.

Claim. If $N \models S$ then $N \preceq M_P$ for some $M \models T$.
Proof of claim. We regard the elements of N as constants naming themselves, and write $L^+(N)$ for the augmentation of L^+ by these (new) constants. Consider the $L^+(N)$-theory

$$\Xi = T \cup \{\sigma^P : \sigma \in \text{eldiag}_L(N)\} \cup \{P(a) : a \in N\}.$$

Here, $\text{eldiag}_L(N)$ (standing for 'elementary diagram') is the set of all $L^+(N)$-sentences of the form $\psi(\bar{a})$, where $\psi(\bar{x})$ is an L-formula, $\bar{a} \in N$, and $N \models \psi(\bar{a})$. We show that Ξ is consistent. If not, then by compactness there are an L-formula $\psi(\bar{x})$ and a tuple $\bar{a} = (a_1, \ldots, a_n) \in N$ with $N \models \psi(\bar{a})$, such that $T \cup \{\psi^P(\bar{a}), P(a_1), \ldots, P(a_n)\}$ is inconsistent. So by the 'lemma on constants' [Hod93, lemma 2.3.2], $T \vdash \forall \bar{x}(P(x_1) \wedge \ldots \wedge P(x_n) \rightarrow \neg \psi^P(\bar{x}))$. That is, $T \vdash (\forall \bar{x} \neg \psi)^P$. By definition of S we have $\forall \bar{x} \neg \psi \in S$, contradicting the assumptions that $N \models S$ and $N \models \psi(\bar{a})$.

So we may take a model M of Ξ. So $M \models T$. T defines a PC'_Δ class, so M_P exists. By relativisation, $M_P \models \text{eldiag}_L(N)$, and it follows (see [Hod93, lemma 2.5.3]) that $N \preceq M_P$ up to isomorphism. The claim is proved.

It is now easy to see that K is the class of all models of S. Certainly, if $M \models T$ then $M_P \models S$. Conversely, let $N \models S$. By the claim, $N \preceq M_P$ for some $M \models T$. Then $M_P \in$ K, and since K is elementary and so closed under elementary equivalence, we obtain $N \in$ K as required. □

We make explicit the import of this for our 2-sorted definitions:

COROLLARY 9.15 *Any pseudo-elementary class closed under ultraroots is elementary, and any pseudo-universal class is elementary and universally axiomatisable — i.e., universal. In each case, an axiomatisation of the class can be obtained effectively from the defining theory.*

Proof. By theorem 9.14 and propositions 9.7 and 9.9. □

290   Chapter 9. Axiomatising pseudo-elementary classes

By Craig's trick [Hod93, exercise 6.1.3], for any (universal) recursively enumerable theory there is an equivalent (universal) recursive theory. So the above result provides recursive (universal) axiomatisations of the pseudo-elementary (respectively, pseudo-universal) classes in example 9.2. As we said, axiomatising such classes by 'simple intuitive' axioms is a central problem of algebraic logic. At least we can now axiomatise them recursively in a uniform way. But the axiomatisation we get from theorem 9.14 is rather obscure: it may not be obvious, given arbitrary σ, whether $T \vdash \sigma^P$ or not. As many algebraic logic classes are known anyway to be varieties, a theorem giving an obscure, albeit recursive, axiomatisation is not telling us all that much new. As Dov Gabbay once said in a similar situation [GabHod$^+$94, p. 201], 'We want to see some axioms and rules'. This is the purpose of the rest of the chapter.

Exercises

1. Let $2 \leq k < \omega$. An (irreflexive symmetric) graph G is said to be *k-colourable* if each node of G can be coloured with one of k colours in such a way that the endpoints of any edge have different colours. Show that the class of k-colourable graphs is *PC*.

2. Show that any PC_Δ class is PC'_Δ.

3. Prove proposition 9.9 and theorem 9.14(2).

4. Assume the hypotheses of theorem 9.14. Let $S = \{\text{L-sentences } \sigma : T \vdash \sigma^P\}$. Show that S axiomatises $\{L\text{-structures } M : \exists N \in \mathsf{K}(N \equiv M)\}$, the elementary closure of K.

5. Let K be a pseudo-elementary class.

   (a) Show that K is closed under ultraproducts.

   (b) Deduce that if K is closed under substructures or ultraroots then it is elementary.

   (See [ChaKei90, exercise 4.1.17, corollary 6.1.16], or theorem 2.32.)

6. Let K be a class of L-structures. Suppose that K and its complement $\{L\text{-structures } M : M \notin \mathsf{K}\}$ are both pseudo-elementary classes. Show that K is elementary and finitely axiomatisable.

7. Show that the class of relation algebras with a finite representation is not pseudo-elementary.

9.4. Model theory of pseudo-elementary classes

8. Let $K = \{M^a\!\restriction_L : M \models U\}$ be a pseudo-elementary class of L-structures, for countable L, U. Show that any structure in K has a countable elementary substructure in K. Deduce that the class of all uncountable sets is pseudo-elementary but not PC^s.

9. Give an example of a PC class K of L-structures, for some countable L, and an L-structure $M \notin K$ such that every countable elementary substructure of M is in K. [Hint: $(\mathbb{R}, <)$.] Deduce that there is a PC class that is not elementary.

10. Show that there is an elementary class of structures for the empty signature that is not PC'. [Cardinalities, diagonalisation.]

11. Show that **RRA** is not a PC class. Do the same for $\mathbf{S\mathfrak{R}aCA}_n$ for finite $n \geq 5$. [You may have to look ahead for this one. Use theorem 18.13 and (e.g.) fact 9.10.]

12. Let I be a set and K_i ($i \in I$) be pseudo-elementary classes in the same signature.

    (a) Show that $\bigcap_{i \in I} K_i$ is also pseudo-elementary. [Add a new sort for each i.]

    (b) If I is finite, show that $\bigcup_{i \in I} K_i$ is pseudo-elementary.
    [If U_i are the L^s-theories, let $U = \{\bigvee_i \sigma_i : \sigma_i \in U_i\}$.]

    (c) Need $\bigcup_{i \in I} K_i$ be pseudo-elementary if I is infinite?

13. Let K be a pseudo-elementary class of similar algebras. Show that $\mathbf{S}K$ and $\mathbf{SP}K$ are pseudo-universal and that $\mathbf{H}K$ is pseudo-elementary.

14. Let K be a class of finite structures in a finite signature L, and assume that K is closed under isomorphism. Show that the following are equivalent:

    - the set of isomorphism types of structures in K is recursively enumerable,
    - K is the finite part of a co-PC' (or equivalently, co-PC^s) class of L-structures. That is, there is a PC' class K' such that $A \in K$ iff $A \notin K'$ for all finite L-structures A.

    [Looking at [Fag74] may help.]

PROBLEM 9.16 (Venema) Does the converse of exercise 5a hold? That is, must a class of structures that is closed under ultraproducts be pseudo-elementary?

9.5 More explicit axioms

We will now obtain explicit axioms for pseudo-elementary and pseudo-universal classes. In this section, we confine our attention to the case of pseudo-universal classes: for these, we can obtain universal axioms. In section 9.6 we will handle the general case.

Let K be a pseudo-universal class defined by the L^s-theory U. Given an arbitrary L-structure $\mathcal{A}$, we try to build a model of U whose **a**-sort is equal to $\mathcal{A}$, by a Henkin-style construction. (The construction builds the **r**-sort.) This is done by having two players, $\forall$ (male) and $\exists$ (female), play a certain game, in which $\forall$ tests $\exists$'s ability to build such a model. We will prove that $\exists$ has a winning strategy in the game iff such a model exists, which is iff $\mathcal{A} \in K$. We will then write L-axioms expressing (roughly) that $\exists$ has a winning strategy in the game played on $\mathcal{A}$, and this way we axiomatise K. Because of the restrictions on U in definition 9.1(3), the axioms will be universal.

9.5.1 The game

From now on we fix a pseudo-universal class $K = \{M^{\mathbf{a}}\!\restriction_L : M \models U\}$ of L-structures, where L^s, U satisfy the conditions of definition 9.1(3), and $U \neq \emptyset$. As usual, we identify the symbols of L with their **a**-sorted copies in L^s, so that $L \subseteq L^s$. We further assume that L, L^s, U are recursive, though until proposition 9.27 we will only need that they are countable. The first step in axiomatising K is to define the game to try to build the **r**-sort of a model of U whose **a**-sort is a fixed L-structure.

DEFINITION 9.17 Let $\mathcal{A}$ be an L-structure.

1. A *pre-network (for $\mathcal{A}$)* is a pair (N, S), where N is a set of nodes (possibly empty) with $N \cap \mathcal{A} = \emptyset$, and S is a set of $L^s(\mathcal{A} \cup N)$-sentences not using '$\to$', in which negations only occur next to atomic subformulas and with no quantifiers $\exists x^{\mathbf{a}}$.

Here, $L^s(\mathcal{A} \cup N)$ is as usual the expansion of the signature L^s by the elements of $\mathcal{A} \cup N$, regarded as (new) constants naming themselves. Elements of $\mathcal{A}$ have sort **a**, and elements of N have sort **r**. Thus, these symbols already have intended interpretations. Note that expressions such as $a = n$ for $a \in \mathcal{A}$, $n \in N$ are not well-formed and cannot be in S.

2. A pre-network (N, S) for $\mathcal{A}$ is said to be a *network* if none of the following hold:

    (a) $\varphi, \neg\varphi \in S$ for some atomic sentence φ.

    (b) There is a quantifier-free $L(\mathcal{A})$-sentence φ (not involving any nodes of N) with $\varphi \in S$ and $\mathcal{A} \models \neg\varphi$.

9.5. More explicit axioms

(c) There is a false equality or inequality in S:
- $a = b \in S$ for some distinct $a, b \in \mathcal{A} \cup N$, or
- $a \neq a \in S$ for some $a \in \mathcal{A} \cup N$.

DEFINITION 9.18 Let $\mathcal{A}$ be an L-structure. We define a game $\Gamma(U, \mathcal{A})$, with $|\mathcal{A}| + \omega$ rounds: for countable $\mathcal{A}$, there are ω rounds, and for uncountable $\mathcal{A}$ there are $|\mathcal{A}|$ rounds. The game is played by $\forall$ and $\exists$ on pre-networks for $\mathcal{A}$. Each round of the game starts with a given pre-network over U, which is modified during the round to leave a new pre-network which is passed on to the next round. Play starts with the empty network, $(\emptyset, \emptyset)$. In each round, the players act as follows. Suppose the pre-network at the start of the round is (N, S).

1. $\forall$ may choose any sentence $\sigma \in U$. The pre-network passed on to the next round is then $(N, S \cup \{\sigma\})$.

2. Or he can pick any function symbol $f \in L^s \setminus L$ (this includes constants as a special case) and any tuple $\bar{x}y$ of distinct variables such that $f(\bar{x}) = y$ is well-formed. The new pre-network becomes $(N, S \cup \{\forall \bar{x} \exists y (f(\bar{x}) = y)\})$; note that this is a pre-network, as by definition 9.1(3), y has sort $\mathbf{r}$.

3. Or he can pick a quantifier-free sentence $\varphi(t) \in S$ and a variable-free (sometimes called 'closed') $L^s(\mathcal{A} \cup N)$-term u, where $\varphi(x)$ is a quantifier-free $L^s(\mathcal{A} \cup N)$-formula, t is a variable-free $L^s(\mathcal{A} \cup N)$-term, and either

   (a) $t = u \in S$, or

   (b) t is an $L(\mathcal{A})$-term (involving no symbols from N or $L^s \setminus L$), $u \in \mathcal{A}$, and $\mathcal{A} \models t = u$.

   The new pre-network is then $(N, S \cup \{\varphi(u)\})$.

4. Or $\forall$ may choose any sentence $\varphi \in S$. What happens next depends on the form of φ.

   (a) If φ is $\psi \wedge \chi$ then the new pre-network existing at the end of the round is either $(N, S \cup \{\psi\})$ or $(N, S \cup \{\chi\})$, at $\forall$'s choice.

   (b) If φ is $\psi \vee \chi$ then the new pre-network is either $(N, S \cup \{\psi\})$ or $(N, S \cup \{\chi\})$, at $\exists$'s choice.

   (c) If φ is $\forall x \psi(x)$, then $\forall$ can choose any element a of $\mathcal{A} \cup N$ of appropriate sort and define the new pre-network to be $(N, S \cup \{\psi(a)\})$.

   (d) If φ is $\exists x \psi(x)$, so that x has sort $\mathbf{r}$, then $\exists$ can choose any element m of N and let the new pre-network be $(N, S \cup \{\psi(m)\})$, or she can choose a new node $m \notin N \cup \mathcal{A}$ and let the new pre-network be $(N \cup \{m\}, S \cup \{\psi(m)\})$.

(e) Otherwise (i.e., if φ is atomic or negated-atomic), nothing happens and the network at the end of the round is just (N, S).

At limit rounds (rounds numbered δ for limit ordinals δ), the pre-network at the start of the round is the union (in the obvious sense) of all the pre-networks played so far. The *outcome* of the play is the pre-network over $\mathcal{A}$ consisting of the union of all pre-networks played during the game. ∃ wins if the outcome is a network, and loses if not.

A strategy for a player in such a game is a set of rules telling the player how to play any round. The strategy is winning if its owner wins any play in which it is used. There is no need to be more formal here.

9.5.2 The game characterises K

Our first goal (of four) is to show that this game characterises the class K. This is done in the following two propositions.

PROPOSITION 9.19 *Let $\mathcal{A}$ be any L-structure. If $\mathcal{A} \in$ K then ∃ has a winning strategy in $\Gamma(U, \mathcal{A})$.*

Proof. Assume that $\mathcal{A} = M^{\mathbf{a}}\!\restriction_L$ for some L^s-structure $M \models U$. ∃ uses M as a guide to winning the game. At each stage, if the current pre-network is (N, S), she maintains the inductive assumption that

$$N \subseteq \mathrm{dom}(M^{\mathbf{r}}) \text{ and } M \models S. \qquad (*)$$

Recall here that the new constants of $L^s(\mathcal{A} \cup N)$ have intended interpretations in $\mathcal{A} \cup N$ and hence in M. So '$M \models S$' makes sense.

$(*)$ is vacuously true initially and is clearly preserved at limit stages. Assume that it is true of (N, S) at the beginning of some round. Let ∀ make a move now. If he adds some $\sigma \in U$ to S, $(*)$ is preserved since $M \models U$. Moves of types 2–3 are easily seen to preserve $(*)$ too. So let him move by picking $\varphi \in S$ in a move of type 4.

(a) If φ is $\psi \wedge \chi$, then $M \models \psi \wedge \chi$, so whichever network ∀ elects to play, $(*)$ is preserved.

(b) If φ is $\psi \vee \chi$, then ∃ plays $(N, S \cup \{\psi\})$ if $M \models \psi$, and $(N, S \cup \{\chi\})$, otherwise. Because $M \models \psi \vee \chi$, $(*)$ is preserved.

(c) If φ is $\forall x \psi(x)$ then clearly, whatever network ∀ plays still satisfies $(*)$.

(d) If φ is $\exists x \psi(x)$, then because x has sort $\mathbf{r}$, $M \models \psi(m)$ for some $m \in M$ of sort $\mathbf{r}$. ∃ chooses such an m, and lets the new network be $(N \cup \{m\}, S \cup \{\psi(m)\})$. This preserves $(*)$.

9.5. More explicit axioms

Let the outcome of play be (N,S). If $\varphi \in S$ is a quantifier-free $L(\mathcal{A})$-sentence, then by $(*)$, $M \models \varphi$, so $\mathcal{A} \models \varphi$, too. Similarly, $(*)$ eliminates the possibility that $\varphi, \neg\varphi \in S$ for some φ or that S contains a false (in)equality. So (N,S) is a network and $\exists$ won the play. We have therefore described a winning strategy for her in this game. □

PROPOSITION 9.20 *Let $\mathcal{A}$ be any L-structure, and assume that $\exists$ has a winning strategy in $\Gamma(U, \mathcal{A})$. Then $\mathcal{A} \in \mathsf{K}$.*

Proof. Assume the hypotheses. Let $\exists$ use her winning strategy in a play of $\Gamma(U, \mathcal{A})$ in which $\forall$ makes every possible move at some stage — as U, L^s are countable and the game lasts $|\mathcal{A}| + \omega$ rounds, he has enough time to do this. (In more detail, if $|\mathcal{A}| + \omega = \kappa$, then the cardinal product $\kappa \cdot \kappa$ is equal to κ. So with each pair (α, β) such that $\alpha, \beta < \kappa$, we may associate some $\gamma_{\alpha\beta} < \kappa$ unique to (α, β). As each round α introduces at most κ new tasks for $\forall$, he can schedule them for future rounds with round numbers of the form $\gamma_{\alpha\beta}$ where $\beta < \kappa$ and $\gamma_{\alpha\beta} > \alpha$. There are κ round numbers of this form, which is enough for him.)

Let (N,S) be the outcome of play; as $\exists$ used a winning strategy, this is a network. We define an L^s-structure M with domain consisting of $\mathcal{A}$ (sort **a**) and N (sort **r**), as follows. (Note that S is therefore a set of $L^s(M)$-sentences.) We define the interpretations of L-symbols by

$$M^{\mathbf{a}}\!\upharpoonright_L = \mathcal{A}.$$

For any $L^s(M)$-sentence α of the form either $R(\bar{a})$ for a relation symbol $R \in L^s \setminus L$ and a tuple $\bar{a} \in M$, or $f(\bar{a}) = b$ for a function symbol $f \in L^s \setminus L$ (possibly a constant) and any $\bar{a}, b \in M$ of appropriate sorts, we define

$$M \models \alpha \iff \alpha \in S.$$

LEMMA 9.21 *This gives a well-defined L^s-structure.*

Proof. The L-symbols are certainly well-defined. Let $f \in L^s \setminus L$ be a function symbol and let $\bar{a} \in M$ be of appropriate sort for f. We must show f^M is defined uniquely on $\bar{a}$: that is, there is a unique $b \in M^{\mathbf{r}}$ with $f(\bar{a}) = b \in S$.

We know $\forall$ played all possible type 2 moves, so that in some round he put $\forall \bar{x} \exists y (f(\bar{x}) = y)$ into S. He would then have made a series of type 4c moves to put $\exists y(f(\bar{a}) = y)$ into S. Then, he would have picked this sentence in a type 4d move at some stage; $\exists$ must have responded so that $f(\bar{a}) = b \in S$ for some $b \in M^{\mathbf{r}}$.

Now let $b, c \in M^{\mathbf{r}}$ be distinct and suppose for contradiction that $f(\bar{a}) = b$ and $f(\bar{a}) = c$ are in S. In some round, $\forall$ would have used a type 3 move, substituting b for $f(\bar{a})$ in $f(\bar{a}) = c$, to put $b = c$ into S. This contradicts the networkhood of (N,S). □

296   *Chapter 9. Axiomatising pseudo-elementary classes*

LEMMA 9.22 *If $\varphi \in S$ is atomic or negated-atomic, then $M \models \varphi$.*

Proof. Assume that $\varphi \in S$ is atomic or negated-atomic. We show $M \models \varphi$ by induction on the total number of function symbols and constants from L^s in φ. If this is zero, then φ has the form $a = b$, $a \neq b$, or $R(\bar{a})$ or $\neg R(\bar{a})$ for $R \in L^s$. As (N,S) is a network, S does not contain false (in)equalities or $L(\mathcal{A})$-sentences, so $M \models \varphi$ if φ is an $L(M)$-sentence. Let $R \in L^s \setminus L$. If φ is $R(\bar{a})$, then $M \models \varphi$ by definition of M. Suppose that φ is $\neg R(\bar{a})$. As (N,S) is a network, $R(\bar{a}) \notin S$. By definition of M, we have $M \not\models R(\bar{a})$, so that $M \models \varphi$.

If some function symbols or constants occur in ψ, then ψ has the form $\psi(f(\bar{a}))$ for some atomic or negated-atomic $L^s(M)$-formula $\psi(x)$ that has fewer function and constant symbols than φ, some function symbol f of L^s (possibly a constant), and some $\bar{a} \in M$. If $f \in L^s \setminus L$, then by lemma 9.21 there is $b \in M$ with $f(\bar{a}) = b \in S$ and $M \models f(\bar{a}) = b$. As $\forall$ made all possible type 3 moves, $\psi(b) \in S$. Alternatively, if $f \in L$, then $\bar{a} \in \mathcal{A}$ and $\mathcal{A} \models f(\bar{a}) = b$ for some $b \in \mathcal{A}$. Again, $M \models f(\bar{a}) = b$, and $\forall$ would have made a type 3 move to place $\psi(b)$ directly into S. So we may assume inductively that $M \models \psi(b)$. That is, $M \models \psi(f(\bar{a}))$, or $M \models \varphi$, as required. □

Since $\forall$ played every possible type 4 moves, it is now straightforward to check by induction on φ that $M \models \varphi$ for every $\varphi \in S$. Since he played every possible type 1 move, we have $U \subseteq S$. So $M \models U$. By definition of M, we have $M^{\mathbf{a}}\restriction_L = \mathcal{A}$. We obtain $\mathcal{A} \in \mathsf{K}$, proving the proposition. □

9.5.3 Short games

Our second goal is to approximate the game $\Gamma(U, \mathcal{A})$ by shorter games. Recall that L, L^s, U are assumed recursive, so we may assume that the variables of L^s of each sort are recursively ordered with order type ω. Fix recursive enumerations $\{\sigma_i : i < \omega\}$ of U and $\{f_i : i < \omega\}$ of the function symbols and constants of $L^s \setminus L$, possibly with repetitions.

DEFINITION 9.23 For an L-structure $\mathcal{A}$, a pre-network (N, S) over $\mathcal{A}$, and $n < \omega$, we let $\Gamma_n(U, \mathcal{A}, N, S)$ be the game played like $\Gamma(U, \mathcal{A})$ but with the following changes:

1. It begins with the pre-network (N, S) instead of $(\emptyset, \emptyset)$.

2. It stops after n rounds. The rounds are numbered $0, 1, \ldots, n-1$. If the pre-network existing after n rounds is a network, $\exists$ wins the play; otherwise, she loses.

3. In a type 1 move in round t (for $t < n$), player $\forall$ must choose a sentence from the set $\{\sigma_i : i < n - t\} \subseteq U$.

9.5. More explicit axioms

4. Similarly, in a type 2 move in round t, $\forall$'s choice is restricted to sentences from the set $\{\forall \bar{x} \exists y (f_i(\bar{x}) = y) : i < n - t\}$, where in each $\forall \bar{x} \exists y (f_i(\bar{x}) = y)$, $\bar{x} y$ is the lexicographically first tuple of distinct variables of L^s of appropriate sorts.

LEMMA 9.24 *Let $\mathcal{A}$ be an L-structure, and (N, S) a pre-network over $\mathcal{A}$.*

1. *For any $n < \omega$, $\exists$ has a winning strategy in $\Gamma_{n+1}(U, \mathcal{A}, N, S)$ iff for every move that $\forall$ may make in round zero of this game, $\exists$ has a response leading to a pre-network (N', S') such that she has a winning strategy in $\Gamma_n(U, \mathcal{A}, N', S')$.*

2. *$\exists$ has a winning strategy in $\Gamma_n(U, \mathcal{A}, N, S)$ for infinitely many $n < \omega$, iff she has a winning strategy in $\Gamma_n(U, \mathcal{A}, N, S)$ for all $n < \omega$.*

Proof. (1) is clear, noting that the numbers in definition 9.23(3,4) behave correctly. For (2), note that if $\exists$ has a winning strategy in the game $\Gamma_n(U, \mathcal{A}, N, S)$ then the same strategy wins $\Gamma_m(U, \mathcal{A}, N, S)$ for her, for any $m < n$. □

PROPOSITION 9.25 *Let $\mathcal{A}$ be any countable L-structure. Then $\exists$ has a winning strategy in $\Gamma(U, \mathcal{A})$ iff she has a winning strategy in $\Gamma_n(U, \mathcal{A}, \emptyset, \emptyset)$ for each $n < \omega$.*

Proof. Left-to-right is trivial. We prove the converse direction essentially by König's tree lemma [Hod93, exercise 5.6.5]. Assume that $\exists$ has a winning strategy in $\Gamma_n(U, \mathcal{A}, \emptyset, \emptyset)$ for all finite n; we will show her how to win $\Gamma(U, \mathcal{A})$. Note that this game has ω rounds. A pre-network (N, S) is said to be a *pole position* if N is finite and $\exists$ has a winning strategy in $\Gamma_n(U, \mathcal{A}, N, S)$ for all $n < \omega$. $\exists$'s strategy in $\Gamma(U, \mathcal{A})$ will be to remain in pole position throughout. As a pole position is certainly a network, if she can do this she will win $\Gamma(U, \mathcal{A})$. We only have to show that she can do it.

By assumption, the initial pre-network $(\emptyset, \emptyset)$ is a pole position. Assume that play in $\Gamma(U, \mathcal{A})$ has reached the pole position (N, S) at the start of some round. We show $\exists$ how to play the round and reach another pole position.

Suppose that $\forall$ makes a type 1 move in $\Gamma(U, \mathcal{A})$ by picking $\sigma_i \in U$. Let $n \geq i$, so that $\forall$ may legally pick σ_i in round 0 of $\Gamma_{n+1}(U, \mathcal{A}, N, S)$. As $\exists$ has a winning strategy in this game, by lemma 9.24 she has a winning strategy in $\Gamma_n(U, \mathcal{A}, N, S \cup \{\sigma_i\})$. This is true for all $n \geq i$, so by the lemma again, it holds for all n: thus, $(N, S \cup \{\sigma_i\})$ is a pole position. The argument for moves of types 2, 3, and the $\wedge$ and $\forall$ cases of type 4 moves is similar, as $\exists$ has no role in these moves.

We now consider type 4 moves in the cases $\vee$ and $\exists$. Let $\forall$ choose $\varphi \vee \psi \in S$ in his move. For each n, as $\exists$ has a winning strategy in $\Gamma_{n+1}(U, \mathcal{A}, N, S)$, she has a winning strategy in either $\Gamma_n(U, \mathcal{A}, N, S \cup \{\varphi\})$ or in $\Gamma_n(U, \mathcal{A}, N, S \cup \{\psi\})$. So clearly, for some $\chi \in \{\varphi, \psi\}$, she has a winning strategy in $\Gamma_n(U, \mathcal{A}, N, S \cup \{\chi\})$

for infinitely many, and hence all, n. So $\exists$ may play $(N, S \cup \{\chi\})$ in $\Gamma(U, \mathcal{A})$ and remain in pole position.

Finally, let $\forall$ choose $\exists x \varphi(x) \in S$. $\exists$ lets him do this as his move in round 0 of each of the games $\Gamma_{n+1}(U, \mathcal{A}, N, S)$ ($n < \omega$), and uses her winning strategy in each case to arrive at position $(N, S)_n$, say; so she has a winning strategy in $\Gamma_n(U, \mathcal{A}, (N, S)_n)$. If there is $m \in N$ such that $(N, S)_n = (N, S \cup \{\varphi(m)\})$ for infinitely many n, then $(N, S \cup \{\varphi(m)\})$ is a pole position and she may play it in the current round of the main game $\Gamma(U, \mathcal{A})$. Otherwise, as N is finite, for infinitely many n we have $(N, S)_n = (N \cup \{m_n\}, S \cup \{\varphi(m_n)\})$ for some new node $m_n \notin N$. Clearly, the result of the game will not depend on the particular choice of new node, so $\exists$ may play the pole position $(N \cup \{m\}, S \cup \{\varphi(m)\})$ for some new node $m \notin N$.

Since $\mathcal{A}$ is countable, $\Gamma(U, \mathcal{A})$ has ω rounds. It follows from the above, by induction on the round number, that $\exists$ can remain in pole position throughout. So she has a winning strategy in this game. □

9.5.4 Axioms for the short games

Our third goal is to write axioms expressing that $\exists$ has a winning strategy in these curtailed games. This is necessarily rather technical, but the underlying idea is straightforward.

DEFINITION 9.26

1. A *syntactic network* is a pair (N, F) where N is a finite set (of 'nodes') and F is a finite set of $L^s(N)$-formulas with free variables all of sort **a**, without using '$\to$', with negations only occurring next to atomic formulas, and with no quantifiers $\exists x^{\mathbf{a}}$. We write $\mathrm{Var}(F)$ for the set of variables occurring free in some formula in F.

2. Given an L-structure $\mathcal{A}$, a syntactic network (N, F), and an assignment $\iota : \mathrm{Var}(F) \to \mathcal{A}$, we write $\iota(F)$ for the set $\{\varphi(\iota(x_1), \ldots, \iota(x_n)) : \varphi(x_1, \ldots, x_n) \in F\}$ of $L^s(\mathcal{A} \cup N)$-sentences. Then $(N, \iota(F))$ is a pre-network over $\mathcal{A}$.

Think of a syntactic network as being obtained from a pre-network over $\mathcal{A}$ by replacing constants of $\mathcal{A}$ in its $L^s(\mathcal{A} \cup N)$-sentences by new variables. A position in $\Gamma(U, \mathcal{A})$ can be represented by a pair consisting of a syntactic network (N, F) and an assignment $\iota : \mathrm{Var}(F) \to \mathcal{A}$.

PROPOSITION 9.27 *For each $n < \omega$ and each syntactic network (N, F), there is a universal first-order L-formula $\eta_n(N, F)$, with free variables among $\mathrm{Var}(F)$, such that for any L-structure $\mathcal{A}$ and assignment $\iota : \mathrm{Var}(F) \to \mathcal{A}$,*

$$\exists \text{ has a winning strategy in } \Gamma_n(U, \mathcal{A}, N, \iota(F)) \iff \mathcal{A}, \iota \models \eta_n(N, F).$$

9.5. More explicit axioms

The formulas $\eta_n(N,F)$ are recursive in n,N,F.

Proof. Cf. proposition 8.2.
The proof is by induction on n. For clarity we try to use x,y,z for L-variables in the formulas η_n, and v,w for arbitrary L^s-variables in the formulas in F (these may occur in the η_n if they have sort **a**). For $n = 0$, plainly $\exists$ has a winning strategy in $\Gamma_0(U,\mathcal{A},N,\iota(F))$ iff $(N,\iota(F))$ is a network. So we let $\eta_0(N,F)$ be the conjunction of the following formulas:

1. Suppose that $\varphi(x_1,\ldots,x_k)$ is an atomic $L^s(N)$-formula, where $x_1,\ldots,x_k$ are distinct variables. Suppose also that $y_1,\ldots,y_k$, $z_1,\ldots,z_k$ are any variables, and that $\varphi(y_1,\ldots,y_k)$, $\neg\varphi(z_1,\ldots,z_k) \in F$. Then we include the conjunct $\bigvee_{1 \le i \le k} y_i \ne z_i$ in $\eta_0(N,F)$.

2. φ, for each quantifier-free L-formula $\varphi \in F$.

3. $\perp$ (falsity), for each formula in F of the form $m = m'$ or $m \ne m$, for any $m \in N$, $m' \in N \setminus \{m\}$. (Formulas in F of the form $x = y$, $x \ne y$ are already covered by the second item.)

It is clear by definition 9.17 that $\mathcal{A}, \iota \models \eta_0(N,F)$ iff $(N,\iota(F))$ is a network over $\mathcal{A}$.

Let $n < \omega$ and assume inductively that we have constructed universal formulas $\eta_n(N,F)$ for all syntactic networks (N,F). As in lemma 9.24, for any syntactic network (N,F) we let $\eta_{n+1}(N,F)$ be the conjunction of the following formulas, representing the kinds of move that $\forall$ may make:

1. $\eta_n(N, F \cup \{\sigma_i\})$ for each $i \le n$. Here, the σ_i are as in definition 9.23.

2. $\eta_n(N, F \cup \{\forall \bar{v} \exists w (f_i(\bar{v}) = w)\})$ for each $i \le n$. Here, $\bar{v}, w$ are as described in definition 9.23.

3. For each quantifier-free $L^s(N)$-formula $\varphi(v,\bar{z})$ and $L^s(N)$-term t such that $\varphi(t,\bar{z}) \in F$, we include the following conjuncts:

    (a) $\eta_n(N, F \cup \{\varphi(u,\bar{z})\})$ whenever u is an $L^s(N)$-term with $t = u \in F$. Note that substituting u for t in φ does not cause a clash of bound variables (there are none).

    (b) In case t is an L-term, we also add the conjunct $\forall y(t = y \to \eta_n(N, F \cup \{\varphi(y,\bar{z})\}))$, where y is some **a**-variable (see below) that does not occur in any formula in F. (There are only finitely many of these conjuncts, because $\varphi(t,\bar{z}) \in F$.)

4. For type 4 moves, we include the following conjuncts of η_{n+1}:

    (a) $\eta_n(N, F \cup \{\varphi\}) \wedge \eta_n(N, F \cup \{\psi\})$ for each formula $\varphi \wedge \psi \in F$.

(b) $\eta_n(N, F \cup \{\varphi\}) \vee \eta_n(N, F \cup \{\psi\})$ for each formula $\varphi \vee \psi \in F$.

(c) (i) $\eta_n(N, F \cup \{\varphi(m, \bar{z})\})$ for each formula $\forall v^{\mathbf{r}} \varphi(v^{\mathbf{r}}, \bar{z}) \in F$ and each $m \in N$.

(ii) $\forall y \, \eta_n(N, F \cup \{\varphi(y, \bar{z})\})$ for each $\forall v^{\mathbf{a}} \varphi(v^{\mathbf{a}}, \bar{z}) \in F$. Here, y is a variable of sort **a** that does not occur in the formulas in F.

(d) $\bigvee_{k \in N} \eta_n(N, F \cup \{\varphi(k, \bar{z})\}) \vee \eta_n(N \cup \{m\}, F \cup \{\varphi(m, \bar{z})\})$ for some arbitrary new node $m \notin N$, for each $\exists v \varphi(v, \bar{z}) \in F$. (Note that by definition 9.26, v has sort **r** here.)

In the definition of η_{n+1}, we should strictly specify various details more carefully so as to make it recursive. For example, the order in which the conjuncts of η_{n+1} are taken can be determined lexicographically. In the cases where a 'new' variable y is taken, we may calculate which conjunct is involved (say the kth) and take y to be the kth **a**-sorted variable in the enumeration that does not occur in F.

So by lemma 9.24, if (N, F) is a syntactic network and $\iota : \mathrm{Var}(F) \to \mathcal{A}$, then $\exists$ has a winning strategy in $\Gamma_{n+1}(U, \mathcal{A}, N, \iota(F))$ iff for every move $\forall$ may make, $\exists$ can respond to reach a position represented by a syntactic network (N', F') and assignment ι' such that she has a winning strategy in $\Gamma_n(U, \mathcal{A}, N', \iota'(F'))$. Inductively, this is iff $\mathcal{A}, \iota' \models \eta_n(N', F')$. It is clear by construction of η_{n+1} that this holds iff $\mathcal{A}, \iota \models \eta_{n+1}(N, F)$.

Let us examine this in a little more detail. Suppose for example that in round 0 of $\Gamma_{n+1}(U, \mathcal{A}, N, \iota(F))$, $\forall$ makes a type 3b move and chooses quantifier-free $\varphi(t(\bar{a}), \bar{c}) \in \iota(F)$, where $\varphi(t(\bar{x}), \bar{z}) \in F$, $t(\bar{x})$ is an L-term, and $\iota(\bar{x}) = \bar{a}$ and $\iota(\bar{z}) = \bar{c}$ are tuples in $\mathcal{A}$. Let $b \in \mathcal{A}$ satisfy $\mathcal{A} \models t(\bar{a}) = b$. Then the new pre-network passed through to round 1 is (N, S'), where $S' = \iota(F) \cup \{\varphi(b, \bar{c})\}$. Let $F' = F \cup \{\varphi(y, \bar{z})\}$ for a new variable y, and let ι' extend ι to y by $\iota'(y) = b$. Then $\iota'(F') = S'$. So the following are equivalent:

- $\exists$ has a winning strategy in $\Gamma_{n+1}(U, \mathcal{A}, N, \iota(F))$ when $\forall$ *makes this particular move in round* 0.

- $\exists$ has a winning strategy in $\Gamma_n(U, \mathcal{A}, N, \iota'(F'))$.

- $\mathcal{A}, \iota' \models \eta_n(N, F')$ (by the inductive hypothesis).

- $\mathcal{A}, \iota \models \forall y (t(\bar{x}) = y \to \eta_n(N, F'))$ (since clearly, ι' is determined by (i) it extends ι and (ii) $\mathcal{A}, \iota' \models t(\bar{x}) = y$).

$\forall y(t(\bar{x}) = y \to \eta_n(N, F'))$ is exactly the conjunct we included in η_{n+1} to cover this kind of move by $\forall$. The other conjuncts handle the other moves he could make. So $\mathcal{A}, \iota \models \eta_{n+1}(N, F)$ iff whatever move $\forall$ makes in round 0, $\exists$ has a winning strategy in Γ_n from the resulting position; by lemma 9.24, this is iff she has a winning strategy in $\Gamma_{n+1}(U, \mathcal{A}, N, \iota(F))$.

The formulas η_n are obviously universal and recursive. □

9.5. More explicit axioms

In particular, the $\forall_1$-sentence $\eta_n(\emptyset,\emptyset)$ is true in $\mathcal{A}$ iff $\exists$ has a winning strategy in the game $\Gamma_n(U,\mathcal{A},\emptyset,\emptyset)$.

9.5.5 The axioms define K

We can now obtain our explicit universal axiomatisation of K: it consists of the sentences $\eta_n(\emptyset,\emptyset)$ (for $n < \omega$). For reference, we restate the hypotheses fully.

THEOREM 9.28 *Let* K *be the pseudo-universal class* $K = \{M^a\!\restriction_L : M \models U\}$ *of L-structures, where U is an L^s-theory as in definition 9.1(3), and L, L^s, U are recursive. Define the L-sentences $\eta_n(\emptyset,\emptyset)$ from U as in proposition 9.27, and define the recursive universal L-theory*

$$T(U) = \{\eta_n(\emptyset,\emptyset) : n < \omega\}.$$

If $\mathcal{A}$ is any L-structure, then $\mathcal{A} \in K$ iff $\mathcal{A} \models T(U)$. So $T(U)$ axiomatises K.

Proof. $T(U)$ is certainly recursively enumerable; it can be seen to be recursive by the argument given in the proof of theorem 8.3.

Let $\mathcal{A} \in K$. Then proposition 9.19 shows that $\exists$ has a winning strategy in $\Gamma(U,\mathcal{A})$, and hence a winning strategy in $\Gamma_n(U,\mathcal{A},\emptyset,\emptyset)$ for each $n < \omega$. By proposition 9.27, $\mathcal{A} \models T(U)$.

Conversely, let $\mathcal{A} \models T(U)$; we will show that $\mathcal{A} \in K$. As L is countable, we may take countable $\mathcal{B} \preceq \mathcal{A}$; then $\mathcal{B} \models T(U)$ as well. By proposition 9.27, $\exists$ has a winning strategy in $\Gamma_n(U,\mathcal{B},\emptyset,\emptyset)$ for all finite n. As $\mathcal{B}$ is countable, by proposition 9.25 $\exists$ has a winning strategy in $\Gamma(U,\mathcal{B})$. By proposition 9.20, $\mathcal{B} \in K$. As K is elementary (by corollary 9.15 or exercise 9.4(5)), and $\mathcal{B} \equiv \mathcal{A}$, we have $\mathcal{A} \in K$ as required. □

REMARK 9.29 By following through this construction for **RRA**, it can be seen that the axiomatisation obtained in this way is basically similar to the one we first constructed in theorem 8.3. The differences are minor, mainly being that the axioms obtained by the general pseudo-elementary construction are rather less 'efficient'. But as we saw in example 9.2, **RRA** can be written as a pseudo-elementary class in a different way, using Monk's theorem that $\mathbf{RRA} = \mathbf{S\mathfrak{R}aCA}_\omega$, and classes such as $\mathbf{RA}_n$ and $\mathbf{S\mathfrak{R}aCA}_n$ can be expressed as pseudo-elementary classes in several different ways using different theories U. The axiomatisations obtained from the different U may be very different from each other. This may be worth further study.

PROBLEM 9.30 *Prove negative results on the relationship between axioms for classes such as* $\mathbf{RA}_n$, $\mathbf{S\mathfrak{R}aCA}_n$, **RRA**, *etc., obtained by expressing them as pseudo-elementary classes in different ways.*

For example, show (if true) that there is no primitive recursive function $f : \omega \to \omega$ such that for all $n < \omega$, the $f(n)$th axiom for **RRA** *obtained from the defining theory U' of examples 9.2(8) logically entails the nth axiom for* **RRA** *obtained using the theory U of examples 9.2(1), and vice versa.*

9.5.6 Varieties and equations

Our fourth and final goal is to convert the universal sentences of $T(U)$, axiomatising K, into equations in the case where K is contained in a discriminator variety. All we have to do is apply theorem 2.60, to obtain:

THEOREM 9.31 *Assume that* D *is a discriminator variety, and* $\mathsf{K} \subseteq \mathsf{D}$ *is a variety and also a pseudo-universal class with defining theory U, say. Then* K *is axiomatised by a recursive set of equations, effectively obtainable from the theory $T(U)$ of theorem 9.28, together with the equations defining* D.

Exercises

1. Extract explicit axioms for **RRA** using the pseudo-elementary approach, and compare them with those obtained in theorem 8.3.

2. Show that there is a recursive function f with the properties of problem 9.30.

9.6 Axiomatising pseudo-elementary classes

A similar but more general version of the argument for pseudo-universal classes provides explicit (but non-universal) recursive axioms for the closure under elementary equivalence of any pseudo-elementary class whose defining theory is recursive. The changes needed in the proof are to definition 9.18, where $\exists$ must be allowed to choose algebra elements in moves of type 4d (this gives rise to existential quantifiers in the formulas η_n), and to proposition 9.25, where saturation or ultraproducts are used (cf. [HirHod97a]).

We now give brief details of the modifications required. Let K be a pseudo-elementary class of L-structures. We may write K in the form $\{M^{\mathbf{a}}\!\upharpoonright_L : M \models U\}$ for some recursive L^s-theory U as in definition 9.1. There is no restriction on the function symbols in $L^s \setminus L$. By pushing negations down next to atomic subformulas, we may suppose that the sentences in U are built from atomic and negated-atomic formulas without using negation; they may have quantifiers $\exists x^{\mathbf{a}}$ over the **a**-sort. So in the definitions of 'pre-network' (definition 9.17) and 'syntactic network' (definition 9.26) we drop the restriction that sentences in S do not involve quantifiers $\exists x^{\mathbf{a}}$ over the **a**-sort. Consequently, in type 4d moves in the game $\Gamma(U, \mathcal{A})$, $\forall$ may

9.6. Axiomatising pseudo-elementary classes

have chosen a sentence $\exists x^{\mathbf{a}} \psi(x)$ from S. If he has done this, we allow and insist that $\exists$ respond by choosing an element $a \in \mathcal{A}$ and playing the pre-network $(N, S \cup \{\psi(a)\})$.

DEFINITION 9.32 We will denote this modified game by $\Gamma^+(U, \mathcal{A})$.

The reader may check that the proofs of propositions 9.19 and 9.20 carry over without change to this new setting. Thus, we have:

PROPOSITION 9.33 *For any L-structure $\mathcal{A}$, $\exists$ has a winning strategy in the game $\Gamma^+(U, \mathcal{A})$ iff $\mathcal{A} \in \mathsf{K}$.*

We may define the truncated versions $\Gamma_n^+(U, \mathcal{A}, N, S)$ of $\Gamma^+(U, \mathcal{A})$ much as before. The analogue of proposition 9.27 still holds:

PROPOSITION 9.34 *For each $n < \omega$ and each syntactic network (N, F), there is a first-order L-formula $\eta_n^+(N, F)$, with free variables among $\mathrm{Var}(F)$, such that for any L-structure $\mathcal{A}$ and assignment $\iota : \mathrm{Var}(F) \to \mathcal{A}$,*

$$\exists \text{ has a winning strategy in } \Gamma_n^+(U, \mathcal{A}, N, \iota(F)) \iff \mathcal{A}, \iota \models \eta_n^+(N, F).$$

The formulas $\eta_n^+(N, F)$ are recursive in n, N, F.

Proof. As before, except that in case 4d of the definition of $\eta_{n+1}^+(N, F)$, for each formula $\exists v^{\mathbf{a}} \varphi(v^{\mathbf{a}}, \bar{z}) \in F$ we add a conjunct $\exists y \eta_n^+(N, F \cup \{\varphi(y, \bar{z})\})$, for a new variable y as before, to reflect the new kind of move that $\exists$ can make. □

However, proposition 9.25 fails in the new situation, at least for infinite algebras (it still holds for finite ones), because $\exists$ may have a choice of infinitely many type 4d moves in a round. So instead, we prove a weaker proposition; it suffices because we are axiomatising the elementary closure of K.

The proof will use saturation, which we first met in definition 3.33. One difficulty that arises in using ordinary saturation is that an ω-saturated structure may be uncountable, while our games Γ^+ work best on countable algebras. One way round this problem is to use elementary chains to extract a suitable countable algebra, and we will do this in chapter 10. However, here we are interested in obtaining recursive axiomatisations, and in the recursive context a more refined notion of saturation is available that avoids the problem altogether. It is called *recursive saturation*. A recursively saturated structure need only realise some types, namely the recursive ones; this keeps its cardinality down.

DEFINITION 9.35 Let L be a recursive first-order language. An L-structure $\mathcal{A}$ is said to be *recursively saturated* if for every tuple $x\bar{y}$ of variables, every recursive set p of L-formulas of the form $\varphi(x, \bar{y})$, and every $|\bar{y}|$-tuple $\bar{a} \in A$, if the type $\{\varphi(x, \bar{a}) : \varphi(x, \bar{y}) \in p\}$ is finitely satisfiable in $\mathcal{A}$ then it is realised in $\mathcal{A}$.

Any countable consistent L-theory (recursive or not) has a countable recursively saturated model [ChaKei90, theorem 2.4.1]. It follows that for any L-structure M, there exists a recursively saturated L-structure elementarily equivalent to M. See, e.g., [Hod93, §10.6], [ChaKei90, §2.4], [Kay91] for information about recursively saturated structures.

PROPOSITION 9.36 *Let $\mathcal{A}$ be any L-structure. Then $\exists$ has a winning strategy in $\Gamma_n^+(U,\mathcal{A},\emptyset,\emptyset)$ for all $n < \omega$ iff there is an L-structure $\mathcal{B} \in \mathsf{K}$ elementarily equivalent to $\mathcal{A}$.*

Proof. Right-to-left is simple. If there is a structure $\mathcal{B}$ as stated, then proposition 9.33 yields that $\exists$ has a winning strategy in $\Gamma^+(U,\mathcal{B})$. It clearly follows that $\exists$ has a winning strategy in $\Gamma_n^+(U,\mathcal{B},\emptyset,\emptyset)$, so (by proposition 9.34) $\mathcal{B} \models \eta_n^+(\emptyset,\emptyset)$, for all $n < \omega$. As $\mathcal{B} \equiv \mathcal{A}$, $\mathcal{A} \models \eta_n^+(\emptyset,\emptyset)$ for all $n < \omega$, too, so by proposition 9.34 again, $\exists$ has a winning strategy in $\Gamma_n^+(U,\mathcal{A},\emptyset,\emptyset)$ for all $n < \omega$.

For the converse, assume that $\exists$ has a winning strategy in $\Gamma_n^+(U,\mathcal{A},\emptyset,\emptyset)$ for all finite n. Take a countable recursively saturated L-structure $\mathcal{B} \equiv \mathcal{A}$. We will show that $\mathcal{B} \in \mathsf{K}$.

Claim. $\exists$ *has a winning strategy in $\Gamma^+(U,\mathcal{B})$.*
Proof of claim. As earlier, we define a *pole position* in this game to be a pre-network (N,S) such that $\exists$ has a winning strategy in the game $\Gamma_n^+(U,\mathcal{B},N,S)$ for infinitely many, or equivalently, all, $n < \omega$. $\exists$'s strategy in $\Gamma^+(U,\mathcal{B})$ will be to remain in pole position; since any pole position is a network, such a strategy is winning.

We show that $\exists$ can remain in pole position. To start, we require that the initial position $(\emptyset,\emptyset)$ is a pole position. By assumption, $\exists$ has a winning strategy in $\Gamma_n^+(U,\mathcal{A},\emptyset,\emptyset)$ for all n. By proposition 9.34, $\mathcal{A} \models \eta_n^+(\emptyset,\emptyset)$ for all n. As $\mathcal{A} \equiv \mathcal{B}$, $\mathcal{B} \models \eta_n^+(\emptyset,\emptyset)$ for all n, too; by the proposition again, it follows that the empty network $(\emptyset,\emptyset)$ is a pole position.

Assume inductively that the pre-network (N,S) existing at the start of the current round of $\Gamma^+(U,\mathcal{B})$ is a pole position. We show that whatever move $\forall$ makes in the current round, $\exists$ can respond in such a way as to reach another pole position. The argument is exactly as in proposition 9.25, except when $\forall$ makes a type 4d move and chooses a $L^s(\mathcal{B} \cup N)$-sentence $\exists v^{\mathbf{a}} \varphi(v,\bar{b}) \in S$, where $\varphi(v,\bar{z})$ is an $L^s(N)$-formula and $\bar{b} \in \mathcal{B}$. In this case, we choose a syntactic network (N,F) with $\exists v^{\mathbf{a}} \varphi(v,\bar{z}) \in F$, and an assignment $\iota : \mathrm{Var}(F) \to \mathcal{B}$ such that $\iota(F) = S$ and $\iota(\bar{z}) = \bar{b}$. This is always possible to do. Let $n < \omega$. By proposition 9.34 and the inductive hypothesis, $\mathcal{B}, \iota \models \eta_{n+1}^+(N,F)$. By the new clause in the definition of η_{n+1}^+, we obtain $\mathcal{B}, \iota \models \exists y \eta_n^+(N, F \cup \{\varphi(y,\bar{z})\})$. Here, y is a new variable not occurring in formulas in F. So there is ι_n extending ι to y and such that $\mathcal{B}, \iota_n \models \eta_n^+(N, F \cup \{\varphi(y,\bar{z})\})$. Clearly, $\mathcal{B}, \iota_n \models \eta_m^+(N, F \cup \{\varphi(y,\bar{z})\})$ for all $m < n$ (cf. the proof of lemma 9.24(2)).

9.6. Axiomatising pseudo-elementary classes

This holds for all $n < \omega$. Now let

$$p = \{\eta_n^+(N, F \cup \{\varphi(y,\bar{z})\}) : n < \omega\},$$

with free variables in $\mathrm{Var}(F \cup \{\varphi(y,\bar{z})\}) = \mathrm{Var}(F) \cup \{y\}$. Let $\iota(p)$ be the type over $\mathrm{rng}(\iota)$ obtained by instantiating each free variable $x \neq y$ of each formula in p by $\iota(x)$. So the formulas in $\iota(p)$ have y as their sole free variable. By the foregoing, $\iota(p)$ is finitely satisfiable in $\mathcal{B}$. As p is evidently recursive and $\mathcal{B}$ is recursively saturated, $\iota(p)$ is realised in $\mathcal{B}$ — there is an assignment $\iota_\omega : \mathrm{Var}(F) \cup \{y\} \to \mathcal{B}$ extending ι and such that $\mathcal{B}, \iota_\omega \models \eta_n^+(N, F \cup \{\varphi(y,\bar{z})\})$ for all n. Thus, if $\exists$ plays the network $(N, \iota_\omega(F \cup \{\varphi(y,\bar{z})\}))$, she will remain in pole position.

As $\mathcal{B}$ is countable, the game $\Gamma^+(U, \mathcal{B})$ has ω rounds. So the above suffices to show that $\exists$ can implement the strategy to remain in pole position throughout. The claim is proved.

By proposition 9.33 and the claim, we obtain $\mathcal{B} \in \mathsf{K}$. Since $\mathcal{B} \equiv \mathcal{A}$, the proof is complete. □

We conclude:

THEOREM 9.37 *Let K be the pseudo-elementary class $\mathsf{K} = \{M^{\mathrm{a}}{\restriction}_L : M \models U\}$ of L-structures, where U is an L^s-theory as in definition 9.1, and L, L^s, U are recursive. Define L-sentences $\eta_n^+(\emptyset,\emptyset)$ from U as in proposition 9.34, and define the recursive L-theory*

$$T^+(U) = \{\eta_n^+(\emptyset,\emptyset) : n < \omega\}.$$

For any L-structure $\mathcal{A}$, we have $\mathcal{A} \models T^+(U)$ iff there is $\mathcal{B} \in \mathsf{K}$ with $\mathcal{A} \equiv \mathcal{B}$. So $T^+(U)$ axiomatises the closure of K under elementary equivalence.

Proof. Let $\mathcal{A} \equiv \mathcal{B} \in \mathsf{K}$. By proposition 9.33, $\exists$ has a winning strategy in $\Gamma(U,\mathcal{B})$, and hence a winning strategy in $\Gamma_n(U, \mathcal{B}, \emptyset, \emptyset)$ for each $n < \omega$. By proposition 9.34, $\mathcal{B} \models T^+(U)$. As $\mathcal{A} \equiv \mathcal{B}$, we have $\mathcal{A} \models T^+(U)$, also.

Conversely, if $\mathcal{A} \models T^+(U)$ then proposition 9.34 yields that $\exists$ has a winning strategy in $\Gamma_n^+(U, \mathcal{A}, \emptyset, \emptyset)$ for all finite n. By proposition 9.36, for some $\mathcal{B}$ we have $\mathcal{A} \equiv \mathcal{B} \in \mathsf{K}$, completing the proof. □

COROLLARY 9.38 *Let K be the pseudo-elementary class $\{M^{\mathrm{a}}{\restriction}_L : M \models U\}$, as above.*

1. *If K is elementary then it is axiomatised by $T^+(U)$.*

2. *For any finite L-structure $\mathcal{A}$, we have $\mathcal{A} \in \mathsf{K}$ iff $\mathcal{A} \models T^+(U)$.*

Proof. Immediate. □

9.7 Generalised Q-operators

We can repeat the idea of section 8.2 in this new situation, to obtain somewhat simpler axioms for pseudo-universal and pseudo-elementary classes but in an expanded signature. We will be brief, as the process is much the same as before, and in any case we already have fact 9.11.

Let K be a pseudo-elementary class defined by the L^s-theory U. For each $n < \omega$ and each syntactic network (N,F), where (recall) F is a set of $L^s(N)$-formulas, enumerate the variables occurring free in formulas in F as $\bar{x}$, say, and introduce a new $|\bar{x}|$-ary relation symbol $Q_n[N,F](\bar{x})$.[1] We can axiomatise K in this expanded signature by the universal closure of the following axioms (taken over all n, N, F), which are obtained directly from the definition of the games Γ, Γ^+:

1. $Q_{n+1}[N,F](\bar{x}) \to Q_n[N, F \cup \{\sigma\}](\bar{x})$, for all $\sigma \in U$.

2. $Q_{n+1}[N,F](\bar{x}) \to Q_n[N, F \cup \{\forall \bar{v} \exists w (f(\bar{v}) = w)\}](\bar{x})$, for any function symbol $f \in L^s \setminus L$ and any tuple $\bar{v}w$ of distinct variables such that $f(\bar{v}) = w$ is well-formed.

3. For any quantifier-free $L^s(N)$-formula $\varphi(v, \bar{z})$ and $L^s(N)$-term t with $\varphi(t, \bar{z}) \in F$,

    (a) for each $L^s(N)$-term u with $t = u \in F$, add the axiom
    $Q_{n+1}[N,F](\bar{x}) \to Q_n[N, F \cup \{\varphi(u, \bar{z})\}](\bar{x})$,

    (b) if t is an L-term, choose a new variable $y \notin \text{rng}(\bar{x})$ and add the axiom
    $t = y \wedge Q_{n+1}[N,F](\bar{x}) \to Q_n[N, F \cup \{\varphi(y, \bar{z})\}](\bar{x}, y)$.

4. For each $\psi \wedge \chi \in F$, add
$Q_{n+1}[N,F](\bar{x}) \to Q_n[N, F \cup \{\psi\}](\bar{x}) \wedge Q_n[N, F \cup \{\chi\}](\bar{x})$.

5. For each $\psi \vee \chi \in F$, add
$Q_{n+1}[N,F](\bar{x}) \to Q_n[N, F \cup \{\psi\}](\bar{x}) \vee Q_n[N, F \cup \{\chi\}](\bar{x})$.

6. For each $\forall x^r \psi(x) \in F$ (here and below, ψ could have more free variables than just x), and any $m \in N$, add $Q_{n+1}[N,F](\bar{x}) \to Q_n[N, F \cup \{\psi(m)\}](\bar{x})$.

7. For each $\forall x^a \psi(x) \in F$, choose a new variable $y \notin \text{rng}(\bar{x})$ and add
$Q_{n+1}[N,F](\bar{x}) \to Q_n[N, F \cup \{\psi(y)\}](\bar{x}, y)$.

8. For each $\exists x^r \psi(x) \in F$, pick a new node $k \notin N$ and add the axiom

$$Q_{n+1}[N,F](\bar{x}) \to Q_n[N \cup \{k\}, F \cup \{\psi(k)\}](\bar{x}) \vee \bigvee_{m \in N} Q_n[N, F \cup \{\psi(m)\}](\bar{x}).$$

---

[1] We are being informal here, as the interpretation of Q_n also depends on the chosen $\bar{x}$, so that we should really write $Q_n[N, F, \bar{x}]$ or proceed as in section 8.2 by enumerating variables in a fixed order. We should also standardise N as (say) an initial segment of ω, in order to obtain a *set* of axioms.

9.7. Generalised Q-operators

9. (This will only arise in the non-pseudo-universal case.)
   For each $\exists x^a \psi(x) \in F$, choose a new variable $y \notin \text{rng}(\bar{x})$ and add the axiom
   $Q_{n+1}[N, F](\bar{x}) \to \exists y Q_n[N, F \cup \{\psi(y)\}](\bar{x}, y)$.

We also need 'pseudo-normality' axioms expressing that under a given assignment of variables $\bar{x}$ to elements of the L-structure $\mathcal{A}$, (N, F) is a network. They are borrowed from the proof of proposition 9.27.

10. Let $\varphi(u_1, \ldots, u_k)$ be an atomic $L^s(N)$-formula, where $u_1, \ldots, u_k$ are distinct variables. Let $y_1, \ldots, y_k, z_1, \ldots, z_k$ be arbitrary variables, and suppose that $\varphi(y_1, \ldots, y_k) \in F$, and $\neg \varphi(z_1, \ldots, z_k) \in F$ (hence the y_i, z_j are all from $\bar{x}$). Choose a tuple $\bar{v}$ of variables of length $|\bar{x}|$ such that for all $i, j < |\bar{x}|$, if $x_i = y_l$ and $x_j = z_l$ for some l with $1 \leq l \leq k$ then $v_i = v_j$. Add the axiom $\neg Q_0[N, F](\bar{v})$.

    Alternatively, add the easier-to-read $\bigwedge_{1 \leq i \leq k} y_i = z_i \to \neg Q_0[N, F](\bar{x})$.

11. $Q_0[N, F](\bar{x}) \to \varphi(\bar{x})$, for each quantifier-free L-formula $\varphi(\bar{x}) \in F$.

12. If there is a formula in F of the form $m \neq m$ or $m = m'$, for any $m \in N$, $m' \in N \setminus \{m\}$, then add $\neg Q_0[N, F](\bar{x})$.

13. $Q_n[\emptyset, \emptyset]$.

Notice that for pseudo-universal classes, the axioms are universal. In this case, we may dispense with the indices n of the $Q_n[N, F]$, by dint of proposition 9.25. In the context of a discriminator variety of BAOs, we may take them to be function symbols instead of relation symbols, coding true by 1 and false by 0 as in section 8.2.2, and rewrite the axioms as equations in the usual way (cf. theorem 2.60). For pseudo-elementary classes, the axioms are $\forall \exists$; one could Skolemise to obtain a universal axiomatisation (in the larger signature of course).

THEOREM 9.39 *Let* K *be a class of L-structures, and let* $\mathcal{A}$ *be an arbitrary L-structure.*

1. *If* K *is pseudo-universal, then* $\mathcal{A}$ *can be expanded to a model of the above axioms (excluding type 9) iff* $\mathcal{A} \in$ K.

2. *If* K *is pseudo-elementary, then* $\mathcal{A}$ *can be expanded to a model of the above axioms (including type 9) iff there is an L-structure* $\mathcal{B}$ *such that* $\mathcal{A} \equiv \mathcal{B} \in$ K.

Proof (sketch). The idea is very similar to the proof of theorem 8.5, so we only give a bare outline of the argument. For part 2, let K be a pseudo-elementary class defined by U. If $\mathcal{A} \equiv \mathcal{B} \in$ K, then by proposition 9.36, $\exists$ has a winning strategy in $\Gamma_n^+(U, \mathcal{A})$ for all $n < \omega$. For $\bar{a} \in \mathcal{A}$, if ι assigns $\bar{x}$ to $\bar{a}$, define $Q_n[N, F](\bar{a})$ to be true iff $\exists$ has a winning strategy in $\Gamma_n^+(U, \mathcal{A}, N, \iota(F))$. Check that the axioms hold. Conversely, assume that $\mathcal{A}$ has an expansion $\mathcal{A}^*$ interpreting the $Q_n[N,F]$ so as to satisfy the axioms. Show by induction on n that for any $\bar{a} \in \mathcal{A}$, any finite n, and any syntactic network (N, F), if $\mathcal{A}^* \models Q_n[N,F](\bar{a})$ and ι assigns $\bar{x}$ to $\bar{a}$, then $\exists$ has a winning strategy in $\Gamma_n^+(U, \mathcal{A}, N, \iota(F))$. Hence $\exists$ has a winning strategy in $\Gamma_n^+(U, \mathcal{A}, \emptyset, \emptyset)$ for all finite n. By proposition 9.36 again, there is $\mathcal{B} \in \mathsf{K}$ with $\mathcal{B} \equiv \mathcal{A}$. Part 1 is proved similarly. □

Chapter 10

Game trees

The preceding chapter showed how to axiomatise certain classes of structures by games. The results proved there are general enough to provide recursive equational axiomatisations for all the varieties of algebras that we meet later in the book. Readers needing to axiomatise a class may use these results directly, or give an *ad hoc* solution in the manner of chapter 8.

But we will also wish to prove negative results, of the form such-and-such a class of algebras is not finitely axiomatisable, even over a given larger class, or sometimes even that the class is not elementary. Results like this can be hard to obtain, and, while it would be possible to do so using the 'pseudo-elementary games' of the preceding chapter, this would greatly complicate and obscure the arguments. To make life easier, we will therefore be defining specific games tailored to the problems in hand. The games guide us in creating the algebras that will serve as counter-examples. Here are some examples:

- games played on 'atomic networks', where the labels on network edges have to be atoms,

- n-pebble network games, where the number of nodes in a network is restricted to n, so network nodes have to be 'picked up and moved' as the game progresses,

- amalgamation games, where ∀ is allowed to pick two networks that have previously been played and that contain as subnetworks isomorphic copies of a common network, and force their amalgamation over the common part,

- games played with 'hypernetworks' — rather like atomic networks, but 'hyperedges' get labels too.

While all these games are formally different, there are strong similarities between them. So in order to avoid too much repetition in the proofs, it is convenient to provide a general setting for games like them, and that is what we do in this chapter. It is worth noting that the 'game trees' of this chapter, while perhaps easier to handle, are closely related to the game Γ of the last chapter: a class of algebras is definable by a 'closed game tree' iff it has the same countable members as some pseudo-elementary class (exercise 10.5(11)).

The programme is as follows.

Section 10.1 Here, we define the generalised games. With only minor modifications, the generalisation we use is that of *Hintikka games*. They are similar to the more 'linear' Keisler games defined in [Hod93, theorem 10.3.7].

Section 10.2 We briefly discuss strategies in these games.

Section 10.3 We show here how the games generalise those of chapter 7.

Section 10.4 Here, we show how to write down first-order formulas expressing that the second player ('$\exists$') has a winning strategy in the games.

Section 10.5 Here, we show how a winning strategy for $\exists$ in a sequence of finite-length games over a sequence of algebras gives her a winning strategy in an infinite-length game over a non-principal ultraproduct of the algebras. Together with an elementary chain construction, we use this to show how to prove non-finite axiomatisability of one class over another. We use ultrapowers to show how to prove that certain classes of algebras are not elementary.

10.1 Trees, and games on them

We begin by defining the games that we will use; we need some terminology for trees first.

DEFINITION 10.1

1. A *tree* is a pair $(T, <)$, where T is a set and $<$ an irreflexive partial ordering on T, such that for any $t \in T$, the set $\widehat{t} = \{u \in T : u < t\}$ is well-ordered by '$<$' (that is, it is linearly ordered with no infinite descending sequences $u_0 > u_1 > \cdots$).

2. As usual, we use T to denote both the tree $(T, <)$ and its underlying set (or domain) T of 'nodes'. For nodes t, u of T, we write $u \leq t$ if $u < t$ or $u = t$.

10.1. Trees, and games on them

3. Let T be a tree, and $t \in T$. We let $\mathrm{ht}_T(t)$, or, when there is no ambiguity, $\mathrm{ht}(t)$ (the 'height of t'), denote the unique ordinal whose order type is $\hat{t}$. (Perhaps 'depth' would be a better term, as we will usually draw trees so that for nodes $t < u$, t lies above u.)

4. If there is a unique node $r \in T$ of height 0, then r is called the *root* of T. All trees we consider will be *rooted* (i.e., they have a root).

5. We let $\mathrm{ht}(T)$ denote the least ordinal α such that T has no elements of height α. As T is a set, such an ordinal exists.

6. If α is an ordinal, we write $T\!\upharpoonright\!\alpha$ for the tree $(T', <\!\upharpoonright_{T'})$, where $T' = \{t \in T : \mathrm{ht}(t) < \alpha\}$ and $<\!\upharpoonright_{T'}$ is the restriction of $<$ to T'.

7. If $t \in T$, we write $T_{\geq t}$ for the tree with nodes $\{u \in T : u \geq t\}$ and with ordering induced by $<$.

8. An *immediate successor* of t in T is a node $u \in T$ with $t < u$ and $\mathrm{ht}(u) = \mathrm{ht}(t) + 1$. In this case t is said to be an *immediate predecessor* of u. We write $\mathrm{Suc}(t, u)$ to denote that u is an immediate successor of t in T. T is said to be *finitely branching* if every node of T has at most finitely many immediate successors (but we do not require a finite bound on the number of immediate successors).

9. A node of T without immediate successors is called a *leaf*.

10. Nodes $t, u \in T$ are said to be *siblings* if $t \neq u$ and there is $v \in T$ such that t, u are both immediate successors of v.

We can now define the games. Unlike in the preceding chapter, we will concentrate on games of length $\leq \omega$, because this is all that we need for applications here, and it is simpler.

DEFINITION 10.2 Let L be a first-order signature. (Typically, it will be the signature $+, -, \cdot, 0, 1, 1', \smile, ;$ of relation algebras.)

1. An L-game tree T is a finitely branching rooted tree of height $\leq \omega$ with each node $t \in T$ labelled by a triple $T(t) = (P_t, \bar{x}_t, \phi_t)$, where:

    - P_t is either $\forall$ or $\exists$ (we may say that t 'belongs to player P_t', or 'is a P_t-node', or that 'P_t owns t'). We require that if $t, u \in T$ are siblings then $P_t = P_u$.
    - $\bar{x}_t$ is a finite tuple of distinct variables.
    - ϕ_t is a first-order L-formula whose free variables all occur in $(\bar{x}_s : s \leq t)$ (this is the tuple of variables obtained by concatenating the tuples $\bar{x}_s$ for all $s \in T$, $s \leq t$, in order of increasing height in T).

Note that some of the variables in $\bar{x}_t$ may occur in $\bar{x}_u$ for $u < t$ — variables may be re-used. Frequently we drop the L and call T simply a *game tree*.

2. Let T be an L-game tree with root r, and for $t \in T$ let $T(t) = (P_t, \bar{x}_t, \phi_t)$. Let $\mathcal{A}$ be an L-structure (typically, a relation algebra) and let v_r be an assignment of the variables of $\bar{x}_r$ to elements of $\mathcal{A}$. We define a game $G(T, \mathcal{A}, v_r)$. In the game, the players $\forall$ and $\exists$ of our earlier acquaintance choose nodes of T, each one being an immediate successor of the last and starting with r, and assign variable symbols to elements of $\mathcal{A}$. If, at some stage of a play of this game, the current node is t and the current variable assignment is v, we say that the current *position* of the play is (t, v). We require that v assign an element of $\mathcal{A}$ to every variable in $\bar{x}_u$ for every node $u \leq t$.

The initial position is (r, v_r).

Suppose that at the start of some round, the current position is (t, v_t).

  (a) If $\mathcal{A}, v_t \models \neg \phi_t$ then player P_t loses the game at this point.

  (b) Otherwise, if t is a leaf of T, then play ends at this point in a win for $\exists$.

  (c) Otherwise, suppose that the immediate successors of t in T belong to player P, say. (Note that they all belong to the same player, by definition of game trees.) P chooses an immediate successor s of t, and an assignment v_s to the variables $(\bar{x}_u : u \leq s)$ that agrees with v_t except perhaps on variables in $\bar{x}_s$ (i.e., $v_s(y) = v_t(y)$ for all variables y in $(\bar{x}_u : u \leq t)$ and not in $\bar{x}_s$). That is, P (re)assigns the variables $\bar{x}_s$. The position now becomes (s, v_s), and the round is complete.

If play goes on for infinitely many moves, then $\exists$ wins.

Thus, in general, position (t, v), with $T(t) = (P_t, \bar{x}_t, \phi_t)$, is arrived at by player P_t choosing the value of $\bar{x}_t$, hopefully making ϕ_t true. Although the initial position is 'given from outside', player P_r will still lose outright if $\mathcal{A}, v_r \models \neg \phi_r$. (If this seems unfair, recall that in the game $G_\omega(N, \mathcal{A})$ of definition 7.12, if N was not a network then $\exists$ lost immediately with no moves being played.)

DEFINITION 10.3

1. Tree operations and definitions apply to game trees T by applying them to the underlying tree. For example:

  (a) We write $t \in T$ to denote that t is a node of the underlying tree of T.

  (b) If $t \in T$ and $0 < n < \omega$, we let $T\!\upharpoonright_n$ denote the game tree based on the tree $T\!\upharpoonright_n$, as defined above, with labelling inherited from T (i.e., $T\!\upharpoonright_n(t) = T(t)$ for all $t \in T\!\upharpoonright_n$).

10.1. Trees, and games on them

(c) We note that $T\upharpoonright_0$ is empty, so not rooted and not strictly a game tree. Rather than requiring that $n > 0$ when using the notation $G(T\upharpoonright_n, \mathcal{A}, v_r)$, we stipulate that $G(\emptyset, \mathcal{A}, v)$ (any $\mathcal{A}, v$) is a trivial game won outright by $\exists$ without any kind of move being played.

(d) If $t \in T$, $T_{\geq t}$ denotes the game tree with underlying tree $T_{\geq t}$ and labelling inherited from T except that $T_{\geq t}(t) = (P_t, \bar{z}, \phi_t)$, where $\bar{z}$ is a tuple of distinct variables consisting of precisely the variables in the tuple $(\bar{x}_u : u \leq t)$ in order of their first occurrence.

(e) If T, U are game trees, we write $T \subseteq U$ if:
- every node of T is a node of U,
- the ordering and labelling on T are induced from U,
- if $t \in T$, $u \in U$, and $U \models u \leq t$, then $u \in T$ ('T is closed downwards in U'),
- if $t, u \in U$ are immediate successors of $v \in U$, and $t \in T$, then $u \in T$ ('T is sibling-closed').

(f) If $T_0 \subseteq T_1 \subseteq \cdots$ is an ω-chain of game trees, $\bigcup_{n<\omega} T_n$ denotes their union: the set of nodes of $\bigcup_{n<\omega} T_n$ is the union of the sets of nodes of the T_n ($n < \omega$), and the ordering and labelling of these nodes are obtained from the orderings and labellings of the nodes of the T_n in the obvious (and well-defined) way. It follows from sibling-closure of the T_n that $\bigcup_{n<\omega} T_n$ is finitely branching, so it can be considered as a game tree in the obvious way.

2. An L-game tree T is said to be *recursive* if L is recursive (definition 9.13) and there exists a Turing machine that will print out a possibly infinite representation of T in some standard form that shows the distinct nodes of T in increasing order of height, their labels, and their mutual relations via '$<$'.

3. A game tree T is said to be *closed* if $\bar{x}_r$ is the empty sequence, where r is the root of T. In this case, we can denote a game on T simply as $G(T, \mathcal{A})$, since the assignment of variables $\bar{x}_r$ is null.

Exercises

1. Let T be a game tree and let $t < s$ in T and $n < \omega$. Show that $(T_{\geq t})_{\geq s} = T_{\geq s}$ and $(T_{\geq t})\upharpoonright_n = (T\upharpoonright_{\mathrm{ht}(t)+n})_{\geq t}$.

2. [König's tree lemma.] Let T be a finitely branching rooted tree of height ω. A node t of T is said to be *live* if $T_{\geq t}$ has height ω. A *branch* of T is a maximal linearly ordered subset of T.

    (a) Show that if $t \in T$ is live then t has a live immediate successor in T.

(b) Deduce that T has an infinite branch. (This needs some form of the axiom of choice.)

3. If T is an L-game tree, show that there is countable $L_0 \subseteq L$ such that T is an L_0-game tree.

4. Show that if T is a recursive game tree and $t \in T$ then $T_{\geq t}$ is also recursive.

10.2 Strategies

We can define a strategy for either player in games of the form $G(T, \mathcal{A}, v)$ in the usual way, as a set of rules telling that player how to play in any situation that can be reached by using the strategy. A strategy is 'winning' if its owner always wins whenever s/he uses it, and 'deterministic' if it gives its owner a unique move in any situation. The strategy's advice in a given position may depend on the entire history of the play so far, though because the conditions for winning do not depend on the history, this dependence can be eliminated.

As we said in chapter 7, it will never be absolutely necessary for us to use a formal definition of the notion of a strategy, and indeed it can be positively harmful to do so, killing the strong and useful intuitions we all have about games. None the less, some readers may disagree, and in any case, since so much of the book depends on games and strategies in them, we had better have a formal definition to hand in case of doubt. The purpose of this section is to provide one. Readers not interested may of course skip it.

Fix a functional signature L, and an L-game tree T with root t_0. For $t \in T$, let $T(t) = (P_t, \bar{x}_t, \phi_t)$. Let $\mathcal{A}$ be an L-structure, and $v_0 : \text{rng}(\bar{x}_{t_0}) \to \mathcal{A}$ an assignment.

Let $\mathcal{P}$ be the set of all pairs (t, v), where $t \in T$ and v is an assignment of the variables in $(\bar{x}_u : u \leq t)$ (and only these variables) to elements of $\mathcal{A}$. We refer to the elements of $\mathcal{P}$ as *positions*, even though some of them may never crop up in the game. A position $(t, v) \in \mathcal{P}$ is said to be *legal* if $\mathcal{A}, v \models \phi_t$, and *illegal* otherwise. We define the binary relation $\sqsubset$ on $\mathcal{P}$ to be the transitive closure of:

$$\{((t,v), (t',v')) \in \mathcal{P} \times \mathcal{P} \ : \ t' \text{ is an immediate successor of } t \text{ in } T, \\ v' \text{ agrees with } v \text{ except perhaps on } \bar{x}_{t'}\}. \quad (10.1)$$

Clearly, $(\mathcal{P}, \sqsubset)$ is a well-founded (with no infinite descending sequences $\pi_0 \sqsupset \pi_1 \sqsupset \cdots$) and irreflexive partial order. (It may not be a tree, since it could have more than one minimal element and there may be more than one $\sqsubset$-ascending chain between two positions.) We often write $\mathcal{P}$ for $(\mathcal{P}, \sqsubset)$, in the usual way. For $\pi, \pi' \in \mathcal{P}$ we write $\pi \sqsubseteq \pi'$ if $\pi \sqsubset \pi'$ or $\pi = \pi'$. If $\pi_1, \pi_2 \in \mathcal{P}$, we say that π_2 is an *immediate successor* of π_1 in $\mathcal{P}$ if (π_1, π_2) is in the relation (10.1) above. We write $\mathcal{U}$ for the set of all legal positions that are not maximal in $\mathcal{P}$.

10.2. Strategies

DEFINITION 10.4

1. A *finished play* of the game $G(T, \mathcal{A}, v_0)$ is a linearly ordered subset $\eta \subseteq \mathcal{P}$ such that:

   (a) $(t_0, v_0) \in \eta$.

   (b) If $\pi \in \eta$, $\pi' \in \mathcal{P}$, and $\pi' \sqsubset \pi$, then $\pi' \in \mathcal{U} \cap \eta$ (in words, η is closed downwards in $\mathcal{P}$ and non-final positions in η are legal).

   (c) η is maximal subject to these conditions.

2. An *unfinished play* of $G(T, \mathcal{A}, v_0)$ is a finite linearly ordered subset η of $\mathcal{U}$ that is closed downwards and such that if $\eta \neq \emptyset$ then $(t_0, v_0) \in \eta$.

3. A *play* of $G(T, \mathcal{A}, v_0)$ is a finished or unfinished play of $G(T, \mathcal{A}, v_0)$.

   If $\eta = \{(t_0, v_0), (t_1, v_1), \ldots\}$ is a play, with $(t_0, v_0) \sqsubset (t_1, v_1) \sqsubset \cdots$, then we view the positions $(t_1, v_1), (t_2, v_2), \ldots$ in η as being successively chosen by $\forall$ or $\exists$ in the game according to the rules of definition 10.2. A finished play η is a win for $\forall$ if it is finite, its maximal element (t, v) (say) is illegal, and t is an $\exists$-node; otherwise, it is a win for $\exists$. An unfinished play encodes the history of the game up to some position at which the game is not yet finished.

4. For $P \in \{\forall, \exists\}$, a *strategy for P in $G(T, \mathcal{A}, v_0)$* is a partial function f from the set of all unfinished plays into $\wp(\mathcal{P})$, such that:

   (a) $\emptyset \in \text{dom}(f)$ and $f(\emptyset) = \{(t_0, v_0)\}$.

   (b) If $\eta \in \text{dom}(f)$, $\eta \neq \emptyset$, (t, v) is the maximal element of η, and the immediate successors of t in T are P-nodes, then $f(\eta)$ is a non-empty set of immediate successors of (t, v) in $\mathcal{P}$.

   (c) If $\eta \in \text{dom}(f)$, $\eta \neq \emptyset$, (t, v) is the maximal element of η, and the immediate successors of t in T belong to the other player than P, then $f(\eta)$ is the set of all immediate successors of (t, v) in $\mathcal{P}$.

   (d) If $\eta \in \text{dom}(f)$, $\pi \in f(\eta)$, and $\eta' = \eta \cup \{\pi\}$ is an unfinished play, then $\eta' \in \text{dom}(f)$.

Now let f be a strategy for player P in the game $G(T, \mathcal{A}, v_0)$.

5. f is said to be *deterministic* if in clause 4b above, $|f(\eta)| = 1$.

6. Let η be a play of $G(T, \mathcal{A}, v_0)$. We say that f is or was *used in the play η* if for every unfinished play $\eta' \subset \eta$, we have $f(\eta') \cap \eta \neq \emptyset$. (It follows from this by induction on $|\eta|$ that $\eta' \in \text{dom}(f)$ for each such η', so the definition makes sense.)

7. f is said to be a *winning strategy* if every play in which f is used is a win for P.

8. A position $(t, v) \in \mathcal{P}$ is said to be a *winning position* for P if P has a winning strategy in the game $G(T_{\geq t}, \mathcal{A}, v)$ — roughly speaking, in the game on T that starts from that position.

In game terms, a strategy f for player P in $G(T, \mathcal{A}, v_0)$ is used by P as follows. If the unfinished play consisting of all positions played so far is η, and player P is to move in the current position, then f is viewed as advising P to move to any position $\pi \in f(\eta)$. If the other player is to move, the new position will be in $f(\eta)$ whatever move is made. Note that the advice given by a strategy in any round depends on the entire history of play so far — i.e., on the unfinished play up to the current position. Note also that any strategy is used in the play $\{(t_0, v_0)\}$. Since the statement that $\exists$ (say) has a winning strategy in $G(T, \mathcal{A}, v)$ can be stated in terms of the existence of a certain partial function as above, readers who are so minded can now mentally remove all mention of games, players, and so on from the rest of the book. See exercise 10.2(1) below.

The notion defined above is the obvious formalisation of our intuitive idea of a strategy as a set of rules telling a player what to do. It is, however, rather cumbersome. The proposition below gives a few alternative formulations of the notion of a *winning strategy for* $\exists$.

PROPOSITION 10.5 *Let* $L, T, \mathcal{A}, t_0, v_0, \mathcal{P}, \mathcal{U}$ *be as above. Then the following are equivalent:*

1. $\exists$ *has a winning strategy in* $G(T, \mathcal{A}, v_0)$.

2. *There is a subset* $\sigma \subseteq \mathcal{P}$ *such that*

    (a) *every finished play* $\eta \subseteq \sigma$ *is a win for* $\exists$,

    (b) $(t_0, v_0) \in \sigma$,

    and for each $(t, v) \in \sigma \cap \mathcal{U}$,

    (c) *if* t *has immediate successors in* T *belonging to* $\exists$, *then* σ *contains some immediate successor (in* $\mathcal{P}$*) of* (t, v),

    (d) *if* t *has immediate successors in* T *belonging to* $\forall$, *then* σ *contains every immediate successor of* (t, v).

3. *There is a total function* $g : \mathcal{U} \to \mathcal{P}$ *such that*

    (a) *for all* $\pi \in \mathcal{U}$, $g(\pi)$ *is an immediate successor of* π *in* $\mathcal{P}$,

10.2. Strategies   317

(b) *if η is a finished play of $G(T,\mathcal{A},v_0)$ such that whenever $(t,v) \in \eta$ and t has immediate successors belonging to $\exists$ then $g((t,v)) \in \eta$, then η is a win for $\exists$.*

Thus, winning strategies can be assumed deterministic and not to depend on the history of the game. (On the other hand, in games of the form $G(T,\mathcal{A},v)$, a fair amount of information about the history may be encoded by the current position.)

Proof. (1) $\Rightarrow$ (2). Let f be a winning strategy for $\exists$ in $G(T,\mathcal{A},v_0)$. Define $\sigma = \bigcup\{\eta : \eta$ is a play of $G(T,\mathcal{A},v_0)$ in which f was used$\}$. We check that σ has the cited properties. Property (b) is clear. For (c,d), let $(t,v) \in \sigma \cap \mathcal{U}$ and pick a play η containing (t,v) and in which f was used. Let $\eta' = \{\pi \in \eta : \pi \sqsubseteq (t,v)\} \subseteq \eta$, and $\eta'' = \{\pi \in \eta : \pi \sqsubset (t,v)\} \subset \eta$. Then $\eta'' \in \text{dom}(f)$, $f(\eta'') \cap \eta \neq \emptyset$, and so $\eta' \in \text{dom}(f)$. Assume that t has immediate successors in T belonging to $\exists$. Then $f(\eta')$ is a non-empty set of immediate successors of (t,v). Let $(t',v') \in f(\eta')$. Then $\eta' \cup \{(t',v')\}$ is a play in which f was used, and so $(t',v') \in \sigma$. The case where t has immediate successors belonging to $\forall$ is similar.

Finally, let $\eta \subseteq \sigma$ be a finished play. If $\exists$ lost this play, then this can only be because η has a maximal element (t,v), say, which is illegal, and t is an $\exists$-node in T. Now $(t,v) \in \sigma$; let ξ be a play in which f was used, with $(t,v) \in \xi$. Clearly, ξ is a finished play which $\exists$ lost, contradicting the assumption that f is a winning strategy for $\exists$.

(2) $\Rightarrow$ (3). If σ is as specified in (2), define $g : \mathcal{U} \to \mathcal{P}$ by:

- If $\pi \in \mathcal{U} \cap \sigma$, then $g(\pi)$ is any element of the set of immediate successors of π in σ.

- If $\pi \in \mathcal{U} \setminus \sigma$, then $g(\pi)$ is an arbitrary immediate successor of π.

Clearly, any finished play satisfying condition (3b) is contained in σ, and so by (2a) is a win for $\exists$.

(3) $\Rightarrow$ (1). Given g, define a map f from the set of unfinished plays into $\wp(\mathcal{P})$ as follows. Define $f(\emptyset) = \{(t_0,v_0)\}$. If η is a non-empty unfinished play with maximal element (t,v), then let $f(\eta)$ be the set of all immediate successors of (t,v) in $\mathcal{P}$, if the immediate successors of t are $\forall$-nodes, and $f(\eta) = \{g((t,v))\}$ otherwise. Then f is a deterministic winning strategy for $\exists$. □

Exercises

1. Formalise the notion of a strategy in $G(T,\mathcal{A},v_0)$ being 'winning' for its owner purely in terms of the structure of $T, \mathcal{P}$, without mentioning games.

2. [cf. Gale–Stewart theorem, [GalSte53].] Let T be a closed L-game tree and $\mathcal{A}$ an L-structure. Assume that the position (t,v) in $G(T,\mathcal{A})$ is legal and not winning for $\forall$.

   (a) If t has immediate successors belonging to $\forall$, show that whatever move $\forall$ makes in the first round of $G(T_{\geq t},\mathcal{A},v)$, the resulting position is not winning for him.

   (b) If t has immediate successors belonging to $\exists$, show that there is a move that $\exists$ can make in the first round of $G(T_{\geq t},\mathcal{A},v)$ that results in a position that is not winning for $\forall$.

   (c) Assume that $\forall$ does not have a winning strategy in $G(T,\mathcal{A})$. Show that $\exists$ can play $G(T,\mathcal{A})$ in such a way that the position is never winning for $\forall$.

   (d) Deduce that either $\forall$ or $\exists$ has a winning strategy in $G(T,\mathcal{A})$ — the game is *determined*.

10.3 Examples

We view games played on game trees as *testing* $\exists$, as follows. Let the signature L be given and let T be an L-game tree with root r and $T(t) = (P_t, \bar{x}_t, \phi_t)$ for $t \in T$, as usual. Let $\mathcal{A}$ be an L-structure and v_r an assignment of the variables $\bar{x}_r$ to elements of $\mathcal{A}$. Suppose that in a play of $G(T, \mathcal{A}, v_r)$, the position (t, v_t) has been reached. Let t have immediate successors owned by $\forall$, say. $\forall$ now has the chance to test $\exists$, by choosing an immediate successor u of t and (re-)assigning the variables $\bar{x}_u$ associated with u. He is restricted to doing this in such a way as to make ϕ_u true; if he does not, his 'test' is illegal and $\exists$ wins by default. If t is a leaf of T, there is no u for him to choose. In this case the testing is over and $\exists$, having survived this far, wins.

Suppose he does this, and that the new position is (u, v_u). Typically, u has immediate successors owned by $\exists$. Now, $\exists$ must 'answer' $\forall$'s challenge by choosing one of them — say, s — and (re-)assigning the variables $\bar{x}_s$ so that ϕ_s holds. If there is no such s, because u is a leaf of T, then the test is not completed and $\exists$ wins by default. But if there is such an s and if she cannot make ϕ_s true, she fails the test and loses the game.

Any successors of s are typically $\forall$'s, and the pattern continues. If $\exists$ survives the testing for infinitely many moves, she passes, and wins the play.

This scheme of game fits all applications later in the book.

10.3. Examples

10.3.1 The game $G_n(I_a, \mathcal{A})$

Consider, for example, the game $G_n(I_a, \mathcal{A})$ of chapter 7, where $n \leq \omega$, $\mathcal{A}$ is a relation algebra, $a \in \mathcal{A} \setminus \{0\}$, and I_a is the 'standard' two-node network with $I_a(0,1) = a$, $I_a(1,0) = 1$, and $I_a(0,0) = I_a(1,1) = 1$'. We can form a game tree T_n, with root r and $\bar{x}_r$ a single variable x, such that for all non-zero $a \in \mathcal{A}$, $\exists$ has a winning strategy in $G_n(I_a, \mathcal{A})$ if and only if she has one in $G(T_n, \mathcal{A}, (w \mapsto a))$.

We do this by defining finite game trees $T_0 \subseteq T_1 \subseteq \cdots \subseteq T_n \subseteq \cdots$ ($n < \omega$) by induction on n, such that for each n, every leaf t of T_n is an $\exists$-node and has an associated term network N_t. (See definition 8.1 for term networks.)

T_0 consists of a single node r, with $T_0(r) = (\forall, w, w \neq 0)$, where w is a variable. We let N_r be the term network with nodes 0, 1, with $N_r(0,1) = w$, $N_r(1,0) = 1$, and $N_r(0,0) = N_r(1,1) = 1$'.

Assume that $n < \omega$ and T_n has been defined, such that every leaf of T_n is an $\exists$-node and has an associated term network. We define $T_{n+1} \supseteq T_n$ as follows.

First, for each leaf $t \in T_n$ with associated term network N_t, and for each pair x, y of nodes of N_t, we insert into T_{n+1} an immediate successor s of t with label $(\forall, (u, v), \top)$, where u, v are distinct variables not occurring in $\bar{x}_{t'}$ for any $t' \leq t$. When $\forall$ selects s and assigns the variables u, v to (say) $a, b \in \mathcal{A}$ in the game-tree game, it corresponds to his choosing the edge (x, y) of N_t and the elements a, b of $\mathcal{A}$ in the game $G(I_a, \mathcal{A})$. There is no restriction in $G(I_a, \mathcal{A})$ on $\forall$'s choice of a and b, so we set $\phi_s = \top$.

For each such s, we then insert two immediate successors s^+, s^- of s. These will be $\exists$-nodes, and we let $N_{s^+} = \text{Acc}(N_t, x, y, u, v)$ and $N_{s^-} = \text{Rej}(N_t, x, y, u, v)$ (see definition 8.1). The labels of s^+, s^- (in T_{n+1}) are defined to be:

$$T_{n+1}(s^+) = (\exists, \emptyset, \rho_0(N_{s^+}))$$
$$T_{n+1}(s^-) = (\exists, \emptyset, \rho_0(N_{s^-})).$$

Here, for a term network N, $\rho_0(N)$ is the formula

$$\bigwedge_{x,y,z \in N} (N(x,x) \leq 1') \wedge ((N(x,z); N(z,y)) \cdot N(x,y) \neq 0)$$

of proposition 8.2, saying that the values of the terms on the term network N yield a legal $\mathcal{A}$-network. When $\exists$ chooses s^+ or s^- in response to $\forall$'s move, it corresponds to her accepting or rejecting (respectively) in the game $G(I_a, \mathcal{A})$.

We let T_{n+1} be the result of extending T_n by these new nodes and labels. See figure 10.1.

Define $T_\omega = \bigcup_{n < \omega} T_n$, a well-defined and recursive game tree. It is clear by construction that any play of $G_n(I_a, \mathcal{A})$ can be viewed as a play of $G(T_n, \mathcal{A}, (w \mapsto a))$, and vice versa. Hence, for any non-zero $a \in \mathcal{A}$ and any $n \leq \omega$, the following are equivalent:

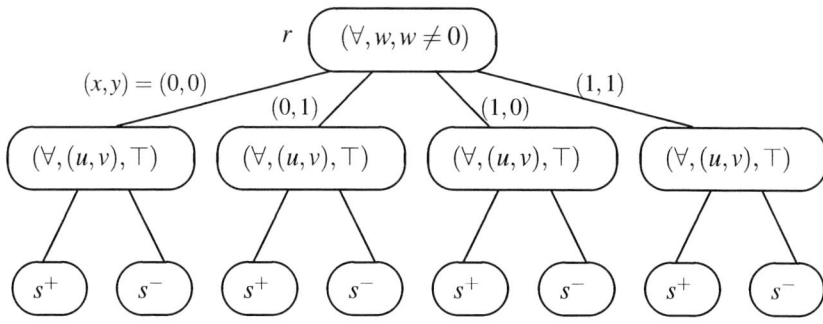

Figure 10.1: The game tree T_1

1. $\exists$ has a winning strategy in $G_n(I_a, \mathcal{A})$;

2. $\exists$ has a winning strategy in $G(T_n, \mathcal{A}, (w \mapsto a))$.

Now let T^+ be the closed game tree formed from T_ω by adding a single new node r^+ beneath the root of T_ω ($r^+ < r$), with $T^+(r^+) = (\exists, \emptyset, \top)$, say ($(\forall, \emptyset, \top)$ would also do). See figure 10.2. By theorem 7.19, for any relation algebra $\mathcal{A}$, $\mathcal{A} \in \mathbf{RRA}$ if and only if $\exists$ has a winning strategy in $G_\omega(I_a, \mathcal{A})$ for all $a \in \mathcal{A} \setminus \{0\}$; and by definition of T^+, it is clear that this is equivalent to $\exists$ having a winning strategy in $G(T^+, \mathcal{A})$.

We emphasise the directness of the correspondence between the two games $G(T_n, \mathcal{A}, (w \mapsto a))$ and $G_n(I_a, \mathcal{A})$. Once this is understood, it will be easy for the reader to construct a suitable game tree corresponding to the games we will introduce in later chapters (or at least, it should be easy to believe that such a tree can be constructed). The following points are also noteworthy:

- Moves in typical 'naturally occurring' games involve (a) choosing elements of an algebra and (b) making one of finitely many choices about some 'physical objects', such as picking two nodes of a network, or choosing whether to add a new node to a network or not. These two aspects are reflected in the game-tree games, where players assign variables to elements of an algebra, and choose one of finitely many immediate successors of the current node.

- Usually, we can easily turn a game tree into a closed game tree, as we did for T_ω above. Clearly, $\exists$ has a winning strategy in $G(T_\omega, \mathcal{A}, (w \mapsto a))$ for all non-zero $a \in \mathcal{A}$ iff she has a winning strategy in $G(T^+, \mathcal{A})$.

- The characterisation of **RRA** above — in terms of a winning strategy in

10.4. Formulas expressing a winning strategy

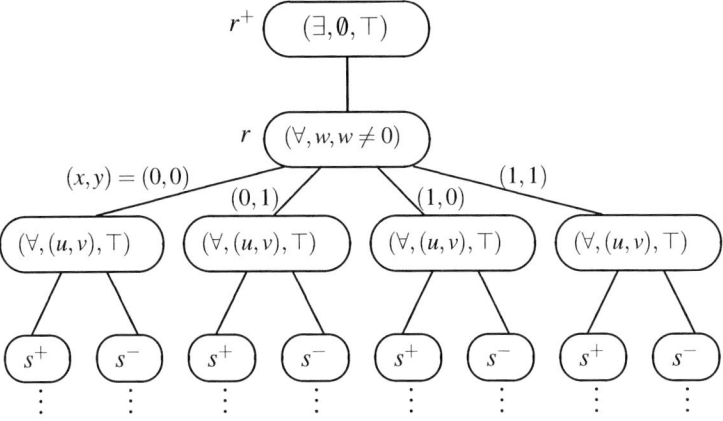

Figure 10.2: The game tree T^+, drawn from the root down

$G(T^+, \mathcal{A})$ — applies only to algebras $\mathcal{A}$ that are known to be relation algebras. The game picks out **RRA** within **RA**. This is a common situation.

- The game $G(T_\omega, \mathcal{A}, (w \mapsto a))$ is clearly *approximated* in some sense by the finite-duration games $G(T_n, \mathcal{A}, (w \mapsto a))$ for $n < \omega$. Again, this is common; we will analyse this situation in more depth in the next two sections.

Exercises

1. Describe a game tree corresponding to the game $G_\omega(N, C)$ for n-dimensional cylindric algebras defined in definition 7.26.

10.4 Formulas expressing a winning strategy

So far, all we have done is to give definitions of games. In this section, we begin to put them to work for us. We will construct formulas expressing that $\exists$ has a winning strategy in games played on game trees. In the next section we will use these to generalise the procedure for extraction of axioms for **RRA** and **RCA**$_\alpha$ introduced in chapter 8.

First, a basic lemma that is rather obvious but encapsulates some crucial facts about winning strategies.

LEMMA 10.6 *Let L be a signature, T an L-game tree with root r and $T(r) = (P_r, \bar{x}_r, \phi_r)$, say. Let $\mathcal{A}$ be an L-structure, and let v assign the variables $\bar{x}_r$ to elements of $\mathcal{A}$.*

1. *If $\exists$ has a winning strategy in $G(T, \mathcal{A}, v)$, then she has a winning strategy in $G(T\!\restriction_n, \mathcal{A}, v)$ for all $n \leq \text{ht}(T)$.*

2. *$\exists$ has a winning strategy in $G(T, \mathcal{A}, v)$ iff:*

    - *if r is an $\forall$-node and $\mathcal{A}, v \models \phi_r$, then one of the three conditions below holds,*
    - *if r is an $\exists$-node, then $\mathcal{A}, v \models \phi_r$ and one of the three conditions below holds,*

   where the three conditions are:

   (a) *r is a leaf of T,*

   (b) *r has immediate successors belonging to $\forall$, and for every move (s, v') that $\forall$ can make in the opening round of $G(T, \mathcal{A}, v)$, $\exists$ has a winning strategy in $G(T_{\geq s}, \mathcal{A}, v')$,*

   (c) *r has immediate successors belonging to $\exists$, and there is a move (s, v') that $\exists$ can make in the opening round of $G(T, \mathcal{A}, v)$ such that she has a winning strategy in $G(T_{\geq s}, \mathcal{A}, v')$.*

Proof. Let $L, T, \mathcal{A}$ be as in the formulation of the lemma. For part 1, suppose that $\exists$ has a winning strategy in $G(T, \mathcal{A}, v)$. Let her use it in the game $G(T\!\restriction_n, \mathcal{A}, v)$, where $n \leq \text{ht}(T)$, for as long as this game lasts. It is clear that she will win it, and hence that she has a winning strategy in it.

Now we prove part 2. Assume first that $\exists$ has a winning strategy in $G(T, \mathcal{A}, v)$. If r is an $\forall$-node and $\mathcal{A}, v \models \phi_r$, then consider a play of $G(T, \mathcal{A}, v)$ in which $\exists$ uses her winning strategy. If r has immediate successors owned by $\forall$, let him begin $G(T, \mathcal{A}, v)$ by making any legal move he likes, to arrive in position (s, v'), say. After his move is made, $\exists$ may proceed with her winning strategy in $G(T, \mathcal{A}, v)$ from the new position (s, v') to win $G(T, \mathcal{A}, v)$. Thus, she has a winning strategy in $G(T_{\geq s}, \mathcal{A}, v')$. If on the other hand r has immediate successors owned by $\exists$, let her use her winning strategy in the opening round of $G(T, \mathcal{A}, v)$. This will cause her to arrive in position (s, v'), say. By continuing with her strategy from this position, she will win $G(T, \mathcal{A}, v)$; so she has a winning strategy in $G(T_{\geq s}, \mathcal{A}, v')$ — namely, 'continue with the winning strategy in progress in $G(T, \mathcal{A}, v)$ from position (s, v')'.

If r is an $\exists$-node, $\exists$ will lose $G(T, \mathcal{A}, v)$ outright if $\mathcal{A}, v \models \neg \phi_r$, contradicting the fact that she has a winning strategy in this game. So $\mathcal{A}, v \models \phi_r$. The rest of the argument is as above.

10.4. Formulas expressing a winning strategy

For the converse, assume the proposed conditions. First suppose that r is an $\forall$-node. If $\mathcal{A}, v \models \neg \phi_r$ then $\exists$ wins $G(T, \mathcal{A}, v)$ outright and so certainly has a winning strategy in it. If $\mathcal{A}, v \models \phi_r$, then by assumption, one of the three conditions (a)–(c) holds. If r is a leaf of T, then $\exists$ wins $G(T, \mathcal{A}, v)$ at this point, so she has a winning strategy in it. If r has immediate successors owned by $\forall$, then let $\forall$ begin $G(T, \mathcal{A}, v)$ by moving to position (s, v'), say. By condition (b), $\exists$ has a winning strategy in $G(T_{\geq s}, \mathcal{A}, v')$. If she uses this strategy in the remainder of $G(T, \mathcal{A}, v)$, she will win it. Thus she has a winning strategy in $G(T_{\geq s}, \mathcal{A}, v')$, namely, 'wait for $\forall$'s move and then pick up a winning strategy in the resulting position'.

If r has immediate successors owned by $\exists$, let her make a move according to condition (c), to position (s, v'), such that she has a winning strategy in the game $G(T_{\geq s}, \mathcal{A}, v')$ starting from that position. She may pick up such a winning strategy and continue with it in the main game $G(T, \mathcal{A}, v)$. This will win the main game for her. So she has a winning strategy in $G(T, \mathcal{A}, v)$, namely, 'move in the opening round to a position (s, v') such that I have a winning strategy in $G(T_{\geq s}, \mathcal{A}, v')$, and then use this winning strategy in the rest of the main game $G(T, \mathcal{A}, v)$'.

Second, assume that r is $\exists$'s node. By assumption, $\mathcal{A}, v \models \phi_r$, so $\exists$ does not lose $G(T, \mathcal{A}, v)$ outright, before any play. By further assumption, one of the three conditions (a)–(c) holds; the rest of the argument is as for when r is an $\forall$-node.

Combining all these cases yields a 'global' winning strategy for $\exists$ in $G(T, \mathcal{A}, v)$ — as required. □

DEFINITION 10.7 Let L be a signature and T a finite L-game tree with root r, and write $T(t) = (P_t, \bar{x}_t, \phi_t)$ for all $t \in T$.

1. For each node $t \in T$ we define formulas $\text{form}_\leq(t)$ and $\text{form}_<(t)$, whose free variables are among those in $(\bar{x}_u : u \leq t)$, by induction on the well-founded order '>' as follows. ('>' is well-founded as T is finite.) Let $t \in T$ and assume inductively that $\text{form}_\leq(s), \text{form}_<(s)$ have been defined for all $s > t$. We let

$$\text{form}_<(t) = \begin{cases} \top, & \text{if } t \text{ is a leaf;} \\ \bigwedge_{\text{Suc}(t,s)} \forall \bar{x}_s \, \text{form}_\leq(s), & \text{if } t \text{ has immediate successors belonging to } \forall; \\ \bigvee_{\text{Suc}(t,s)} \exists \bar{x}_s \, \text{form}_\leq(s), & \text{if } t \text{ has immediate successors belonging to } \exists. \end{cases}$$

$$\text{form}_\leq(t) = \begin{cases} \phi_t \to \text{form}_<(t), & \text{if } t \text{ is an } \forall\text{-node;} \\ \phi_t \wedge \text{form}_<(t), & \text{if } t \text{ is an } \exists\text{-node.} \end{cases}$$

2. We now define θ_T to be the formula $\text{form}_\leq(r)$, with free variables among $\bar{x}_r$. For the empty tree T (see definition 10.3(1c)), we define $\theta_T = \top$.

Clearly, the formulas form$_\leq(t)$ and θ_T are constructible effectively from (a description of) T. If T is closed, θ_T is a sentence.

LEMMA 10.8 *Let L be a signature and T a finite L-game tree with root r. Let $T(t) = (P_t, \bar{x}_t, \phi_t)$, for $t \in T$. Then for any L-structure $\mathcal{A}$ and assignment v_r of the variables $\bar{x}_r$ to elements of $\mathcal{A}$, we have $\mathcal{A}, v_r \models \theta_T(\bar{x}_r)$ if and only if $\exists$ has a winning strategy in the game $G(T, \mathcal{A}, v_r)$.*

Proof. If T is empty, the result is trivial. Assume not; let $t \in T$, and let v_t be an assignment of the variables in $(\bar{x}_u : u \leq t)$ to elements of $\mathcal{A}$. We prove by induction over '$>$' that

- $\mathcal{A}, v_t \models$ form$_\leq(t)$ iff $\exists$ has a winning strategy in $G(T_{\geq t}, \mathcal{A}, v_t)$.

Note that v_t is defined on all free variables of form$_\leq(t)$. Suppose inductively that we have proved the equivalence for all nodes $s > t$ in T. We prove it for t. By the definition of form$_<(t)$ and the inductive hypothesis, the following hold:

(a') If t is a leaf of T, then $\mathcal{A}, v_t \models$ form$_<(t)$.

(b') If t has immediate successors belonging to $\forall$, then we have $\mathcal{A}, v_t \models$ form$_<(t)$ iff $\mathcal{A}, v_t \models \bigwedge_{\text{Suc}(t,s)} \forall \bar{x}_s$ form$_\leq(s)$, iff for every move (s, v') that $\forall$ makes in the opening round of $G(T_{\geq t}, \mathcal{A}, v_t)$, $\exists$ has a winning strategy in $G(T_{\geq s}, \mathcal{A}, v')$.

(c') If t has immediate successors belonging to $\exists$, then we have $\mathcal{A}, v_t \models$ form$_<(t)$ iff $\mathcal{A}, v_t \models \bigvee_{\text{Suc}(t,s)} \exists \bar{x}_s$ form$_\leq(s)$, iff there is a move (s, v') that $\exists$ can make in the opening round of the game $G(T_{\geq t}, \mathcal{A}, v_t)$ to give her a winning strategy in $G(T_{\geq s}, \mathcal{A}, v')$.

So we see that $\mathcal{A}, v_t \models$ form$_<(t)$ iff one of the conditions (a)–(c) of lemma 10.6(2) holds for the game $G(T_{\geq t}, \mathcal{A}, v_t)$. This lemma now yields that $\exists$ has a winning strategy in $G(T_{\geq t}, \mathcal{A}, v_t)$ iff:

1. if t is an $\forall$-node and $\mathcal{A}, v_t \models \phi_t$, then $\mathcal{A}, v_t \models$ form$_<(t)$, and

2. if t is an $\exists$-node, then $\mathcal{A}, v_t \models \phi_t$ and $\mathcal{A}, v_t \models$ form$_<(t)$.

So by definition of form$_\leq(t)$, $\exists$ has a winning strategy in $G(T_{\geq t}, \mathcal{A}, v_t)$ iff $\mathcal{A}, v_t \models$ form$_\leq(t)$. This completes the induction.

Taking $t = r$, we obtain that $\mathcal{A}, v_r \models \theta_T(\bar{x}_r)$ if and only if $\exists$ has a winning strategy in $G(T, \mathcal{A}, v_r)$. □

DEFINITION 10.9 Let L be a signature, and T a closed L-game tree. We let Θ_T be the first-order L-theory $\{\theta_{T \restriction n} : n < \omega\}$ (see definition 10.7).

THEOREM 10.10 *Let L be a signature, T a closed L-game tree, and $\mathcal{A}$ any L-structure. Then $\mathcal{A} \models \Theta_T$ iff $\exists$ has a winning strategy in $G(T \restriction_n, \mathcal{A})$ for all $n < \omega$.*

Proof. Immediate from lemma 10.8. □

We end this section with a useful definition.

DEFINITION 10.11 Closed L-game trees T, U are said to be *equivalent* if for any L-structure $\mathcal{A}$, $\exists$ has a winning strategy in $G(T, \mathcal{A})$ iff she has a winning strategy in $G(U, \mathcal{A})$.

Exercises

1. A finite game tree T is said to be *universal* if for any $s \in T$,

   (a) if $T(s) = (\exists, \bar{x}_s, \phi_s)$ then the sequence $\bar{x}_s$ of variables is empty and ϕ_s is universal, and

   (b) if $T(s) = (\forall, \bar{x}_s, \phi_s)$ then ϕ_s is existential.

   Show that for any such tree, θ_T is a universal formula.

   Find conditions on T for θ_T to be existential.

2. Let L be a countable signature and let σ be an L-sentence. Find a closed L-game tree T such that for any L-structure $\mathcal{A}$, $\exists$ has a winning strategy in $G(T, \mathcal{A})$ iff $\mathcal{A} \models \sigma$. Do the same, replacing σ by a first-order L-theory Σ.

3. If T is a recursive L-game tree, show that $\{\theta_{T \upharpoonright n} : n < \omega\}$ is a recursive L-theory.

4. Show that for any game tree T, there exists an equivalent game tree U (definition 10.11) such that for all $u \in U$, with $U(u) = (P_u, \bar{x}_u, \phi_u)$, ϕ_u is atomic.

10.5 Games and non-finite axiomatisability

We will want to use game trees and games played on them to prove results of the form: class K of algebras is not finitely axiomatisable, perhaps over some given larger class K′ ⊇ K. We now explain in outline how to do it. The detailed proofs will follow afterwards.

1. The first step in the process is to find an (infinite) closed game tree T that is 'sound for K over K′' in the sense that for any countable algebra $\mathcal{A} \in$ K′, if $\exists$ has a winning strategy in $G(T, \mathcal{A})$ then $\mathcal{A} \in$ K. This will involve designing a specific 'sound' game, and then showing (usually easily) that its rules can be written as a closed game tree, say T. We did this for the case K = **RRA**, K′ = **RA** in chapter 7 and section 10.3.1, and we'll see several more examples later in the book.

2. We then try to find a sequence of algebras $\mathcal{A}_n \in \mathsf{K}' \setminus \mathsf{K}$ ($n < \omega$) such that $\exists$ has a winning strategy in $G(T\restriction n, \mathcal{A}_n)$ for each n. This can be hard to do; part IV of the book discusses some methods.

    Frequently, a converse to (1) holds: if $\mathcal{A} \in \mathsf{K}$ then $\exists$ has a winning strategy in $G(T, \mathcal{A})$. Then, showing that $\mathcal{A}_n \notin \mathsf{K}$ can be done by proving that $\forall$ has a winning strategy in $G(T, \mathcal{A}_n)$.

3. We can then show that K is not finitely axiomatisable over K'. We prove this by constructing some countable algebra $\mathcal{A}_\omega$ which is the 'limit', in two senses, of the $\mathcal{A}_n$. First, $\exists$ should have a winning strategy in the full-strength game $G(T, \mathcal{A}_\omega)$. Second, any first-order sentence true in $\mathcal{A}_\omega$ should also be true in infinitely many $\mathcal{A}_n$.

    $\mathcal{A}_\omega$ can be built using ω-saturation or ultraproducts, and then using elementary chains to get countability. The method using saturation is left to the exercises; here, we concentrate on ultraproducts. We consider an ultraproduct $\prod_D \mathcal{A}_n = \prod_{n<\omega} \mathcal{A}_n / D$ of the $\mathcal{A}_n$, where D is a non-principal ultrafilter over ω. $\exists$'s winning strategies in the games $G(T\restriction n, \mathcal{A}_n)$ (each n) can be used to construct a winning strategy for her in the infinite game $G(T, \prod_D \mathcal{A}_n)$ by 'playing along the ultraproduct' — see theorem 10.12. We cannot conclude, though, that $\prod_D \mathcal{A}_n \in \mathsf{K}$ because it may not be countable. So we use elementary chains to obtain a countable elementary subalgebra $\mathcal{A}_\omega \preceq \prod_D \mathcal{A}_n$ such that $\exists$ still has a winning strategy in the game $G(T, \mathcal{A}_\omega)$ (proposition 10.13). It now follows easily that K is not finitely axiomatisable over K'.

We may also wish to show that K is not elementary — it is not the class of models of any first-order theory. For this, it is only necessary to ensure additionally that all the $\mathcal{A}_n$ in (2) above are the same algebra. Again, part IV of the book will give some techniques for doing this.

10.5.1 Ultraproducts and games

Here, we prove that if T is a closed L-game tree, $\mathcal{A}_n$ ($n < \omega$) are L-structures, and $\exists$ has a winning strategy in $G(T\restriction n, \mathcal{A}_n)$ for all $n < \omega$, then she has a winning strategy in $G(T, \mathcal{B})$, where $\mathcal{B}$ is a non-principal ultraproduct of the form $\prod_D \mathcal{A}_n$ for any non-principal ultrafilter D over ω.

THEOREM 10.12 *Let L be a signature and T a closed L-game tree. For each $n < \omega$, let $\mathcal{A}_n$ be an L-structure such that $\exists$ has a winning strategy in the game $G(T\restriction n, \mathcal{A}_n)$. Let $\mathcal{B}$ be any non-principal ultraproduct $\prod_D \mathcal{A}_n$ (where D is a non-principal ultrafilter over ω). Then $\exists$ has a winning strategy in the game $G(T, \mathcal{B})$.*

Proof. Cf. [Hod93, theorem 10.3.7]. For each $n < \omega$, fix a winning strategy σ_n for $\exists$ in $G(T\restriction n, \mathcal{A}_n)$. The underlying idea in the proof is as follows. $\forall$ and $\exists$ will

10.5. Games and non-finite axiomatisability

play $G(T, \mathcal{B})$, and to help her win, $\exists$ will simultaneously play a 'private' game $G(T\restriction n, \mathcal{A}_n)$ for each $n < \omega$, using her winning strategy σ_n. At each stage, she will ensure that the play of $G(T, \mathcal{B})$ so far is approximated by a 'large' (i.e., in the ultrafilter) set of private games. In effect, their ultraproduct will be the play of $G(T, \mathcal{B})$.

In more detail, her approach is to preserve condition (10.2) below. In a play of $G(T, \mathcal{B})$, let the position at the start of some round be (t, v) — recall that $t \in T$ is the current node and v is the current assignment to the variables used in the path from the root to t. For each of the finitely many variables $x \in \mathrm{dom}(v)$, pick an element $\alpha_x \in \prod_{n<\omega} \mathcal{A}_n$ such that the element α_x/D of $\mathcal{B}$ corresponding to α_x is $v(x)$. For $n < \omega$, let v_n be the variable assignment, with the same domain as v, defined by $v_n(x) = \alpha_x(n) \in \mathcal{A}_n$ for each $x \in \mathrm{dom}(v)$. Then $\exists$ attempts to preserve the condition that

$$S_{t,v} \stackrel{\mathrm{def}}{=} \left\{ n : \begin{array}{l} \mathrm{ht}_T(t) < n < \omega, \\ \exists \text{ has a winning strategy} \\ \text{in } G((T\restriction n)_{\geq t}, \mathcal{A}_n, v_n) \end{array} \right\} \in D. \qquad (10.2)$$

Observe that (a) by assumption, (10.2) is true at the initial position $(r, \emptyset)$, where r is the root of T; and (b) although the definition of $S_{t,v}$ depends on the choice of representatives α_x of the equivalence classes $v(x) \in \mathcal{B}$, and hence on the choice of v_n, (10.2) holds independently of this choice.

Assume inductively that (10.2) holds for (t, v) and some choice of v_n. We show that $\exists$ can preserve it to the next round of $G(T, \mathcal{B})$ if there is one. For $t \in T$, we let $T(t) = (P_t, \bar{x}_t, \phi_t)$, as usual.

The round is played by first evaluating whether $\mathcal{B}, v \models \phi_t$ holds. If not, the game is over and there is nothing to show. So assume that $\mathcal{B}, v \models \phi_t$. By Łoś's theorem, there is $\Phi \in D$ such that $\mathcal{A}_n, v_n \models \phi_t$ for every $n \in \Phi$. If t is a leaf of T then again, the game is over, so we may assume not. There are now two cases.

$\forall$ is to move. First, suppose that t has immediate successors in T belonging to $\forall$, so that $\forall$ is to move in $G(T, \mathcal{B})$. He does so by picking an immediate successor t' of t in T and an assignment v' into $\mathcal{B}$ that agrees with v except perhaps on $\bar{x}_{t'}$. For each $x \in \mathrm{rng}(\bar{x}_{t'})$, we choose a representative $\alpha'_x \in \prod_{n<\omega} \mathcal{A}_n$ such that $v'(x) = \alpha'_x/D$. For each $n < \omega$, let v'_n be the same as v_n except that $v'_n(x) = \alpha'_x(n)$ for every $x \in \mathrm{rng}(\bar{x}_{t'})$.

We now take arbitrary $n \in S_{t,v} \cap \Phi$ with $n > \mathrm{ht}(t')$. Since D is non-principal, there are (infinitely many) such n. Then $\forall$ can legally play (t', v'_n) from the initial position (t, v_n) in $G((T\restriction n)_{\geq t}, \mathcal{A}_n, v_n)$. For, since $n > \mathrm{ht}(t')$, it is clear that $t' \in (T\restriction n)_{\geq t}$. Since $n \in \Phi$, we have $\mathcal{A}_n, v_n \models \phi_t$, so this game does not end in its initial position without any rounds being played. By definition, v'_n agrees with v_n except perhaps on $\bar{x}_{t'}$.

So let $\forall$ play (t', v'_n) in the first round of $G((T\restriction n)_{\geq t}, \mathcal{A}_n, v_n)$. By (10.2), $\exists$ has a winning strategy in this game, so she must still have a winning strategy from the new position (cf. lemma 10.6). That is, she has a winning strategy in $G((T\restriction n)_{\geq t'}, \mathcal{A}_n, v'_n)$. We have $v'(x) = \langle v'_n(x) : n < \omega \rangle / D$ for all $x \in \mathrm{dom}(v')$, so we may define the set $S_{t',v'}$ using the v'_n. With this definition, we see that $n \in S_{t',v'}$.

We have shown that $(S_{t,v} \cap \Phi) \setminus \{\mathrm{ht}(t')\} \subseteq S_{t',v'}$. Since the left-hand set is certainly in D, (10.2) is re-established.

∃ is to move Alternatively, suppose that t has immediate successors in T belonging to ∃. As before, for each $n \in S_{t,v} \cap \Phi$, $n > \mathrm{ht}(t')$, ∃ is to make a move in the initial position in $G((T\restriction n)_{\geq t}, \mathcal{A}_n, v_n)$. By (10.2), she has a winning strategy in this game. Let her use it to move from the initial position (t, v_n), and let the resulting position be (t'_n, v'_n). As T is finitely branching, there are only finitely many immediate successors of t in T. So as D is an ultrafilter, there is an immediate successor t' of t such that

$$S' = \{n \in S_{t,v} \cap \Phi : n > \mathrm{ht}(t'),\ t'_n = t'\} \in D.$$

We now show ∃ how to respond to ∀'s move in the main game $G(T, \mathcal{B})$. She will pick the immediate successor t' of t, but she must also select a new assignment agreeing with v except perhaps on $\bar{x}_{t'}$. To do this, let X be the domain of the assignments v'_n for any $n \in S'$ (X does not depend on the choice of n). For $n \in \omega \setminus S'$, let v'_n be an arbitrary assignment of the variables in X to elements of $\mathcal{A}_n$. For each $x \in X$ define $\alpha'_x \in \prod_{n<\omega} \mathcal{A}_n$ by: for each $n < \omega$, $\alpha'_x(n) = v'_n(x)$. We may now define an assignment $v' : X \to \mathcal{B}$ by $v'(x) = \alpha'_x / D$, for $x \in X$. Because $S' \in D$, v' as so defined does not depend on the v'_n for $n \notin S'$. Moreover, for any variable $x \in X \setminus \mathrm{rng}(\bar{x}_s)$, we have $\alpha'_x(n) = v'_n(x) = v_n(x) = \alpha_x(n)$ for all $n \in S'$. As $S' \in D$, we have $v'(x) = \alpha'_x / D = \alpha_x / D = v(x)$. So v' agrees with v on all variables not in $\mathrm{rng}(\bar{x}_s)$.

Thus, in the main game $G(T, \mathcal{B})$, it is legal for ∃ to play t' and the assignment v', so taking her to the new position (t', v'). We let her do so, and check that (10.2) still holds. Because $S' \in D$, it suffices to show that with the given choice of the v_n we have $S' \subseteq S_{t',v'}$. If $n \in S'$, then certainly $n > \mathrm{ht}(t')$. Moreover, because ∃ used a winning strategy in $G((T\restriction n)_{\geq t}, \mathcal{A}_n, v_n)$ to make the move to $(t'_n, v'_n) = (t', v'_n)$, she has a winning strategy in $G((T\restriction n)_{\geq t'}, \mathcal{A}_n, v'_n)$ (cf. lemma 10.6). Thus, $n \in S_{t',v'}$, as required.

We have shown that (10.2) holds initially, and that if it holds in a given position then ∃ can play one round of $G(T, \mathcal{B})$ and preserve it. It only remains to check that while (10.2) holds, ∃ has not lost the game, and so she eventually wins. We note that ∃ can only lose a play of $G(T, \mathcal{B})$ in a position (t, v) where $T(t) = (\exists, \bar{x}_t, \phi_t)$ and $\mathcal{B}, v \models \neg \phi_t$. If this happens, then choose assignments v_n as above. By Łoś's theorem, $\Phi = \{n < \omega : \mathcal{A}_n, v_n \models \neg \phi_t\} \in D$. By (10.2), $S_{t,v} \in D$, too, so there is $n \in \Phi \cap S_{t,v}$. Since ∃ has a winning strategy in $G((T\restriction n)_{\geq t}, \mathcal{A}_n, v_n)$, we must have $\mathcal{A}_n, v_n \models \phi_t$. This is a contradiction, and we conclude that ∃ never loses. □

10.5. Games and non-finite axiomatisability

10.5.2 Countable, elementary subalgebra

Theorem 10.12 gave a winning strategy for $\exists$ in the infinite game $G(T,\mathcal{B})$, under certain conditions. Typically these games will be designed so that, for countable structures $\mathcal{A}$, a winning strategy for $\exists$ in $G(T,\mathcal{A})$ implies that $\mathcal{A} \in \mathsf{K}$, for some class of structures K. The remaining problem is that the ultraproduct $\mathcal{B}$ of theorem 10.12 is likely to be uncountable. No problem, though, we can fix that.

PROPOSITION 10.13 *Let L be a countable signature and T be a closed L-game tree. Suppose that $\mathcal{B}$ is an L-structure, and that $\exists$ has a winning strategy in $G(T,\mathcal{B})$. Then there is a countable elementary substructure $\mathcal{C} \preceq \mathcal{B}$ such that $\exists$ has a winning strategy in $G(T,\mathcal{C})$.*

Proof. First we fix some conventions for the proof. For $t \in T$, let $T(t) = (P_t, \bar{x}_t, \phi_t)$ as usual. Because $\exists$ wins $G(T,\mathcal{B})$ if she never loses at any finite stage, we can assume without loss of generality that the moves recommended to $\exists$ by her winning strategy in $G(T,\mathcal{B})$ in a given position depend only on that position and not on how it was arrived at (the 'history'). We can also assume that her strategy is *deterministic,* in that it always gives $\exists$ a unique move to make in any position. See proposition 10.5 for more details. These assumptions are made for simplicity, and are not essential for the proof below.

We define an elementary chain $C_0 \preceq C_1 \preceq \cdots \preceq \mathcal{B}$ of countable elementary substructures of $\mathcal{B}$, by induction. We let C_0 be any countable elementary substructure of $\mathcal{B}$; by the downward Löwenheim–Skolem–Tarski theorem [Hod93, corollary 3.1.5], [ChaKei90, theorem 3.1.6], we can find such a C_0. If $n < \omega$ and C_n has been defined, we consider plays of $G(T,\mathcal{B})$ in which $\forall$ only assigns variables to elements of C_n and $\exists$ uses her winning strategy. Let X be the set of all elements of $\mathcal{B}$ chosen by $\exists$ as assignments to variables during such plays. It is easily seen that X is countable. We let C_{n+1} be a countable elementary substructure of $\mathcal{B}$ containing C_n and X; again, existence follows from the downward Löwenheim–Skolem–Tarski theorem.

To see that X is countable in a more formal way, write $\mathcal{P}$ for the set of all positions in $G(T,\mathcal{B})$, as in section 10.2. Define $\mathcal{P}(m) \subseteq \mathcal{P}$, for $m < \omega$, to be the set of all possible positions at the start of the mth round of plays of $G(T,\mathcal{B})$ in which $\forall$ only chooses elements of C_n and $\exists$ uses her winning strategy. We may define $\mathcal{P}(m)$ formally by induction on $m < \omega$. Clearly, $\mathcal{P}(0) = \{(t_0, \emptyset)\}$, where t_0 is the root of T. If $\mathcal{P}(m)$ is defined, then $\mathcal{P}(m+1)$ is evidently the smallest subset of $\mathcal{P}$ such that for every $(t,v) \in \mathcal{P}(m)$,

- if t is the immediate predecessor of an $\forall$-node in T, t' is an immediate successor in T of t, v' is an assignment of the variables $\bar{x}_u$ ($u \leq t'$) to elements of $\mathcal{B}$, v' agrees with v except perhaps on $\bar{x}_{t'}$, and $v'(x) \in C_n$ for every $x \in \mathrm{rng}(\bar{x}_{t'})$, then $(t', v') \in \mathcal{P}(m+1)$,

- if t is the immediate predecessor in T of an $\exists$-node, and $\exists$'s deterministic winning strategy in $G(T, \mathcal{B})$ directs her in position (t, v) to move to position (t', v'), then $(t', v') \in \mathcal{P}(m+1)$.

A simple induction shows that each $\mathcal{P}(m)$ is countable. But plainly,

$$X = \{v(x) : (t, v) \in \bigcup_{m<\omega} \mathcal{P}(m), t \text{ an } \exists\text{-node}, x \in \text{rng}(\bar{x}_t)\},$$

so X is also countable.

Now define $C = \bigcup_{n<\omega} C_n$. By the elementary chain theorem (cf. [ChaKei90, 3.1.9], [Hod93, 2.5.2], exercise 10 below), $C \preceq \mathcal{B}$, and C is countable.

We claim that $\exists$ has a winning strategy in $G(T, C)$. To prove the claim, we show that $\exists$'s winning strategy in $G(T, \mathcal{B})$ restricts to a winning strategy in the game $G(T, C)$. More formally, $\exists$ can play $G(T, C)$ in such a way that if the successive positions taken up are $(t_0, v_0), (t_1, v_1), \ldots$, then $(t_0, v_0), (t_1, v_1), \ldots$ is also a play of $G(T, \mathcal{B})$ in which $\exists$ uses her winning strategy. It will follow that $\exists$ does not lose $G(T, C)$ at any finite stage. For, this could only happen if she arrived in a position (t_n, v_n) such that she owns t_n and $C, v_n \models \neg \phi_{t_n}$. But $C \preceq \mathcal{B}$, so that $\mathcal{B}, v_n \models \neg \phi_{t_n}$, too. This means that $\exists$ loses the play $(t_0, v_0), (t_1, v_1), \ldots, (t_n, v_n)$ of $G(T, \mathcal{B})$, contradicting the fact that she is using a winning strategy in this game. Hence, she wins $G(T, C)$. So she has a winning strategy in $G(T, C)$ — namely, to play in this way.

So assume inductively that the play of $G(T, C)$ so far is $(t_0, v_0), (t_1, v_1), \ldots, (t, v_t)$, for some $t \in T$, and that this is the initial part of a play of $G(T, \mathcal{B})$ in which $\exists$ is using her winning strategy. This is true by the original assumption in the initial position (t_0, v_0). We show how to preserve it for one more move.

If t is a leaf or $C, v_t \models \neg \phi_t$, then the game $G(T, C)$ is over and we are done. Assume not. Then as $C, v_t \models \phi_t$ and $C \preceq \mathcal{B}$, we have $\mathcal{B}, v_t \models \phi_t$, too. So $G(T, \mathcal{B})$ is not over, either, and $\exists$'s strategy in it continues in operation. If t is the immediate predecessor of an $\forall$-node, then $\forall$ will move in $G(T, C)$ to a position (s, v_s), say. Thus, the play so far becomes $(t_0, v_0), \ldots, (t, v_t), (s, v_s)$ and this is still the initial part of a play of $G(T, \mathcal{B})$ in which $\exists$ is using her winning strategy.

Assume now that t is the immediate predecessor of an $\exists$-node. Inductively, $(t_0, v_0), \ldots, (t, v_t)$ is the initial part of a play of $G(T, \mathcal{B})$ in which $\exists$ is using her winning strategy; and it is now her move. Let her apply her strategy, and suppose that it advises her to move to position (s, v_s). Since $\forall$ has only made finitely many assignments thus far during the play, all being to elements of C, there is $n < \omega$ such that all of his assignments have been to elements of C_n. By construction of the C_n, v_s assigns variables to elements of C_{n+1}, and therefore to elements of C. So she can legally continue $G(T, C)$ by moving to position (s, v_s). The play so far will then be $(t_0, v_0), \ldots, (t, v_t), (s, v_s)$ — still the initial part of a play of $G(T, \mathcal{B})$ in which $\exists$ is using her winning strategy.

10.5. Games and non-finite axiomatisability

The claim is now proved, and with it, the proposition. □

10.5.3 Non-finite axiomatisability

We will be applying the above two results — theorem 10.12 and proposition 10.13 — to prove that various classes of algebras are not finitely axiomatisable or not elementary. The following theorem shows how.

THEOREM 10.14 *Let L be a countable signature, T a closed L-game tree, and K a class of L-structures. Let U be a first-order L-theory. Assume that if $\mathcal{A}$ is a countable model of U and $\exists$ has a winning strategy in $G(T, \mathcal{A})$ then $\mathcal{A} \in \mathsf{K}$.*

1. *Suppose that K' is a class of L-structures containing K, and that for each $n < \omega$, there is an L-structure $\mathcal{A}_n \in \mathsf{K}' \setminus \mathsf{K}$ with $\mathcal{A}_n \models U$ and such that $\exists$ has a winning strategy in $G(T{\restriction}n, \mathcal{A}_n)$. Then K is not finitely axiomatisable over K' in first-order logic with signature L.*

2. *Let $\mathcal{A} \models U$, and suppose that $\exists$ has a winning strategy in $G(T{\restriction}n, \mathcal{A})$ for every finite n. Then there is $C \in \mathsf{K}$ elementarily equivalent to $\mathcal{A}$. Hence, if $\mathcal{A} \notin \mathsf{K}$ then K is not closed under elementary equivalence, and so is not an elementary class.*

Proof. Let D be any non-principal ultrafilter over ω. Since $\exists$ has a winning strategy in each game $G(T{\restriction}n, \mathcal{A}_n)$ $(n < \omega)$, we know that $\exists$ has a winning strategy in the game $G(T, \prod_D \mathcal{A}_n)$, by theorem 10.12. Łoś' theorem tells us that $\prod_D \mathcal{A}_n \models U$. By proposition 10.13, there is a countable elementary subalgebra $C \preceq \prod_D \mathcal{A}_n$ such that $\exists$ has a winning strategy in $G(T, C)$. Since C is elementarily equivalent to $\prod_D \mathcal{A}_n$, we know that $C \models U$. By the condition in the theorem, $C \in \mathsf{K}$.

Now for the first part, assume for contradiction that K is finitely axiomatisable over K', so that there is a first-order L-sentence σ such that for any L-structure $\mathcal{A}$, $\mathcal{A} \in \mathsf{K}$ iff $\mathcal{A} \in \mathsf{K}'$ and $\mathcal{A} \models \sigma$. Then $C \models \sigma$ and so $\prod_D \mathcal{A}_n \models \sigma$. By Łoś' theorem, $\{n : \mathcal{A}_n \models \sigma\} \in D$. In particular, this set is non-empty. But $\mathcal{A}_n \in \mathsf{K}' \setminus \mathsf{K}$ implies $\mathcal{A}_n \not\models \sigma$ for all $n < \omega$, a contradiction.

For the second part, we have $\mathcal{A} \preceq \prod_D \mathcal{A} \succeq C$ by corollary 2.30, so $\mathcal{A} \equiv C$ and, as above, $C \in \mathsf{K}$. If $\mathcal{A} \notin \mathsf{K}$ then K is not closed under elementary equivalence, so K is not an elementary class. □

Exercises

In the exercises, L is a countable signature and T a closed L-game tree. We say that T defines a class K of L-structures over a larger one, K', if whenever $\mathcal{A} \in \mathsf{K}'$ is countable, $\mathcal{A} \in \mathsf{K}$ iff $\exists$ has a winning strategy in $G(T, \mathcal{A})$.

1. Let $\mathcal{A}$ be an L-structure. Assume any one of the following:

(a) for every ∃-node $t \in T$, $\bar{x}_t$ is the empty sequence of variables

(b) $\mathcal{A}$ is finite

(c) $\mathcal{A}$ is ω-saturated

(d) L, T are recursive and $\mathcal{A}$ is recursively saturated.

Show that the following are equivalent:

(i) ∃ has a winning strategy in $G(T, \mathcal{A})$.

(ii) ∃ has a winning strategy in $G(T\!\restriction\! n, \mathcal{A})$ for every finite n.

2. Suppose that for every ∃-node $t \in T$, $\bar{x}_t$ is the empty sequence of variables. Assume that $\mathsf{K} \subseteq \mathsf{K}'$ are classes of L-structures, K' is axiomatised by Ξ, and T defines K over K'. Show that for any countable L-structure $\mathcal{A}$, $\mathcal{A} \in \mathsf{K}$ iff $\mathcal{A} \models \Xi \cup \Theta_T$.

3. Prove theorem 10.14 without using ultraproducts, using compactness and saturation instead. So in the first part of the theorem suppose, for contradiction, that K is axiomatised by a single formula σ over K'. Use compactness to show that $U \cup \Theta_T \cup \{\neg\sigma\}$ is consistent and then consider an ω-saturated model of this theory. Use proposition 10.13 to obtain a contradiction.

4. Does proposition 10.13 still hold if we replace '∃' by '∀'?

5. For classes $\mathsf{K} \subseteq \mathsf{K}'$ of L-structures, let T, U be L-game trees such that T defines K over K', and U defines $\mathsf{K}' \setminus \mathsf{K}$ over K'. If K' is elementary, show that there is a finite game tree V that defines K over K'. Show that this need not hold if K' is not elementary. [Compactness; finite linear orders.]

6. If K is a class of L-structures that is defined by some closed game tree over the class of all L-structures, must the complement of K be so definable?

7. For classes $\mathsf{K} \subseteq \mathsf{K}' \subseteq \mathsf{K}''$ of L-structures, suppose that there are closed game trees T, T' such that T defines K over K' and T' defines K' over K''. Find a game tree that defines K over K''. Generalise to a descending ω-chain of classes.

8. For each $n < \omega$ let K_n be a class of L-structures and T_n a closed game tree defining K_n over the class of all L-structures. Find closed game trees defining

- $\bigcap_{n<\omega} \mathsf{K}_n$,
- $\bigcup_{m<n} \mathsf{K}_m$, for each $n < \omega$,

over the class of all L-structures.

10.5. Games and non-finite axiomatisability

9. Let T be a closed L-game tree defining some class K, over the class of all L-structures.

    (a) Define the rank of a position in a game $G(T, \mathcal{A})$ (cf. definition 7.4).

    (b) Show that $\exists$ has a winning strategy in $G(T, \mathcal{A})$ iff the initial position has rank ∞.

    (c) Let K have rank α if α is the least ordinal such that for any L-structure $\mathcal{A}$, if the rank of the initial position in $G(T, \mathcal{A})$ is $\geq \alpha$ then $\mathcal{A} \in$ K. (Not every K need have a rank.) Show that if T is universal (see exercise 10.4(1)) then K has rank $\leq \omega$.

10. The elementary chain theorem as in [ChaKei90, 3.1.9], [Hod93, 2.5.2] states that if $C_0 \preceq C_1 \preceq \cdots$ are structures then the structure $C \stackrel{\text{def}}{=} \bigcup_{i<\omega} C_i$ satisfies $C \succeq C_i$ for all $i < \omega$. The application of this in proposition 10.13 requires that if $\mathcal{B}$ is a structure and $C_0 \preceq C_1 \preceq \cdots$ are elementary substructures of $\mathcal{B}$, then $\bigcup_{i<\omega} C_i \preceq \mathcal{B}$. Derive this from the elementary chain theorem.

11. Let K be a class of L-structures. Show the following to be equivalent:

    - there exists a closed L-game tree T that defines K over the class of all L-structures,

    - K has the same countable members as some pseudo-elementary class.

    [For $\Downarrow$, encode a winning strategy for $\exists$ in $G(T,\mathbf{a}$-sort). For example, you might use new $|\bar{x}_t|$-ary relation symbols E_t ($t \in T$) and arrange that for any assignment $v_t : \bar{x}_t \to \mathcal{A}$, (t, v_t) is a winning position for $\exists$ in $G(T, \mathcal{A})$ iff $\mathcal{A}, v_t \models E_t(\bar{x}_t)$. For $\Uparrow$, find a tree version of the game Γ^+ of section 9.6.]

Chapter 11

Atomic networks

11.1 Introduction

So far, we have considered arbitrary relation algebras. In chapter 7, we defined a game that characterised when a relation algebra is representable, and this was used in chapter 8 to construct axioms for representability. Now, after a two-chapter interlude on games, we return to relation algebras and begin to shift our attention to the *atomic* ones, the focus of almost all of the rest of the book. This chapter serves as a gentle introduction to their peculiarities.

Let us recall what we know about atomic relation algebras. Any finite relation algebra is atomic. There are infinite relation algebras that are atomic, but not all of them are, and indeed, there exist relation algebras with no atoms at all. We have seen (definition 3.21) that the atoms of an atomic relation algebra can be endowed with intrinsic structure induced from the algebra; we wrote down axioms that were satisfied by the resulting 'atom structure' (see definition 3.22), and showed conversely that any atomic L_{RA}-BAO whose atom structure satisfies these axioms is an atomic relation algebra (lemma 3.24). For finite atom structures, this relation algebra is unique (up to isomorphism over the atom structure). Many of the examples of relation algebras in chapter 4 were presented via their atom structures.

Ordinary representations This is all possible because **RA** is a Sahlqvist variety and so is well represented by its atom structures. But we are also interested in **RRA**. It would be attractive to reduce the study of atomic representable relation algebras entirely to their atom structures, simply because they are so much easier to work with (after all, a relation algebra with n atoms has all of 2^n elements). We already discussed this approach for boolean algebras with operators in section 2.7.5.

As a first test question, we ask:

1. Are there (first-order) conditions on the atom structure of an atomic relation algebra equivalent to the representability of the relation algebra?

For finite algebras, this question has a positive answer, by general BAO considerations: a finite relation algebra is representable iff its atom structure is in At**RRA**, which by theorem 2.84 is an elementary class. But in general, there is more than one relation algebra with a given atom structure. In chapter 14 and again in section 17.7, we will construct two atomic relation algebras with the same atom structure, only one of them being representable! This will have the consequence that **RRA** is not closed under completions and is not Sahlqvist-axiomatisable. In terms of section 2.7.5, it means that Str**RRA** $\subset$ At**RRA**. The answer to question 1 for arbitrary atomic relation algebras is therefore 'no': representability of an atomic relation algebra is not determined by its atom structure.

Atom structures generating RRA? In section 2.7.5, we saw that for a variety V of BAOs, even when StrV $\subset$ AtV, there is still the possibility of finding an elementary class K of atom structures that generates V. This is only possible when V is canonical, but all known canonical varieties of BAOs have such a K, and indeed, one satisfying CstV $\subseteq$ K $\subseteq$ StrV, where CstV = $\{At \mathcal{A}^+ : \mathcal{A} \in V\}$. Moreover, V = **S**$\mathfrak{Cm}$K in this case. Since **RRA** is canonical (theorem 3.36), we should to be able to find such a K for it, which might then serve as a useful counterpart of **RRA** on the atom structure level.

We cannot take K = Str**RRA**: in chapter 14, we will show that Str**RRA** is not elementary. Nor can we take K to be the elementary closure of Str**RRA**: we do not know whether this class is elementary, but even if it is, it generates a strictly larger class than **RRA**.

Complete representations To find a suitable K, it turns out to help if we shift our attention to *complete* representations (or equivalently, atomic representations — we saw in theorem 2.21 that a representation of a boolean algebra is complete iff it is atomic, and this applies to relation algebras too since any representation of a relation algebra is *inter alia* a representation of its underlying boolean algebra). Only atomic algebras can have complete representations. If we wish to work with representations of relation algebras via their atom structures, complete representations are more appropriate than ordinary ones, because they are determined in a simple way by the interpretations of the atoms. So we ask a second question:

2. Are there (first-order) conditions on the atom structure of an atomic relation algebra equivalent to the relation algebra's having a *complete* representation?

11.1. Introduction

Unlike for ordinary representations, whether an (atomic) relation algebra has a complete representation is determined by its atom structure (exercise 3.4(4)). So if we drop the 'first-order' requirement, the answer to question 2 is 'yes'.

But it turns out that complete representations are harder to capture than ordinary ones. We will see in section 17.2 that having a complete representation is not an elementary property of a relation algebra (i.e., no first-order theory characterises it). Consequently, the class of atom structures of completely representable relation algebras is not elementary. So if we do require a first-order characterisation, the answer to question 2 in general is 'no'. It follows from this, and is important to realise, that *not every representable atomic relation algebra has a complete representation* (we will see examples in sections 14.4 and 17.2).

Lyndon conditions Nonetheless, we are able to obtain a characterisation of complete representability for countable atomic relation algebras by infinite-length games ('atomic' versions of the games of chapter 7). We know that such games have finite-length approximations and the existence of a winning strategy for the second player in them is expressible by first-order sentences. We loosely call these sentences the *Lyndon conditions,* in honour of Lyndon's [Lyn50]. This seminal paper gave certain first-order conditions on an atomic relation algebra that amount to characterising when $\exists$ has a winning strategy in certain games, and showed that a finite relation algebra satisfies the conditions if and only if it is representable. For more details, see section 11.3.2. The conditions given in [Lyn50] are equivalent to the ones we obtain from the games.

The games theorems of chapter 10 can now be deployed to show that an atomic relation algebra $\mathcal{A}$ satisfies the Lyndon conditions iff it is elementarily equivalent to a completely representable relation algebra. It follows that if $\mathcal{A}$ has a complete representation then $\mathcal{A}$ satisfies the Lyndon conditions, and if $\mathcal{A}$ satisfies the Lyndon conditions then it is representable. For finite and canonical embedding algebras, all three statements are equivalent; in general, neither implication reverses. Now it will follow from our definition of the game that the Lyndon conditions can equally be viewed as first-order conditions on atom structures. So we see that

1. if $\mathcal{A} \in$ **RRA** then At $\mathcal{A}^+$ satisfies the Lyndon conditions,

2. if A is a relation algebra atom structure that satisfies the Lyndon conditions, then $\mathfrak{Cm} A \in$ **RRA**.

Hence, the class '**LCAS**' of relation algebra atom structures that satisfy the Lyndon conditions is an elementary class with Cst**RRA** $\subseteq$ **LCAS** $\subseteq$ Str**RRA**, as sought above. Further, **RRA** = S $\mathfrak{Cm}$ **LCAS**.

Usefulness of the Lyndon conditions Can we really replace the study of **RRA** by the study of relation algebra atom structures satisfying the Lyndon conditions?

To a large extent, yes. The Lyndon conditions are always sufficient for an atomic relation algebra to be representable, and for finite relation algebras and canonical extensions of relation algebras, they characterise representability. We can always check representability of a relation algebra by seeing if the Lyndon conditions hold in its canonical extension. There are other elementary classes of atom structures that generate **RRA** (see, e.g., exercise 3.4(6)), but not all of them have this property.

The usefulness of the Lyndon conditions is further enhanced by the fact that they have a transparent game-theoretic meaning. In practice, it is the game that is most useful in checking representability: we will use it all the time in parts IV and V of the book. We will also modify the game, for example by imposing a limit on the number of nodes in an atomic network played in the game, in order to characterise new classes of algebras and representations. In chapter 12, we introduce various kinds of 'bases' for atomic relation algebras. Bases were first defined by Maddux [Madd78b, Madd83, Madd89b] essentially as sets of atomic networks with certain closure properties. An atomic relation algebra has a given type of basis iff ∃ has a winning strategy in a certain, modified atomic game. In chapter 13, we use bases to determine when a relation algebra has certain kinds of relativised representations.

But the Lyndon conditions have their limits. They are not always the most appropriate means of studying non-completely representable relation algebras. When we come to consider Str **RRA** more closely, in chapter 14, we will use the game of chapter 7 in preference — though by exercise 14.2(7), the Lyndon conditions could be used instead.

Outline of chapter

This chapter occupies a central place in the book, because we will be working mostly with atomic relation algebras from now on. We will study them using games. We begin in section 11.2 by adapting the games of chapter 7 to work with atomic relation algebras, or their atom structures. In section 11.3, we discuss other views of the 'atomic games', including their equivalence to Lyndon's earlier approach and the corresponding first-order conditions (the Lyndon conditions). In section 11.4, we develop the connections with complete representations. The general game setting of the preceding chapter then applies straightforwardly, and in section 11.5 we read off some facts that will be useful in chapter 17 when we prove non-finite axiomatisability results.

11.2 Atomic networks and games

Let $\mathcal{A}$ be an atomic relation algebra.

11.2. Atomic networks and games

DEFINITION 11.1 An $\mathcal{A}$-network N is said to be *atomic* if $N(x,y)$ is an atom of $\mathcal{A}$ for every $x, y \in N$.

LEMMA 11.2 *Let N be an atomic $\mathcal{A}$-network, for some atomic relation algebra $\mathcal{A}$, and let $x,y,z \in N$. Then*

1. $N(x,y) = N(y,x)^{\smile}$,
2. $N(x,y) \leq (N(x,z) ; N(z,y))$,
3. $N(x,x) = 1' \cdot (N(x,y) ; N(y,x))$.

Proof. Left to the reader (cf. lemma 5.10 and exercise 7.1(5)). □

DEFINITION 11.3 Let N be an atomic $\mathcal{A}$-network. We define a new 'atomic' game, $G_n^a(N, \mathcal{A})$, of length $n \leq \omega$. In it, the players $\forall, \exists$ build a chain $N = N_0 \subseteq N_1 \subseteq \cdots \subseteq N_t \subseteq N_{t+1} \subseteq \cdots$ $(t < n)$ of atomic $\mathcal{A}$-networks. In the tth round, for $t < n$, if N_t was the last network to be constructed,

- $\forall$ chooses $x, y \in N_t$ and atoms $a, b \in \mathcal{A}$ such that $a ; b \geq N_t(x,y)$.

- $\exists$ responds, if she can, with an atomic network $N_{t+1} \supseteq N_t$ such that there is $z \in N_{t+1}$ with $N_{t+1}(x,z) = a$ and $N_{t+1}(z,y) = b$. We may assume (without affecting who may have a winning strategy) that N_{t+1} has at most one extra node beyond those in N_t.

$\exists$ wins the play if she never gets stuck — if she can always find a suitable network to respond with. So roughly, in each round $\forall$ demands a certain 'triangle' (at least if $x \neq y$) to be added to the network and $\exists$ must comply, or lose.

The game $G_n^a(\mathcal{A})$ is a slight variation of this. In an extra, unnumbered preliminary round, $\forall$ chooses an atom of $\mathcal{A}$, and $\exists$ responds with an atomic network N with a among its labels (i.e., N contains nodes x,y with $N(x,y) = a$). We assume without loss of generality that the nodes of N are just x and y. The game then continues as $G_n^a(N, \mathcal{A})$.

It is evident that the rules of the game $G_n^a(\mathcal{A})$ depend only on the atom structure At $\mathcal{A}$ of $\mathcal{A}$ (definition 3.21). Consequently, if $\mathcal{S}$ is a relation algebra atom structure, we may write $G_n^a(\mathcal{S})$ instead of the more accurate $G_n^a(\mathcal{A})$ for some atomic relation algebra $\mathcal{A}$ with At $\mathcal{A} = \mathcal{S}$ (for example, $\mathcal{A} = \mathfrak{Cm}\,\mathcal{S}$).

Exercises

1. Show that the two simplifying assumptions in the definition of G^a are legitimate. That is, show that

(a) ∃ has a winning strategy in a version of $G_n^a(N, \mathcal{A})$ where there is no restriction on how many nodes her networks N_{t+1} have, iff she has a winning strategy in the official game where N_{t+1} has at most one more node than N_t (for each $t < n$),

(b) ∃ has a winning strategy in a version of $G_n^a(\mathcal{A})$ where her response to ∀'s choice of atom a in the preliminary round can be any atomic network N such that $N(x,y) = a$ for some $x, y \in N$, iff she has a winning strategy in the official game $G_n^a(\mathcal{A})$ where the nodes of N are precisely $\{x, y\}$.

11.3 Alternative views of the game

Three other views of the game G^a, relating it to other work, are of interest here.

11.3.1 Relation to the game G_n of chapter 7

The reader may be recalling the two kinds of move (accepting and rejecting) that ∃ could make in the game G_n of definition 7.12 that characterised **RRA**, and wondering why here there is only one. In fact, there is no real difference between the two games in this regard. The game G^a is essentially the earlier game, except that we require ∀ always to choose atoms of the algebra and ∃ always to play atomic networks. Imagine such a game. In some round, where the current (atomic) network is N, say, suppose that ∀ moves by choosing nodes x, y of N and atoms a, b of the algebra $\mathcal{A}$. ∃ must respond with an atomic network $N' \supseteq N$ with the property that $N'(x, y)$ is either $N(x, y) \cdot (a; b)$ ('accept') or $N(x, y) \cdot -(a; b)$ ('reject'). Since $N(x, y)$ is an atom, one of these two elements of $\mathcal{A}$ is zero and ∃ would lose immediately if she played a network with such a label. So whether she should accept or reject is determined by ∀'s move — she has no choice. But if she rejects, because $N(x, y) \cdot (a; b) = 0$, then $N(x, y) \cdot -(a; b) = N(x, y)$, so it is legal for her to respond to ∀'s move with simply the old network, N. Since ∃ has such a straightforward response — obviously her best response as it commits her to nothing beyond what she is already — we see no point in allowing ∀ to make such a move. So we require that ∀'s move x, y, a, b always satisfies $N(x, y) \leq a; b$. Then ∃ must always accept, so the rejection option is never needed. We arrive at the game G^a of definition 11.3 above.

Nonetheless, requiring that ∀ always chooses atoms and ∃ always plays atomic networks makes a substantial difference. For example, whenever ∃ plays a network properly extending the existing one, she must choose atoms of the algebra to label all but two of the new edges. This means that the first-order sentences expressing that ∃ has a winning strategy will no longer be universal.

11.3.2 Lyndon conditions

The game G^a is easily seen to be equivalent to an older, celebrated characterisation of the finite representable relation algebras, due to Lyndon. In [Lyn50], he gave an infinite (recursive) set of first-order sentences in the language of relation algebras, which we will call the *Lyndon conditions*. Here they are, with very minor changes (see exercise 4 below). First, some notation. We use variables of the form x_{ij} for distinct $i, j < \omega$. Roughly, x_{ij} corresponds to the value of a pre-network label $N(i,j)$. For $a, b, n < \omega$, we define the formula

$$v^n_{a,b} = \bigwedge_{i \in n \setminus \{a,b\}} \left(x_{in} \leq \left((x_{ia} ; x_{an}) \cdot (x_{ib} ; x_{bn}) \cdot \prod_{j < i,\, j \neq a,b} x_{ij} ; x_{jn} \right) \right).$$

We now define the formulas $\lambda^n_{\alpha,\beta}$ that will be used to form the Lyndon conditions. α, β denote elements of $^k\omega$, for some $k < \omega$, and the definition is by induction on k. We let ε denote the unique empty sequence in $^0\omega$; for a finite sequence $\alpha = (\alpha_0, \ldots, \alpha_{k-1}) \in {}^k\omega$ and $a < \omega$, we let $a\alpha$ denote the sequence $(a, \alpha_0, \ldots, \alpha_{k-1}) \in {}^{k+1}\omega$. All quantification in the $\lambda^n_{\alpha,\beta}$ is implicitly relativised to atoms of a relation algebra, and for any variable x_{ij} introduced by a quantifier, another variable x_{ji} is also implicitly introduced, with value $\check{x}_{ij}$. Thus, for example, $\exists x_{23} \varphi$ is really $\exists x_{23} x_{32} (\text{atom}(x_{23}) \wedge x_{32} = \check{x}_{23} \wedge \varphi)$, where $\text{atom}(x)$ is $\forall y (y < x \leftrightarrow y = 0)$ as we have seen before. The $\lambda^n_{\alpha,\beta}$ are defined as follows:

$$\lambda^n_{\varepsilon,\varepsilon} = \top,$$
$$\lambda^n_{a\alpha,b\beta} = \forall x_{an} x_{nb} \left(x_{ab} \leq x_{an} ; x_{nb} \rightarrow \exists_{i \in n \setminus \{a,b\}} x_{in} \left(v^n_{a,b} \wedge \lambda^{n+1}_{\alpha,\beta} \right) \right).$$

For $k < \omega$, let $\Sigma_k = \{\alpha \in {}^k\omega : \alpha_i \leq i+2 \text{ for all } i < k\}$. We can now define the set LC of Lyndon conditions:

$$\text{LC} = \left\{ \forall x_{01} \lambda^2_{0\alpha, 1\beta}(x_{01}) \; : \; (\alpha, \beta) \in \bigcup_{k < \omega} \Sigma_k \times \Sigma_k \right\}.$$

We let **LC** denote the class of all atomic relation algebras $\mathcal{A}$ such that $\mathcal{A} \models \text{LC}$.

Game view of Lyndon conditions The Lyndon conditions have a natural game-theoretic interpretation: they correspond to a slightly modified form of the game $G^a_k(\mathcal{A})$. There are three changes.

1. In the preliminary round, $\exists$ must respond to $\forall$'s choice of atom with an atomic network N_0 over $\mathcal{A}$ with precisely two nodes: 0, 1, say. In round 0 of the main game, $\forall$ must pick the nodes 0 and 1 of N_0, in that order.

2. From then on, play proceeds similarly to $G_k^a(\mathcal{A})$, except that each successive network N_{t+1} of the play is an atomic $\mathcal{A}$-network with *exactly* one more node than the preceding N_t. $\exists$ must always add a new node, even if a suitable node already exists in N_t. The nodes of N_t can therefore be taken to be $\{0, 1, \ldots, t+1\}$.

3. The last and most interesting modification is that $\forall$ must declare his choice of edges in advance, before play starts. (There is no such restriction on his choice of atoms.) He can do this because the preceding alterations mean that the nodes and edges of each network produced during play (though not their labels) are known in advance.

Each Lyndon condition

$$\forall x_{01} \, \lambda^2_{0\alpha, 1\beta}(x_{01}) \;=\; \forall x_{01} x_{02} x_{21}(x_{01} \leq x_{02}; x_{21} \to \lambda^3_{\alpha,\beta}),$$

where $\alpha, \beta \in \Sigma_k$, expresses that $\exists$ has a winning strategy in the $(k+1)$-round game described above, when $\forall$ promises to choose the edge (α_t, β_t) of N_t in round $t+1$ of the modified version of $G_k^a(N_0, \mathcal{A})$. This is easily seen: for example, the combined effect of the subformulas $v^n_{a,b}$ is to express that each N_t is a network, and reflexive labels $N_t(i,i)$ can be inferred uniquely from the irreflexive labels using lemma 5.10. Therefore, the Lyndon conditions *en masse* express that $\exists$ has a winning strategy in every such game (of each finite length).

This modified game is sufficiently close to the original G^a to let us prove:

THEOREM 11.4 *Let $\mathcal{A}$ be any atomic relation algebra. The following are equivalent.*

1. *$\mathcal{A}$ satisfies the Lyndon conditions.*

2. *$\exists$ has a winning strategy in the games $G_n^a(\mathcal{A})$ for all $n < \omega$.*

Proof. Since the modifications mainly serve to restrict $\forall$, it is clear that if $\exists$ can win all $G_n^a(\mathcal{A})$ then she can win the modified games, so that $\mathcal{A}$ satisfies the Lyndon conditions. Thus, $2 \Rightarrow 1$.

$1 \Rightarrow 2$ is shown by straightforward game-theoretic arguments. For example, let N be an atomic network with three nodes. Suppose that $\forall$ declares that he will choose his edges in n 'tranches', in the following manner. In the first tranche, $\forall$ will choose all the edges of N, one by one in some arbitrary order. Let N_1 be the network that results at the end of this tranche. In his second tranche, $\forall$ will choose all edges of N_1. This process continues for n tranches altogether. Then if $\exists$ has a winning strategy in this game, she also has a winning strategy in $G_n^a(N, \mathcal{A})$. We will leave the details of the proof as an exercise for the interested reader. □

11.3. Alternative views of the game

Discussion   Although Lyndon only defined the conditions in the context of simple atomic relation algebras, they make sense in any atomic relation algebra. He proved that any finite (simple) relation algebra satisfying the conditions is representable. His proof in [Lyn50] works by building a representation in a step-by-step fashion and has a strong game-theoretic flavour, especially p. 711 and the 'graphical interpretation' on p. 712, though games are not explicitly mentioned. Lyndon then gave an example of a finite relation algebra, with 56 atoms, that did not satisfy the Lyndon conditions and was therefore not representable. This was the first example of a non-representable relation algebra.

[Lyn50] mistakenly claimed that the Lyndon conditions held in any representable (simple, atomic) relation algebra. The error lay in assuming that every representation of an atomic relation algebra must be atomic. What his argument does show, as he pointed out in [Lyn56], is that any completely representable[1] atomic relation algebra satisfies the Lyndon conditions. Since any finite representable relation algebra is completely representable, this means that the Lyndon conditions hold in a finite relation algebra iff it is representable. Thus, the Lyndon conditions characterise the representable relation algebras within the class of finite relation algebras.

[Lyn56] also gave the first correct axiomatisation of **RRA**, by an approach that is in some ways similar to the one given in chapter 8. See exercise 7.6(6) for a corresponding game.

11.3.3   Game tree view

Our third view of the games G^a (definition 11.3) is by game trees, as in chapter 10. We can easily construct a game tree (see definition 10.2) for the G^a. We let L be the signature of relation algebras.

THEOREM 11.5 *There is a closed L-game tree T^a so that for any atomic relation algebra $\mathcal{A}$ and any $n \leq \omega$, the game $G^a_n(\mathcal{A})$ is equivalent to the game $G(T^a\!\restriction_{3+2n}, \mathcal{A})$ in the sense that $\exists$ has a winning strategy in one if and only if she has one in the other. [Note: $3 + 2\omega = \omega$.]*

Proof. For each $n < \omega$, we define a finite closed game tree T_n of depth $3 + 2n$ such that for any atomic relation algebra $\mathcal{A}$, $\exists$ has a winning strategy in $G^a_n(\mathcal{A})$ if and only if she has one in $G(T_n, \mathcal{A})$. We do this by induction on n, such that for each n, every leaf t of T_n is an $\exists$-node and has an associated term network N_t. This is as

---
[1] Following Dana Scott, Lyndon's term was 'strongly representable'.

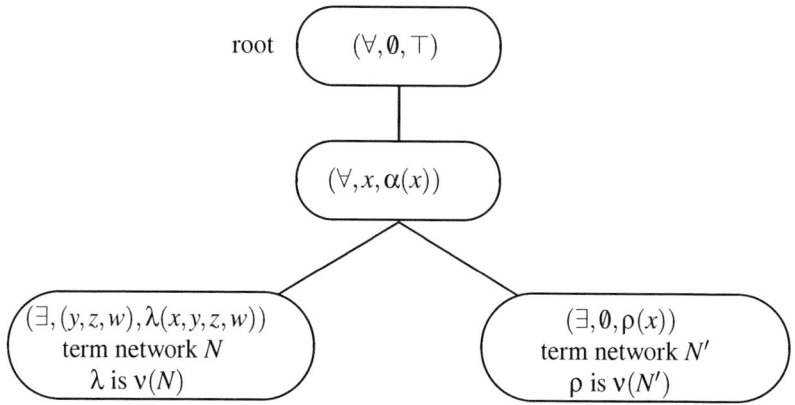

Figure 11.1: The game tree T_0 for G_0^a

in section 10.3.1. For a term network N, we define the formula

$$v(N) \stackrel{def}{=} \bigwedge_{x,y \in N} \alpha(N(x,y)) \wedge (N(x,x) \le \text{`}1\text{'})$$
$$\wedge \bigwedge_{x,y,z \in N} ((N(x,y);N(y,z)) \cdot N(x,z) \ne 0),$$

based on that of proposition 8.2 and saying that the values of the terms on the term network N yield a legal atomic $\mathcal{A}$-network. Here and below, $\alpha(v)$ is the formula $\forall u(u < v \leftrightarrow u = 0)$, expressing that v is an atom.

We can now start building the trees. T_0 consists of four nodes, as shown in figure 11.1. The term network N associated with the bottom left node has two points, $0, 1$, say, with $N(0,1) = x$, $N(1,0) = y$, $N(0,0) = z$, and $N(1,1) = w$; and the formula $\lambda(x,y,z,w)$ is $v(N)$. The term network N' associated with the bottom right-hand node has only one node, say 0, and satisfies $N'(0,0) = x$. The formula $\rho(x)$ is $v(N')$.

This tree represents the game $G_0^a(\mathcal{A})$ in the following way. When $\forall$ chooses an atom of $\mathcal{A}$ in the preliminary round of $G_0^a(\mathcal{A})$, it corresponds to his moving from the top node in figure 11.1 to the middle one, and choosing an assignment to the variable x in the middle node. That node, together with the chosen value of x, becomes the new position in the tree game. $\exists$ must now respond to $\forall$ in $G_0^a(\mathcal{A})$ by selecting an atomic network with x among its labels. Her choice of network here is determined by whether it has one or two nodes (the game $G(T_0, \mathcal{A})$ reflects this by allowing her to pick the right or left leaf of T_0), and, if it has two nodes,

11.3. Alternative views of the game

what its other labels are (this corresponds to her choice of atoms for y,z,w in the game $G(T_0, \mathcal{A})$ if she chooses the left leaf). Note how the players' choices are constrained to be atoms, by the formula α.

Assume that $n < \omega$ and T_n has been defined, the inductive hypothesis still holding. For $t \in T_n$ we write $T_n(t) = (P_t, \bar{x}_t, \phi_t)$. We define $T_{n+1} \supseteq T_n$ as follows.

First, for each leaf $t \in T_n$ with associated term network N_t, and for each pair x,y of nodes of N_t, we insert into T_{n+1} an immediate successor s of t with label

$$(\forall, (u,v), \alpha(u) \wedge \alpha(v) \wedge u;v \geq N_t(x,y)),$$

where u,v are distinct variables not occurring in $\bar{x}_{t'}$ for any $t' \leq t$. In the formula, $N_t(x,y)$ is the term labelling the edge (x,y) of the term network N_t. When $\forall$ selects s and assigns the variables u,v to (say) $a,b \in \mathcal{A}$ in the game-tree game, it corresponds to his choosing the edge (x,y) of N_t and the elements a,b of $\mathcal{A}$ in the game G^a. If either a or b is not an atom of $\mathcal{A}$, or $a;b$ does not lie above the atom that is associated with $N_t(x,y)$ by the current assignment of variables in the branch to t, then $\alpha(a) \wedge \alpha(b) \wedge u;v \geq N_t(x,y)$ fails and $\forall$ loses the tree game.

We then give each such s two immediate successors: s^- and s^+. If $\exists$ chooses s^-, it expresses that $\exists$ responded to $\forall$'s move in the game $G^a_n(\mathcal{A})$ by playing the same network as was already there — so it should already contain a suitable node z. So we let the label of s^- in T_{n+1} be $(\exists, \emptyset, \bigvee_{z \in N_t}(N_t(x,z) = u \wedge N_t(z,y) = v))$. s^- is an $\exists$-node and we let its term network be N_t.

The other successor s^+ is for when $\exists$ decides to extend the network. It will also be an $\exists$-node, so we need a term network for it. This network, N_{s^+}, has as nodes those of N_t plus a new one, say z. The edges of N_{s^+} that are edges of N_t carry the same labels as in N_t. The new edges $(x,z), (z,y)$ are labelled with the variables u, v, respectively. All other edges of N_{s^+} are labelled with new distinct variables. Let the set of these new variables be $\{w_0, \ldots, w_{k-1}\}$. We let the label of s^+ in T_{n+1} be $(\exists, (w_0, \ldots, w_{k-1}), \nu(N_{s^+}))$.

We let T_{n+1} be the result of extending T_n by these new nodes and labels. Thus, all nodes of T_n are nodes of T_{n+1} and have the same $<$-relations holding between them in T_{n+1}, and the same labels, as in T_n. This completes the induction.

Then we define $T_\omega = \bigcup_{n<\omega} T_n$ — a well-defined and recursive closed game tree. It is clear by construction that for any $n \leq \omega$, any play of $G^a_n(\mathcal{A})$ can be viewed as a play of $G(T_n, \mathcal{A})$, and vice versa. Hence, $\exists$ has a winning strategy in $G^a_n(\mathcal{A})$ iff she has a winning strategy in $G(T_n, \mathcal{A})$. So we may let T^a be T_ω — the obvious fact that $T_n = T_\omega \lceil 3+2n$, for each n, finishes the proof. □

Exercises

1. [Cf. exercise 12.4(1) later.] Let $\mathcal{A}$ be an atomic non-associative algebra. Show that $\mathcal{A}$ is associative (and hence a relation algebra) if and only if $\exists$ has

a winning strategy in $G_2^a(\mathcal{A})$, the game where $\forall$ can force the construction of a four-node atomic network (note that $G_n^a(\mathcal{A})$ is well-defined for any atomic non-associative algebra).

2. Let $\mathcal{A}$ be an atomic relation algebra, let $n \leq \omega$, and suppose that $\exists$ has a winning strategy in $G_n^a(\mathcal{A})$. Show that she has a winning strategy in the game $G_n(\mathcal{A})$ of definition 7.12.

3. Find a recursive function $f : \omega \times \omega \to \omega$ such that for any finite relation algebra $\mathcal{A}$ and $n < \omega$, if $\exists$ has a winning strategy in the game $G_{f(|\mathcal{A}|,n)}(\mathcal{A})$ of definition 7.12 then she has a winning strategy in $G_n^a(\mathcal{A})$.

4. How do the Lyndon conditions given in section 11.3.2 differ from Lyndon's original definition in [Lyn50]?

5. Check that the Lyndon conditions do express that $\exists$ has a winning strategy in the modified version of $G_k^a(\mathcal{A})$ as described in section 11.3.2.

6. Fill in the details of the proof of theorem 11.4.

7. Strengthen theorem 11.4 by finding a function $f : \omega \to \omega$ such that for any atomic relation algebra $\mathcal{A}$ and $n < \omega$, if $\exists$ has a winning strategy in the modified (Lyndon-style) game of length $f(n)$ then she has a winning strategy in $G_n^a(\mathcal{A})$.

8. Let X be any finite set and let v_{xy} $(x, y \in X)$ be variables. The formula $\alpha(v)$, defining atoms, is as above. Define formulas φ_n^X, with free variables v_{xy} for $x, y \in X$, by induction on $n \geq 0$. Let

$$\varphi_0^X \stackrel{\text{def}}{=} \bigwedge_{x,y \in X} \alpha(v_{xy}) \wedge \bigwedge_{x \in X} (v_{xx} \leq 1') \wedge \bigwedge_{x,y,z \in X} (v_{xz}; v_{zy} \geq v_{xy}).$$

Given φ_n^X, pick $z \notin X$ (for definiteness, $z = X$), let $Z = X \cup \{z\}$, and let

$$\varphi_{n+1}^X \stackrel{\text{def}}{=} \bigwedge_{x,y \in X} \forall ab \Big(v_{xy} \leq a; b \to \exists_{w \in X} v_{wz}, v_{zz}, v_{zw} (v_{xz} \leq a \wedge v_{zy} \leq b \wedge \varphi_n^Z) \Big).$$

Let σ_n be $\forall v_{xy}(\alpha(v_{xy}) \to \exists v_{yx} v_{xx} v_{yy} \varphi_n^{\{x,y\}})$ (any distinct x, y). Prove that $\theta_{T^a \restriction 3+2n}$ (see definition 10.7) and σ_n are equivalent, for all finite n.

PROBLEM 11.6 Is there a recursive function $f : \omega \to \omega$ such that for any finite relation algebra $\mathcal{A}$ and any $n < \omega$, if $\exists$ has a winning strategy in $G_{f(n)}(\mathcal{A})$ of definition 7.12 then she has a winning strategy in the game $G_n^a(\mathcal{A})$? (Cf. exercise 8.1(1). The examples of section 14.4 will show that the answer is negative if we replace 'finite relation algebra' by 'atomic relation algebra'.)

11.4 Atomic games and complete representations

We will now justify our interest in the games G^a. Their main application is given by the following theorem.

THEOREM 11.7 *Let $\mathcal{A}$ be an atomic relation algebra.*

1. *If $\mathcal{A}$ has a complete representation then $\exists$ has a winning strategy in $G^a_\omega(\mathcal{A})$.*

2. *If $\mathcal{A}$ is countable (or, more generally, an atomic relation algebra with countably many atoms), then the converse holds.*

Proof. This is much as in theorem 7.19.

1. If h is a complete representation of $\mathcal{A}$, then $\exists$ can use h as a guide to her moves, and win $G^a_\omega(\mathcal{A})$. Let the base of h be U. Any finite subset $N \subseteq U$ defines an atomic $\mathcal{A}$-network, also called N, whose set of nodes is N and whose labelling is given by: $N(x,y)$ is the unique atom $a \in \mathcal{A}$ with $(x,y) \in h(a)$, for each $x,y \in N$. Notice that a exists because h is a complete, and hence by corollary 2.21 an atomic, representation. It is easy to check that, so defined, N is an atomic network over $\mathcal{A}$. $\exists$ ensures that each network N_t she plays is defined in this way.

   For $t = 0$, let $\forall$ choose the atom a. $\exists$ chooses $u,v \in U$ with $(u,v) \in h(a)$ — such u and v exist because h is a representation of $\mathcal{A}$. She defines the atomic network N_0 by stating that its nodes are u,v.

   Given the inductive hypothesis ($N_t \subseteq U$) at round t, if $\forall$ chooses $x,y \in N_t$ and atoms a,b with $N_t(x,y) \leq a;b$, then $(x,y) \in h(a;b)$, so there is $z \in U$ with $(x,z) \in h(a)$ and $(z,y) \in h(b)$. $\exists$ selects such a z and forms N_{t+1} by stating that its nodes are the set of nodes of N_t together with z. This strategy allows her to win.

2. Conversely, assume that At$\mathcal{A}$ is countable and that $\exists$ has a winning strategy for $G^a_\omega(\mathcal{A})$. Let $a \in \text{At}(\mathcal{A})$ be arbitrary, and consider a play of the game in which she uses this strategy, and $\forall$ begins by playing a and continues by making every possible move x,y,b,c (where x,y are nodes constructed during play, and b,c are atoms of $\mathcal{A}$) in some round of the game. He can do this because $\mathcal{A}$ has countably many atoms. Exactly as in theorem 7.19, we obtain a homomorphism $h_a : \mathcal{A} \to \wp(U_a \times U_a)$, where U_a is the quotient of the set of nodes constructed during play by the equivalence relation $\sim$ defined as in the earlier theorem. (For nodes x,y, we have $x \sim y$ iff the label on (x,y) is ≤ 1', and for $r \in \mathcal{A}$, $h_a(r) = \{(x/\sim, y/\sim) : \text{the label on } (x,y) \text{ is } \leq r\}$.) Finally, we define h by $h(r) = \bigcup \{h_a(r) : a \in \text{At}\mathcal{A}\}$ for $r \in \mathcal{A}$. Then h is a representation of $\mathcal{A}$.

But h will also be *complete*, because the networks played in $G_\omega^a(\mathcal{A})$ are always atomic. For let $(x,y) \in h(1)$. Clearly, $(x,y) \in h_a(1)$ for some atom a. Consider the play of $G_\omega^a(\mathcal{A})$ used to create h_a, and choose $t < \omega$ so that $x,y \in N_t$. Then $N_t(x,y)$ is an atom b of $\mathcal{A}$, and we clearly have $(x,y) \in h(b)$. Thus, h is an atomic representation, so, by corollary 2.21, complete.

□

REMARK 11.8 Theorem 11.7 works best for algebras having countably many atoms. What about algebras with uncountably many atoms? Then theorem 11.7(2) may fail. It is still true that ∃ will have a winning strategy in the game $G_\omega^a(\mathcal{A})$ for any completely representable relation algebra $\mathcal{A}$, but (see theorem 17.25) it is not true that such a winning strategy is sufficient for $\mathcal{A}$ to have a complete representation. A theorem that we could prove is that an atomic relation algebra $\mathcal{A}$ has a complete representation if and only if ∃ has a winning strategy in the game $G_\kappa^a(\mathcal{A})$ for some infinite cardinal $\kappa \geq |\mathrm{At}(\mathcal{A})|$, but we leave the details of defining this game and proving the result to exercise 3 below.

REMARK 11.9 An alternative is to consider the completely representable relation algebras as a pseudo-elementary class, with defining theory U as given in example 9.2(2). For countable atomic relation algebras $\mathcal{A}$, the game $\Gamma^+(U,\mathcal{A})$ of definition 9.32 boils down to the game $G_\omega^a(\mathcal{A})$ of definition 11.3.

Exercises

1. Let $\mathcal{A}$ be an atomic relation algebra with an atom f such that for all non-identity atoms $x,y \in \mathrm{At}(\mathcal{A})$, the triple (f,x,y) is consistent (such an atom is called a *flexible* atom). Prove that ∃ has a winning strategy in the game $G_\omega^a(\mathcal{A})$. Describe her winning strategy in this game.

2. It is easy to find finite, representable relation algebras with only infinite representations. The representations of the 3-atom *point algebra* of chapter 4 are essentially dense linear orders without endpoints (exercise 4.4(3)). Is it possible to find a finite, representable, *symmetric* relation algebra with only infinite representations?

   Maddux found that the question can be solved positively with a sort of 'Monk algebra' $\mathcal{A}$ — or perhaps, if it's not disrespectful, an anti-Monk algebra. (Monk algebras were discussed in chapter 1 and will be developed further in part IV.) $\mathcal{A}$ is the first of an infinite sequence of algebras suggested to Maddux by Tarski and described in [Madd78b, pp 65–66]. The proof that ∃ has a winning strategy in $G_\omega^a(\mathcal{A})$, and hence that $\mathcal{A}$ is representable, is an excellent primer for defining and checking winning strategies and gives a flavour of the kinds of strategies we'll use in part IV of the book.

11.5. Axioms for complete representability?

Consider the following relation algebra atom structure S with four atoms, 1',r,b,g (red, blue, and green), all self-converse, where $R_{1'} = \{1'\}$ and where the forbidden triples of atoms are the permutations of $(1',x,y)$ (any atoms $x \neq y$) and (r,b,g). So here, instead of forbidding monochromatic triangles, as with Monk algebras, we forbid the trichromatic triangles. Let $\mathcal{A}$ be the complex algebra $\mathfrak{Cm}\,S$.

(a) Check that S is a relation algebra atom structure in the sense of definition 3.22.

(b) Show that $\mathcal{A}$ has no finite representation. For this, show by induction on $n < \omega$ that in the base of any representation h of $\mathcal{A}$, there are points $x_0,\ldots,x_{n-1}$ such that for each $i < n$ there is $c_i \in \{r,b,g\}$ with $(x_j,x_i) \in h(c_i)$ for all $j < i$, and $c_i \neq c_{i-1}$ for $0 < i < n$.

(c) Show that $\mathcal{A}$ is representable. For this, carefully define a winning strategy for $\exists$ in $G_\omega^a(\mathcal{A})$.

3. Define a game $G_\kappa^a(\mathcal{A})$ for an arbitrary atomic relation algebra $\mathcal{A}$ and cardinal κ. Show that $\mathcal{A}$ has a complete representation iff $\exists$ has a winning strategy in this game, for a suitable κ (which κ?). See remark 11.8.

11.5 Axioms for complete representability?

We can now obtain a first-order characterisation of the elementary closure of the completely representable relation algebras — namely, the Lyndon conditions. We will combine it with theorems 11.4–11.7 to obtain theorem 11.10 below.

THEOREM 11.10 *Let T^a be the closed game tree of theorem 11.5. For any atomic relation algebra $\mathcal{A}$, the following are equivalent:*

(a) $\mathcal{A} \models \Theta_{T^a}$ *(see definition 10.9),*

(b) $\exists$ *has a winning strategy in $G_n^a(\mathcal{A})$ for all finite n,*

(c) $\mathcal{A}$ *satisfies the Lyndon conditions,*

(d) *there is a completely representable relation algebra C elementarily equivalent to $\mathcal{A}$.*

Hence,

1. For any atomic relation algebra $\mathcal{A}$, if $\mathcal{A}$ is completely representable then $\mathcal{A}$ satisfies the Lyndon conditions, and if $\mathcal{A}$ satisfies the Lyndon conditions then $\mathcal{A}$ is representable.

2. If $\mathcal{A}$ is a finite relation algebra, then $\mathcal{A}$ is representable iff $\mathcal{A}$ is completely representable, iff $\exists$ has a winning strategy in $G^a_\omega(\mathcal{A})$, iff $\mathcal{A}$ satisfies the Lyndon conditions.

Proof. By theorem 10.10, $\mathcal{A} \models \Theta_{T^a}$ iff $\exists$ has a winning strategy in $G(T^a{\restriction}n, \mathcal{A})$ for all finite n. It follows from theorem 11.5 that parts (a) and (b) are equivalent. By theorem 11.4, (b) holds iff $\mathcal{A}$ satisfies the Lyndon conditions. This proves the equivalences (a)–(c).

We prove $(b) \Rightarrow (d)$ by theorem 10.14. Let 'L' of that theorem be the signature of relation algebras, 'T' the game tree T^a, 'K' the class of completely representable relation algebras, and 'U' the axioms defining atomic relation algebras. By theorems 11.5 and 11.7, the hypothesis of the theorem applies. By (b) and theorem 11.5, $\exists$ has a winning strategy in $G(T^a{\restriction}n, \mathcal{A})$ for all finite n, so by theorem 10.14(2), there is $C \in \mathsf{K}$ with $C \equiv \mathcal{A}$, proving (d).

Now we prove $(d) \Rightarrow (a)$. If there is such a C, then by theorem 11.7, $\exists$ has a winning strategy in $G^a_\omega(C)$; it follows that (b), and hence also (a), holds for C. But $\mathcal{A} \equiv C \models \Theta_{T^a}$ implies $\mathcal{A} \models \Theta_{T^a}$, as required. Thus, all four conditions are equivalent.

For part 1 of the conclusions, if the atomic relation algebra $\mathcal{A}$ is completely representable then since $\mathcal{A} \equiv \mathcal{A}$, by $(d) \Rightarrow (c)$ we see that $\mathcal{A}$ satisfies the Lyndon conditions. And $\mathcal{A}$'s satisfying the Lyndon conditions implies, by $(c) \Rightarrow (d)$, the existence of a completely representable $C \equiv \mathcal{A}$. Since $C \in \mathbf{RRA}$ and $\mathbf{RRA}$ is (a variety and so) closed under elementary equivalence, we have $\mathcal{A} \in \mathbf{RRA}$, too.

For part 2, if $\mathcal{A}$ is representable then as any representation of a finite relation algebra is complete (corollary 2.21), $\mathcal{A}$ is completely representable. By theorem 11.7, this implies that $\exists$ has a winning strategy in $G^a_\omega(\mathcal{A})$, which implies by $(b) \Rightarrow (c)$ that $\mathcal{A}$ satisfies the Lyndon conditions, which implies by part 1 that $\mathcal{A}$ is representable. □

REMARK 11.11 We will see in chapter 14 that both converses to part 1 of the theorem fail.

Part 2 gives Lyndon's result of [Lyn50], mentioned in section 11.3.2. We will use this characterisation of the finite representable relation algebras frequently in parts IV and V of the book.

We will see in theorem 17.6 that the completely representable relation algebras do not form an elementary class, and axiomatising its elementary closure is the best we can do.

Exercises

1. Let $\mathcal{A}$ be any relation algebra. Prove that $\mathcal{A}$ is representable iff $\exists$ has a winning strategy in the game $G^a_\omega(\mathcal{A}^+)$, iff she has one in each game $G^a_n(\mathcal{A}^+)$ for

11.5. Axioms for complete representability? 351

$n < \omega$. (Recall that $\mathcal{A}^+$ is the canonical extension of $\mathcal{A}$; it is atomic.) Hence, $\mathcal{A}$ is representable iff $\mathcal{A}^+$ is representable, iff $\mathcal{A}^+$ satisfies the Lyndon conditions.

2. Use the Lyndon conditions and fact 2.86 to deduce Monk's theorem (theorem 3.36) that the variety **RRA** is canonical.

3. This exercise extends exercise 4.5(2) and proves Monk's important result [Mon64] that **RRA** is not finitely axiomatisable. (See exercise 14.2(9), remark 15.13, and section 17.1 for other proofs.) We use the notation of the earlier exercise: $\mathcal{A}_n$ continues to denote the Lyndon algebra with $n+2$ atoms (including 1'), for finite $n \geq 2$. Infinitely many $\mathcal{A}_n$ are representable and infinitely many are not.

   (a) For $2 \leq n < \omega$, show that the Lyndon algebra $\mathcal{A}_n$ is interpretable, uniformly in n, in the L-structure $(\mathrm{bool}(\mathcal{A}_n), 1'^{\mathcal{A}_n})$, where L is the signature of boolean algebras augmented with a constant 1'.

   (b) For each sentence σ of the language of relation algebras, let σ' denote an L-sentence constructed via the interpretation above, such that whenever $2 \leq n < \omega$ we have $(\mathrm{bool}(\mathcal{A}_n), 1'^{\mathcal{A}_n}) \models \sigma'$ iff $\mathcal{A}_n \models \sigma$. Let $\mathsf{LC}' = \{\lambda' : \lambda \in \mathsf{LC}\}$. Let ρ be the conjunction of the axioms defining **RA**. Let α be an L-sentence defining the class of atomic boolean algebras and stating that 1' is an atom.

   Assume for contradiction that σ is a first-order sentence defining **RRA**. Define the following L-theories:
   - $R = \mathsf{LC}' \cup \{\alpha, \rho', \exists_{\geq n} x(x=x) : n < \omega\}$,
   - $N = \{\alpha, \rho', \neg \sigma', \exists_{\geq n} x(x=x) : n < \omega\}$.

   Show that R, N are consistent.

   (c) Take a countable model $\mathcal{A}' \models R$, and let $\mathcal{A}$ be the structure resulting from applying the interpretation above to $\mathcal{A}'$. Show that $\mathcal{A}$ is an atomic relation algebra that satisfies the Lyndon conditions.

   (d) Take a countable model $\mathcal{B}' \models N$ and define $\mathcal{B}$ from $\mathcal{B}'$ as above. Show that $\mathcal{B}$ is an atomic non-representable relation algebra with $\mathrm{At}\,\mathcal{B} \cong \mathrm{At}\,\mathcal{A}$. Deduce that $\mathcal{B}$ satisfies the Lyndon conditions.

   (e) Conclude that **RRA** is not finitely axiomatisable in first-order logic.

4. This exercise is to reproduce the proof of [Jón91] that **RRA** is not axiomatisable by any set of equations using finitely many variables.

   (a) Let $2 \leq n < \omega$. Let B be the domain of a maximal proper subalgebra of $\mathrm{bool}(\mathcal{A}_n)$, and suppose that 1' $\in B$. Show that B is the domain of a subalgebra $\mathcal{B}$ of $\mathcal{A}_n$. [Check carefully that B is closed under composition ';' in $\mathcal{A}_n$.]

(b) Show that $\mathcal{B}$ as in part 4a embeds into $\mathcal{A}_m$ for any $m \geq n$. Deduce that every proper subalgebra of $\mathcal{A}_n$ is representable.

(c) Use part 4a to show that for all $k < \omega$ there is $n_0 < \omega$ such that for all $n > n_0$, $\mathcal{A}_n$ is not generated by any subset with at most k elements.

(d) Assume that **RRA** is axiomatised by a set of equations written with only k variables, for some finite k. Show that an arbitrary relation algebra $\mathcal{A}$ is representable iff every k-generated subalgebra of $\mathcal{A}$ (i.e., one generated by a subset of $\mathcal{A}$ of size at most k) is representable.

(e) Conclude that **RRA** is not axiomatisable by k-variable equations.

5. Show that **RRA** is not axiomatisable by a set of prenex universal sentences using finitely many variables.

6. Show that Jónsson's theorem (exercise 4) implies Monk's theorem (exercise 3). [Compactness.]

Part III

Approximations

Introduction to approximations 355

A substantial fraction of the remainder of the book will be concerned with finite-dimensional approximations to **RRA**. The story will take some time to develop. Here, we give a short introduction to the three main approaches to approximations. In chapter 12, we will develop the elementary properties of 'bases', perhaps the most concrete of them. The other two — relation algebra reducts and relativised representations — will be discussed in chapter 13, where all three approaches will be proven equivalent. Hierarchy results will be proved in chapters 15 and 17, undecidability results in chapter 18, and finite model property results in chapter 19.

Approaches to approximating representability

When Tarski originally set down his equations defining relation algebras (definition 3.8), it must have been hoped that they would characterise the isomorphism types of proper relation algebras — that is, that they would axiomatise **RRA**. This was not to be. Lyndon exhibited the first non-representable relation algebra in [Lyn50], a paper also noteworthy for introducing a 'point-by-point' approach to building representations of relation algebras which we presented in chapter 11 in terms of the game $G_n^a(\mathcal{A})$. Later examples of Lyndon's using projective planes (section 4.5) led to Monk's famous proof ([Mon64], to be proved in section 17.1) that **RRA** is not finitely axiomatisable in first-order logic.

We can view these examples as demonstrating that a relation algebra can contain 'contradictions'. These are invisible in small parts (with at most four points) of a representation of the algebra, which are well controlled by the axioms governing the basic operations (associativity of composition, etc.); but they show up when we try to build, in Lyndon's way, larger fragments of a representation and can destroy the possibility of a representation existing at all. For example, the proof in chapter 4 that McKenzie's relation algebra is not representable required a 5-point fragment (see figure 4.1). The size of fragment needed to detect 'contradictions' in a relation algebra can be arbitrarily large, and Monk used this to prove that no one first-order sentence can control them, so yielding the non-finite axiomatisability of **RRA**. Monk extended his result to cylindric algebras, the higher-arity analogue of relation algebras, in [Mon69].

Monk's results were very influential. One reaction to them was to try to control the behaviour of larger fragments of a representation. We may try to characterise those relation algebras where we have no problems in building up to n points of a potential representation, for some finite n. (We say 'potential' because of course the relation algebra may turn out not to have a representation. We will return to this point below.) As n increases, the classes of such algebras approach **RRA**. We can ask whether the classes are finitely axiomatisable, whether they form a strict hierarchy as n increases, or whether they reach **RRA** at some finite n. At any rate, they should be useful approximations to **RRA**, and shed light on it.

Relational and cylindric bases

Two striking ways of doing this were found. In a series of publications dating from the late 1970s, Maddux used *n-dimensional relational and cylindric bases,* which we will discuss in chapter 12. Recall (definition 11.2) that an atomic network describes which elements of $\mathcal{A}$ hold between each pair of its nodes; the coherency conditions in the definition ensure that the description respects the relation algebra operations. We may think of an atomic network as the isomorphism type of an n-point fragment of a potential complete representation of $\mathcal{A}$. An n-dimensional basis for an atomic relation algebra $\mathcal{A}$ is a set of n-point atomic networks over $\mathcal{A}$, with certain closure properties; the difference between relational and cylindric bases is in the closure properties that are required.

The class $\mathbf{RA}_n$ denotes those relation algebras that have an n-dimensional relational basis [Madd83], or embed in an algebra that does. For such a relation algebra, we can use the basis to build a (complete) representation of it up to n points in Lyndon's manner; and we can even keep on going, except that for every new point we add, we must forget some other point in order to limit the total size of the fragment to n. The relational basis always shows us how to continue, because of its closure properties. Essentially, it encodes a winning strategy for $\exists$ in an 'n-pebble' version of the game G_ω^a of definition 11.3, in which networks are restricted to having n nodes.

If the algebra has a cylindric basis [Madd78b, Madd89b], we can do even more, in that the different ways that the basis dictates to add points are guaranteed to be compatible with each other, again so long as no more than n points are simultaneously considered. This compatibility property is a very strong 'Church–Rosser condition' obtained from an amalgamation closure property in the definition of cylindric bases. It corresponds closely to homogeneous representations, and also to commutativity of quantifiers in first-order logic: the equivalence of $\exists x \exists y \varphi$ and $\exists y \exists x \varphi$. Cylindric bases also encode winning strategies, but in a new kind of game in which $\forall$ is also allowed to demand the amalgamation of two previously-played atomic networks, provided they are 'compatible' and their amalgam has no more than n nodes.

The fact that bases 'are' winning strategies in games makes them very useful to us. Their combinatorial properties are accessible via the games, and their existence can be approximated by truncating these games.

Maddux introduced relational bases in [Madd83] in the context of n-variable proof theory, to characterise models in which sequents are valid if and only if they are provable in a certain sequent calculus using at most n variables. Cylindric bases are used when the proof theory has an axiom for commuting quantifiers.

Introduction to approximations 357

Relation algebra reducts of cylindric algebras

The second way to control large fragments of a potential representation of a relation algebra is due to Henkin, Monk, and Tarski. If the relation algebra is the relation algebra reduct (definition 5.40) of an n-dimensional cylindric algebra, or is a subalgebra of such a reduct, then we can arrange that fragments of a potential representation of up to n points are controlled by the cylindric algebra axioms of definition 5.16, just as smaller fragments (size four or less) are controlled by the axioms of relation algebras. The class of relation algebras that arise in this way is of course $\mathbf{S\Re aCA}_n$ (definition 5.45).

On the face of it, the definition of $\mathbf{S\Re aCA}_n$ is more abstract than that of $\mathbf{RA}_n$. For example, given even a finite relation algebra, it is not immediately clear how one might determine whether it is in $\mathbf{S\Re aCA}_n$ for given n. (In fact, for $n \geq 5$ this problem is undecidable: see theorem 18.13.) Maddux made progress towards a more concrete characterisation, by proving in [Madd78b, Madd89b] that any subalgebra of a relation algebra with an n-dimensional cylindric basis is in $\mathbf{S\Re aCA}_n$. This is a useful sufficient condition for membership of $\mathbf{S\Re aCA}_n$. He established it by showing that an n-dimensional cylindric basis can be viewed as the atom structure of an n-dimensional cylindric algebra. The original relation algebra can be recovered from this cylindric algebra by taking its relation algebra reduct, and possibly then taking a subalgebra of that. Unfortunately, whether the condition is also necessary — whether every algebra in $\mathbf{S\Re aCA}_n$ is a subalgebra of a relation algebra with an n-dimensional cylindric basis — remains an open question. So we cannot at present characterise $\mathbf{S\Re aCA}_n$ by n-dimensional cylindric bases.

Nonetheless, we will prove in chapter 13 that $\mathbf{S\Re aCA}_n$ can be equivalently defined in the same way as $\mathbf{RA}_n$, using not cylindric bases but *n-dimensional hyperbases*. Hyperbases generalise cylindric bases by using *hypernetworks* instead of atomic networks. Hypernetworks contain more information than simply the relation algebra structure on n points of a potential representation. Each sequence of up to n points in an n-dimensional hypernetwork carries a 'hyperlabel'. The idea is that the points of the hypernetwork are controlled by an element of an n-dimensional cylindric algebra, and the additional hyperlabelling structure describes this n-ary relation more fully than the merely binary relation algebra structure on the points can do. We believe that this extra information is essential to capture $\mathbf{S\Re aCA}_n$ and cannot be eliminated, but further research is needed.

Relativised representations

How can we formalise the notion of 'potential representation' of a possibly non-representable relation algebra? The answer is to use the relativised representations of definition 5.1. In chapter 5, we saw a natural way of defining a relativised representation for a relation algebra, or more generally a non-associative algebra,

$\mathcal{A}$. We dropped the requirement that the representation be fully classical (i.e., the original unrelativised notion of representation), and instead allowed relativised interpretations of all the relation algebra operators. The unit (the interpretation in the representation of the top element 1) is a reflexive, symmetric relation on the base set M of the representation, and all operations, in particular negation '$-$' and composition ';', are relativised to the unit. So a pair (x,y) of elements of M stands in the relation $r;s$ (where $r,s \in \mathcal{A}$) if there is $z \in M$ with (x,z) standing in the relation r and (z,y) in the relation s, *so long as (x,y) is in the unit*. The interpretation of '$-$' is defined similarly; the other operations are unaffected by relativisation since 1^M is reflexive and symmetric. Maddux used this kind of relativised representation to characterise weakly associative algebras (section 5.1 and theorem 7.5).

We now want to insist that 'on the small scale', the relativised representation does appear to be classical. We may then vary the scale up to which classical behaviour is required. If we only demand that things appear classical when considering at most three points of the base of the relativised representation, then we capture semi-associativity and hence the class **SA**. If we require classical behaviour of four-point fragments, we capture associativity and hence ordinary relation algebras. Indeed, these are alternative ways of defining **SA** and **RA**. Requiring classical behaviour for fragments of up to five points takes us to a strictly smaller class than **RA**, up to six takes us to a still smaller one, and so on. The 'limit' or intersection of these classes is **RRA**. Thus, we obtain a whole series of 'approximations' to, or perhaps 'n-variable analogues' of, **RRA**.

n-square relativised representations In fact, there is more than one way of requiring classical behaviour of fragments of a relativised representation up to a given size. The basic requirement is that the relativisation (of negation and relational composition) is only visible when fragments of the representation above a certain size (say, n points) are considered. The n-square relativised representations of definition 5.7 are like this. Recall from definition 5.7 that a subset of a relativised representation M is a *clique* if all pairs of elements of it lie in the unit. We defined M to be *n-square* if for any clique $C \subseteq M$ of size less than n, if $x,y \in C$ and (x,y) lies in the relation $r;s$ then there is a clique $C' \supseteq C$ in M containing a point z realising this composition, so that (x,z) lies in the relation r, and (z,y) in s. An n-square representation is 'locally classical', in that if we look at it through a moveable 'window' only big enough to show n points, we will never discover using the relation algebra operations that it is relativised.

n-flat relativised representations A stronger requirement than n-squareness involves commuting quantifiers. To explain it, we will use 'clique-relativised' semantics. By definition 5.1, a relativised representation M of a non-associative algebra $\mathcal{A}$ is a model of a certain first-order theory $R_{\mathcal{A}}$ in the signature $\mathcal{L}(\mathcal{A})$ of

Introduction to approximations 359

which each element of $\mathcal{A}$ is a binary relation symbol. Any n-variable formula φ in the first-order language in signature $\mathcal{L}(\mathcal{A})$, written with variables $x_0, \ldots, x_{n-1}$, say, can be naturally interpreted in M by relativising quantifiers of φ to cliques. In the *clique-relativised semantics* of φ, the range of every assignment of variables to elements of M used in the inductive evaluation of φ should be a clique. Equivalently, we can syntactically relativise all quantifiers in φ to $\bigwedge_{i,j<n} 1(x_i, x_j)$, and evaluate the resulting formula classically in M.

We may or may not choose to require that the quantifiers in formulas behave classically in the clique-relativised semantics: in particular, that any n-variable formula of the form $\exists x_i \exists x_j \varphi$ is equivalent to $\exists x_j \exists x_i \varphi$ (for $i, j < n$). A relativised representation satisfying this very natural requirement is said to be *n-flat*. Any n-flat relativised representation is n-square, but not always conversely. Considering n-flat relativised representations gets us closer to a classical representation.

In order to handle complete relativised representations it is helpful to use a stronger, infinitary logic, $\mathcal{L}(\mathcal{A})^n_{\infty\omega}$, in which arbitrary conjunctions and disjunctions of sets of formulas are allowed. A relativised representation in which any infinitary formula of the form $\exists x_i \exists x_j \varphi$ is equivalent to $\exists x_j \exists x_i \varphi$ (for $i, j < n$) is said to be *infinitarily n-flat*.

One may ask if n-flatness has a more concrete characterisation, like the 'clique addition property' used to define n-square. The answer is in a limited sense 'yes': we will define a notion of *n-smooth* relativised representation which has an 'amalgamation' property, but to make this work we will have to expand the representation by further relations, essentially stating when two tuples agree in the clique-relativised semantics on all n-variable formulas.

Both opting in to commuting quantifiers, and opting out, have been considered in the literature, and we will examine both options here in a uniform way.

Unifying the three approaches

So we have three approaches to approximating **RRA** — by bases (**RA**$_n$), relation algebra reducts (**S$\mathfrak{R}$aCA**$_n$), and relativised representations. In chapter 13, we will unify all of them. We show that for all finite $n \geq 4$:

- **S$\mathfrak{R}$aD**$_n$ = **S$\mathfrak{R}$aG**$_n$ = **RA**$_n$ is the class of all relation algebras with an n-square relativised representation,

- **S$\mathfrak{R}$aCA**$_n$ = **S$\mathfrak{R}$a(CA**$_n \cap$ **D**$_n$) = **S$\mathfrak{R}$a(CA**$_n \cap$ **G**$_n$) is the class of subalgebras of relation algebras with an n-dimensional hyperbasis, and the class of relation algebras with an n-flat (and infinitarily n-flat, and n-smooth) relativised representation.

360   *Introduction to approximations*

Thus, in each case, the same class can be equivalently characterised in terms of bases, relation algebra reducts, and relativised representations.

The 3-dimensional case   Above, we required that $n \geq 4$. The case $n = 3$ is rather special, because $\mathbf{S\mathfrak{R}aG_3} = \mathbf{WA} \supset \mathbf{RA}_3$ (cf. lemma 5.54) and $\mathbf{S\mathfrak{R}aCA}_3 \not\subseteq \mathbf{NA}$. The proof that a relation algebra reduct has a basis requires $n \geq 4$. But all is not lost: we can show that $\mathbf{RA}_3 = \mathbf{SA} = \mathbf{S\mathfrak{R}a(G_3 \cap CA_3)}$, this being the class of non-associative algebras with a 3-square and/or 3-flat and/or 3-smooth relativised representation.

The payoff

We give some illustrations of the benefits of having this three-way characterisation of the classes $\mathbf{RA}_n$ and $\mathbf{S\mathfrak{R}aCA}_n$.

Representation theory for RA   Since $\mathbf{RA} = \mathbf{RA}_4 = \mathbf{S\mathfrak{R}aCA}_4$, a corollary of our results is that the class $\mathbf{RA}$ of relation algebras is precisely the class of non-associative algebras with 4-square (or equivalently, 4-flat or 4-smooth) relativised representations. Indeed, we will see in chapter 19 that the class of finite relation algebras is precisely the class of all finite non-associative algebras with finite 4-square (or equivalently, finite 4-flat or 4-smooth) relativised representations. The representation theory developed here characterises arbitrary relation algebras, and may therefore be useful for studying relation algebras without classical representations.

Bases and representability   Using the representation theory developed here, we can give a direct argument (proposition 13.48) showing that the $\mathbf{RA}_n$ (and the $\mathbf{S\mathfrak{R}aCA}_n$) converge to $\mathbf{RRA}$. A similar argument using relational bases appears in [Madd91b, p. 112].

Canonicity   We can also prove that the varieties $\mathbf{RA}_n$ ($4 \leq n < \omega$) are canonical by generalising Monk's theorem for $\mathbf{RRA}$ (theorem 3.36) to n-square (and n-flat) relativised representations.

Axiomatising $\mathbf{RA}_n$ and $\mathbf{S\mathfrak{R}aCA}_n$ It is easily seen that the varieties $\mathbf{RA}_n$ and $\mathbf{S\mathfrak{R}aCA}_n$ ($5 \leq n < \omega$) are pseudo-universal classes (definition 9.1) by each of their characterisations. Being contained in $\mathbf{RA}$, they are discriminator varieties. So we can use the techniques of chapter 9 to give equational axioms for them *in three different ways,* corresponding to our three kinds of characterisation. We do this in outline in theorem 13.55. Though the axiomatisations are infinite, and for $n \geq 5$ necessarily so, they are explicit and recursive. It may be interesting to compare the axiomatisations obtained from the three different characterisations of the classes.

Introduction to approximations 361

A hierarchy of approximations

In part IV we will be better able to assess the merits of $\mathbf{RA}_n$ and $\mathbf{S\mathfrak{R}aCA}_n$ as approximations to $\mathbf{RRA}$. The following diagram and text summarises some of what we will find.

$$
\begin{array}{ccccccccc}
\mathbf{NA} & & \mathbf{RA}_3 & & \mathbf{RA}_4 & \supset & \mathbf{RA}_5 & \supset \cdots \supset & \mathbf{RRA} \\
\cup^f & & \| & & \| & & \| & & \\
\mathbf{S\mathfrak{R}aG}_3 & & \mathbf{S\mathfrak{R}a(CA}_3\cap\mathbf{G}_3) & & \mathbf{S\mathfrak{R}aG}_4 & & \mathbf{S\mathfrak{R}aG}_5 & \cdots & \\
\| & & \| & & \| & & & & \| \\
\mathbf{WA} & \supset^f & \mathbf{SA} & \supset^f & \mathbf{RA} & \cup & & \cdots & \\
& & \| & & \| & & & & \\
& & \mathbf{S\mathfrak{R}a(CA}_3\cap\mathbf{G}_3) & & \mathbf{S\mathfrak{R}aCA}_4 & \supset & \mathbf{S\mathfrak{R}aCA}_5 & \supset \cdots \supset & \mathbf{RRA}
\end{array}
$$

Inclusions between classes

All classes shown in the diagram are canonical conjugated varieties, all except $\mathbf{NA}$ and $\mathbf{WA}$ are discriminator varieties, all inclusions are strict, and all except the $\supset^f$s are not finitely axiomatisable (i.e., the smaller class, which may be $\mathbf{RRA}$, is not finitely axiomatisable over the larger). Remark 15.13 will show that $\mathbf{RA}_m \not\subseteq \mathbf{S\mathfrak{R}aCA}_n$ for all finite $m, n \geq 5$, so no more inclusions other than those derivable by transitivity can be added to the diagram above. Also, proposition 13.48 will show (combining work of Monk and Maddux) that

$$\bigcap_{3\leq n<\omega} \mathbf{RA}_n = \bigcap_{3\leq n<\omega} \mathbf{S\mathfrak{R}aCA}_n = \mathbf{RRA}.$$

So by exercise 1 below, all inclusions derived by transitivity from the non-finitely axiomatisable ones shown in the diagram are also non-finitely axiomatisable. For example, $\mathbf{RRA}$ is not finitely axiomatisable over $\mathbf{RA}_5$.

Lemma 5.54 and exercise 5.7(4) showed that $\mathbf{S\mathfrak{R}aG}_3 = \mathbf{WA}$. The other equalities in the diagram will be proved in corollaries 13.47 and 13.50.

It is well known that $\mathbf{SA} \supset \mathbf{RA}$. The inclusions $\mathbf{RA}_4 \supset \mathbf{RA}_5 \supset \cdots$ were proven strict in [Madd92]. Infinitely many of the inclusions $\mathbf{S\mathfrak{R}aCA}_4 \supseteq \mathbf{S\mathfrak{R}aCA}_5 \supseteq \cdots$ were shown to be strict in [Madd89b], and in [HirHod+02c] and chapter 15 it is shown that they are all strict. The non-finite axiomatisability results are proved in [HirHod00, HirHod01a] and chapters 15 and 17.

For $n \geq 6$, $\mathbf{RA}_n$ and $\mathbf{S\mathfrak{R}aCA}_n$ are not closed under completions (see section 17.7) and are therefore not Sahlqvist varieties (theorem 2.96). The analogous questions for $n = 5$ are open.

362   Introduction to approximations

In some ways, however, $\mathbf{RA}_n$ is better behaved than $\mathbf{S\Re aCA}_n$:

- $\mathbf{RA}_3 = \mathbf{SA}$, but as we saw in fact 5.46, $\mathbf{S\Re aCA}_3 \not\subseteq \mathbf{NA}$. (This is why we use $\mathbf{S\Re a}(\mathbf{CA}_3 \cap \mathbf{G}_3)$ in the bottom line of the diagram above. The irregularity caused is mostly only visual, since $\mathbf{S\Re aCA}_n = \mathbf{S\Re a}(\mathbf{CA}_n \cap \mathbf{G}_n)$ for finite $n \geq 4$.)

- Given finite $n \geq 5$, it is decidable (in polynomial time) whether a finite relation algebra $\mathcal{A}$ is in $\mathbf{RA}_n$ (corollary 12.32), but undecidable whether $\mathcal{A} \in \mathbf{S\Re aCA}_n$ (theorem 18.13).

- For any finite relation algebra $\mathcal{A}$, we have $\mathcal{A} \in \mathbf{RA}_n$ iff $\mathcal{A}$ has a finite n-square relativised representation (theorem 19.18). This 'finite model property' fails for $\mathbf{S\Re aCA}_n$ and n-flat relativised representations (proposition 19.19), although by theorem 19.20 it does hold for the class $\mathbf{S\Re a}\{C \in \mathbf{CA}_n : C \text{ finite}\}$.

- [Ste00, SteVen98] characterised $\mathbf{RA}_n$ as the class of subalgebras of relation algebra reducts of 'Q_n-algebras', where Q_n is an n^2-ary operator symbol similar to those defined by Jónsson in [Jón91], and those in section 8.2. The Q_n-algebras form a finitely axiomatisable Sahlqvist variety (see exercise 12.4(12) below). We do not know of a similar subreduct characterisation of $\mathbf{S\Re aCA}_n$.

Approximations or analogues?

The conclusion is that $\mathbf{RA}_n$ and $\mathbf{S\Re aCA}_n$ do not approximate $\mathbf{RRA}$ very closely: for any finite $n \geq 5$, $\mathbf{RRA}$ and even $\mathbf{RA}_{n+1}$ are 'infinitely far away' from $\mathbf{RA}_n$ in terms of axioms required to make up the difference, and similarly for $\mathbf{S\Re aCA}_n$. Even $\mathbf{S\Re aCA}_n$ is infinitely far from $\mathbf{RA}_n$. Perhaps these classes are better seen as *finite-dimensional analogues* of $\mathbf{RRA}$. (The 'dimension' here ('n') is connected to the number of variables used in proofs in first-order logic; we will draw out this connection later.) They are non-finitely axiomatisable canonical varieties with many of the properties of $\mathbf{RRA}$, and they have a workable representation theory.

Exercises

1. Let K_n ($n \leq \omega$) be elementary classes of relation algebras (say), with $\mathsf{K}_0 \supseteq \mathsf{K}_1 \supseteq \cdots$ and $\mathsf{K}_\omega = \bigcap_{n<\omega} \mathsf{K}_n$. Assume that for each $n < \omega$, K_{n+1} is not finitely axiomatisable over K_n — that is, there is no first-order L_{RA}-sentence σ such that $\mathsf{K}_{n+1} = \{\mathcal{A} \in \mathsf{K}_n : \mathcal{A} \models \sigma\}$.

    (a) Prove that for any $n < m < \omega$, K_m is not finitely axiomatisable over K_n.
    (b) Prove that K_ω is not finitely axiomatisable over K_n for any $n < \omega$.

Chapter 12

Relational, cylindric, and hyperbases

We have a lot of work to do on approximations to **RRA**. We begin gently in this chapter, by developing the basic theory of relational bases and hyperbases. We also explain how they are connected to Maddux's cylindric bases, and how cylindric bases give rise to homogeneous representations.

12.1 Hypernetworks

We start by generalising the atomic networks introduced in chapter 11, by allowing them to carry higher-arity information. Many of our results apply not just to relation algebras, but more generally to non-associative algebras (definition 5.2) as well. So until further notice, $\mathcal{A}$ will be an *atomic non-associative algebra* (unless otherwise stated). For this section, the dimension n with $2 \leq n \leq \omega$ is fixed.

12.1.1 Definition of hypernetworks

DEFINITION 12.1 Let Λ be a non-empty set.

An *n-dimensional Λ-hypernetwork N over $\mathcal{A}$* is a map $N : {}^{\leq n}n \to \Lambda \cup \operatorname{At}\mathcal{A}$ such that $N(\bar{x}) \in \operatorname{At}\mathcal{A}$ if $|\bar{x}| = 2$, and $N(\bar{x}) \in \Lambda$ if $|\bar{x}| \neq 2$, for any $\bar{x} \in {}^{\leq n}n$, and with the following properties:

1. $N(x,x) \leq 1$' (or more strictly, $N(\bar{x}) \leq 1$', where $\bar{x} = (x,x) \in {}^2 n$)

2. $N(x,y) \leq N(x,z) ; N(z,y)$

for all $x,y,z < n$, and

3. If $\bar{x}, \bar{y} \in {}^{\leq n}n$, $|\bar{x}| = |\bar{y}|$, and $N(x_i, y_i) \leq 1'$ for all $i < |\bar{x}|$, then $N(\bar{x}) = N(\bar{y})$.

We write $H_n(\mathcal{A}, \Lambda)$ for the set of all n-dimensional Λ-hypernetworks over $\mathcal{A}$.

We will drop $\Lambda, \mathcal{A}, n$ when the context is clear or we are not interested in what they are, and simply say that N is a hypernetwork. In the context of definition 12.1, x, y, z are called *nodes* of the hypernetwork N; a sequence $\bar{x} \in {}^{\leq n}n$ with $|\bar{x}| = 2$ is called an *edge*, and if $|\bar{x}| \neq 2$ it is called a *hyperedge*. The set Λ is the set of *hyperlabels*. So we are thinking of an n-dimensional hypernetwork N as a directed labelled hypergraph with set of nodes n.

Discussion of hyperlabels

REMARK 12.2 As already suggested, the hyperlabels allow encoding more of the context of a tuple of points in a relation algebra representation than is captured by the relations of the relation algebra, which are only binary. This may explain the intuition behind condition 3 of definition 12.1. Suppose that we 'embed' a hypernetwork N into a representation by a map ι — ι maps the nodes of N to elements of the base of the representation, in such a way that for all nodes x, y, the relation $N(x, y)$ holds on the arc $(\iota(x), \iota(y))$ of the representation. Two distinct tuples $\bar{x}, \bar{y}$ of a hypernetwork N may be necessarily equal when embedded in this way, because $N(x_i, y_i) \leq 1'$ for all $i < |\bar{x}| = |\bar{y}|$. In that case, the information they carry should be the same, so we require that they have the same hyperlabel.

Actually, condition 3 of definition 12.1 follows from the others when $|\bar{x}| = |\bar{y}| = 2$. For in this case, if $N(x_0, y_0), N(x_1, y_1) \leq 1'$ then

$$\begin{aligned}
N(y_0, y_1) \;\leq\; N(y_0, x_1); N(x_1, y_1) \;&\leq\; (N(y_0, x_0); N(x_0, x_1)); N(x_1, y_1) \\
&\leq\; (1'; N(x_0, x_1)); 1' \;=\; N(x_0, x_1).
\end{aligned}$$

The converse inequality, $N(x_0, x_1) \leq N(y_0, y_1)$, is proved similarly, and follows anyway as edge labels are atoms of $\mathcal{A}$. Hence, $N(x_0, x_1) = N(y_0, y_1)$.

The reader may wonder why we do not label edges (of length 2) with hyperlabels as well as atoms. There are places in chapter 13 (lemmas 13.35 and 13.44), where it would help if we did — but only in the case $n = 4$, and this case can always be handled by direct means. So rather than define a hypernetwork to be a pair of maps (N, η), where $N : {}^2 n \to \text{At}\,\mathcal{A}$ and $\eta : {}^{\leq n}n \to \Lambda$, to keep the notation simple we prefer to adopt the definition above.

It will be seen that the important hyperlabels are the ones labelling the hyperedges of length $n - 2$.

REMARK 12.3 The structure obtained from a hypernetwork by discarding its hyperlabels is clearly an atomic network as defined in definition 11.1 (except for the technical quibble that we only defined networks for relation algebras, not for non-associative algebras). It should be clear that if we take Λ to be a singleton set, no

12.1. Hypernetworks

real information is carried by the hyperlabels. In that case, a hypernetwork is effectively an atomic network, and we will usually regard such hypernetworks simply as atomic networks. Thus, hypernetworks generalise ordinary atomic networks (at least if we restrict consideration to atomic networks with nodes n).

We have the following hypernetwork version of lemma 11.2; the proof is the same as before.

LEMMA 12.4 *Let N be an n-dimensional hypernetwork and let $x, y < n$. Then $N(x, y) = N(y, x)^{\smile}$.*

Proof. By property 2 of definition 12.1, $N(x,y) ; N(y,x) \geq N(x,x)$, so by property 1, $(N(x,y) ; N(y,x)) \cdot 1' \neq 0$. By the Peircean law in $\mathcal{A}$ (lemma 5.9), we obtain $N(x,y) \cdot N(y,x)^{\smile} \neq 0$. Now $N(x,y)$ is an atom of $\mathcal{A}$, and by lemma 3.11, $N(y,x)^{\smile}$ is also an atom. We obtain $N(x,y) = N(y,x)^{\smile}$, as required. □

12.1.2 Comparing and altering hypernetworks

The following is a basic way to compare two hypernetworks.

DEFINITION 12.5 Let M, N be n-dimensional hypernetworks.

1. For $x < n$, we write $M \equiv_x N$ if $M(\bar{y}) = N(\bar{y})$ for all $\bar{y} \in {}^{\leq n}(n \setminus \{x\})$.

2. More generally, if $x_0, \ldots, x_{k-1} < n$, we write $M \equiv_{x_0,\ldots,x_{k-1}} N$ if $M(\bar{y}) = N(\bar{y})$ for all $\bar{y} \in {}^{\leq n}(n \setminus \{x_0, \ldots, x_{k-1}\})$. And if $S \subseteq n$, then $M \equiv_S N$ means that $M(\bar{y}) = N(\bar{y})$ for all $\bar{y} \in {}^{\leq n}(n \setminus S)$.

In Maddux's terminology, $M \equiv_x N$ means that M and N 'agree off of x'. Obviously, $\equiv_x$ and $\equiv_{x_0,\ldots,x_{k-1}}$ are equivalence relations on hypernetworks.

We can create new hypernetworks from old ones using transformations on n.

DEFINITION 12.6 *If N is an n-dimensional Λ-hypernetwork over $\mathcal{A}$, and $\tau : n \to n$ is any map, then $N\tau$ denotes the n-dimensional Λ-hypernetwork over $\mathcal{A}$ with labellings defined by*

$$N\tau(\bar{x}) = N(\tau(\bar{x})), \quad \text{for all } \bar{x} \in {}^{\leq n}n.$$

Recall that $\tau(\bar{x})$ is $(\tau(x_0), \ldots, \tau(x_{l-1}))$, where $|\bar{x}| = l < \omega$, and $\tau(\bar{x}) = \tau \circ \bar{x} \in {}^{\omega}n$ for $\bar{x} \in {}^{\omega}n$.

We can think of $N\tau$ as a hypernetwork that 'embeds' into N in the model-theoretic sense — the embedding is τ and it takes a node x of $N\tau$ to the node $\tau(x)$ of N. This embedding need not be one-one, of course, but it does preserve all labels.

We now justify the claim in the definition that $N\tau$ is indeed a hypernetwork.

LEMMA 12.7 *Let N be an n-dimensional Λ-hypernetwork over $\mathcal{A}$, and $\tau : n \to n$ be a map. Then $N\tau$ is also a Λ-hypernetwork over $\mathcal{A}$.*

Proof. We make the necessary checks. Certainly, $N\tau(\bar{x}) \in \mathrm{At}\,\mathcal{A}$ if $|\bar{x}| = 2$, and $N\tau(\bar{x}) \in \Lambda$, otherwise. Let $x, y, z < n$. Then

$$\begin{aligned} N\tau(x,x) &= N(\tau(x),\tau(x)) \leq 1\text{'}, \\ N\tau(x,y) &= N(\tau(x),\tau(y)) \\ &\leq N(\tau(x),\tau(z)) ; N(\tau(z),\tau(y)) \\ &= N\tau(x,z) ; N\tau(z,y). \end{aligned}$$

Finally, let $\bar{x}, \bar{y} \in {}^l n$ for some $l \leq n$, and assume that $N\tau(x_i, y_i) \leq 1\text{'}$ for each $i < l$. So $N(\tau(\bar{x})_i, \tau(\bar{y})_i) = N(\tau(x_i), \tau(y_i)) = N\tau(x_i, y_i) \leq 1\text{'}$ for each $i < l$. As N is a hypernetwork, $N(\tau(\bar{x})) = N(\tau(\bar{y}))$, so by definition, $N\tau(\bar{x}) = N\tau(\bar{y})$. □

Recall from definition 5.22 that the 'substitution' map $[x/y] : n \to n$, for $x, y < n$, is defined to be the identity on $n \setminus \{x\}$ and such that $[x/y](x) = y$. The following lemma characterises the hypernetwork $N[x/y]$.

LEMMA 12.8 *Let M, N be n-dimensional hypernetworks over $\mathcal{A}$, and let $x, y < n$ be distinct. Then $M = N[x/y]$ iff $M \equiv_x N$ and $M(x, y) \leq 1\text{'}$.*

Proof. Left-to-right is clear. For the converse, suppose that $M \equiv_x N$ and $M(x,y) \leq 1\text{'}$. For brevity, we write $z' = [x/y](z)$ for each $z < n$. Let $\bar{z} \in {}^{\leq n}n$, and define $\bar{z}' = [x/y](\bar{z})$. Then $N[x/y](\bar{z}) = N(\bar{z}') = M(\bar{z}')$, since $\bar{z}' \in {}^{\leq n}(n \setminus \{x\})$. Now because $M(x,y) \leq 1\text{'}$, it follows that $M(z'_i, z_i) \leq 1\text{'}$ for each $i < |\bar{z}|$. So $M(\bar{z}) = M(\bar{z}') = N[x/y](\bar{z})$, for all $\bar{z}$. That is, $M = N[x/y]$. □

Exercises

1. Let N be an n-dimensional hypernetwork, and $\sigma, \tau : n \to n$.

   (a) Show that $(N\sigma)\tau = N(\sigma \circ \tau)$.

   (b) Show that if $x < n$ and $\sigma(y) = \tau(y)$ for all $y < n$, $y \neq x$, then $N\sigma \equiv_x N\tau$.

   (c) More generally, let $S \subseteq n$, and assume that $N(\tau(s), \sigma(s)) \leq 1\text{'}$ for all $s \in S$. Prove that $N\tau \equiv_{n \setminus S} N\sigma$.

12.2 Relational bases and hyperbases

Now we will formally define the notions of relational basis and hyperbasis. We do this only for hypernetworks, in order to save work and avoid proliferating definitions: these two are the only kinds of basis we need. In section 12.6 we will

12.2. Relational bases and hyperbases

discuss the connections to Maddux's original definitions, one of which was somewhat different. For the present, we restrict to finite-dimensional bases, leaving some results about ω-dimensional bases to section 17.5.

The atomic non-associative algebra $\mathcal{A}$ remains fixed. The dimension n will be finite, with $n \geq 2$ for relational bases and $n \geq 3$ for hyperbases. We also fix a non-empty set Λ.

12.2.1 Relational bases

DEFINITION 12.9 Let $2 \leq n < \omega$. An *n-dimensional relational Λ-basis* for $\mathcal{A}$ is a set $\mathcal{R}$ of n-dimensional Λ-hypernetworks over $\mathcal{A}$, such that:

1. For all $a \in \operatorname{At}\mathcal{A}$ there is $N \in \mathcal{R}$ such that $N(0,1) = a$.

2. For all $N \in \mathcal{R}$, all $x, y, z < n$ with $z \neq x, y$, and for all $a, b \in \operatorname{At}\mathcal{A}$ such that $N(x,y) \leq a\,;b$, there is $M \in \mathcal{R}$ with $M \equiv_z N$, $M(x,z) = a$, and $M(z,y) = b$.

   We call this the *triangle addition property* — roughly, given any hypernetwork in a relational basis, we can attach to it any consistent triangle and still remain in the basis.

We say that $\mathcal{A}$ *has an n-dimensional relational basis* if for some set Λ as above, there exists an n-dimensional relational Λ-basis for $\mathcal{A}$. We let **RB**$_n$ denote the class of all atomic non-associative algebras that have an n-dimensional relational basis.

It is easily proved that for relational bases, Λ is irrelevant, so that really we are dealing with atomic networks.

LEMMA 12.10 *Suppose that $\mathcal{A}$ has an n-dimensional relational Λ-basis, where Λ is some set. Then for any non-empty Λ', $\mathcal{A}$ has an n-dimensional relational Λ'-basis.*

Proof. If $\mathcal{R}$ is an n-dimensional relational Λ-basis for $\mathcal{A}$, then pick any function $f : \Lambda \to \Lambda'$ (for example, a constant function) and let $\mathcal{R}'$ be obtained from $\mathcal{R}$ by redefining the hyperlabels of hypernetworks in $\mathcal{R}$ to be their f-images in Λ'. Clearly, $\mathcal{R}'$ is a relational Λ'-basis. $\square$

The proof shows that we can blur the information carried by the hyperlabels without affecting relational basis-hood. In particular, if $\mathcal{A}$ has any relational basis then it has a relational $\{\lambda\}$-basis consisting essentially of atomic networks. So we will often take a relational Λ-basis to consist of atomic networks, not hypernetworks, and refer to it simply as a *relational basis*. The hyperedges and hyperlabels play no role at all. However, for the hyperbases to be defined next, Λ cannot be dispensed with in this way.

12.2.2 Hyperbases

DEFINITION 12.11 Let $3 \leq n < \omega$. An *n-dimensional Λ-hyperbasis* for $\mathcal{A}$ is an *n*-dimensional relational Λ-basis $\mathcal{H}$ for $\mathcal{A}$ with the additional property that:

3. For all $M, N \in \mathcal{H}$ and $x, y < n$ with $M \equiv_{xy} N$, there is $L \in \mathcal{H}$ such that $M \equiv_x L \equiv_y N$.

    We call this the *amalgamation property:* see figure 12.1. We may assume that $x \neq y$ here, since if $x = y$ we may take $L = M$.

We say that $\mathcal{A}$ *has an n-dimensional hyperbasis* if for some set Λ as above, there exists an *n*-dimensional Λ-hyperbasis for $\mathcal{A}$.

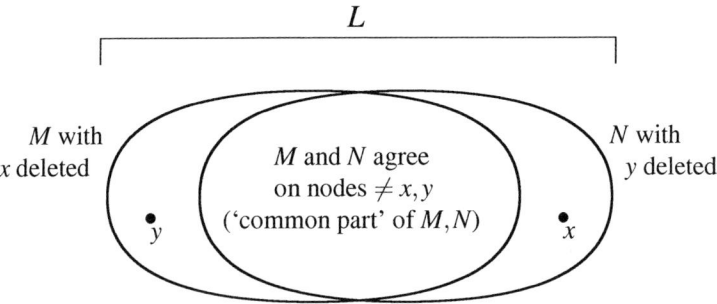

Figure 12.1: Amalgamation property in a hyperbasis

We do not permit $n = 2$ in definition 12.11 because that would require amalgamating hypernetworks with no 'common part' ($N \equiv_{01} M$ for all 2-dimensional hypernetworks M, N). Such amalgamation is not always possible for non-simple relation algebras, even representable ones. Results such as proposition 12.22 below fail for $m = 2$. We conclude that a 2-dimensional hyperbasis would be a pathological notion.

There is no analogue of lemma 12.10 for hyperbases. Changing the hyperlabels of hypernetworks in a hyperbasis may introduce new instances of hypernetworks N, M with $M \equiv_{xy} N$, for which no witnesses L to the amalgamation property can be found in the original hyperbasis.

The property of having a hyperbasis is in general stronger than that of having a relational basis. We will see in remark 15.13 that for all $n \geq 5$, there is a finite relation algebra with an *n*-dimensional relational basis but no *n*-dimensional hyperbasis.

12.3 Elementary properties of bases

Here we derive some simple consequences of the definitions given above. $\mathcal{A}$ remains a fixed atomic non-associative algebra, and we also fix the dimension $n < \omega$ with $n \geq 2$ for relational bases and $n \geq 3$ for hyperbases.

12.3.1 Symmetric bases

First we investigate closure of bases under substitutions and permutations. The following fact will be very useful here and in the next chapter. Let α be any ordinal. Recall from section 5.4.2 that a (total) map $f : \alpha \to \alpha$ is *finitary* if $\{i < \alpha : f(i) \neq i\}$ is finite.

FACT 12.12 *Any finitary $f : \alpha \to \alpha$ that is not a permutation of α is either of the form $[i/j]$ (for some $i, j < \alpha$) or a composition of finitely many maps of this form.*

Proving the fact is a not-too-difficult exercise (exercise 4 below). It can be derived from [Jón62]. [How78, theorem 1] proves it for finite α, and [Tho93, corollary 1.2] for arbitrary α.

LEMMA 12.13 *Let $\mathcal{R}$ be an n-dimensional relational basis for $\mathcal{A}$. Then $\mathcal{R}$ is closed under substitutions: if $N \in \mathcal{R}$ and $x, y < n$ then $N[x/y] \in \mathcal{R}$. Hence, if $\tau : n \to n$ is not a permutation then $N \in \mathcal{R}$ implies $N\tau \in \mathcal{R}$.*

Proof. If $x = y$, the result is trivial, so assume otherwise. By definition 12.1, $N(y, y) \leq N(y, y) ; N(y, y)$. By the triangle addition property, we can find some $M \in \mathcal{R}$ with $M \equiv_x N$ and $M(y, x) = M(x, y) = N(y, y)$. In particular, $M(x, y) \leq 1$'. By lemma 12.8, $M = N[x/y]$.

The last part is immediate by fact 12.12 and induction. □

DEFINITION 12.14 *An n-dimensional relational basis $\mathcal{R}$ is said to be symmetric if whenever $N \in \mathcal{R}$ and $\sigma : n \to n$ then $N\sigma \in \mathcal{R}$.*

Of course, this definition applies to hyperbases as a special case. We say that an atomic non-associative algebra $\mathcal{A}$ *has a symmetric n-dimensional relational basis (or hyperbasis)* if there exists a symmetric n-dimensional relational Λ-basis (respectively, a symmetric n-dimensional Λ-hyperbasis) for $\mathcal{A}$, for some Λ.

Already, for any n-dimensional relational basis $\mathcal{R}$, if $N \in \mathcal{R}$ and $\tau : n \to n$ is not a permutation then by lemma 12.13, $N\tau \in \mathcal{R}$. We will now show how to make fully symmetric bases, closed even under permutations.

LEMMA 12.15 *Let $\mathcal{R}$ be an n-dimensional relational Λ-basis for $\mathcal{A}$, and define*

$$\overline{\mathcal{R}} = \{N\sigma : N \in \mathcal{R},\ \sigma : n \to n\},$$

370   Chapter 12. Relational, cylindric, and hyperbases

the symmetric closure of $\mathcal{R}$. Then $\overline{\mathcal{R}}$ is a symmetric n-dimensional relational Λ-basis for $\mathcal{A}$. Further, if $\mathcal{R}$ is a hyperbasis then so is $\overline{\mathcal{R}}$.

Proof. By lemma 12.7, the elements of $\overline{\mathcal{R}}$ are bona fide hypernetworks, and $\overline{\mathcal{R}}$ is certainly symmetric (see exercise 12.1(1a)).

We check the defining properties of relational bases. For any atom $a \in \mathcal{A}$, there is $N \in \overline{\mathcal{R}}$ with $N(0,1) = a$, because $\overline{\mathcal{R}} \supseteq \mathcal{R}$. For the triangle addition property, let $N \in \overline{\mathcal{R}}$ and $\sigma : n \to n$, let $x,y,z < n$ with $z \neq x,y$, and let $a,b \in \mathrm{At}\,\mathcal{A}$ with $N\sigma(x,y) \leq a;b$. We seek $P \in \overline{\mathcal{R}}$ with $P \equiv_z N\sigma$, $P(x,z) = a$, and $P(z,y) = b$.

Let $\tau = \sigma \circ [z/x]$. It is clear that τ is not a permutation of n, so by lemma 12.13, $N\tau \in \mathcal{R}$. Since clearly $N\tau \equiv_z N\sigma$ (see lemma 12.8 or exercise 12.1(1b)), we have $N\tau(x,y) \leq a;b$. As $\mathcal{R}$ is a relational basis, there is $P \in \mathcal{R}$ with $P \equiv_z N\tau$, $P(x,z) = a$, and $P(z,y) = b$. Plainly, $P \in \overline{\mathcal{R}}$ and $P \equiv_z N\sigma$, so P is as required.

Now suppose that $\mathcal{R}$ is a hyperbasis. We check that $\overline{\mathcal{R}}$ is also a hyperbasis. Let $N, M \in \overline{\mathcal{R}}$, $\sigma, \tau : n \to n$, and assume that $N\sigma \equiv_{xy} M\tau$ for some distinct $x,y < n$. We require $N\sigma \equiv_i P \equiv_j M\tau$ for some $P \in \overline{\mathcal{R}}$. By lemma 12.8/exercise 12.1(1b) again, we have

$$N(\sigma \circ [x/y]) \equiv_x N\sigma \equiv_{xy} M\tau \equiv_y M(\tau \circ [y/x]),$$

so that $N(\sigma \circ [x/y]) \equiv_{xy} M(\tau \circ [y/x])$. Again, $\sigma \circ [x/y]$ is not a permutation of n, so by lemma 12.13, $N(\sigma \circ [x/y]) \in \mathcal{R}$. Similarly, $M(\tau \circ [y/x]) \in \mathcal{R}$. As $\mathcal{R}$ is a hyperbasis, there is $P \in \mathcal{R}$ with $N(\sigma \circ [x/y]) \equiv_x P \equiv_y M(\tau \circ [y/x])$. Hence, $P \in \overline{\mathcal{R}}$ and $N\sigma \equiv_x P \equiv_y M\tau$, as required.   □

12.3.2 Interpolation in hyperbases

The remainder of this section mostly concerns properties specific to hyperbases. The definition of hyperbasis allows us to find an 'interpolant' L between two hypernetworks M,N that differ on two nodes only. The next lemma allows us to find a sequence of interpolants in a hyperbasis between two hypernetworks that may differ on many nodes.

LEMMA 12.16 *Let $\mathcal{H}$ be an n-dimensional hyperbasis for $\mathcal{A}$. Suppose that the hypernetworks $M,N \in \mathcal{H}$ satisfy $M \equiv_{\bar{x}} N$, for some sequence $\bar{x} = (x_0, \ldots, x_{l-1}) \in {}^{<\omega}n$ such that $\mathrm{rng}(\bar{x}) \neq n$. Then there exist hypernetworks $N_i \in \mathcal{H}$ for $i \leq l$ such that*

$$M = N_0 \equiv_{x_0} N_1 \equiv_{x_1} N_2 \equiv_{x_2} \cdots \equiv_{x_{l-2}} N_{l-1} \equiv_{x_{l-1}} N_l = N.$$

Proof. By induction on $|\bar{x}|$. For $|\bar{x}| \leq 2$, the result holds by definition of a hyperbasis. So pick finite $l > 2$, let $\bar{x} \in {}^l n$ be such that $\mathrm{rng}(\bar{x}) \neq n$, and suppose

12.3. Elementary properties of bases

the lemma holds for all shorter sequences. Pick $k \in n \setminus \mathrm{rng}(\bar{x})$. Suppose that $M, N \in \mathcal{H}$ and $M \equiv_{\bar{x}} N$. By lemma 12.13, $M[x_0/k], N[x_0/k] \in \mathcal{H}$, and clearly, we have $M[x_0/k] \equiv_{(x_1, x_2, \ldots, x_{l-1})} N[x_0/k]$. So inductively, there are hypernetworks $N_i \in \mathcal{H}$ for $1 \leq i \leq l$ such that

$$M \equiv_{x_0} M[x_0/k] = N_1 \equiv_{x_1} N_2 \equiv_{x_2} \cdots$$
$$\cdots \equiv_{x_{l-2}} N_{l-1} \equiv_{x_{l-1}} N_l = N[x_0/k] \equiv_{x_0} N.$$

Finally we must reorder these equivalences. From the last two equivalences we see that $N_{l-1} \equiv_{x_0, x_{l-1}} N$. From the definition of a hyperbasis, there exists $N'_{l-1} \in \mathcal{H}$ with $N_{l-1} \equiv_{x_0} N'_{l-1} \equiv_{x_{l-1}} N$. Now, we have $N_{l-2} \equiv_{x_{l-2}} N_{l-1} \equiv_{x_0} N'_{l-1}$, so as before, there is $N'_{l-2} \in \mathcal{H}$ with $N_{l-2} \equiv_{x_0} N'_{l-2} \equiv_{x_{l-2}} N'_{l-1}$.

$$
\begin{array}{ccccccccc}
M & \equiv_{x_0} & N_1 & \equiv_{x_1} & N_2 & \equiv_{x_2} & \cdots & \equiv_{x_{l-2}} & N_{l-1} & \equiv_{x_{l-1}} & N_l \\
& & \|\|_{x_0} & & \|\|_{x_0} & & \cdots & & \|\|_{x_0} & & \|\|_{x_0} \\
& & N'_1 & \equiv_{x_1} & N'_2 & \equiv_{x_2} & \cdots & \equiv_{x_{l-2}} & N'_{l-1} & \equiv_{x_{l-1}} & N
\end{array}
$$

Proceeding to fill in the bottom row of the diagram above in this way, from right to left, we obtain hypernetworks $N'_i \in \mathcal{H}$ for $1 \leq i < l$ as shown. Since the first two equivalences give $M \equiv_{x_0} N'_1$, we obtain our result. □

12.3.3 From hyperbasis to cylindric algebra

A set of hypernetworks can be made into the atom structure of a cylindric-type algebra, as follows.

DEFINITION 12.17 Let $\mathcal{H}$ be a set of n-dimensional hypernetworks over $\mathcal{A}$. We define an algebra $\mathfrak{Ca}(\mathcal{H})$ of the signature of n-dimensional cylindric algebras. The domain of $\mathfrak{Ca}(\mathcal{H})$ is the full power set $\wp(\mathcal{H})$. The boolean operations are defined as expected (as complement and union of sets). For $i, j < n$, we define the diagonal

$$\mathsf{d}_{ij} = \{N \in \mathcal{H} : N(i, j) \leq 1'\},$$

and for $i < n$ we define the cylindrifier c_i by

$$\mathsf{c}_i S = \{N \in \mathcal{H} : \exists M \in S(N \equiv_i M)\}, \text{ for } S \subseteq \mathcal{H}.$$

Equivalently, we can regard $\mathcal{H}$ as a cylindric-type atom structure (see definition 2.62), via

- $R_{\mathsf{c}_i}(N, M)$ holds iff $N \equiv_i M$,

- $R_{\mathsf{d}_{ij}}(N)$ holds iff $N(i,j) \leq 1$',

and define $\mathfrak{Ca}\,\mathcal{H}$ to be the complex algebra $\mathfrak{Cm}\,\mathcal{H}$ as in section 2.7.2.

The next proposition relates hyperbases to relation algebra reducts (see definition 5.40 for these).

PROPOSITION 12.18 *Let $\mathcal{H}$ be an n-dimensional hyperbasis for $\mathcal{A}$. Then*

1. $\mathfrak{Ca}(\mathcal{H}) \in \mathbf{CA}_n$,

2. $\mathcal{A}$ embeds into $\mathfrak{Ra}(\mathfrak{Ca}(\mathcal{H}))$ via $a \mapsto \{N \in \mathcal{H} : N(0,1) \leq a\}$.

For a proof, borrow the idea of [Madd89b, theorem 10] where the same is proved for cylindric bases, or try the similar exercise 14 below. This is one way to prove a result in the next chapter — that any subalgebra of a relation algebra with an n-dimensional hyperbasis is in $\mathbf{S\mathfrak{Ra}CA}_n$ — though we will actually prove it via theorem 13.46 parts 7 and 8. Proposition 12.18 will be useful in chapter 15.

12.3.4 Reducing the dimension of a relational basis

We can decrease the dimension of a relational basis in a well behaved way, by restricting its hypernetworks to a smaller set of nodes.

DEFINITION 12.19 *Let $2 \leq m \leq n$.*

1. If N is an n-dimensional Λ-hypernetwork over $\mathcal{A}$, then $N{\upharpoonright}_m$ denotes the restriction of the map N to $^{\leq m}m$. Clearly, $N{\upharpoonright}_m$ is an m-dimensional Λ-hypernetwork.

2. If $\mathcal{M}$ is an n-dimensional relational basis for $\mathcal{A}$, we write $\mathcal{M}{\upharpoonright}_m$ for $\{N{\upharpoonright}_m : N \in \mathcal{M}\}$.

We now prove that $\mathcal{M}{\upharpoonright}_m$ is an m-dimensional relational basis for $\mathcal{A}$.

LEMMA 12.20 *Let Λ be a non-empty set, and let $2 \leq m \leq n$. If $\mathcal{M}$ is an n-dimensional relational Λ-basis over $\mathcal{A}$, then $\mathcal{M}{\upharpoonright}_m$ is an m-dimensional relational Λ-basis over $\mathcal{A}$.*

Proof. We check the conditions for relational bases. If $a \in \mathsf{At}\,\mathcal{A}$ then there is $N \in \mathcal{M}$ with $N(0,1) = a$, so $N{\upharpoonright}_m(0,1) = a$ and condition 1 holds. To check the triangle addition property, let $N{\upharpoonright}_m$ with $N \in \mathcal{M}$, $x,y,z < m$ with $z \neq x,y$, and $a,b \in \mathsf{At}\,\mathcal{A}$ with $N{\upharpoonright}_m(x,y) \leq a;b$. We know $N(x,y) \leq a;b$, and since $\mathcal{M}$ is a relational basis, there is $N' \equiv_z N \in \mathcal{M}$ with $N'(x,z) = a$ and $N'(z,y) = b$. The required m-dimensional hypernetwork witnessing the triangle addition property is $N'{\upharpoonright}_m$. So $\mathcal{M}{\upharpoonright}_m$ is an m-dimensional relational basis for $\mathcal{A}$. □

12.3. Elementary properties of bases

12.3.5 Reducing the dimension of a hyperbasis

We now show how to restrict a hyperbasis. This is another technique that will be useful in chapter 15. However, if we simply restrict an n-dimensional hypernetwork N via $N \mapsto N\!\restriction_m$ for $m < n$, as for relational bases, we run into the difficulty that for $\bar{x} \in {}^n m$, the hyperlabel $N(\bar{x})$ is not available in $N\!\restriction_m$ because $\bar{x}$ is too long.[1] We would like to keep it, in order to preserve amalgamation properties, so we simply generalise the earlier definitions to allow long hyperedges of length up to n.

DEFINITION 12.21 *Let $3 \leq m \leq n \leq k < \omega$ and let Λ be a non-empty set.*

1. *An n-wide m-dimensional Λ-hypernetwork over $\mathcal{A}$ is a map $N : {}^{\leq n}m \to \Lambda \cup \mathrm{At}\,\mathcal{A}$ with the properties of definition 12.1 (extended to longer hyperedges in the obvious way). $H_m^n(\mathcal{A}, \Lambda)$ will denote the set of all such objects.*

2. *For n-wide m-dimensional Λ-hypernetworks N, M, and $\bar{x} \in {}^{<\omega}m$, we define $N \equiv_{\bar{x}} M$ if $N(\bar{y}) = M(\bar{y})$ for all $\bar{y} \in {}^{\leq n}(m \setminus \mathrm{rng}(\bar{x}))$.*

3. *An n-wide m-dimensional Λ-hyperbasis for $\mathcal{A}$ is a set $\mathcal{H} \subseteq H_m^n(\mathcal{A}, \Lambda)$ with the properties of definition 12.11. So if $n = m$, $\mathcal{H}$ is just a hyperbasis.*

4. *For a k-wide n-dimensional hypernetwork N, we let $N\!\restriction_m^k$ be the restriction of the map N to ${}^{\leq k}m$. So we restrict to the first m nodes but keep all hyperlabels on them. Clearly, $N\!\restriction_m^k \in H_m^k(\mathcal{A}, \Lambda)$.*

5. *For $\mathcal{H} \subseteq H_n^k(\mathcal{A}, \Lambda)$, we let $\mathcal{H}\!\restriction_m^k = \{N\!\restriction_m^k : N \in \mathcal{H}\}$. We also define the CA_n-type algebra $\mathfrak{Ca}(\mathcal{H})$ as in definition 12.17.*

Earlier results such as lemma 12.16 generalise to cover wide hyperbases. Proposition 12.18 also generalises, to show that if $\mathcal{H}$ is a k-wide n-dimensional Λ-hyperbasis for $\mathcal{A}$ then $\mathfrak{Ca}\,\mathcal{H} \in \mathbf{CA}_n$ and $\mathcal{A}$ embeds into $\mathfrak{Ra}\,\mathfrak{Ca}\,\mathcal{H}$ via the same map as before (see exercise 2). We can also generalise lemma 12.20 to hyperbases, and handle neat reducts:

PROPOSITION 12.22 *Let $3 \leq m \leq n \leq k < \omega$ be given, and let $\mathcal{H}$ be a k-wide n-dimensional hyperbasis over $\mathcal{A}$. Then $\mathcal{H}\!\restriction_m^k$ is a k-wide m-dimensional hyperbasis, and $\mathfrak{Ca}(\mathcal{H}\!\restriction_m^k) \cong \mathfrak{Nr}_m(\mathfrak{Ca}(\mathcal{H}))$ (see definition 5.40 for the neat reduct operation $\mathfrak{Nr}$).*

Proof. The proof of lemma 12.20 shows that $\mathcal{H}\!\restriction_m^k$ is an m-dimensional relational basis. It remains to check amalgamation. Let $N, M \in \mathcal{H}$ and $x, y < m$ satisfy $N\!\restriction_m^k \equiv_{xy} M\!\restriction_m^k$. Then if $\bar{z} \in {}^{\leq k}(m \setminus \{x, y\})$, we have $N\!\restriction_m^k(\bar{z}) = M\!\restriction_m^k(\bar{z})$ — i.e., $N(\bar{z}) = M(\bar{z})$. Hence, $N \equiv_{(x,y,m,m+1,\ldots,n-1)} M$. Since $m \geq 3$, we see that $\{x, y, m, m+1, \ldots, n-1\} \neq n$. So (the generalised) lemma 12.16 applies to yield

---

[1][HirHod01a, proposition 14] is wrong on this point.

$K, L \in \mathcal{H}$ with $N \equiv_x K \equiv_y L \equiv_{(m,m+1,\ldots,n-1)} M$. Then $K\!\restriction_m^k \in \mathcal{H}\!\restriction_m^k$ and $N\!\restriction_m^k \equiv_x K\!\restriction_m^k \equiv_y L\!\restriction_m^k = M\!\restriction_m^k$. Thus, $\mathcal{H}\!\restriction_m^k$ is a hyperbasis.

We now define a map $h : \mathfrak{Ca}(\mathcal{H}\!\restriction_m^k) \to \mathfrak{Nr}_m(\mathfrak{Ca}(\mathcal{H}))$ by

$$h(S) = \{N \in \mathcal{H} : N\!\restriction_m^k \in S\}, \text{ for } S \subseteq \mathcal{H}\!\restriction_m^k.$$

It should be clear that $h(S)$ is indeed in the neat reduct: if $N \in h(S)$, $m \leq i < n$, and $N' \in \mathcal{H}$ with $N' \equiv_i N$, then $N'\!\restriction_m^k = N\!\restriction_m^k \in S$, so $N' \in h(S)$. Hence, $c_i(h(S)) = h(S)$, and so $h(S) \in \mathfrak{Nr}_m(\mathfrak{Ca}(\mathcal{H}))$.

We show h is an isomorphism. Let $S, S' \subseteq \mathcal{H}\!\restriction_m^k$. Clearly, $h(S \cup S') = h(S) \cup h(S')$, $h(\mathcal{H}\!\restriction_m^k \setminus S) = \mathcal{H} \setminus h(S)$, and $h(S) = \emptyset$ iff $S = \emptyset$. It follows that h preserves the boolean operations and is one-one. For surjectivity, let $X \in \mathfrak{Nr}_m \mathfrak{Ca}\,\mathcal{H}$; define $S = \{N\!\restriction_m^k : N \in X\}$. Plainly, $X \subseteq h(S)$. Conversely, if $M \in h(S)$ then $M\!\restriction_m^k = N\!\restriction_m^k$ for some $N \in X$. Clearly, $M \equiv_{(m,m+1,\ldots,n-1)} N$. By lemma 12.16 again, there are $L_m, L_{m+1}, \ldots, L_{n-2} \in \mathcal{H}$ with $M \equiv_m L_m \equiv_{m+1} L_{m+1} \cdots L_{n-2} \equiv_{n-1} N$. So $M \in c_m c_{m+1} \cdots c_{n-1} X = X$. Hence, $h(S) = X$, proving that h is onto.

For the diagonals, let $i, j < m$. Let $\mathsf{d}_{ij}^{\mathfrak{Ca}(\mathcal{H}\!\restriction_m^k)}$ and $\mathsf{d}_{ij}^{\mathfrak{Ca}(\mathcal{H})}$ denote the ijth diagonals of $\mathfrak{Ca}(\mathcal{H}\!\restriction_m^k)$ and $\mathfrak{Ca}(\mathcal{H})$, respectively. Then

$$\begin{aligned} h(\mathsf{d}_{ij}^{\mathfrak{Ca}(\mathcal{H}\!\restriction_m^k)}) &= \{N \in \mathcal{H} : N\!\restriction_m^k(i,j) \leq 1'\} \\ &= \{N \in \mathcal{H} : N(i,j) \leq 1'\} \\ &= \mathsf{d}_{ij}^{\mathfrak{Ca}(\mathcal{H})}. \end{aligned}$$

Finally, for the cylindrifiers, let $S \subseteq \mathcal{H}\!\restriction_m^k$ and $i < m$. We need to show that $h(\mathsf{c}_i^{\mathfrak{Ca}(\mathcal{H}\!\restriction_m^k)}(S)) = \mathsf{c}_i^{\mathfrak{Ca}(\mathcal{H})}(h(S))$. To do this, we require that for any $N \in \mathcal{H}$,

$$N\!\restriction_m^k \equiv_i P \text{ for some } P \in S \iff N \equiv_i Q \text{ for some } Q \in \mathcal{H} \text{ with } Q\!\restriction_m^k \in S.$$

So let $N \in \mathcal{H}$. Right-to-left is straightforward: if $Q \in \mathcal{H}$, $N \equiv_i Q$, and $Q\!\restriction_m^k \in S$, then $N\!\restriction_m^k \equiv_i Q\!\restriction_m^k \in S$. For the converse, assume that $N\!\restriction_m^k \equiv_i P$ for some $P \in S$. As $S \subseteq \mathcal{H}\!\restriction_m^k$, we may take $Q' \in \mathcal{H}$ satisfying $Q'\!\restriction_m^k = P$. Then, $N \equiv_{(i,m,m+1,\ldots,n-1)} Q'$. By lemma 12.16, there is $Q \in \mathcal{H}$ such that $N \equiv_i Q \equiv_{(m,m+1,\ldots,n-1)} Q'$. So $N \equiv_i Q$ and $Q\!\restriction_m^k = Q'\!\restriction_m^k = P \in S$, as required.

Thus, h is a cylindric algebra isomorphism. □

REMARK 12.23 In the case where $|\Lambda| = 1$, hyperlabels carry no information. Proposition 12.22 thus yields that if $\mathcal{H}$ is an n-dimensional $\{\lambda\}$-hyperbasis over $\mathcal{A}$, for some λ, then $\mathcal{H}\!\restriction_m$ is an m-dimensional hyperbasis and $\mathfrak{Ca}(\mathcal{H}\!\restriction_m) \cong \mathfrak{Nr}_m \mathfrak{Ca}\,\mathcal{H}$.

Exercises

1. Show, for $3 \leq m \leq n \leq k < \omega$ and any atomic non-associative algebra $\mathcal{A}$ and non-empty set Λ, that $H_m^k(\mathcal{A}, \Lambda) = H_n^k(\mathcal{A}, \Lambda)\!\restriction_m^k$.

12.3. Elementary properties of bases

2. Generalise lemma 12.16 and proposition 12.18 to wide hyperbases.

3. Show that if an atomic non-associative algebra has a wide hyperbasis then it has an ordinary one. [Encode overlong hyperlabels in shorter ones.]

4. (Cf. fact 12.12.) Let α be an ordinal, and let $\tau : \alpha \to \alpha$ be a finitary non-bijective map. Show that τ is a product of substitutions: i.e., $\tau = [i_0/j_0] \circ \cdots \circ [i_k/j_k]$ for some $k < \omega$ and $i_0, \ldots, i_k, j_0, \ldots, j_k < \alpha$.

5. Show that any relation algebra with a complete representation has an n-dimensional hyperbasis for all finite $n \geq 3$. [If the base of the complete representation is M, let $\Lambda = {}^{\leq n}M$. Cf. theorems 12.37 and 12.41 below.]

6. Let $n \geq 5$, let $\mathcal{A}, \mathcal{B}$ be non-associative algebras, and suppose $\mathcal{A} \subseteq^c \mathcal{B}$ (see definition 2.15). Show that if $\mathcal{B}$ is atomic and has an n-dimensional relational basis, then $\mathcal{A}$ is also atomic and has an n-dimensional relational basis. Repeat for hyperbases. [For amalgamation, encode edge labellings in hyperedge labels.]

7. Let $\mathcal{H}$ be an n-dimensional Λ-hyperbasis for the atomic non-associative algebra $\mathcal{A}$, where $n \geq 3$. Show that if $\mathfrak{Ca}\,\mathcal{H} \in \mathbf{RCA}_n$ then $\mathcal{A} \in \mathbf{RRA}$. Prove that the converse holds for finite $\mathcal{A}$, $|\Lambda| = 1$, and $n = 3$ (cf. [Mon61b]).

8. Show that in the triangle addition property (as in definition 12.9(2)) for n-dimensional relational bases of *semi-associative algebras,* where $n \geq 3$, we can add the restriction that $x \neq y$ and obtain an equivalent definition. [Cf. lemma 13.26.]

9. Let $\mathcal{A}$ be an atomic semi-associative algebra. For finite $n \geq 3$, show that $\mathcal{A}$ has an n-dimensional relational basis iff there is a set $\mathcal{S}$ of n-dimensional Λ-hypernetworks over $\mathcal{A}$ such that:

    (a) For all $a \in \mathrm{At}\,\mathcal{A}$ there is $N \in \mathcal{S}$ such that $N(0,1) = a$.
    (b) For all $N \in \mathcal{S}$ and for all $a, b \in \mathrm{At}\,\mathcal{A}$ such that $N(0,1) \leq a;b$, there is $M \in \mathcal{S}$ with $M \equiv_2 N$, $M(0,2) = a$, and $M(2,1) = b$.

    [Hint for '$\Leftarrow$': lemma 13.26.]

10. One might surmise that an atomic non-associative algebra $\mathcal{A}$ is weakly associative iff it has a 2-dimensional relational basis. Prove this is false by showing that any atomic non-associative algebra has a 2-dimensional relational basis.

11. [Madd91b, theorem 35] Let $\mathcal{A}$ be an atomic non-associative algebra. Show that the following are equivalent:

(a) $\mathcal{A}$ is semi-associative ($\mathcal{A} \in \mathbf{SA}$),

(b) $\mathcal{A}$ has a 3-dimensional relational basis,

(c) $\mathcal{A}$ has a 3-dimensional $\{\lambda\}$-hyperbasis.

[Consider the set of all 3-dimensional $\{\lambda\}$-hypernetworks; use lemma 5.10 and exercise 5.1(10).]

12. [[Madd83, theorem 5] plus exercise 11] Let $\mathcal{A}$ be an atomic non-associative algebra. Show that the following are equivalent:

    (a) $\mathcal{A}$ is a relation algebra ($\mathcal{A} \in \mathbf{RA}$),

    (b) $\mathcal{A}$ has a 4-dimensional relational basis,

    (c) $\mathcal{A}$ has a 4-dimensional $\{\lambda\}$-hyperbasis.

13. Show that for $n = 3, 4$, any n-dimensional relational $\{\lambda\}$-basis of an atomic non-associative algebra is actually an n-dimensional $\{\lambda\}$-hyperbasis, for arbitrary λ.

14. Let $\mathcal{A}$ be an atomic non-associative algebra and let $3 \leq n < \omega$. Show that if $\mathcal{A}$ has an n-dimensional relational basis then $\mathcal{A} \in \mathbf{S\mathfrak{R}\mathfrak{a}G}_n$. In addition, show that if $\mathcal{A}$ is finite and has an n-dimensional relational basis then there are finite $C \in \mathbf{G}_n$ and a relation algebra reduct embedding from $\mathcal{A}$ into C. (We do not write $\mathcal{A} \in \mathbf{S\mathfrak{R}\mathfrak{a}}\{C \in \mathbf{G}_n : C \text{ is finite}\}$ because we promised in remark 5.50 not to define such a class.) [As in proposition 12.18, turn the basis into the atom structure of an algebra. Use the axioms for $\mathbf{G}_n$ given in fact 5.38. Cf. exercise 5.7(4).]

12.4 Games

We now justify our earlier claim that bases 'are' winning strategies for games.

12.4.1 Game for relational bases

We first define a game that tests the existence of an n-dimensional relational basis. As we remarked earlier, the hyperlabels are not important here, so the game will be played over atomic networks. We will also assume here that the set of nodes of an atomic network can be arbitrary, not necessarily of the form n for some $n \leq \omega$. This freer approach, matching the original definition 7.1, will help in chapter 16.

DEFINITION 12.24 Let $\mathcal{A}$ be an atomic non-associative algebra, and let $2 \leq n \leq \omega$ and $r \leq \omega$. The game $G_r^n(\mathcal{A})$ is played over atomic $\mathcal{A}$-networks by our players $\forall$ and $\exists$, and has r rounds.

12.4. Games

1. In round 0, $\forall$ picks an atom $a \in \text{At}\,\mathcal{A}$. $\exists$ must respond with an atomic $\mathcal{A}$-network N_0 with set of nodes $\{x,y\}$, say, (possibly $x = y$), and such that $N_0(x,y) = a$.

2. In any round t with $0 < t < r$, the current network being N_{t-1}, $\forall$ moves as follows. First, if $|N_{t-1}| = n$ then he chooses some node $z \in N_{t-1}$ to delete, and defines N to be the subnetwork of N_{t-1} resulting from deleting node z. (See definition 7.3 for subnetworks.) If $|N_{t-1}| < n$ then he sets $N = N_{t-1}$.

   Now he chooses nodes $x, y \in N$ and atoms $a, b \in \text{At}\,\mathcal{A}$ such that $N(x,y) \leq a\,;b$. $\exists$ must respond with an atomic $\mathcal{A}$-network $N_t \supseteq N$ with $|N_t| \leq n$ and containing a node z such that $N_t(x,z) = a$ and $N_t(z,y) = b$.

As usual, $\exists$ wins if she responds legally to $\forall$'s move in each round $t < r$, and she loses otherwise. By convention, $\exists$ wins the 0-round game $G_0^n(\mathcal{A})$ by default.

This game is very similar to the atomic network game $G_r^a(\mathcal{A})$ of definition 11.3 (when $\mathcal{A}$ is a relation algebra and $n = \omega$ they are essentially the same), but now $\mathcal{A}$ can be non-associative, and the networks are bounded in size by n. It is analogous to what in modal logic and finite model theory is called an 'n-pebble game'. This is why $\forall$ has to delete a node when the size of the next network threatens to exceed n. Apart from the potential non-associativity of $\mathcal{A}$, the game $G_r^\omega(\mathcal{A})$ is the same as the game $G_r^a(\mathcal{A})$ of definition 11.3. Here, we are mainly interested in finite dimensions.

The following proposition shows the correspondence between the game G_r^n and relational bases. It is proved by viewing a relational basis as a winning strategy, and vice versa.

PROPOSITION 12.25 *Let $\mathcal{A}$ be an atomic non-associative algebra, and let $2 \leq n < \omega$. Then $\mathcal{A}$ has an n-dimensional relational basis iff $\exists$ has a winning strategy in $G_\omega^n(\mathcal{A})$.*

Proof. If $\mathcal{R}$ is an n-dimensional relational basis for $\mathcal{A}$, then $\exists$ can win $G_\omega^n(\mathcal{A})$ by always playing a subnetwork of a network in $\mathcal{R}$. In round 0, when $\forall$ plays the atom $a \in \mathcal{A}$, say, she may choose $N \in \mathcal{R}$ with $N(0,1) = a$ and play $N\!\restriction_{\{0,1\}}$. In round $t > 0$, if the current network is $N_{t-1} \subseteq M \in \mathcal{R}$, then however $\forall$ defines N, we have $N \subseteq M$ and $|N| < n$, so there is $z < n$ with $z \notin N$. Let $\forall$ pick $x, y \in N$ and atoms a, b of $\mathcal{A}$ with $a\,;b \geq N(x,y)$. So $a\,;b \geq M(x,y)$, and by the triangle addition property for $\mathcal{R}$ there is $M' \in \mathcal{R}$ with $M' \equiv_z M$, $M'(x,z) = a$, and $M'(z,y) = b$. $\exists$ may respond to $\forall$'s move with the restriction of M' to $\text{nodes}(N) \cup \{z\}$. Clearly, she always plays a valid atomic $\mathcal{A}$-network, so she never loses and hence wins the game.

Conversely, assume that $\exists$ has a winning strategy in $G_\omega^n(\mathcal{A})$. For any atomic network N with $|N| \leq n$, let the game $G_\omega^n(N, \mathcal{A})$ be defined as for $G_\omega^n(\mathcal{A})$ except that in round 0, $\forall$ does nothing and $\exists$ simply plays $N_0 = N$. Let $\mathcal{R}$ be the set of all atomic $\mathcal{A}$-networks M with set of nodes n such that $\exists$ has a winning strategy in $G_\omega^n(M, \mathcal{A})$. We claim that $\mathcal{R}$ is an n-dimensional relational basis for $\mathcal{A}$.

To see this, first let $a \in \mathrm{At}\,\mathcal{A}$. We seek $N \in \mathcal{R}$ with $N(0,1) = a$. Let $\forall$ play a in round 0 of $G_\omega^n(\mathcal{A})$, and let $\exists$ respond with her winning strategy by playing N_0, with nodes $\{x, y\}$. Let $f : n \to \{x, y\}$ be given by $f(0) = x$ and $f(i) = y$ for $0 < i < n$. Define M with nodes n by $M(i, j) = N_0(f(i), f(j))$, for $i, j < n$. Now $\exists$ clearly has a winning strategy in $G_\omega^n(N_0, \mathcal{A})$, namely, 'continue with the winning strategy in progress in $G_\omega^n(\mathcal{A})$'. It is now easily checked that $M \in \mathcal{R}$ (cf. exercise 2 below), and obviously $M(0,1) = a$.

Next, we check that $\mathcal{R}$ has the triangle addition property. Let $M \in \mathcal{R}$, $x, y, z < n$, $z \neq x, y$, and $a, b \in \mathrm{At}\,\mathcal{A}$ with $a; b \geq M(x, y)$. Now $\exists$ has a winning strategy in $G_\omega^n(M, \mathcal{A})$. Let $\forall$ play round 1 of this game by deleting z from M, leaving N, say, and playing x, y, a, b. $\exists$ will respond with a network $N_1 \supseteq N$ containing a node v such that $N_1(x, v) = a$ and $N_1(v, y) = b$. Again, she clearly has a winning strategy in $G_\omega^n(N_1, \mathcal{A})$. If $v \notin N$ then without loss of generality we may assume that $v = z$, and so N_1 satisfies $N_1 \in \mathcal{R}$, $N_1 \equiv_z N$, $N_1(x, z) = a$, $N_1(z, y) = b$, as required. If however $v \in N$, define $f : n \to \mathrm{nodes}(N)$ by $f(i) = i$ if $i \neq z$, and $f(z) = v$. Define M' by $M'(i, j) = N(f(i), f(j))$, for $i, j < n$. Then as in the previous case, $M' \in \mathcal{R}$ is as required. □

12.4.2 Game for hyperbases

Now we define a stronger game, for hyperbases. For countable $\mathcal{A}$, a winning strategy for $\exists$ in this game will ensure that $\mathcal{A}$ has an n-dimensional Λ-hyperbasis, and so by proposition 12.18 that $\mathcal{A} \in \mathbf{S}\mathfrak{R}\mathfrak{a}\mathbf{CA}_n$. But the game is designed to be stronger than that: a winning strategy for $\exists$ over the countable algebra $\mathcal{A}$, using a countable set of labels Λ, actually yields an n-dimensional Λ-hyperbasis $\mathcal{H}$ containing extensions of all n-wide m-dimensional Λ-hypernetworks. It follows by proposition 12.22 that $\mathfrak{Ca} H_m^n(\mathcal{A}, \Lambda) \in \mathfrak{Nr}_m \mathbf{CA}_n$. This additional strength will be required in chapter 15, when we prove the non-finite axiomatisability of $\mathbf{S}\mathfrak{Nr}_m \mathbf{CA}_{n+1}$ over $\mathbf{S}\mathfrak{Nr}_m \mathbf{CA}_n$ for $3 \leq m < n$.

In the hyperbasis game, the number of nodes of the networks played is fixed at n.

DEFINITION 12.26 Let $3 \leq m \leq n < \omega$ and $r \leq \omega$, let $\mathcal{A}$ be an atomic non-associative algebra, and let Λ be any non-empty set. We define a game $G_r^{m,n}(\mathcal{A}, \Lambda)$ (or we may just write $G_r^{m,n}(\mathcal{A})$ if Λ is given by the context), played by $\forall$ and $\exists$ on n-dimensional Λ-hypernetworks. The rounds are numbered $0, 1, \ldots, t, \ldots$ for $t < r$, and in each round, $\exists$ produces an n-dimensional Λ-hypernetwork N_t, say, in response to some action by $\forall$. In round t:

12.4. Games

1. ∀ may pick any n-wide m-dimensional Λ-hypernetwork M over $\mathcal{A}$ — i.e., he can freely choose all edge labels and hyperlabels of M subject to the result being a wide hypernetwork. We call such a move an m-*dimensional move* and denote it by (M). ∃ must respond with an n-dimensional hypernetwork N_t such that $N_t \restriction_m^n = M$.

2. ∀ may pick a previously played hypernetwork N_u (for some $u < t$), nodes $x, y, z < n$ with $z \neq x, y$, and atoms $a, b \in \operatorname{At}\mathcal{A}$ such that $N_u(x, y) \leq a; b$. We call such a move by ∀ a *triangle move* and denote it by (N_u, x, y, z, a, b). ∃ must respond with a hypernetwork N_t such that $N_t \equiv_z N_u$, $N_t(x, z) = a$, and $N_t(z, y) = b$.

3. Or he may choose (M, N, x, y), where $M, N \in \{N_u : u < t\}$ and $x, y < n$ are distinct, such that $M \equiv_{xy} N$. Such a move by ∀ is called an *amalgamation move*, denoted (M, N, x, y). ∃ must then respond with a hypernetwork N_t satisfying $N \equiv_x N_t \equiv_y M$. See figure 12.2; here we write $N - x$ for the 'hypernetwork' resulting from N by deleting node x, and define $M - y$ similarly.

∃ wins the play if she never gets stuck in any round. As a convention, we say that ∃ wins the game $G_0^{m,n}(\mathcal{A}, \Lambda)$.

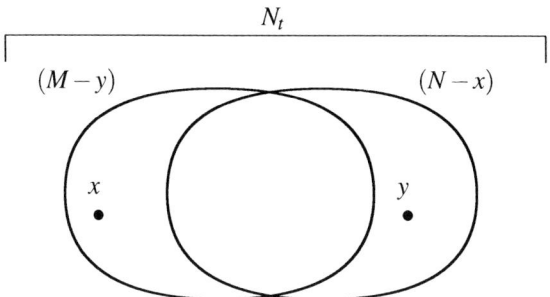

Figure 12.2: An amalgamation move

Recall from definition 12.21 that $H_m^n(\mathcal{A}, \Lambda)$ denotes the set of all n-wide m-dimensional Λ-hypernetworks over $\mathcal{A}$.

PROPOSITION 12.27 *Let $3 \leq m \leq n < \omega$, let $\mathcal{A}$ be any atomic non-associative algebra, and let Λ be a non-empty set.*

1. *If $\mathcal{A}$ has an n-dimensional Λ-hyperbasis $\mathcal{H}$ such that $\mathcal{H}\restriction_m^n = H_m^n(\mathcal{A}, \Lambda)$, then ∃ has a winning strategy in $G_\omega^{m,n}(\mathcal{A}, \Lambda)$.*

2. *If $\mathcal{A}$ has countably many atoms, Λ is countable, and ∃ has a winning strategy in $G_\omega^{m,n}(\mathcal{A}, \Lambda)$ then $\mathcal{A}$ has an n-dimensional Λ-hyperbasis $\mathcal{H}$ with $\mathcal{H}\restriction_m^n = H_m^n(\mathcal{A}, \Lambda)$. Hence, $\mathcal{A} \in \mathbf{S\mathfrak{R}aCA}_n$ and $\mathfrak{Ca}\, H_m^n(\mathcal{A}, \Lambda) \in \mathfrak{Nr}_m \mathbf{CA}_n$.*

Proof. If $\mathcal{A}$ has an n-dimensional Λ-hyperbasis $\mathcal{H}$ containing extensions of arbitrary n-wide m-dimensional Λ-hypernetworks, then $\exists$'s strategy is to always respond with a hypernetwork from $\mathcal{H}$.

For the second part, assume that $\mathcal{A}$ has countably many atoms, Λ is countable, and $\exists$ has a winning strategy in $G_\omega^{m,n}(\mathcal{A},\Lambda)$. Suppose that during a play of $G_\omega^{m,n}(\mathcal{A},\Lambda)$, $\forall$ plays (i) the m-dimensional move (M), for each n-wide m-dimensional Λ-hypernetwork M, (ii) the triangle move (N,x,y,z,a,b) for each hypernetwork N that occurs in this play of the game and all legitimate x,y,z,a,b, and (iii) the amalgamation move (M,N,x,y) for all hypernetworks M,N that are played and all distinct $x,y < n$ such that $M \equiv_{xy} N$. Since there are only countably many atoms in $\mathcal{A}$ and countably many labels in Λ, it is possible for $\forall$ to schedule all these moves in a play of the game. Let $\mathcal{H} = \{N : N$ occurs in the play$\}$. Clearly, $\mathcal{H}$ is an n-dimensional Λ-hyperbasis for $\mathcal{A}$ with $\mathcal{H}\restriction_m^n = H_m^n(\mathcal{A},\Lambda)$.

Now, since $\mathcal{A}$ has an n-dimensional hyperbasis $\mathcal{H}$, $\mathfrak{Ca}\,\mathcal{H} \in \mathbf{CA}_n$ and $\mathcal{A} \in \mathbf{S\mathfrak{Ra}CA}_n$ follow by proposition 12.18. Also, using proposition 12.22, we have $\mathfrak{Ca}\,H_m^n(\mathcal{A},\Lambda) = \mathfrak{Ca}(\mathcal{H}\restriction_m^n) \cong \mathfrak{Nr}_m\,\mathfrak{Ca}\,\mathcal{H} \in \mathfrak{Nr}_m\mathbf{CA}_n$, as required. $\square$

12.4.3 Expressing the games by game trees

We now show how to obtain equivalent games to $G_r^n, G_r^{m,n}$ from game trees (see definition 10.1). Later, this will allow us to use the non-finite axiomatisability results from chapter 10.

The tree for G_ω^n is constructed in much the same way as the tree for G_r^a in theorem 11.5. We leave the details as exercise 4 below.

To express $G_\omega^{m,n}(\mathcal{A},\Lambda)$ by a tree game, we consider the game to be played on a structure $\mathfrak{M}$ larger than just the algebra $\mathcal{A}$. This is for convenience and because it is the form of game required in chapter 15. Exercise 6 below asks for a tree expressing the game when viewed as played on $\mathcal{A}$ alone.

Fix m,n with $3 \leq m < n < \omega$, and a non-empty set Λ. The aim is to build a game tree $T(m,n)$ such that $\exists$ has a winning strategy in $G_\omega^{m,n}(\mathcal{A},\Lambda)$ iff she has a winning strategy in $G(T(m,n),\mathfrak{M})$. The structure $\mathfrak{M}$ will have five sorts, $\mathfrak{M}^1, \ldots, \mathfrak{M}^5$. We will denote the signature of $\mathfrak{M}$ by L.

1. $\mathfrak{M}^1$ is the atomic non-associative algebra $\mathcal{A}$. L will include the symbols of the signature of non-associative algebras, restricted to sort 1 and interpreted in $\mathcal{A}$ in the natural way.

2. $\mathfrak{M}^2$ is Λ.

3. $\mathfrak{M}^3$ is $H_m^n(\mathcal{A},\Lambda)$. Each wide hypernetwork is represented by a single element of this sort. L contains unary function symbols f_{ij} : sort 3 $\to$ sort 1, for $i,j < m$, to express the labels on the edge (i,j) of hypernetworks. For

12.4. Games

each $N \in \mathfrak{M}^3$, $f_{ij}^{\mathfrak{M}}(N)$ will be the atom $N(i,j)$ of $\mathcal{A}$. There are similar function symbols $f_{\bar{a}}$: sort 3 $\to$ sort 2 (for each $\bar{a} \in {}^{\leq n}m$ with $|\bar{a}| \neq 2$) to express the hyperlabels. For $N \in \mathfrak{M}^3$, $f_{\bar{a}}^{\mathfrak{M}}(N)$ will be the hyperlabel $N(\bar{a})$.

4. $\mathfrak{M}^4$ is $H_n(\mathcal{A}, \Lambda)$. Similar function symbols to the above are included in L: they are g_{ij} : sort 4 $\to$ sort 1 for all $i, j < n$, and $g_{\bar{a}}$: sort 4 $\to$ sort 2 for all $\bar{a} \in {}^{\leq n}n$ with $|\bar{a}| \neq 2$.

5. $\mathfrak{M}^5$ is any atomic BAO C of the similarity type of m-dimensional cylindric algebras, such that the atom structure At C is isomorphic to At $\mathfrak{Ca}(H_m^n(\mathcal{A}, \Lambda))$ — i.e., to At $\mathfrak{Ca}\,\mathfrak{M}^3$. For example, C could be $\mathfrak{Ca}(H_m^n(\mathcal{A}, \Lambda))$ itself. C will play no role in the game, but it will be useful in section 15.6. L contains the symbols of the signature of $\mathbf{CA}_m$, restricted to $\mathfrak{M}^5$ and interpreted in the natural way, and also a function symbol ι : sort 3 $\to$ sort 5. $\iota^{\mathfrak{M}}$ maps each $N \in H_m^n(\mathcal{A}, \Lambda)$ to whichever atom of C corresponds by isomorphism to the atom $\{N\}$ of $\mathfrak{Ca}(H_m^n(\mathcal{A}, \Lambda))$.

We denote $\mathfrak{M}$ in full by $\mathfrak{M}(\mathcal{A}, m, n, \Lambda, C)$. Note that the following relations are definable in $\mathfrak{M}$ by appropriate first-order formulas:

- $N \upharpoonright_m^n = P$, for $N \in \mathfrak{M}^4 = H_n(\mathcal{A}, \Lambda)$ and $P \in \mathfrak{M}^3 = H_m^n(\mathcal{A}, \Lambda)$. This is expressible by

$$\bigwedge_{\bar{a} \in {}^{\leq n}m} g_{\bar{a}}(N) = f_{\bar{a}}(P).$$

- $N \equiv_{ij} P$, for $N, P \in \mathfrak{M}^4 = H_n(\mathcal{A}, \Lambda)$ and $i, j < n$. This is expressible by

$$\bigwedge_{\bar{a} \in {}^{\leq n}(n \setminus \{i,j\})} g_{\bar{a}}(N) = g_{\bar{a}}(P).$$

Formulas expressing $\equiv_i$ on $\mathfrak{M}^4$ and the relations $\equiv_i, \equiv_{ij}$ on $\mathfrak{M}^3$ are defined similarly.

REMARK 12.28 There are first-order L-sentences expressing the following:

1. $\mathfrak{M}^3 = H_m^n(\mathfrak{M}^1, \mathfrak{M}^2)$. This can be said by the conjunction of sentences saying that each $x \in \mathfrak{M}^3$ is a wide hypernetwork:

   - $\forall x \bigwedge_{i < m} f_{ii}(x) \leq 1'$
   - $\forall x \bigwedge_{i,j,k < m} f_{ij}(x) \leq f_{ik}(x) ; f_{kj}(x),$
   - $\forall x \bigwedge_{\bar{a}, \bar{b} \in {}^l m} \left(\left(\bigwedge_{i < l} f_{a_i b_i}(x) \leq 1' \right) \to f_{\bar{a}}(x) = f_{\bar{b}}(x) \right)$ for all $l \leq n$,

382   Chapter 12. Relational, cylindric, and hyperbases

together with one more sentence expressing that every n-wide m-dimensional $\mathfrak{M}^2$-hypernetwork over $\mathfrak{M}^1$ occurs in $\mathfrak{M}^3$. We leave the construction of this last formula to exercise 5.

2. $\mathfrak{M}^4 = H_n(\mathfrak{M}^1, \mathfrak{M}^2)$. This is similar to (1).

3. $\mathfrak{M}^5$ is an atomic m-dimensional cylindric-type BAO and ι is an isomorphism witnessing $\mathrm{At}(\mathfrak{M}^5) \cong \mathrm{At}\,\mathfrak{Ca}(\mathfrak{M}^3)$.

    This is straightforward: we can include the axioms defining normality and additivity of the c_i ($i < m$), it is easy to express atomicity, and the atom structure $\mathrm{At}\,\mathfrak{Ca}(\mathfrak{M}^3)$ is easily definable from $\mathfrak{M}^3 = H_m^n(\mathfrak{M}^1, \mathfrak{M}^2)$.

LEMMA 12.29 *For each m, n with $3 \le m < n < \omega$, there is a closed game tree $T(m,n)$ (see definition 10.2) such that for any atomic non-associative algebra $\mathcal{A}$, any non-empty set Λ, any atomic m-dimensional cylindric-type BAO C with atom structure isomorphic to $\mathrm{At}\,\mathfrak{Ca}(H_m^n(\mathcal{A}, \Lambda))$, and any $r \le \omega$, $\exists$ has a winning strategy in $G_r^{m,n}(\mathcal{A}, \Lambda)$ iff she has a winning strategy in $G(T(m,n) \restriction_{1+2r}, \mathfrak{M}(\mathcal{A}, m, n, \Lambda, C))$.*

Here, $1 + 2r$ is evaluated using ordinal sum and product, so that $1 + 2\omega = \omega$.

Proof. We write T for $T(m,n)$. The root of the tree, τ_0, has labelling $T(\tau_0) = (\exists, \emptyset, \top)$. On any branch of T, we will arrange that the nodes belong to $\exists$ and $\forall$ alternately: nodes of even height will belong to $\exists$. A node $\tau \neq \tau_0$ belonging to $\exists$ will correspond to round $\rho(\tau) = ht(\tau)/2 - 1$ in the game $G_\omega^{m,n}(\mathcal{A}, \Lambda)$, and will have a variable $v_{\rho(\tau)}$ of sort 4 associated with it. The idea will be that if v_t is assigned to $N_t \in H_n(\mathcal{A}, \Lambda)$ for each $t \le \rho(\tau)$ by the current assignment, this means that the play of $G_\omega^{m,n}(\mathcal{A}, \Lambda)$ up to round $\rho(\tau)$ is $N_0, N_1, \ldots, N_{\rho(\tau)-1}$, and that $\exists$ chooses the hypernetwork $N_{\rho(\tau)}$ in round $\rho(\tau)$.

For each $\exists$-node τ, including the root, T contains the following immediate successors of τ.

m-dimensional moves There is an immediate successor τ_1 of τ with labelling defined by

$$T(\tau_1) = (\forall, x, \top),$$

where x is a new variable of sort 3, not associated with any node on the branch from the root up to τ. T also contains an immediate successor τ_2 of τ_1 labelled

$$T(\tau_2) = \bigl(\exists,\ v_{\rho(\tau)+1},\ v_{\rho(\tau)+1}\restriction_m^n = x\bigr).$$

12.4. Games

Triangle moves For each node $\upsilon \leq \tau$ belonging to $\exists$ and each $i, j, k < n$ with $k \neq i, j$, there is an immediate successor τ' of τ with labelling

$$T(\tau') = \big(\forall,\ \langle x, y\rangle,\ \alpha(x) \wedge \alpha(y) \wedge x; y \geq g_{ij}(v_{\rho(\upsilon)})\big).$$

Here, x and y are new variables of sort 1, and $\alpha(x), \alpha(y)$ are formulas stating that x, y are atoms of $\mathfrak{M}^1$. There is an immediate successor τ'' of τ' labelled as follows, where $v = v_{\rho(\tau)+1}$:

$$T(\tau'') = \big(\exists,\ v,\ v \equiv_k v_{\rho(\upsilon)} \wedge g_{ik}(v) = x \wedge g_{kj}(v) = y\big).$$

Amalgamation moves For each pair $\upsilon, \sigma \leq \tau$ of distinct nodes belonging to $\exists$, and each distinct $i, j < n$, there is an immediate successor τ^* of τ with labelling

$$T(\tau^*) = \big(\forall, \emptyset, v_{\rho(\upsilon)} \equiv_{ij} v_{\rho(\sigma)}\big).$$

There is then an immediate successor τ^+ of τ^* labelled as follows, where $v = v_{\rho(\tau)+1}$:

$$T(\tau^+) = \big(\exists,\ v,\ v_{\rho(\upsilon)} \equiv_i v \wedge v \equiv_j v_{\rho(\sigma)}\big).$$

The definition of $T(m, n)$ follows that of $G_r^{m,n}$. $\exists$-nodes of height $2, 4, 6, \ldots$ in T correspond to rounds $0, 1, 2, \ldots$ of $G_r^{m,n}$, respectively. For example, plays of $G_3^{m,n}$, which have rounds 0, 1, 2 only, correspond to plays of $G(T, \mathfrak{M})$ using nodes of height up to 6 — that is, the nodes of $T\!\restriction_7$. So it should be clear that for all $r \leq \omega$, $\exists$ has a winning strategy in $G_r^{m,n}(\mathcal{A}, \Lambda)$ iff she has a winning strategy in $G(T\!\restriction_{1+2r}, \mathfrak{M}(\mathcal{A}, m, n, \Lambda, C))$. □

Exercises

1. [Cf. exercise 11.3(1).] Prove that for any atomic non-associative algebra $\mathcal{A}$,

   (a) if $3 \leq p \leq \omega$, $\exists$ has a winning strategy in $G_2^p(\mathcal{A})$ iff $\mathcal{A} \in \mathbf{SA}$,

   (b) if $4 \leq p \leq \omega$, $\exists$ has a winning strategy in $G_3^p(\mathcal{A})$ iff $\mathcal{A} \in \mathbf{RA}$.

2. In the proof of proposition 12.25, check that if N is an atomic $\mathcal{A}$-network with at most n nodes, M is a set, $f : M \to N$ is a map, and we define a pre-network with nodes M by $M(x, y) = N(f(x), f(y))$ for all $x, y \in M$, then

   - M is an atomic $\mathcal{A}$-network,
   - if $\exists$ has a winning strategy in $G_\omega^n(N, \mathcal{A})$ and $|M| \leq n$, then $\exists$ also has a winning strategy in $G_\omega^n(M, \mathcal{A})$.

3. Let $\mathcal{A}$ be a finite non-associative algebra, and let $2 \leq n < \omega$. Show that $\mathcal{A}$ has an n-dimensional relational basis iff $\exists$ has a winning strategy in $G_r^n(\mathcal{A})$ for all finite r.

4. Construct a game tree equivalent to the game G_r^n for relational bases.

5. Completing remark 12.28, write L-sentences expressing the properties that every n-wide m-dimensional $\mathfrak{M}^2$-hypernetwork over $\mathfrak{M}^1$ occurs in $\mathfrak{M}^3$, and C is an atomic cylindric-type BAO with At $C \cong$ At $\mathfrak{Ca}(H_m^n(\mathcal{A}, \Lambda))$.

6. Construct a game tree T for the game $G_\omega^{m,n}(\mathcal{A}, \Lambda)$ viewed as being played on $\mathcal{A}$ alone — so $G_r^{m,n}(\mathcal{A}, \Lambda)$ is 'equivalent' to $G(T\restriction_{1+2r}, \mathcal{A})$ for all $r \leq \omega$ — where (a) Λ is a fixed *infinite* set, (b) Λ is a fixed *finite* set.

[The only significance of hyperlabels is whether two of them are equal or not. So in a play of $G_r^{m,n}(\mathcal{A}, \Lambda)$ for given finite r, the hyperlabels needed by the players can without loss of generality be taken from an arbitrary finite set whose size can be calculated from r in advance. The way these finitely many hyperlabels are handled by $\forall$ and $\exists$ in the game can be encoded in the tree structure. The edge labels of n-dimensional hypernetworks can be similarly encoded using n^2 variables taking values in $\mathcal{A}$.]

12.5 The variety RA$_n$

We can now define this.

DEFINITION 12.30 [Madd83, Madd91b] For finite $n \geq 2$, **RA**$_n$ is the class of all non-associative algebras $\mathcal{A}$ such that $\mathcal{A}$ is a subalgebra of some complete atomic non-associative algebra that has an n-dimensional relational basis.

We concentrate on finite dimensions, as although **RA**$_n$ can be defined for infinite n, it is then identical to **RRA**. It follows from exercise 12.3(10) that **RA**$_2$ = **NA**, so the case $n = 2$ is also of little interest. It is easily seen (exercise 3 below) that completeness plays no role in definition 12.30, and so **RA**$_n$ = **S**(**RB**$_n$) (see definition 12.9 for **RB**$_n$).

RA$_n$ is defined in terms of relational bases. We know of no analogous definition for any other kind of basis in the literature. We could invent some notation such as **S**(**HB**$_n$) for the class of all non-associative algebras that embed in an atomic non-associative algebra with an n-dimensional hyperbasis, but for $n \geq 4$ this class turns out to be **S$\mathfrak{R}\mathfrak{a}$CA**$_n$ (theorem 13.46), so we do not do so.

Let $\mathcal{A}$ be a non-associative algebra. Clearly, if $\mathcal{A}^+$ has an n-dimensional relational basis, then $\mathcal{A} \in$ **RA**$_n$. It turns out that the converse also holds:

$$\text{if } \mathcal{A} \in \mathbf{RA}_n \text{ then } \mathcal{A}^+ \text{ has an } n\text{-dimensional relational basis.} \quad (12.1)$$

12.5. The variety $\mathbf{RA}_n$

We will prove this in corollary 13.47 below; it can also be proved directly (exercise 5 below). The following two lemmas assume (12.1).

PROPOSITION 12.31 [Madd83, theorems 8, 9] $\mathbf{RA}_n$ *is a canonical variety, for each $n \geq 3$.*

Proof. The proof is taken from [Madd83]. By Birkhoff's theorem 2.45, to show that $\mathbf{RA}_n$ is a variety it is sufficient to show that it is closed under subalgebras, direct products, and homomorphic images. Closure under subalgebras is obvious. For products, let $\mathcal{A}_i \subseteq \mathcal{B}_i$ ($i \in I$), and suppose that each $\mathcal{B}_i$ is complete and atomic and has an n-dimensional relational basis $\mathcal{M}_i$. We may suppose that the $\mathcal{B}_i$ are pairwise disjoint. Let C be the complex algebra over the atom structure consisting of the disjoint union of the atom structures of the $\mathcal{B}_i$. C is complete and atomic. Check that $\prod_{i \in I} \mathcal{A}_i \subseteq C$, and that $\bigcup_{i \in I} \mathcal{M}_i$ is an n-dimensional relational basis for C.

Now let $\mathcal{A} \in \mathbf{RA}_n$ and $h : \mathcal{A} \to \mathcal{B}$ be a surjective homomorphism. We show that $\mathcal{B} \in \mathbf{RA}_n$ too; several checks in the proof are left for exercise 4 below.

Let $K = \{a \in \mathcal{A} : h(a) = 0\}$, the kernel of h, and let $z = -\sum K$, the meet being evaluated in $\mathcal{A}^+$. Let $\mathcal{D}$ be the relativisation $\mathcal{A}^+\!\upharpoonright_z$ of $\mathcal{A}^+$ to z, as defined in the proof of proposition 2.58. $\mathcal{D}$ is clearly complete and atomic. Define $g : \mathcal{A}^+ \to \mathcal{D}$ by $g(a) = a \cdot z$. It can be checked that g is a homomorphism of $\mathcal{A}^+$ onto $\mathcal{D}$. Hence, $\mathcal{D} \in \mathbf{NA}$, being a homomorphic image of $\mathcal{A}^+ \in \mathbf{NA}$. Moreover, for all $a \in \mathcal{A}$, we have $g(a) = 0$ iff $a \in K$. It follows that $\mathcal{B}$ embeds in $\mathcal{D}$, via $b \mapsto g(a)$ for any $a \in h^{-1}[b]$ (where $b \in \mathcal{B}$) — this is well-defined and one-one.

By the result (12.1) that we are assuming, $\mathcal{A}^+$ has an n-dimensional relational basis, say $\mathcal{M}$. It can be checked that $\{N \in \mathcal{M} : N(x,y) \in \mathcal{D} \text{ for all } x,y < n\}$ is an n-dimensional relational basis for $\mathcal{D}$. Since $\mathcal{B}$ is up to isomorphism a subalgebra of $\mathcal{D}$, it follows from the definition of $\mathbf{RA}_n$ that $\mathcal{B} \in \mathbf{RA}_n$ as required.

Canonicity of $\mathbf{RA}_n$ says that if $\mathcal{A} \in \mathbf{RA}_n$ then $\mathcal{A}^+ \in \mathbf{RA}_n$. This is clear, as by our assumption (12.1), $\mathcal{A}^+$ actually has an n-dimensional relational basis, and it is certainly complete and atomic. As we said, we will prove (12.1) later; the proof will not use the current proposition! □

For $n \geq 4$, $\mathbf{RA}_n$ is contained in $\mathbf{RA}$ (exercise 10 below) and so is a conjugated discriminator variety.

COROLLARY 12.32 *Let $3 \leq n < \omega$.*

1. *A finite non-associative algebra has an n-dimensional relational basis if and only if it is in $\mathbf{RA}_n$.*

2. *It is algorithmically decidable (in PTIME) whether a finite non-associative algebra is in $\mathbf{RA}_n$ or not.*

Proof. The first part follows immediately from (12.1), since $\mathcal{A} \cong \mathcal{A}^+$ if $\mathcal{A}$ is finite. A non-deterministic algorithm can now decide whether a finite algebra is in $\mathbf{RA}_n$ by 'guessing' an n-dimensional relational basis and checking that it is one. For fixed n, a deterministic algorithm can do the job in polynomial time by a method of Pratt [Pra79]. Let $\mathcal{A}$ be given, with $|\mathcal{A}| = 2^m$, so that $\mathcal{A}$ has m atoms. The algorithm starts by calculating the atom structure At $\mathcal{A}$, which can be done in time $O(|\mathcal{A}|^3)$. It then writes down $\mathcal{M} = \{$all n-dimensional atomic $\mathcal{A}$-networks$\}$. Clearly, $|\mathcal{M}| \leq m^{n^2}$, and it follows that $\mathcal{M}$ can be constructed in polynomial time in $|\mathcal{A}|$. The algorithm then checks that each network $N \in \mathcal{M}$ does not violate the triangle addition property. This involves considering all 'triangles' (x,y,z,a,b) for $x,y,z < n$, $z \neq x,y$, and $a,b \in$ At $\mathcal{A}$; there are at most $n^3 m^2$ of these. For each triangle, the algorithm tests that if $N(x,y) \leq a;b$, there is $P \in \mathcal{M}$ with $P \equiv_z N$, $P(x,z) = a$, and $P(z,y) = b$. This takes time $O(|\mathcal{M}|)$. It is done for every atomic network in $\mathcal{M}$, so each pass through $\mathcal{M}$ takes time $O(|\mathcal{M}|^2 n^3 m^2)$. If a network fails the test, it is deleted from $\mathcal{M}$ and the process starts again. The number of passes required before termination is clearly at most the original size of $\mathcal{M}$, so the total time required is $O(|\mathcal{M}|^3 n^3 m^2)$. On termination, a final check is made that for each atom $a \in \mathcal{A}$ there is $N \in \mathcal{M}$ with $N(0,1) \leq a$; this takes time at most $m \cdot |\mathcal{M}| \leq m^{n^2+1}$. If the check succeeds, the algorithm announces that $\mathcal{A} \in \mathbf{RA}_n$; otherwise it responds that $\mathcal{A} \notin \mathbf{RA}_n$. This can easily be seen to be correct: if the algorithm accepts then the final value of $\mathcal{M}$ is a relational basis, while conversely, any relational basis for $\mathcal{A}$ is contained in $\mathcal{M}$ at each stage, so if $\mathcal{A}$ has any relational basis at all then the algorithm will find one and accept. The total time taken is of order a polynomial in $|\mathcal{A}|$. □

PROBLEM 12.33 Investigate the parallel complexity of determining whether a finite non-associative algebra is in $\mathbf{RA}_n$. Investigate the complexity of determining, for a finite relation algebra atom structure $\mathcal{S}$, whether $\mathfrak{Cm}\,\mathcal{S} \in \mathbf{RA}_n$.

Corollary 12.32(1) remains true for $\mathbf{S\mathfrak{R}aCA}_n$ and hyperbases, as will be seen in theorem 13.46. Corollary 12.32(2), on the other hand, fails for $\mathbf{S\mathfrak{R}aCA}_n$: in theorem 18.13, we show that for $n \geq 5$, it is undecidable whether an arbitrary finite relation algebra is in $\mathbf{S\mathfrak{R}aCA}_n$.

Exercises

1. Show that $\mathbf{RA}_3 \supseteq \mathbf{RA}_4 \supseteq \mathbf{RA}_5 \supseteq \cdots$.

2. Show (e.g., using games) that McKenzie's 4-atom non-representable relation algebra $\mathcal{K}$ (described in section 4.4) is not in $\mathbf{RA}_5$.

3. Show that the requirement 'complete' can be omitted from definition 12.30 without changing its meaning.

12.5. The variety $\mathbf{RA}_n$

4. Check the following for the proof of proposition 12.31 (see [Madd83, theorem 9] if stuck):

    (a) $\Sigma K = \Sigma K; 1 = 1; \Sigma K$.

    (b) $z = z; 1 = 1; z$.

    (c) For all $x, y \in \mathcal{A}^+$, $x; y \cdot z = (x \cdot (z; \breve{y})); (y \cdot (\breve{x}; z)) \cdot z$.

    (d) $(x; y) \cdot z = ((x \cdot z); (y \cdot z)) \cdot z$ for all $x, y \in \mathcal{A}^+$.

    (e) g is a homomorphism : $\mathcal{A}^+ \to \mathcal{D}$.

    (f) For all $a \in \mathcal{A}$, $g(a) = 0$ iff $h(a) = 0$. [This will need BPI (fact 2.8) or similar.] Hence, the map $b \mapsto g(a)$, for any $a \in \mathcal{A}$ such that $h(a) = b$, is a well-defined embedding : $\mathcal{B} \to \mathcal{D}$.

    (g) $\{N \in \mathcal{M} : N(x,y) \in \mathcal{D}$ for all $x, y < n\}$ is an n-dimensional relational basis for $\mathcal{D}$.

5. Let $\mathcal{A}$ be a non-associative algebra and $3 \leq n < \omega$. Show that $\mathcal{A} \in \mathbf{RA}_n$ if and only if $\mathcal{A}^+$ has an n-dimensional relational basis. [Use ω-saturation on a structure encoding a relational basis; cf. theorem 3.36. This proves (12.1).]

6. Prove that $\mathbf{RRA} \subseteq \mathbf{RA}_n$ for all $n \geq 3$. [Cf. exercise 12.6(2) later.]

7. [Venema] Fix $3 \leq n < \omega$. Find an existential second-order sentence defining the class of all atomic relation-type algebras with an n-dimensional relational basis. Can you find a similar sentence in a suitable fixed-point logic (see, e.g., [EbbFlu99])?

8. Take finite $n \geq 5$. Show that for any relation algebra $\mathcal{A}$, $\mathcal{A} \in \mathbf{RA}_n$ iff there is an atomic algebra $\mathcal{B} \supseteq \mathcal{A}$ with an n-dimensional relational basis and with $|\mathcal{B}| = |\mathcal{A}|$. Deduce that $\mathbf{RA}_n$ is a *PC* class (definition 9.6) — i.e., it is definable by an existential second-order sentence.

9. Show that for $3 \leq n < \omega$, the problem of whether a finite non-associative algebra is in $\mathbf{RA}_n$ can be solved in polylogarithmic space.

10. Extending exercises 12.3(11, 12), prove that the following classes are equal:

    - **RA**
    - $\mathbf{S}\{\mathcal{A} \in \mathbf{NA} : \mathcal{A}$ atomic and has a 4-dimensional hyperbasis$\}$
    - $\mathbf{S}\{\mathcal{A} \in \mathbf{NA} : \mathcal{A}$ atomic with a 4-dimensional relational basis$\} = \mathbf{RA}_4$.

    Do the same for **SA** and 3-dimensional bases. [Deal with atomic algebras, then use the canonical embedding construction. Cf. [Madd83, theorem 6] and [Madd91b, theorem 35].]

11. Use exercise 10 and earlier results to prove that $\mathbf{RA} = \mathbf{S\Re aCA}_4$.

12. Let $n \geq 3$ be finite. Using the ideas of sections 8.2 and 9.7 (Jónsson Q-operators), find a finite set of equations in an expanded signature such that $\mathbf{RA}_n$ is the class of all subalgebras of the reduct to the signature of relation algebras of algebras satisfying the equations. [Cf. [SteVen98].]

13. [Maddux; see also problem 21 in chapter 21.] Let $n < \omega$. Prove that 'almost all' finite non-associative algebras belong to $\mathbf{RA}_n$ in the following sense. Let $\mathbf{NA}(k)$ be the number of non-isomorphic finite non-associative algebras with at most k atoms and let $\mathbf{RA}_n(k)$ be the number of non-isomorphic finite non-associative algebras in $\mathbf{RA}_n$, with at most k atoms. Prove that $\lim_{k \to \infty} \frac{\mathbf{RA}_n(k)}{\mathbf{NA}(k)} = 1$. [Hint: First consider non-associative algebras $\mathcal{A}$ in which the identity is an atom. $\mathcal{A}$ is determined by the set of forbidden triples of atoms. Let each triple of diversity atoms, and its five other Peircean transforms, be a forbidden triple with probability one half. Let B be the set of all atomic $\mathcal{A}$-networks over the nodes $0, 1, \ldots, n-1$. You have to prove that B is a relational basis for $\mathcal{A}$ with probability greater than some function of k whose limit, as k tends to infinity, is one. Then consider the general case.]

14. The following quasi-equation states a property of 'quasi-projections'. It is (roughly) one of Jónsson's axioms for weakly representable relation algebras, given in [Jón59] and discussed a little in section 5.2.

$$\pi(p,q) \wedge \breve{v}; x \cdot w; \breve{y} \leq \breve{p}; q \to v; w \cdot x; y = (v; \breve{p} \cdot x; \breve{q}); (p; w \cdot q; y), \quad (12.2)$$

where $\pi(p,q)$ denotes

$$\breve{p}; p \leq 1' \wedge \breve{q}; q \leq 1' \wedge p; 1 = q; 1 \wedge p; \breve{p} \cdot q; \breve{q} \leq 1' \wedge \breve{p}; 1; q = \breve{p}; q,$$

saying that p, q form a 'full direct product'. Prove that (12.2) is valid in $\mathbf{RA}_5$. [The 'unsharpness problem' was to determine whether it is valid in $\mathbf{RA}$. Maddux solved it negatively in [Madd95], using a 58-atom algebra.]

12.6 Maddux's bases

In this section we take a short detour to discuss bases as defined by their inventor, Roger Maddux, and compare them with the bases already introduced. In the next section, we connect cylindric bases to homogeneous representations of relation algebras.

In Maddux's terminology, bases are certain sets of 'basic matrices', these being atomic networks with a standardised set of nodes. He defined two kinds of basis:

12.6. Maddux's bases

relational bases, and cylindrical or cylindric bases. Below, except where otherwise stated, $\mathcal{A}$ continues to be a fixed atomic non-associative algebra, and n is the fixed dimension with $n \geq 2$ for relational bases and $n \geq 3$ for cylindric bases.

12.6.1 Relational and cylindric bases

DEFINITION 12.34 [Madd83, Madd89b] An *n-dimensional relational basis* for $\mathcal{A}$ is a set $\mathcal{M}$ of atomic $\mathcal{A}$-networks N with nodes$(N) = n$, such that:

(R$_0$) For every atom $a \in \text{At}\,\mathcal{A}$ there is $N \in \mathcal{M}$ such that $N(0,1) = a$.

(R$_1$) If $N \in \mathcal{M}$, $x, y < n$, $a, b \in \text{At}\,\mathcal{A}$, $N(x,y) \leq a\,;b$, and $x, y \neq z < n$, then there is some $M \in \mathcal{M}$ such that $M \equiv_z N$, $M(x,z) = a$, and $M(z,y) = b$.

This is identical to our definition of relational basis (definition 12.9) except that it uses atomic networks instead of hypernetworks. We already said (remark 12.3) that for relational bases, hyperlabels are of no significance, and that we can equally take the hypernetworks in a relational basis to be atomic networks. So our use of the same term 'relational basis' for both here will cause no problem.

The definition of cylindric basis, next, is significantly different from the 'basis' definitions we have seen so far.

DEFINITION 12.35 [Madd78b], [Madd89b, definition 4]   An *n-dimensional cylindric basis* for $\mathcal{A}$ is a set $\mathcal{M}$ of n-dimensional atomic networks satisfying:

(C$_0$) If $a, b, c \in \text{At}\,\mathcal{A}$ and $a \leq b\,;c$, then there is $N \in \mathcal{M}$ with $N(0,1) = a$, $N(0,2) = b$, and $N(2,1) = c$.

(C$_1$) If $M, N \in \mathcal{M}$, $x, y < n$, $x \neq y$, and $M \equiv_{xy} N$, then there is some $L \in \mathcal{M}$ such that $M \equiv_x L \equiv_y N$.

(C$_2$) If $N \in \mathcal{M}$ and $x, y < n$ then $N[x/y] \in \mathcal{M}$.

In [Madd83] relational bases were defined for any atomic semi-associative algebra and for any $3 \leq n \leq \omega$. It was shown [Madd83, theorem 4] that any atomic semi-associative algebra has a 3-dimensional relational basis, (theorem 5) that an atomic semi-associative algebra has a 4-dimensional relational basis iff it is a relation algebra, and (theorem 6) that $\mathbf{RA}_\omega = \mathbf{RRA}$.

In [Madd89b, definitions 3, 4], any atomic non-associative algebra was allowed for relational and cylindric bases, and n was any ordinal with $n \geq 2$ for relational bases and $n \geq 3$ for cylindric bases. It was shown [Madd89b, theorem 10] that any atomic weakly associative algebra with an n-dimensional cylindric basis is in $\mathbf{S\mathfrak{R}aCA}_n$.

In [Madd91b], any atomic non-associative algebra was again allowed for relational bases (cylindric bases were not used in this paper), and n was any ordinal with $n \geq 2$. It was shown in [Madd91b, theorem 35] that an atomic non-associative algebra is semi-associative iff it has a 3-dimensional relational basis. We may conclude that $\mathbf{RA}_3 = \mathbf{SA}$ and $\mathbf{RA}_4 = \mathbf{RA}$.

12.6.2 Comparing cylindric bases with hyperbases

Here, we show that any cylindric basis of a weakly associative algebra is a hyperbasis (and hence, a relational basis). The assumption of weak associativity is essential here: see exercise 5 below. Then we show that the converse fails — as one would expect, given the presence of hyperlabels.

The idea of the proof of the first statement is well known (see, e.g., [Madd89b, p. 954]) but seems not to have appeared in print. The proof assumes that $n \geq 4$; see exercise 4 below for $n = 3$.

LEMMA 12.36 *Let $\mathcal{A}$ be an atomic weakly associative algebra, let $n \geq 4$, and let $\mathcal{M}$ be a set of atomic networks over $\mathcal{A}$ with set of nodes n. We regard each network in $\mathcal{M}$ as a $\{\lambda\}$-hypernetwork in the usual way (see remark 12.3). The following are equivalent.*

1. *$\mathcal{M}$ is an n-dimensional $\{\lambda\}$-hyperbasis, for some λ.*

2. *$\mathcal{M}$ is an n-dimensional cylindric basis.*

Proof. Assume (1). We show that $\mathcal{M}$ is a cylindric basis. (C$_0$) holds because if $a, b, c \in \operatorname{At} \mathcal{A}$ with $a \leq b; c$, then there is $N \in \mathcal{M}$ with $N(0,1) = a$, and (by the triangle addition property) $M \in \mathcal{M}$ with $M \equiv_2 N$, $M(0,2) = b$, and $M(2,1) = c$. So M is as required for (C$_0$). (C$_1$) is clear, and (C$_2$) was shown in lemma 12.13.

For the converse, assume that $\mathcal{M}$ is a cylindric basis for $\mathcal{A}$; we show that $\mathcal{M}$ is a hyperbasis. If $a \in \operatorname{At} \mathcal{A}$, then we must find $N \in \mathcal{M}$ with $N(0,1) \leq a$. But $a \leq 1;1$, so by complete additivity of ';' in $\mathcal{A}$, there are atoms $b, c \in \mathcal{A}$ with $a \leq b; c$. By (C$_0$), there is $N \in \mathcal{M}$ with $N(0,1) \leq a$, etc. — N is as required.

The amalgamation property is immediate, by (C$_1$); we pass to the triangle addition property. Let $x, y, z < n$ with $z \neq x, y$, and suppose that $a, b \in \operatorname{At} \mathcal{A}$, $N \in \mathcal{M}$, and $N(x,y) \leq a; b$. We wish to find $P \in \mathcal{M}$ with $P \equiv_z N$, $P(x,z) = a$, and $P(z,y) = b$.

Claim. There is $Q \in \mathcal{M}$ with $Q(x,y) = N(x,y)$, $Q(x,z) = a$, and $Q(z,y) = b$.
Proof of claim. By (C$_0$), there is $M \in \mathcal{M}$ with $M(0,1) = N(x,y)$, $M(0,2) = a$, and $M(2,1) = b$. First assume that $x \neq y$; we will deal with the other case in a moment. As $n \geq 4$ and x, y, z are distinct, by fact 12.12 we may choose $\sigma : n \to n$, a product of substitutions, such that $\sigma(x) = 0$, $\sigma(y) = 1$, and $\sigma(z) = 2$. By repeated use of (C$_2$), we see that $Q \stackrel{\text{def}}{=} M\sigma \in \mathcal{M}$. Then

12.6. Maddux's bases

- $Q(x,y) = M\sigma(x,y) = M(0,1) = N(x,y)$,
- $Q(x,z) = M(0,2) = a$,
- $Q(z,y) = M(2,1) = b$.

If on the other hand $x = y$, then $N(x,y) \leq 1$'. Hence, by lemmas 11.2 and 3.11, $M(1,0) = M(0,1)\breve{} = N(x,y)\breve{} \leq \breve{1}' = 1$'. Choose a product $\sigma : n \to n$ of substitutions with $\sigma(x) = \sigma(y) = 0$ and $\sigma(z) = 2$. Then $Q \stackrel{\text{def}}{=} M\sigma \in \mathcal{M}$, and:

- $Q(x,y) = M\sigma(x,y) = M(0,0) \leq M(0,1); M(1,0) \leq N(x,y); 1' = N(x,y)$. As both are atoms, we obtain $Q(x,y) = N(x,y)$.
- $Q(x,z) = M(0,2) = a$.
- $Q(z,y) = M(2,0) \leq M(2,1); M(1,0) \leq b; 1' = b$. As both are atoms, we have $Q(z,y) = b$.

This proves the claim.

Take Q as in the claim, and enumerate $n \setminus \{x,y,z\}$ as $\{t_1 \ldots, t_l\}$. Then, using that $\mathcal{A}$ is weakly associative, we can show that

$$N \equiv_{z,t_1,\ldots,t_l} Q.$$

Indeed, the claim gives us $N(x,y) = Q(x,y)$, and by lemma 11.2, $N(y,x) = N(x,y)\breve{}$ $= Q(x,y)\breve{} = Q(y,x)$. We plainly have $N(x,x) \leq N(x,y); N(y,x) \cdot 1$', and as in weakly associative algebras the right-hand side is an atom (lemma 5.10), we have equality: $N(x,x) = N(x,y); N(y,x) \cdot 1$'. Similarly, $Q(x,y); Q(y,x) \cdot 1' = Q(x,x)$. Hence,

$$N(x,x) = N(x,y); N(y,x) \cdot 1' = Q(x,y); Q(y,x) \cdot 1' = Q(x,x).$$

The proof that $N(y,y) = Q(y,y)$ is similar. So $N \equiv_{z,t_1,\ldots,t_l} Q$ is established.

The proof of lemma 12.16, with axiom (C$_2$) playing the role of lemma 12.13, now shows that there is $P \in \mathcal{M}$ with $N \equiv_z P \equiv_{t_1,\ldots,t_l} Q$, and the second equivalence here gives $P(x,z) = Q(x,z) = a$, and $P(z,y) = Q(z,y) = b$. So P is as required. □

So in dimension at least 4 (and also 3, by exercise 4 below), any cylindric basis of an atomic weakly associative algebra 'is' a relational basis and in fact a $\{\lambda\}$-hyperbasis for any λ. We now show that the converse of this can fail when the set Λ of hyperlabels is non-trivial.

THEOREM 12.37 *There exists a (finite) relation algebra with an n-dimensional hyperbasis for all finite $n \geq 3$, but with no 5-dimensional cylindric basis (and hence, by exercise 1 below, no n-dimensional cylindric basis for any $n \geq 5$).*

392   Chapter 12. Relational, cylindric, and hyperbases

Proof. Take p to be a whole number, at least 2. Recall from section 4.5 that the Lyndon algebra $\mathcal{A}_p$ is a finite relation algebra with $p+2$ atoms, say $1', a_0, \ldots, a_p$. It is defined by:

- $a_i; a_i = a_i + 1'$ if $p \geq 3$, and $a_i; a_i = 1'$ if $p = 2$.
- $a_i; a_j = \sum_{k \neq i,j} a_k$ if $i \neq j$,
- (necessarily) $\breve{a}_i = a_i$,

where $i, j, k \leq p$. On arbitrary elements of $\mathcal{A}$, ';' can be calculated from this using additivity of ';'. So can '˘': we have $\breve{r} = r$ for all $r \in \mathcal{A}_p$.

It can be shown that any given Lyndon algebra $\mathcal{A}_p$ is representable iff there is a projective plane of order p. For prime p, there is such a plane, so infinitely many $\mathcal{A}_p$ are representable [Lyn61, theorem 1].

Now choose $p \geq 4$ such that $\mathcal{A}_p$ is representable. As $\mathcal{A}_p$ is finite, it is completely representable, so by exercise 12.2(5) it has an n-dimensional hyperbasis for all $n \geq 3$. In more detail, let M be a representation of $\mathcal{A}_p$, regarded as a model of the theory $T_{\mathcal{A}_p}$ of definition 3.30. We can assume that $M \models \forall xy\, 1(x,y)$, since $\mathcal{A}_p$ is simple. Let $n \geq 3$. For each n-tuple $\bar{a} = (a_0, \ldots, a_{n-1}) \in {}^n M$, define a hypernetwork $N_{\bar{a}}$ over $\mathcal{A}_p$, by: $N_{\bar{a}}(x,y)$ is the unique atom $\alpha \in \mathcal{A}_p$ such that $M \models \alpha(a_x, a_y)$, for $x, y < n$; and for $\bar{x} = (x_0, \ldots, x_{l-1}) \in {}^{\leq n} n$ with $l \neq 2$, $N_{\bar{a}}(\bar{x}) = (a_{x_0}, \ldots, a_{x_{l-1}})$. It is easily checked that $\{N_{\bar{a}} : \bar{a} \in {}^n M\}$ is an n-dimensional hyperbasis for $\mathcal{A}_p$.

It remains to show that $\mathcal{A}_p$ does not have a 5-dimensional cylindric basis. Assume for contradiction that $\mathcal{M}$ is such a basis. For convenience, let a, b, c, d, e be distinct diversity atoms ($\neq 1'$) of $\mathcal{A}_p$. By property (C$_0$) of the definition of cylindric basis, there are atomic networks in $\mathcal{M}$ with the triangle on nodes 0, 1, 2 labelled by atoms (a,b,c), (a,c,d), and (a,c,e). Since $\mathcal{M}$ is closed under substitutions (C$_2$), there are $M_3, M_4, N_3, N_4 \in \mathcal{M}$ with (see figure 12.3):

- $N_3(0,1) = a$, $N_3(0,3) = b$, $N_3(3,1) = c$, and $N_3(1,4) = 1'$,
- $M_3(1,2) = a$, $M_3(1,3) = c$, $M_3(2,3) = d$, and $M_3(1,4) = 1'$,
- $N_4(0,1) = a$, $N_4(0,4) = b$, $N_4(4,1) = c$, and $N_4(1,3) = 1'$,
- $M_4(1,2) = a$, $M_4(1,4) = c$, $M_4(2,4) = e$, and $M_4(1,3) = 1'$.

Now $M_3 \equiv_{02} N_3$ and $M_4 \equiv_{02} N_4$, so by (C$_1$) there are $P_3, P_4 \in \mathcal{M}$ with $M_3 \equiv_0 P_3 \equiv_2 N_3$ and $M_4 \equiv_0 P_4 \equiv_2 N_4$. Notice that

$$\begin{aligned} P_3(0,2) &\leq (P_3(0,1); P_3(1,2)) \cdot (P_3(0,3); P_3(3,2)) \\ &= (a;a) \cdot (b;d) = a, \end{aligned}$$

so $P_3(0,2) = a$. Similarly, $P_4(0,2) = a$. See figure 12.4.

12.6. Maddux's bases

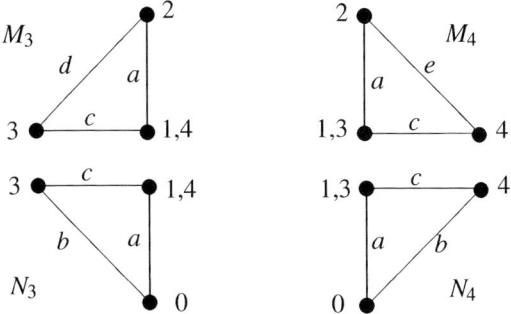

Figure 12.3: The networks N_3, N_4, M_3, M_4

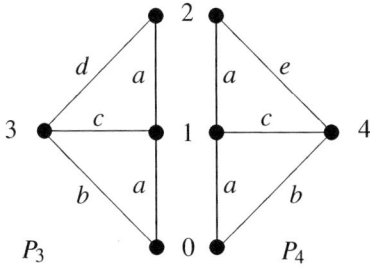

Figure 12.4: The networks P_3, P_4

Hence, $P_3 \equiv_{34} P_4$. By (C$_1$) again, there is $P \in \mathcal{M}$ with $P_3 \equiv_4 P \equiv_3 P_4$. But now,

$$\begin{aligned} P(3,4) &\leq \prod_{i=0,1,2} P(3,i)\,;P(i,4) \\ &= \prod_{i=0,1,2} P_3(3,i)\,;P_4(i,4) \\ &= (b\,;b)\cdot(c\,;c)\cdot(d\,;e) = 0, \end{aligned}$$

a contradiction. □

However, if $\mathcal{A}$ is a representable relation algebra, then $\mathcal{A}$ is a *subalgebra* of a relation algebra with an *n*-dimensional cylindric basis, for all $n \geq 4$. See exercise 2 below. We do not know if this is true 'level-by-level':

PROBLEM 12.38 *For arbitrary finite $n \geq 5$, does every atomic relation algebra with an n-dimensional hyperbasis embed in an atomic relation algebra with an n-dimensional cylindric basis?*

Exercises

1. If $3 \leq m \leq n < \omega$, show that any atomic non-associative algebra with an n-dimensional cylindric basis has an m-dimensional cylindric basis.

2. Let $\mathcal{A} \in \mathbf{RRA}$. Prove that $\mathcal{A} \subseteq \mathcal{B}$ for some atomic relation algebra $\mathcal{B}$ that has an n-dimensional cylindric basis for every $n \geq 3$. [Consider the full proper relation algebra over the base of a representation of $\mathcal{A}$.]

3. Generalising lemma 12.36 slightly, show that for finite $n \geq 4$, any set of n-dimensional Λ-hypernetworks over an atomic weakly associative algebra $\mathcal{A}$ that satisfies conditions (C$_0$–C$_2$) of definition 12.35 is an n-dimensional Λ-hyperbasis for $\mathcal{A}$.

4. Let $\mathcal{A}$ be an atomic weakly associative algebra, let $n \in \{3,4\}$, and let $\mathcal{M}$ be a set of atomic $\mathcal{A}$-networks with set of nodes n. Prove that the following are equivalent:

    (a) $\mathcal{M}$ is an n-dimensional relational basis for $\mathcal{A}$,

    (b) $\mathcal{M}$ is an n-dimensional cylindric basis for $\mathcal{A}$,

    (c) $\mathcal{M}$ is an n-dimensional $\{\lambda\}$-hyperbasis for $\mathcal{A}$, where we regard an atomic $\mathcal{A}$-network as a $\{\lambda\}$-hypernetwork over $\mathcal{A}$ in the usual way.

    Deduce from exercise 12.2(11) that an atomic weakly associative algebra is semi-associative iff it has a 3-dimensional cylindric basis, and is a relation algebra iff it has a 4-dimensional cylindric basis. (Exercise 5 shows that this is false for non-associative algebras.)

5. This exercise is to show that for non-associative algebras, in arbitrarily high dimension, not every cylindric basis is a relational basis. It constructs an atomic non-associative algebra $\mathcal{B}$ with an n-dimensional cylindric basis for all $n \geq 3$, but with no n-dimensional relational basis for any $n \geq 3$ (since $\mathcal{B}$ is not even weakly associative). The 'reason' why this can happen is that networks do not start to work properly until weak associativity is assumed. For an atomic network N over a weakly associative algebra, lemma 5.10 shows that $N(0,1)$ determines $N\!\upharpoonright_{\{0,1\}}$. For arbitrary non-associative algebras, this is no longer true.

    Let $\mathcal{A}$ be an atomic relation algebra satisfying 1' < 1. Define an atomic non-associative algebra $\mathcal{B}$ by 'splitting identity atoms of $\mathcal{A}$', as follows. For each identity atom $e \in \mathcal{A}$ (with $e \leq 1'$), introduce new distinct red and green copies e^r, e^g of e, and for each diversity atom $a \in \mathcal{A}$ (with $a \cdot 1' = 0$), let $a^r = a^g = a$. The set of atoms of $\mathcal{B}$ is now defined to be $\{a^r, a^g : a \in \mathrm{At}\mathcal{A}\}$. The identity atoms are e^r, e^g for all $e \in \mathrm{At}\mathcal{A}$ with $e \leq 1'$, and conversion

12.7. Cylindric bases and homogeneous representations

is defined on atoms of $\mathcal{B}$ by $\breve{a}^r = (\breve{a})^r$ for $a \in \operatorname{At}\mathcal{A}$, and similarly for a^g. Finally, the set of consistent triples of atoms of $\mathcal{B}$ (see definition 3.25) is defined to be $\{(a^r,b^r,c^r),(a^g,b^g,c^g) : (a,b,c) \text{ a consistent triple of atoms of } \mathcal{A}\}$. We now define $\mathcal{B}$ to be the complex algebra over this atom structure.

(a) Show that $\mathcal{B} \in \mathbf{NA} \setminus \mathbf{WA}$. Deduce that $\mathcal{B}$ has no n-dimensional relational basis for any $n \geq 3$.

Now let $n \geq 3$, and suppose that $\mathcal{A}$ has an n-dimensional cylindric basis, say $\mathcal{M}$. (E.g., take $\mathcal{A}$ to be the '$\mathcal{B}$' in exercise 2.) For any atomic $\mathcal{A}$-network N, let N^g be the $\mathcal{B}$-pre-network given by $N^g(x,y) = N(x,y)^g$, for $x,y \in N$. Define N^r similarly. Let $\mathcal{M}' = \{N^r, N^g : N \in \mathcal{M}\}$.

(b) Show that N^r, N^g are atomic $\mathcal{B}$-networks, for each $N \in \mathcal{M}$.

(c) Prove that $\mathcal{M}'$ is an n-dimensional cylindric basis for $\mathcal{B}$.

(d) Show explicitly that $\mathcal{M}'$ is not a relational basis.

PROBLEM 12.39 *Investigate the complexity of deciding, for fixed $n \geq 3$, whether a finite weakly associative algebra has an n-dimensional cylindric basis. [It is clearly in NP; is it NP-complete?]*

12.7 Cylindric bases and homogeneous representations

DEFINITION 12.40 A (classical) representation of a relation algebra $\mathcal{A}$ can be equivalently regarded as a model of the theory $T_\mathcal{A}$ of definition 3.30. Such a representation is said to be *homogeneous* if it is 'ultrahomogeneous' in the model-theoretic sense: every partial isomorphism of M with finite domain extends to an automorphism of M.

For finite relation algebras, the following amalgamation argument, in the style of [Fra54], shows that cylindric bases and homogeneous representations 'coincide'. See exercises 13.6(1) and 19.6(4) for related results.

THEOREM 12.41 *Let $\mathcal{A}$ be a finite relation algebra. $\mathcal{A}$ has a homogeneous representation if and only if $\mathcal{A}$ has an n-dimensional cylindric basis for all finite $n \geq 3$.*

Proof (sketch). If M is a homogeneous representation of $\mathcal{A}$, then let $n \geq 3$ and consider the set of all atomic networks N with set of nodes n that 'embed' in M (that is, there is a map $\nu : n \to M$ such that $M \models [N(i,j)](\nu(i),\nu(j))$ for all $i,j < n$). This can be checked to be an n-dimensional cylindric basis for $\mathcal{A}$; homogeneity of M is used to prove the amalgamation property.

To prove the converse, suppose that $\mathcal{A}$ has an n-dimensional cylindric basis $\mathcal{M}_n$ for all finite $n \geq 3$. By lemmas 12.15 and 12.36, we may suppose that each $\mathcal{M}_n$ is symmetric (closed under permutations). By remark 12.23 and lemma 12.36, $\mathcal{M}_n \restriction_m$ is a symmetric m-dimensional cylindric basis for $\mathcal{A}$ whenever $3 \leq m < n$.

Now there are only finitely many m-dimensional atomic networks over $\mathcal{A}$, for any finite m. So by König's tree lemma (exercise 10.1(2)), we may suppose without loss of generality that $\mathcal{M}_n \restriction_m = \mathcal{M}_m$ whenever $3 \leq m < n < \omega$. We will construct a homogeneous representation of $\mathcal{A}$ by playing a suitable game. The following definition and lemma will show that $\exists$ has a winning strategy in the game.

DEFINITION 12.42 Let N be an atomic network over the atomic relation algebra $\mathcal{A}$. A *partial isomorphism* f of N is a partial one-one map $f : \mathrm{nodes}(N) \to \mathrm{nodes}(N)$ such that for all $i, j \in \mathrm{dom}(f)$ we have $N(i, j) = N(f(i), f(j))$.

LEMMA 12.43 *Let $\mathcal{A}$ be a finite relation algebra, and let $\mathcal{M}_n$ be a symmetric n-dimensional cylindric basis for $\mathcal{A}$ such that $\mathcal{M}_n \restriction_m = \mathcal{M}_m$ for $3 \leq m \leq n < \omega$. Let $3 \leq t < \omega$ and $N \in \mathcal{M}_t$.*

1. *Let $i, j < t$, $r, s \in \mathcal{A}$, and assume that $r;s \geq N(i,j)$. There is $N^+ \in \mathcal{M}_{t+1}$ with $N \subseteq N^+$ and $N^+(i,t) \leq r$, $N^+(t,j) \leq s$.*

2. *If $i < t$ and f is a partial isomorphism of N, then there are $N^+ \in \mathcal{M}_{t+1}$ and a partial isomorphism f^+ of N^+ such that $N^+ \supseteq N$, $i \in \mathrm{dom}(f^+)$, and $f^+ \supseteq f$.*

Proof.

1. Since $\mathcal{M}_t = \mathcal{M}_{t+1} \restriction_t$, there is some $N' \in \mathcal{M}_{t+1}$ with $N' \restriction_{\{0,\dots,t-1\}} = N$. As $\mathcal{M}_{t+1}$ is a cylindric basis, by lemma 12.36 there is $N^+ \in \mathcal{M}_{t+1}$ with $N^+ \equiv_t N'$ and $N^+(i,t) \leq r$, $N^+(t,j) \leq s$.

2. We may assume that $f \neq \emptyset$ and $i \notin \mathrm{dom} f$. Again, take $N' \in \mathcal{M}_{t+1}$ with $N' \restriction_{\{0,\dots,t-1\}} = N$. Let $g : t+1 \to t+1$ be any total map extending f^{-1} and with $g(t) = i$. Since $\mathcal{M}_{t+1}$ is symmetric, $N'g \in \mathcal{M}_{t+1}$ (see definition 12.6 for $N'g$). Also, $(N'g)(f(x), f(y)) = N'(g(f(x)), g(f(y))) = N'(x,y)$, for $x, y \in \mathrm{dom}(f)$. So since f is a partial isomorphism, for all $x, y \in \mathrm{dom}(f)$ we have

$$N'(f(x), f(y)) = N'(x,y) = (N'g)(f(x), f(y)).$$

Thus, if $\bar{x}$ is a tuple with $\mathrm{rng}(\bar{x}) = t \setminus \mathrm{rng}(f)$ then $N' \equiv_{t\bar{x}} N'g$ — the two networks agree on all labels *except* perhaps those involving nodes outside $\mathrm{rng}(f)$. Since $\mathrm{rng}(f) \neq \emptyset$ and $\mathcal{M}_{t+1}$ is a cylindric basis, by lemmas 12.36 and 12.16 there is $N^+ \in \mathcal{M}_{t+1}$ such that

$$N' \equiv_t N^+ \equiv_{\bar{x}} N'g.$$

12.7. Cylindric bases and homogeneous representations

Since $N'\restriction_{\{0,\ldots,t-1\}} = N$ and $N' \equiv_t N^+$, we get $N^+\restriction_t = N$. And since $N^+ \equiv_{\bar{x}} N'g$, for all $x \in \mathrm{dom}(f)$ we have $N^+(t, f(x)) = (N'g)(t, f(x)) = N'(i, x) = N^+(i, x)$. We can show $N^+(f(x), t) = N^+(x, i)$, $N^+(t, t) = N^+(i, i)$, and $N^+(f(x), f(x)) = N^+(x, x)$ similarly. So we see that $f^+ = f \cup \{(i, t)\}$ is a partial isomorphism of N^+ extending f and defined on i (it is one-one since $i \notin \mathrm{dom}\, f$ and $t \notin \mathrm{rng}\, f$).

□

Two players, ∀ (male) and ∃ (female), now play a game to build a homogeneous representation of $\mathcal{A}$. The game has ω rounds, numbered $2, 3, \ldots, t, \ldots$ ($2 \leq t < \omega$). We start at 2 because the $\mathcal{M}_t$ start at $t = 3$. In round 2, ∀ picks non-zero $r_0 \in \mathcal{A}$, and ∃ responds with some $N_3 \in \mathcal{M}_3$ with $N_3(0, 1) \leq r_0$; she can find such a network because $\mathcal{M}_3$ is a cylindric basis. In each subsequent round $t \geq 3$, if the current network is $N_t \in \mathcal{M}_t$, then ∀ can make one of two types of move:

Triangle move: He chooses $i, j < t$ and $r, s \in \mathcal{A}$ with $N_t(i, j) \leq r; s$. ∃ must respond with some $N_{t+1} \in \mathcal{M}_{t+1}$ extending N_t such that $N_{t+1}(i, t) \leq r$ and $N_{t+1}(t, j) \leq s$. Lemma 12.43(1) shows that she can do this.

Partial isomorphism move: Or he picks a partial isomorphism $f : N_t \to N_t$ and a node $i \in \mathrm{nodes}(N_t)$. ∃ must respond with some $N_{t+1} \in \mathcal{M}_{t+1}$ extending N_t such that there is a partial isomorphism $f^+ : N_{t+1} \to N_{t+1}$ extending f and with $i \in \mathrm{dom}(f^+)$. She can do this, by lemma 12.43(2).

So ∃ has a winning strategy in this game.

Consider a play $N_3, N_4, \ldots$ of the game in which ∀ plays $r_0 \in \mathcal{A} \setminus \{0\}$ and then makes every possible move in some round. That is, if $N_t(i, j) \leq r; s$ for some t, i, j, r, s, then he plays a triangle move i, j, r, s in some round $u \geq t$; and if f is any partial isomorphism of N_t and $i \in \mathrm{nodes}(N_t)$ then ∀ plays the partial isomorphism move f, i in some round $u \geq t$ (note that f is a partial isomorphism of N_u and $i \in \mathrm{nodes}(N_u)$). He can do all this because $\mathcal{A}$ and each network N_t is finite, and he makes countably many moves during the game. Since $N_t \subseteq N_u$ for $3 \leq t \leq u < \omega$, the play has a well-defined limit N_ω, a network of dimension ω, where for each $i, j < \omega$, $N_\omega(i, j) = N_k(i, j)$ for any $k > 2, i, j$. N_ω has the following properties:

1. $N_\omega(0, 1) \leq r_0$ (by ∀'s first move).

2. N_ω is a network: $N_\omega(i, i) \leq 1$' and $N_\omega(i, j) \leq N_\omega(i, k); N_\omega(k, j)$ for all $i, j, k < \omega$ (because N_ω is the 'limit' of networks).

3. If $N_\omega(i, j) \leq r; s$ for some $i, j < \omega$ and $r, s \in \mathcal{A}$, there is $k < \omega$ with $N_\omega(i, k) \leq r$ and $N_\omega(k, j) \leq s$. (Take $k > i, j$ such that ∀ played the triangle move i, j, r, s in round k.)

4. If $f : N_\omega \to N_\omega$ is a finite partial isomorphism and $i < \omega$, then f extends to a partial isomorphism defined on i. (Find $t < \omega$ such that $\mathrm{dom}(f) \cup \mathrm{rng}(f) \subseteq N_t$ and $i \in N_t$. If $i \notin \mathrm{dom}(f)$, there is $u \geq t$ where $\forall$ played the partial isomorphism move f, i in round u.)

5. Any finite partial isomorphism of N_ω is induced by a partial isomorphism that is actually a permutation of ω. (As N_ω is countable, repeated 'back-and-forth' application of property 4 shows this.)

Define a binary relation $\sim$ on ω by $i \sim j$ iff $N_\omega(i,j) \leq 1$'. It is easily checked (cf. lemma 7.22) that $\sim$ is an equivalence relation, and indeed a *congruence*, in that if $i \sim i'$ and $j \sim j'$ then $N_\omega(i,j) = N_\omega(i',j')$. Write $\omega/\sim$ for the set of equivalence classes, and $i^\sim$ for the equivalence class of i ($i < \omega$). Let M_{r_0} be the $\mathcal{L}(\mathcal{A})$-structure with domain $\omega/\sim$ given by

$$M_{r_0} \models r(i^\sim, j^\sim) \text{ iff } N_\omega(i,j) \leq r,$$

for $r \in \mathcal{A}$. (We make explicit the dependence on $\forall$'s first move r_0 here.) There is such a structure M_{r_0} for each non-zero $r_0 \in \mathcal{A}$. As is easily seen, property 5 ensures that each M_{r_0} is homogeneous.

Let $r \approx s$ in $\mathrm{At}\,\mathcal{A}$ iff $r \leq 1; s; 1$. Then $\approx$ is an equivalence relation on $\mathrm{At}\,\mathcal{A}$, and, as can be checked, $r \approx s$ iff $M_r \models \exists xy\, s(x,y)$, for all atoms r,s. Take a set E of representatives for the $\approx$-classes, and let M be the disjoint union of the structures M_e for $e \in E$. Then $M \models \exists xy\, r(x,y)$ for all non-zero $r \in \mathcal{A}$. It can be checked routinely that M is a representation of $\mathcal{A}$; see theorem 7.19 for similar work. We check that it is homogeneous. Let θ be any non-empty finite partial isomorphism of M, and take $x \in \mathrm{dom}(\theta) \cap M_e$ for arbitrary e. There is $a \in \mathrm{At}\,\mathcal{A}$ such that $M \models a(x,x)$. If $\theta(x) \in M_{e'}$, then since $M \models a(\theta(x), \theta(x))$, we have $e \approx a \approx e'$, whence $e = e'$. So θ is a union of partial isomorphisms of the M_e. Each one extends to an automorphism of M_e, and their union is an extension of θ to an automorphism of M. So M is homogeneous. □

Exercises

1. Prove that any relation algebra $\mathcal{A}$ with a complete homogeneous representation has an n-dimensional cylindric basis for all finite $n \geq 3$.

2. Show that for any group $\mathcal{G}$, the algebra $\mathfrak{Cm}\,\mathcal{G}$ (definition 4.3) has an n-dimensional cylindric basis for all finite $n \geq 3$.

3. Let $\mathcal{A} \in \mathbf{RRA}$. Prove that $\mathcal{A} \subseteq \mathcal{B}$ for some atomic relation algebra $\mathcal{B}$ that has a homogeneous representation. [Cf. exercises 9.3(4) and 12.6(2).]

Chapter 13

Approximations to RRA

In the introduction to part III, we discussed three approaches to approximating **RRA**: by bases, relation algebra reducts, and relativised representations. In this chapter, we develop the kinds of 'locally well behaved' relativised representation we have in mind: n-square, n-flat and n-smooth relativised representations. We will strengthen the notion of an n-flat relativised representation by insisting that quantifiers commute not just for ordinary first-order formulas but also for formulas in an infinitary logic. We'll call relativised representations with this property *infinitarily n-flat*. We then prove the equivalence of the three approaches, the chief goal of this chapter. The result of all this work appears in theorem 13.46 way below. The case $n = 3$ is handled separately, after that. The chapter ends with remarks on finite versions of this result, and recursive axiomatisations of $\mathbf{RA}_n$ and $\mathbf{S\Re aCA}_n$.

13.1 Representation theory

We begin by reviewing the main points about relativised representations from earlier chapters. Let $\mathcal{A}$ be a non-associative algebra. Recall from definition 5.1 that $L(\mathcal{A})$ is the first-order language in a signature consisting of one binary relation symbol for each element of $\mathcal{A}$, and that $R_\mathcal{A}$ is the $L(\mathcal{A})$-theory consisting of the following axioms:

$$\forall xy[1'^{\mathcal{A}}(x,y) \leftrightarrow x = y]$$
$$\forall xy[r(x,y) \leftrightarrow s(x,y) \vee t(x,y)] \quad \text{for all } r,s,t \in \mathcal{A} \text{ with } r = s+t$$
$$\forall xy[1^{\mathcal{A}}(x,y) \rightarrow (r(x,y) \leftrightarrow \neg s(x,y))] \quad \text{for all } r,s \in \mathcal{A} \text{ with } r = -s$$
$$\forall xy[r(x,y) \leftrightarrow s(y,x)] \quad \text{for all } r,s \in \mathcal{A} \text{ with } r = \breve{s}$$
$$\forall xy[1^{\mathcal{A}}(x,y) \rightarrow (r(x,y) \leftrightarrow \exists z(s(x,z) \wedge t(z,y)))] \quad \text{for all } r,s,t \in \mathcal{A} \text{ with } r = s;t$$
$$\exists xy\, r(x,y) \quad \text{for all } r \in \mathcal{A} \text{ with } r \neq 0.$$

We will usually drop the index '$\mathcal{A}$' in $\mathcal{L}(\mathcal{A})$-formulas. A *relativised representation* of $\mathcal{A}$ is a model of $R_\mathcal{A}$. A *complete relativised representation* of $\mathcal{A}$ is a model M of $R_\mathcal{A}$ such that for any set S of elements of $\mathcal{A}$ whose supremum $\sum S$ exists in $\mathcal{A}$, and any $x,y \in M$, $M \models (\sum S)(x,y)$ iff $M \models s(x,y)$ for some $s \in S$. Any representation of a finite algebra is complete. By lemma 5.3, a relativised representation of $\mathcal{A}$ is a representation of the boolean reduct of $\mathcal{A}$. So by the De Morgan laws, a relativised representation M is complete iff whenever $x,y \in M$, $S \subseteq \mathcal{A}$, and $\prod S$ exists in $\mathcal{A}$, we have $M \models (\prod S)(x,y)$ iff $M \models s(x,y)$ for all $s \in S$. By theorem 2.21, M is a complete relativised representation iff for every $x,y \in M$ with $M \models 1(x,y)$, there is an atom $a \in \mathcal{A}$ such that $M \models a(x,y)$. Hence, any non-associative algebra with a complete relativised representation is atomic.

13.1.1 Relativised semantics for $\mathcal{L}(\mathcal{A})$

We saw in theorem 7.5 that **WA** can be characterised by relativised representations — a relation-type BAO has a relativised representation if and only if it is a weakly associative algebra. But by imposing additional conditions on relativised representations, we can define smaller classes of algebras than **WA**. A convenient and general way to do this is by using n-variable first-order $\mathcal{L}(\mathcal{A})$-formulas in a relativised interpretation over a relativised representation. For handling complete relativised representations it is useful to use n-variable formulas in an infinitary language $\mathcal{L}(\mathcal{A})^n_{\infty\omega}$.

Until the end of section 13.1, we fix a non-associative algebra $\mathcal{A}$, and n with $3 \leq n < \omega$. The following definition is as in section 2.2.

DEFINITION 13.1 We define the *n-variable infinitary language* $\mathcal{L}(\mathcal{A})^n_{\infty\omega}$. The variables are $x_0, \ldots, x_{n-1}$, and the set of atomic formulas is $\{r(x_i, x_j) : r \in \mathcal{A}, i,j < n\} \cup \{x_i = x_j : i,j < n\}$. The class of formulas of $\mathcal{L}(\mathcal{A})^n_{\infty\omega}$ is defined to be the smallest class containing all the atomic formulas, closed under negations, quantification and arbitrary conjunctions over sets of formulas. Formally, if $\varphi \in \mathcal{L}(\mathcal{A})^n_{\infty\omega}$, $i < n$ and $\Phi \subseteq \mathcal{L}(\mathcal{A})^n_{\infty\omega}$ is a set, then $\neg\varphi, \exists x_i \varphi, \bigwedge \Phi \in \mathcal{L}(\mathcal{A})^n_{\infty\omega}$.

We write $\bigvee \Phi$ as an abbreviation for $\neg \bigwedge \{\neg \varphi : \varphi \in \Phi\}$. $\varphi \wedge \psi$ abbreviates $\bigwedge \{\varphi, \psi\}$, and other abbreviations such as $\to$ are defined as usual. The first-order fragment of $\mathcal{L}(\mathcal{A})^n_{\infty\omega}$ is denoted by $\mathcal{L}(\mathcal{A})^n$.

DEFINITION 13.2 Let M be a relativised representation of $\mathcal{A}$.

1. A *clique* in M is a subset $C \subseteq M$ such that $M \models 1(x,y)$ for all $x,y \in C$.

2. We write $C^n(M)$ for the set $\{\bar{a} \in {}^n M : \text{rng}(\bar{a}) \text{ is a clique in } M\}$.

3. When we write $\bar{a} \in C^n(M)$, it will be implicit that $\bar{a} = (a_0, \ldots, a_{n-1})$.

13.1. Representation theory

4. For tuples $\bar{a} = (a_0,\ldots,a_{n-1}), \bar{b} = (b_0,\ldots,b_{n-1})$ in $C^n(M)$ and $i_0,\ldots,i_{k-1} < n$, we write $\bar{a} \equiv_{i_0,\ldots,i_{k-1}} \bar{b}$ if $a_j = b_j$ for all $j < n$ with $j \notin \{i_0,\ldots,i_{k-1}\}$.

$C^n(M)$ is obviously closed under substitutions: if $(a_0,\ldots,a_{n-1}) \in C^n(M)$ and $i,j < n$ then $(a_0,\ldots,a_{i-1},a_j,a_{i+1},\ldots,a_{n-1}) \in C^n(M)$. It also satisfies a stronger property: if $\bar{a} \in C^n(M)$ and $\pi : n \to n$ is any map, then $\bar{a} \circ \pi = (a_{\pi(0)},\ldots,a_{\pi(n-1)}) \in C^n(M)$. (Substitution-closure covers the case when $\pi = [i/j]$. Recall here (definition 5.22) that the map $[i/j] : n \to n$, for $i,j < n$, is defined to be the identity on $n \setminus \{i\}$ and such that $[i/j](i) = j$. The notation $[i/j]$ depends implicitly on n: if we change n, we get a different map. But here, n is fixed.)

DEFINITION 13.3 Let M be a relativised representation of $\mathcal{A}$. We define the *(n-dimensional) clique-relativised semantics* $M \models_C \varphi(\bar{a})$, for $\varphi \in L(\mathcal{A})^n_{\infty\omega}$ and $\bar{a} \in C^n(M)$, as follows.

- If φ is $r(x_i, x_j)$ for $r \in \mathcal{A}$ and $i,j < n$, then $M \models_C \varphi(\bar{a})$ iff $M \models r(a_i, a_j)$.
- If φ is $x_i = x_j$ for some $i,j < n$, then $M \models_C \varphi(\bar{a})$ iff $a_i = a_j$.
- $M \models_C \neg \varphi(\bar{a})$ iff $M \not\models_C \varphi(\bar{a})$.
- For any set Φ of $L(\mathcal{A})^n_{\infty\omega}$-formulas, $M \models_C \bigwedge \Phi(\bar{a})$ iff $M \models_C \varphi(\bar{a})$ for all $\varphi \in \Phi$.
- For $i < n$, $M \models_C \exists x_i \varphi(\bar{a})$ iff $M \models_C \varphi(\bar{b})$ for some $\bar{b} \in C^n(M)$ with $\bar{b} \equiv_i \bar{a}$.

Note that $\models_C$ depends implicitly on n. (It hardly seems necessary to introduce notation such as $\models_C^n$, since n will usually be fixed by the context.) As we will see in section 19.2.3, the clique-relativised semantics is related to the 'clique-guarded fragment' of first-order logic.

13.1.2 Square relativised representations

We now recall from definition 5.7 the first special kind of relativised representation. The definition here is equivalent as we are only dealing with finite n.

DEFINITION 13.4 Let M be a relativised representation of $\mathcal{A}$. M is said to be *n-square* if whenever $\bar{a} \in C^n(M)$, $r,s \in \mathcal{A}$, $i,j,k < n$, $k \neq i,j$, and $M \models (r;s)(a_i, a_j)$, then there is $\bar{b} \in C^n(M)$ with $\bar{b} \equiv_k \bar{a}$ such that $M \models r(b_i, b_k)$ and $M \models s(b_k, b_j)$.

LEMMA 13.5 *Let $\mathcal{B}$ be a non-associative algebra with $\mathcal{B} \supseteq \mathcal{A}$. If $\mathcal{B}$ has an n-square relativised representation, then so does $\mathcal{A}$.*

Proof. Let M^+ be an n-square relativised representation of $\mathcal{B}$, and let M be the reduct of M^+ to the language $L(\mathcal{A})$. As $M^+ \models R_{\mathcal{B}}$, and plainly $R_{\mathcal{A}} \subseteq R_{\mathcal{B}}$, we have $M \models R_{\mathcal{A}}$, so M is a relativised representation of $\mathcal{A}$. The n-squareness of M with respect to $\mathcal{A}$ is immediate from the given n-squareness of M^+ with respect to $\mathcal{B}$. □

13.1.3 Flat relativised representations

We now introduce a second, stronger kind of relativised representation.

DEFINITION 13.6 Let M be a relativised representation of $\mathcal{A}$. We say that M is *infinitarily n-flat* if for all $\varphi \in L(\mathcal{A})^n_{\infty\omega}$, all $\bar{a} \in C^n(M)$, and all $i, j < n$, we have

$$M \models_C (\exists x_i \exists x_j \varphi \leftrightarrow \exists x_j \exists x_i \varphi)(\bar{a}). \tag{13.1}$$

A weaker condition is the following: M is *n-flat* if (13.1) holds for all first-order formulas $\varphi \in L(\mathcal{A})^n$.

That is, up to $\models_C$-equivalence, quantifiers commute. We will now establish some basic properties of n-flat and infinitarily n-flat relativised representations, showing that flatness is a natural, powerful, and well behaved concept. Similar lemmas exist in [Madd89b]; cf. also lemma 12.16.

LEMMA 13.7 *Let $\mathcal{A} \subseteq \mathcal{B}$ be non-associative algebras. If $\mathcal{B}$ has an infinitarily n-flat relativised representation then so does $\mathcal{A}$. If $\mathcal{B}$ has an n-flat relativised representation then so does $\mathcal{A}$.*

Proof. As in lemma 13.5, let M^+ be an infinitarily n-flat relativised representation of $\mathcal{B}$, and let M be its reduct to $L(\mathcal{A})$. Then M is a relativised representation of $\mathcal{A}$; it is clearly infinitarily n-flat, since $C^n(M) = C^n(M^+)$ and $L(\mathcal{A})^n_{\infty\omega} \subseteq L(\mathcal{B})^n_{\infty\omega}$. The (non-infinitary) n-flat case is similar. □

LEMMA 13.8 *Let M be an infinitarily n-flat (respectively n-flat) relativised representation of $\mathcal{A}$, let $i_0, \ldots, i_{k-1} < n$ for some $k < n$, and let $\varphi \in L(\mathcal{A})^n_{\infty\omega}$ (respectively $\varphi \in L(\mathcal{A})^n$) and $\bar{a} \in C^n(M)$. Then $M \models_C (\exists x_{i_0} \ldots \exists x_{i_{k-1}} \varphi)(\bar{a})$ iff $M \models_C \varphi(\bar{b})$ for some $\bar{b} \in C^n(M)$ with $\bar{a} \equiv_{i_0, \ldots, i_{k-1}} \bar{b}$.*

Proof. '$\Rightarrow$' is clear. We prove '$\Leftarrow$' by induction on k. If $k = 0$, it is trivial, and if $k = 1$, it holds by definition of $\models_C$. Let $k > 1$, and assume the result for smaller k. Since $k < n$, there is $j \in n \setminus \{i_0, \ldots, i_{k-1}\}$. Then $a_j = b_j$. Let $\bar{a}' \in C^n(M)$ be the result of replacing $a_{i_{k-1}}$ by a_j in $\bar{a}$. Define $\bar{b}'$ similarly. Clearly, $\bar{a} \equiv_{i_{k-1}} \bar{a}' \equiv_{i_0, \ldots, i_{k-2}} \bar{b}' \equiv_{i_{k-1}} \bar{b}$. So if $M \models_C \varphi(\bar{b})$, then by three applications of the inductive hypothesis we obtain $M \models_C (\exists x_{i_{k-1}} \varphi)(\bar{b}')$, $M \models_C (\exists x_{i_0} \ldots \exists x_{i_{k-1}} \varphi)(\bar{a}')$, and $M \models_C (\exists x_{i_{k-1}} (\exists x_{i_0} \ldots \exists x_{i_{k-1}} \varphi))(\bar{a})$. Now M is infinitarily n-flat, so by the commutativity of existential quantifiers and a straightforward induction on k, we obtain $M \models_C (\exists x_{i_0} \ldots \exists x_{i_{k-1}} \varphi)(\bar{a})$, as required.

The non-infinitary case is entirely similar. □

Now we prove that free variables of $L(\mathcal{A})^n_{\infty\omega}$-formulas behave as we would hope. Cf. [Madd89b, lemma 20]. Bear in mind that variables can be 're-used' in n-variable formulas, so that x_0 occurs both free and bound in $r(x_0, x_1) \wedge \exists x_0 s(x_0, x_1)$, for example.

13.1. Representation theory

LEMMA 13.9 *Let M be an infinitarily n-flat relativised representation of $\mathcal{A}$ (respectively, an n-flat relativised representation of $\mathcal{A}$), let $\varphi \in L(\mathcal{A})^n_{\infty\omega}$ ($\varphi \in L(\mathcal{A})^n$), and let x_i (for some $i < n$) be a variable that does not occur free in φ. Then $M \models_C (\varphi \leftrightarrow \exists x_i \varphi)(\bar{a})$ for all $\bar{a} \in C^n(M)$. Equivalently, if $\bar{a}, \bar{b} \in C^n(M)$ and $\bar{a} \equiv_i \bar{b}$ then $M \models_C \varphi(\bar{a})$ iff $M \models_C \varphi(\bar{b})$.*

Hence, if $i_0, \ldots, i_{k-1} < n$ for some $k < n$ and none of $x_{i_0}, \ldots, x_{i_{k-1}}$ are free in φ, then whenever $\bar{a} \equiv_{i_0,\ldots,i_{k-1}} \bar{b}$ in $C^n(M)$, we have $M \models_C \varphi(\bar{a})$ iff $M \models_C \varphi(\bar{b})$.

Proof. We treat the infinitary case; the non-infinitary case is similar. We show by induction on φ that if x_i is not free in φ, $\bar{a}, \bar{b} \in C^n(M)$, and $\bar{a} \equiv_i \bar{b}$, then $M \models_C \varphi(\bar{a})$ iff $M \models_C \varphi(\bar{b})$. If φ is atomic, this is trivial, and the cases of negation and conjunction are also straightforward. Assume the result for φ and consider $\exists x_j \varphi$, assuming that x_i is not free in $\exists x_j \varphi$. If $j = i$, the result follows from the fact that $\equiv_i$ is an equivalence relation on $C^n(M)$. So assume that $j \neq i$. We let $\bar{a} \equiv_i \bar{b}$ and $M \models_C \exists x_j \varphi(\bar{a})$, and check that $M \models_C \exists x_j \varphi(\bar{b})$, also. (The converse is similar.) Plainly, $M \models_C \exists x_i \exists x_j \varphi(\bar{b})$. By infinitary n-flatness, $M \models_C \exists x_j \exists x_i \varphi(\bar{b})$. So there are $\bar{c}, \bar{d} \in C^n(M)$ with $\bar{b} \equiv_j \bar{c} \equiv_i \bar{d}$ and $M \models_C \varphi(\bar{d})$. Now as $i \neq j$, x_i is not free in φ. So by the inductive hypothesis, $M \models_C \varphi(\bar{c})$. Hence, $M \models_C \exists x_j \varphi(\bar{b})$, as required.

The last part follows by a straightforward induction on k, using lemma 13.8. □

We can use these results to prove that n-flatness is at least as strong a property as n-squareness.

LEMMA 13.10 *Any n-flat relativised representation of $\mathcal{A}$ is n-square.*

Proof. Let M be an n-flat relativised representation of $\mathcal{A}$, let $\bar{a} \in C^n(M)$, $i, j, k < n$ with $k \neq i, j$, and $r, s \in \mathcal{A}$. Suppose that $M \models (r;s)(a_i, a_j)$. We seek $\bar{b} \in C^n(M)$ with $\bar{b} \equiv_k \bar{a}$, $M \models r(b_i, b_k)$, and $M \models s(b_k, b_j)$.

As $M \models (r;s)(a_i, a_j)$, there is $c \in M$ with $M \models r(a_i, c) \wedge s(c, a_j)$. Let $\bar{c}$ be the n-tuple given by $c_i = a_i$, $c_j = a_j$, and $c_l = c$ for $l < n$, $l \neq i, j$. As 1^M is a reflexive, symmetric relation on M (lemma 5.3), we see that $\{a_i, a_j, c\}$ is a clique in M, so that $\bar{c} \in C^n(M)$.

Let χ be the $L(\mathcal{A})^n$-formula $r(x_i, x_k) \wedge s(x_k, x_j)$. Clearly, $M \models_C \chi(\bar{c})$, so $M \models_C \exists x_k \chi(\bar{c})$. But also, $\bar{c} \equiv_{n \setminus \{i,j\}} \bar{a}$, and no x_l for $l \neq i, j$ is free in $\exists x_k \chi$. So by the last part of lemma 13.9, $M \models_C \exists x_k \chi(\bar{a})$. Thus, there is $\bar{b} \in C^n(M)$ with $\bar{b} \equiv_k \bar{a}$ and $M \models_C \chi(\bar{b})$. By definition of χ, $\bar{b}$ is as required. □

13.1.4 Smooth relativised representations

These form an alternative approach to flat relativised representations. Essentially, we make an infinitarily n-flat relativised representation n-smooth by dropping explicit mention of the formulas φ of $\mathcal{L}(\mathcal{A})^n_{\infty\omega}$, and stating instead by means of equivalence relations which n-tuples of a relativised representation agree on all these formulas with respect to $\models_C$. We can axiomatise the properties required for quantifiers to commute, by stating that the equivalence relations should have certain 'n-back-and-forth' properties. The reader may consult [DawLin+95, EbbFlu99] for similar work in finite model theory, showing that the equivalence relations can be taken to be definable in fixed-point logic. n-smooth representations have the disadvantage (over infinitarily n-flat ones) that one must expand a relativised representation M by adding further relations, but the advantage that the infinitely many conditions $M \models_C (\exists x_i \exists x_j \varphi \leftrightarrow \exists x_j \exists x_i \varphi)(\bar{a})$, for all formulas φ, reduce to finitely many. We will see that $\mathcal{A}$ has an n-smooth relativised representation iff it has an n-flat one, though getting from one to the other will require saturation.

NOTATION 13.11 *If $\bar{x}, \bar{y}$ are m-tuples, we write $(\bar{x} \mapsto \bar{y})$ for $\{(x_i, y_i) : i < m\}$; this may or may not be a well-defined map. The concatenation of tuples $\bar{x}, \bar{y}$ is denoted $\bar{x}\bar{y}$.*

DEFINITION 13.12 Let M be a relativised representation of $\mathcal{A}$. M is said to be n-smooth if it is n-square[1] and there is an equivalence relation E^m on $C^m(M)$ for each $m = 1, 2, \ldots, n$, the E^m together satisfying:

- if $0 < l, m \leq n$, $(\bar{x}, \bar{y}) \in E^m$, and $\theta : l \to m$, then
  $(\bar{x} \circ \theta, \bar{y} \circ \theta) = ((x_{\theta(0)}, \ldots, x_{\theta(l-1)}), (y_{\theta(0)}, \ldots, y_{\theta(l-1)})) \in E^l$,

- for any $(\bar{x}, \bar{y}) \in E^m$, $r \in \mathcal{A}$, and $i, j < m$, if $M \models r(x_i, x_j)$ then $M \models r(y_i, y_j)$ (i.e., $(\bar{x} \mapsto \bar{y})$ is a well-defined partial isomorphism of M), and

- if $(\bar{x}, \bar{y}) \in E^{n-2}$ and $\text{rng}(\bar{x}x)$ and $\text{rng}(\bar{y}y)$ are cliques in M, then there exists a point $z \in M$ such that $(\bar{x}x, \bar{y}z) \in E^{n-1}$ and $\text{rng}(\bar{y}yz)$ is a clique.

REMARK 13.13 $\Theta = \{(\bar{x} \mapsto \bar{y}) : (\bar{x}, \bar{y}) \in \bigcup_{m \leq n} E^m\}$ is a certain kind of n-back-and-forth system of partial isomorphisms of M. Whenever $\theta \in \Theta$ is a map and $0 < |\text{dom}(\theta)| \leq n - 2$, θ can be extended within Θ to be defined on a new point a, so long as $\text{dom}(\theta) \cup \{a\}$ is a clique. Moreover, the extension can be chosen so that its range extends to a clique containing some other new point b, so long as $\text{rng}(\theta) \cup \{b\}$ was already a clique. Such n-back-and-forth systems offer an alternative definition of n-smooth.

LEMMA 13.14 *Let $\mathcal{A} \subseteq \mathcal{B}$ be non-associative algebras. If $\mathcal{B}$ has an n-smooth relativised representation then so does $\mathcal{A}$.*

Proof. Let M be an n-smooth relativised representation of $\mathcal{B}$. Consider the reduct of M to $\mathcal{L}(\mathcal{A})$. By lemma 13.5, this reduct is an n-square relativised representation of $\mathcal{A}$, and the equivalence relation properties defining n-smooth relativised representations of $\mathcal{A}$ are easily seen to hold in it. □

---
[1] This condition is not needed, as it follows from the one below; but it is easier to add it explicitly.

13.1. Representation theory

13.1.5 Links between the notions

For convenience, we round up some facts about these relativised representations.

PROPOSITION 13.15 *Let $3 \leq n < \omega$. Below, relativised representations are of an arbitrary non-associative algebra.*

1. *Any n-flat relativised representation is n-square [lemma 13.10].*

2. *Any infinitarily n-flat relativised representation is n-flat [trivial].*

3. *Any infinitarily n-flat relativised representation is n-smooth [exercise 13.5(1)].*

4. *Any finite (or more generally, ω-saturated) n-flat relativised representation is infinitarily n-flat [exercise 8 below].*

13.1.6 Elementary view

It is easily seen that n-square, n-flat, and n-smooth relativised representations can be characterised by first-order theories, although infinitarily n-flat relativised representations may not.

DEFINITION 13.16

1. We can extend the theory $R_{\mathcal{A}}$ to an $L(\mathcal{A})$-theory $\mathrm{Sq}^n(\mathcal{A})$ whose models (if any) are precisely the n-square relativised representations of $\mathcal{A}$, by adding axioms

    $$\forall \bar{x}(\mathrm{Clique}(\bar{x}) \wedge (r;s)(x_i,x_j) \to \exists x_k(\mathrm{Clique}(\bar{x}) \wedge r(x_i,x_k) \wedge s(x_k,x_j))),$$

    for all $i,j,k < n$ with $k \neq i,j$, and all $r,s \in \mathcal{A}$. In these axioms, $\bar{x}$ is an n-tuple $(x_0, \ldots, x_{n-1})$ of distinct variables, and $\mathrm{Clique}(\bar{x})$ is the formula $\bigwedge_{l,m<n} 1(x_l,x_m)$ where $|\bar{x}| = n$.

2. Similarly, we can extend $R_{\mathcal{A}}$ to an $L(\mathcal{A})$-theory $\mathrm{Fl}^n(\mathcal{A})$ whose models are the n-flat relativised representations of $\mathcal{A}$. Recall how to relativise quantifiers of an $L(\mathcal{A})^n$-formula φ to $C^n(M)$. The result, φ^{cl}, say, is defined by induction: $\varphi^{cl} = \varphi$ for atomic φ, $(\neg \varphi)^{cl} = \neg \varphi^{cl}$, $(\varphi \wedge \psi)^{cl} = \varphi^{cl} \wedge \psi^{cl}$, and $(\exists x_i \varphi)^{cl} = \exists x_i(\mathrm{Clique}(\bar{x}) \wedge \varphi^{cl})$. Then, for any relativised representation M of $\mathcal{A}$ and any $\bar{a} \in C^n(M)$, we have $M \models \varphi^{cl}(\bar{a})$ iff $M \models_C \varphi(\bar{a})$, for all $\varphi \in L(\mathcal{A})^n$.

    Now, the theory $\mathrm{Fl}^n(\mathcal{A})$ is obtained by adding axioms

    $$\forall \bar{x}\bigl(\mathrm{Clique}(\bar{x}) \to (\exists x_i \exists x_j \varphi \leftrightarrow \exists x_j \exists x_i \varphi)^{cl}\bigr),$$

    to $R_{\mathcal{A}}$, for all $\varphi \in L(\mathcal{A})^n$ and all $i,j < n$.

3. We can also extend the language $L(\mathcal{A})$ to the language $L(\mathcal{A},E)$ by adding new $2m$-ary predicate symbols E^m, for $0 < m \leq n$, and extend the theory $\mathrm{Sq}^n(\mathcal{A})$ to an $L(\mathcal{A},E)$-theory $\mathrm{Sm}^n(\mathcal{A})$ whose models are precisely the n-smooth relativised representations of $\mathcal{A}$. So, for example, the statement 'E^m is an equivalence relation on $C^m(M)$' can be written as the $L(\mathcal{A},E)$-sentence

$$\forall \bar{x},\bar{y},\bar{z}\,[(\mathrm{Clique}(\bar{x}) \leftrightarrow E^m(\bar{x},\bar{x})) \wedge (E^m(\bar{x},\bar{y}) \wedge E^m(\bar{y},\bar{z}) \to E^m(\bar{z},\bar{x}))],$$

where $\bar{x},\bar{y},\bar{z}$ are sequences of m distinct variables. The other axioms are derived from definition 13.12 above in the obvious way:

$$\forall \bar{x}\bar{y}(E^m(\bar{x},\bar{y}) \to E^l(\bar{x}\circ\theta,\bar{y}\circ\theta)) \quad \text{for } l,m < n,\ \theta : l \to m,$$
$$\forall \bar{x}\bar{y}(E^m(\bar{x},\bar{y}) \wedge r(x_i,x_j) \to r(y_i,y_j)) \quad \text{for } r \in \mathcal{A},$$
$$\forall \bar{x}\bar{y}(E^{n-2}(\bar{x},\bar{y}) \wedge \mathrm{Clique}(\bar{x}x) \wedge \mathrm{Clique}(\bar{y}y)$$
$$\to \exists z(E^{n-1}(\bar{x}x,\bar{y}z) \wedge \mathrm{Clique}(\bar{y}yz))).$$

The following is now clear.

PROPOSITION 13.17 *Let $\mathcal{A}$ be a non-associative algebra, let M be a $L(\mathcal{A})$-structure, and let $3 \leq n < \omega$.*

1. *M is an n-square relativised representation of $\mathcal{A}$ iff $M \models \mathrm{Sq}^n(\mathcal{A})$,*

2. *M is an n-flat relativised representation of $\mathcal{A}$ iff $M \models \mathrm{Fl}^n(\mathcal{A})$,*

3. *M is an n-smooth relativised representation of $\mathcal{A}$ iff M expands to a model of $\mathrm{Sm}^n(\mathcal{A})$.*

We can use this to prove an analogue for relativised representations of Monk's theorem 3.36.

COROLLARY 13.18 *Let $\mathcal{A}$ be an non-associative algebra and let $3 \leq n < \omega$.*

1. *If $\mathcal{A}$ has an n-square relativised representation then $\mathcal{A}^+$ has a complete n-square relativised representation.*

2. *If $\mathcal{A}$ has an n-smooth relativised representation then $\mathcal{A}^+$ has a complete n-smooth relativised representation.*

Proof (sketch). Suppose that $\mathcal{A}$ has an n-square relativised representation. So the theory $\mathrm{Sq}^n(\mathcal{A})$ is consistent. Let M be an ω-saturated model of it. The proof of theorem 3.36 adapts to show that M is essentially a complete n-square relativised representation of $\mathcal{A}^+$ — the proof of n-squareness is similar to the proof that ';' is respected. Similarly, if $\mathcal{A}$ has an n-smooth relativised representation then an ω-saturated model M of $\mathrm{Sm}^n(\mathcal{A})$ is (essentially) a complete n-smooth relativised representation of $\mathcal{A}^+$. Here, we note that any partial isomorphism of M is also a partial isomorphism of 'M regarded as a relativised representation of $\mathcal{A}^+$'. □

For an analogous result for flat relativised representations, see exercise 9 below.

13.1. Representation theory

Exercises

In the exercises, n is finite and satisfies $n \geq 3$ unless otherwise stated.

1. Show that any n-square relativised representation (of a non-associative algebra) is also m-square for any $3 \leq m < n$. Repeat for n-flat and infinitarily n-flat.

2. Let $\mathcal{A}$ be a non-associative algebra and M be a relativised representation of $\mathcal{A}$. Show that the following are equivalent:

    (a) M is n-square.
    (b) $M \models_C ([r;s](x_i,x_j) \leftrightarrow \exists x_k (r(x_i,x_k) \wedge s(x_k,x_j)))(\bar{a})$ for all $\bar{a} \in C^n(M)$, $i,j,k < n$ with $k \neq i, j$, and $r,s \in \mathcal{A}$.
    (c) For every clique $C \subseteq M$ with $|C| < n$, for every $r,s \in \mathcal{A}$, and every $x,y \in C$ with $M \models (r;s)(x,y)$, there is $z \in M$ with $M \models r(x,z) \wedge s(z,y)$ such that $C \cup \{z\}$ is a clique.

3. Let $\mathcal{A}$ be a non-associative algebra. For a permutation σ of n and a formula $\varphi \in L(\mathcal{A})^n_{\infty\omega}$, let φ^σ be the result of replacing the variable x_i in φ by $x_{\sigma(i)}$, simultaneously for each $i < n$. Given an infinitarily n-flat relativised representation M of $\mathcal{A}$, show that for all $\varphi \in L(\mathcal{A})^n_{\infty\omega}$, all permutations $\sigma : n \to n$, and all $\bar{a} \in C^n(M)$, we have $M \models_C \varphi^\sigma(\bar{a})$ iff $M \models_C \varphi(\bar{a} \circ \sigma)$.

4. Let $\mathcal{A}$ be any non-associative algebra. Show, for any formula $\varphi \in L(\mathcal{A})^3$, that there is a quantifier-free formula φ^* of $L(\mathcal{A})^3$ such that in any relativised representation M of $\mathcal{A}$ and for any $\bar{a} \in C^3(M)$, we have $M \models_C (\varphi \leftrightarrow \varphi^*)(\bar{a})$. [See the proof of theorem 3.32.]

5. Show that any representation of a representable relation algebra $\mathcal{A}$, when regarded as a model of the theory $T_{\mathcal{A}}$ of definition 3.30, is (a) n-square, (b) infinitarily n-flat, (c) n-smooth, for all finite $n \geq 3$.

6. Complete the proof of corollary 13.18.

7. Let M be a relativised representation of a non-associative algebra $\mathcal{A}$. Show that M is infinitarily n-flat iff there is an equivalence relation $\sim$ on $C^n(M)$ such that:

    - if $\bar{a}, \bar{b} \in C^n(M)$ and $\bar{a} \sim \bar{b}$, then $M \models r(a_i, a_j) \leftrightarrow r(b_i, b_j)$ for all $r \in \mathcal{A}$ and $i, j < n$,
    - if $\bar{a}, \bar{b} \in C^n(M)$, $\bar{a} \sim \bar{b}$, $i < n$, and $\bar{a} \equiv_i \bar{a}'$, then there is $\bar{b}' \in C^n(M)$ with $\bar{a}' \sim \bar{b}' \equiv_i \bar{b}$ (so $\sim$ is a kind of back-and-forth system),
    - if $\bar{a}, \bar{b}, \bar{c} \in C^n(M)$, $i, j < n$, and $\bar{a} \equiv_i \bar{b} \equiv_j \bar{c}$, then there are $\bar{b}', \bar{c}' \in C^n(M)$ with $\bar{a} \equiv_j \bar{b}' \equiv_i \bar{c}' \sim \bar{c}$.

[If stuck, see lemmas 13.41 and 13.42 later on.]

8. Show that any finite (or more generally, ω-saturated) n-flat relativised representation of a non-associative algebra is infinitarily n-flat.

9. Use saturation to show that if $\mathcal{A}$ has an n-flat relativised representation then $\mathcal{A}^+$ has a complete infinitarily n-flat relativised representation.

10. Show that any relativised representation satisfying the 'equivalence relation' conditions for smoothness in definition 13.12 is already n-square.

11. Show that the class of all non-associative algebras that have an n-flat relativised representation is a variety. Do the same for n-square relativised representations. [Imitate the argument for **RRA**.]

12. [For set theorists] Show that the formulas of $L(\mathcal{A})^n_{\infty\omega}$ do form a (proper) class.

13.2 From relativised representations to relation algebra reducts

We now prove the first main result of the chapter. It takes us from relativised representations to relation algebra reducts of various cylindric-type algebras. We have seen the idea before, in exercise 5.7(4).

Before we proceed, we make a remark on notation. For much of this chapter we will be using several first-order signatures all at once: the signatures L_{RA} of relation algebras, the signatures of cylindric algebras of various dimensions, and the signature $L(\mathcal{A})$ of relativised representations. These overlap, since both relation algebras and cylindric algebras have the boolean symbols. We will generally write these symbols in the same way; their home signature may be identified from the context.

If M is an L-structure and s a symbol in L, the conventions of section 2.2.2 dictate that we write s^M for the interpretation of s in M. This can lead to excesses: for example, if M is a relativised representation of a non-associative algebra $\mathcal{A}$, the unit of M should strictly be written $(1^{\mathcal{A}})^M$. As earlier, we will sometimes drop superfixes, leaving the context to disambiguate the meaning. So for the constant 1 of L_{RA}, for example, since $1^{\mathcal{A}}$ is both an element of the algebra $\mathcal{A}$ and a binary relation symbol of $L(\mathcal{A})$, dropping the superfix $^{\mathcal{A}}$ here means that 1 may have any of three different meanings, to be determined by the context. It is not hard to do this: for example, in $1(x,y)$, the '(x,y)' shows that 1 means the relation symbol $1^{\mathcal{A}} \in L(\mathcal{A})$.

13.2. From relativised representations to relation algebra reducts

DEFINITION 13.19 Let M be a relativised representation of a non-associative algebra $\mathcal{A}$, and let $3 \leq n < \omega$. For $\varphi \in L(\mathcal{A})^n_{\infty\omega}$, write
$$\varphi^C = \{\bar{a} \in C^n(M) : M \models_C \varphi(\bar{a})\}.$$

Let $\mathcal{D} = \mathcal{D}(M)$ be the following $\mathbf{CA}_n$-type algebra (its signature is the signature $\{+, -, 0, 1, \mathsf{d}_{ij}, \mathsf{c}_i : i, j < n\}$ of cylindric algebras).

- The domain of $\mathcal{D}$ is the set $\{\varphi^C : \varphi \in L(\mathcal{A})^n_{\infty\omega}\}$ of all sets definable by infinitary n-variable formulas in the clique-relativised semantics. (Though $L(\mathcal{A})^n_{\infty\omega}$ is a proper class, $\mathrm{dom}(\mathcal{D})$ is a set because M is a set.)
- 0 is interpreted in $\mathcal{D}$ as $(0(x_0, x_1))^C$ (i.e., as $\emptyset$), and 1 is interpreted as $(1(x_0, x_1))^C$ (i.e., as $C^n(M)$).
- $+$ and $-$ are interpreted in $\mathcal{D}$ as follows: $\varphi^C + \psi^C = \varphi^C \cup \psi^C = (\varphi \vee \psi)^C$, and $-(\varphi^C) = C^n(M) \setminus \varphi^C = (\neg\varphi)^C$ (this is plainly well-defined).
- d_{ij} is interpreted as $(x_i = x_j)^C$ (i.e., as $\{\bar{a} \in C^n(M) : a_i = a_j\}$).
- c_i is interpreted by: $\mathsf{c}_i(\varphi^C) = (\exists x_i \varphi)^C$ (this is well-defined).

We also define $\mathcal{D}^0 = \mathcal{D}^0(M)$ to be the subalgebra of $\mathcal{D}$ with domain $\{\varphi^C : \varphi \in L(\mathcal{A})^n\}$.

THEOREM 13.20 *Let $3 \leq n < \omega$ and let M be an n-square relativised representation of a non-associative algebra $\mathcal{A}$.*

1. *$\mathcal{D}(M), \mathcal{D}^0(M) \in \mathbf{G}_n$ (see definition 5.35).*
2. *If M is n-flat then $\mathcal{D}^0(M) \in \mathbf{CA}_n$ (see definition 5.16).*
3. *If M is infinitarily n-flat, then $\mathcal{D}(M) \in \mathbf{CA}_n$ and $\mathcal{D}(M)$ is atomic.*
4. *There are relation algebra reduct embeddings (see definition 5.49) from $\mathcal{A}$ into $\mathcal{D}(M)$ and $\mathcal{D}^0(M)$. If M is a complete relativised representation, these embeddings are complete (see definition 2.15).*

Hence, $\mathcal{A} \in \mathbf{S\mathfrak{R}aG}_n$; if M is n-flat then $\mathcal{A} \in \mathbf{S\mathfrak{R}a}(\mathbf{CA}_n \cap \mathbf{G}_n)$; and if M is complete and infinitarily n-flat then $\mathcal{A} \subseteq^c \mathfrak{R}a\mathcal{D}$ for some atomic $\mathcal{D} \in \mathbf{CA}_n \cap \mathbf{G}_n$.

Proof. We write $\mathcal{D}$ for $\mathcal{D}(M)$, and $\mathcal{D}^0$ for $\mathcal{D}^0(M)$. Assume that M is n-square. Clearly, $\mathcal{D}$ is a subalgebra of the cylindric relativised set algebra ($\mathbf{Crs}_n$) with domain $\wp(C^n(M))$, so $\mathcal{D} \in \mathbf{Crs}_n$ itself. The unit $C^n(M)$ of $\mathcal{D}$ is obviously closed under all maps $: n \to n$: that is, if $\bar{a} \in C^n(M)$ then $(a_{\sigma(0)}, \ldots, a_{\sigma(n-1)}) \in C^n(M)$ for all maps $\sigma: n \to n$. So $\mathcal{D} \in \mathbf{G}_n$. Similarly, or because $\mathbf{G}_n = \mathbf{SG}_n$, $\mathcal{D}^0(M) \in \mathbf{G}_n$. This proves part 1. For part 2, suppose that M is n-flat. By definition of n-flatness, we have $\mathcal{D}^0 \models \forall x(\mathsf{c}_i \mathsf{c}_j x = \mathsf{c}_j \mathsf{c}_i x)$ for all $i, j < n$. As $\mathcal{D}^0 \in \mathbf{G}_n$, we obtain $\mathcal{D}^0 \in \mathbf{CA}_n$ by proposition 5.39(5).

For part 3, assume that M is infinitarily n-flat. The proof that $\mathcal{D} \in \mathbf{CA}_n$ is the same as for $\mathcal{D}^0$, immediately above. We show that $\mathcal{D}$ is atomic. Let $\varphi^C \in \mathcal{D} \setminus \{0\}$. Choose $\bar{a} \in \varphi^C$ and consider $\tau = \bigwedge \{\psi \in L(\mathcal{A})^n_{\infty\omega} : M \models_C \psi(\bar{a})\}$. This conjunction is over a proper class, but because M is a set we can replace it by an M-equivalent conjunction over a set. So essentially, $\tau \in L(\mathcal{A})^n_{\infty\omega}$. Then τ^C is an atom of $\mathcal{D}$ beneath φ^C.

We prove part 4 for $\mathcal{D}$; the proof for $\mathcal{D}^0$ is similar. We show that the map $\iota: \mathcal{A} \to \mathcal{D}$ given by

$$\iota(r) = r(x_0, x_1)^C \in \mathcal{D}, \quad \text{for } r \in \mathcal{A},$$

is a relation algebra reduct embedding from $\mathcal{A}$ into $\mathcal{D}$. First, we check that $\iota(r)$ is 2-dimensional for all r. So let $r \in \mathcal{A}$ and $2 \leq i < n$; we must show that $r(x_0, x_1)^C = \mathsf{c}_i(r(x_0, x_1)^C) = (\exists x_i r(x_0, x_1))^C$. This is easily checked. Let $\bar{a} \in C^n(M)$. It is trivial that if $M \models_C r(x_0, x_1)(\bar{a})$ then $M \models_C (\exists x_i r(x_0, x_1))(\bar{a})$, because $\bar{a} \equiv_i \bar{a}$. Assume $M \models_C (\exists x_i r(x_0, x_1))(\bar{a})$. So $M \models_C r(x_0, x_1)(\bar{b})$ for some $\bar{b} \in C^n(M)$ with $\bar{b} \equiv_i \bar{a}$. As $r(x_0, x_1)$ is an atomic formula, we have $M \models r(b_0, b_1)$. But $i \geq 2$, so $a_0 = b_0$ and $a_1 = b_1$; thus, $M \models r(a_0, a_1)$. By definition of $\models_C$ for atomic formulas, $M \models_C r(x_0, x_1)(\bar{a})$, as required.

It is easy to check that ι preserves the boolean operations. If $r, s \in \mathcal{A}$, we require $\iota(r + s) = \iota(r) \cup \iota(s)$. Let $\bar{a} \in C^n(M)$. Then $\bar{a} \in \iota(r + s)$ iff $M \models_C ((r + s)(x_0, x_1))(\bar{a})$, iff $M \models (r + s)(a_0, a_1)$ as $(r + s)(x_0, x_1)$ is atomic. This is iff $M \models r(a_0, a_1)$ or $M \models s(a_0, a_1)$, as M is a relativised representation of $\mathcal{A}$. Reversing the argument, this is iff $\bar{a} \in \iota(r) \cup \iota(s)$, as required. For negation, let $r \in \mathcal{A}$; we require $\iota(-r) = C^n(M) \setminus \iota(r)$. If $\bar{a} \in C^n(M)$, then $M \models 1(a_0, a_1)$; so by the $-$-axiom of the theory $R_\mathcal{A}$ defining relativised representations, $M \models (-r)(a_0, a_1) \leftrightarrow \neg r(a_0, a_1)$. That is, $M \models_C (\neg r(x_0, x_1) \leftrightarrow -r(x_0, x_1))(\bar{a})$. So $\bar{a} \notin \iota(r)$ iff $\bar{a} \in \iota(-r)$, as required. It now follows that $\iota(0) = \emptyset$ and $\iota(1) = C^n(M)$.

Now, to show that ι is one-one, we only need check that if $r \in \mathcal{A} \setminus \{0\}$ then $\iota(r) \neq 0$ in $\mathcal{D}$. But M is a relativised representation of $\mathcal{A}$, so there are $a, b \in M$ with $M \models r(a, b)$. By the $+$-axiom of $R_\mathcal{A}$, $M \models 1(a, b)$. It follows by lemma 5.3 that $\{a, b\}$ is a clique in M, so there is certainly some $\bar{a} \in C^n(M)$ with $a_0 = a, a_1 = b$. Hence, $M \models_C r(x_0, x_1)(\bar{a})$ and $\bar{a} \in \iota(r)$, which is therefore non-empty as required.

Next, we check that ι preserves the relation algebra operations. By the identity axiom of $R_\mathcal{A}$, $M \models \forall xy(1\text{'}(x, y) \leftrightarrow x = y)$. So for all $\bar{a} \in C^n(M)$, we have $\bar{a} \in \iota(1\text{'})$ iff $M \models_C 1\text{'}(x_0, x_1)(\bar{a})$ iff $M \models 1\text{'}(a_0, a_1)$, iff $a_0 = a_1$, iff $\bar{a} \in \mathsf{d}_{01}$. So $\iota(1\text{'}) = \mathsf{d}_{01}$.

For the other operations, it will help to use notation 5.52 again, writing $\bar{a}^i_j$ for the tuple $(a_0, \ldots, a_{i-1}, a_j, a_{i+1}, \ldots)$. Note that if $i, j < n$ and $\bar{a} \in C^n(M)$, then $\bar{a}^i_j \in C^n(M)$, and $\bar{a} \in \mathsf{s}^i_j x$ iff $\bar{a}^i_j \in x$, for all $x \in \mathcal{D}$.

13.2. From relativised representations to relation algebra reducts

Now we check conversion. Let $r \in \mathcal{A}$; we need to prove that $\iota(\breve{r}) = s_0^2 s_1^0 s_2^1 \iota(r)$. Let $\bar{a} \in C^n(M)$. Then $\bar{a} \in s_0^2 s_1^0 s_2^1 \iota(r)$ iff $\bar{a}_{012}^{201} \in \iota(r)$. But we have $\bar{a}_{012}^{201} = (a_1, a_0, a_0, a_3, \ldots, a_{n-1})$ (see notation 5.52), so this is iff $M \models r(a_1, a_0)$. By the conversion axiom of $R_\mathcal{A}$, this is iff $M \models \breve{r}(a_0, a_1)$ — i.e., iff $\bar{a} \in \iota(\breve{r})$, as required.

Finally, we check composition. We require $\iota(r;s) = c_2(s_2^1 \iota(r) \cdot s_2^0 \iota(s))$ for all $r, s \in \mathcal{A}$. Fix such r, s. First, let $\bar{a} \in c_2(s_2^1 \iota(r) \cdot s_2^0 \iota(s))$. So there is $\bar{b} \in C^n(M)$ with $\bar{b} \equiv_2 \bar{a}$ and $\bar{b} \in s_2^1 \iota(r) \cdot s_2^0 \iota(s)$. Such a $\bar{b}$ satisfies $\bar{b}_2^1 = (b_0, b_2, b_2, \ldots) \in \iota(r)$ and $\bar{b}_2^0 = (b_2, b_1, b_2, \ldots) \in \iota(s)$, so $M \models r(b_0, b_2) \land s(b_2, b_1)$. That is, $M \models r(a_0, b_2) \land s(b_2, a_1)$. But $M \models 1(a_0, a_1)$, because $\bar{a} \in C^n(M)$. As M is a relativised representation of $\mathcal{A}$, we have $M \models (r;s)(a_0, a_1)$, so that $\bar{a} \in \iota(r;s)$.

Conversely, let $\bar{a} \in \iota(r;s)$, so that $M \models (r;s)(a_0, a_1)$. By the final defining property of n-squareness, there is $\bar{b} \in C^n(M)$ with $\bar{b} \equiv_2 \bar{a}$, $M \models r(b_0, b_2)$, and $M \models s(b_2, b_1)$. So clearly, $\bar{b} \in s_2^1 \iota(r) \cdot s_2^0 \iota(s)$, as above, and $\bar{a} \in c_2(s_2^1 \iota(r) \cdot s_2^0 \iota(s))$.

Thus ι is a relation algebra reduct embedding from $\mathcal{A}$ into $\mathcal{D}$. Finally we check that it is a complete embedding, under the assumption that M is a complete relativised representation. By corollary 2.22, $\mathcal{A}$ is atomic. Let $\varphi \in L(\mathcal{A})_{\infty\omega}^n$ be such that $\varphi^C \neq 0$. By lemma 2.17, it suffices to find an atom α of $\mathcal{A}$ such that $\varphi^C \cdot \iota(\alpha) \neq 0$. Pick $\bar{a} \in \varphi^C$. Now M is a complete relativised representation and $\bar{a} \in C^n(M)$, so by theorem 2.21 there is $\alpha \in \text{At}(\mathcal{A})$ such that $M \models \alpha(a_0, a_1)$. Then $\bar{a} \in \alpha(x_0, x_1)^C = \iota(\alpha)$, so $\iota(\alpha) \cdot \varphi^C \neq 0$, as required. □

Exercises

1. The reader may have observed that only the following apparently weak form of n-squareness is needed in the proof of theorem 13.20. Let $\mathcal{A}$ be a non-associative algebra and let $n \geq 3$ be finite. We say that a relativised representation M of $\mathcal{A}$ is $(n - \frac{1}{2})$-square if for all $\bar{a} \in C^n(M)$ and all $r, s \in \mathcal{A}$, if $M \models r;s(a_0, a_1)$ then there is $\bar{b} \in C^n(M)$ with $\bar{b} \equiv_2 \bar{a}$, $M \models r(b_0, b_2)$, and $M \models s(b_2, b_1)$. That is, we restrict to $(i, j, k) = (0, 1, 2)$ in definition 13.4. In particular, $i \neq j$.

   (a) Show that if $\mathcal{A}$ has an $(n - \frac{1}{2})$-square relativised representation then $\mathcal{A} \in \mathbf{S\mathfrak{R}aG}_n$.

   (b) Show that any relativised representation of $\mathcal{A}$ is $2\frac{1}{2}$-square. Deduce as in exercise 5.7(4) that $\mathbf{WA} \subseteq \mathbf{S\mathfrak{R}aG}_3$.

   (c) Show that any n-square relativised representation is $(n - \frac{1}{2})$-square, and that any $(n + \frac{1}{2})$-square relativised representation is n-square.

   (d) Show that if $\mathcal{A} \in \mathbf{SA}$ then any $(n - \frac{1}{2})$-square relativised representation of $\mathcal{A}$ is n-square. [Cf. lemma 13.26 below.]

It turns out that any non-associative algebra with a 3-square relativised representation is semi-associative (theorem 13.49). Hence, $(n-\frac{1}{2})$-squareness is actually weaker than n-squareness only for non-semi-associative algebras when $n=3$.

2. Let M be an n-flat relativised representation of a non-associative algebra $\mathcal{A}$. Assume that for all $S \subseteq \mathcal{D}^0(M)$ such that $\sum S$ exists in $\mathcal{D}^0(M)$, we have $\sum S = \bigcup S$. Show that M is infinitarily n-flat.

13.3 From reducts to relational bases

Next, we show that whenever $n \geq 4$, $\mathcal{D} \in \mathbf{D}_n$ (see definition 5.35), $\mathcal{D}$ is atomic, $\mathcal{A}$ is a non-associative algebra, and $\iota : \mathcal{A} \to \mathcal{D}$ is a complete relation algebra reduct embedding (see definition 5.49), then $\mathcal{A}$ has an n-dimensional relational basis. (We must not try to prove that $\mathfrak{Ra}\mathcal{D}$ has a basis because $\mathfrak{Ra}\mathcal{D}$ *is in general not defined for* $\mathcal{D} \in \mathbf{D}_n$ — recall our discussion in remark 5.50.) In section 13.4, we will prove an analogous result for $\mathbf{CA}_n$ and hyperbases. Non-atomic algebras will be handled later using these results.

THEOREM 13.21 *Let $n \geq 4$ be finite and let $\mathcal{D} \in \mathbf{D}_n$ be atomic. Let $\mathcal{A}$ be a non-associative algebra and $\iota : \mathcal{A} \to \mathcal{D}$ a complete relation algebra reduct embedding. Then $\mathcal{A}$ is atomic and has an n-dimensional relational basis.*

Proof. We may assume that $\mathrm{dom}(\mathcal{A}) \subseteq \mathrm{dom}(\mathcal{D})$ and ι is the inclusion map. Then $\mathrm{bool}(\mathcal{A}) \subseteq^c \mathrm{bool}(\mathcal{D})$, so by lemma 2.16 $\mathcal{A}$ is atomic. By proposition 5.37, we may also assume that for each n-tuple $\bar{a} \in 1^{\mathcal{D}}$ there is an atom γ of $\mathcal{D}$ with $\bar{a} \in \gamma$. By lemma 2.17, there is an atom α of $\mathcal{A}$ with $\gamma \cdot \alpha \neq 0$. Hence, $\gamma \leq \alpha$. So $\bar{a} \in \alpha$, and α is clearly the unique atom of $\mathcal{A}$ containing $\bar{a}$.

This allows us to define an atomic network from $\bar{a}$.

DEFINITION 13.22 For each $\bar{a} \in 1^{\mathcal{D}}$, we define a labelled graph $N_{\bar{a}}$ with nodes n, with labelling defined by: for each $x,y < n$, $N_{\bar{a}}(x,y)$ is the unique atom of $\mathcal{A}$ containing $(a_x, a_y, a_x, \ldots, a_x)$.

Because $n \geq 4$, the map $\theta : n \to n$ given by $\theta(1) = y$, $\theta(i) = x$ for $i \in n \setminus \{1\}$ is not injective, so by fact 12.12 it is a product of substitutions. Since $\mathcal{D} \in \mathbf{D}_n$, $\bar{a} \circ \theta = (a_x, a_y, a_x, \ldots, a_x) \in 1^{\mathcal{D}}$. So by the above, $N_{\bar{a}}(x,y)$ is well-defined. But further, the definition of relation algebra reduct embedding yields $c_i r = r$ for all $r \in \mathcal{A}$ and $2 \leq i < n$. This allows us to prove that only the first two entries of tuples matter for determining $N_{\bar{a}}(x,y)$.

LEMMA 13.23 *Let $\bar{a} = (a_0, \ldots, a_{n-1}) \in 1^{\mathcal{D}}$, let $x, y < n$, and suppose that $\bar{b} = (a_x, a_y, b_2, \ldots, b_{n-1}) \in 1^{\mathcal{D}}$. Then for all atoms r of $\mathcal{A}$, $\bar{b} \in r$ iff $r = N_{\bar{a}}(x,y)$.*

13.3. From reducts to relational bases

Proof. We have $r = c_2 r$, $\bar{b} \in 1^{\mathcal{D}}$, $(a_x, a_y, a_x, b_3, \ldots, b_{n-1}) \in 1^{\mathcal{D}}$ since $\mathcal{D} \in \mathbf{D}_n$, and $\bar{b} \equiv_2 (a_x, a_y, a_x, b_3, \ldots, b_{n-1})$. So

$$\bar{b} \in r \iff (a_x, a_y, a_x, b_3, \ldots, b_{n-1}) \in r.$$

Repeating this argument for entries $3, \ldots, n-1$ yields

$$\bar{b} \in r \iff (a_x, a_y, a_x, \ldots, a_x) \in r,$$

and the right-hand side here is equivalent to $r = N_{\bar{a}}(x, y)$. □

LEMMA 13.24 *For each $\bar{a} \in 1^{\mathcal{D}}$, $N_{\bar{a}}$ is an atomic $\mathcal{A}$-network.*

Proof. First we check that reflexive edges have labels beneath the identity. Let $x < n$. We know that $(a_x, a_x, \ldots, a_x) \in N_{\bar{a}}(x, x)$, by definition of $N_{\bar{a}}$. But $(a_x, \ldots, a_x) \in d_{01}^{\mathcal{D}}$, so $N_{\bar{a}}(x, x) \cdot d_{01}^{\mathcal{D}} \neq 0$. Since $d_{01}^{\mathcal{D}} = 1'^{\mathcal{A}} \in \mathcal{A}$ and $N_{\bar{a}}(x, x) \in \mathrm{At}\,\mathcal{A}$, we see that $N_{\bar{a}}(x, x) \leq 1'^{\mathcal{A}}$ as required.

Now we check triangle consistency. Let $x, y, z < n$. We require that $N_{\bar{a}}(x, y) \leq N_{\bar{a}}(x, z) ; N_{\bar{a}}(z, y)$. Consider the n-tuple $(a_x, a_y, a_z, a_x, \ldots, a_x)$. Since $n \geq 4$, this is obtainable from $\bar{a}$ by substitutions, so it is in $1^{\mathcal{D}}$. Then by lemma 13.23,

$$(a_x, a_y, a_z, a_x, \ldots, a_x) \in N_{\bar{a}}(x, y).$$

Similarly, $(a_x, a_z, a_z, a_x, \ldots, a_x) \in N_{\bar{a}}(x, z) \cdot d_{12}$, so

$$(a_x, a_y, a_z, a_x, \ldots, a_x) \in c_1(N_{\bar{a}}(x, z) \cdot d_{12}) = s_2^1 N_{\bar{a}}(x, z).$$

Similarly, $(a_x, a_y, a_z, a_x, \ldots, a_x) \in s_2^0 N_{\bar{a}}(z, y)$. Hence,

$$\begin{aligned}(a_x, a_y, a_z, a_x, \ldots, a_x) &\in c_2(s_2^1 N_{\bar{a}}(x, z) \cdot s_2^0 N_{\bar{a}}(z, y)) \\ &= N_{\bar{a}}(x, z) ; N_{\bar{a}}(z, y).\end{aligned}$$

So $N_{\bar{a}}(x, y) \cdot (N_{\bar{a}}(x, z) ; N_{\bar{a}}(z, y)) \neq 0$. It follows that $N_{\bar{a}}(x, y) \leq N_{\bar{a}}(x, z) ; N_{\bar{a}}(z, y)$, since the left-hand side is an atom of $\mathcal{A}$ and the right-hand side is in $\mathcal{A}$.

Thus, $N_{\bar{a}}$ is an atomic network. □

Now let $\mathcal{M}^- = \{N_{\bar{a}} : \bar{a} \in 1^M\}$.

LEMMA 13.25 $\mathcal{M}^-$ *has the following properties:*

1. *For any atom $r \in \mathrm{At}\,\mathcal{A}$, there is $N \in \mathcal{M}^-$ with $N(0, 1) = r$.*

2. *If $N \in \mathcal{M}^-$, $r, s \in \mathrm{At}\,\mathcal{A}$ and $N(0, 1) \leq r ; s$ then there is $P \in \mathcal{M}^-$ with $P \equiv_2 N$, $P(0, 2) = r$ and $P(2, 1) = s$ (this is the triangle addition property restricted to $0, 1, 2$).*

Proof. First, let $r \in \text{At}\,\mathcal{A}$. Since $r \neq 0$, there is $\bar{a} \in 1^{\mathcal{D}}$ with $\bar{a} \in r$. By lemma 13.23, $N_{\bar{a}}(0,1) = r$.

For the second part, let $N_{\bar{a}} \in \mathcal{M}^-$ and $r, s \in \text{At}\,\mathcal{A}$ and suppose that $N_{\bar{a}}(0,1) \leq r; s = c_2(s_2^1 r \cdot s_2^0 s)$. By lemma 13.23, $\bar{a} \in N_{\bar{a}}(0,1)$. So for some z in the base set of $\mathcal{D}$, we have $\bar{b} \stackrel{\text{def}}{=} (a_0, a_1, z, a_3, \ldots, a_{n-1}) \in s_2^1 r \cdot s_2^0 s$. Thus,

$$(a_0, z, z, a_3, \ldots, a_{n-1}) \in r, \tag{13.2}$$
$$(z, a_1, z, a_3, \ldots, a_{n-1}) \in s. \tag{13.3}$$

Since $\bar{b}$ and $\bar{a}$ are identical, except that $b_2 = z$ which may differ from a_2, it follows that $N_{\bar{b}} \equiv_2 N_{\bar{a}}$. By (13.2) and lemma 13.23, $N_{\bar{b}}(0,2) = r$. Similarly, using (13.3) we can obtain $N_{\bar{b}}(2,1) = s$. □

So we have the triangle addition property 'restricted to nodes $(0,1,2)$' and we seek a set of atomic networks with the unrestricted triangle addition property. This would be very easy to achieve if the unit of $\mathcal{D}$ were closed under all maps, not just substitutions — i.e., if $\mathcal{D} \in \mathbf{G}_n$ — but we only know that $\mathcal{D} \in \mathbf{D}_n$. The following set of atomic networks allows us to deal with this. Let

$$\mathcal{M} = \{N : N\sigma \in \mathcal{M}^- \text{ for all non-injective maps } \sigma : n \to n\}.$$

LEMMA 13.26 *$\mathcal{M}$ is a (symmetric) n-dimensional relational basis for $\mathcal{A}$.*

Proof. (Cf. exercise 12.3(9).) We check first that the elements of $\mathcal{M}$ are atomic $\mathcal{A}$-networks. So let $N \in \mathcal{M}$. For the identity condition, let $x < n$ and consider the constant map $\sigma : i \mapsto x$, $i < n$. This is non-injective, so $N\sigma \in \mathcal{M}^-$. Hence by lemma 13.24, $N(x,x) = (N\sigma)(x,x) \leq 1'$. For triangle consistency let $x, y, z < n$ and let $\tau : n \to n$ be defined by $\tau(x) = x$, $\tau(y) = y$, $\tau(z) = z$ and $\tau(w) = x$ for all $w \in n \setminus \{x, y, z\}$. Since $n \geq 4$, this is non-injective. So again by lemma 13.24,

$$N(x,y) = (N\tau)(x,y)$$
$$\leq (N\tau)(x,z); (N\tau)(z,y) = N(x,z); N(z,y).$$

For symmetry, let $N \in \mathcal{M}$ and let $\pi : n \to n$ be any map. For any non-injective map $\sigma : n \to n$, $(N\pi)\sigma = N(\pi\sigma) \in \mathcal{M}^-$, since $\pi\sigma$ is also non-injective. So $N\pi \in \mathcal{M}$.

Next we show that $\mathcal{M}^- \subseteq \mathcal{M}$. Let $N_{\bar{a}} \in \mathcal{M}^-$, for some $\bar{a} \in 1^{\mathcal{D}}$ and let $\sigma : n \to n$ be non-injective. Then $N_{\bar{a}}\sigma = N_{\bar{a}\circ\sigma} \in \mathcal{M}^-$ (see exercise 1), and thus $N_{\bar{a}} \in \mathcal{M}$.

Now we check that $\mathcal{M}$ is a relational basis. The atom property of definition 12.9 holds for $\mathcal{M}$ because it does for $\mathcal{M}^-$, and $\mathcal{M}^- \subseteq \mathcal{M}$. We check the triangle addition property. Let $N \in \mathcal{M}$ and $x, y, z < n$, $z \notin \{x, y\}$ and $r, s \in \text{At}\,\mathcal{A}$, and suppose $N(x,y) \leq r; s$. We seek $P \in \mathcal{M}$ with $P \equiv_z N$ and with $P(x,z) = r$ and $P(z,y) = s$.

13.3. From reducts to relational bases

First, we assume that $x \neq y$. Let $\tau : n \setminus \{2\} \to n \setminus \{z\}$ be any bijection such that $\tau(0) = x$ and $\tau(1) = y$. Let $\tau^+ = \tau \cup \{(2,x)\}$; τ^+ is clearly non-injective, so $N\tau^+ \in \mathcal{M}^-$. $N(x,y) \leq r;s$ implies $N\tau^+(0,1) \leq r;s$ so by lemma 13.25 there is $Q \in \mathcal{M}^-$ with $Q \equiv_2 N\tau^+$ and $Q(0,2) = r$, $Q(2,1) = s$. Now let $\sigma = \tau^{-1} \cup \{(z,2)\}$. The network we require is $Q\sigma$. By the above, $Q\sigma \in \mathcal{M}$. $(Q\sigma)(x,z) = Q(\sigma(x),\sigma(z)) = Q(0,2) = r$ and $(Q\sigma)(z,y) = Q(\sigma(z),\sigma(y)) = Q(2,1) = s$. And if $u,v < n$ and $z \notin \{u,v\}$ then we have

$$Q\sigma(u,v) = Q(\sigma(u),\sigma(v)) = Q(\tau^{-1}(u),\tau^{-1}(v)) = N\tau^+(\tau^{-1}(u),\tau^{-1}(v)) = N(u,v).$$

Hence $N \equiv_z Q\sigma$, as required.

Now assume that $x = y$. Note that $0 < N(x,x) \leq (r;s) \cdot 1'$, and as r,s are atoms, it follows by the Peircean law that $\breve{r} = s$. Choose $w < n$, $w \notin \{x,z\}$. Now as $n \geq 4$, by lemma 5.53 we have $\mathcal{A} \in \mathbf{S\mathfrak{R}aD}_n \subseteq \mathbf{RA} \subseteq \mathbf{SA}$. So

$$N(x,w) \leq N(x,x);N(x,w)$$
$$\leq (r;s);N(x,w) \leq (r;1);1 =_{\mathbf{SA}} r;1.$$

By complete additivity of ';' in $\mathcal{A}$, we may choose an atom $t \in \mathcal{A}$ with $N(x,w) \leq r;t$. By the case for $x \neq y$, proved above, there is $M \in \mathcal{M}$, $M \equiv_z N$, $M(x,z) = r$, and $M(z,w) = t$. By lemma 11.2, $M(z,y) = M(z,x) = \breve{r} = s$, so M is as required. □

This completes the proof of theorem 13.21. □

Exercises

1. In the notation of this section, consider n-tuples $\bar{a} \in 1^{\mathcal{D}}$ as maps from n into the base of $\mathcal{D}$, via $i \mapsto a_i$, so that we can compose them with maps $\tau : n \to n$. Thus, $\bar{a} \circ \tau = (a_{\tau(0)}, \ldots, a_{\tau(n-1)})$.

    (a) Prove that if $\bar{a}, \bar{b} \in 1^{\mathcal{D}}$, $z < n$, and $\bar{a} \equiv_z \bar{b}$, then $N_{\bar{a}} \equiv_z N_{\bar{b}}$.

    (b) Prove that if $\bar{a} \in 1^{\mathcal{D}}$ and $\sigma : n \to n$ is not injective, then $\bar{a} \circ \sigma \in 1^{\mathcal{D}}$ and $N_{\bar{a}}\sigma = N_{\bar{a} \circ \sigma}$. [Use fact 12.12.]

    (c) Deduce that $\mathcal{M}^-$ above is closed under substitutions: if $N \in \mathcal{M}^-$ and $x,y < n$ then $N[x/y] \in \mathcal{M}^-$.

2. In the notation of this section, show directly (without using the argument of lemma 13.26) that if $n \geq 4$ and $\mathcal{D} \in \mathbf{G}_n$ then $\mathcal{A}$ has a symmetric n-dimensional relational basis. Repeat for $n = 3$, assuming additionally that $\mathcal{A} \in \mathbf{SA}$.

13.4 From reducts to hyperbases

We will now prove the analogue of theorem 13.21 for relation algebra reducts of cylindric algebras, and hyperbases:

THEOREM 13.27 *Suppose that $4 \leq n < \omega$, $C \in \mathbf{CA}_n$ is atomic, and $\mathcal{A} \subseteq^c \mathfrak{Ra}C$. Then $\mathcal{A}$ is atomic and has an n-dimensional hyperbasis.*

13.4.1 Preliminary results on substitutions

We will need some consequences of the results of section 5.4.2 on substitutions in cylindric algebras, which the reader may wish to review. Some of the consequences hold for arbitrary $\mathbf{CA}_\alpha$, but we only need them and prove them for $\mathbf{CA}_n$ for finite $n \geq 4$. Fix such an n. Unless otherwise stated, all s-c-words (definition 5.21) will be of the signature of $\mathbf{CA}_n$.

LEMMA 13.28 *Let $m \leq n$, and let w be an s-c-word with $m \subseteq \mathrm{dom}(\widehat{w})$. Let $C \in \mathbf{CA}_n$. Then the map $x \mapsto wx$, for $x \in \mathfrak{Nr}_m C$, is a complete homomorphism from the boolean reduct $\mathrm{bool}(\mathfrak{Nr}_m C)$ into $\mathrm{bool}(C)$.*

Proof. Let $x, y \in \mathfrak{Nr}_m C$. Since substitutions and cylindrifications preserve 0 and 1, it is clear that $w0 = 0$ and $w1 = 1$. Now suppose that $S \subseteq C$ and $\sum^{\mathfrak{Nr}_m C} S$ exists. By lemma 5.42, $\sum^{\mathfrak{Nr}_m C} S = \sum^C S$. By facts 5.17 and 5.20(2), substitutions and cylindrifications are completely additive operators on cylindric algebras. So by induction on $|w|$, we obtain $w(\sum^C S) = \sum^C \{ws : s \in S\}$. Hence, $w(\sum^{\mathfrak{Nr}_m C} S) = \sum^C \{ws : s \in S\}$ as required.

We check that $w(-x) = -wx$, for $x \in \mathfrak{Nr}_m C$. By repeated application of lemma 5.33, we can assume that $w = u\rho_{\Gamma_0} \ldots \rho_{\Gamma_{k-1}}$, for some s-word u and some sets $\Gamma_0, \ldots, \Gamma_{k-1} \subseteq n$. Let $\Gamma = \bigcup_{i<k} \Gamma_i$. Clearly, $n \setminus \Gamma = \mathrm{dom}(\widehat{w}) \supseteq m$. So for any $\mathsf{c}_i, \mathsf{s}_j^i$ occurring in $\rho_{\Gamma_0} \ldots \rho_{\Gamma_{k-1}}$, we have $i, j \in \Gamma$, so $i, j \geq m$. Hence, $\mathsf{c}_i x = x$, and by fact 5.20, $\mathsf{s}_j^i x = \mathsf{s}_j^i \mathsf{c}_i x = \mathsf{c}_i x = x$. Similarly, $\mathsf{c}_i - x = -x$ (fact 5.17(5)) and $\mathsf{s}_j^i - x = -x$ (fact 5.20). It follows that $wx = ux$ and $w(-x) = u(-x)$. But by fact 5.20 again, $\mathsf{s}_j^i(-x) = -\mathsf{s}_j^i x$ for any $i, j < n$ and any $x \in C$. So by induction on $|u|$, we obtain $w(-x) = u(-x) = -ux = -wx$, as required. □

LEMMA 13.29 *For any partial non-surjective map $f : n \to n$, there exists an s-c-word w of the form uv, where u is an s-word and v is a c-word, with $\widehat{w} = f$.*

Proof. See definition 5.23 for $\widehat{w}$. If $f = \emptyset$, we may take $w = \mathsf{c}_0 \ldots \mathsf{c}_{n-1}$. Assume that $f \neq \emptyset$. Let $n \setminus \mathrm{dom}(f) = \{i_0, \ldots, i_{k-1}\}$, where $k < n$, take $j \in \mathrm{rng}(f)$, and consider the total map $f^+ = f \cup \{(i_l, j) : l < k\} : n \to n$. Now $\mathrm{rng}(f^+) = \mathrm{rng}(f)$, so by fact 12.12, f^+ is a product of substitutions. It follows by definition 5.23 that there is an s-word u with $\widehat{u} = f^+$. Then if $w = u\mathsf{c}_{i_0} \ldots \mathsf{c}_{i_{k-1}}$, we have $\widehat{w} = f^+ \restriction_{\mathrm{dom}(f)} = f$. □

13.4. From reducts to hyperbases

NOTATION 13.30 If $\bar{a} \in {}^{<n-1}n$, we write $\mathsf{s}_{\bar{a}}$ for an arbitrarily-chosen s-c-word w of the form uv with u an s-word, v a c-word, and $\widehat{w} = \bar{a}$. Such a word exists by lemma 13.29. By example 5.24, any two such words are $\sim$-equivalent, and since also $|n \setminus \mathrm{rng}(\bar{a})| \geq 2$, they are congruent (theorem 5.25). So the value of $\mathsf{s}_{\bar{a}}x$ for $x \in C \in \mathsf{CA}_n$ is independent of the choice of w representing $\mathsf{s}_{\bar{a}}$. As a slight abbreviation, we may write $\mathsf{s}_{a_0 a_1 \ldots a_{l-1}}$ for $\mathsf{s}_{\bar{a}}$, where $\bar{a} = (a_0, \ldots, a_{l-1})$. In this way, we can write s_{ijk} instead of, say, $\mathsf{s}_{(i,j,k)}$. It will help to remember that $\widehat{\mathsf{s}_{ijk}}$ is the partial map from n to n that takes 0 to i, 1 to j, 2 to k, and is undefined on larger numbers.

The definition of composition in the relation algebra reduct of a cylindric algebra is in terms of the indices 0, 1, and 2. These indices can be 'moved', using substitutions.

LEMMA 13.31 *Let $C \in \mathsf{CA}_n$, and let $i, j, k < n$ with $k \neq i, j$. Then in C we have $\mathsf{s}_{ij}(r;s) = \mathsf{c}_k(\mathsf{s}_{ik}r \cdot \mathsf{s}_{kj}s)$ for all $r, s \in \mathfrak{Ra}C$. Here, $r;s$ is evaluated in $\mathfrak{Ra}C$.*

Proof. Remember that $n \geq 4$. First, as $k \neq i, j$, it can easily be checked (using exercise 5.4(6)) that $\mathsf{c}_k \mathsf{s}_{ijk} \sim \mathsf{s}_{ij}\mathsf{c}_2$, and clearly $\widehat{\mathsf{c}_k \mathsf{s}_{ijk}}$ has range of size at most $n-2$. We obtain $\mathsf{c}_k \mathsf{s}_{ijk} \simeq \mathsf{s}_{ij}\mathsf{c}_2$ by theorem 5.25. Similarly, $\mathsf{s}_{ik} \simeq \mathsf{s}_{ijk}\mathsf{s}_2^1 \mathsf{c}_2$ and $\mathsf{s}_{kj} \simeq \mathsf{s}_{ijk}\mathsf{s}_2^0 \mathsf{c}_2$.

Second, if $3 \leq i < n$ then by fact 5.20, $\mathsf{c}_i \mathsf{s}_2^1 r = \mathsf{s}_2^1 \mathsf{c}_i r = \mathsf{s}_2^1 r$ (as $r \in \mathfrak{Nr}_2 C$), and similarly, $\mathsf{c}_i \mathsf{s}_2^0 s = \mathsf{s}_2^0 \mathsf{c}_i s = \mathsf{s}_2^0 s$. So $\mathsf{s}_2^1 r, \mathsf{s}_2^0 s \in \mathfrak{Nr}_3 C$.

Third, since $3 \subseteq \mathrm{dom}(\widehat{\mathsf{s}_{ijk}})$, lemma 13.28 implies that the map $x \mapsto \mathsf{s}_{ijk}x$ is a boolean homomorphism from $\mathrm{bool}(\mathfrak{Nr}_3 C)$ to $\mathrm{bool}(C)$. So $\mathsf{s}_{ijk}(\mathsf{s}_2^1 r \cdot \mathsf{s}_2^0 s) = \mathsf{s}_{ijk}\mathsf{s}_2^1 r \cdot \mathsf{s}_{ijk}\mathsf{s}_2^0 s$. Now we obtain

$$\begin{array}{rll} \mathsf{s}_{ij}(r;s) & = & \mathsf{s}_{ij}\mathsf{c}_2(\mathsf{s}_2^1 r \cdot \mathsf{s}_2^0 s) \quad \text{composition in } \mathfrak{Ra}C \\ & = & \mathsf{c}_k \mathsf{s}_{ijk}(\mathsf{s}_2^1 r \cdot \mathsf{s}_2^0 s) \quad \text{as } \mathsf{c}_k \mathsf{s}_{ijk} \simeq \mathsf{s}_{ij}\mathsf{c}_2 \\ & = & \mathsf{c}_k(\mathsf{s}_{ijk}\mathsf{s}_2^1 r \cdot \mathsf{s}_{ijk}\mathsf{s}_2^0 s) \quad \text{by the above} \\ & = & \mathsf{c}_k(\mathsf{s}_{ijk}\mathsf{s}_2^1 \mathsf{c}_2 r \cdot \mathsf{s}_{ijk}\mathsf{s}_2^0 \mathsf{c}_2 s) \quad \text{as } r, s \in \mathfrak{Nr}_2 C \\ & = & \mathsf{c}_k(\mathsf{s}_{ik}r \cdot \mathsf{s}_{kj}s) \quad \text{as } \mathsf{s}_{ijk}\mathsf{s}_2^1 \mathsf{c}_2 \simeq \mathsf{s}_{ik} \\ & & \quad\quad\quad\quad\quad\quad\quad \text{and } \mathsf{s}_{ijk}\mathsf{s}_2^0 \mathsf{c}_2 \simeq \mathsf{s}_{kj}, \end{array}$$

as required. □

13.4.2 Finding the hyperbasis

We will now prove theorem 13.27. Assume that $\mathcal{A} \subseteq^c \mathfrak{Ra}C$ for some atomic $C \in \mathsf{CA}_n$ and some n with $4 \leq n < \omega$. By lemma 5.42, $\mathfrak{Ra}C$ is atomic, so by lemma 2.16, $\mathcal{A}$ is atomic too. We will show that $\mathcal{A}$ has an n-dimensional hyperbasis.

The case $n = 4$

First we dispatch the case $n = 4$. By theorem 5.44, $\mathfrak{Ra}\,C \in \mathbf{RA}$. Hence, $\mathcal{A} \in \mathbf{RA}$. So by exercise 12.2(12), $\mathcal{A}$ has a 4-dimensional hyperbasis.

The case $n \geq 5$

From now on, we assume that $n \geq 5$. (All but one of the results below go through unchanged if $n = 4$; and even the one (lemma 13.35) that requires $n \geq 5$ can be generalised to cover the case $n = 4$ at the cost of complicating the definition of hypernetwork in the way discussed in section 12.1.)

DEFINITION 13.32 Recall from lemma 5.42 that if $m < n$ then $\mathfrak{Nr}_m C$ is atomic. Let $\Lambda = \bigcup_{m<n-1} \mathrm{At}(\mathfrak{Nr}_m C)$, and let $\lambda \in \Lambda$ be arbitrary.

For each atom x of C, we define what we will see to be an n-dimensional Λ-hypernetwork N_x over $\mathcal{A}$, as follows. Let $\bar{a} \in {}^{\leq n}n$. Then:

- if $|\bar{a}| = 2$, $N_x(\bar{a})$ is defined to be the unique atom $r \in \mathrm{At}\,\mathcal{A}$ such that $x \leq \mathsf{s}_{\bar{a}} r$,

- if $2 \neq |\bar{a}| < n-1$, $N_x(\bar{a})$ is defined to be the unique atom $r \in \mathfrak{Nr}_{|\bar{a}|} C$ such that $x \leq \mathsf{s}_{\bar{a}} r$,

- otherwise (i.e., if $|\bar{a}| \in \{n-1, n\}$), we let $N_x(\bar{a}) = \lambda$.

We check that this is well-defined. By lemma 13.28, the map $\theta : r \mapsto \mathsf{s}_{\bar{a}} r$ is a complete boolean homomorphism from $\mathrm{bool}(\mathfrak{Nr}_{|\bar{a}|} C)$ into $\mathrm{bool}(C)$. Hence, we have $\sum \{\mathsf{s}_{\bar{a}} r : r \in \mathrm{At}\,\mathfrak{Nr}_{|\bar{a}|} C\} = 1^C \geq x$. So as x is an atom of C, there is a unique atom $r \in \mathfrak{Nr}_{|\bar{a}|} C$ with $x \leq \mathsf{s}_{\bar{a}} r$. When $|\bar{a}| = 2$, since $\mathrm{bool}(\mathcal{A}) \subseteq^c \mathrm{bool}(\mathfrak{Nr}_2 C)$, $\theta\!\restriction_{\mathcal{A}}$ is a complete boolean homomorphism from $\mathrm{bool}(\mathcal{A})$ into $\mathrm{bool}(C)$. So again, there is a unique atom $r \in \mathcal{A}$ with $x \leq \mathsf{s}_{\bar{a}} r$. Notice that

$$\begin{aligned} x \leq \mathsf{s}_{ij} r &\iff N_x(i,j) \leq r, &\text{for all } r \in \mathcal{A}, \\ x \leq \mathsf{s}_{\bar{a}} r &\iff N_x(\bar{a}) \leq r, &\text{for all } r \in \mathfrak{Nr}_{|\bar{a}|} C, \end{aligned} \quad (13.4)$$

for all $i, j < n$ and $\bar{a} \in {}^{<n-1}n$ with $|\bar{a}| \neq 2$.

LEMMA 13.33 *Let $x \in \mathrm{At}\,C$. Then N_x is an n-dimensional Λ-hypernetwork over $\mathcal{A}$.*

Proof. Bear in mind that $n \geq 4$. We first check that $\mathcal{A} \models N_x(i,i) \leq 1$' for each $i < n$. Since $1'^{\mathcal{A}} = \mathsf{d}_{01}^C$, by (13.4) it suffices to show that $C \models x \leq \mathsf{s}_{ii} \mathsf{d}_{01}$. By lemma 5.27(2), $\mathsf{s}_{ii} \mathsf{d}_{01} = \mathsf{d}_{ii} = 1$ in C, and we are done.

Next, we let $i, j, k < n$ and check that $N_x(i,k); N_x(k,j) \geq N_x(i,j)$ in $\mathcal{A}$. Let $N_x(i,k) = r$ and $N_x(k,j) = s$, where $r, s \in \mathrm{At}\,\mathcal{A}$. Then $x \leq \mathsf{s}_{ik} r \cdot \mathsf{s}_{kj} s \leq \mathsf{c}_k(\mathsf{s}_{ik} r \cdot \mathsf{s}_{kj} s) = \mathsf{s}_{ij}(r;s)$ by lemma 13.31. By (13.4), $N_x(i,j) \leq r;s$ as required.

13.4. From reducts to hyperbases

Finally, we have to check that if $\bar{a}, \bar{b} \in {}^{\leq n}n$ are of equal length l, say, and satisfy $N_x(a_i, b_i) \leq 1$' for each $i < l$, then $N_x(\bar{a}) = N_x(\bar{b})$. If $l \in \{n-1, n\}$ then $N_x(\bar{a}) = N_x(\bar{b}) = \lambda$, so suppose $l < n-1$. By (13.4) and lemma 5.27(2) again, the condition is equivalent to $x \leq \mathsf{d}_{a_i b_i}$ for each $i < l$.

Let $d(\bar{a}, \bar{b}) = |\{i < l : a_i \neq b_i\}|$. The proof is by induction on $d(\bar{a}, \bar{b})$. If $d(\bar{a}, \bar{b}) = 0$, then $\bar{a} = \bar{b}$ and there is nothing to prove. Assume that $d(\bar{a}, \bar{b}) = 1$. Let $i < l$ be the index with $a_i \neq b_i$. Now $|\mathrm{rng}(\bar{a}) \cup \mathrm{rng}(\bar{b})| \leq n-1$, so we may choose $j < n$ with $j \notin \mathrm{rng}(\bar{a}) \cup \mathrm{rng}(\bar{b})$. Let $\bar{c} \in {}^l n$ be given by $\bar{c} \equiv_i \bar{a}$, $c_i = j$. Then $\widehat{\mathsf{s}_{\bar{a}}} = (\mathsf{s}^j_{a_i} \mathsf{s}_{\bar{c}})\widehat{}$ and $\widehat{\mathsf{s}_{\bar{b}}} = (\mathsf{s}^j_{b_i} \mathsf{s}_{\bar{c}})\widehat{}$; and it follows from exercise 5.4(6) that $\mathsf{s}_{\bar{a}} \sim \mathsf{s}^j_{a_i} \mathsf{s}_{\bar{c}}$ and $\mathsf{s}_{\bar{b}} \sim \mathsf{s}^j_{b_i} \mathsf{s}_{\bar{c}}$. So by theorem 5.25, $\mathsf{s}_{\bar{a}} \simeq \mathsf{s}^j_{a_i} \mathsf{s}_{\bar{c}}$ and $\mathsf{s}_{\bar{b}} \simeq \mathsf{s}^j_{b_i} \mathsf{s}_{\bar{c}}$.

Now we show $N_x(\bar{a}) = N_x(\bar{b})$. Let $N_x(\bar{a}) = r$. Then $x \leq \mathsf{s}_{\bar{a}} r$. Also, by assumption, $x \leq \mathsf{d}_{a_i b_i}$. So

$$
\begin{aligned}
x &\leq \mathsf{d}_{a_i b_i} \cdot \mathsf{s}_{\bar{a}} r & \text{by assumption} \\
&= \mathsf{d}_{a_i b_i} \cdot \mathsf{s}^j_{a_i} \mathsf{s}_{\bar{c}} r & \text{as } \mathsf{s}_{\bar{a}} \simeq \mathsf{s}^j_{a_i} \mathsf{s}_{\bar{c}} \\
&= \mathsf{d}_{a_i b_i} \cdot \mathsf{s}^j_{b_i} \mathsf{s}_{\bar{c}} r & \text{by fact 5.20(4)} \\
&\leq \mathsf{s}^j_{b_i} \mathsf{s}_{\bar{c}} r = \mathsf{s}_{\bar{b}} r & \text{as } \mathsf{s}^j_{b_i} \mathsf{s}_{\bar{c}} \simeq \mathsf{s}_{\bar{b}}.
\end{aligned}
$$

Hence, $N_x(\bar{b}) = r$, as required.

Now let $d(\bar{a}, \bar{b}) \geq 2$ and assume the result for smaller d. Choose $i < l$ with $a_i \neq b_i$, and let $\bar{c} \in {}^l n$ be given by $\bar{c} \equiv_i \bar{a}$, $c_i = b_i$. Then clearly, $d(\bar{a}, \bar{c}), d(\bar{c}, \bar{b}) < d(\bar{a}, \bar{b})$, and $x \leq \mathsf{d}_{a_j c_j}$, $x \leq \mathsf{d}_{c_j b_j}$ for all $j < l$. By the inductive hypothesis, $N_x(\bar{a}) = N_x(\bar{c}) = N_x(\bar{b})$, which completes the induction and the proof. □

The next two lemmas relate atoms to hypernetworks. The second is in some way the converse of the first.

LEMMA 13.34 *Let $x, y \in \mathrm{At}\, C$, $i < n$, and suppose that $\mathsf{c}_i x = \mathsf{c}_i y$. Then $N_x \equiv_i N_y$.*

Proof. Let $\bar{a} \in {}^{\leq n}(n \setminus \{i\})$. We show $N_x(\bar{a}) = N_y(\bar{a})$. We may assume that $|\bar{a}| < n - 1$. Let $N_x(\bar{a}) = r$, so that $x \leq \mathsf{s}_{\bar{a}} r$. Then $y \leq \mathsf{c}_i x \leq \mathsf{c}_i \mathsf{s}_{\bar{a}} r$. But $i \notin \mathrm{rng}(\bar{a})$, so $\mathsf{c}_i \mathsf{s}_{\bar{a}} \simeq \mathsf{s}_{\bar{a}}$ by lemma 5.33. Hence, $\mathsf{c}_i \mathsf{s}_{\bar{a}} r = \mathsf{s}_{\bar{a}} r$, so that $y \leq \mathsf{s}_{\bar{a}} r$ and $N_y(\bar{a}) = r$. □

LEMMA 13.35 *Let $x, y \in \mathrm{At}\, C$, let $i, j < n$ be distinct, and suppose that $N_x \equiv_{ij} N_y$. Then $\mathsf{c}_i \mathsf{c}_j x = \mathsf{c}_i \mathsf{c}_j y$.*

Proof. Pick $\bar{a} \in {}^{n-2}n$ with $\mathrm{rng}(\bar{a}) = n \setminus \{i, j\}$. Then $N_x(\bar{a}) = N_y(\bar{a})$.

Clearly, $\bar{a}$ is one-one, so it has an inverse, a one-one map $f : n \setminus \{i, j\} \to n - 2$. Let w be the concatenation of an s-word and the c-word $\mathsf{c}_i \mathsf{c}_j$, with $\widehat{w} = f$. Such a w exists by lemma 13.29.

Now we will use some of the material of section 5.4. Clearly, $\mathrm{rng}(\widehat{w}) = n - 2$ and $\bar{a} \circ \widehat{w} = \mathrm{Id}_{n \setminus \{i, j\}}$. By lemma 5.27(1), $\widehat{\mathsf{s}_{\bar{a}} w} = \bar{a} \circ \widehat{w} = \widehat{\mathsf{c}_i \mathsf{c}_j}$. Also, $\widehat{\mathsf{s}_{\bar{a}} w}$ is one-one,

420   Chapter 13. Approximations to **RRA**

so by exercise 5.4(3) we obtain $(s_{\bar{a}}w)^* = Id_n = (c_i c_j)^*$. Hence, $s_{\bar{a}}w \sim c_i c_j$. As $|\mathrm{rng}(\widehat{c_i c_j})| = n - 2$, we obtain $s_{\bar{a}}w \simeq c_i c_j$ by theorem 5.25. Thus,

$$x \leq c_i c_j x = s_{\bar{a}} w x.$$

Now $n-2, n-1 \notin \mathrm{rng}(\widehat{w})$, so by lemma 5.33, $c_{n-2}w \simeq c_{n-1}w \simeq w$. Hence,

$$wx \in \mathfrak{Nr}_{n-2} C.$$

Since $n - 2 \neq 2$ (here, we use our assumption that $n \geq 5$ for the only time), the above and (13.4) yield $N_x(\bar{a}) \leq wx$. So $y \leq s_{\bar{a}} N_y(\bar{a}) = s_{\bar{a}} N_x(\bar{a}) \leq s_{\bar{a}} wx = c_i c_j x$. Because x, y are atoms of C, we obtain $c_i c_j x = c_i c_j y$ by fact 5.17. □

PROPOSITION 13.36 $\mathcal{H} = \{N_x : x \in \mathrm{At}\, C\}$ *is an n-dimensional* Λ-*hyperbasis for* $\mathcal{A}$.

Proof. We check the three conditions for hyperbases (definition 12.11). For any atom a of $\mathcal{A}$, pick an atom x of C with $x \leq a$. Then $x \leq a = s_{01} a$, so by (13.4), $N_x(0, 1) \leq a$. Since these are atoms of $\mathcal{A}$, it follows that $N_x(0, 1) = a$, as required.

Next we check the triangle addition property for $\mathcal{H}$. Let $N_x \in \mathcal{H}$, $i, j < n$, $k \in n \setminus \{i, j\}$, and $r, s \in \mathrm{At}\,\mathcal{A}$, and assume that $N_x(i, j) \leq r; s$. So by (13.4), $x \leq s_{ij}(r;s) = c_k(s_{ik} r \cdot s_{kj} s)$ (by lemma 13.31). By complete additivity of c_k (see fact 5.17), there is an atom y of C with $x \leq c_k y$ and $y \leq s_{ik} r \cdot s_{kj} s$. By fact 5.17, $c_k x = c_k y$, so by lemma 13.34, $N_y \equiv_k N_x$. Also, since $y \leq s_{ik} r$ we have $N_y(i, k) = r$ by definition of N_y, and similarly, $N_y(k, j) = s$.

Finally we check the amalgamation property. Let $N_x, N_y \in \mathcal{H}$, let $i, j < n$ be distinct, and suppose that $N_x \equiv_{ij} N_y$. By lemma 13.35, $c_i c_j x = c_i c_j y$. Hence, by additivity of cylindrifications, there is an atom $z \in C$ with $c_i x = c_i z$ and $c_j z = c_j y$. Then $N_z \in \mathcal{H}$, and by lemma 13.34, $N_x \equiv_i N_z \equiv_j N_y$. □

This completes the proof of theorem 13.27.

Exercises

1. Check that $s_{ik} \simeq s_{ijk} s_2^1 c_2$ and $s_{kj} \simeq s_{ijk} s_2^0 c_2$ in lemma 13.31, and that $s_{\bar{a}} \sim s_{a_i}^j s_{\bar{c}}$ and $s_{\bar{b}} \sim s_{b_i}^j s_{\bar{c}}$ in lemma 13.33.

2. Using s-c-words, find explicit expressions for the values of $N_x(\bar{a})$ when $2 \neq |\bar{a}| < n - 1$ (definition 13.32).

13.5 From bases to relativised representations

Now we can complete the triangle by passing from bases to relativised representations. This is quite easy to do, by treating a basis as a 'saturated set of mosaics' in the sense of Németi [Ném95, VenMar98]. It can be 'spun out' into a relativised representation as follows ([Madd89a] gives a related approach).

PROPOSITION 13.37 *Let $\mathcal{A}$ be an atomic non-associative algebra, and let $3 \leq n < \omega$. Let $\mathcal{H}$ be a symmetric n-dimensional relational Λ-basis for $\mathcal{A}$. Then there is a non-empty labelled directed graph M, and a labelling of some n-tuples of M, with the following properties.*

1. *The set of edges forms a reflexive and symmetric binary relation on M.*

2. *Each directed edge (x,y) of M is labelled by a unique atom of $\mathcal{A}$, written $M(x,y)$.*

3. *If (x,y) is an edge of M, then $M(x,y) \leq 1$' iff $x = y$.*

4. *Say (as usual in graph theory) that a subset C of M is a clique if (x,y) is an edge of M for all $x, y \in C$. Then for any clique $\{x_0, \ldots, x_{n-1}\} \subseteq M$, there is unique $N \in \mathcal{H}$ such that the n-tuple $(x_0, \ldots, x_{n-1})$ is labelled by N. We write this labelling as $M(x_0, \ldots, x_{n-1}) = N$.*

5. *If $x_0, \ldots, x_{n-1} \in M$ and $M(x_0, \ldots, x_{n-1}) = N \in \mathcal{H}$, then for all $i, j < n$, (x_i, x_j) is an edge of M and $M(x_i, x_j) = N(i, j) \in \text{At}\mathcal{A}$.*

6. *If $x_0, \ldots, x_{n-1} \in M$, $M(x_0, \ldots, x_{n-1}) = N$, and $\sigma : n \to n$, then $M(x_{\sigma(0)}, \ldots, x_{\sigma(n-1)}) = N\sigma$.*

7. *Suppose that $\{x_0, \ldots, x_{n-1}\}$ is a clique in M, $k < n$, and $N \in \mathcal{H}$. Then $M(x_0, \ldots, x_{n-1}) \equiv_k N$ iff there is $y \in M$ such that $M(x_0, \ldots, x_{k-1}, y, x_{k+1}, \ldots, x_{n-1}) = N$.*

8. *For every $N \in \mathcal{H}$, there are $x_0, \ldots, x_{n-1} \in M$ with $M(x_0, \ldots, x_{n-1}) = N$.*

Proof. We will build a chain of possibly uncountable labelled, directed hypergraphs M_t ($t < \omega$); they will not be complete hypergraphs. Their union will essentially be the M we seek. Each M_t will have edges (ordered pairs) labelled by atoms of $\mathcal{A}$, and hyperedges (l-tuples for $l \leq n$, $l \neq 2$) labelled by elements of Λ (the labelling by hypernetworks required in (4) will be added later). No non-edges or non-hyperedges are labelled.

We will require (inductively) that for each $t < \omega$, M_t satisfies conditions 1–3 of the proposition (with 'M_t' replacing 'M' in each), and (χ) below. For such a graph M_t, and a hypernetwork $N \in \mathcal{H}$, a map $\nu : N \to M_t$ (formally, a map from n into $\text{dom}(M_t)$) is said to be an *embedding* if

- for all $i, j < n$, $(\nu(i), \nu(j))$ is an edge of M_t and $M_t(\nu(i), \nu(j)) = N(i,j)$,

- whenever $\bar{a} \in {}^{\leq n}n$ with $|\bar{a}| \neq 2$, then $\nu(\bar{a})$ is a hyperedge of M_t and is labelled with $N(\bar{a})$.

Note that despite their name, embeddings need not be one-to-one, but they do preserve atoms under 1'. We further require of M_t:

(χ) Any clique in M_t is contained in $\mathrm{rng}(\nu)$ for some $N \in \mathcal{H}$ and some embedding $\nu : N \to M_t$.

We start by building M_0. Let N be a hypernetwork and let $S \subseteq n$. We say that the labelled hypergraph $N{\upharpoonright}_S$ induced by N on S is *strict* if for all distinct $i, j \in S$ we have $N(i,j) \cdot 1' = 0$. $N{\upharpoonright}_S$ is *maximal strict* if it is strict and for all $S \subset T \subseteq n$, $N{\upharpoonright}_T$ is not strict. Let M_0 be the disjoint union of all maximal strict labelled graphs $N{\upharpoonright}_S$, where $N \in \mathcal{H}$ and $S \subseteq n$, and the edge and hyperedge labelling is induced from N. Thus, M_0 satisfies requirements 1–3 above. If $N{\upharpoonright}_S$ is maximal strict, then for all $i < n$ there is unique $s_i \in S$ such that $N(i, s_i) \leq 1'$. The map $\nu = \{(i, s_i) : i < n\}$ is an embedding of N onto $N{\upharpoonright}_S$. So M_0 satisfies requirement (χ), too.

Assume inductively that M_t is defined for some $t < \omega$. Then we define the extension M_{t+1} of M_t so that for every quadruple of the form (N, ν, k, N'), where $N, N' \in \mathcal{H}$, $k < n$, $N \equiv_k N'$, and $\nu : N \to M_t$ is an embedding, the restriction $\nu{\upharpoonright}_{n \setminus \{k\}}$ of ν extends to an embedding $\nu' : N' \to M_{t+1}$. We do this as follows.

- If $N'(k, i) \leq 1'$ for some $i \neq k$, then $N' = N[k/i]$ by lemma 12.8, and we may (must) set $\nu'(k) = \nu(i)$. This is well-defined if there are several such i, and is an embedding. No change to M_t is made for these (N, ν, k, N').

- For each other (N, ν, k, N'), we adjoin a new node $\pi = \pi_{(t, N, \nu, k, N')}$ to M_t. We add just the following new edges: $(\pi, \nu(i)), (\nu(i), \pi)$ for each $i \in n \setminus \{k\}$, and (π, π).

- The new edges are labelled by atoms of $\mathcal{A}$ as follows: (π, π) is labelled by $N'(k, k)$, $(\pi, \nu(i))$ by $N'(k, i)$, and $(\nu(i), \pi)$ by $N'(i, k)$. This is well-defined.

- We may extend $\nu{\upharpoonright}_{n \setminus \{k\}}$ to ν' defined on k, by setting $\nu'(k) = \pi$.

- We also add a new hyperedge $\nu'(\bar{a})$ for every $\bar{a} \in {}^{\leq n}n$ of length $\neq 2$ with $k \in \mathrm{rng}(\bar{a})$. We label it by $N'(\bar{a})$. This is well-defined.

  Because $N' \equiv_k N$, ν' is an embedding $: N' \to M_{t+1}$.

- For distinct (N, ν, k, N'), the new points $\pi_{(t, N, \nu, k, N')}$ are (defined to be) distinct.

- M_{t+1} will consist of M_t, with its old labels, edges, and hyperedges, together with all these new points, edges, hyperedges, and labels.

13.5. From bases to relativised representations

It is easy to check that the properties 1–3 and χ above are preserved by these actions. For (χ), note that any clique in M_{t+1} is either a clique in M_t, for which we have the result inductively, or else it contains a new point $\pi_{(t,N,\mathsf{v},k,N')}$, in which case it is contained in rng(v'), since the only edges involving $\pi_{(t,N,\mathsf{v},k,N')}$ lie in this set.

We define a labelled hypergraph M as follows. The set of nodes of M is the union of the sets of nodes of the M_t (for all $t < \omega$). For $x, y \in M$, (x,y) is defined to be an edge of M iff it is an edge of some M_t ($t < \omega$), and its label $M(x,y)$ is defined to be $M_t(x,y)$; this is well-defined, and satisfies properties 1–3 of the proposition.

The hyperedges will be the n-tuples $(x_0, \ldots, x_{n-1})$ of elements of M such that $\{x_0, \ldots, x_{n-1}\}$ is a clique in M. For each such tuple, we choose any $t < \omega$ such that $\{x_0, \ldots, x_{n-1}\} \subseteq M_t$, and let $M(x_0, \ldots, x_{n-1})$ be the unique $N \in \mathcal{H}$ such that there is an embedding $\mathsf{v} : N \to M_t$ with $\mathsf{v}(i) = x_i$ for all $i < n$. Uniqueness of N is clear from the definition of 'embedding'. To confirm existence of N, first note that by (χ), there is $N \in \mathcal{H}$ and an embedding $\mathsf{v} : N \to M_t$ with $\{x_0, \ldots, x_{n-1}\} \subseteq$ rng(v). So we may choose $\tau : n \to n$ such that $x_i = \mathsf{v}(\tau(i))$ for each $i < n$. As $\mathcal{H}$ is symmetric, $N\tau \in \mathcal{H}$; and clearly, $\mathsf{v} \circ \tau$ is an embedding : $N\tau \to M_t$. Since $x_i = \mathsf{v} \circ \tau(i)$ for each $i < n$, we may let $M(x_0, \ldots, x_{n-1}) = N\tau$. Since $M(x_0, \ldots, x_{n-1})$ clearly does not depend upon the chosen t, it is well-defined and satisfies property 4. The same argument shows that property 6 holds. By construction, M satisfies properties 5 and 7, and by definition of M_0 it satisfies property 8. $\square$

THEOREM 13.38 *Let $\mathcal{A}$ be an atomic non-associative algebra, and let $3 \leq n < \omega$.*

1. *Suppose that $\mathcal{A}$ has an n-dimensional relational basis. Then $\mathcal{A}$ has a complete n-square relativised representation.*

2. *If $\mathcal{A}$ has an n-dimensional hyperbasis, then it has a complete relativised representation which is both infinitarily n-flat and n-smooth.*

Proof. Let $\mathcal{H}$ be an n-dimensional relational basis for $\mathcal{A}$. By lemma 12.15, we may suppose that $\mathcal{H}$ is symmetric. Let M satisfy the properties of proposition 13.37. Define M as an $\mathcal{L}(\mathcal{A})$-structure by:

$$M \models r(x,y) \iff (x,y) \text{ is an edge of } M \text{ and } M(x,y) \leq r,$$

for each $r \in \mathcal{A}$ and $x, y \in M$. We will show that M satisfies the conclusion of the theorem. Below, 'property xxx' will mean part xxx of proposition 13.37.

LEMMA 13.39 *M is a complete relativised representation of $\mathcal{A}$.*

Proof. We check that the axioms in the theory $R_\mathcal{A}$ of definition 5.1 (recalled also in section 13.1) all hold in M. To see that $M \models \forall xy(1'(x,y) \leftrightarrow x = y)$, use properties 1 and 3 (of proposition 13.37). The boolean clauses follow easily from the definition of the $\mathcal{L}(\mathcal{A})$-structure on M, using the fact that $M(x,y)$ is an atom of $\mathcal{A}$.

We check the 'converse' axiom $\forall xy(r(x,y) \leftrightarrow \breve{r}(y,x))$, for $r \in \mathcal{A}$. Suppose that $M \models r(x,y)$. By property 1, $\{x,y\}$ is a clique in M. Let $M(x,y,\ldots,y) = N \in \mathcal{H}$ (property 4). Then by property 5 and lemma 12.4, using that the map $r \mapsto \breve{r}$ preserves $\leq$ (lemma 3.11), we have

$$\breve{r} \geq M(x,y)^{\smile} = N(0,1)^{\smile} = N(1,0) = M(y,x).$$

Hence, $M \models \breve{r}(y,x)$. The other direction is similar.

Now consider the composition axiom. Let $x, y \in M$ with $M \models 1(x,y)$. First, suppose that $M \models r(x,z) \wedge s(z,y)$ for some $z \in M$ and $r, s \in \mathcal{A}$. We require $M \models [r;s](x,y)$. By property 1, it follows that $\{x,y,z\}$ is a clique in M, so by property 4, we obtain $M(x,y,z,z,\ldots,z) = N$ for some $N \in \mathcal{H}$. Then by property 5,

$$M(x,y) = N(0,1) \leq N(0,2); N(2,1) = M(x,z); M(z,y) \leq r;s,$$

so $M \models [r;s](x,y)$.

Conversely, suppose that $M \models [r;s](x,y)$. Since (x,y) is an edge of M, properties 1 and 4 yield $M(x,y,y,\ldots,y) = N$ for some $N \in \mathcal{H}$. Then $N(0,1) = M(x,y) \leq r;s$. As $\mathcal{H}$ is a relational basis, by the triangle addition property there is $P \in \mathcal{H}$ with $N \equiv_2 P$, $P(0,2) \leq r$, and $P(2,1) \leq s$. By property 7, there is $z \in M$ with $M(x,y,z,y,y,\ldots,y) = P$. Then $M(x,z) = P(0,2) \leq r$ and $M(z,y) = P(2,1) \leq s$, so $M \models r(x,z) \wedge s(z,y)$ as required.

Lastly we check the axiom $\exists xy\, r(x,y)$, for $r \neq 0$ in $\mathcal{A}$. Choose an atom $a \leq r$. As $\mathcal{H}$ is a relational basis for $\mathcal{A}$, there is $N \in \mathcal{H}$ with $N(0,1) = a$. By property 8, there are $x_0, \ldots, x_{n-1} \in M$ with $M(x_0, \ldots, x_{n-1}) = N$. Clearly, $M(x_0, x_1) = N(0,1) = a \leq r$ so that $M \models r(x,y)$ as required.

Since every edge of M is labelled by an atom, we see by lemma 5.3 and theorem 2.21 it is *complete,* respecting all meets and joins that exist in $\mathcal{A}$. □

So M is a relativised representation of $\mathcal{A}$. Note that $M \models 1(x,y)$ iff (x,y) is an edge of M. Hence, the notion of clique given above coincides with that of definition 13.2. Recall that $C^n(M) = \{\bar{a} \in {}^n M : \mathrm{rng}(\bar{a})$ is a clique in $M\}$.

LEMMA 13.40 *M is n-square.*

Proof. This is similar to the proof of lemma 13.39. If $\bar{x} = (x_0, \ldots, x_{n-1}) \in C^n(M)$, $i, j, k < n$ with $k \neq i, j$, $r, s \in \mathcal{A}$, and $M \models (r;s)(x_i, x_j)$, we require that there is $\bar{y} \in C^n(M)$ with $\bar{y} \equiv_k \bar{x}$, $M \models r(y_i, y_k)$ and $M \models s(y_k, y_j)$. So take such $\bar{a}, r, s$. By property 4 (of proposition 13.37), $M(\bar{x}) = N$ for some $N \in \mathcal{H}$. Thus, $N(i,j) = M(x_i, x_j) \leq r;s$. As $\mathcal{H}$ is a relational basis, there is $P \in \mathcal{H}$, $P \equiv_k N$, $P(i,k) \leq r$, and $P(k,j) \leq s$. By property 7, there is an n-tuple $\bar{y}$ of elements of M with $\bar{y} \equiv_k \bar{x}$ and $M(\bar{y}) = P$. Property 5 of the proposition now yields $\bar{y} \in C^n(M)$, $M(y_i, y_k) = P(i,k) \leq r$, and $M(y_k, y_j) = P(k,j) \leq s$, as required. □

13.5. From bases to relativised representations

We now deal with flatness.

LEMMA 13.41 *Let $\bar{a}, \bar{b} \in C^n(M)$ satisfy $M(\bar{a}) = M(\bar{b})$. Then $M \models_C \varphi(\bar{a}) \leftrightarrow \varphi(\bar{b})$, for all $\varphi \in L(\mathcal{A})^n_{\infty\omega}$.*

Proof. By induction on φ. Let $M(\bar{a}) = M(\bar{b}) = N \in \mathcal{H}$. If φ is atomic, of the form $r(x_i, x_j)$, we have $M \models_C r(x_i, x_j)(\bar{a})$ iff $M \models r(a_i, a_j)$, iff $M(a_i, a_j) \leq r$, iff $M(b_i, b_j) \leq r$ (since by property 5 of proposition 13.37, $M(a_i, a_j) = N(i, j) = M(b_i, b_j)$), iff $M \models r(b_i, b_j)$, iff $M \models_C r(x_i, x_j)(\bar{b})$. If φ is $x_i = x_j$ then φ is equivalent in M to $1'(x_i, x_j)$, so the preceding case gives the result here. The boolean cases are easy: $M \models_C \neg\varphi(\bar{a}) \iff M \not\models_C \varphi(\bar{a}) \iff_{I.H.} M \not\models_C \varphi(\bar{b}) \iff M \models_C \neg\varphi(\bar{b})$; and for any set $\Phi \subseteq L(\mathcal{A})^n_{\infty\omega}$ we have $M \models_C \bigwedge \Phi(\bar{a})$ iff $M \models_C \varphi(\bar{a})$ for all $\varphi \in \Phi$, iff $M \models_C \varphi(\bar{b})$ for all $\varphi \in \Phi$, by induction hypothesis, iff $M \models_C \bigwedge \Phi(\bar{b})$.

Consider $\exists x_i \varphi$. If $M \models_C \exists x_i \varphi(\bar{a})$, then there is $\bar{a}' \in C^n(M)$ with $\bar{a}' \equiv_i \bar{a}$ and $M \models_C \varphi(\bar{a}')$. By property 4, there is $N' \in \mathcal{H}$ with $M(\bar{a}') = N'$. By property 7, $N' \equiv_i N$, and the same property applied to $\bar{b}$ in the other direction shows that there is $\bar{b}' \equiv_i \bar{b}$ with $M(\bar{b}') = N'$. By property 5, $\bar{b}' \in C^n(M)$. By the inductive hypothesis, $M \models_C \varphi(\bar{b}')$. But $\bar{b}' \equiv_i \bar{b}$, so $M \models_C \exists x_i \varphi(\bar{b})$. The converse is similar. $\square$

Essentially we showed that the set of all partial maps on M of the form $\bar{a} \mapsto \bar{b}$, where $\bar{a}, \bar{b} \in C^n(M)$ and $M(\bar{a}) = M(\bar{b})$, forms an '*n*-back-and-forth system' of partial isomorphisms of M (cf. [Hod93, §3.2]). Flatness and smoothness of the relativised representation follows from this when $\mathcal{H}$ is a hyperbasis:

LEMMA 13.42 *If $\mathcal{H}$ is a hyperbasis then M is infinitarily n-flat and n-smooth.*

Proof. Suppose that $\mathcal{H}$ is a hyperbasis. Let $M \models_C \exists x_i \exists x_j \varphi(\bar{a})$ for some $\bar{a} \in C^n(M)$, $\varphi \in L(\mathcal{A})^n_{\infty\omega}$, and distinct $i, j < n$. So there are $\bar{b}, \bar{c} \in C^n(M)$ with $\bar{a} \equiv_i \bar{b} \equiv_j \bar{c}$ and $M \models_C \varphi(\bar{c})$. As $\bar{a}, \bar{b}, \bar{c} \in C^n(M)$, property 4 provides $N, P, Q \in \mathcal{H}$ with $M(\bar{a}) = N, M(\bar{b}) = P$, and $M(\bar{c}) = Q$. By property 7, $N \equiv_i P \equiv_j Q$ so that $N \equiv_{ij} Q$. As $\mathcal{H}$ is a hyperbasis, there is $R \in \mathcal{H}$ with $N \equiv_j R \equiv_i Q$. By the 'existential' part of property 7, there are $\bar{d}, \bar{e} \in C^n(M)$ with $\bar{a} \equiv_j \bar{d} \equiv_i \bar{e}$, $M(\bar{d}) = R$, and $M(\bar{e}) = Q$. Now, $M \models_C \varphi(\bar{c})$ and lemma 13.41 yield $M \models_C \varphi(\bar{e})$. Hence, $M \models_C \exists x_i \varphi(\bar{d})$ and $M \models_C \exists x_j \exists x_i \varphi(\bar{a})$, as required for infinitary *n*-flatness.

For *n*-smoothness we need to define equivalence relations E^m on $C^m(M)$, for $0 < m \leq n$, and we then need to check the conditions of definition 13.12. For $0 < m \leq n$ and $\bar{a} \in C^m(M)$, let $\bar{a}^\circ = (a_0, a_1 \ldots, a_{m-1}, a_0, \ldots, a_0) \in C^n(M)$, where we have $n - m$ copies of a_0 after $a_0, a_1 \ldots, a_{m-1}$. Now, for $\bar{a}, \bar{b} \in C^m(M)$, we let

$$E^m(\bar{a}, \bar{b}) \iff M(\bar{a}^\circ) = M(\bar{b}^\circ).$$

Clearly, for $0 < m \leq n$, E^m is an equivalence relation on $C^m(M)$. If $(\bar{a}, \bar{b}) \in E^m$, $0 < l \leq n$ and $\theta : l \to m$, extend θ to $\theta^\circ : n \to n$ by $\theta^\circ(l) = \cdots = \theta^\circ(n-1) = \theta(0)$.

Then by property 6,

$$M((\bar{a} \circ \theta)^\circ) = M(\bar{a}^\circ \circ \theta^\circ) = M(\bar{a}^\circ)\theta^\circ,$$
$$M((\bar{b} \circ \theta)^\circ) = M(\bar{b}^\circ \circ \theta^\circ) = M(\bar{b}^\circ)\theta^\circ.$$

Since $M(\bar{a}^\circ) = M(\bar{b}^\circ)$, we have $M(\bar{a}^\circ)\theta^\circ = M(\bar{b}^\circ)\theta^\circ$, so $(\bar{a} \circ \theta, \bar{b} \circ \theta) \in E^l$. Property 5 ensures that $(\bar{a}, \bar{b}) \in E^m$ implies $\bar{a} \mapsto \bar{b}$ is a well-defined partial isomorphism.

To check the last condition in definition 13.12, let $\bar{a}, \bar{b} \in C^{n-2}(M)$ with $(\bar{a}, \bar{b}) \in E^{n-2}$, and let $a, b \in M$ be such that $\mathrm{rng}(\bar{a}a)$ and $\mathrm{rng}(\bar{b}b)$ are cliques. We seek $c \in M$ such that $\mathrm{rng}(\bar{b}bc)$ is a clique and $(\bar{a}a, \bar{b}c) \in E^{n-1}$. Well, $(\bar{a}, \bar{b}) \in E^{n-2}$ implies $M(\bar{a}a_0a_0) = M(\bar{b}b_0b_0)$. So by property 7,

$$\begin{aligned} M(\bar{a}aa_0) &\equiv_{n-2} M(\bar{a}a_0a_0) \\ &= M(\bar{b}b_0b_0) \\ &\equiv_{n-1} M(\bar{b}b_0b). \end{aligned}$$

Since these hypernetworks belong to the hyperbasis $\mathcal{H}$, there is $N \in \mathcal{H}$ such that

$$M(\bar{a}aa_0) \equiv_{n-1} N \equiv_{n-2} M(\bar{b}b_0b).$$

By property 7, there is $c \in M$ such that $\mathrm{rng}(\bar{b}cb)$ is a clique and $M(\bar{b}cb) = N$. It remains to show that $(\bar{a}a, \bar{b}c) \in E^{n-1}$. On the one hand, $M(\bar{a}aa_0) \equiv_{n-1} N$ and (by property 5) $M(\bar{a}aa_0)(0, n-1) \leq 1'$. So by lemma 12.8, $M(\bar{a}aa_0) = N[n-1/0]$. On the other hand, by property 6 we have

$$M(\bar{b}cb_0) = M(\bar{b}cb)[n-1/0] = N[n-1/0] = M(\bar{a}aa_0).$$

Hence $(\bar{a}a, \bar{b}c) \in E^{n-1}$, as required. □

This completes the proof of theorem 13.38. □

REMARK 13.43 The proof of theorem 13.38 only required that M satisfy the properties of proposition 13.37, and it constructed an infinitarily n-flat relativised representation with base M. Thus, if we could find a finite labelled graph M satisfying properties 1–8 then $\mathcal{A}$ would have a finite n-square or infinitarily n-flat and n-smooth relativised representation. We will do this later in theorem 19.20.

Exercises

1. For any finite $n \geq 3$, show that any infinitarily n-flat relativised representation of a non-associative algebra is n-smooth.

13.6 From smooth to hyperbasis

Before we tie all this together we need to go from a complete smooth relativised representation to a hyperbasis.

LEMMA 13.44 *Let $3 \leq n < \omega$. If $\mathcal{A}$ is an atomic non-associative algebra with a complete n-smooth relativised representation, then $\mathcal{A}$ has an n-dimensional hyperbasis.*

Proof. First we deal with the case $n = 4$. Then $\mathcal{A}$ has a complete 4-smooth relativised representation, which is by definition 4-square. By considering the set of 4-dimensional atomic networks that embed into M, it can be shown that $\mathcal{A}$ has a 4-dimensional relational basis. By exercise 12.3(12), $\mathcal{A}$ has an n-dimensional hyperbasis.

Now assume $n \neq 4$. Let M be a complete n-smooth relativised representation of $\mathcal{A}$. For $m \leq n$, E^m is an equivalence relation on $C^m(M)$. So the union $E = \bigcup_{0 < m \leq n} E^m$ is an equivalence relation on $\bigcup_{0 < m \leq n} C^m(M)$. Let Λ be the set of E-equivalence classes. For any n-dimensional Λ-hypernetwork N over $\mathcal{A}$ and map $\nu : n \to M$, we say that ν is an *embedding* of N into M if

1. for all $r \in \mathrm{At}(\mathcal{A})$ and $i, j < n$, we have $N(i,j) = r$ iff $M \models r(\nu(i), \nu(j))$, and

2. for any $\bar{a} \in {}^{\leq n}n$ with $|\bar{a}| \neq 2$, $\nu(\bar{a})$ is a member of the equivalence class $N(\bar{a}) \in \Lambda$.

We let $\mathcal{H}$ be the set of all n-dimensional Λ-hypernetworks over $\mathcal{A}$ that embed into M, and check that $\mathcal{H}$ is a hyperbasis. The first two properties of hyperbases are easy to verify, using n-squareness. We check the 'amalgamation' condition for $\mathcal{H}$. So take $P, Q \in \mathcal{H}$ and distinct $i, j < n$ with $P \equiv_{ij} Q$. We seek a hypernetwork $R \in \mathcal{H}$ with $P \equiv_i R \equiv_j Q$. As usual, we may assume $i \neq j$, else we could take $R = P$.

Let $\bar{b}$ be any $(n-2)$-tuple enumerating $n \setminus \{i, j\}$. Since $P, Q \in \mathcal{H}$, there are embeddings $\pi : P \to M$ and $\psi : Q \to M$. Since $P \equiv_{ij} Q$ and $n \neq 4$, we know that $P(\bar{b}) = Q(\bar{b}) \in \Lambda$, so that $(\pi(\bar{b}), \psi(\bar{b})) \in E$. Now $\mathrm{rng}(\pi\restriction_{n \setminus \{i\}})$ and $\mathrm{rng}(\psi\restriction_{n \setminus \{j\}})$ are both cliques, so since M is n-smooth, there is a point $z \in M$ such that $\mathrm{rng}(\psi\restriction_{n \setminus \{j\}}) \cup \{z\}$ is a clique and

$$((\pi(b_0), \ldots, \pi(b_{n-3}), \pi(j)), (\psi(b_0), \ldots, \psi(b_{n-3}), z)) \in E.$$

Let $\rho : n \to M$ be defined by: $\rho(k) = \psi(k)$ for $k \neq j$, and $\rho(j) = z$. By the first part of the definition of n-smooth (definition 13.12), for any $\bar{a} \in {}^{\leq n}(n \setminus \{i\})$ we have $(\pi(\bar{a}), \rho(\bar{a})) \in E$.

We use ρ to define the required hypernetwork $R \in \mathcal{H}$ in the obvious way: for $k, l < n$, $R(k, l)$ is the atom α of $\mathcal{A}$ satisfying $M \models \alpha(\rho(k), \rho(l))$, and for $\bar{a} \in {}^{\leq n}n$

of length $\neq 2$, $R(\bar{a})$ is the E-class of $\rho(\bar{a})$. Since ψ and ρ agree on all points except perhaps j, it follows that $R \equiv_j Q$. For any $\bar{a} \in {}^{\leq n}(n \setminus \{i\})$, because $E(\pi(\bar{a}), \rho(\bar{a}))$, we have $P(\bar{a}) = R(\bar{a})$. Hence $P \equiv_i R$, as required. □

Exercises

1. Consider the following definition (cf. definition 13.12 above). For finite $n \geq 3$, a relativised representation M of the non-associative algebra $\mathcal{A}$ is said to be n-homogeneous if

   - it is n-square, and
   - if $\theta : M \to M$ is a partial map such that:

     (a) θ is a partial isomorphism of M — i.e., for all $r \in \mathcal{A}$ and $x, y \in \text{dom}(\theta)$, we have $M \models r(x,y) \leftrightarrow r(\theta(x), \theta(y))$,

     (b) $0 < |\theta| \leq n-2$, and

     (c) if $x, y \in M$ are such that $\text{dom}(\theta) \cup \{x\}$, $\text{rng}(\theta) \cup \{y\}$ are cliques,

     then there is $x' \in M$ such that $\theta \cup \{(x, x')\}$ is a partial isomorphism and $\text{rng}(\theta) \cup \{x', y\}$ is a clique.

   (a) Show that any n-homogeneous relativised representation is n-smooth.

   (b) Prove that an atomic weakly associative algebra has an n-dimensional cylindric basis iff it has a complete n-homogeneous relativised representation. [Cf. theorem 12.41. See exercise 19.6(4) for a 'finite version' of this exercise.]

13.7 Summary and discussion

We can now tie up our results. We begin with the results on atomic non-associative algebras, and then we consider arbitrary non-associative algebras. This is for $n \geq 4$; afterwards, we tackle the three-dimensional case. Finally we touch on finite versions of our results, a topic to be continued later in the book, and show how to obtain recursive axiomatisations of the classes described in this chapter.

13.7.1 Atomic non-associative algebras

Combining the results so far:

THEOREM 13.45 *Let $4 \leq n < \omega$ and let $\mathcal{A}$ be a non-associative algebra. The following are equivalent.*

1. *$\mathcal{A}$ is atomic and has an n-dimensional relational basis.*

13.7. Summary and discussion

2. $\mathcal{A}$ has a complete n-square relativised representation.

3. There is a complete relation algebra reduct embedding from $\mathcal{A}$ into some atomic $C \in \mathbf{G}_n$.

4. There is a complete relation algebra reduct embedding from $\mathcal{A}$ into some atomic $C \in \mathbf{D}_n$.

The following are also equivalent (to each other, not the above):

5. $\mathcal{A}$ is atomic and has an n-dimensional hyperbasis.

6. $\mathcal{A}$ has a complete infinitarily n-flat relativised representation.

7. $\mathcal{A} \subseteq^c \mathfrak{Ra}C$ for some atomic $C \in \mathbf{CA}_n \cap \mathbf{G}_n$ (see definition 2.15 for $\subseteq^c$).

8. $\mathcal{A} \subseteq^c \mathfrak{Ra}C$ for some atomic $C \in \mathbf{CA}_n \cap \mathbf{D}_n$.

9. $\mathcal{A} \subseteq^c \mathfrak{Ra}C$ for some atomic $C \in \mathbf{CA}_n$.

10. $\mathcal{A}$ has a complete n-smooth relativised representation.

When $n = 4$, all 10 parts are equivalent.

Proof (sketch). All conditions imply that $\mathcal{A}$ is atomic: for conditions 2, 6, and 10, use corollary 2.22, and for conditions 3, 4, and 7–9 use lemma 2.16. When $n = 4$, exercise 12.3(12) shows that parts 1 and 5 are equivalent. The following diagram summarises the proof.

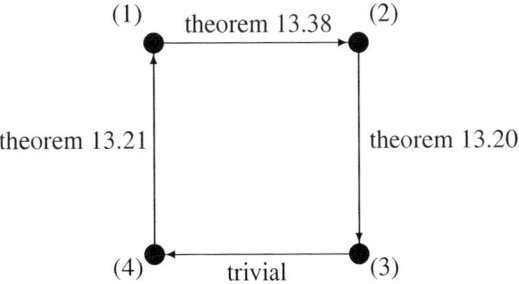

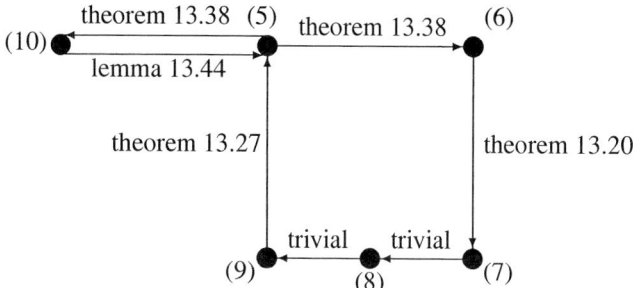

13.7.2 Arbitrary non-associative algebras

We now apply our results about complete relativised representations to relativised representations of arbitrary non-associative algebras.

THEOREM 13.46 *Let $\mathcal{A}$ be a non-associative algebra, and let $4 \leq n < \omega$. The following are equivalent.*

1. *$\mathcal{A}$ has an n-square relativised representation.*

2. *$\mathcal{A} \in S\mathfrak{R}\mathfrak{a}\mathbf{G}_n$.*

3. *$\mathcal{A} \in S\mathfrak{R}\mathfrak{a}\mathbf{D}_n$.*

4. *$\mathcal{A}^+$ has an n-dimensional relational basis.*

5. *$\mathcal{A} \in \mathbf{RA}_n$.*

Further, the following are equivalent (to each other, not to the above):

6. *$\mathcal{A}$ has an infinitarily n-flat relativised representation.*

7. *$\mathcal{A}$ has an n-flat relativised representation.*

8. *$\mathcal{A} \in S\mathfrak{R}\mathfrak{a}(\mathbf{CA}_n \cap \mathbf{G}_n)$.*

9. *$\mathcal{A} \in S\mathfrak{R}\mathfrak{a}(\mathbf{CA}_n \cap \mathbf{D}_n)$.*

10. *$\mathcal{A} \in S\mathfrak{R}\mathfrak{a}\mathbf{CA}_n$.*

11. *$\mathcal{A}^+$ has an n-dimensional hyperbasis.*

12. *There is an embedding from $\mathcal{A}$ into some atomic non-associative algebra that has an n-dimensional hyperbasis.*

13.7. Summary and discussion

13. $\mathcal{A}$ has an n-smooth relativised representation.

For $n = 4$, all 13 parts are equivalent.

Proof. We prove 1–5 and then 6–13 cyclically.

$1 \Rightarrow 2$ This can be proved in two ways. By corollary 13.18, $\mathcal{A}^+$ has a complete n-square relativised representation, and by theorem 13.45, $\mathcal{A}^+ \in \mathbf{S\Re aG}_n$ for some $C \in \mathbf{G}_n$. Thus, $\mathcal{A} \in \mathbf{S\Re aG}_n$. Alternatively, just use theorem 13.20.

$2 \Rightarrow 3$ Trivial.

$3 \Rightarrow 4$ Suppose that there is a relation algebra reduct embedding from $\mathcal{A}$ into some $\mathcal{B} \in \mathbf{D}_n$. By theorem 5.51, there is a complete relation algebra reduct embedding from $\mathcal{A}^+$ into $\mathcal{B}^+$; by fact 5.38, $\mathcal{B}^+ \in \mathbf{D}_n$ up to isomorphism. By theorem 13.45, $\mathcal{A}^+$ has an n-dimensional relational basis.

$4 \Rightarrow 5$ Trivial, by definition of $\mathbf{RA}_n$.

$5 \Rightarrow 1$ If $\mathcal{A} \in \mathbf{RA}_n$, then $\mathcal{A} \subseteq \mathcal{B}$ for some atomic non-associative algebra $\mathcal{B}$ with an n-dimensional relational basis. By theorem 13.45, $\mathcal{B}$ has a complete n-square relativised representation. So by lemma 13.5, $\mathcal{A}$ has an n-square relativised representation too.

$6 \Rightarrow 7$ Trivial, since $L(\mathcal{A})^n \subseteq L(\mathcal{A})^n_{\infty\omega}$.

$7 \Rightarrow 8$ By theorem 13.20, if M is an n-flat relativised representation of $\mathcal{A}$ then $\mathcal{D}^0(M) \in \mathbf{CA}_n \cap \mathbf{G}_n$ and $\mathcal{A} \subseteq \Re\mathfrak{a}(\mathcal{D}^0(M))$.

$8 \Rightarrow 9, 9 \Rightarrow 10$ Trivial.

$10 \Rightarrow 11$ By theorem 5.43, if $\mathcal{A} \subseteq C \in \mathbf{CA}_n$ then $\mathcal{A}^+ \subseteq^c \Re\mathfrak{a} C^+$, and $C^+ \in \mathbf{CA}_n$ since $\mathbf{CA}_n$ is canonical. By theorem 13.45, $\mathcal{A}^+$ has an n-dimensional hyperbasis.

$11 \Rightarrow 12$ Trivial.

$12 \Rightarrow 13$ Let $\mathcal{A} \subseteq \mathcal{B}$ for some atomic $\mathcal{B}$ with an n-dimensional hyperbasis. By theorem 13.45, $\mathcal{B}$ has a complete n-smooth relativised representation, and hence by lemma 13.14, $\mathcal{A}$ has an n-smooth relativised representation.

$13 \Rightarrow 6$ If $\mathcal{A}$ has an n-smooth relativised representation then by corollary 13.18, $\mathcal{A}^+$ has a complete n-smooth relativised representation. By theorem 13.45, $\mathcal{A}^+$ has an infinitarily n-flat relativised representation, and so (lemma 13.7) $\mathcal{A}$ has an infinitarily n-flat relativised representation.

Maddux [Madd83, theorem 5] proved that $\mathbf{RA}_4 = \mathbf{RA}$. Monk proved that every relation algebra is in $\mathbf{S\mathfrak{R}aCA}_3$ [Mon61b, Th. 9.10]; Maddux improved this to show that every relation algebra is in $\mathbf{S\mathfrak{R}aCA}_4$ [Madd78b, Ch. 10,Th. (21)], [HenMon+85, Th. 5.3.17]. The converse inclusion, $\mathbf{S\mathfrak{R}aCA}_4 \subseteq \mathbf{RA}$, was proved by Henkin and Tarski (see theorem 5.44). Thus, $\mathbf{RA} = \mathbf{S\mathfrak{R}aCA}_4$. So parts (5) and (10) are equivalent when $n = 4$, and so all 13 parts are equivalent in this case. □

We now make explicit some corollaries of this result.

COROLLARY 13.47 Let $4 \leq n < \omega$ and let $\mathcal{A}$ be a non-associative algebra.

1. $\mathbf{S\mathfrak{R}aD}_n = \mathbf{S\mathfrak{R}aG}_n = \mathbf{RA}_n$.

2. $\mathbf{S\mathfrak{R}aCA}_n = \mathbf{S\mathfrak{R}a}(\mathbf{CA}_n \cap \mathbf{D}_n) = \mathbf{S\mathfrak{R}a}(\mathbf{CA}_n \cap \mathbf{G}_n)$.

3. $\mathbf{RA} = \mathbf{RA}_4 = \mathbf{S\mathfrak{R}aD}_4 = \mathbf{S\mathfrak{R}aG}_4 = \mathbf{S\mathfrak{R}aCA}_4 = \mathbf{S\mathfrak{R}a}(\mathbf{CA}_4 \cap \mathbf{D}_4)$
   $= \mathbf{S\mathfrak{R}a}(\mathbf{CA}_4 \cap \mathbf{G}_4)$. Also, $\mathcal{A}$ is a relation algebra iff $\mathcal{A}$ has 4-square, 4-flat, infinitarily 4-flat, and 4-smooth relativised representations, iff $\mathcal{A}^+$ has a 4-dimensional relational basis and hyperbasis.

4. If $\mathcal{A} \in \mathbf{RA}_n$ then $\mathcal{A}^+ \in \mathbf{RA}_n$, and indeed, $\mathcal{A}^+$ has an n-dimensional relational basis. So $\mathbf{RA}_n$ is a canonical variety.

5. If $\mathcal{A}$ has an n-flat relativised representation, then $\mathcal{A}^+$ has a complete n-flat relativised representation. Analogous facts hold for infinitarily n-flat, n-smooth, and n-square relativised representations. [The last two were proved in corollary 13.18. Cf. Monk's theorem 3.36 for $\mathbf{RRA}$.]

6. $\mathbf{S\mathfrak{R}aCA}_n \subseteq \mathbf{RA}_n$.

Proof. Most is immediate from theorem 13.46. We note that the last part of part 4 (that $\mathbf{RA}_n$ is a variety) follows from proposition 12.31. This proposition assumed the result (†) that if $\mathcal{A} \in \mathbf{RA}_n$ then $\mathcal{A}^+$ has an n-dimensional relational basis, which the first half of part 4 proves. Part 6 holds because any hyperbasis is a relational basis. □

We saw in proposition 5.48 that $\mathbf{S\mathfrak{R}aCA}_n$ is a canonical variety. We will see in chapter 15 that for $n \geq 5$, $\mathbf{S\mathfrak{R}aCA}_n \subset \mathbf{RA}_n$, and indeed that $\mathbf{S\mathfrak{R}aCA}_n$ is not finitely axiomatisable over $\mathbf{RA}_n$. So conditions 1–5 of theorem 13.46 imply conditions 6–13 only if $n = 4$.

PROPOSITION 13.48 (Monk, Maddux)

$$\bigcap_{3 \leq n < \omega} \mathbf{RA}_n = \bigcap_{3 \leq n < \omega} \mathbf{S\mathfrak{R}aCA}_n = \mathbf{S\mathfrak{R}aCA}_\alpha = \mathbf{RRA} \quad \textit{(all ordinals } \alpha \geq \omega\textit{)}.$$

13.7. Summary and discussion

Proof (sketch). Suppose that a simple countable non-associative algebra has an n-square relativised representation for every finite $n \geq 3$. A compactness argument applied to proposition 13.17 will show that it has a single relativised representation which is ω-square (n-square for every finite n — see definition 5.7). A standard step-by-step construction will now build a chain of larger and larger finite cliques in this representation resolving more and more compositional defects, so that their union will have no defects. Being a clique, the union will be a classical representation of the algebra. Since a classical representation is evidently ω-square, we see that $\bigcap_n \mathbf{RA}_n$ and $\mathbf{RRA}$ coincide on countable simple algebras. As they are both discriminator varieties of relation algebras, it follows by theorem 2.44 and corollary 2.56 that they are equal. So $\bigcap_{3 \leq n < \omega} \mathbf{RA}_n = \mathbf{RRA}$.

By corollary 13.47, $\mathbf{S\Re aCA}_n \subseteq \mathbf{RA}_n$. Though this inclusion fails for $n = 3$, by exercise 5.6(1) we have

$$\mathbf{S\Re aCA}_\alpha \subseteq \mathbf{S\Re aCA}_\beta \text{ for any ordinals } \alpha \geq \beta \geq 3, \tag{13.5}$$

so $\bigcap_{3 \leq n < \omega} \mathbf{S\Re aCA}_n = \bigcap_{4 \leq n < \omega} \mathbf{S\Re aCA}_n$. It is easy to show (see exercise 5.6(2)) that

$$\mathbf{RRA} \subseteq \mathbf{S\Re aCA}_\alpha \text{ for any ordinal } \alpha \geq 3. \tag{13.6}$$

So we obtain:

$$\mathbf{RRA} \subseteq \bigcap_{3 \leq n < \omega} \mathbf{S\Re aCA}_n \subseteq \bigcap_{4 \leq n < \omega} \mathbf{RA}_n = \mathbf{RRA}.$$

Now for infinite α, we obtain $\mathbf{RRA} \subseteq \mathbf{S\Re aCA}_\alpha \subseteq \bigcap_{3 \leq n < \omega} \mathbf{S\Re aCA}_n = \mathbf{RRA}$ by (13.6) and (13.5), completing the proof. □

See [Madd91b, p. 112] for a similar argument with relational bases, and [Mon61b, theorems 4.1, 9.11, 9.12] for analogous results for neat reducts.

13.7.3 Three-dimensional version of theorem 13.46

In dimension 3, we have the following:

THEOREM 13.49 *Let $\mathcal{A}$ be a non-associative algebra. The following are equivalent.*

1. *$\mathcal{A} \in \mathbf{SA}$.*

2. *$\mathcal{A}^+$ has a 3-dimensional relational basis.*

3. *$\mathcal{A} \in \mathbf{RA}_3$ — i.e., $\mathcal{A}$ embeds in some atomic non-associative algebra with a 3-dimensional relational basis.*

4. $\mathcal{A}$ has a 3-square relativised representation.

5. $\mathcal{A}^+$ has a 3-dimensional hyperbasis.

6. $\mathcal{A}$ embeds in some atomic non-associative algebra that has a 3-dimensional hyperbasis.

7. $\mathcal{A}$ has a infinitarily 3-flat relativised representation.

8. $\mathcal{A}$ has a 3-flat relativised representation.

9. $\mathcal{A}$ has a 3-smooth relativised representation.

10. $\mathcal{A} \in \mathbf{S\mathfrak{R}a}(\mathbf{CA}_3 \cap \mathbf{G}_3)$.

Proof. The following diagram summarises the proof.

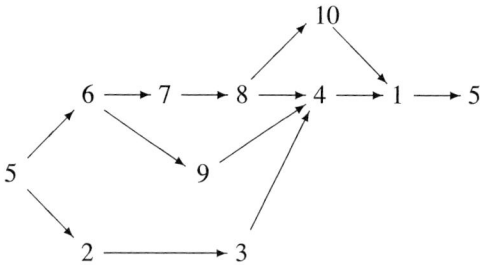

$5 \Rightarrow 2$, $2 \Rightarrow 3$, $5 \Rightarrow 6$, $7 \Rightarrow 8$, $9 \Rightarrow 4$ Trivial.

$3 \Rightarrow 4$ Use theorem 13.38 and lemma 13.5.

$6 \Rightarrow 7$ Use theorem 13.38 and lemma 13.7.

$8 \Rightarrow 4$ Use lemma 13.10.

$4 \Rightarrow 1$ Exercise.

$1 \Rightarrow 5$ By theorem 5.8, if $\mathcal{A} \in \mathbf{SA}$ then $\mathcal{A}^+ \in \mathbf{SA}$. As $\mathcal{A}^+$ is atomic, by exercise 12.3(11) it has a 3-dimensional hyperbasis.

$6 \Rightarrow 9$ By theorem 13.38 and lemma 13.14.

$8 \Rightarrow 10$ If $\mathcal{A}$ has a 3-flat relativised representation, theorem 13.20 applies to show that $\mathcal{A} \in \mathbf{S\mathfrak{R}a}(\mathbf{CA}_3 \cap \mathbf{G}_3)$.

$10 \Rightarrow 1$ This follows from theorems 6.7 and 6.8, which show that for $C \in \mathbf{CA}_3$, we have $\mathfrak{R}aC \in \mathbf{SA}$ iff $C \in \mathbf{G}_3$. Alternatively, use exercise 5.7(5).

□

13.7. Summary and discussion

The following is immediate:

COROLLARY 13.50 $\mathbf{SA} = \mathbf{RA}_3 = \mathbf{S\mathfrak{R}a}(\mathbf{CA}_3 \cap \mathbf{G}_3)$.

It may be worth recalling here that $\mathbf{S\mathfrak{R}aG}_3 = \mathbf{WA}$ (see lemma 5.54 and exercise 5.7(4)), and $\mathbf{S\mathfrak{R}aCA}_3 \not\subseteq \mathbf{NA}$. In problem 5.56 we asked how $\mathbf{S\mathfrak{R}aD}_3$ and $\mathbf{S\mathfrak{R}a}(\mathbf{CA}_3 \cap \mathbf{D}_3)$ are related to the usual suspects.

13.7.4 Finite versions of theorem 13.46 (first part)

We now ask (roughly speaking) whether we can put 'finite' everywhere in theorem 13.46. To begin, consider the first five conditions of the theorem, to do with relational bases. We have:

THEOREM 13.51 *Let $\mathcal{A}$ be a finite non-associative algebra and let $n \geq 4$ be finite. Then the following are equivalent:*

1. *There is a relation algebra reduct embedding from $\mathcal{A}$ into some finite $C \in \mathbf{G}_n$.*

2. *There is a relation algebra reduct embedding from $\mathcal{A}$ into some finite $C \in \mathbf{D}_n$.*

3. *$\mathcal{A} = \mathcal{A}^+$ has a (finite) n-dimensional relational basis.*

4. *$\mathcal{A} \in \mathbf{RA}_n$.*

5. *$\mathcal{A}$ embeds in some finite non-associative algebra that has an n-dimensional al relational basis.*

Proof. $1 \Rightarrow 2$ is clear, and $2 \Rightarrow 3$ follows from theorem 13.46, since if a finite non-associative algebra has an n-dimensional relational basis then the basis must obviously be finite. $3 \Rightarrow 5 \Rightarrow 4 \Rightarrow 3$ are immediate from the definition of $\mathbf{RA}_n$ and theorem 13.46. For $3 \Rightarrow 1$, see exercise 12.3(14). □

By theorem 13.46, if $\mathcal{A}$ has a finite n-square relativised representation then condition (4) above holds. Later, in theorem 19.18, we will show that '$\mathcal{A}$ has a finite n-square relativised representation' can in fact be added as another equivalent condition in theorem 13.51.

13.7.5 Finite versions of theorem 13.46 (second part)

Now we consider finitising the $\mathbf{S\mathfrak{R}aCA}_n$ side of theorem 13.46. Here, things are not so straightforward, because a finite relation algebra or non-associative algebra could in principle have an infinite n-dimensional hyperbasis but no finite one. There are only finitely many n-dimensional atomic networks over a given finite relation algebra, so any n-dimensional relational basis must be finite; but because of the presence of hyperlabels, this may not be so for a hyperbasis. Indeed, in proposition 19.19 we will show that for $n \geq 5$, if a finite relation algebra $\mathcal{A}$ has an n-dimensional hyperbasis, it does *not* follow that $\mathcal{A}$ has a *finite* n-dimensional hyperbasis. Nor need $\mathcal{A}$ have a finite n-flat relativised representation.

Nonetheless, the subclass of finite non-associative algebras with a finite hyperbasis is well behaved and satisfies its own version of theorem 13.46. In chapter 19, we will prove that the following are equivalent for a finite non-associative algebra $\mathcal{A}$:

1. $\mathcal{A}$ has a finite n-flat (or infinitarily n-flat, or n-smooth) relativised representation;

2. $\mathcal{A} \subseteq \mathfrak{Ra}C$ for some finite $C \in \mathbf{CA}_n$;

3. $\mathcal{A}$ has a finite n-dimensional hyperbasis.

Inspection of the proof of theorem 13.46 shows that $1 \Rightarrow 2$ and $2 \Rightarrow 3$ hold; we will prove $3 \Rightarrow 1$ in theorem 19.20.

PROBLEM 13.52 Must any n-smooth relativised representation of an arbitrary non-associative algebra be infinitarily n-flat? Or even n-flat? (For a converse, see exercise 13.5(1).)

PROBLEM 13.53 Is every n-flat relativised representation of a non-associative algebra also infinitarily n-flat?

PROBLEM 13.54 For finite $n \geq 5$, is there an atomic non-associative algebra with a complete n-flat relativised representation but with no complete infinitarily n-flat relativised representation? (By theorem 13.46, if 'complete' is omitted then the answer is 'no'. Hence, any algebra forming a counter-example must be infinite.)

Exercises

1. Show directly that if $\mathcal{A}$ is a finite non-associative algebra with a finite n-square relativised representation then $\mathcal{A}$ has a finite n-dimensional relational basis. Do the same for n-flat relativised representations and hyperbases.

2. State and prove a version of theorem 13.49 for complete relativised representations, and of theorem 13.51 for $n = 3$.

3. Formulate a notion of '$2\frac{1}{2}$-dimensional relational basis' such that **WA** is the class of all subalgebras of atomic non-associative algebras $\mathcal{A}$ with such a basis.

13.8. Equational axioms for **RA**$_n$ and **S**$\mathfrak{R}$**aCA**$_n$

4. Given finite $n \geq 3$, a finite non-associative algebra $\mathcal{A}$, and a finite $L(\mathcal{A})$-structure M, prove that it is decidable whether

   (a) M is a relativised representation of $\mathcal{A}$,

   (b) M is an n-square relativised representation of $\mathcal{A}$,

   (c) M is an n-flat relativised representation of $\mathcal{A}$. [This is harder: there are infinitely many formulas in $L(\mathcal{A})^n$ of the form $\exists x_i \exists x_j \varphi \leftrightarrow \exists x_j \exists x_i \varphi$. Exercise 13.1(7) may help.]

13.8 Equational axioms for **RA**$_n$ and **S**$\mathfrak{R}$**aCA**$_n$

In this section, we use theorem 13.46 to give recursive equational axiomatisations of the classes **RA**$_n$ and **S**$\mathfrak{R}$**aCA**$_n$, for each finite $n \geq 5$. First, we note that **RA**$_n$ and **S**$\mathfrak{R}$**aCA**$_n$ are pseudo-universal classes (definition 9.1). By theorem 13.46, this can be seen in any of three ways, by expressing that the given algebra $\mathcal{A}$ is the first, **a**-sort of a structure M such that one of the following holds:

1. The second, **n**-sort of M consists of an algebra C in $\mathbf{D}_n, \mathbf{G}_n$ (for **RA**$_n$), or **CA**$_n$ (for **S**$\mathfrak{R}$**aCA**$_n$). This condition is expressed by including axioms for $\mathbf{D}_n, \mathbf{G}_n$, or **CA**$_n$, or in the first two cases by including a third sort **b**, say, an $(n+1)$-ary relation $\text{holds}(\mathbf{n}, \mathbf{b}, \ldots, \mathbf{b})$, and axioms stating directly that if $\hat{c} = \{(b_0, \ldots, b_{n-1}) : M \models \text{holds}(c, b_0, \ldots, b_{n-1})\}$ for $c \in C$, then $\{\hat{c} : c \in C\}$ is a relativised set algebra isomorphic to C, with unit closed under substitutions or permutations. Further, one of the functions of M is a relation algebra reduct embedding $: \mathcal{A} \to C$.

2. The **n**-sort of M is an n-square (for **RA**$_n$) or n-flat/smooth (for **S**$\mathfrak{R}$**aCA**$_n$) relativised representation of $\mathcal{A}$ (see definition 13.16 for suitable axioms);

3. The **n**-sort of M is in two halves, a non-associative algebra in which $\mathcal{A}$ embeds by a function of M, and an n-dimensional relational basis (for **RA**$_n$) or hyperbasis (for **S**$\mathfrak{R}$**aCA**$_n$) of this larger algebra (cf. section 12.4.3). For **S**$\mathfrak{R}$**aCA**$_n$, the set of hyperlabels will form an additional sort of M.

We leave the details as exercises for the reader. Using theorem 9.28, we can now obtain recursive universal axiomatisations of **RA**$_n$ and **S**$\mathfrak{R}$**aCA**$_n$, for $n \geq 5$ (these classes are equal to **RA** for $n = 4$, and so are finitely axiomatisable). For any relation algebra $\mathcal{A}$, we have

- $\mathcal{A} \models \{\chi_k^n : k < \omega\}$ iff $\mathcal{A} \in \mathbf{RA}_n$,

- $\mathcal{A} \models \{\mu_k^n : k < \omega\}$ iff $\mathcal{A} \in \mathbf{S}\mathfrak{R}\mathbf{aCA}_n$.

We now replace the universal sentences χ_k^n, μ_k^n by equations. Recall that **RA** is a discriminator variety: the discriminator term is $1; x; 1$ (see definition 2.52 and theorem 3.19). Further, **RA**$_n$ and **S$\Re$aCA**$_n$ are varieties contained in **RA**. So we may apply theorem 2.60 to obtain equations $\varepsilon(\chi_k^n)$ and $\varepsilon(\mu_l^n)$ defining **RA**$_n$ and **S$\Re$aCA**$_n$, respectively.

THEOREM 13.55 *For each finite $n \geq 5$:*

1. **RA**$_n$ *is axiomatised by the equations defining* **RA** *together with the equations* $\varepsilon(\chi_k^n)$ *for $k < \omega$.*

2. **S$\Re$aCA**$_n$ *is axiomatised by the equations defining* **RA** *together with the equations* $\varepsilon(\mu_k^n)$ *for $k < \omega$.*

Part IV

Constructing Relation Algebras

Introduction to the constructions 441

The constructions

In this part of the book, we will construct a number of relation algebras that fail to have certain kinds of representation. In chapter 4, we saw a finite, four-atom relation algebra designed by McKenzie that was not representable. We introduced the Lyndon algebras, and saw that infinitely many of them were not representable. Also, some of the representable Lyndon algebras failed to have homogeneous representations (by the proof of theorem 12.37, and theorem 12.41). But constructing these examples is clearly quite delicate. We are looking for more general methods of building relation algebras that have or do not have certain kinds of representation.

In part II, we defined various games played with networks over a relation algebra corresponding to different kinds of representation. For various values $m < n \leq \omega$, we now wish to provide relation algebras for which $\exists$ can win such a game if there are only m rounds to play, but for which she has no winning strategy if there are n rounds in the game. These constructions will allow us to prove various non-finite axiomatisability results quite easily. If we can construct a relation algebra for which $\exists$ has a winning strategy in the game of length n, for any $n < \omega$, but for which she has no winning strategy in the ω-length game, then we will be able to show that the class of relation algebras defined by the ω-length game is not even elementary. A similar argument applies to atom structures: for example, if we can find a sequence of relation algebra atom structures such that $\exists$ can win all finite-length games on the complex algebras over them but not on the complex algebra over an ultraproduct of them, this will show that the relation algebra atom structures whose complex algebras are representable relation algebras is not an elementary class.

Monk algebras In chapter 14, we consider the classes of *strongly* and *weakly representable atom structures* — those relation algebra atom structures such that every (respectively, some) atomic relation algebra with that atom structure is representable. We prove that the class of strongly representable atom structures is not elementary.

To do this, we use a variant of *Monk algebras* (see section 4.4). In a Monk algebra, the non-identity atoms are given colours, and composition is essentially defined by stating that 'monochromatic triples of atoms' are inconsistent, though we may wish to include additional rules about consistent triples of atoms to suit the particular task in hand. If a Monk algebra has many more atoms than colours, it follows from Ramsey's theorem that any representation of it would contain a monochromatic triangle, so the algebra is not representable. Algebras of this form were used by Monk in [Mon69] to prove that the representable cylindric algebras cannot be defined by finitely many axioms.

To show that the class of strongly representable atom structures is not elementary, we use a kind of Monk algebra (we will still call it a Monk algebra, for simplicity) in which the rule about monochromatic triples of atoms is relaxed slightly, together with a powerful graph-theoretic result of Erdös, to construct a sequence of strongly representable atom structures with an ultraproduct that is not strongly representable. The result now follows, since all elementary classes are closed under ultraproducts. Moreover, since we know from chapter 2 that the class of weakly representable atom structures is elementary, it follows that there exist weakly but not strongly representable relation algebra atom structures. This has the striking consequence that there exist two atomic relation algebras with the same atom structure, only one of which is representable.

Also in chapter 14 we will describe another construction, of atomic relation algebras whose atom structures are strongly representable but do not satisfy the Lyndon conditions. It is due to Maddux and is related to an earlier construction of Lyndon.

In chapter 15, we consider the classes $\mathbf{S\mathfrak{R}aCA}_n$ (for $4 \leq n < \omega$). For a countable atomic relation algebra $\mathcal{A}$, if $\exists$ has a winning strategy in the 'n-pebble game' $G_\omega^{m,n}(\mathcal{A}, \Lambda)$ defined in chapter 12, then $\mathcal{A}$ has an n-dimensional hyperbasis and so belongs to $\mathbf{S\mathfrak{R}aCA}_n$. Then we define a collection of finite Monk algebras $\mathfrak{A}(n,r)$ (for $4 \leq n < \omega$, $0 < r < \omega$). We show, for $3 \leq m < n < \omega$ and infinite Λ, that $\exists$ can win the games $G_\omega^{m,n}(\mathfrak{A}(n,r), \Lambda)$ and $G_r^{m,n+1}(\mathfrak{A}(n,r), \Lambda)$, but that $\mathfrak{A}(n,r) \notin \mathbf{S\mathfrak{R}aCA}_{n+1}$. It follows that $\mathbf{S\mathfrak{R}aCA}_{n+1}$ cannot be finitely axiomatised, even relative to $\mathbf{S\mathfrak{R}aCA}_n$, and that $\mathbf{S\mathfrak{N}r}_m\mathbf{CA}_{n+1}$ cannot be finitely axiomatised relative to $\mathbf{S\mathfrak{N}r}_m\mathbf{CA}_n$, for $3 \leq m < n < \omega$.

Rainbow construction For some non-finite axiomatisability results like the ones just mentioned, it suffices to find algebras $\mathcal{A}_r$ ($r < \omega$) such that $\exists$ has a winning strategy in the game of length r but not in the game of length ω. Monk algebras are well-suited to this role. But there are other cases where more accurate control over who possesses a winning strategy is required.

A construction that gives us great control over who wins games of various lengths is the *rainbow construction,* introduced in chapter 16. The rainbow construction is motivated by games, not by Ramsey's theorem. It reduces the difficult problem of finding a relation algebra where $\exists$ has a winning strategy in a game of a certain length but no winning strategy in longer games to the much easier problem of finding two structures A, B in a binary relational language such that $\exists$ can win short Ehrenfeucht–Fraïssé 'forth' games played on the pair of structures (A, B) but not longer games. More precisely, from these two structures A and B we construct a 'rainbow relation algebra' $\mathcal{A}_{A,B}$ and show that

$$\exists \text{ has a winning strategy in } G_{1+r}^{2+p}(\mathcal{A}_{A,B})$$
$$\iff \exists \text{ has a winning strategy in } EF_r^p(A, B),$$

Introduction to the constructions 443

where G_{1+r}^{2+p} is the $(2+p)$-pebble, $(1+r)$-round atomic network game of chapter 12, $p, r \leq \omega$, and $EF_r^p(A,B)$ is a kind of p-pebble, r-round Ehrenfeucht–Fraïssé 'forth' game from A to B (see theorem 16.5).

The rainbow construction therefore allows fine control over both the length of game that $\exists$ can win, and (more importantly) the number of pebbles that she can cope with. This is one of its key advantages over Monk algebras. Typically, with Monk algebras it is very hard to calculate the exact maximum r such that $\exists$ can win the r-round game. More importantly, to prove that $\forall$ can defeat her in an n-pebble game, we usually take n to be a large number determined by some Ramsey calculation; an exact estimate of how large n needs to be is often unavailable, and anyway we can't control it very well. The construction and proofs of chapter 15, discussed above, are exceptions to this. For the Monk algebras of that chapter, by extensive use of the amalgamation property for hyperbases we can show that $\forall$ has a winning strategy in the hyperbasis game with $n+1$ pebbles but not with n. But for relational bases, which need not have the amalgamation property, we cannot see how to limit the number of pebbles to $n+1$ and still give $\forall$ a winning strategy. So the rainbow construction is particularly useful for obtaining a non-finite axiomatisability result for $\mathbf{RA}_{n+1}$ over $\mathbf{RA}_n$.

The transparency of working out who has a winning strategy in various games played on rainbow algebras has further benefits than simply being able to calculate who will win. It allows rather sophisticated constructions — e.g., the one in chapter 18 in part V, proving that the problem of whether a finite relation algebra is representable is undecidable. We cannot see how such a result could be proved with Monk algebras. On the other hand, in our experience, theorems proved using Monk algebras can also be proved using the rainbow construction, and indeed, we first found the results of chapters 14 and 15 by using it. The rainbow construction constitutes a general method for building interesting relation algebras that is arguably easier to use than Monk algebras (once one is familiar with it). It also generalises easily to cylindric algebras — see [HirHod97a, Hodk97, HodMik00].

In chapter 17, we'll see a number of applications of the rainbow construction, including a proof of Monk's result on the non-finite axiomatisability of $\mathbf{RRA}$, and the further result of Jónsson that $\mathbf{RRA}$ cannot be axiomatised by equations using finitely many variables. We'll also construct a single relation algebra such that $\exists$ has a winning strategy in all finite-length atomic games over the algebra, but not in the infinitely long game. It follows that the class of completely representable relation algebras is not elementary (section 17.2). We use the rainbow construction again to show that the weakly representable relation algebras cannot be axiomatised finitely, and that $\mathbf{RA}_n$ and $\mathbf{S\mathfrak{R}aCA}_n$ are not closed under completions for $n \geq 6$.

Chapter 14

Strongly representable relation algebra atom structures

14.1 Introduction

Representability is difficult to characterise for relation algebras. As Monk proved, the representable relation algebras cannot be defined by any first-order sentence, and we'll see later that for finite relation algebras, representability is undecidable. On the other hand, representability is trivially characterised for boolean algebras: every field of sets is a boolean algebra, and by Stone's theorem, every boolean algebra can be represented as a field of sets.

Since the relation algebra operations are completely additive, an atomic relation algebra is completely determined by (a) its atom structure (definitions 2.62 and 3.22) together with (b) its boolean structure. For the boolean structure, we need to know which suprema of sets of atoms are present in the algebra and which are not.

The fact that representability is so difficult to pin down for relation algebras but so easy with boolean algebras, together with the informal equation 'relation algebra = boolean algebra + atom structure' just established, might lead one to conclude that the atom structure alone determines whether or not an atomic relation algebra is representable. However, in this chapter we'll show that this is false: there are two atomic relation algebras with the same atom structure, one representable, the other, not.

So we can distinguish two kinds of 'representability of an atom structure':

Chapter 14. Strongly representable relation algebra atom structures

DEFINITION 14.1 (cf. definition 2.74.) Let S be a relation algebra atom structure.

1. S is said to be *weakly representable* if there exists a representable atomic relation algebra with atom structure S.

2. S is said to be *strongly representable* if every atomic relation algebra with atom structure S is representable.

3. **WRAS** and **SRAS** denote, respectively, the classes of weakly and strongly representable relation algebra atom structures.

We have the following basic lemma.

LEMMA 14.2 *Let S be a relation algebra atom structure.*

1. *S is strongly representable iff its complex algebra $\mathfrak{Cm}\, S$ (definition 2.65) is representable.*

2. *S is weakly representable iff the subalgebra $\tau(S)$ of $\mathfrak{Cm}\, S$ generated by the atoms of $\mathfrak{Cm}\, S$ is representable. (We call $\tau(S)$ the* term algebra *over S, because every element of it is the value of some relation algebra term with variables assigned to atoms of S.)*

Proof. Since clearly $\mathrm{At}\,\mathfrak{Cm}\, S = \mathrm{At}\,\tau(S) \cong S$, '$\Rightarrow$' of (1) and '$\Leftarrow$' of (2) are trivial. For the converses, note that by proposition 2.66 and complete additivity of the relation algebra operations (theorem 3.14), any atomic relation algebra $\mathcal{A}$ with atom structure S embeds into $\mathfrak{Cm}\, S$, via

$$\theta : x \mapsto \{a \in S : a \leq x\} \in \mathfrak{Cm}\, S, \text{ for } x \in \mathcal{A}.$$

So we may regard $\mathcal{A}$ as a subalgebra of $\mathfrak{Cm}\, S$. Since **RRA** is closed under subalgebras, we have:

1. if $\mathfrak{Cm}\, S$ is representable, then $\mathcal{A}$ is also representable,

2. if $\mathcal{A}$ is representable, then since $\tau(S)$ is the smallest subalgebra of $\mathfrak{Cm}\, S$ containing S, it is a subalgebra of $\mathcal{A}$ and so is also representable. □

It follows that in terms of definitions 2.74 and 2.89, we have

$$\begin{aligned}\mathbf{WRAS} &= \mathrm{At}\,\mathbf{RRA},\\ \mathbf{SRAS} &= \mathrm{Str}\,\mathbf{RRA}.\end{aligned}$$

Theorem 2.84 therefore showed that **WRAS** is elementary and can be axiomatised explicitly from an equational axiomatisation of **RRA** such as the one given in

14.2. **SRAS** *is not an elementary class*

chapter 8. Exercise 4.5(2) showed using Lyndon algebras that neither **WRAS** nor **SRAS** is finitely axiomatisable.

In this chapter, we investigate these classes further. In section 14.2 we show that **SRAS** is not elementary. We use a kind of Monk algebra, together with graphs of high chromatic number and girth. In section 14.3 we consider the relation between these classes and others: in particular, the class of atom structures of completely representable relation algebras (**CRAS**; see definition 3.35), and its elementary closure **LCAS**, the class of atom structures that satisfy the Lyndon conditions. In section 14.4 we outline a different construction, due to Maddux, of relation algebra atom structures that are strongly representable but do not satisfy the Lyndon conditions.

14.2 SRAS is not an elementary class

Our aim is to prove:

THEOREM 14.3 **SRAS** *is not closed under ultraproducts and hence is not elementary.*

This will be proved in section 14.2.3. The construction makes use of a result of Erdös [Erd59], which provides finite graphs of arbitrarily high chromatic number and girth. By taking disjoint unions of them, we obtain an infinite graph Γ_r, for each $r < \omega$, that cannot be coloured by any finite number of colours but whose 'small' subgraphs of size $< r$ have no cycles and so can be coloured by just two colours. From such a graph Γ we construct an atom structure $\alpha(\Gamma)$. The fact that Γ_r cannot be coloured is enough to prove that $\alpha(\Gamma_r)$ is strongly representable. The fact that small subgraphs of Γ_r can be two-coloured is enough to show that a non-principal ultraproduct of the Γ_r is two-colourable, and this suffices to show that the ultraproduct of the atom structures $\alpha(\Gamma_r)$ is not strongly representable.

As is often the case, the use of ultraproducts can be easily replaced by first-order compactness.

14.2.1 Graphs and colourings

DEFINITION 14.4 Let $\Gamma = (V, E)$ be an undirected graph (V is the set of vertices or nodes, and E is an irreflexive, symmetric binary relation on V), and let C be a non-empty set (of 'colours').

1. A subset $X \subseteq V$ is said to be an *independent set* if $(x, y) \notin E$ for all $x, y \in X$.

2. A function $f : V \to C$ is called a *C-colouring* of Γ if $(v, w) \in E \Rightarrow f(v) \neq f(w)$.

3. The *chromatic number* of Γ, denoted $\chi(\Gamma)$, is the size of the smallest finite set C such that there exists a C-colouring of Γ, if such a set C exists, and $\chi(\Gamma) = \infty$ otherwise.

4. A *cycle* in Γ is a finite sequence $\gamma = \langle v_0, v_1, \ldots, v_{k-1} \rangle$ of distinct nodes (some $k \geq 3$) such that $(v_0, v_1), \ldots, (v_{k-2}, v_{k-1}), (v_{k-1}, v_0) \in E$. The *length* of such a cycle is k.

5. The *girth* of Γ, denoted $g(\Gamma)$, is the length of the shortest cycle in Γ if Γ has any cycles, and $g(\Gamma) = \infty$ if not.

So a colouring assigns colours to nodes in such a way that any two nodes connected by an edge get different colours. Note that $f : V \to C$ is a C-colouring of Γ iff $f^{-1}[c]$ is an independent set, for all $c \in C$. Thus, a C-colouring of Γ is essentially a partition of V into at most $|C|$ independent sets. Exercise 1 below connects colourings to cycles.

THEOREM 14.5 (Erdős, [Erd59]) *For all $r < \omega$ there is a finite graph G_r with $\chi(G_r) > r$ and $g(G_r) > r$.*

A recent presentation of Erdős' proof of this theorem can be found in [Die97, theorem 11.2.2]. Erdős' proof was radical and was one of the earliest results using the probabilistic method. The idea is to construct a graph G on n vertices by randomly choosing whether or not to include an edge between two distinct vertices, the probability of including it being p, and repeating this random choice independently for each pair of distinct vertices in the graph. For $p > \frac{6 \ln n}{n}$ the probability that G contains an independent set of size $\frac{n}{2r}$ tends to zero as n tends to infinity.

If $p < n^{\frac{1}{r}-1}$ then although there may be cycles of length r or less ('short' cycles) in G, we find that the probability that there are at least $\frac{n}{2}$ short cycles also tends to zero as n tends to infinity. For large n, it is possible to find a value of p with $\frac{6 \ln n}{n} < p < n^{\frac{1}{r}-1}$. With such a value for p, for sufficiently large n, the probability that G has no independent set of size $\frac{n}{2r}$ and fewer than $\frac{n}{2}$ short cycles is strictly positive. It follows that a graph G with both properties exists.

From each of the short cycles of such a graph G we delete a single node to obtain a graph H. Clearly, H has no short cycles, so $g(H) > r$. Also, $|H| \geq \frac{n}{2}$ and H has no independent set of size $\frac{n}{2r}$. But for any r-colouring of H there must be a set of at least $\frac{|H|}{r} \geq \frac{n}{2r}$ nodes, all with the same colour. There are no independent sets in H of this size, so there cannot be such an r-colouring after all. Hence $\chi(H) > r$.

DEFINITION 14.6 For $r < \omega$, let Γ_r be the disjoint union $\bigcup_{r \leq s < \omega} G_s$, where G_s is a graph, as in theorem 14.5, with $\chi(G_s) > s$ and $g(G_s) > s$. So if $G_s = (V_s, E_s)$ for each s, the V_s being pairwise disjoint, then $\Gamma_r = (\bigcup_{r \leq s < \omega} V_s, \bigcup_{r \leq s < \omega} E_s)$.

14.2. SRAS is not an elementary class

COROLLARY 14.7 *For all $r < \omega$, Γ_r is an infinite graph with $g(\Gamma_r) > r$ and $\chi(\Gamma_r) = \infty$. Also, if D is any non-principal ultrafilter over ω then $\chi(\prod_D \Gamma_r) = 2$.*

Proof. The first part of the corollary is clear, from the definition of Γ_r. For the second part, since the girth of Γ_r is more than r, Γ_r does not satisfy the following sentence σ_i:

$$\exists x_0 \ldots \exists x_{i-1} \left(\bigwedge_{j<k<i} x_j \neq x_k \wedge \bigwedge_{j<i-1} E(x_j, x_{j+1}) \wedge E(x_{i-1}, x_0) \right)$$

for any i with $3 \leq i \leq r$. Hence $\Gamma_r \models \neg \sigma_i$, for $3 \leq i \leq r$.

So let Γ be any non-principal ultraproduct of the Γ_r. By Łoś theorem, $\Gamma \models \neg \sigma_i$ for all finite $i \geq 3$. Thus Γ has no cycles, which implies (see exercise 1 below) that Γ can be coloured with just two colours. Clearly, Γ has edges and hence has no 1-colouring, completing the proof. □

COROLLARY 14.8 *The class $\mathbf{N} = \{\Gamma : \chi(\Gamma) = \infty\}$ is not elementary.*

Proof. By corollary 14.7, $\Gamma_r \in \mathbf{N}$ for $r < \omega$, but $\prod_D \Gamma_r \notin \mathbf{N}$. This shows that $\mathbf{N}$ is not closed under ultraproducts and so by theorem 2.32 cannot be elementary. □

REMARK 14.9 It is easily seen (exercise 2 below) that $\mathbf{N}$ is closed under elementary equivalence, so (equivalently) is closed under ultraroots and ultrapowers. The complement $\{$graphs $\Gamma : \chi(\Gamma) < \infty\}$ of $\mathbf{N}$ is easily seen not to be closed under ultraproducts (e.g., of complete graphs of arbitrarily large finite size) so is not elementary either.

14.2.2 The construction

As usual with structures, we identify notationally a graph with its set of nodes (its domain).

DEFINITION 14.10 Let Γ be a graph. We define a relation algebra atom structure $\alpha(\Gamma)$ with atoms 1' (the sole identity atom) and r_i, b_i, g_i for $i \in \Gamma$. The non-identity atoms r_i, b_i, g_i are regarded as red, blue, and green, respectively. All atoms are self-converse. To define composition on atoms, we list the *forbidden triples* of atoms (see definition 3.25). They are:

$$(1', a, b), (a, 1', b), (a, b, 1') \quad \text{where } a \neq b, \qquad \text{(I)}$$
$$(r_i, r_j, r_k), (b_i, b_j, b_k), (g_i, g_j, g_k) \quad \text{where } \{i, j, k\} \subseteq \Gamma \text{ is independent.} \qquad \text{(II)}$$

Thus, all polychromatic and some monochromatic triples of atoms are consistent. It is easily checked that $\alpha(\Gamma)$ meets the conditions of being a relation algebra atom structure (definition 3.22). For example, the consistent triples are closed under permutations, and hence under Peircean transforms. Therefore, by lemma 3.24, $\mathfrak{Cm}(\alpha(\Gamma))$ is a relation algebra — a kind of Monk algebra.

DEFINITION 14.11 For a graph Γ and $X \subseteq \Gamma$, we define $\mathsf{R}_X, \mathsf{B}_X, \mathsf{G}_X \in \mathfrak{Cm}(\alpha(\Gamma))$ to be the sets of all red, blue, and green atoms, respectively, with indices in X: that is, $\mathsf{R}_X = \{r_i : i \in X\}$, etc. A non-zero element s of $\mathfrak{Cm}(\alpha(\Gamma))$ is said to be *monochromatic* if $s \leq 1$', $s \leq \mathsf{R}_\Gamma$, $s \leq \mathsf{B}_\Gamma$, or $s \leq \mathsf{G}_\Gamma$.

THEOREM 14.12 *If Γ is infinite and $\chi(\Gamma) < \infty$ then $\alpha(\Gamma)$ is not strongly representable.*

Proof. We show that $\mathfrak{Cm}(\alpha(\Gamma))$ is not representable. Assume for contradiction that $h : \mathfrak{Cm}(\alpha(\Gamma)) \to \mathcal{A}$ is an embedding from the complex algebra $\mathfrak{Cm}(\alpha(\Gamma))$ into a proper relation algebra $\mathcal{A}$ with base set X. Each $h(a)$ ($a \in \mathfrak{Cm}(\alpha(\Gamma))$) is a binary relation on X, and h respects the relation algebra operations. In particular,

(1) if $a, b \in \mathfrak{Cm}(\alpha(\Gamma))$ then $h(a+b) = h(a) \cup h(b)$,

(2) if $x, y \in X$ and $(x, y) \in h(1')$ then $x = y$,

(3) if $x_0, x_1, x_2 \in X$, $a, b, c \in \mathfrak{Cm}(\alpha(\Gamma))$, $(x_0, x_1) \in h(a)$, $(x_1, x_2) \in h(b)$, and $(x_0, x_2) \in h(c)$, then $(a;b) \cdot c \neq 0$.

We know that Γ has a finite colouring, so partition the nodes of Γ into sets C_j ($j < n$ for some finite n) such that there are no edges within any C_j. Let $\Pi = \{1', \mathsf{R}_{C_j}, \mathsf{B}_{C_j}, \mathsf{G}_{C_j} : j < n\}$. Then $\sum \Pi = 1$ in $\mathfrak{Cm}(\alpha(\Gamma))$. As Π is finite, repeated use of (1) shows that for any $x, y \in X$ there is $\mathsf{P} \in \Pi$ with $(x, y) \in h(\mathsf{P})$.

Since $\mathfrak{Cm}(\alpha(\Gamma))$ is infinite, X must also be infinite. By Ramsey's theorem (see the end of section 2.1), there are distinct $x_i \in X$ ($i < \omega$) and $\mathsf{P} \in \Pi$ such that $(x_i, x_j) \in h(\mathsf{P})$ for all $i < j < \omega$. By (2) above, $\mathsf{P} \neq 1'$. By (3), $(\mathsf{P};\mathsf{P}) \cdot \mathsf{P} \neq 0$. But because the C_j are independent sets and P is monochromatic, it is easily checked using rule (II) of the definition of $\alpha(\Gamma)$ that $(\mathsf{P};\mathsf{P}) \cdot \mathsf{P} = 0$. This contradiction completes the proof that $\mathfrak{Cm}(\alpha(\Gamma))$ is not representable. □

THEOREM 14.13 *If $\chi(\Gamma) = \infty$ then $\alpha(\Gamma)$ is strongly representable.*

Proof. We show that $\alpha(\Gamma)$ is strongly representable by showing that for all finite n and all non-zero $a_0 \in \mathfrak{Cm}(\alpha(\Gamma))$, $\forall$ does not have a winning strategy in $G_n(I_{a_0}, \mathfrak{Cm}(\alpha(\Gamma)))$. Here, G_n is the n-round representation game of definition 7.12, and I_{a_0} is the network with two nodes, x, y, with labelling $I_{a_0}(x, y) = a_0$, etc., as

14.2. SRAS is not an elementary class

defined in definition 7.18. Since finite-length games are determined (or by exercise 7.5(5)), $\exists$ has a winning strategy in each of these games, so by theorem 7.19, $\mathfrak{Cm}(\alpha(\Gamma))$ is representable.

Write α for $\alpha(\Gamma)$. Suppose for contradiction that $\forall$ has a winning strategy in some game $G_n(I_{a_0}, \mathfrak{Cm}\,\alpha)$ ($n < \omega$, $a_0 \neq 0$). By lemma 7.17, there is a finite set S of pairs of elements of $\mathfrak{Cm}\,\alpha$ such that his winning strategy only ever directs him to choose pairs of elements in S.

We define a finite boolean subalgebra $\mathcal{B}$ of the boolean reduct bool($\mathfrak{Cm}\,\alpha$) of $\mathfrak{Cm}\,\alpha$, in three steps. Start with the finite set

$$B_0 = \{1', a_0, \mathsf{R}_\Gamma, \mathsf{B}_\Gamma, \mathsf{G}_\Gamma\} \cup \{a, b, a;b : (a,b) \in W\}.$$

First, let B_1 be the finite boolean subalgebra of bool($\mathfrak{Cm}\,\alpha$) generated by B_0 (the set of all unions of intersections of members of B_0 and complements of members of B_0). Second, let $B_2 = B_1 \cup \{\mathsf{R}_X, \mathsf{B}_X, \mathsf{G}_X : \mathsf{R}_X \in B_1$ or $\mathsf{B}_X \in B_1$ or $\mathsf{G}_X \in B_1\}$ (copy monochromatic elements to all colours). Finally, let $\mathcal{B}$ be the boolean subalgebra of bool($\mathfrak{Cm}\,\alpha$) generated by B_2. Observe that:

- $1', a_0 \in \mathcal{B}$, and if $(a,b) \in S$ then $a, b, (a;b) \in \mathcal{B}$,
- each atom of $\mathcal{B}$ is monochromatic,
- for any $X \subseteq \Gamma$, the following are equivalent: $\mathsf{R}_X \in \mathcal{B}$, $\mathsf{B}_X \in \mathcal{B}$, $\mathsf{G}_X \in \mathcal{B}$.

A subset $X \subseteq \Gamma$ is said to be *small* if $\mathsf{R}_X \in \mathrm{At}\,\mathcal{B}$. Clearly, the small sets form a finite partition of Γ. Since there is no finite colouring of Γ, there must be some small set $X \subseteq \Gamma$ such that (j,k) is an edge of Γ for some $j, k \in X$. Then by rule (II) of the definition of $\alpha(\Gamma)$, (r_i, r_j, r_k), (b_i, r_j, r_k), and (g_i, r_j, r_k) are consistent triples of atoms of α for any $i \in \Gamma$. Also, $(1', r_j, r_j)$ is consistent. The same holds with b_j, b_k, and g_j, g_k. So by definition of complex algebras, we obtain

$$\mathsf{C}_X ; \mathsf{C}_X = 1 \text{ in } \mathfrak{Cm}\,\alpha, \text{ whenever C is R, B, or G.} \tag{14.1}$$

To obtain a contradiction, we will show how $\exists$ can win the game $G_n(I_{a_0}, \mathfrak{Cm}\,\alpha)$ when $\forall$ uses his supposedly winning strategy. As usual, we let N_t denote the pre-network in play after t rounds of the game, for each $t \leq n$. To help her win, $\exists$ will maintain an auxiliary $\mathfrak{Cm}\,\alpha$-network M_t ($t \leq n$) satisfying the following conditions:

- $M_t(e) \in \mathrm{At}\,\mathcal{B}$ for each edge e of M_t,
- if $x, y \in M_t$ then $M_t(x,y) = M_t(y,x)$, and $M_t(x,y) = 1'$ if and only if $x = y$,
- there is a map $' : \mathrm{nodes}(N_t) \to \mathrm{nodes}(M_t)$ such that $M_t(x', y') \leq N_t(x,y)$ for all $x, y \in \mathrm{nodes}(N_t)$.

The initial network is $N_0 = I_{a_0}$, with two nodes, x and y. If $1' \leq a_0$, $\exists$ defines M_0 to be the network with a single node, say x, with $M_0(x,x) = 1'$; she defines $x' = y' = x$. Otherwise, since $1' \in \operatorname{At} \mathfrak{Cm}\,\alpha$, we have $a_0 \leq -1'$. $\exists$ chooses an atom a of $\mathcal{B}$ with $a \leq a_0$, and defines M_0 to have nodes x, y only, with $M_0(x,y) = a$, $M_0(y,x) = \breve{a} = a \in \operatorname{At} \mathcal{B}$, and $M_0(x,x) = M_0(y,y) = 1'$. She lets $x' = x$ and $y' = y$. This all meets the conditions above.

Let $t < n$ and assume inductively that $\exists$ has defined M_t for N_t. We will show how $\exists$ can respond in the next, tth round of the game and how she can construct a new network M_{t+1} meeting the above conditions with respect to N_{t+1}. $\forall$ plays the round by choosing nodes $x, y \in N_t$ and elements $a, b \in \mathfrak{Cm}\,\alpha$ according to his strategy. Note that $a, b, a; b \in \mathcal{B}$. $\exists$ uses the label $M_t(x', y')$ to calculate her response. If $M_t(x', y') \leq -(a;b)$, then she rejects his move and plays N_{t+1} as specified in definition 7.12, with the same nodes as N_t. She defines $M_{t+1} = M_t$. This preserves the requirements on M, N.

Otherwise, since $M_t(x', y') \in \operatorname{At} \mathcal{B}$, we have $M_t(x', y') \leq a;b$. $\exists$ then accepts $\forall$'s move by playing a labelled graph N_{t+1} with a new node z and labelling it as specified in definition 7.12. She now faces the problem of defining M_{t+1} meeting the conditions above.

If there is already a node $p \in M_t$ with $M_t(x', p) \leq a$ and $M_t(p, y') \leq b$, then $\exists$ can set $M_{t+1} = M_t$ and $z' = p$. This meets the required conditions on N_{t+1}, M_{t+1}.

Otherwise, since $a;b \geq M_t(x', y') \in \operatorname{At} \mathcal{B}$ and $a, b \in \mathcal{B}$, by additivity of composition there must be $a^-, b^- \in \operatorname{At} \mathcal{B}$ with $a^- \leq a$, $b^- \leq b$, and $(a^-;b^-) \cdot M_t(x', y') \neq 0$. $\exists$ chooses such a^-, b^-; it can be checked using standard relation algebra properties that $a^-, b^- \neq 1'$, and that if $x' = y'$ then $a^- = b^-$. Note that a^-, b^- are monochromatic. Choose C from R, B, G of different colour from the colours of (atoms in) a^-, b^-. $\exists$ now refines M_t to $M_{t+1} \supseteq M_t$ by adding a new node, defining z' to be that new node, and labelling edges of M_{t+1} with atoms of $\mathcal{B}$ as follows:

- $M_{t+1}(x', z') = M_{t+1}(z', x') = a^-$,
  $M_{t+1}(z', z') = 1'$,
  $M_{t+1}(z', y') = M_{t+1}(y', z') = b^-$.

  This is well-defined if $x' = y'$, as $a^- = b^-$ in that case.

- For $p \in \operatorname{nodes}(M_t) \setminus \{x', y'\}$, $\exists$ labels $M_{t+1}(z', p) = M_{t+1}(p, z') = \mathsf{C}_X$, where X is as in (14.1).

We claim that every triangle (p, q, r) of nodes of the labelled graph M_{t+1} is consistent — i.e., $M_{t+1}(p, q) \cdot \bigl(M_{t+1}(p, r); M_{t+1}(r, q)\bigr) \neq 0$ in $\mathfrak{Cm}\,\alpha$. If $p, q, r \in M_t$, this is clear inductively, if $|\{p, q, r\}| < 3$ it is trivial, and by definition of α and $\mathfrak{Cm}\,\alpha$ (or by exercise 7.1(4)), the order of p, q, r is not significant, so we need only consider triangles of distinct nodes of the form (p, q, z').

1. We already know (x', y', z') to be consistent, since $(a^-;b^-) \cdot M_t(x', y') \neq 0$.

14.2. SRAS is not an elementary class

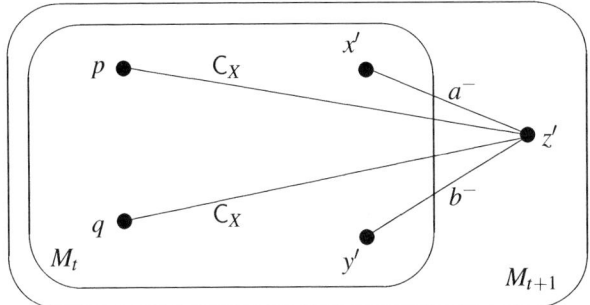

Figure 14.1: Labels in M_{t+1}

2. The labels in any triangle of the form (x', q, z') with $q \neq x', y'$ are c, C_X, a^- for some $c \in \text{At}\,\mathcal{B} \setminus \{1'\}$. Because C is not the colour of a^-, the triangle must be consistent. The case of (y', q, z') is similar.

3. Finally, if (p, q, z') is a triangle with $p, q \in M_t \setminus \{x, y\}$, we have $M_{t+1}(p, q) = c$ for some $c \in \text{At}\,\mathcal{B}$, and $M_{t+1}(p, z') = M_{t+1}(q, z') = C_X$. See figure 14.1. By (14.1) above, $c \leq C_X ; C_X$, so (p, q, z') is consistent.

This proves the claim.

By the claim, M_{t+1} is a network. We have arranged that every edge of M_{t+1} is labelled by an element of At $\mathcal{B}$ beneath the corresponding label in N_{t+1}. Thus, the inductive conditions on M_t are maintained for another round.

It follows that each labelled graph N_t ($t \leq n$) is a network. To see this, let $x, y, z \in N_t$. Then

$$N_t(x, z) ; N_t(z, y) \cdot N_t(x, y) \geq M_t(x', z') ; M_t(z', y') \cdot M_t(x', y') \neq 0.$$

Thus, $\exists$ never loses, which is a contradiction. $\square$

14.2.3 SRAS is not elementary

We can now prove theorem 14.3.

Proof. For $r < \omega$ let Γ_r be an infinite graph, as in definition 14.6, with $g(\Gamma_r) > r$ and $\chi(\Gamma_r) = \infty$ (corollary 14.7). Then the atom structure $\alpha(\Gamma_r)$ is strongly representable by theorem 14.13. Consider a non-principal ultraproduct $\prod_D \alpha(\Gamma_r)$. It is easily seen that this is isomorphic to $\alpha(\prod_D \Gamma_r)$. But, by corollary 14.7, $\prod_D \Gamma_r$ is 2-colourable, so by theorem 14.12, $\alpha(\prod_D \Gamma_r)$ is not strongly representable. Hence **SRAS** is not closed under ultraproducts, and so by theorem 2.32 is not elementary. $\square$

Exercises

1. Let Γ be a graph. Show that Γ has a 2-colouring iff it has no cycles of odd length.

2. Prove that the class **N** of non-finitely colourable graphs defined in corollary 14.8 is closed under elementary equivalence.

3. Prove theorem 14.3 and corollary 14.8 using compactness instead of ultraproducts.

4. In the context of the proof of theorem 14.13 above, check that $a^-, b^- \neq 1'$, and that if $x' = y'$ then $a^- = b^-$.

5. Prove that $\prod_D \alpha(\Gamma_r) \cong \alpha(\prod_D \Gamma_r)$ (in the proof of theorem 14.3 above).

6. Let Γ be a graph. Show that the following are equivalent:

   (a) Γ is not n-colourable for any finite n,

   (b) the boolean algebra $(\wp(\Gamma), \emptyset, \Gamma, \cup, \setminus)$ has a non-principal ultrafilter that does not contain any independent subset of Γ.

7. Prove theorem 14.13 using the game $G_\omega^a(\mathfrak{Cm}(\alpha(\Gamma))^+)$ and exercise 6.

8. Let Γ be an infinite graph and $\mathcal{A}$ a subalgebra of $\mathfrak{Cm}(\alpha(\Gamma))$ such that $\alpha(\Gamma) \subseteq \mathcal{A}$ and $\mathsf{R}_\Gamma, \mathsf{B}_\Gamma, \mathsf{G}_\Gamma \in \mathcal{A}$. Suppose that whenever $E_0, \ldots, E_n$ is a finite partition of Γ and $\mathsf{R}_{E_i} \in \mathcal{A}$ for all $i \leq n$, then some E_i is not independent, and similarly for G, B instead of R. Show that $\mathcal{A}$ is representable.

9. Use theorems 14.12 and 14.13 and ultraproducts to prove Monk's theorem [Mon64] that **RRA** is not finitely axiomatisable.

14.3 Consequences of the theorem

Here we sketch some corollaries of theorem 14.3.

14.3.1 Closure properties

These are straightforward to prove.

COROLLARY 14.14 **SRAS** *is not closed under elementary equivalence, ultrapowers, nor ultraproducts.*

14.3. Consequences of the theorem

Proof. Recall that **SRAS** = Str **RRA**. By proposition 2.90 [Gol89, theorem 3.8.4], **SRAS** is elementary iff it is closed under elementary equivalence, iff it is closed under ultrapowers, iff it is closed under ultraproducts. By theorem 14.3, **SRAS** has none of these closure properties. □

COROLLARY 14.15 **SRAS** ⊂ **WRAS**. *Hence,*

1. *there exist two atomic relation algebras with the same atom structure, only one of which is representable,*

2. **RRA** *is not closed under completions, and is not atom-canonical,*

3. *there exists a non-representable relation algebra with a dense representable subalgebra,*

4. *[Ven97b]* **RRA** *is not Sahlqvist-axiomatisable.*

Proof. By the definitions, **SRAS** ⊆ **WRAS**. Since **WRAS** = At **RRA** is elementary (theorem 2.84), and by theorem 14.3, **SRAS** is not, it follows that the inclusion is strict. (See exercise 8 below for a specific atom structure in **WRAS** \ **SRAS**.)

1. A trivial consequence of **SRAS** ⊂ **WRAS**.

2. Let $S \in$ **WRAS** \ **SRAS**. Since S is weakly representable, there exists a representable atomic relation algebra $\mathcal{A}$ with At $\mathcal{A} = S$. Since S is not strongly representable, it follows from lemma 14.2 that $\mathfrak{Cm}\,S$ is not representable. By remark 2.67, $\mathfrak{Cm}\,S$ is the completion of $\mathcal{A}$, and clearly, At $\mathfrak{Cm}\,S =$ At $\mathcal{A} = S$.

3. Since a relation algebra is by definition dense in its completion, this follows from part 2.

4. That **RRA** is not a Sahlqvist variety now follows from Venema's theorem 2.96.
□

14.3.2 Related classes

We now examine the connections between **SRAS** and **WRAS** and other classes of atom structures.

DEFINITION 14.16 We define three more classes of atom structures.

1. **CRAS** denotes the class of relation algebra atom structures S such that some atomic relation algebra with atom structure S has a complete representation. We will call such atom structures *completely representable*.

2. Recall the Lyndon conditions, discussed in section 11.3.2. Since all quantifiers in the Lyndon conditions are relativised to atoms, the conditions can be equally regarded as conditions on an atom structure, and indeed they can be translated into 'equivalent' first-order sentences in the signature of relation algebra atom structures. If $\mathcal{A}, \mathcal{B}$ are atomic relation algebras with the same atom structure, $\mathcal{A}$ satisfies the Lyndon conditions iff $\mathcal{B}$ does. We will therefore say even-handedly that an atom structure or a relation algebra satisfies the Lyndon conditions, whichever is more convenient.

LCAS denotes the class of all relation algebra atom structures satisfying the Lyndon conditions. Equivalently, by theorem 11.4, **LCAS** denotes the class of all relation algebra atom structures S such that $\exists$ has a winning strategy in $G_n^a(\mathfrak{Cm}\,S)$ for each $n < \omega$.

3. For a relation algebra atom structure S, the *first-order algebra over* S is the subalgebra of $\mathfrak{Cm}\,S$ consisting of all sets of atoms that are first-order definable with parameters in S — that is, sets of the form $\{a \in S : S \models \varphi(a, \bar{b})\}$ for some first-order formula $\varphi(x, \bar{y})$ of the signature of S and some tuple $\bar{b}$ of elements of S. By exercise 2.7(20), these sets indeed form a subalgebra of $\mathfrak{Cm}\,S$.

FOAS denotes the class of relation algebra atom structures whose first-order algebra is representable.[1]

We will prove that

THEOREM 14.17 **CRAS** $\subset$ **LCAS** $\subset$ **SRAS** $\subset$ **FOAS** $\subset$ **WRAS**, *the elementary classes being underlined.*

Proof (sketch). It is convenient to state this theorem here, though one part of the proof will rely on a result to be proved later, in chapter 17. We will use a few facts about first-order interpretations: see section 2.1 and [Hod93] for details of these.

First, we show that **CRAS** $\subseteq$ **LCAS** $\subseteq$ **SRAS** $\subseteq$ **FOAS** $\subseteq$ **WRAS**. For the first two inclusions we use theorem 11.10(1), which states that any completely representable atomic relation algebra satisfies the Lyndon conditions, and that any atomic relation algebra satisfying the Lyndon conditions is representable. The first inclusion is now immediate. The second inclusion, **LCAS** $\subseteq$ **SRAS**, also follows: if $S \in$ **LCAS** then $\mathfrak{Cm}\,S$ satisfies the Lyndon conditions, so is representable; by lemma 14.2, $S \in$ **SRAS**. The inclusions **SRAS** $\subseteq$ **FOAS** $\subseteq$ **WRAS** are immediate from the definitions.

Next we prove that the inclusions are proper. We have **CRAS** $\subset$ **LCAS** because **LCAS** is by definition elementary, while by corollary 17.7 later on, **CRAS**

---
[1] We thank Rob Goldblatt for suggesting this as an example of an elementary class lying strictly between **SRAS** and **WRAS**.

14.3. Consequences of the theorem

is not. The inclusion **LCAS** $\subset$ **SRAS** is proper because **LCAS** is elementary while by theorem 14.3, **SRAS** is not. For an explicit example of an atom structure in **SRAS** \ **LCAS**, take a graph Γ with $\chi(\Gamma) = \infty$ and $g(\Gamma) > 15$ (cf. corollary 14.7). Theorem 14.13 shows that $\alpha(\Gamma) \in$ **SRAS**. But it follows from theorem 11.4 and the claim below that $\alpha(\Gamma)$ fails all but finitely many Lyndon conditions. So $\alpha(\Gamma) \in$ **SRAS** \ **LCAS**.

Claim. Let n be large enough so that any 3-colouring of the edges of a complete graph of size n must contain six nodes such that all the edges between the six have the same colour. (Such an n exists by Ramsey's theorem.) Then $\exists$ has no winning strategy in the game $G_n^a(\mathfrak{Cm}(\alpha(\Gamma)))$ of definition 11.3.

Proof of claim. Assume for contradiction that she has, and let $\forall$ play simply so as to create a strict atomic network N with n nodes. Choose a set X of six nodes of N such that the colour of $N(x,y)$ for distinct $x,y \in X$ is constant — say, r. For distinct $x, y \in X$, let $v(x,y) \in \Gamma$ be the node such that $N(x,y) = N(y,x) = r_{v(x,y)}$. Let Δ be the induced subgraph of Γ with nodes $\{v(x,y) : x,y \in X, x \neq y\}$. Since $|\Delta| \leq 15$ and $g(\Gamma) > 15$, Δ is 2-colourable and we can partition its nodes into independent sets D_0, D_1. By exercise 7 below, any 2-colouring of the edges of a complete graph of size 6 has a monochromatic triangle, so there are distinct $x,y,z \in X$ and $d < 2$ such that $v(x,y), v(y,z), v(x,z) \in D_d$. So $\{v(x,y), v(y,z), v(z,x)\}$ is an independent set in Γ. But then, $(N(x,y), N(y,z), N(z,x)) = (r_{v(x,y)}, r_{v(y,z)}, r_{v(z,x)})$ is inconsistent and $\exists$ lost the game, a contradiction. This proves the claim.

In section 14.4 we'll give another construction (due to Maddux) of a strongly representable atom structure that fails to satisfy the Lyndon conditions, showing again that **LCAS** $\subset$ **SRAS**.

FOAS can be shown to be elementary by Venema's argument in theorem 2.84 [Ven97a], or by using theorem 2.32. Since **SRAS** is not elementary, the inclusion **SRAS** $\subset$ **FOAS** is proper. Indeed, regarding $\mathbb{Z}$ as a graph whose edges are all pairs of consecutive nodes, an Ehrenfeucht–Fraïssé game argument will establish that the atom structure $\alpha(\mathbb{Z})$ is in **FOAS** \ **SRAS**. See also exercise 8 below.

To see that the inclusion **FOAS** $\subset$ **WRAS** is proper, let Γ consist of ω disjoint copies of the three-node graph with nodes 0, 1, 2, and edges 0–1 and 1–2 only. As Γ has a first-order-definable 2-colouring (nodes can be coloured according to their degree: 1 or 2), and Γ is interpretable in $\alpha(\Gamma)$, the argument of theorem 14.12 shows that the first-order algebra over $\alpha(\Gamma)$ is not representable. Hence, $\alpha(\Gamma) \notin$ **FOAS**. However, it can be checked that the term algebra over $\alpha(\Gamma)$ (see lemma 14.2) consists of all finite and cofinite subsets of $\alpha(\Gamma)$, which can be shown to be representable by exercise 14.2(8). So $\alpha(\Gamma) \in$ **WRAS**.

The proof above shows that the only elementary classes in the formulation of the theorem are those underlined. $\square$

REMARK 14.18 If S is a finite relation algebra atom structure, then $S \in$ **WRAS** iff $S \in$ **CRAS**. So membership of all five classes in theorem 14.17 is equivalent for finite atom structures.

PROBLEM 14.19 If S, S' are elementarily equivalent relation algebra atom structures, must the term algebras of S and S' also be elementarily equivalent?

PROBLEM 14.20 For which $\alpha \geq 3$, if any, is $\operatorname{Str}\mathbf{RCA}_\alpha$ elementary?

Exercises

1. Check remark 14.18. Generalise it to atom structures S of the form $\operatorname{At} \mathcal{A}^+$ for a relation algebra $\mathcal{A}$.

2. Show that a relation algebra atom structure S is in **CRAS** iff every atomic relation algebra $\mathcal{A}$ with $\operatorname{At} \mathcal{A} = S$ is completely representable. (Cf. exercise 3.4(4).)

3. Show that the class of weakly but non-strongly representable atom structures is not elementary.

4. Show that **SRAS** is closed under ultraroots, and that **CRAS** is closed under ultraproducts.

5. [Venema] Show directly that **WRAS** is closed under ultraproducts and ultraroots (and hence elementary). Show also that it is pseudo-elementary (with a transparent defining theory), and hence can be explicitly axiomatised (though the method of theorem 2.84 is much more direct).

6. Show that if S, S' are $L_{\infty\omega}$-equivalent relation algebra atom structures, then the term algebras of S and S' are $L_{\infty\omega}$-equivalent. See [Hod93] for information about $L_{\infty\omega}$.

7. Show that any 2-colouring of the *edges* of the complete graph of size 6 has a monochromatic triangle.

8. As in the proof of theorem 14.17, regard $\mathbb{Z}$ as a graph with set of nodes $\mathbb{Z}$ and set of edges $\{(n, n+1) : n \in \mathbb{Z}\}$. Prove that $\alpha(\mathbb{Z})$ is weakly but not strongly representable. [The first half is harder: you will have to calculate the term algebra over $\alpha(\mathbb{Z})$. Exercise 14.2(8) may then help.]

9. Check the details of the proof above that **SRAS** $\subset$ **FOAS** $\subset$ **WRAS**.

14.4 Maddux's construction

An earlier construction of strongly but not completely representable atom structures was given some time ago by Maddux [Madd78b, examples 23, p. 154ff]. Below, we have simplified them slightly. The main results are presented only in outline, through structured exercises. For $6 \leq q < \omega$, these atom structures $\mathcal{X}_q$ have the following properties:

- $\mathcal{X}_q$ is an infinite atom structure.
- $\mathcal{X}_q$ is strongly representable.
- However, $\forall$ has a winning strategy in $G_q^a(\mathcal{X}_q)$, (recall from definition 11.3 that we write $G_q^a(\mathcal{X}_q)$ instead of $G_q^a(\mathfrak{Cm}\,\mathcal{X}_q)$, as the game depends only on the atom structure). So $\mathcal{X}_q$ fails all but finitely many Lyndon conditions.
- Consequently, although it is strongly representable, there is no completely representable relation algebra with atom structure equal to, or even elementarily equivalent to, $\mathcal{X}_q$ (see theorem 11.10).

Thus, the 'Maddux algebra' $\mathfrak{Cm}\,\mathcal{X}_q$ is representable but fails infinitely many Lyndon conditions. The algebras $\mathfrak{Cm}\,\mathcal{X}_q$ were not the first-constructed examples with this property: a rather complicated algebra defined in [Lyn50] is representable but fails some Lyndon conditions, though the implications of this were not clear at the time of its publication. At any rate, the $\mathcal{X}_q$ show that $\exists$'s having a winning strategy in the games $G^a(\mathcal{X}_q)$ is not a necessary condition for $\mathfrak{Cm}\,\mathcal{X}_q$ to be representable.

Nonetheless, a simple way to show that the $\mathfrak{Cm}\,\mathcal{X}_q$ are in fact representable is by exhibiting winning strategies for $\exists$ in the games $G_n^a(\mathfrak{Cm}\,\mathcal{X}_q^+)$ on the canonical embedding algebra of $\mathfrak{Cm}\,\mathcal{X}_q$, for all finite n. In fact, we may as well deal with $G_\omega^a(\mathfrak{Cm}\,\mathcal{X}_q^+)$: exercise 11.5(1) shows that $\mathfrak{Cm}\,\mathcal{X}_q$ is representable iff $\exists$ has such a strategy.

14.4.1 The atom structures

Fix q with $6 \leq q < \omega$. The atoms of $\mathcal{X}_q$ are:

1. $e_i\ (i < q)$
2. $d_{ij}^{lr}\ (r < \omega,\ i,j,l < q,\ l \neq i,j)$.

The identity relation R_1, is defined to be $\{e_i : i < q\}$. The d_{ij}^{lr} will be the diversity atoms (not in R_1,). i and j are called the start and end indices, respectively, l is called the middle index, and r the rank. In e_i, i is called both the start and the end index.

We must also specify conversion:

3. each e_i is self-converse ($\breve{e}_i = e_i$)
4. $(d_{ij}^{lr})^\smile = d_{ji}^{lr}$.

Finally, we specify composition by listing the consistent triples. All Peircean transforms of the following triples of atoms are consistent, no other triples of atoms are consistent:

5. $(e_i, e_i, e_i), (d_{ij}^{lr}, e_j, d_{ji}^{lr})$
6. $(d_{ij}^{lr}, d_{jk}^{ms}, d_{ki}^{nt})$, where $r > s, t$ and $l \neq k$.

Since the start and end indices in consistent triples must match, we can define the 'induced index' of any node in an atomic network. This index will be the end index of the label of any incoming arc and the start index of the label of any outgoing arc, including the reflexive arc. The crucial point about the definition of composition is that a triple of the form

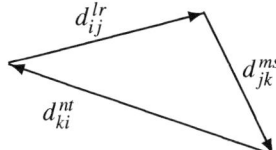

is consistent iff there is a d-atom, say d_{ij}^{lr}, among the three that has maximum rank (strictly larger than the others) and its middle index l is different from the induced index, k, of the node opposite.

REMARK 14.21 In the atoms d_{ij}^{lr}, we required that $l \neq i, j$. This requirement can be dropped at the cost of replacing G_{q-1}^a by G_{3q-1}^a in exercise 2 below.

Exercises

1. Show that $\mathcal{X}_q$ is a relation algebra atom structure (see definition 3.22). The key point is to check the associativity axiom, which can be done by showing that $\exists$ has a winning strategy in $G_4^a(\mathcal{X}_q)$.

2. Show, for $q \geq 6$, that $\mathcal{X}_q$ fails some of the Lyndon conditions. This can be done by demonstrating a winning strategy for $\forall$ in the $(q-1)$-round atomic game $G_{q-1}^a(\mathcal{X}_q)$. Hint: let $\forall$ use his q moves to ensure that exactly one node is generated of every possible induced index $< q$. Let the final labelled atomic graph of the play be N. Then consider an edge of N labelled by a diversity atom of the form d_{ij}^{lr}, with maximal rank r. Let $z \in N$ have induced index l; z exists because $\forall$ played as he did. Note that x, y, z are distinct. Consider the triangle x, y, z (figure 14.2).

14.4. Maddux's construction

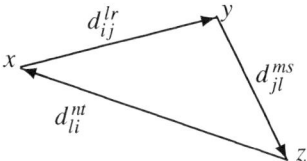

Figure 14.2: An inconsistent triangle in N

14.4.2 $\mathcal{X}_q$ is strongly representable

It remains to prove:

LEMMA 14.22 $\mathfrak{Cm}\,\mathcal{X}_q$ *is representable.*

Proof. We will prove the lemma by showing that $\exists$ has a winning strategy in the atomic network game $G_\omega^a((\mathfrak{Cm}\,\mathcal{X}_q)^+)$. Here, $(\mathfrak{Cm}\,\mathcal{X}_q)^+$ is the canonical extension of $\mathfrak{Cm}\,\mathcal{X}_q$. By exercise 11.5(1), it follows that $\mathfrak{Cm}\,\mathcal{X}_q$ is representable.

Recall that $(\mathfrak{Cm}\,\mathcal{X}_q)^+$ is the complete atomic relation algebra with atom structure $\mathrm{At}((\mathfrak{Cm}\,\mathcal{X}_q)^+)$ obtained from the set of all ultrafilters of $\mathfrak{Cm}\,\mathcal{X}_q$. So the game $G_\omega^a((\mathfrak{Cm}\,\mathcal{X}_q)^+)$ is played on networks in which the edges are labelled by ultrafilters of $\mathfrak{Cm}\,\mathcal{X}_q$.

The ultrafilters of $\mathfrak{Cm}\,\mathcal{X}_q$ For $i,j,l < q$ with $l \neq i,j$, and $X \subseteq \mathcal{X}_q$ (i.e., $X \in \mathfrak{Cm}\,\mathcal{X}_q$), define

$$X(i,j,l) = \{r < \omega : d_{ij}^{lr} \in X\}.$$

Consider an ultrafilter φ of $\mathfrak{Cm}\,\mathcal{X}_q$. If φ is principal, we can regard it as an atom of $\mathfrak{Cm}\,\mathcal{X}_q$, i.e. $\varphi \in \mathcal{X}_q$. In any event, using exercise 2.3(12) we see that either it contains 1' and hence contains some identity atom e_i, or else it contains exactly one of the finitely many sets $D_{ij}^l = \{d_{ij}^{lr} : r < \omega\}$, for $i,j,l < q$, $l \neq i,j$, in which case

$$\varphi^* = \{X(i,j,l) : X \in \varphi\}$$

is an ultrafilter on ω (that is, an ultrafilter of the boolean algebra $\mathcal{P}(\omega)$ of exercise 2.3(1)). Conversely, for any $i,j,l < q$ with $l \neq i,j$, any ultrafilter f on ω gives rise to the ultrafilter

$$f_{ij}^l = \{X \subseteq \mathcal{X}_q : X(i,j,l) \in f\}$$

of $\mathfrak{Cm}\,\mathcal{X}_q$ containing D^l_{ij}. Clearly, if $D^l_{ij} \in \varphi$ then $(\varphi^*)^l_{ij} = \varphi$, and $(f^l_{ij})^* = f$, so the correspondence is one-one. We conclude that the ultrafilters of $\mathfrak{Cm}\,\mathcal{X}_q$ are precisely the principal ones (generated by) e_i for $i < q$, and those of the form f^l_{ij} for $i,j,l < q$, $l \neq i,j$, f an ultrafilter on ω.

The atom structure of $\mathfrak{Cm}\,\mathcal{X}_q^+$ We now consider which triples of ultrafilters of $\mathfrak{Cm}\,\mathcal{X}_q$ form a consistent triple of atoms of $\mathfrak{Cm}\,\mathcal{X}_q^+$.

Exercises

1. Use lemma 3.26 and the definition of consistent triples of atoms in $\mathcal{X}_q$ to show that whenever $i,j,k,l,m,n < q$ satisfy $l \neq i,j$, $m \neq j,k$, $n \neq i,k$, and f,g,h are ultrafilters on ω, f is non-principal, and $l \neq k$, then the triple of atoms $(f^l_{ij}, g^m_{jk}, h^n_{ki})$ is consistent in $\mathfrak{Cm}\,\mathcal{X}_q^+$.

2. Describe a winning strategy for $\exists$ in $G^a_\omega(\mathfrak{Cm}\,\mathcal{X}_q^+)$. Explain why this strategy should win. [Use the previous exercise to find suitable ultrafilters for $\exists$ to label edges with. The restriction $q \geq 6$ will be needed when choosing suitable middle indices.]

From the last exercise, we see that $\mathfrak{Cm}\,\mathcal{X}_q^+$ and hence $\mathfrak{Cm}\,\mathcal{X}_q$ are representable, as required. □

Chapter 15

Non-finite axiomatisability of $S\mathfrak{R}aCA_{n+1}$ over $S\mathfrak{R}aCA_n$

15.1 Outline of chapter

Here, we continue our investigation into the connections between cylindric algebras of various finite dimensions, and relation algebras. In chapter 12, we generalised Maddux's definition of a cylindric basis by introducing the notion of an n-dimensional hyperbasis for an atomic relation algebra. This was used in chapter 13 to characterise $S\mathfrak{R}aCA_n$ for finite $n \geq 4$ by hyperbases and by n-flat and n-smooth relativised representations. We know by proposition 5.48 that $S\mathfrak{R}aCA_n$ is a variety. By example 9.2(8), it is pseudo-universal, a fact used to obtain an equational, recursive axiomatisation of $S\mathfrak{R}aCA_n$ in theorem 13.55. Although we did not include it, and we will not use it here, a similar recursive axiomatisation could be obtained for the class $S\mathfrak{Nr}_mCA_n$ of subalgebras of m-dimensional neat reducts of n-dimensional cylindric algebras, for $3 \leq m \leq n < \omega$, and perhaps a relativised semantics for the algebras in this class.

The classes $S\mathfrak{R}aCA_n$ and $S\mathfrak{Nr}_mCA_n$ are important approximations to **RRA** and **RCA**$_m$, and have long been studied. In this chapter, we are concerned with the way they vary with the dimension. By exercise 5.6(1), we have

$$S\mathfrak{R}aCA_3 \supseteq S\mathfrak{R}aCA_4 \supseteq S\mathfrak{R}aCA_5 \supseteq \cdots.$$

In the same way, it can be shown that

$$CA_m = S\mathfrak{Nr}_mCA_m \supseteq S\mathfrak{Nr}_mCA_{m+1} \supseteq S\mathfrak{Nr}_mCA_{m+2} \supseteq \cdots.$$

In this chapter, we use the results of chapter 13 to prove:

THEOREM 15.1

1. Each of the inclusions $\mathbf{S\Re aCA}_3 \supset \mathbf{S\Re aCA}_4 \supset \cdots$ is strict.

2. All but the first inclusion above cannot be finitely axiomatised. I.e., for $4 \leq n < \omega$, there is no first-order sentence σ of the language of relation algebras such that for all $\mathcal{A} \in \mathbf{S\Re aCA}_n$ we have $\mathcal{A} \models \sigma \iff \mathcal{A} \in \mathbf{S\Re aCA}_{n+1}$.

Let $3 \leq m < \omega$.

3. Each of the inclusions $\mathbf{CA}_m = \mathbf{S\Re t}_m \mathbf{CA}_m \supset \mathbf{S\Re t}_m \mathbf{CA}_{m+1} \supset \cdots$ is strict.

4. All but the first inclusion in (3) cannot be finitely axiomatised.

Consider the first part of the theorem first. It has been known for some time that some of the inclusions are strict. The first inclusion is certainly strict, because if $\mathfrak{C}$ is a 3-dimensional cylindric algebra, $\Re a\mathfrak{C}$ is not necessarily associative (see fact 5.46), whereas if $\mathfrak{C}$ is a 4-dimensional cylindric algebra then $\Re a\mathfrak{C}$ is a relation algebra and so is definitely associative (see theorem 5.44 or [Madd91a, pp 377–378]; in fact, exercise 12.5(11) showed that $\mathbf{S\Re aCA}_4 = \mathbf{RA}$, the class of all relation algebras). That $\mathbf{S\Re aCA}_4 \supset \mathbf{S\Re aCA}_5$ was also known: see [TarGiv87, p. 93]. Maddux proved in [Madd91c] that $\mathbf{S\Re aCA}_n \neq \mathbf{S\Re aCA}_{3n-7}$ for all finite $n \geq 4$, so that infinitely many of the inclusions are strict.

Part 1 of the theorem was first proved in [HirHod+02d]. Here, we prove it as a consequence of the proof of part 2. A corollary of part 1 is that for $3 \leq n < \omega$, there exists a first-order 3-variable formula ϕ_n such that $\vdash_{n+1} \phi_n$ but $\nvdash_n \phi_n$, for a certain n-variable proof system $\vdash_n$ of [TarGiv87]. See [HirHod+02c] for an explicit construction of such a ϕ_n. A similar result was proved in [Gor00]. We will return to this in section 15.7.

Recall from theorem 13.46 that $\mathcal{A} \in \mathbf{S\Re aCA}_n$ if and only if $\mathcal{A}^+$ has an n-dimensional hyperbasis. To prove part 2 of theorem 15.1, we will use the two-player game $G_r^{m,n}(\mathcal{A}, \Lambda)$ of definition 12.26. This game constructs an approximation to an n-dimensional Λ-hyperbasis for an atomic relation algebra $\mathcal{A}$ and tests its closure properties (amalgamation, etc) r times. For countable $\mathcal{A}$, if it survives ω tests it is a genuine hyperbasis.

Then, for each $n \geq 4$, we will construct finite 'Monk' relation algebras $\mathfrak{A}(n,r)$ ($1 \leq r < \omega$) which have an approximation to an $(n+1)$-dimensional hyperbasis that survives r tests of this kind, but which have no genuine hyperbasis. So the $\mathfrak{A}(n,r)$ are never actually in $\mathbf{S\Re aCA}_{n+1}$ (for finite algebras, this is guaranteed by the lack of a hyperbasis), but as r increases, they get closer and closer to it. It follows that an ultraproduct of them *is* in $\mathbf{S\Re aCA}_{n+1}$, so by Łoś' theorem, $\mathbf{S\Re aCA}_{n+1}$ is not finitely axiomatisable. We will even show that $\mathfrak{A}(n,r) \in \mathbf{S\Re aCA}_n \cap \mathbf{RA}_{n+1}$ for all r, so that $\mathbf{S\Re aCA}_{n+1}$ is not finitely axiomatisable over $\mathbf{S\Re aCA}_n$, nor over $\mathbf{RA}_{n+1}$.

15.1. Outline of chapter

For part 3 of theorem 15.1, note that Monk proved [Mon69, theorem 1.11] that $\mathbf{S\mathfrak{N}r}_m\mathbf{CA}_n \neq \mathbf{RCA}_m$ whenever $3 \leq m < n < \omega$; since $\mathbf{RCA}_m = \bigcap_{m<n<\omega} \mathbf{S\mathfrak{N}r}_m\mathbf{CA}_n$, this implies that infinitely many of the inclusions in (3) are strict. Andréka proved [And90a] that $\mathbf{S\mathfrak{N}r}_m\mathbf{CA}_n \neq \mathbf{S\mathfrak{N}r}_m\mathbf{CA}_{n+1}$ whenever $3 \leq m \leq n < 2m < \omega$. We could prove part 3 as a fairly easy corollary of part 1 (see exercise 1), but we cannot see how to deduce part 4 from part 2 so instead we focus on the stronger part 4 of the theorem.

Some existing results in the literature are closely related to part 4. [And97a, theorem 2] proves that for $n \geq m+2$, $\mathbf{S\mathfrak{N}r}_m\mathbf{CA}_n$ is not finitely axiomatisable and cannot be axiomatised at all with a set of prenex universal sentences using only a bounded number of variables. (Of course, this does not imply even part 3 of theorem 15.1, as the $\mathbf{S\mathfrak{N}r}_m\mathbf{CA}_n$ for various n might still be equal.) Sayed Ahmed proves in [Say01] that $\mathfrak{N}r_m\mathbf{CA}_n$ is not elementary, for $1 < m < n$; this is not implied by theorem 15.1. Here, we prove part 4: for $3 \leq m < n < \omega$, there is no first-order sentence $\rho_{m,n}$ such that for all $C \in \mathbf{S\mathfrak{N}r}_m\mathbf{CA}_n$, we have $C \models \rho_{m,n} \iff C \in \mathbf{S\mathfrak{N}r}_m\mathbf{CA}_{n+1}$. Interestingly, the inequality $m < n$ is indispensable here: Andréka shows that $\mathbf{S\mathfrak{N}r}_m\mathbf{CA}_{m+1}$ *is* finitely axiomatisable [And90a]. We use the restriction $m < n$ only once in our proof.

Our proof is again by games. The reader may recall that the game $G_r^{m,n}$ is designed to test a rather stronger property than the simple existence of a hyperbasis: it tests the existence of an n-dimensional hyperbasis $\mathcal{H}$ such that any n-wide m-dimensional hypernetwork M (see definition 12.21) extends to a hypernetwork in $\mathcal{H}$. (Here we assume a fixed set of hyperlabels: say, ω.) This is helpful in obtaining the non-finite axiomatisability result about neat reducts. For $3 \leq m \leq n < \omega$, $0 < r < \omega$, we can define an m-dimensional cylindric algebra C_r by taking as atom structure the set of all $(n+1)$-wide m-dimensional hypernetworks over $\mathfrak{A}(n,r)$. Much as for $\mathbf{S\mathfrak{R}a CA}_n$ in part 2, we show that each $C_r \in \mathbf{S\mathfrak{N}r}_m\mathbf{CA}_n \setminus \mathbf{S\mathfrak{N}r}_m\mathbf{CA}_{n+1}$, but for $m < n$, a non-principal ultraproduct of the C_r is in $\mathbf{S\mathfrak{N}r}_m\mathbf{CA}_{n+1}$. It follows, for $3 \leq m < n < \omega$, that $\mathbf{S\mathfrak{N}r}_m\mathbf{CA}_{n+1}$ is not finitely axiomatisable over $\mathbf{S\mathfrak{N}r}_m\mathbf{CA}_n$.

In section 15.7, we discuss an application of part 4 in finite variable proof theory. In [HenMon+85, AndNém+01], a certain natural Hilbert system $\vdash_{m,n}$ is given, to prove m-variable formulas in an m-ary relational signature. Proofs in $\vdash_{m,n}$ can use up to n variables. With some restrictions, an m-variable formula φ corresponds in a natural way to a $\mathbf{CA}_m$-term $\widehat{\varphi}$, and it is shown that $\vdash_{m,n} \varphi$ iff $\mathbf{S\mathfrak{N}r}_m\mathbf{CA}_n \models \widehat{\varphi} = 1$. Because $\mathbf{S\mathfrak{N}r}_m\mathbf{CA}_{n+1} \subset \mathbf{S\mathfrak{N}r}_m\mathbf{CA}_n$, it follows that $\vdash_{m,n+1}$ is strictly stronger than $\vdash_{m,n}$. Because for $3 \leq m < n < \omega$, $\mathbf{S\mathfrak{N}r}_m\mathbf{CA}_{n+1}$ is not finitely axiomatisable over $\mathbf{S\mathfrak{N}r}_m\mathbf{CA}_n$, it follows that for any finite set Σ of m-variable schemata, the system '$\Sigma \vdash_{m,n}$' with all m-variable instances of schemata in Σ as additional axioms of $\vdash_{m,n}$ cannot prove the same theorems as $\vdash_{m,n+1}$. See theorem 15.17.

Exercises

1. [HirHod+02c] Show that part 1 of theorem 15.1 implies part 3.

15.2 The algebras $\mathfrak{A}(n,r)$ and C_r

DEFINITION 15.2 Let $3 \leq n < \omega$ and $r < \omega$, and let Ω satisfy $n, r \ll \Omega < \omega$ — say, $\Omega = (nr)^{nr}$. $\mathfrak{A}(n,r)$ is the finite non-associative algebra (a kind of Monk algebra) defined as follows. As it is finite, it is isomorphic to the complex algebra over its atom structure, so it suffices to specify this atom structure.

- The atoms of $\mathfrak{A}(n,r)$ are 1' (the sole identity atom) and $a^k(i,j)$, for each $i < n-1$, $j < r$, and $k < \Omega$.

- All atoms are self-converse.

- It remains to list the *forbidden triples* (a,b,c) of atoms of $\mathfrak{A}(n,r)$ — those such that $a \cdot (b;c) = 0$. This defines composition: for $x, y \in \mathfrak{A}(n,r)$ we have

$$x; y = \{a \in \operatorname{At}\mathfrak{A}(n,r) : \exists b, c \in \operatorname{At}\mathfrak{A}(n,r),\ b \leq x,\ c \leq y, \\ (a,b,c) \text{ is not inconsistent}\}.$$

Any permutation of the triple $(1', s, t)$ will certainly be inconsistent unless $t = s$. Also, all permutations of the following triples are inconsistent:

$$(a^k(i,j), a^{k'}(i,j), a^{k''}(i,j')),$$

if $j \leq j' < r$, where $i < n-1$ and $k, k', k'' < \Omega$ are arbitrary. That is, a triple of a-atoms with all i-indices the same and two js the same must (for consistency) have the third j strictly less than the other two.

All other triples of atoms are consistent.

This completes the definition of $\mathfrak{A}(n,r)$. It is trivial to check that the atom structure just defined satisfies all the provisions of definition 3.22 except perhaps the associativity axiom, so that $\mathfrak{A}(n,r) \in \mathbf{NA}$. We'll see in corollary 15.6 that for $n \geq 4$, $\mathfrak{A}(n,r)$ is a relation algebra. Of course, $\mathfrak{A}(n,0)$ has no atoms except 1' and so is the trivial representable relation algebra with domain $\{0,1\}$, where $1' = 1$. The superscripts k don't play much of a role yet, but to prove that $\mathfrak{A}(n,r) \notin \mathbf{S\mathfrak{R}aCA}_{n+1}$ (theorem 15.8) we will need 'many' atoms.

Note that as $\mathfrak{A}(n,r)$ is finite, it is isomorphic to $\mathfrak{A}(n,r)^+$, its canonical extension. As abbreviations, we write

$$a(i,j) = \sum_{k<\Omega} a^k(i,j), \quad a(i) = \sum_{j<r} a(i,j), \quad a^k = \sum_{\substack{i<n-1 \\ j<r}} a^k(i,j).$$

(All suprema exist because the algebra is finite.)

We now define an associated cylindric-type algebra C_r. Recall from definition 12.21 the notion of an n-wide m-dimensional ω-hypernetwork over $\mathfrak{A}(n,r)$, the set $H_m^n(\mathfrak{A}(n,r),\omega)$ of all such hypernetworks, and the construction of the $\mathbf{CA}_m$-type algebra $\mathfrak{Ca}\,\mathcal{H}$ for a set $\mathcal{H}$ of wide m-dimensional hypernetworks.

DEFINITION 15.3 For $3 \le m \le n$, we define an infinite algebra C_r of the signature of $\mathbf{CA}_m$ by[1]

$$C_r = \mathfrak{Ca}(H_m^{n+1}(\mathfrak{A}(n,r),\omega)).$$

(To avoid cluttering the notation, m and n here are determined by context.)

15.3 $\mathfrak{A}(n,r) \in \mathbf{S\mathfrak{R}aCA}_n$

Here, we prove that $\mathfrak{A}(n,r)$ has an n-dimensional hyperbasis (definition 12.11), and so is in $\mathbf{S\mathfrak{R}aCA}_n$. Informally, the proof is based on the following idea.

We prove that the set of *all* $((n+1)$-wide) n-dimensional Λ-hypernetworks (for any non-empty Λ) forms a hyperbasis. The key points to check for this are the triangle addition and amalgamation properties, and they raise similar problems. Consider the latter: for any two n-dimensional hypernetworks M, N with $M \equiv_{xy} N$ (some $x, y < n$), we have to find a hypernetwork L with $M \equiv_x L \equiv_y N$. These conditions determine the label on every edge and hyperedge of L except those involving both x and y. Since the case $x = y$ is trivial (let $L = N$), suppose $x \ne y$. The labels on hyperedges involving x and y are not too difficult to define (while maintaining the consistency conditions in the definition of hypernetwork). The critical part is to find an atom for the edge-label $L(x,y)$. Except in trivial cases, we select an atom $a^0(i,0)$ for some $i < n - 1$ so that all triangles (x,y,z) for $z < n$, $z \ne x, y$ are consistent in L. In particular, to ensure consistency, we avoid choosing $a^0(i,0)$ if $N(x,z) \le a(i)$ and $M(z,y) \le a(i)$, for some $z \in n \setminus \{x,y\}$. There are only $n-2$ nodes $z \in n \setminus x,y$, so we can always choose $i < n - 1$ so that $L(x,y) = a^0(i,0)$ avoids these inconsistencies.

NOTATION 15.4 For a complete directed graph N with edges labelled by atoms of some non-associative algebra, and tuples $\bar{x}, \bar{y}$ of nodes of N, we write $\bar{x} \sim_N \bar{y}$ if $|\bar{x}| = |\bar{y}| = l$, say, and $N(x_i, y_i) \le 1$' for all $i < l$.

THEOREM 15.5 *Let $3 \le n < \omega$, $r < \omega$, and let Λ be an arbitrary non-empty set. Then the set $H_n^{n+1}(\mathfrak{A}(n,r),\Lambda)$ (or $\mathcal{H}$ for short) of all $(n+1)$-wide n-dimensional Λ-hypernetworks over $\mathfrak{A}(n,r)$ is an $(n+1)$-wide n-dimensional Λ-hyperbasis for $\mathfrak{A}(n,r)$. Hence, $\mathfrak{A}(n,r) \in \mathbf{S\mathfrak{R}aCA}_n$.*

---
[1] The corresponding definition in [HirHod01a] should be modified to use wide hypernetworks.

468   Chapter 15. Non-finite axiomatisability of $S\mathfrak{R}aCA_{n+1}$ over $S\mathfrak{R}aCA_n$

Proof. We show that $\mathcal{H}$ is a hyperbasis. If $r = 0$, $\mathfrak{A}(n,r)$ is trivial and all edge labels on any hypernetwork are 1'; under these circumstances it is easily seen that $\mathcal{H}$ is a hyperbasis. So we may suppose $r > 0$.

First, let $a \in \text{At}\mathfrak{A}(n,r)$. Let N_a be the $(n+1)$-wide hypernetwork with edge labelling defined as follows: $N_a(x,1) = N_a(1,x) = a$ for $x < n$, $x \neq 1$, and all other edges are labelled by the identity 1'. Let the hyperedges of N_a be labelled by an arbitrary element $\lambda_0 \in \Lambda$. N_a is easily seen to be a hypernetwork, and so $N_a \in \mathcal{H}$. Thus, $\mathcal{H}$ satisfies the first condition of the definition of a hyperbasis.

Second, we check that $\mathcal{H}$ has the triangle addition property. Let $N \in \mathcal{H}$, $x,y,z < n$, $z \neq x,y$, and $a,b \in \text{At}\mathfrak{A}(n,r)$, and suppose that $N(x,y) \leq a;b$. We seek a hypernetwork $M \equiv_z N$ with $M(x,z) = a$ and $M(z,y) = b$. If $a = 1$' then let $M = N[z/x]$ (cf. lemma 12.8). Evidently, $M \equiv_z N$, $M(x,z) = 1' = a$, and $M(z,y) = N(x,y) \leq 1';b = b$, so that $M(z,y) = b$ as $M(z,y)$ is an atom. Otherwise, if $b = 1'$ then let $M = N[z/y]$; we see that $M(x,z) = a$ and $M(z,y) = b$ in a similar way.

So assume that $a,b \neq 1'$. We define M as follows. The condition $M \equiv_z N$ already defines all labels of edges and hyperedges of M not involving z; and we define $M(z,z) = 1'$. We must define the labels on edges (w,z) in M, for $w < n$ with $w \neq z$ (hyperlabels will be dealt with later). Because atoms of $\mathfrak{A}(n,r)$ are self-converse, by lemma 12.4 the converse edge (z,w) must have the same label as (w,z) in M, so we need only define one of them. These labels are defined one at a time, as follows.

First, we let $M(x,z) = a$ and $M(y,z) = b$. This is well-defined, as if $x = y$ then $N(x,y) = 1' \leq a;b$ so that by definition 15.2, $a = b$.

We continue through the remaining edges (w,z) in some arbitrary order, as follows. Let (w,z) be the next edge to label. If $N(w,v) \leq 1'$ for some $v < n$, $v \neq z$ such that (v,z) has already been labelled, then we have no choice but to let $N(w,z) = N(v,z)$. This is well-defined if there is more than one such v. If not, we let $M(w,z) = a^0(i,0)$, for some $i < n-1$ to be determined next. (We use our assumption that $r > 0$ here.) We choose i to be the least number such that there is no already-labelled edge (v_i,z) with $M(w,v_i), M(v_i,z) \leq a(i)$. Since there are only $n-2$ nodes v_i in the hypernetwork different to w and z, while there are $n-1$ possible values of i to choose from, it will always be possible to find such an i. It can be seen that for all $v,w \in n \setminus \{z\}$, the triple $(M(v,w), M(w,z), M(z,v))$ is consistent. Thus, we may define all the edge labels in M while avoiding any inconsistency.

We still have to define the hyperlabels of hyperedges involving z in M. This is easily done: we choose arbitrary $\lambda_0 \in \Lambda$ and set $M(\bar{x}) = \lambda_0$ for all $\bar{x} \in {}^{\leq n+1}n$ with $|\bar{x}| \neq 2$ and $z \in rng(\bar{x})$. For this to yield a valid hypernetwork, we require in terms of notation 15.4 that if $\bar{x}, \bar{y} \in {}^{\leq n+1}n$ and $\bar{x} \sim_M \bar{y}$ then $M(\bar{x}) = M(\bar{y})$. To see this, observe that for all $w < n$, $M(w,z) = 1'$ iff $w = z$. Consequently, if $\bar{x}, \bar{y} \in {}^{\leq n+1}n$

15.4. $\mathfrak{A}(n,r) \notin \mathbf{S\mathfrak{R}aCA}_{n+1}$ 469

satisfy $\bar{x} \sim_M \bar{y}$, then $z \in rng(\bar{x})$ iff $z \in rng(\bar{y})$. So if $z \notin rng(\bar{x})$ then since $M \equiv_z N$ and N is a hypernetwork, we have $M(\bar{x}) = N(\bar{x}) = N(\bar{y}) = M(\bar{y})$. If $z \in rng(\bar{x})$, then assuming $|\bar{x}| \neq 2$, $M(\bar{x}) = \lambda_0 = M(\bar{y})$, and the case where $|\bar{x}| = 2$ is covered by $\exists$'s choice of atoms labelling the edges $M(w,z)$ (indeed, by remark 12.2 it follows from consistency of the edge labelling). So $M \in \mathcal{H}$.

It remains to check the amalgamation property. Suppose $M, N \in \mathcal{H}$ and $x, y < n$ are such that $M \equiv_{xy} N$. We seek a hypernetwork $L \in \mathcal{H}$ such that $M \equiv_x L \equiv_y N$. If $x = y$ then $L = M$ will do, so we may assume not. The requirement $M \equiv_x L \equiv_y N$ uniquely determines all the edge labels and hyperlabels of L except those involving both x and y. Even these are determined if $M(y,z) \leq 1'$ or $N(x,z) \leq 1'$ for some $z \in n \setminus \{x,y\}$, for then, by lemma 12.8, we must have $L = N[y/z]$ or $L = M[x/z]$, respectively (this can be checked to be well-defined if both occur, or if there are several such z). So we may assume not. The labels of (x,y), (y,x) are now defined to be $a^0(i,0)$ for some $i < n-1$ such that there is no node $z < n$, $z \neq x, y$, with $L(x,z), L(z,y) \leq a(i)$. There are at most $n-2$ such z, so exactly as in the previous case, there will always be such a value i. As before, we let $M(\bar{x}) = \lambda_0$ for all $\bar{x} \in {}^{\leq n+1}n$ with $|\bar{x}| \neq 2$ and $x, y \in rng(\bar{x})$. If $\bar{x}, \bar{y} \in {}^{\leq n+1}n$ and $\{x,y\} \subseteq rng(\bar{x})$ but $\{x,y\} \not\subseteq rng(\bar{y})$ then $\bar{x} \not\sim_M \bar{y}$, so the hyperlabelling is consistent and M is a hypernetwork.

Thus, $\mathcal{H}$ is an $(n+1)$-wide n-dimensional hyperbasis for $\mathfrak{A}(n,r)$. By proposition 12.18, generalised to wide hyperbases in exercise 12.3(2), we have $\mathfrak{Ca}\,\mathcal{H} \in \mathbf{CA}_n$ and $\mathfrak{A}(n,r) \subseteq \mathfrak{RaCa}\,\mathcal{H}$ up to isomorphism. We conclude as required that $\mathfrak{A}(n,r) \in \mathbf{S\mathfrak{R}aCA}_n$. □

Since it is known (theorem 5.44) that $\mathbf{S\mathfrak{R}aCA}_n \subseteq \mathbf{RA}$ for $n \geq 4$, we obtain:

COROLLARY 15.6 *For $4 \leq n < \omega$, $r < \omega$, we have $\mathfrak{A}(n,r) \in \mathbf{RA}$.*

COROLLARY 15.7 *Let $3 \leq m \leq n < \omega$ and $r < \omega$. Then $C_r \in \mathfrak{Nr}_m\mathbf{CA}_n$.*

Proof. As we saw in theorem 15.5, $\mathcal{H} = H_n^{n+1}(\mathfrak{A}(n,r), \omega)$ is a wide n-dimensional ω-hyperbasis and $\mathfrak{Ca}(\mathcal{H}) \in \mathbf{CA}_n$. By exercise 12.3(1), $H_m^{n+1}(\mathfrak{A}(n,r), \omega) = \mathcal{H}\restriction_m^{n+1}$. So by proposition 12.22,

$$C_r = \mathfrak{Ca}(H_m^{n+1}(\mathfrak{A}(n,r), \omega)) = \mathfrak{Ca}(\mathcal{H}\restriction_m^{n+1}) \cong \mathfrak{Nr}_m\mathfrak{Ca}\,\mathcal{H} \in \mathfrak{Nr}_m\mathbf{CA}_n.$$

□

15.4 $\mathfrak{A}(n,r) \notin \mathbf{S\mathfrak{R}aCA}_{n+1}$

We prove that $\mathfrak{A}(n,r)$ does not have an $(n+1)$-dimensional hyperbasis. The sketched argument at the start of section 15.3 which showed how we can amalgamate two n-dimensional hypernetworks and label the new edge with the atom

$a^0(i,0)$, for some $i < n - 1$, will not work here in $n + 1$ dimensions. The reason is that if M, N are $n + 1$-dimensional hypernetworks, $x \neq y < (n + 1)$ and $M \equiv_{xy} N$, then we cannot be sure to find some $i < n - 1$ so that for all $z \in (n + 1) \setminus \{x, y\}$ either $N(x, z) \not\leq a(i)$ or $M(z, y) \not\leq a(i)$ because now there are $n - 1$ nodes $z \in (n+1) \setminus \{x, y\}$ so potentially $n - 1$ values of i to avoid. To show that $\mathfrak{A}(n, r) \notin \mathbf{S\mathfrak{R}aCA}_{n+1}$ we'll suppose instead that $\mathfrak{A}(n, r)$ does have an $(n + 1)$-dimensional hyperbasis and deduce that this 'bad situation' does indeed arise — a contradiction.

In very rough outline, we prove our result as follows. We assume for contradiction that there is an $(n + 1)$-dimensional hyperbasis for $\mathfrak{A}(n, r)$. We show that any hyperbasis must contain a sequence of large sets S_t of hypernetworks in the hyperbasis, which get closer and closer to an impossibility as t gets bigger. We show that there are $x \neq y < n + 1$ and you can pick two hypernetworks L_0, M' in the hyperbasis (L_0 will be a fixed hypernetwork in S_t and M' will be obtained from any other hypernetwork in S_t by a substitution) such that $L_0 \equiv_{xy} M'$, and for each $z \in (n + 1) \setminus \{x, y\}$ there are $i < n - 1$ and $j < r$ (which depend on t and z only) such that $M'(x, z), L_0(z, y) \leq a(i, j)$. So there is an amalgam N in the hyperbasis with $M' \equiv_x N \equiv_y L_0$ and if the label $N(x, y)$ is below $a(i)$ then it must be below $a(i, j')$ for some $j' < j$. The set of hypernetworks S_{t+1} will consist of *some* hypernetworks N obtained like this. We select the subset S_{t+1} from the set of all possible N obtained like this in such a way that there will be a fixed x and y for S_{t+1} and a uniform i and j for each suitable z, as above, in the next iteration S_{t+1}. The amalgamation ensures that either there is a wider range of values i as z ranges over $(n + 1) \setminus \{x, y\}$ corresponding to S_{t+1}, or for one such z we have the same i, but a strictly lower value for j. Eventually, all of the $n - 1$ possible values of i will be witnessed by one of the z's. Then, as t continues to increase, we get closer and closer to a contradiction because the value for j cannot be forced below zero.

Let's try to make this precise.

THEOREM 15.8 *For $3 \leq n < \omega$ and $0 < r < \omega$, we have $\mathfrak{A}(n, r) \notin \mathbf{S\mathfrak{R}aCA}_{n+1}$.*

Proof. By theorem 13.46 parts 10 and 11, and the finiteness of $\mathfrak{A}(n, r)$, it is enough to establish that $\mathfrak{A}(n, r)$ ($\cong \mathfrak{A}(n, r)^+$) has no $(n + 1)$-dimensional hyperbasis. So suppose for contradiction that $\mathcal{H}$ is such a hyperbasis. From this, we will show that for $t = 0, 1, \ldots, nr$ there is a 'large' set $S_t \subseteq \mathcal{H}$, where the hypernetworks in S_t are subject to certain constraints which depend on t and get tighter as t increases. This will lead to a contradiction when we take a sufficiently large value of t.

We begin by constructing S_0. Since $\mathcal{H}$ is a hyperbasis, there is a hypernetwork $N \in \mathcal{H}$ with $N(0, 1) = a^0(0, 0)$. Letting $\tau : (n + 1) \to (n + 1)$ be the map given by $\tau(1) = 1$, $\tau(i) = 0$ (all $i \leq n$ with $i \neq 1$), we see by lemma 12.13 that $N\tau \in \mathcal{H}$. For each $k < \Omega$ we have $a^k(0, 0); a^0(1, 0) \geq a^0(0, 0)$ (check that this is not one of the inconsistent triples). So by the triangle addition property for $\mathcal{H}$, there is a hypernetwork $M_k \in \mathcal{H}$ with $N\tau \equiv_2 M_k$, $M_k(0,2) = a^k(0,0)$, and $M_k(2,1) = a^0(1,0)$. See figure 15.1.

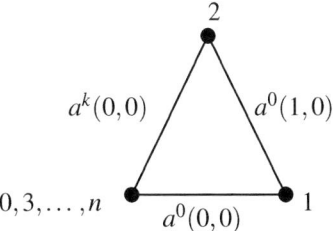

Figure 15.1: The hypernetwork M_k

Observe that $M_k \equiv_2 M_l$ for $k,l < \Omega$, but if $M_k(0,2), M_l(0,2) \le a^m$ (some $m < \Omega$) then as $M_k(0,2) \le a^k$, $M_l(0,2) \le a^l$, and the a^k are pairwise disjoint in $\mathfrak{A}(n,r)$, we have $k = l \,(= m)$. Let $S_0 = \{M_k : k < \Omega\}$.

Inductively, for $0 \le t < nr$, suppose there is a set $S_t \subseteq \mathcal{H}$ of hypernetworks with

1. $|S_t| \ge (nr)^{nr-t}$,

an integer s_t with $1 \le s_t \le n$, and functions $I_t : (n+1) \setminus \{s_t\} \to (n-1)$, $J_t : (n+1) \setminus \{s_t\} \to r$, such that for any $L, M \in S_t$:

2. $L \equiv_{s_t} M$,

3. if $L(0, s_t), M(0, s_t) \le a^k$ (any $k < \Omega$) then $L = M$,

4. $I_t(0) = J_t(0) = 0$,

5. $L(x, s_t) \le a(I_t(x), J_t(x))$, for all $x \le n$, $x \ne s_t$.

See figure 15.2. This is true for $t = 0$ if we let $s_0 = 2$, $I_0(x) = 0$ for $x \le n$, $x \ne 1, 2$, $I_0(1) = 1$, and $J_0(x) = 0$ for all $x \le n$ with $x \ne 2$. See figure 15.1.

DEFINITION 15.9 Let $1 \le s \le n$, $D = (n+1) \setminus \{s\}$, $I : D \to (n-1)$, and $J : D \to r$, as above. For $i < n-1$, define the *index of i with respect to I, J* by

$$\text{Ind}(i, I, J) = \min\{J(x) : x \in D, I(x) = i\} - r < 0$$

if the minimum is taken over a non-empty set, and let $\text{Ind}(i, I, J) = 0$ otherwise[2]. Thus, $0 \ge \text{Ind}(i, I, J) \ge -r$, for each $i < n-1$. Then define the *rank* of (I, J):

$$\text{rk}(I, J) = \sum_{i < n-1} \text{Ind}(i, I, J).$$

---

[2] Or as Roger Maddux puts it, "For an *i*-trip you pay your lowest ticket price; if you've not got a ticket you pay the maximum possible plus one."

Chapter 15. Non-finite axiomatisability of $S\mathfrak{R}aCA_{n+1}$ over $S\mathfrak{R}aCA_n$

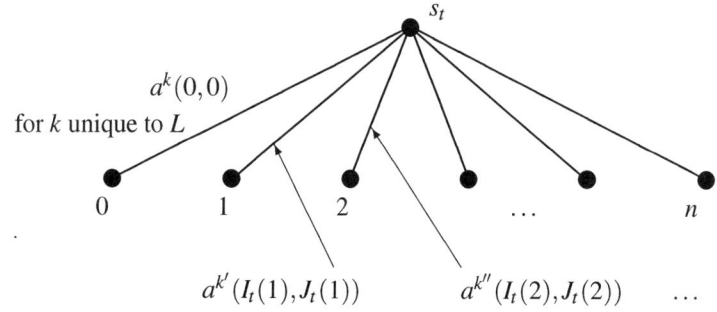

Figure 15.2: A hypernetwork $L \in S_t$

So $0 \geq \mathrm{rk}(I,J) \geq -(n-1)r$.

We have one more inductive assumption.

6. $\mathrm{rk}(I_t, J_t) < -t$.

Observe that $\mathrm{Ind}(0, I_0, J_0) = \mathrm{Ind}(1, I_0, J_0) = -r$, and $\mathrm{Ind}(i, I_0, J_0) = 0$ for $i > 1$. Hence, $\mathrm{rk}(I_0, J_0) = -2r < 0$. So all the hypotheses are true for $t = 0$.

Assuming that S_t, s_t, I_t, J_t exist with properties 1 to 6, let us see if we can find $S_{t+1}, s_{t+1}, I_{t+1}, J_{t+1}$ also with these properties.

Clearly, $I_t : (n+1) \setminus \{s_t\} \to (n-1)$ cannot be injective. Let $p, q \in (n+1) \setminus \{s_t\}$ be distinct such that $I_t(p) = I_t(q)$. We can suppose that $J_t(p) \geq J_t(q)$, and because $J_t(0) = 0$, we can suppose further that $p > 0$. *We will let $s_{t+1} = p$.* Observe that p does not contribute to the 'min' in the expression for $\mathrm{Ind}(i, I_t, J_t)$ for $i = I_t(p)$, and certainly not for other i, so that for all $i < n - 1$,

$$\mathrm{Ind}(i, I_t, J_t) = \begin{cases} \min(K_i) - r, & \text{if } K_i \neq \emptyset; \\ 0, & \text{otherwise,} \end{cases} \quad (15.1)$$

where

$$K_i = \{J_t(x) : x \leq n, x \notin \{s_t, s_{t+1}\}, I_t(x) = i\}. \quad (15.2)$$

Note that $1 \leq s_{t+1} \leq n$ and $s_{t+1} \neq s_t$.

Fix $L_0 \in S_t$. For each hypernetwork $M \in S_t \setminus \{L_0\}$, we have seen (lemma 12.13) that $M[s_{t+1}/s_t] \in \mathcal{H}$. Since $M \equiv_{s_t} L_0$, we have $M[s_{t+1}/s_t] \equiv_{s_t, s_{t+1}} L_0$. By the amalgamation property of hyperbases, there must be $M' \in \mathcal{H}$ with

$$M[s_{t+1}/s_t] \equiv_{s_t} M' \equiv_{s_{t+1}} L_0. \quad (15.3)$$

15.4. $\mathfrak{A}(n,r) \not\subseteq S\mathfrak{R}\mathfrak{a}\mathbf{CA}_{n+1}$

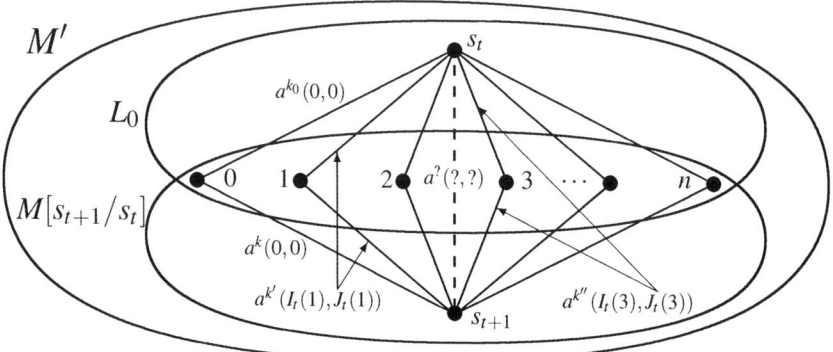

Figure 15.3: The hypernetwork M'

See figure 15.3.

Choose such an M' for each $M \in S_t \setminus \{L_0\}$, and let $S'_t = \{M' : M \in S_t \setminus \{L_0\}\} \subseteq \mathcal{H}$. The required set S_{t+1} will be a subset of S'_t to be defined shortly. First, we show that properties 2 and 3 hold for S'_t and hence for any subset of it. Property 2 holds because $M' \equiv_{s_{t+1}} L_0$ for all $M' \in S'_t$. For property 3, recalling that $s_t > 0$, observe that $M'(0, s_{t+1}) = M[s_{t+1}/s_t](0, s_{t+1}) = M(0, s_t)$ for any $M' \in S'_t$. If $M'_1(0, s_{t+1}), M'_2(0, s_{t+1}) \leq a^k$ for some $M_1, M_2 \in S_t \setminus \{L_0\}$, $k < \Omega$, then $M_1(0, s_t), M_2(0, s_t) \leq a^k$. Since $M_1, M_2 \in S_t$, property 3 for S_t yields $M_1 = M_2$, so $M'_1 = M'_2$. Note that by the same argument, $M'_1 = M'_2 \Rightarrow M_1 = M_2$, so by property 1 for S_t,

$$|S'_t| = |S_t| - 1. \tag{15.4}$$

We will find functions $I_{t+1} : (n+1) \setminus \{s_{t+1}\} \to (n-1)$, $J_{t+1} : (n+1) \setminus \{s_{t+1}\} \to r$ so that properties 1, 4, and 5 remain true for $S_{t+1}, s_{t+1}, I_{t+1}, J_{t+1}$, for some suitable $S_{t+1} \subseteq S'_t$. The domain of I_{t+1} (and J_{t+1}) is the same as that of I_t except that it includes s_t instead of s_{t+1}. We define I_{t+1} to agree with I_t on all points in both domains:

$$\begin{array}{l} I_{t+1}(x) = I_t(x) \\ \text{for all } x \in \mathrm{dom}(I_t) \cap \mathrm{dom}(I_{t+1}) = (n+1) \setminus \{s_t, s_{t+1}\}, \end{array} \tag{15.5}$$

and similarly for J_{t+1}. Now, only the values of $I_{t+1}(s_t), J_{t+1}(s_t)$ remain undecided: they will be settled below.

Property 4 for I_{t+1}, J_{t+1} is obvious, as $0 \in \mathrm{dom}(I_t) \cap \mathrm{dom}(I_{t+1})$. The case where $x \neq s_t$ in property 5 will automatically hold, because for any $M \in S_t \setminus \{L_0\}$, $M'(x, s_{t+1}) = M[s_{t+1}/s_t](x, s_{t+1}) = M(x, s_t) \leq a(I_t(x), J_t(x)) = a(I_{t+1}(x), J_{t+1}(x))$ by (15.3), (15.5), and property 5 for S_t.

474   Chapter 15. Non-finite axiomatisability of $S\mathfrak{R}aCA_{n+1}$ over $S\mathfrak{R}aCA_n$

But the case $x = s_t$ in property 5 is not covered, as we do not know much about $M'(s_t, s_{t+1})$ for $M \in S_t \setminus \{L_0\}$. At least, we can show that it is disjoint from 1'. For this, note that $M'(0, s_t) = L_0(0, s_t)$, and $M'(0, s_{t+1}) = M[s_{t+1}/s_t](0, s_{t+1}) = M(0, s_t)$. Since $L_0 \neq M$, property 3 implies that $L_0(0, s_t) \neq M(0, s_t)$. Hence, $M'(0, s_t) \neq M'(0, s_{t+1})$, so because M' is a hypernetwork, $M'(s_t, s_{t+1}) \neq 1'$.

Thus, for each $M' \in S'_t$ we have $M'(s_t, s_{t+1}) \leq a(i, j)$ for some i and j which depend on M'. The number of possible values of (i, j) is at most $(n-1)r$. Using (15.4), property 1 for S_t, and the pigeon-hole principle, we can find a subset S_{t+1} of S'_t with

$$|S_{t+1}| \geq \frac{|S'_t|}{(n-1)r} \geq \frac{(nr)^{nr-t} - 1}{(n-1)r} \geq (nr)^{nr-(t+1)},$$

such that there are fixed $i_0 < n-1$ and $j_0 < r$ with $M'(s_t, s_{t+1}) \leq a(i_0, j_0)$ for all $M' \in S_{t+1}$. We now complete the definition of the functions I_{t+1} and J_{t+1} by: $I_{t+1}(s_t) = i_0$ and $J_{t+1}(s_t) = j_0$.

We have now defined $S_{t+1}, s_{t+1}, I_{t+1}, J_{t+1}$. By construction, property 5 holds even for the case $x = s_t$. Since we have chosen S_{t+1} to be 'large', property 1 holds, too.

Finally, we check property 6. As this property holds for t, it is enough to prove that $\mathrm{rk}(I_{t+1}, J_{t+1}) < \mathrm{rk}(I_t, J_t)$. We do this by showing that $\mathrm{Ind}(i, I_{t+1}, J_{t+1}) \leq \mathrm{Ind}(i, I_t, J_t)$ for all $i < n-1$, and then that $\mathrm{Ind}(i_0, I_{t+1}, J_{t+1}) < \mathrm{Ind}(i_0, I_t, J_t)$. (This can also be seen informally by inspecting figure 15.3.)

Recall the definition of Ind:

$$\begin{aligned}\mathrm{Ind}(i, I_{t+1}, J_{t+1}) \\ = \min\{J_{t+1}(x) : x \leq n, x \neq s_{t+1}, I_{t+1}(x) = i\} - r\end{aligned} \quad (15.6)$$

if this set is non-empty, and 0 otherwise. By (15.5), we can replace I_t, J_t by I_{t+1}, J_{t+1} in (15.2), without changing K_i. So $K_i = \{J_{t+1}(x) : x \leq n, x \notin \{s_t, s_{t+1}\}, I_{t+1}(x) = i\}$, and by (15.6),

$$\begin{aligned}\mathrm{Ind}(i, I_{t+1}, J_{t+1}) \\ = \min(K_i \cup \{J_{t+1}(s_t) : I_{t+1}(s_t) = i\}) - r\end{aligned} \quad (15.7)$$

if the set is non-empty, and 0, otherwise. Comparing (15.1) and (15.7), it is plain that $\mathrm{Ind}(i, I_{t+1}, J_{t+1}) \leq \mathrm{Ind}(i, I_t, J_t)$ for all $i < n-1$, since the minimum in (15.7) is taken over at least as large a set.

We now prove that $\mathrm{Ind}(i_0, I_{t+1}, J_{t+1}) < \mathrm{Ind}(i_0, I_t, J_t)$. We defined $I_{t+1}(s_t) = i_0$ and $J_{t+1}(s_t) = j_0$, so (15.7) reduces to

$$\mathrm{Ind}(i_0, I_{t+1}, J_{t+1}) = \min(K_{i_0} \cup \{j_0\}) - r. \quad (15.8)$$

Comparing (15.8) and (15.1), it is clear that if $K_{i_0} = \emptyset$ then $\mathrm{Ind}(i_0, I_{t+1}, J_{t+1}) < \mathrm{Ind}(i_0, I_t, J_t)$. If $K_{i_0} \neq \emptyset$, we need $j_0 < \min(K_{i_0})$. So take an arbitrary element

15.4. $\mathfrak{A}(n,r) \notin \mathbf{S\mathfrak{R}aCA}_{n+1}$ 475

of K_{i_0}: it has the form $J_t(x)$, where $x \leq n$, $x \neq s_t, s_{t+1}$, and $I_t(x) = i_0$. We show $j_0 < J_t(x)$. Now by property 1, $S_{t+1} \neq \emptyset$. Let $M' \in S_{t+1}$, where $M \in S_t \setminus \{L_0\}$. Then

$$M'(x, s_t) = L_0(x, s_t) \leq a(I_t(x), J_t(x)) = a(i_0, J_t(x)),$$
$$M'(x, s_{t+1}) = M[s_{t+1}/s_t](x, s_{t+1})$$
$$= M(x, s_t) \leq a(I_t(x), J_t(x)) = a(i_0, J_t(x)),$$
$$M'(s_t, s_{t+1}) \leq a(i_0, j_0).$$

As M' is a hypernetwork, this implies that $[a(i_0, J_t(x)); a(i_0, J_t(x))] \cdot a(i_0, j_0) \neq 0$. From the definition of composition in $\mathfrak{A}(n,r)$, this gives $j_0 < J_t(x)$, as required.

Thus, for $t < nr$, S_t, s, I_t, J_t exist with the listed properties. Now, taking $t = (n-1)r$ would give a set S_t of size at least $(nr)^{nr-(n-1)r}$ — certainly non-empty — where the rank of all the hypernetworks would be constrained by $\mathrm{rk}(I_t, J_t) < -(n-1)r$. But by definition 15.9, the minimum possible rank is just $-(n-1)r$. From this contradiction we deduce that the hyperbasis $\mathcal{H}$ does not exist. Hence, $\mathfrak{A}(n,r) \notin \mathbf{S\mathfrak{R}aCA}_{n+1}$. □

Recall that $C_r = \mathfrak{Ca}(H_m^{n+1}(\mathfrak{A}(n,r), \omega))$ for $r < \omega$, where $H_m^{n+1}(\mathfrak{A}(n,r), \omega)$ is the set of all $(n+1)$-wide m-dimensional ω-hypernetworks over $\mathfrak{A}(n,r)$. We have seen (corollary 15.7) that $C_r \in \mathfrak{Nr}_m\mathbf{CA}_n$.

COROLLARY 15.10 *If $3 \leq m \leq n < \omega$ and $0 < r < \omega$, then $C_r \notin \mathbf{S\mathfrak{N}r}_m\mathbf{CA}_{n+1}$.*

Proof. Suppose for contradiction that $C_r \in \mathbf{S\mathfrak{N}r}_m\mathbf{CA}_{n+1}$. Then we have $\mathfrak{Ra}(C_r) \in \mathfrak{RaS\mathfrak{N}r}_m\mathbf{CA}_{n+1} \subseteq \mathbf{S\mathfrak{R}aCA}_{n+1}$. But it is easy to check (cf. proposition 12.18) that $\mathfrak{A}(n,r) \subseteq \mathfrak{Ra}(C_r)$ via the embedding $a \mapsto \{M \in H_m^{n+1}(\mathfrak{A}(n,r), \omega) : M(0,1) \leq a\}$. So $\mathfrak{A}(n,r) \in \mathbf{S\mathfrak{R}aCA}_{n+1}$, contrary to theorem 15.8. □

COROLLARY 15.11 *Each inclusion in theorem 15.1 parts 1 and 3 is strict — i.e., for all finite $n \geq 3$ we have $\mathbf{S\mathfrak{R}aCA}_n \neq \mathbf{S\mathfrak{R}aCA}_{n+1}$, and for $3 \leq m \leq n$ we have $\mathbf{S\mathfrak{N}r}_m\mathbf{CA}_n \neq \mathbf{S\mathfrak{N}r}_m\mathbf{CA}_{n+1}$.*

Proof. For $n \geq 3$ and $0 < r < \omega$ we have $\mathfrak{A}(n,r) \in \mathbf{S\mathfrak{R}aCA}_n \setminus \mathbf{S\mathfrak{R}aCA}_{n+1}$ (theorems 15.5 and 15.8), and for $3 \leq m \leq n < \omega$ and $0 < r < \omega$ we have $C_r \in \mathbf{S\mathfrak{N}r}_m\mathbf{CA}_n \setminus \mathbf{S\mathfrak{N}r}_m\mathbf{CA}_{n+1}$ (corollaries 15.7 and 15.10). □

This proves parts 1 and 3 of theorem 15.1.

Exercises

1. Let $\mathbf{S\mathfrak{R}aCA}_n^f$ denote $\mathbf{S\mathfrak{R}a}\{C \in \mathbf{CA}_n : C \text{ is finite}\}$. Show that $\mathbf{S\mathfrak{R}aCA}_{n+1}^f \subset \mathbf{S\mathfrak{R}aCA}_n^f$ for all finite $n \geq 4$.

15.5 ∃ can win $G_r^{m,n+1}(\mathfrak{A}(n,r), \Lambda)$

Here we provide a winning strategy for ∃ in the game $G_r^{m,n+1}(\mathfrak{A}(n,r), \Lambda)$ of definition 12.26. We will use it to prove parts 2 and 4 of theorem 15.1.

Recall, from the informal discussion at the start of section 15.4 or the proof of theorem 15.8, that we can use the amalgamation condition for hyperbases to show that if $\mathfrak{A}(n,r)$ does have an $(n+1)$-dimensional hyperbasis then there must be a sequence of subsets S_t of the hyperbasis in which the 'ranks' of the hypernetworks get lower and lower, until we arrive at a rank which is impossibly low, giving a contradiction. This shows that $\mathfrak{A}(n,r)$ cannot have an $(n+1)$-dimensional hyperbasis.

That means that ∀ can win an ω-length game which tests the existence of a hyperbasis over $\mathfrak{A}(n,r)$. Essentially, this is because an amalgamation move can force down the rank of the hypernetworks by forcing a lower value of j as the label of the new edge until ∀ wins. But he can only force the rank down by one. We'll use this to show that ∃ can survive at least r moves in such a game. In fact we use the game $G_r^{m,n+1}$ where ∀ can play arbitrary $(n+1)$-wide m-dimensional hypernetworks. We show that ∃ can still win this r-round game. The m-dimensional moves will be important for part 4 of theorem 15.1, which involves m-dimensional neat reducts. Of course, as is often the case, turning this sketched idea into a formal strategy for ∃ in the game and then proving that this strategy is a winning one is rather complicated, as there are many possible ways that ∀ can play and her strategy must be capable of dealing with any of them.

THEOREM 15.12 *Let $3 \leq m < n < \omega$ and $r < \omega$, and let Λ be an infinite set. Then ∃ has a winning strategy in $G_r^{m,n+1}(\mathfrak{A}(n,r), \Lambda)$.*

Proof. Certainly, ∃ has a winning strategy in $G_0^{m,n+1}(\mathfrak{A}(n,0), \Lambda)$, as no rounds are played. So we suppose $r \geq 1$. All hypernetworks mentioned will be Λ-hypernetworks and, in the main, they will be $(n+1)$-dimensional with nodes $\{0, 1, \ldots, n\}$. In terms of notation 15.4, requirement 3 in the definition of hypernetworks says that $\bar{x} \sim_N \bar{y} \Rightarrow N(\bar{x}) = N(\bar{y})$. Observe that the relation $\sim_N$ is determined by the edge labelling part of N alone.

We define a strategy for ∃ in $G_r^{m,n+1}(\mathfrak{A}(n,r), \Lambda)$. Let the play so far be $N_0, N_1, \ldots, N_{t-1}$. In round t, whatever type of move ∀ makes, ∃ has to choose a suitable hypernetwork N_t. We will specify below how ∃ labels the edges of N_t. This determines the relation $\sim_{N_t}$. But having done this, there is an easy way to specify the labels on hyperedges of N_t, which ∃ uses on many occasions. We will call this method the *default labelling*. It is as follows.

Default labelling Some of the labels of N_t may be determined by ∀'s move. For example, if he plays an amalgamation move (L, M, x, y) in round t then ∃ must

15.5. ∃ can win $G_r^{m,n+1}(\mathfrak{A}(n,r),\Lambda)$ 477

reply with a hypernetwork N_t such that $L \equiv_x N_t \equiv_y M$. So if $\bar{x} \in {}^{\leq n+1}((n+1) \setminus \{x\})$ then ∃ must set $N_t(\bar{x}) = L(\bar{x})$, etc. The label on such an edge, and any $\sim_{N_t}$-equivalent edge, is determined directly by ∀'s move. ∃ labels all the remaining hyperedges of N_t one at a time in some arbitrary order, as follows. If $\bar{x}$ is the next hyperedge to be labelled, then:

1. If $\bar{x} \sim_{N_t} \bar{y}$ for some hyperedge $\bar{y}$ that is already labelled in N_t, she lets $N_t(\bar{x}) = N_t(\bar{y})$.

2. Otherwise, she lets $N_t(\bar{x})$ be some new label that is not the label of any hyperedge that is already labelled in N_t nor is it the label of any hyperedge in any hypernetwork N_u for $u < t$. We do need Λ to be infinite or at any rate fairly large for this to be possible. *In this case we say that ∃ chose the label $N_t(\bar{x})$ in round t.*

Thus, if ∃ chose the label $N_t(\bar{x})$ in round t then for all $s \leq t$ and for all $\bar{y} \in {}^{\leq n+1}(n+1)$ we have $N_t(\bar{x}) = N_s(\bar{y}) \iff (s=t) \wedge (\bar{x} \sim_{N_t} \bar{y})$. We will see that hyperlabels $N_t(\bar{x})$ chosen by ∃ act as 'timestamps', determining in which round they were defined as well as in which $\sim_{N_t}$-class they lie.

∃'s strategy in round t. We refer to definition 12.26 for the rules of the game $G_r^{m,n+1}(\mathfrak{A}(n,r),\Lambda)$. There are three kinds of move in the game: m-dimensional moves, triangle addition moves, and amalgamation moves. We deal with each in turn.

Suppose that ∀ makes an m-dimensional move in round t, choosing an $(n+1)$-wide m-dimensional hypernetwork M over $\mathfrak{A}(n,r)$. ∃ lets N_t be an $(n+1)$-dimensional hypernetwork such that $N_t\!\restriction_m^{n+1} = M$ and $N_t(j,0) = $ '1' for each j with $m \leq j < n+1$. It is easily seen that there exists a (unique) hypernetwork N_t with this property. Observe that because $m < n$,

there do not exist $i_0, \ldots, i_{n-1} < n+1$ such that the atoms $N_t(i_{n-1}, i_j)$ (for $j = 0, \ldots, n-2$) are pairwise distinct and not equal to '1'. (15.9)

This is the only kind of move that ∀ can make in $G_1^{m,n+1}(\mathfrak{A}(n,1))$, so the proof is complete for that case. From now on, we suppose $r \geq 2$.

Suppose that ∀ makes a triangle move: say (N_u, x, y, z, a, b) for $u < t$, $x, y, z \leq n$, $z \notin \{x, y\}$, and $a, b \in \text{At}\,\mathfrak{A}(n,r)$ with $N_u(x,y) \leq a;b$. ∃ must find $N_t \equiv_z N_u$ with $N_t(x,z) = a$ and $N_t(z,y) = b$. If $a = $ '1' then she must let $N_t = N_u[z/x]$, and if $b = $ '1' she must let $N_t = N_u[z/y]$. This is well-defined if $a = b = $ '1'.

Assume now that $a, b \neq $ '1'. ∃ first labels the edges of N_t. Her technique is similar to that of theorem 15.5, but as the dimension is now $n+1$, there is an extra node to worry about. Since she must arrange that $N_t \equiv_z N_u$, the only edges she has

478   *Chapter 15. Non-finite axiomatisability of* $S\mathfrak{R}aCA_{n+1}$ *over* $S\mathfrak{R}aCA_n$

to label are those involving z. She enumerates the n nodes w with $w \leq n$, $w \neq z$ in some fashion, as w_l ($l < n$), say, with x, y first in the enumeration. Then she labels the edges (w_l, z) one by one. (The converse edges are given the same labels, as usual.) She lets $N_t(x, z) = a$ and $N_t(y, z) = b$; this is well-defined if $x = y$, since then, $a = b$. She continues as follows. Let (w_l, z) be the next edge to label.

- If $N_u(w_l, w_{l'}) = 1$' for some $l' < l$, so that the edge $(w_{l'}, z)$ has already been labelled, she must define $N_t(w_l, z) = N_t(w_{l'}, z)$; this is well-defined.

- Otherwise, if $l < n - 1$, she lets $N_t(w_l, z) = a^0(i, r - 1)$ for some $i < n - 1$ such that every previously labelled edge $(w_{l'}, z)$ for $l' < l$, including (x, z) and (y, z), satisfies $N_t(w_{l'}, z) \not\leq a(i)$. At this stage, she has labelled at most $n - 2$ previous edges, and there are $n - 1$ possible values of i to choose from, so this can be done.

- Finally, $\exists$ lets $N_t(w_{n-1}, z) = a^0(i, j)$ for some $i < n - 1$ such that no already-labelled edge $(w_{l'}, z)$ *except perhaps* (w_{n-2}, z), has label beneath $a(i)$ (again there are enough values of i to find a solution), and with $j = r - 1$ *unless* $N_u(w_{n-1}, w_{n-2}), N_t(w_{n-2}, z) \leq a(i, r - 1)$, *in which case* $j = r - 2 \geq 0$. By choice of i, the triangle (w_{n-1}, w_l, z) of N_t is consistent, for $l < n - 2$. Consider the triangle (w_{n-1}, w_{n-2}, z).

  1. If $N_u(w_{n-1}, w_{n-2}), N_t(w_{n-2}, z) \leq a(i, r - 1)$, then $\exists$ chose $N_t(w_{n-1}, z) = a^0(i, r - 2)$, so (w_{n-1}, w_{n-2}, z) is consistent.

  2. Otherwise, if $N_u(w_{n-1}, w_{n-2}), N_t(w_{n-2}, z) \leq a(i)$, then as $n \geq 4$ we have $w_{n-2} \neq x, y$, so that $\exists$ chose $N_t(w_{n-2}, z) = a^0(i, r - 1)$ a moment ago. So as case 1 fails, we must have $N_u(w_{n-1}, w_{n-2}) \leq a(i, j)$ for some $j < r - 1$. But $\exists$ chose $N_t(w_{n-1}, z) = a^0(i, r - 1)$, so again, (w_{n-1}, w_{n-2}, z) is consistent.

  3. Otherwise, $N_u(w_{n-1}, w_{n-2}) \not\leq a(i)$ or $N_t(w_{n-2}, z) \not\leq a(i)$, so consistency is clear.

$\exists$ then sets $N_t(z, z) = 1$', and uses the default labelling for hyperedges. Labels of hyperedges not involving z are determined directly by N_u, and other hyperedges are labelled with a unique new label as described above. It is plain that the resulting N_t is a hypernetwork.

REMARK 15.13 The proof so far covers m-dimensional and triangle moves by $\forall$ when $r \geq 2$, and does not depend on t or Λ. Effectively, we have proved that $\exists$ has a winning strategy in the game $G_\omega^{n+1}(\mathfrak{A}(n, r))$ of definition 12.24. So by proposition 12.25, if $r \geq 2$ then $\mathfrak{A}(n, r)$ has an $(n + 1)$-dimensional relational basis and so is in $\mathbf{RA}_{n+1}$, but by theorem 15.8 is not in $S\mathfrak{R}aCA_{n+1}$. So $S\mathfrak{R}aCA_n \subset \mathbf{RA}_n$ for all finite $n \geq 5$ (since by corollary 13.47, $S\mathfrak{R}aCA_n \subseteq \mathbf{RA}_n$).

15.5. ∃ can win $G_r^{m,n+1}(\mathfrak{A}(n,r),\Lambda)$

Moreover, for any finite $k \geq 3$ and $n \geq 4$, if r is large enough (say, $r \geq k^2$) then $\mathfrak{A}(n,r) \in \mathbf{RA}_k$. This can be proved by providing a winning strategy for ∃ in $G_\omega^k(\mathfrak{A}(n,r))$, as follows. We consider only triangle moves here. ∃ may respond to triangle moves (N,x,y,z,a,b) by ∀ much as in the proof above, but instead of choosing labels of the form $a^0(i,r-1)$ or $a^0(i,r-2)$ to label the edges (w,z), where $w < k$, $w \neq x,y,z$, she uses $a^0(i,j_w)$, where

- $i < n-1$ and $a,b \not\leq a(i)$ (this is to ensure consistency of the triangles (x,z,w) and (y,z,w); we need $n \geq 4$ here),

- the numbers j_w ($w \in k \setminus \{x,y,z\}$) are distinct elements of

$$\{j < r : \neg \exists u,v \in k \setminus \{z\}(N(u,v) \leq a(i,j))\}.$$

For large enough r, this set has size at least $|k \setminus \{x,y,z\}|$ and so it is possible to find such j_w.

Then in any triangle involving z and with edges labelled by $a^k(i,j)$, $a^{k'}(i,j')$, and $a^{k''}(i,j'')$, the indices j, j', j'' are distinct and so the triangle is consistent.

This has several interesting consequences:

1. By theorem 15.8, the algebra $\mathfrak{A}(n-1,k^2)$ witnesses $\mathbf{RA}_k \not\subseteq \mathbf{S\Re aCA}_n$, for all finite $k,n \geq 5$.

2. Since the $\mathbf{RA}_k$ are all elementary classes, any non-principal ultraproduct $\prod_{r<\omega} \mathfrak{A}(n,r)/D$ of the $\mathfrak{A}(n,r)$ over ω is in $\bigcap_{3 \leq k < \omega} \mathbf{RA}_k = \mathbf{RRA}$ (by proposition 13.48). Since $\mathbf{RRA} \subseteq \mathbf{S\Re aCA}_{n+1}$, this already proves part 2 of theorem 15.1. However, we prefer to proceed with a direct proof since the method is interesting in its own right and also gives information about C_r, leading to part 4 of the theorem. See exercise 2 below for an alternative method.

3. $\mathfrak{A}(n,r) \in \mathbf{RA}_{n+1} \cap \mathbf{S\Re aCA}_n \setminus \mathbf{S\Re aCA}_{n+1}$, and yet $\prod_{r<\omega} \mathfrak{A}(n,r)/D \in \mathbf{RRA}$. This implies that

   - for all finite $n \geq 5$, the two inclusions $\mathbf{S\Re aCA}_n \subset \mathbf{RA}_n$ and $\mathbf{S\Re aCA}_n \subset \mathbf{S\Re aCA}_{n-1} \cap \mathbf{RA}_n$ are not finitely axiomatisable,

   - for all finite $n \geq 5$, $\mathbf{RRA}$ is not finitely axiomatisable over $\mathbf{RA}_n$ or $\mathbf{S\Re aCA}_n$.

This implies Monk's theorem [Mon64] that $\mathbf{RRA}$ is not finitely axiomatisable.

PROBLEM 15.14 *For precisely which k,r ($r \geq k > n \geq 4$) does $\mathfrak{A}(n,r)$ have a k-dimensional relational basis?*

∃'s **strategy in round t (continued)** Suppose finally that ∀ plays an amalgamation move, say (M, N, x, y), in round t, for $M, N \in \{N_u : u < t\}$ with $M \equiv_{xy} N$ for distinct $x, y < n + 1$. See figure 15.4. ∃'s strategy should select a hypernetwork N_t

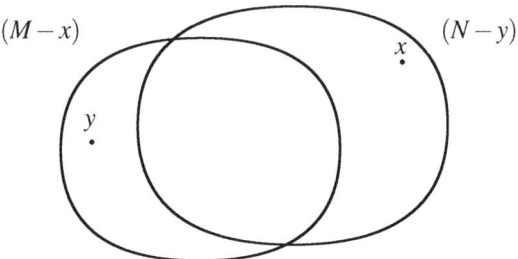

Figure 15.4: ∀'s move in round t

such that $M \equiv_x N_t \equiv_y N$.

∃'s strategy (for rounds when ∀ makes an amalgamation move) is as follows. There are two cases, according to whether or not $M - x$ and $N - y$ (see figure 15.4) embed in the sense of definition 12.6 into a hypernetwork N_u played earlier, the two embeddings agreeing up to $\sim_{N_u}$-equivalence on the 'common part' of M, N.

1. If there are $u < t$ and maps $\xi_1, \xi_2 : (n+1) \to (n+1)$ such that

   - $M \equiv_x N_u \xi_1$ and $N \equiv_y N_u \xi_2$ (see definition 12.6),
   - $N_u(\xi_1(z), \xi_2(z)) = 1$' for all $z \in (n+1) \setminus \{x, y\}$,

   then she chooses any such ξ_1, ξ_2, u and uses them to define N_t. More precisely, let $\xi : n+1 \to n+1$ be defined by $\xi(x) = \xi_2(x)$, and $\xi(z) = \xi_1(z)$ if $z \ne x$. Then N_t is defined to be $N_u \xi$. By lemma 12.7, this is a hypernetwork. It is easy to check that $M \equiv_x N_t \equiv_y N$.

2. Otherwise, she selects an atom $a^0(i, j)$ to label (x, y) such that:

   (a) Every triangle of N_t of the form (x, y, z) is consistent — i.e., there is no $z \le n$ with $z \ne x, y$ such that the triple $(a^0(i, j), M(y, z), N(z, x))$ violates the rules defining $\mathfrak{A}(n, r)$ given at the start of section 15.2.

   (b) $r - 1 - t \le j < r$.

   As earlier, the labels of edges of N_t not involving both x, y are already determined by the condition $M \equiv_x N_t \equiv_y N$.

   ∃ then chooses hyperlabels using the default labelling in the way described before. Notice that if there is $z \le n$ with $z \ne x, y$ such that $N(x, z) = 1$', then

15.5. ∃ can win $G_r^{m,n+1}(\mathfrak{A}(n,r),\Lambda)$

case 1 can be used, by taking $u < t$ with $N_u = M$ and letting $\xi_1 = \text{Id}_{(n+1)}$ (the identity map on $n+1$) and $\xi_2 = [x/z]$. The situation if $M(y,z) = 1$' for some z is similar. So neither is the case. Hence, the default labelling will choose new labels for all hyperedges involving both x and y.

This completes the definition of ∃'s strategy for amalgamation moves. We check that it can always be implemented. We assume inductively that she has successfully implemented it, and the other strategies for m-dimensional and triangle moves given above, in all rounds $u < t$. Let ∀ make the amalgamation move (M,N,x,y) as above in round t. First off, note that $t > 0$, else ∀ could not find M,N. If $t = 1$, then $M = N = N_0$, so ∃ can respond using case 1 of her strategy (with $u = 0$, $\xi_1 = \xi_2 = \text{Id}_{n+1}$). So we may assume that $t \geq 2$. As already mentioned, if there is $z \in (n+1) \setminus \{x,y\}$ with $N(x,z) \leq 1$' or $M(y,z) \leq 1$', then case 1 applies; so we may assume not.

Suppose that case 2 of ∃'s strategy cannot be used, because no suitable atom $a^0(i,j)$ for any $i < n-1$ with $j \geq r-1-t$ can be found to label $N_t(x,y)$. We will show that case 1 of the strategy can be used in this situation.

By inapplicability of case 2, for each $i < n-1$ there is $z_i \in (n+1) \setminus \{x,y\}$ such that the triple $(a^0(i,r-1-t), M(y,z_i), N(z_i,x))$ is inconsistent. Recall that the only inconsistent triples are those consisting of 1' and two distinct atoms, or triples of three $a^k(i,j)$ atoms with all three is the same and two js the same and no bigger than ($\leq$) the third j. We assumed above that $N(x,z_i), M(z_i,y) \not\leq 1$'. So inconsistency of $(a^0(i,r-1-t), M(y,z_i), N(z_i,x))$ implies that $M(y,z_i), N(z_i,x) \leq a(i)$. Since the $a(i')$ ($i' < n-1$) are pairwise disjoint in $\mathfrak{A}(n,r)$, the $z_{i'}$ must be pairwise distinct. So *for any* $j \geq r-1-t$, the triple $(a^0(i,j), M(y,z_{i'}), N(z_{i'},x))$ is certainly consistent if $i' \neq i$. But by cardinalities, $n+1 \setminus \{x,y\} = \{z_i : i < n-1\}$, so $(a^0(i,j), M(y,z), N(z,x))$ is consistent for all $z \in (n+1) \setminus \{x,y,z_i\}$. We conclude that for every $j \geq r-1-t$, the triple $(a^0(i,j), M(y,z_i), N(z_i,x))$ is inconsistent. Now if $N(x,z_i) \leq a(i,j)$ and $M(z_i,y) \leq a(i,j')$ for distinct $j,j' < r$, then $(a^0(i,r-1), M(y,z_i), N(z_i,x))$ is consistent. So we must have $M(y,z_i), N(z_i,x) \leq a(i,j_i)$ for some single $j_i < r$; and $j_i \leq r-1-t$, else $(a^0(i,r-1-t), M(y,z_i), N(z_i,x))$ is consistent.

So we have arrived at the following relatively simple situation: for each $i < n-1$, there are $z_i \in (n+1) \setminus \{x,y\}$ and $j_i \leq r-1-t$ such that $M(y,z_i), N(z_i,x) \leq a(i,j_i)$. The z_i are pairwise distinct, and $\{x,y,z_0,\ldots,z_{n-2}\} = n+1$. See figure 15.5.

Choose least possible $u_1 < t$ such that there is a map $\xi_1 : (n+1) \to (n+1)$ with $M \equiv_x N_{u_1}\xi_1$. Since $M \in \{N_u : u < t\}$, such a u_1 exists. Similarly, choose $\xi_2 : (n+1) \to (n+1)$ such that $N \equiv_y N_{u_2}\xi_2$ for least possible $u_2 < t$. Define $\bar{z} = (z_0,\ldots,z_{n-2}) \in {}^{\leq n+1}(n+1)$. We will show that $u_1 = u_2$ and $\xi_1(\bar{z}) \sim \xi_2(\bar{z})$ in N_{u_1}, so that case 1 of ∃'s strategy applies.

Consider u_1. We aim now to show that ∃ *chose the label on the hyperedge* $\xi_1(\bar{z})$ *of* N_{u_1} *in round* u_1 *of the game, using the default labelling*. To do this, we

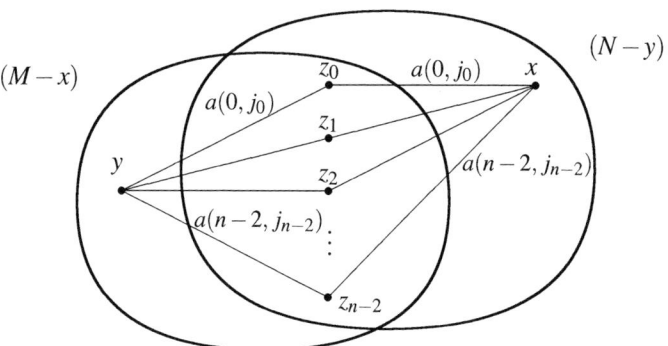

Figure 15.5: $\forall$'s move in round t in more detail

deal in turn with each kind of move that $\forall$ may have made in that round.

m-dimensional moves There exist $n-1$ distinctly labelled edges in M incident with a common node: to wit, $(y,z_0),\dots,(y,z_{n-2})$. None of these edges is labelled by 1' (see figure 15.5). As x is not involved in these edges and $M \equiv_x N_{u_1}\xi_1$, the same holds for $N_{u_1}\xi_1$ and hence for N_{u_1}. By $\exists$'s earlier strategy (see (15.9) above) and since $m \leq n-1$, this cannot be so if she was responding to an m-dimensional move by $\forall$ in round u_1. So he did not make such a move. *The assumption $m < n$ is vital here, but only here.*

Triangle moves Assume that in round u_1, $\forall$ made the triangle move (N_v, x', y', z', a, b). So $v < u_1$. It follows by minimality of u_1 that $z' \in rng(\xi_1 \restriction_{(n+1)\setminus\{x\}})$. (For if not, then since $N_v \equiv_{z'} N_{u_1}$, we have $N_v\xi_1 \equiv_x N_{u_1}\xi_1 \equiv_x M$ and we could have taken $u_1 = v$.)

We claim that $z' \neq \xi_1(y)$. For suppose otherwise. Recall that $t \geq 2$. We know (by $M \equiv_x N_{u_1}\xi_1$ and figure 15.5) that for each $i < n-1$,

$$N_{u_1}(z', \xi_1(z_i)) = N_{u_1}\xi_1(y, z_i) = M(y, z_i) \leq a(i, j_i), \qquad (15.10)$$
$$\text{where } j_i \leq r-1-t \leq r-3.$$

The $a(i, j_i)$ for $i < n-1$ are pairwise disjoint, and $n \geq 4$, so there are at least three different values of $N_{u_1}(z', \xi_1(z_i))$ as i varies. Now by minimality of u_1, we have $N_{u_1} \neq N_v[z'/x']$ and $N_{u_1} \neq N_v[z'/y']$ (else $M \equiv_x N_v[z'/x']\xi_1$, etc). By examining $\exists$'s strategy in triangle moves, we see that in round u_1 $\exists$ must have chosen all but at most two of the edge labels $N_{u_1}(z', w)$ for $w \neq z'$. So she chose at least one of the edge labels $N_{u_1}(z', \xi_1(z_i))$. But her strategy only chooses atoms $a^0(i,j)$ for $j = r-1$ or $j = r-2$. As $j_i \leq r-3$, this contradicts (15.10), proving the claim.

15.5. ∃ can win $G_r^{m,n+1}(\mathfrak{A}(n,r), \Lambda)$ 483

Hence, $z' = \xi_1(z_i)$ for some $i < n-1$. Looking again at the default labelling method shows that ∃ chose the label $N_{u_1}(\xi_1(\bar{z}))$, as the 'new node' z' of N_{u_1} is one of the nodes in this hyperedge.

Amalgamation moves Suppose now that ∀ made an amalgamation move in round u_1 — say, the move (M', N', x', y'). By minimality of u_1, we see that $x', y' \in rng(\xi_1 \restriction_{(n+1) \setminus \{x\}})$, and that ∃ used case 2 of her amalgamation strategy in round u_1. Hence,

$$N_{u_1}(x', y') \leq a(i, j) \text{ for some } i < n - 1 \text{ and } j \geq r - 1 - u_1. \quad (15.11)$$

We claim that $x', y' \in rng(\xi_1(\bar{z}))$.

If not, then as $x', y' \in rng(\xi_1 \restriction_{(n+1) \setminus \{x\}})$, we have $\xi_1(y) \in \{x', y'\}$. Let $i < n - 1$ be such that $\{x', y'\} = \{\xi_1(y), \xi_1(z_i)\}$. We know (by $M \equiv_x N_{u_1} \xi_1$ and figure 15.5 again) that

$$N_{u_1}(x', y') = N_{u_1}(\xi_1(y), \xi_1(z_i))$$
$$= N_{u_1} \xi_1(y, z_i) = M(y, z_i) \leq a(i, j_i).$$

So by (15.11), $j = j_i$. But by the rest of (15.11), $j_i \leq r - 1 - t < r - 1 - u_1 \leq j$, a contradiction. This proves the claim.

By this claim, the $(n-1)$-tuple $\xi_1(\bar{z})$ was labelled by ∃ in round u_1 of the game, using case 2 of her strategy, using the default labelling.

We have now proved that ∃ chose the label on $N_{u_1}(\xi_1(\bar{z}))$ in round u_1 using the default labelling. Note that this label is equal to $M(\bar{z})$, since $M \equiv_x N_{u_1} \xi_1$. Pursuing the same argument for N, u_2, ξ_2 shows that ∃ also chose the label on the hyperedge $\xi_2(\bar{z})$ in round u_2 of the game. We have $N_{u_2}(\xi_2(\bar{z})) = N(\bar{z})$.

But $M(\bar{z}) = N(\bar{z})$, since $M \equiv_{xy} N$. So $N_{u_1}(\xi_1(\bar{z})) = N_{u_2}(\xi_2(\bar{z}))$. Examination of ∃'s default labelling strategy shows that she never chooses the same label for hyperedges in different rounds. Hence, $u_1 = u_2$.

Moreover, she never uses the same label for two hyperedges $\bar{x}, \bar{y}$ in N_{u_1} unless $\bar{x} \sim_{N_{u_1}} \bar{y}$ — i.e., $N_{u_1}(x_i, y_i) = 1'$ for all $i < |\bar{x}| = |\bar{y}|$. Hence, $N_{u_1}(\xi_1(z_i), \xi_2(z_i)) = 1'$ for all $i < n - 1$. So, taking the 'u' of case 1 of ∃'s strategy to be $u_1 = u_2$, we see that she can use this case in the current round, t. □

Exercises

1. For $m < d < \omega$, a d-dimensional relational basis $\mathcal{R}$ for an atomic relation algebra $\mathcal{A}$ is said to have the *m-dimensional network addition property* if whenever N, N' are m-dimensional atomic $\mathcal{A}$-networks, $k < m$, $N \equiv_k N'$, $M \in \mathcal{R}$, and $\nu : m \to d$ is an embedding of N into M (i.e., $N(i,j) = M(\nu(i), \nu(j))$

for all $i,j < m$), then there is $M' \in \mathcal{R}$ with $M' \equiv_{\nu(k)} M$ and such that ν is an embedding of N' into M'.

Show that the 3-dimensional network addition property is equivalent to the triangle addition property.

For $3 \leq m < n$, $d < \omega$, and $r \geq d^2$, show that $H_d(\mathfrak{A}(n,r),\{\lambda\})$ is a d-dimensional relational basis with the m-dimensional network addition property.

2. For C_r as in definition 15.3, where $3 \leq m < n < \omega$ and $r < \omega$, show that any non-principal ultraproduct $\prod_{r<\omega} C_r/D$ is in $\mathbf{RCA}_m$. [Find a representation of $\prod_{r<\omega} \mathfrak{A}(n,r)/D$ (or some countable elementary subalgebra of it — cf. proposition 10.13) that embeds all m-dimensional hypernetworks in a way respecting $\equiv_i$ for $i < m$. You will need $m < n$.] Deduce part 4 of theorem 15.1, and that $\mathbf{RCA}_m$ is not finitely axiomatisable over $\mathbf{S\mathfrak{N}r}_m\mathbf{CA}_n$.

15.6 Non-finite axiomatisability

Let us recall and summarise our results so far. Fix $3 \leq m < n < \omega$. We use ω as our infinite set of hyperlabels. By theorems 15.5 and 15.8, for $r > 0$ we have

$$\mathfrak{A}(n,r) \in \mathbf{S\mathfrak{R}aCA}_n \setminus \mathbf{S\mathfrak{R}aCA}_{n+1}, \tag{15.12}$$

and by corollaries 15.7 and 15.10,

$$C_r \stackrel{\text{def}}{=} \mathfrak{Ca}(H_m^{n+1}(\mathfrak{A}(n,r),\omega)) \in \mathbf{S\mathfrak{N}r}_m\mathbf{CA}_n \setminus \mathbf{S\mathfrak{N}r}_m\mathbf{CA}_{n+1}. \tag{15.13}$$

Write T for the game tree $T(m,n+1)$ for $G_r^{m,n+1}$ constructed in lemma 12.29. Recall from that lemma that $\mathfrak{M}_r = \mathfrak{M}(\mathfrak{A}(n,r),m,n+1,\omega,C_r)$ is a 5-sorted structure with sorts $\mathfrak{A}(n,r)$, ω, $H_m^{n+1}(\mathfrak{A}(n,r),\omega)$, $H_{n+1}(\mathfrak{A}(n,r),\omega)$, and C_r. In the lemma, C_r is required to be (and it clearly is) an atomic m-dimensional cylindric-type BAO with atom structure isomorphic to $\operatorname{At}\mathfrak{Ca}(H_m^{n+1}(\mathfrak{A}(n,r),\omega))$. By theorem 15.12 and lemma 12.29,

$$\exists \text{ has a winning strategy in } G(T\restriction_{1+2r},\mathfrak{M}_r). \tag{15.14}$$

We can now apply results of chapter 10 to prove theorem 15.1. It is only necessary to prove parts 2 and 4 of theorem 15.1, since parts 1 and 3 follow from these and in any case were already proved in corollary 15.11. Our proof is based on ultraproducts and Łoś' theorem. We fix a non-principal ultrafilter D over $\omega \setminus \{0\}$. All ultraproducts here will be of the form $\prod_{0<r<\omega} X_r/D$, for structures X_r which also depend on the fixed m,n. We write $\prod_D X_r$ as shorthand for $\prod_{0<r<\omega} X_r/D$.

15.6. Non-finite axiomatisability

By (15.14), ∃ has a winning strategy in $G(T\restriction_r, \mathfrak{M}_r)$ for all finite $r > 0$. So by theorem 10.12, ∃ has a winning strategy in $G(T, \prod_D \mathfrak{M}_r)$. By proposition 10.13, which clearly holds for many-sorted structures, there is a countable elementary substructure $\mathfrak{N} \preceq \prod_D \mathfrak{M}_r$ such that ∃ has a winning strategy in $G(T, \mathfrak{N})$.

Now remark 12.28 showed that the links between the sorts of $\mathfrak{M}$ can be expressed in the language of $\mathfrak{M}$. Also, the C_r are m-dimensional cylindric algebras. Using Łoś' theorem and $\mathfrak{N} \preceq \prod_D \mathfrak{M}_r$, it follows that $\mathfrak{N}$ is of the form $\mathfrak{M}(\mathcal{B}, m, n+1, \Lambda, C)$ for some countable atomic non-associative algebra $\mathcal{B}$, countable set Λ, and countable atomic m-dimensional cylindric algebra C with At $C \cong$ At $\mathfrak{Ca}(H_m^{n+1}(\mathcal{B}, \Lambda))$. Furthermore,

$$\mathcal{B} \preceq \prod_D \mathfrak{A}(n,r) \text{ and } C \preceq \prod_D C_r. \tag{15.15}$$

So ∃ has a winning strategy in $G(T, \mathfrak{M}(\mathcal{B}, m, n+1, \Lambda, C))$. By lemma 12.29, she also has a winning strategy in $G_\omega^{m,n+1}(\mathcal{B}, \Lambda)$. So by proposition 12.27, $\mathcal{B} \in \mathbf{S\mathfrak{Ra}CA}_{n+1}$ and $\mathfrak{Ca}(H_m^{n+1}(\mathcal{B}, \Lambda)) \in \mathfrak{N}\mathfrak{r}_m \mathbf{CA}_{n+1}$. Now C is a cylindric algebra, so (fact 5.17) is a completely additive BAO. So by proposition 2.66, C embeds into $\mathfrak{Ca}(H_m^{n+1}(\mathcal{B}, \Lambda))$. We thus obtain

$$\mathcal{B} \in \mathbf{S\mathfrak{Ra}CA}_{n+1} \text{ and } C \in \mathbf{S\mathfrak{N}\mathfrak{r}}_m \mathbf{CA}_{n+1}. \tag{15.16}$$

Actually, both algebras are representable (remark 15.13 and exercise 15.5(2)).

Proof of theorem 15.1(2) Suppose for contradiction that there is a single first-order sentence σ of the language of relation algebras such that for any algebra $\mathcal{A} \in \mathbf{S\mathfrak{Ra}CA}_n$, we have $\mathcal{A} \in \mathbf{S\mathfrak{Ra}CA}_{n+1}$ iff $\mathcal{A} \models \sigma$. So, by (15.12), $\mathfrak{A}(n,r) \not\models \sigma$, for each $0 < r < \omega$. By (15.16), $\mathcal{B} \in \mathbf{S\mathfrak{Ra}CA}_{n+1}$, so $\mathcal{B} \models \sigma$. By (15.15), $\prod_D \mathfrak{A}(n,r) \models \sigma$ too. But then Łoś' theorem tells us that $\{r : \mathfrak{A}(n,r) \models \sigma\} \in D$, contradicting the fact that $\mathfrak{A}(n,r) \not\models \sigma$ for all non-zero $r < \omega$.

Proof of theorem 15.1(4) The proof that $\mathbf{S\mathfrak{N}\mathfrak{r}}_m \mathbf{CA}_{n+1}$ is not finitely axiomatisable over $\mathbf{S\mathfrak{N}\mathfrak{r}}_m \mathbf{CA}_n$ is similar to that of part 2, using (15.13) instead of (15.12).

This completes the proof of theorem 15.1. We note by exercise 1 of the introduction to part III (p. 362) and exercise 5.6(3) that $\mathbf{S\mathfrak{Ra}CA}_l$ is not finitely axiomatisable over $\mathbf{S\mathfrak{Ra}CA}_n$, whenever $4 \leq n < l \leq \omega$. Nor is $\mathbf{RRA} = \bigcap_{n \leq l < \omega} \mathbf{S\mathfrak{Ra}CA}_l$ finitely axiomatisable over $\mathbf{S\mathfrak{Ra}CA}_n$. Similar facts hold for neat reducts.

Exercises

1. For finite $n \geq 5$, let $\mathcal{A}$ be an infinite atomic relation algebra defined similarly to $\mathfrak{A}(n,r)$, but with atoms $a^k(i,j)$ for $i < n-2$, $j < \omega$ and $k < \omega_1$ (ω_1 is the first uncountable ordinal).

486   Chapter 15. Non-finite axiomatisability of $S\mathcal{R}aCA_{n+1}$ over $S\mathcal{R}aCA_n$

(a) Show that $\mathcal{A}$ has no n-dimensional hyperbasis.
(b) Show that $\exists$ has a winning strategy in $G_r^{m,n}(\mathcal{A},\omega)$ for any $r < \omega$ and $m < n-1$.
(c) Deduce that the class of atomic relation algebras with a complete infinitarily n-flat relativised representation (or an n-dimensional hyperbasis, or satisfying any of conditions 5–10 of theorem 13.45) is not elementary.

15.7 Proof theory

We can easily apply theorem 15.1 to proof theory with finitely many variables. Because this area is already well represented in the research literature (see, for example, [Madd83, Madd89b, HenMon+85, TarGiv87, Giv91, AndNém+01]), we do not go into great detail.

In [AndNém+01, section 7], a Hilbert system for proving m-variable formulas with n-variable proofs (for $m \leq n < \omega$) is given. All formulas are in a fixed countably infinite signature $\mathcal{R}$ consisting of m-ary relation symbols. The logic L^n consists of all such formulas written with variables $v_0, \ldots, v_{n-1}$ only, where all atomic formulas are equalities or of the form $R(v_0, \ldots, v_{m-1})$ for $R \in \mathcal{R}$. That is, the order of variables in non-equality atomic formulas is fixed. The axioms of $\vdash_{m,n}$ are as follows, where $i,j,k < n$ and φ, ψ are L^n-formulas:

1. all propositional tautologies,
2. $\forall v_i(\varphi \to \psi) \to (\forall v_i \varphi \to \forall v_i \psi)$,
3. $\forall v_i \varphi \to \varphi$,
4. $\forall v_i \forall v_j \varphi \to \forall v_j \forall v_i \varphi$,
5. $\forall v_i \varphi \to \forall v_i \forall v_i \varphi$,
6. $\exists v_i \varphi \to \forall v_i \exists v_i \varphi$,
7. $v_i = v_i$,
8. $v_i = v_j \to (v_i = v_k \to v_j = v_k)$,
9. $\exists v_i(v_i = v_i)$,
10. $v_i = v_j \to \forall v_k(v_i = v_j)$ if $k \notin \{i,j\}$,
11. $v_i = v_j \to (\varphi \to \forall v_i(v_i = v_j \to \varphi))$ if $i \neq j$,
12. $R(v_0, \ldots, v_{m-1}) \to \forall v_i R(v_0, \ldots, v_{m-1})$ if $R \in \mathcal{R}$ and $i \geq m$.

15.7. Proof theory

The rules of inference are modus ponens (infer ψ from φ and $\varphi \to \psi$) and universal generalisation (infer $\forall v_i \varphi$ from φ).

DEFINITION 15.15 Let Σ be any set of L^n-formulas, and φ an L^m-formula. We write $\Sigma \vdash_{m,n} \varphi$ if any set S of L^n-formulas containing Σ and all axioms above and closed under the inference rules also contains φ.

Now we identify elements of $\mathcal{R}$ with first-order variables of the language of $\mathbf{CA}_m$. For an L^m-formula φ, define a $\mathbf{CA}_m$-term $\widehat{\varphi}$ by induction: $R(v_0, \ldots, v_{m-1})\widehat{} = R$ for $R \in \mathcal{R}$, $\widehat{v_i = v_j} = \mathsf{d}_{ij}$, $\widehat{\neg \varphi} = -\widehat{\varphi}$, $\widehat{\varphi \wedge \psi} = \widehat{\varphi} \cdot \widehat{\psi}$, $\widehat{\exists v_i \varphi} = \mathsf{c}_i \widehat{\varphi}$. We quote a result from [AndNém$^+$01, section 7] relating $\vdash_{m,n}$ to neat reducts:

FACT 15.16 *For any L^m-formula φ, we have $\vdash_{m,n} \varphi$ iff $\mathbf{S}\mathfrak{Nr}_m\mathbf{CA}_n \models \widehat{\varphi} = 1$.*

For a proof, see [HenMon$^+$85, 4.3.25]. Note here that the variables (from $\mathcal{R}$) in the equation $\widehat{\varphi} = 1$ are implicitly universally quantified.

Schemata An *m-schema* is an L^m-formula $\sigma(R_1, \ldots, R_k)$, where $R_1, \ldots, R_k \in \mathcal{R}$ are the relation symbols occurring in σ. Example:

$$\forall v_i R(v_0, \ldots, v_{m-1}) \to \forall v_i \forall v_i R(v_0, \ldots, v_{m-1}),$$

where $i < m$, $R \in \mathcal{R}$. An *m-instance* of σ is a formula of the form $\sigma(\chi_1, \ldots, \chi_k)$, where $\chi_1, \ldots, \chi_k$ are L^m-formulas and each atomic subformula $R_l(v_0, \ldots, v_{m-1})$ of σ has been replaced by χ_l ($l = 1, \ldots, k$). Example: $\forall v_i \varphi \to \forall v_i \forall v_i \varphi$, where φ is any L^m-formula. Note that any *m*-instance of an *m*-schema is an L^m-formula.

THEOREM 15.17 *Let $3 \leq m < n < \omega$. There is no finite set of m-schemata whose set Σ of m-instances satisfies*

$$\Sigma \vdash_{m,n} \varphi \iff \vdash_{m,n+1} \varphi$$

for all L^m-formulas φ.

Proof. It is enough to prove that there is no single *m*-schema with the above property. Assume that $\sigma(R_1, \ldots, R_k)$ is such a schema. We can assume without loss of generality that σ is a sentence (see exercise 2 below). We will show that the equation $\widehat{\sigma} = 1$ axiomatises $\mathbf{S}\mathfrak{Nr}_m\mathbf{CA}_{n+1}$ over $\mathbf{S}\mathfrak{Nr}_m\mathbf{CA}_n$, contradicting theorem 15.1.

It is easy to show that $\widehat{\sigma} = 1$ is valid in $\mathbf{S}\mathfrak{Nr}_m\mathbf{CA}_{n+1}$. Clearly, σ is an *m*-instance of itself, so $\sigma \in \Sigma$ and consequently $\Sigma \vdash_{m,n} \sigma$. By assumption, $\vdash_{m,n+1} \sigma$, and by fact 15.16, $\mathbf{S}\mathfrak{Nr}_m\mathbf{CA}_{n+1} \models \widehat{\sigma} = 1$, as required.

For the converse, let $C \in \mathbf{S}\mathfrak{Nr}_m\mathbf{CA}_n$ satisfy $C \models \widehat{\sigma} = 1$, the variables of $\widehat{\sigma}$ being implicitly universally quantified here. We show $C \in \mathbf{S}\mathfrak{Nr}_m\mathbf{CA}_{n+1}$.

As $\mathfrak{SNr}_m\mathbf{CA}_{n+1}$ is a variety [Mon61a], it suffices to show that C satisfies any equation valid in $\mathfrak{SNr}_m\mathbf{CA}_{n+1}$. So take such an equation. We may assume it to have the form $s = 1$, where the variables of the $\mathbf{CA}_m$-term s come from $\mathcal{R}$; then clearly, there is an L^m-formula φ with $\widehat{\varphi} = s$. We require $C \models \widehat{\varphi} = 1$.

Now $\mathfrak{SNr}_m\mathbf{CA}_{n+1} \models \widehat{\varphi} = 1$ implies by fact 15.16 that $\vdash_{m,n+1} \varphi$, so by assumption, $\Sigma \vdash_{m,n} \varphi$. Hilbert systems are compact (see exercise 1 below), so there is finite $\Sigma_0 \subseteq \Sigma$ with $\Sigma_0 \vdash_{m,n} \varphi$. Let $\varphi' = (\bigwedge \Sigma_0) \to \varphi$, an L^m-formula. Then by exercise 3 below, $\vdash_{m,n} \varphi'$, from which fact 15.16 yields $\mathfrak{SNr}_m\mathbf{CA}_n \models \widehat{\varphi'} = 1$. So

$$C \models \widehat{\varphi} + \left(-\prod_{\psi \in \Sigma_0} \widehat{\psi}\right) = 1. \tag{15.17}$$

Consider a formula $\psi = \sigma(\chi_0, \ldots, \chi_k) \in \Sigma_0$. It is evident that $\sigma(\chi_0, \ldots, \chi_k)\widehat{}$ is the result of substituting the terms $\widehat{\chi}_1, \ldots, \widehat{\chi}_k$ for the variables $R_1, \ldots, R_k$ in $\widehat{\sigma}$. So $C \models \widehat{\sigma} = 1$ implies $C \models \sigma(\chi_0, \ldots, \chi_k)\widehat{} = 1$ — since we implicitly universally quantify $\mathcal{R}$-variables on both sides here, the latter is a special case of the former. That is, $C \models \widehat{\psi} = 1$ for every $\psi \in \Sigma_0$. By (15.17), $C \models \widehat{\varphi} = 1$, as required. □

See [Gor00] for a similar result.

PROBLEM 15.18 *As we said at the beginning of this chapter, [And90a] showed that $\mathfrak{SNr}_n\mathbf{CA}_{n+1}$ is a finitely axiomatisable class. Does theorem 15.17 hold when $m = n$?*

Exercises

In the exercises, $3 \leq m < n < \omega$ and an L^m-formula φ are fixed.

1. Let Σ be a set of L^m-formulas. Prove that $\Sigma \vdash_{m,n} \varphi$ iff there are $k < \omega$ and L^n-formulas α_i ($i \leq k$) such that $\alpha_k = \varphi$ and for each $i \leq k$, one of the following holds:

   - α_i is an axiom,
   - $\alpha_i \in \Sigma$,
   - there are $j, j' < i$ such that $\alpha_{j'} = \alpha_j \to \alpha_i$,
   - there are $j < i$ and $l < n$ such that $\alpha_i = \forall v_l \alpha_j$.

   (The sequence $(\alpha_0, \ldots, \alpha_k)$ is called a *proof of φ from Σ*.)
   Deduce that $\Sigma \vdash_{m,n} \varphi$ iff $\Sigma_0 \vdash_{m,n} \varphi$ for some finite $\Sigma_0 \subseteq \Sigma$ ('compactness').

2. For any set Σ of L^m-formulas and $i < m$, show that $\Sigma \vdash_{m,n} \varphi$ iff $\{\forall v_i \sigma : \sigma \in \Sigma\} \vdash_{m,n} \varphi$. Deduce that for any L^m-formula σ whose set of m-instances is Σ, there is an L^m-sentence σ', whose set of m-instances is Σ', such that $\Sigma \vdash_{m,n} \varphi$ iff $\Sigma' \vdash_{m,n} \varphi$ for all L^m-formulas φ.

3. Let Σ be a finite set of L^m-sentences. Prove that $\Sigma \vdash_{m,n} \varphi$ iff $\vdash_{m,n} (\bigwedge \Sigma) \to \varphi$. [For $\Rightarrow$, let $\alpha_0, \ldots, \alpha_k$ be a proof of φ from Σ and show by induction on $i \leq k$ that $\vdash_{m,n} \bigwedge \Sigma \to \alpha_i$.] Must the equivalence hold if Σ is a set of formulas (not necessarily sentences)?

Chapter 16

The rainbow construction for relation algebras

Here we define the rainbow construction. The purpose of this construction, for various types of games, is to produce relation algebras where $\exists$ has a winning strategy in games of certain (shorter) lengths but not in games of longer lengths. Using this construction we can prove various results about the theory of the representation class corresponding to the game. The problem of constructing a relation algebra such that $\exists$ has winning strategies in games of certain lengths but not other lengths, is a hard one. The point of the rainbow construction is to reduce this problem to that of finding two relational structures A, B where $\exists$ can win an Ehrenfeucht–Fraïssé-style game over A, B provided the game does not go on too long, but she loses for longer games.

We will first define a slight variation of a well known Ehrenfeucht–Fraïssé 'forth' game. We will call it $EF_r^p(A,B)$, where $p, r \leq \omega$ are ordinals. It is played between two ordinary model-theoretic structures A, B in a relational signature. Then we will construct an atomic non-associative 'rainbow algebra' $\mathcal{A}_{A,B}$ from arbitrary A, B in a binary (relational) signature, and show (in theorem 16.5) that a winning strategy for $\exists$ in $EF_r^p(A,B)$ is equivalent to one in the game $G_{1+r}^{2+p}(\mathcal{A}_{A,B})$ of definition 12.24. The significance of this game is that for an atomic non-associative algebra $\mathcal{A}$, a winning strategy for $\exists$ in $G_\omega^n(\mathcal{A})$ is equivalent to $\mathcal{A}$'s having an n-dimensional relational basis. So we can use the 'equivalence' of $EF_r^p(A,B)$ and $G_{1+r}^{2+p}(\mathcal{A}_{A,B})$ to solve a variety of problems concerning relational bases, in the next chapter. It gives us freedom to create algebras on which $\exists$ can win G_r^p for certain p, r but not others, just by varying the structures A, B. It will then be an easy matter to prove the theorems we want, by choosing A, B carefully.

16.1 Ehrenfeucht–Fraïssé 'forth' games

The game EF is a slight modification of a standard Ehrenfeucht–Fraïssé forth' game, which we therefore define first.

16.1.1 The standard Ehrenfeucht–Fraïssé game

The notion of 'homomorphism' was explained in section 2.1. A partial homomorphism is similar, but it need not be defined on the whole of a structure. Formally:

DEFINITION 16.1 Let L be a relational signature (with no function or constant symbols). Let A, B be L-structures. A *partial homomorphism* from A to B is a partial map $f : A \to B$ such that for every k-ary relation symbol $P \in L$ and $a_1, \ldots, a_k \in \text{dom}(f)$, if $A \models P(a_1, \ldots, a_k)$ then $B \models P(f(a_1), \ldots, f(a_k))$.

DEFINITION 16.2 Let L be a relational signature. Let A, B be L-structures, and let $p, r \leq \omega$. The game $\mathbf{EF}_r^p(A, B)$ is defined as follows. It is a game played by $\forall, \exists$, with r rounds and p pairs of 'pebbles'. At each stage of the game, some of the pebble pairs may be in play, the remainder being left on the side for later use. If a pair is in play, then one pebble of the pair is placed on an element of A, and the other on an element of B. The game has r rounds, numbered $0, 1, \ldots, i, \ldots$ for $i < r$. Initially there are no pebbles in play. In each round:

- if he wishes, $\forall$ may remove one or more pairs of pebbles in play, placing them at the side. (It is convenient to allow him to do this, even though it is evidently not in general to his advantage.)

- $\forall$ then picks up one pebble from a pair not in play, and places it on an element of A. (If all pebbles are in play at the beginning of the round, $\forall$ will first have to exercise his option to remove a pair.)

- $\exists$ then places the other pebble of that pair on an element of B.

That completes the round. $\forall$ wins at this point if the relation $R \subseteq A \times B$ defined by the pebble pairs is not a well-defined partial homomorphism from A to B. So formally, let R consist of all pairs (a, b) such that $a \in A$ and $b \in B$ are elements covered by pebbles in the same pair. $\exists$ loses unless R defines a partial homomorphism from A to B. If $\exists$ does not lose in this way, the game goes on for another round. If $\exists$ never loses in any round, then she wins the play.

EXAMPLE 16.3 As a simple (but useful) example, let A, B be finite complete undirected graphs without loops (reflexive edges). If $|A| > |B|$, then $\forall$ has a winning strategy in $\mathbf{EF}_r^p(A, B)$ for any $p, r > |B|$. In each round $0, 1, \ldots, |B|$, he places a new pebble on a new element of A. The edge relation in B is irreflexive, so to

16.1. Ehrenfeucht–Fraïssé 'forth' games

avoid losing, $\exists$ must respond by placing the other pebble of the pair on an unused element of B. After $|B|$ rounds there will be no such element, and she will lose at the next round. On the other hand, if $p \leq |B|$ then $\exists$ has a winning strategy for $\mathbf{EF}^p_\omega(A,B)$, because with only at most $|B|$ pebble pairs, there always will be an unused element of B.

REMARK 16.4 It is clear (see exercise 1 below) that if $\forall$ has a winning strategy in $\mathbf{EF}^p_r(A,B)$, then he has a winning strategy in which he only removes a pair of pebbles from play if he has to in order to create a spare pebble pair, and that he only removes at most one pair of pebbles per round.

16.1.2 The modified Ehrenfeucht–Fraïssé game

The definition of the Ehrenfeucht–Fraïssé game $\mathbf{EF}^p_r(A,B)$, above, is quite standard. But for simplicity in what follows, we prefer to use a minor modification of it, which we denote by $EF^p_r(A,B)$. In this game, $\forall$ has to ensure that at each round there are at least two pebbles on distinct elements of A. So we require that $\forall$ *makes his first two moves together* and they must be distinct; and that if at any later time the number of distinct elements of A covered by pebbles falls below two, then $\exists$ wins immediately and the game is terminated at that point. The effects of this are:

- If any of p, r, $|A|$ are less than two, we get a degenerate game in which $\forall$ cannot make a legal move and so he is certain to lose. $\exists$ has a winning strategy in any such game.

- If $p=2$ or $r=2$, $\forall$ can only hope to win in his combined first two rounds, for at that point (if $r>2$) he must pick up a pebble on A. For most games of this form, $\exists$ has a winning strategy — all that she requires is that for any distinct $a, a' \in A$, there exist $b, b' \in B$, not necessarily distinct, such that the partial map $f : A \to B$ given by $f(a) = b$ and $f(a') = b'$ (i.e., $f = \{(a,b), (a',b')\}$) is a partial homomorphism from A to B.

Thus, in the combined round 0 and round 1, $\forall$ is required to place two pebbles on two distinct points $a, a' \in A$. $\exists$ then has to place the two corresponding pebbles on points $b, b' \in B$ such that $\{(a,b), (a',b')\}$ is a partial homomorphism from A to B. If she cannot manage to do this then she loses immediately.

The play then continues as in $\mathbf{EF}^p_r(A,B)$, except that if at any stage $\forall$ exercises his option of removing a pair of pebbles, and as a result, either there are no pebbles left in play or all his remaining pebbles occupy a single point $a \in A$, then $\forall$ loses the play and the game is terminated.

Clearly, if $\forall$ has a winning strategy in this game then he has a winning strategy in which he never places two pebbles on the same point $a \in A$ — and vice versa.

See exercise 2 below. So we assume that the game rules additionally state that he always plays this way, whether or not he has a winning strategy. If he ever places a pebble on top of another (in A), he loses. All other rules of $EF_r^p(A,B)$ remain unchanged from $\mathbf{EF}_r^p(A,B)$.

Exercises

1. Show that if $\forall$ has a winning strategy in $\mathbf{EF}_r^p(A,B)$ then he has a winning strategy in which he only removes at most one pair of pebbles from play in any round, and then only if he has to in order to create a spare pebble pair to bring back into play in the main part of the round.

2. Show that $\forall$ has a winning strategy in the game $EF_r^p(A,B)$ as first defined iff he has a winning strategy in which he never places two pebbles on the same element of A. Don't forget to consider the case where $|A| < p$.

3. If L and r are finite and A,B are L-structures, show that $\exists$ has a winning strategy in $\mathbf{EF}_r^p(A,B)$ if and only if $A \models \sigma \Rightarrow B \models \sigma$ for every positive existential L-sentence σ written with at most p variables, possibly re-used, and of quantifier depth at most r. [In example 16.3 above, a relevant sentence is $\exists x_0,\ldots,x_{|B|} \bigwedge_{i<j\leq|B|} E(x_i,x_j)$, where E is the edge relation (playing the role of $\neq$). A similar result for arbitrary L and $r = \omega$ can also be proved [Bar77].]

4. Show that if $\exists$ has a winning strategy in the standard game $\mathbf{EF}_r^p(A,B)$ then she has one in the modified game $EF_r^p(A,B)$ too. Find r, L, and L-structures A,B such that $\exists$ has a winning strategy in $EF_r^2(A,B)$ but not $\mathbf{EF}_r^2(A,B)$.

16.2 The rainbow algebra $\mathcal{A}_{A,B}$

Until the end of this chapter, we fix structures A, B in a signature L consisting of only at most unary and binary relation symbols. We will define an atomic non-associative algebra $\mathcal{A}_{A,B}$, by listing its atoms, stating their converses, and then defining composition on the atoms. We can then define $\mathcal{A}_{A,B}$ by taking the power set of this set of atoms, and define conversion and composition by an infinitary distribution rule: that is, $\mathcal{A}_{A,B}$ will be the complex algebra over the atom structure we define. (Any subalgebra with the same atoms will also do.) The reader should not confuse the structure A with the domain of the algebra $\mathcal{A}_{A,B}$.

We will regard the atoms as coloured. (This gave rise to the name 'rainbow construction', which was coined by Yde Venema.) The identity, '1', is an atom with no colour. Nonidentity atoms are of five different colours: green, white, yellow,

16.2. The rainbow algebra $\mathcal{A}_{A,B}$

black, and red, and here they are:

$$
\begin{aligned}
\text{Green} &= \{g_a : a \in A\} \\
\text{White} &= \{w\} \cup \{w_S : S \subseteq A, |S| \leq 2\} \\
\text{Yellow} &= \{y\} \\
\text{Black} &= \{b\} \\
\text{Red} &= \{r_{bb'} : b, b' \in B\}.
\end{aligned}
$$

The atoms shown are all distinct. Note that if A and B are finite, there are finitely many atoms and $\mathcal{A}_{A,B}$ will be finite.

All atoms are self-converse ($\check{x} = x$) except the red atoms, for which we have $(r_{bb'})^\vee = r_{b'b}$.

Composition in an atomic non-associative algebra may be defined by listing all triples of atoms (x, y, z) such that $x; y \geq \check{z}$ (see definition 3.25). These are the *consistent triples* of atoms. If (x, y, z) is a consistent triple, then so are its Peircean transforms (y, z, x), (z, x, y), $(\check{x}, \check{z}, \check{y})$, $(\check{y}, \check{x}, \check{z})$, and $(\check{z}, \check{y}, \check{x})$. A triple is *forbidden* if it is not consistent. We define the forbidden triples of atoms of $\mathcal{A}_{A,B}$ to consist of all Peircean transforms of the following, where the indices a, a' range over A and the indices $b, b', b_1, b_1', b_2, b_2', b_3, b_3'$ range over B.

$$(1', s, t) \quad \text{unless } s = \check{t} \text{ (any atoms } s, t) \tag{16.1}$$

$$(g, g', g''), (g, g', W) \quad \text{any green atoms } g, g', g'' \text{ and any white atom } W \tag{16.2}$$

$$(y, y, y), (y, y, b) \tag{16.3}$$

$$(g_a, y, w_S) \quad \text{unless } a \in S \tag{16.4}$$

$$(g_a, g_{a'}, r_{b'b}) \quad \text{unless } \{(a, b), (a', b')\} \text{ is a partial homomorphism} \tag{16.5}$$

$$(r_{b_1 b_1'}, r_{b_2 b_2'}, r_{b_3 b_3'}) \quad \text{unless } b_1' = b_2, \, b_2' = b_3, \text{ and } b_3' = b_1. \tag{16.6}$$

We will refer to these rules as 'rules 16.1 to 16.6'. This completes the definition of $\mathcal{A}_{A,B}$. The atom structure just defined satisfies all provisions of definition 3.22 except perhaps the associativity axiom, and so $\mathcal{A}_{A,B}$ is a non-associative algebra. (Indeed, by exercise 16.4(2), $\mathcal{A}_{A,B} \in \mathbf{SA}$.)

Our aim is to prove:

THEOREM 16.5 (rainbow theorem) *Let A, B be structures in a relational signature L consisting of only at most unary and binary relation symbols. Let $p, r \leq \omega$. Then $\exists$ has a winning strategy in the game $G_{1+r}^{2+p}(\mathcal{A}_{A,B})$ if and only if she has a winning strategy in the game $EF_r^p(A, B)$.*

Here, '+' denotes ordinal sum, so that $2 + \omega = \omega$, for example.

REMARK 16.6 The cases where $p < 2$ or $r < 2$ are degenerate and we make no use of them later. By exercise 12.4(1), $\exists$ has a winning strategy in $G_3^4(\mathcal{A}_{A,B})$ if and only if $\mathcal{A}_{A,B}$ is associative (and hence a relation algebra). By the rainbow theorem, if $\exists$ has a winning strategy in $EF_2^2(A,B)$ then $\mathcal{A}_{A,B}$ will be a relation algebra. The game $EF_2^2(A,B)$ is near-degenerate and in almost all cases — certainly in the cases that concern us in chapter 17 — $\exists$ will have a winning strategy in it, so the algebra $\mathcal{A}_{A,B}$ will in fact be a relation algebra.

Theorem 16.5 will follow from corollary 16.9 and proposition 16.13 below, which prove the directions '$\Rightarrow$' and '$\Leftarrow$', respectively.

16.3 How $\forall$ can win $G(\mathcal{A}_{A,B})$

We refer to definition 12.24 for the rules of the game G_r^n. In the remainder of the chapter, L, A, and B will be as stated in theorem 16.5, and when we say 'network' we will mean an atomic $\mathcal{A}_{A,B}$-network whose set of nodes is arbitrary, not necessarily an ordinal $\leq \omega$. Recall (from just after definition 7.1) that a network N is *strict* if for all $x, y \in N$, if $N(x,y) = 1$' then $x = y$.

First, a definition and an observation.

DEFINITION 16.7 A *red clique* is a possibly empty network R such that for any distinct elements m, n of R, the label $R(m, n)$ is a red atom.

The observation is that given any red clique R with at least two nodes, each node $n \in R$ can be given a unique index $\beta(n, R) \in B$ such that if $m, n \in R$ are distinct then $N(m, n) = r_{\beta(m, R), \beta(n, R)}$. (Explicitly, we may let $\beta(n, R) = b$, where $N(n, m) = r_{bb'}$ for some $m \in R$ and $b' \in B$; rule 16.6 shows that this is well-defined.) When the context is clear, we will refer to the index of the node n simply as $\beta(n)$.

PROPOSITION 16.8 *Suppose that $p, r \leq \omega$ and that $\forall$ has a winning strategy in $EF_r^p(A, B)$. Then he has a winning strategy in $G_{1+r}^{2+p}(\mathcal{A}_{A,B})$.*

Proof. We have seen that if p, r, or $|A|$ is less than 2 then $EF_r^p(A, B)$ is degenerate and $\forall$ cannot win. So assume that $p, r, |A| \geq 2$.

Let $\forall$ start a play of $G_{1+r}^{2+p}(\mathcal{A}_{A,B})$ with the following sequence of moves. In round 0, he picks the atom w. $\exists$ must respond with a network N_0, consisting (without loss of generality) of just two nodes, c, c', with $c \neq c'$ and $N_0(c, c') = $ w. In subsequent rounds, $\forall$ never deletes these two nodes.

Next, $\forall$ starts a private play of $EF_r^p(A, B)$ to help him choose his subsequent moves in $G_{1+r}^{2+p}(\mathcal{A}_{A,B})$. Suppose his winning strategy for $EF_r^p(A, B)$ tells him to place two pebbles on the (distinct) elements $a, a' \in A$ in the combined round 0 and round 1. Then in round 1 of $G_{1+r}^{2+p}(\mathcal{A}_{A,B})$ he chooses the edge (c, c') and the atoms

16.3. How ∀ can win $G(\mathcal{A}_{A,B})$

g_a, y. Note that (g_a, y, w) is a consistent triple of atoms according to the definition of $\mathcal{A}_{A,B}$, so $g_a ; y \geq w = N_0(c, c')$ and this is a legal move. In round 2, he chooses the same edge (c, c') and the atoms $g_{a'}, y$. Since $a \neq a'$, after two rounds this forces ∃ to construct a strict network N_2 with distinct nodes c, c', n, n'. She has to choose a label for the edge (n_0, n_1) in N_2. This label cannot be 1', green, or white (by rules 16.1 and 16.2 applied to the triangle (n, n', c)), nor yellow or black (by rule 16.3 on triangle (n, n', c')). So it must be red — say, $r_{bb'}$, for some $b, b' \in B$. See figure 16.1. In ∀'s private play of $EF_r^p(A, B)$, he now lets ∃ place two pebbles

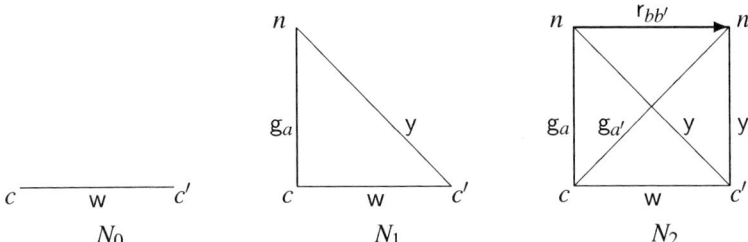

Figure 16.1: ∀'s first three moves

on b and b', corresponding to his pebbles on a, a', respectively.[1]

We can now finish off the case $p = 2$. We are assuming that ∀ has a winning strategy in $EF_r^p(A, B)$; but when $p = 2$, by removing a pair of pebbles he would leave only one pair in play and according to the rules he would lose at this point. Hence we must assume that he has won $EF_r^p(A, B)$ already after rounds 0 and 1, and he never has to remove a pair of pebbles. This can only mean that the map $\{(a, b), (a', b')\}$ is not a partial homomorphism from A to B. By rule 16.5, the triangle (c, n, n') is inconsistently labelled in N_2 and ∃ has lost $G_{1+r}^{2+p}(\mathcal{A}_{A,B})$.

So assume that $p \geq 3$. In round t $(3 \leq t < 1+r)$, suppose inductively that the current network N_{t-1} has nodes $\{c, c', n_0, \ldots n_{q-1}\}$ for some $q < \omega$ with $2 \leq q \leq p$, and that for each $j < q$, $N_{t-1}(c', n_j) = y$ and $N_{t-1}(c, n_j) = g_{a_j}$ for some $a_j \in A$, the a_j being pairwise distinct. Thus, N_{t-1} is a strict network. As above, the labels on the edges (n_j, n_k) must be red (for distinct $j, k < q$), so the subnetwork of N_{t-1} with nodes $\{n_j : j < q\}$ forms a red clique of size at least two. Hence, each node n_j has an index $\beta(n_j) \in B$. As part of an inductive hypothesis, suppose also that at the start of round $t-1$ of $EF_r^p(A, B)$, ∃ has not lost yet, ∀ is still using his winning strategy, and the situation corresponds to the situation in round t of $G_{1+r}^{2+p}(\mathcal{A}_{A,B})$. That is, there are pairs of pebbles on $(a_j, \beta(n_j))$ for each $j < q$.

---

[1] This shows why we modified the Ehrenfeucht–Fraïssé game. The nodes in a red clique do not have indices until there are at least two of them. In the play of $EF_r^p(A, B)$, ∃ does not have to decide where to place her pebbles in B until there are two distinct pebbles in A.

By exercise 16.1(1), we can suppose that ∀ only removes a single pair of pebbles and only when he has to. If the number q of pairs of pebbles is already p, then ∀'s strategy will direct him to remove a pair of pebbles, say the pair on $(a_j, \beta(n_j))$ for some $j < q$. In this case there must be $2 + p$ nodes in the current network N_{t-1}, and ∀ removes the corresponding node n_j in the play of $G_{1+r}^{2+p}(\mathcal{A}_{A,B})$. Since $p \geq 3$, removing the pair $(a_j, \beta(n_j))$ still leaves at least two distinct pairs of pebbles, so the play of $EF_r^p(A,B)$ continues. Similarly, removing n_j still leaves at least two nodes in the red clique, so they retain their indices.

Suppose that ∀'s winning strategy for $EF_r^p(A,B)$ now tells him to place a pebble on $a \in A$, say. In the play of $G_{1+r}^{2+p}(\mathcal{A}_{A,B})$ he picks the edge (c, c') — as always — and the atoms g_a, y. That completes his move.

By the rules of $EF_r^p(A,B)$, a is distinct from all other points in A covered by pebbles, so this forces ∃ to add a new node n to the network. As we have seen previously, n must be part of a red clique of size at least two. So n has an index $\beta(n) \in B$. In ∀'s private play of $EF_r^p(A,B)$, he lets ∃ place her corresponding pebble on $\beta(n)$.

In this way our inductive assumptions are maintained for another round.

Because ∀ has a winning strategy in $EF_r^p(A,B)$, there will come a time when ∀ places a pebble on $a \in A$ but there is nowhere in B for ∃ to place the other pebble while maintaining a partial homomorphism. At this point, no matter how ∃ has played in $G_{1+r}^{2+p}(\mathcal{A}_{A,B})$, the map $\{(a_j, \beta(n_j)) : j < q\}$ is not a partial homomorphism. Because the signature of A and B contains only unary and binary relations, there must exist $j < j' < q$ such that $\{(a_j, \beta(n_j)), (a_{j'}, \beta(n_{j'}))\}$ is not a partial homomorphism. But this means that the triangle $(n_j, n_{j'}, c)$ is not consistent with rule 16.5, and so ∃ has lost in the game $G_{1+r}^{2+p}(\mathcal{A}_{A,B})$. □

COROLLARY 16.9 *Suppose that $p, r \leq \omega$ and that ∃ has a winning strategy in $G_{1+r}^{2+p}(\mathcal{A}_{A,B})$. Then she has a winning strategy in $EF_r^p(A,B)$.*

Proof. Assume that ∃ has a winning strategy in $G_{1+r}^{2+p}(\mathcal{A}_{A,B})$. Clearly, ∀ cannot also have a winning strategy in this game. By proposition 16.8, ∀ has no winning strategy in $EF_r^p(A,B)$. Let ∃ play $EF_r^p(A,B)$ according to the strategy: 'do not allow ∀ to get into a position from which he has a winning strategy'. The initial (empty) position satisfies this condition; and if in the current position, ∀ has no winning strategy, then whatever move ∀ makes, ∃ must obviously have a response to create a position from which he has no winning strategy. So she can implement this strategy throughout the game. She never loses (a lost position would certainly be one at which ∀ has a winning strategy), so this must be a winning strategy for her. (We have simply proved that the game $EF_r^p(A,B)$ is determined; cf. [GalSte53]. We saw determinacy of representation-building games in chapter 7; the idea here is similar but applied to back-and-forth games.) □

16.3. How ∀ can win $G(\mathcal{A}_{A,B})$

Discussion

In the proof of proposition 16.8, ∀ in effect forced ∃ to attempt the game $EF_r^p(A,B)$ — and he proceeded to defeat her. He did it using 'his' green atoms, forcing ∃ to use 'her' red atoms to build a red clique on a 'base' (c,c').

DEFINITION 16.10 Let (c,c') be any edge in a strict network N (with $c \neq c'$). Define the *red clique based on* (c,c') *in N,* denoted $R_N(c,c')$, to be the subnetwork of N with nodes

$$\{n \in N : N(c,n) \text{ is green and } N(c',n) \text{ is yellow}\}.$$

$R_N(c,c')$ may be empty, but by rules 16.1–16.3 it is certainly a red clique.

In proposition 16.13 below, we are going to prove the converse of proposition 16.8 — that if ∃ has a winning strategy in $EF_r^p(A,B)$ then she has one in $G_{1+r}^{2+p}(\mathcal{A}_{A,B})$. The idea behind the proof is not hard: if ∀ plays the same strategy used in proposition 16.8 then ∃ can win by using a winning strategy in $EF_r^p(A,B)$. The proof is complicated somewhat, because we cannot assume that ∀ plays in this way. We have to provide a winning ∃-strategy to cover every eventuality. But it turns out that ∀'s best strategy is more or less as described in proposition 16.8, in that the only way he could win would be to defeat ∃ on red cliques. So we will focus on red cliques in the proof.

∃ will associate a private Ehrenfeucht–Fraïssé game $EF_r^p(A,B)$ with each red clique $R_N(c,c')$ of size at least two, where N is the current network, $c,c' \in N$ are distinct, and $N(c,c') \neq w_S$ (for any S). These games will help her win the main game $G_{1+r}^{2+p}(\mathcal{A}_{A,B})$.

REMARK 16.11 The case where $N(c,c') = w_S$ (some S) represents a degenerate clique. In a sense, such a clique is 'frozen', in that ∀ can only force a red clique of size $|S| \leq 2$ based on the edge (c,c'). From this position he can make no further progress.

These atoms w_S and the freezing of cliques form a critical part of the rainbow construction. To see why they are necessary, consider the possible course of events in a play of G_ω^ω on an algebra $\mathcal{A}'$ similar to $\mathcal{A}_{A,B}$ but without w_S atoms and with rule 16.4 of the definition of $\mathcal{A}_{A,B}$ deleted. For definiteness, we let A be the complete (undirected loop-free) graph on $\{0,1,2\}$, and B be the disjoint union of A and the complete graph on $\{3,4\}$. See figure 16.2.

∃ has a winning strategy in $EF_\omega^\omega(A,B)$, namely, 'copy ∀'s moves'. By proposition 16.13 below, she has a winning strategy in $G_\omega^\omega(\mathcal{A}_{A,B})$. However, the following strategy in $G_\omega^\omega(\mathcal{A}')$ is winning for ∀: in fact, it wins $G_4^5(\mathcal{A}')$ for him. In his first three moves, by choosing the atom r_{34} and then playing so as to create c, c', ∀ can create the third network N_2 shown in figure 16.3. ∃ has to label the edge (c,c')

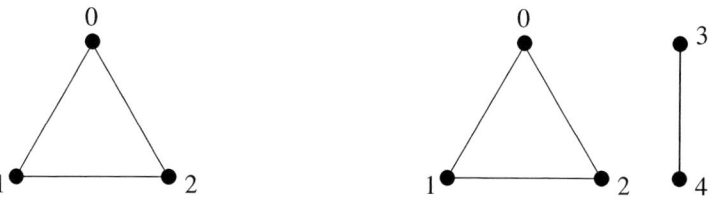

Figure 16.2: The graphs A and B

in N_2. Whatever she chooses, $\forall$ then chooses the nodes c, c' and atoms g_2, y, as shown on the right of figure 16.3. This move would be illegal if $\exists$ had been able to label (c, c') with $w_{\{0,1\}}$ (as per her strategy in proposition 16.13). Here, it is legal, and $\exists$ now faces a clique-labelling problem similar to that in proposition 16.8. She has to label the edges (n_0, n_2) and (n_1, n_2) in the next network N_3, necessarily with red atoms r_{3b}, r_{4b} (respectively) for some $b \in B$ such that $(3, b)$ and $(4, b)$ are both edges of B (because $(0, 2)$ and $(1, 2)$ are edges of A). There is no such b, and $\exists$ loses.

What went wrong is that $\forall$ was able to choose the first red edge in a red clique that he then forced $\exists$ to build, as she did in proposition 16.8. The atoms w_S are introduced to stop him doing this, by 'freezing' any clique that he tries to initiate, so that $\exists$ can choose all the edges in every red clique with more than two elements. The introduction of these atoms is entirely motivated by games. We will see similar atoms (u and v) in chapter 18.

16.4 How $\exists$ can win $G(\mathcal{A}_{A,B})$

Before we show how $\exists$ can win $G_{1+r}^{2+p}(\mathcal{A}_{A,B})$, we introduce a harmless assumption which applies for the rest of the chapter. We will use this assumption later.

ASSUMPTION 16.12 In a play of some $G_r^p(\mathcal{A}_{A,B})$, suppose that in some round, t, $\forall$ picks an edge (m, n) from the current network N (N is either N_{t-1} or what remains of it after $\forall$ has removed a node), and atoms $a, b \in \mathcal{A}_{A,B}$ with $a; b \geq N(m, n)$. If there is already a node $l \in N$ such that $N(m, l) = a$ and $N(l, n) = b$, then $\forall$ wasted the round because $\exists$ does not need to extend the network N for her move — she may set $N_t = N$. There is no advantage to $\forall$ in making such a move, and we assume that he never does so. If he gets stuck as a result of this requirement, then he loses.

PROPOSITION 16.13 Suppose that $p, r \leq \omega$ and that $\exists$ has a winning strategy in $EF_r^p(A, B)$. Then she also has a winning strategy in $G_{1+r}^{2+p}(\mathcal{A}_{A,B})$.

16.4. How ∃ can win $G(\mathcal{A}_{A,B})$

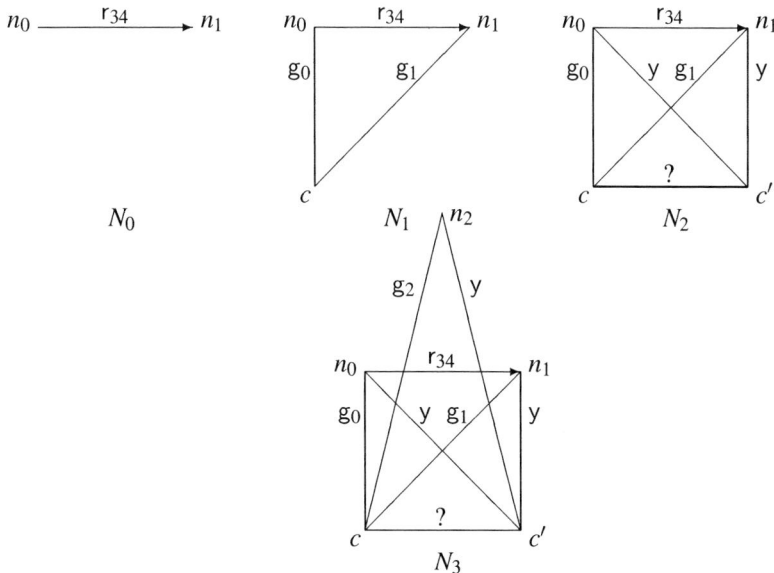

Figure 16.3: ∀'s successive moves in $G_4^5(\mathcal{A}')$

Proof. First let us deal with the case where either p or r is less than 2. By considering the rules of the game $G_{1+r}^{2+p}(\mathcal{A}_{A,B})$ we see that we will never get a network with more than three nodes. There is only one case where ∃ has to choose the label on an edge of such a network, and that is when we have a two-node network N with nodes n_0, n_1, say, and ∀ picks a reflexive edge, say (n_0, n_0), and a pair of atoms $a, \breve{a}$. ∃ has to add a new node n_2 and choose the label on the edge (n_2, n_1) (by lemma 11.2, this determines the label on the edge (n_1, n_2)). But she can always do this, because whatever the values of $a, N(n_0, n_1)$ are, we have $s = \breve{a}; N(n_0, n_1) \neq 0$; this may be checked using the definition of $\mathcal{A}_{A,B}$. ∃ may choose any atom $e \leq s$ to label (n_2, n_1). To be specific, we let her take $e = b$ unless $a = N(n_0, n_1) = y$, in which case she objects that ∀ should not have made such a move according to assumption 16.12, or $N(n_0, n_1) = 1$', in which case she lets $e = \breve{a}$. Thus, ∃ certainly has a winning strategy if either p or r is less than two, though the game is rather degenerate.

So assume that $p, r \geq 2$, and assume the hypotheses of the proposition. In each round of $G^{2+p}_{1+r}(\mathcal{A}_{A,B})$, $\exists$ plays using the following strategy.

In round 0, let $\forall$ play the atom a. If $a = 1'$, $\exists$ responds with the network N_0 with a single node, 0, with $N_0(0,0) = 1'$. If not, she plays the network N_0 with two nodes, 0, 1, say, with $N_0(0,0) = N_0(1,1) = 1'$, $N_0(0,1) = a$, and $N_0(1,0) = \breve{a}$. It is easily checked that N_0 is a strict atomic $\mathcal{A}_{A,B}$-network.

At the start of the tth round ($1 \le t < 1+r$), let the current network be N_{t-1}. We are going to describe how $\exists$ responds to a move of $\forall$ in this round, and to do this, we will again assume inductively that the network N_{t-1} has certain properties. We will check that these properties are preserved by $\exists$'s actions.

Some terminology will be very useful. $\forall$ may first choose to delete a node from N_{t-1}, leaving a subnetwork N; if he does not, let $N = N_{t-1}$. If $\forall$ then selects $(e, e') \in N$ and atoms $\varepsilon, \varepsilon'$, and if $\exists$ makes a proper extension $N_t \supset N$ with a new node v with $N_t(e, v) = \varepsilon$ and $N_t(v, e') = \varepsilon'$, then we say that the labels on the edges (e, v), (v, e), (e', v), (v, e'), and (v, v) of N_t are *chosen by* $\forall$. All labels on other edges involving the new node v are said to be *chosen by* $\exists$. Labels of edges not involving the new node v will have been chosen by $\forall$ or $\exists$ in earlier rounds on the same principles. All labels on edges of the initial network N_0 are chosen by $\forall$.

Inductive assumptions on N_{t-1}

I. In the play so far, $\exists$ has never chosen green or yellow atoms as labels for edges, nor the white atom w.

II. N_{t-1} is strict — if $m, n \in N_{t-1}$ are distinct then $N_{t-1}(m, n) \ne 1'$.

III. For each ordered pair (c, c'), where c, c' are distinct nodes of N_{t-1} with $|R_{N_{t-1}}(c, c')| \ge 2$ (see definition 16.10 for the notation) and $N_{t-1}(c, c') \ne w_S$ (any S), there is an associated private play of $EF_r^p(A, B)$, denoted $EF_r^p[c, c']$, in which $\exists$ is using her winning strategy. Further, there is a one-one correspondence between the pairs of pebbles currently in play in $EF_r^p[c, c']$ and the nodes of $R_{N_{t-1}}(c, c')$, such that for each such pair, if the elements of A, B covered by the pebbles in the pair are $a \in A, b \in B$, then the corresponding node $n \in R_{N_{t-1}}(c, c')$ satisfies

- $N_{t-1}(c, n) = g_a$,
- $\beta(n, R_{N_{t-1}}(c, c')) = b$.

IV. There are at least as many rounds remaining in the play of any private game $EF_r^p[c, c']$ as there are rounds remaining in the main game $G^{2+p}_{1+r}(\mathcal{A}_{A,B})$.

V. N_{t-1} is a network.

16.4. How $\exists$ can win $G(\mathcal{A}_{A,B})$

We will show that $\exists$ has a strategy that can maintain all these assumptions to the next round. Our main aim is to prove inductive assumption V — this will show that $\exists$ has a winning strategy and prove the proposition. We need assumptions III and IV in order that we can define $\exists$'s strategy at all.

These properties hold trivially in N_0. Assuming they hold in N_{t-1}, we must make sure they hold in N_t.

$\forall$'s move   Let $\forall$ begin his move now. Suppose that he removes some node x from N_{t-1} to create a network $N \subseteq N_{t-1}$. At this point $\exists$ adjusts the private games so that the inductive assumptions still hold in N. This is mostly trivial to do, as follows. For any $x' \in N_{t-1}$, there may be an associated play $EF_r^p[x,x']$. If so, there are no winners in this game nor any survivors: this play is abandoned. Similarly, plays $EF_r^p[x',x]$ are also abandoned.

For each $c,c' \in N_{t-1}$ such that $x \in R_{N_{t-1}}(c,c')$, if there is an associated play $EF_r^p[c,c']$ then she gets $\forall$ to exercise his option (see definition 16.2) to remove the pair of pebbles associated with x in the play $EF_r^p[c,c']$. If $|R_N(c,c')| < 2$ (because removing x from the red clique based on (c,c') brought the size of the clique below 2) then removing the associated pair of pebbles in $EF_r^p[c,c']$ leaves only one pair of pebbles in play. According to the rules, $\exists$ wins and this play is terminated.

If $\forall$ does not remove a node from N_{t-1}, we define N to be N_{t-1}. In either case, we have arranged that our inductive assumptions hold in N.

Then $\forall$ picks the edge (e,e') (say) from N, and the pair of atoms $\varepsilon, \varepsilon' \in \mathcal{A}_{A,B}$. So the triple $(\varepsilon, \varepsilon', N(e',e))$ is consistent with the original rules of the definition of $\mathcal{A}_{A,B}$.

$\exists$'s response   By assumption 16.12, there is not already a node $z \in N$ such that $N(e,z) = \varepsilon$ and $N(z,e') = \varepsilon'$. It follows that $\varepsilon, \varepsilon' \neq 1'$. Now $\exists$ forms N_t by adding to N a new node, v, defining $N_t(v,v) = 1'$, $N_t(e,v) = \varepsilon$, $N_t(v,e') = \varepsilon'$, and, necessarily, $N_t(v,e) = \breve{\varepsilon}$ and $N_t(e',v) = \breve{\varepsilon}'$. This is well-defined if $e = e'$, for then, the triple $(\varepsilon, \varepsilon', 1')$ is consistent, so $\varepsilon = \breve{\varepsilon}'$ by rule 16.1 of the definition of $\mathcal{A}_{A,B}$.

Now $\exists$ has to choose labels for the new edges $(v,x), (x,v)$ (for $x \in N \setminus \{e, e'\}$), in such a way that the resulting graph forms an atomic network $N_t \supseteq N$ meeting the inductive assumptions above. We know by lemma 11.2 that the label $N_t(v,x)$ must always be the converse of $N_t(x,v)$, so she need only specify one of them. We will first state which labels $\exists$ chooses and make any definitions needed to keep the inductive conditions, and then we will check that these inductive conditions are indeed re-established. In particular, we will check that $\exists$'s choices result in a consistent network N_t.

When choosing a label for (v,x), $\exists$'s first concern is that the triangles (v,x,e) and (v,x,e') are rendered consistent in N_t. The labels on (x,e) and (e,v) are already

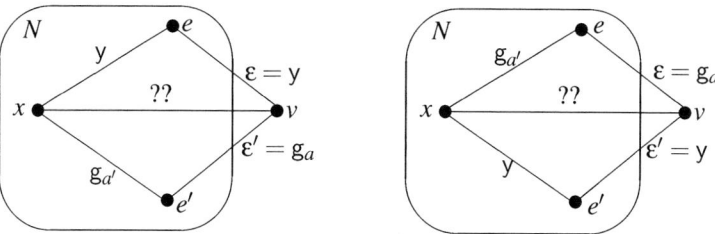

Figure 16.4: ∃ must adjoin v to the red clique containing x

fixed, and she certainly cannot select a label for (v,x) that causes (v,x,e) to violate the rules, or N_t would not be a network. The same goes for (v,x,e').

First, she defines $N_t(v,x)$ for all those x satisfying conditions 1 or 2 below.

1. If there is no 'green path' from v to x via e or e' — that is, it is not true that ε and $N(e,x)$ are green or that ε' and $N(e',x)$ are green — then ∃ can choose a white atom for $N_t(v,x)$. She picks w_S, where

$$S = \{a \in A : g_a \in \{\varepsilon, \varepsilon', N(e,x), N(e',x)\}\}.$$

(By the case assumption, S has size at most two, so the definition makes sense.)

2. Otherwise, if there is no yellow path from v to x via e or e' — that is, it is not true that $\varepsilon = N(e,x) = y$ or that $\varepsilon' = N(e',x) = y$ — then ∃ defines $N_t(v,x) = b$.

Now she simultaneously considers all remaining $x \in N \setminus \{e,e'\}$, if any. For each such x, there are both green and yellow paths from v to x via e or e', so one of the situations shown in figure 16.4 applies — either the left or the right, and the same one for each x, because $\varepsilon, \varepsilon'$ are fixed for this round. The indices a, a' of the green atoms shown can be any distinct (by assumption 16.12) elements of A. Assume that the second case shown (right) is the one that holds; the other situation is dealt with similarly (just swap e and e'). Thus, the set of x that ∃ now considers is:

$$R_N(e,e') = \{x \in N : N(e,x) \text{ is green and } N(e',x) = y\}.$$

We can suppose that $R_N(e,e') \neq \emptyset$, for otherwise there are no labels to choose here. So we must have $e \neq e'$. ∃ labels (x,v) for every $x \in R_N(e,e')$ according to the following three cases:

16.4. How ∃ can win $G(\mathcal{A}_{A,B})$

3. $N(e,e') = w_S$ for some $S \subseteq A, |S| \leq 2$.

   Take $x \in R_N(e,e')$ and let $N(x,e) = g_{a'}$, as in the second diagram in figure 16.4. ∃ starts a new game $EF_r^p(A,B)$ by getting ∀ to place two pebbles on $a, a' \in A$. As $a \neq a'$, this is legal. She uses her winning strategy in this game to place two corresponding pebbles on $b, b' \in B$. Then she labels the edge (x,v) with the atom $r_{b'b}$.

4. $N(e,e') \neq w_S$ (any S) and $|R_N(e,e')| = 1$.

   ∃ proceeds exactly as for case 3. Additionally, inductive hypothesis III requires that a private play $EF_r^p[e,e']$ of $EF_r^p(A,B)$ be defined. She defines it to be the play that she just started — it is well-defined because $|R_N(e,e')| = 1$.

5. Otherwise: $N(e,e') \neq w_S$ for any S, and $|R_N(e,e')| \geq 2$.

   In this case, by our inductive assumptions, there is already an Ehrenfeucht–Fraïssé game $EF_r^p[e,e']$ in progress. Since $|N| < 2 + p$ and $e, e' \notin R_N(e,e')$, we have $|R_N(e,e')| < p$, so fewer than p pairs of pebbles are in play in this game. ∃ picks up a spare pebble pair and, playing the role of ∀, places one of the pebbles in the pair on $a \in A$ (recall that $\varepsilon = g_a$). She uses her winning strategy to respond by placing the other one on $b \in B$, say. Then for each $x \in R_N(e,e')$, she labels the edge (x,v) with the atom $r_{\beta(x),b}$; here, $\beta(x)$ is of course $\beta(x, R_N(e,e'))$.

This completes the definition of N_t and of the strategy for ∃ in $G_{1+r}^{2+p}(\mathcal{A}_{A,B})$.

Inductive assumptions are still true

Clearly ∃ never chose green or yellow atoms nor w, 1'. This is enough to show that N_t is strict. So the inductive assumptions I and II hold for N_t.

Let us check inductive assumption III now. Let $R_{N_t}(c,c')$ be a red clique with $N_t(c,c') \neq w_S$ (any S) and $|R_{N_t}(c,c')| \geq 2$. So $c \neq c'$. We have to check that there is an associated game $EF_r^p[c,c']$. If $R_{N_t}(c,c') \cup \{c,c'\} \subseteq N$, we already know that there is a suitable game in progress because the inductive assumptions hold in N. So we can suppose otherwise.

Now $c, c' \in N$. For assume not — say, $c \notin N$. Then c is the new node added in the current round (i.e., $c = v$), and at most two edges incident with c can have labels chosen by ∀. Since $|R_{N_t}(c,c')| \geq 2$, there must be at least two distinct nodes $n, m \in R_{N_t}(c,c')$ with $(c,n), (c,m)$ labelled with green atoms and $(c',n), (c',m)$ labelled by y. So ∃ must have chosen the labels on all other edges incident with c: in particular, she chose the label on (c,c'). The only green edges incident with c are (c,n) and (c,m); and since (n,c') and (m,c') are yellow, there is no 'green path' from c to c'. Hence, according to ∃'s strategy (case 1), she would have chosen $N_t(c,c') = w_S$ for some S. But we are assuming that $N_t(c,c') \neq w_S$. This is a contradiction, and we obtain $c \in N$. The proof that $c' \in N$ is similar.

So we can suppose that $R_{N_t}(c,c') \not\subseteq N$ and $c,c' \in N$. Thus, the new node v lies in $R_{N_t}(c,c')$. In the definition of $\exists$'s strategy, this situation is covered by cases 4 and 5: where $\forall$ chose the nodes $c,c' \in N$ with $N(c,c') \neq w_S$, $\exists$ has to choose a red atom to label an edge (x,v), and where $|R_N(c,c')|$ is 1 or more than 1 respectively. In these two cases, her strategy was designed precisely to maintain inductive assumption III.

Inductive assumption IV is easy. Until round 2 is complete in $G_{1+r}^{2+p}(\mathcal{A}_{A,B})$ (when there are four nodes in the network N), there can be no red clique $R_N(c,c')$ of size two or more. At the end of round 2, it may be that such a clique comes into being and that a game $EF_r^p[c,c']$ is started. This would involve playing the combined rounds 0 and 1 of this game. That still leaves $r-2$ rounds left of $EF_r^p[c,c']$ (or ω rounds if $r = \omega$) which is the same as the number of rounds $1+r-3$ remaining in $G_{1+r}^{2+p}(\mathcal{A}_{A,B})$ (or ω if $r = \omega$). So at this stage, the inductive assumption is true. In each subsequent round of $G_{1+r}^{2+p}(\mathcal{A}_{A,B})$, at most one round of any existing game $EF_r^p[c,c']$ is played, and therefore this inductive assumption will remain true. Games resulting from cliques formed after round 2 have even more rounds remaining, so the inductive assumption holds for them, too.

Inductive assumption V: N_t is a network For the resulting N_t to be a network, it is sufficient that every triangle (x,x',v), for distinct $x,x' \in N$, is consistent: i.e., $\bigl(N_t(x,x'), N_t(x',v), N_t(v,x)\bigr)$ is a consistent triple of atoms of $\mathcal{A}_{A,B}$. There are three kinds of triangle of this form: the triangle (e,e',v) (if $e \neq e'$), which $\forall$ chose and so is guaranteed consistent; 'primary' triangles, which involve e or e' and one other node of N, so that $\exists$ had to define only one of the labels on the edges of the triangle; and finally 'secondary' ones involving neither of e,e', in which she had to define two labels. See figure 16.5.

$\exists$'s first concern has been that all primary triangles created were consistent, and it is mostly evident that she was successful. Rules 16.2 and 16.3 are not violated in primary triangles, since her strategy never selects a green or yellow atom to label an edge (v,x), and only chooses a white (respectively, black) label when there is no green (respectively, yellow) path from v to x. Her indices for white atoms were chosen precisely to conform with rule 16.4 in primary triangles.

$\exists$ only chooses a red atom if the two atoms chosen by $\forall$ were green and yellow. Thus, rule 16.6 cannot be broken on primary triangles. Rule 16.5 does require a little checking. Suppose there is a primary triangle (v,w,x) (for some $w \in \{e,e'\}$, $x \in N \setminus \{e,e'\}$), two of whose edges are green and one red. Now $\exists$ chose the label on (v,x). As she never chooses green atoms to label edges, this must be the red

16.4. How ∃ can win $G(\mathcal{A}_{A,B})$

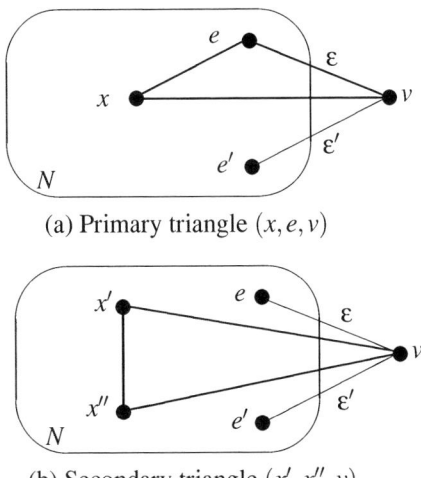

(a) Primary triangle (x, e, v)

(b) Secondary triangle (x', x'', v)

Figure 16.5: The triangles to check

edge. We therefore have

$$N_t(w,v) = g_a,\ N_t(w,x) = g_{a'},\ N_t(v,x) = r_{b,b'},$$

for some $a, a' \in A$, $b, b' \in B$. For her to choose a red label on $N_t(v,x)$ according to her strategy, ∀ must have selected the edge $(w, w') \in \{(e, e'), (e', e)\}$ and a green and a yellow atom for his move. In this case, ∃'s strategy ensures that $\{(a,b), (a',b')\}$ is part of a position in $EF_r^p[w,w']$ in which she is using her winning strategy (see cases 3–5 of ∃'s strategy), so that the map $\{(a,b), (a',b')\}$ is a partial homomorphism from A to B. Hence, the triangle (v, w, x) does not violate rule 16.5.

Moreover, all secondary triangles are consistent. We show this as follows. Note first that any secondary triangle has two edges coloured by ∃, hence two edges labelled by white, black, or red atoms. Inspection of the original definition of consistent triples of atoms of $\mathcal{A}_{A,B}$ shows that in such a case, inconsistency is only possible if all three edges are red, rule 16.6 being the rule at risk.

So let the new node be v as usual, and let (x', x'', v) be a secondary triangle with each edge labelled by a red atom. ∃ chose two red atoms as labels for (v, x') and (v, x''). She would only do this if ∀ picked the edge (e, e') (say) and the atoms $\varepsilon = g_a$ (for some $a \in A$) and $\varepsilon' = y$ (the other case, where he picked (e', e), is symmetrical), $N(e, x') = g_{a'}$, $N(e, x'') = g_{a''}$ (some $a', a'' \in A$), and $N(e', x') = N(e', x'') = y$. See figure 16.6.

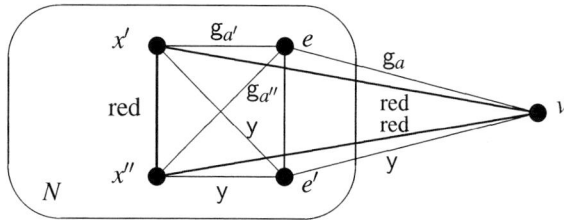

Figure 16.6: The secondary triangle x', x'', v

Claim. $N(e, e') \neq \mathsf{w}_S$ for any $S \subseteq A$.

Proof of claim. Suppose for contradiction that $N(e, e') = \mathsf{w}_S$ for some $S \subseteq A$. By rule 16.4 and the consistency of the triangles (e, e', x'), (e, e', x''), and (e, e', v), we know that $\{a, a', a''\} \subseteq S$. Since $|S| \leq 2$, two of a, a', a'' must be equal to each other.

If a is equal to either a' or a'' then $\forall$ has broken assumption 16.12 by making a move which does not require a proper extension to the network. So suppose otherwise, so that $a' = a''$. We'll see that this also involves a violation of assumption 16.12. Consider the subnetwork of N with nodes x', x'', e, e' (see figure 16.6). Which of these four nodes was added most recently in the play of the game? Whichever it is, it should be incident with at most two edges within the subnetwork that were labelled by $\forall$; the rest must have been labelled by $\exists$. If it was x', then $\forall$ must have labelled the edge (x', e) (by $g_{a'}$) and the edge (x', e') (by y), as $\exists$ never chooses these colours. But this breaks assumption 16.12, because in the round when x' was created, $\exists$ did not actually need to add the node x' as there was already the node x'' available. Similarly, x'' cannot have been the most recently added node (because x' was already there). If the most recent node was e, then in the round when e was added, $\forall$ labelled the edges (x', e) and (x'', e) with the atom $g_{a'}$. Consequently, $\exists$ must have chosen the label on (e, e'). This is also true if e' was the last node to be built, by similar reasoning. In either case, her strategy would have directed her to choose the atom $\mathsf{w}_{\{a'\}}$ to label (e, e'), as there was no green path from e to e' and the two yellow-green paths both involved the atoms y and $g_{a'}$. Thus, $N(e, e') = \mathsf{w}_{\{a'\}}$. But then, by the consistency of $\forall$'s triangle (e, e', v), we must have $a \in S$; so $a = a'$, contradicting our assumption. This proves the claim.

So $N(e, e') \neq \mathsf{w}_S$ (any S). Since $x', x'', v \in R_{N_t}(e, e')$, $\exists$ used part 5 of her strategy to assign a label to the edges (x', v) and (x'', v) of N_t — she will have defined $N_t(x', v) = \mathsf{r}_{\beta(x'), b}$ and $N_t(x'', v) = \mathsf{r}_{\beta(x''), b}$, for some $b \in B$ determined by her strategy in $EF_r^p[e, e']$. Here, $\beta(x')$ in full is $\beta(x', R_N(e, e'))$, and similarly for $\beta(x'')$. By definition of the indices $\beta(x'), \beta(x'')$, we have $N_t(x',x'') = N(x',x'') = r_{\beta(x'),\beta(x'')}$. Thus, rule 16.6 is not violated in triangle (x',x'',v).

We see that the triangle chosen by $\forall$, the primary triangles, and the secondary triangles are all consistent with rules 16.1 to 16.6. This shows that N_t is a network. We conclude that the given strategy for $\exists$ is winning in $G^{2+p}_{1+r}(\mathcal{A}_{A,B})$, as required. □

Corollary 16.9 and proposition 16.13 yield theorem 16.5.

16.5 Modifications to the rainbow algebra

For some applications, such as the one in section 17.7 to come, it is helpful to adjust the rainbow algebras $\mathcal{A}_{A,B}$ slightly. The crucial point to note in doing this is that the green atoms of an algebra $\mathcal{A}_{A,B}$ essentially belong to $\forall$, and the red ones to $\exists$. $\forall$ used the 'excess' of green atoms over red ones to defeat $\exists$ in the game $G^{2+p}_{1+r}(\mathcal{A}_{A,B})$ in proposition 16.8. Similarly, $\exists$ used the sufficiency of red atoms to win $G^{2+p}_{1+r}(\mathcal{A}_{A,B})$ in proposition 16.13. So informally we have:

1. If $\forall$ has a winning strategy in $G^{2+p}_{1+r}(\mathcal{A}_{A,B})$, and we add further green atoms and/or delete red atoms from $\mathcal{A}_{A,B}$, then he still has a winning strategy in $G^{2+p}_{1+r}(\mathcal{A}_{A,B})$. (We do not add white atoms w_S when we add green atoms.)

2. If $\exists$ has a winning strategy in $G^{2+p}_{1+r}(\mathcal{A}_{A,B})$, and we add further red atoms and/or delete green atoms from $\mathcal{A}_{A,B}$, then she still has a winning strategy in $G^{2+p}_{1+r}(\mathcal{A}_{A,B})$. (We do not delete any white atoms w_S for $S \subseteq A$, $|S| = 2$, when we delete green atoms.)

To formalise these statements, we would have to define precisely what is meant by a new 'red' or 'green' atom. This can be done by sensible modifications of the rules 16.1–16.6 in the definition of $\mathcal{A}_{A,B}$. Here, we will only use the following easy-to-state formal version, involving the red atoms. We use '⊆' to denote substructure in the conventional model-theoretic sense.

PROPOSITION 16.14 *Let L, A, B, p, r be as in theorem 16.5, and let $\mathcal{A}_{A,B}$ be as defined in section 16.2.*

1. *Let S be an atom structure with $S \subseteq \text{At}\,\mathcal{A}_{A,B}$ and containing every non-red atom of $\text{At}\,\mathcal{A}_{A,B}$. Suppose that $\forall$ has a winning strategy in $G^{2+p}_{1+r}(\mathcal{A}_{A,B})$. Then he also has a winning strategy in $G^{2+p}_{1+r}(\mathfrak{Cm}\,S)$.*

2. *Let B' be an L-structure with $B' \supseteq B$, and let $\mathcal{R}$ be an atom structure with $\text{At}\,\mathcal{A}_{A,B} \subseteq \mathcal{R} \subseteq \text{At}\,\mathcal{A}_{A,B'}$. (Here, we identify the non-red atoms of $\mathcal{A}_{A,B}$ with*

the corresponding atoms of $\mathcal{A}_{A,B'}$.) Assume that $\exists$ has a winning strategy in $G_{1+r}^{2+p}(\mathcal{A}_{A,B})$. Then she also has a winning strategy in $G_{1+r}^{2+p}(\mathfrak{Cm}\,\mathcal{R})$.

Proof. For part 1, assume that $\forall$ has a winning strategy in $G_{1+r}^{2+p}(\mathcal{A}_{A,B})$. So by theorem 16.5, he has a winning strategy in $EF_r^p(A,B)$. Now, the proof of proposition 16.8 gives a winning strategy for him in $G_{1+r}^{2+p}(\mathcal{A}_{A,B})$ that utilises only white, yellow, and green atoms. It never asks him to choose a red atom. Since $\mathcal{S}$ has all the non-red atoms of $\mathcal{A}_{A,B}$, he can use exactly this strategy in $G_{1+r}^{2+p}(\mathfrak{Cm}\,\mathcal{S})$. Any play of $G_{1+r}^{2+p}(\mathfrak{Cm}\,\mathcal{S})$ in which he uses the strategy is also a play of $G_{1+r}^{2+p}(\mathcal{A}_{A,B})$ in which he uses it. The resulting play is therefore a win for $\forall$ in $G_{1+r}^{2+p}(\mathcal{A}_{A,B})$. Since $\mathcal{S} \subseteq \mathrm{At}\,\mathcal{A}_{A,B}$, it is clear that the play is also a win for him in $G_{1+r}^{2+p}(\mathfrak{Cm}\,\mathcal{S})$.

The proof of the second part is similar. If $\exists$ has a winning strategy in the game $G_{1+r}^{2+p}(\mathcal{A}_{A,B})$, then as above, she has a winning strategy in $EF_r^p(A,B)$. This same strategy evidently wins $EF_r^p(A,B')$ for her, and it never chooses an element of $B' \setminus B$. The proof of proposition 16.13 now shows that she has a winning strategy σ in $G_{1+r}^{2+p}(\mathcal{A}_{A,B'})$ that only ever directs her to choose atoms in $\mathcal{A}_{A,B}$, and hence in $\mathcal{R}$. So she may use σ in the game $G_{1+r}^{2+p}(\mathfrak{Cm}\,\mathcal{R})$. Any play of this game in which she uses σ is a play of $G_{1+r}^{2+p}(\mathcal{A}_{A,B'})$ in which σ was used. Hence it is a win for $\exists$ in $G_{1+r}^{2+p}(\mathcal{A}_{A,B'})$. So clearly, it is a win for her in the game $G_{1+r}^{2+p}(\mathfrak{Cm}\,\mathcal{R})$. This shows that σ is a winning strategy for her in this game. □

Exercises

1. Let A, B be structures in a relational signature with only unary and binary relation symbols. Verify that the rainbow algebra $\mathcal{A}_{A,B}$ is a non-associative algebra.

2. Find a winning strategy for $\exists$ in the game $G_\omega^3(\mathcal{A}_{A,B})$. Deduce, using proposition 12.25 that $\mathcal{A}_{A,B}$ has a 3-dimensional relational basis and use exercise 12.3(11) to conclude that $\mathcal{A}_{A,B} \in \mathbf{SA}$.

3. The definition of EF_r^p is in terms of partial homomorphisms. Define games $EF1_r^p, EFi_r^p$ analogously, using one-one partial homomorphisms, and partial isomorphisms (partial maps preserving all quantifier-free formulas), respectively. Show that for any L, A, B as in theorem 16.5, there are non-associative algebras $\mathcal{A}1_{A,B}, \mathcal{A}i_{A,B}$ such that $\exists$ has a winning strategy in $EF1_r^p(A,B)$ (respectively, in $EFi_r^p(A,B)$) iff she has one in $G_{1+r}^{2+p}(\mathcal{A}1_{A,B})$ (respectively, in $G_{1+r}^{2+p}(\mathcal{A}i_{A,B})$). [Reduce to the homomorphism case by adjusting L, A, B.]

4. What happens to theorem 16.5 if L contains constants as well as relation symbols?

16.5. Modifications to the rainbow algebra

5. Let A, B be L-structures for some binary signature L, and assume that $|A| \geq 2$. Prove that $\mathcal{A}_{A,B}$ has a complete representation iff there is a homomorphism $f : A \to B$.

6. Generalise the atomic network game transfinitely to a game $G_\lambda^\kappa(\mathcal{A})$ with λ rounds over κ-dimensional atomic networks. Pay particular attention to the definitions at rounds indexed by limit ordinals. Also generalise the Ehrenfeucht–Fraïssé game to $EF_\lambda^\kappa(A, B)$ and show that the 'rainbow' theorem 16.5 still holds transfinitely: $\exists$ has a winning strategy in $G_{1+\lambda}^{2+\kappa}(\mathcal{A}_{A,B})$ iff she has a winning strategy in $EF_\lambda^\kappa(A, B)$, for any ordinals κ, λ.

Chapter 17

Applying the rainbow construction

With the aid of the rainbow theorem from the preceding chapter, we can now easily reel off a string of results on various classes of relation algebras. In this chapter, we will prove:

1. **RRA** is not finitely axiomatisable in first-order logic [Mon64], nor by equations using finitely many variables [Jón91], nor by prenex universal sentences using finitely many variables. (For other proofs of these results, see exercises 11.5(3–5) and remark 15.13.)

2. The class **WRAS** of weakly representable atom structures is not finitely axiomatisable in first-order logic. (For another proof, see exercise 4.5(2).) The class **LCAS** of atom structures satisfying the Lyndon conditions is not finitely axiomatisable either.

3. The class **CRA** of completely representable relation algebras is not elementary.

4. For finite $n \geq 5$, the variety $\mathbf{RA}_n$ is not finitely axiomatisable over $\mathbf{RA}_{n-1}$, and the class '$\mathbf{CRA}_n$' of relation algebras with complete n-square relativised representations is not elementary.

5. The class **wRRA** of weakly representable relation algebras (with representations respecting the operations $\cdot$, 0, 1', $\smile$, and ;) is not finitely axiomatisable in first-order logic [Hai87, Hai91, HodMik00].

6. **RRA**, $\mathbf{S\mathfrak{Ra}CA}_n$, and $\mathbf{RA}_n$, for $n \geq 6$, are not closed under completions, and are not atom-canonical. (See corollary 14.15 for another proof for **RRA**.)

The ease with which these results are proved is illustrated by the simplicity (even triviality) of the structures A, B that will be used as input to the rainbow theorem: they will mostly be complete graphs, paths from directed graphs, linear orders, or combinations of these. (But do not despair: in chapter 18, richer structures will be coded into a different kind of rainbow algebra.)

The following notation will be useful throughout.

NOTATION 17.1 *For a cardinal $\kappa > 0$, let K_κ be a complete undirected graph without loops (irreflexive edges), with set of nodes κ. We regard these graphs as structures in the signature consisting of a single binary relation symbol, interpreted as the edge relation.*

17.1 Non-finite axiomatisability of RRA

First, we prove (again) the celebrated theorem of Monk [Mon64] that **RRA** is not finitely axiomatisable. (For other proofs, see exercise 11.5(3) and remark 15.13.) The hard work has already been done, and what remains of the proof is very simple. For each $1 \leq n < \omega$, let

$$\mathcal{A}_n = \mathcal{A}_{K_{n+3}, K_{n+2}},$$

where $\mathcal{A}_{K_{n+3}, K_{n+2}}$ is as defined in section 16.2. As for all $n < \omega$, $\exists$ has a winning strategy in $EF_2^2(K_{n+3}, K_{n+2})$, we see by remark 16.6 that $\mathcal{A}_n$ is a (finite) relation algebra.

LEMMA 17.2 *For each $n < \omega$, $\forall$ has a winning strategy in the game $G_\omega^a(\mathcal{A}_n)$, but $\exists$ has a winning strategy in $G_n^a(\mathcal{A}_n)$.*

Proof. Recall that G_n^a (for $n \leq \omega$) is effectively the same game as G_n^ω. So by theorem 16.5, we only need show that $\forall$ has a winning strategy in the game $EF_\omega^\omega(K_{n+3}, K_{n+2})$. This is easily seen (cf. example 16.3). $\forall$ uses his first $n+3$ moves to pebble the whole of K_{n+3}. When responding, $\exists$ cannot help but pebble an element of K_{n+2} twice. Since K_{n+2} has no reflexive edges, once she has done so, the partial map from K_{n+3} to K_{n+2} defined by the pebble positions cannot be a partial homomorphism. So $\forall$ has won.

On the other hand, it is clear that this approach takes $\forall$ $n+3$ rounds in the worst case. In the first n rounds, $\exists$ may always find an unpebbled node of K_{n+2} to use; so she can survive n rounds without losing. By theorem 16.5, $\exists$ has a winning strategy in $G_n^a(\mathcal{A}_n)$. □

We obtain:

17.2. Complete representations

THEOREM 17.3 (Monk, 1964) *There is no finite set of first-order sentences S such that for any algebra $\mathcal{A}$ of the signature of relation algebras, $\mathcal{A} \models S$ if and only if $\mathcal{A} \in$ **RRA**.*

Proof. By theorem 11.10, for any atomic relation algebra $\mathcal{A}$, if $\exists$ has a winning strategy in $G^a_\omega(\mathcal{A})$ then $\mathcal{A} \in$ **RRA**; for finite algebras the converse holds too. So by lemma 17.2, $\mathcal{A}_n \notin$ **RRA**. Now by theorem 11.5, G^a_ω has an equivalent game tree. Lemma 17.2 and theorem 10.12 now imply that $\exists$ has a winning strategy in $G^a_\omega(\prod_D \mathcal{A}_n)$ for any non-principal ultrafilter D over ω. So $\prod_D \mathcal{A}_n \in$ **RRA**.

Now if **RRA** were finitely axiomatised by a first-order sentence $\sigma = \bigwedge S$, say, then $\prod_D \mathcal{A}_n \models \sigma$. It follows from Łoś' theorem (cf. exercise 2.4(6)) that $\mathcal{A}_n \models \sigma$ for infinitely many $n < \omega$, contradicting $\mathcal{A}_n \notin$ **RRA**. □

We also obtain the following:

PROPOSITION 17.4 *The following classes of relation algebra atom structures cannot be defined by finitely many first-order axioms.*

1. *The class* **WRAS** *of weakly representable atom structures (definition 14.1).*

2. *The class* **LCAS** *of atom structures satisfying the Lyndon conditions (definition 14.16).*

Proof. We saw in lemma 17.2 that $\forall$ has a winning strategy in $G^a_\omega(\mathcal{A}_n)$. As $\mathcal{A}_n$ is finite, by theorem 11.10(2) it is not representable, and $\text{At}\,\mathcal{A}_n \notin$ **WRAS**. This holds for each $n < \omega$. But $\exists$ has a winning strategy in $G^a_n(\mathcal{A}_n)$. It follows by theorems 10.12 and 11.4 that for any non-principal ultrafilter D over ω, the ultraproduct $\prod_D \mathcal{A}_n$ satisfies all the Lyndon conditions. So $\prod_D \text{At}\,\mathcal{A}_n \cong \text{At}(\prod_D \mathcal{A}_n) \in$ **LCAS**. Since by theorem 14.17, **LCAS** $\subseteq$ **WRAS**, an application of Łoś' theorem gives the result. □

It may be helpful here to recall from theorem 14.17 that the class **CRAS** of atom structures of relation algebras with complete representations, and the class **SRAS** of strongly representable atom structures, are not elementary.

Exercises

1. Prove that the class **FOAS** of definition 14.16 is not finitely axiomatisable.

17.2 Complete representations

A very similar example can be used to show that **CRA**, the class of relation algebras that have a complete representation, is not an elementary class. This kind of question has been of interest at least since Lyndon's 1956 paper [Lyn56]: a remark on its last page states that **CRA** is not axiomatisable by universal sentences.

Continuing to use notation 17.1, we let K be the disjoint union of the graphs K_n ($1 \leq n < \omega$), and $\mathcal{B} = \mathcal{A}_{K_\omega, K}$. $\mathcal{B}$ is an infinite atomic relation algebra.

LEMMA 17.5 $\forall$ *has a winning strategy in* $G_\omega^a(\mathcal{B})$. $\exists$ *has a winning strategy in* $G_n^a(\mathcal{B})$ *for all* $n < \omega$.

Proof. $\exists$ has a winning strategy in $EF_n^\omega(K_\omega, K)$ for all finite n, as she may always respond to $\forall$'s move by placing a pebble in $K_n \subseteq K$. There are only n rounds, so $\forall$ will never defeat her. But $\forall$ can win $EF_\omega^\omega(K_\omega, K)$, by placing successive pebbles on distinct elements of K_ω. At the outset, $\exists$ must choose which K_n in K to respond in; and from then on in the game she must stick with this same K_n. However large n is, it will be finite; after n rounds, she will have pebbled every element of this copy, and she will lose in the next round.

The lemma now follows from theorem 16.5. □

We deduce:

THEOREM 17.6 **CRA** *is not an elementary class.*

Proof. Since $\exists$ has no winning strategy in $G_\omega^a(\mathcal{B})$, it follows by theorem 11.7 that $\mathcal{B} \notin$ **CRA**. On the other hand, since by theorem 11.5 G_ω^a has an equivalent game tree, lemma 17.5 and theorem 10.12 show that $\exists$ does have a winning strategy in $G_\omega^a(\prod_D \mathcal{B})$ for any non-principal ultrafilter D over ω. Proposition 10.13 now yields a countable elementary subalgebra $C \preceq \prod_D \mathcal{B}$ such that $\exists$ has a winning strategy in $G_\omega^a(C)$. C is obviously atomic. By theorem 11.7 again, $C \in$ **CRA**. Since $C \equiv \mathcal{B}$, we see that **CRA** is not closed under elementary equivalence and so is not an elementary class. □

Recall from definition 14.16 that **CRAS** is the class of atom structures of relation algebras with complete representations.

COROLLARY 17.7 **CRAS** *is not elementary.*

Proof. By theorem 17.6, there are atomic relation algebras $\mathcal{A}, \mathcal{B}$ with $\mathcal{A} \equiv \mathcal{B}$, $\mathcal{A} \in$ **CRA**, and $\mathcal{B} \notin$ **CRA**. So certainly, At $\mathcal{A} \in$ **CRAS**. Note that *no* atomic relation algebra with atom structure At $\mathcal{B}$ is completely representable (see exercise 14.3(2)). So At $\mathcal{B} \notin$ **CRAS**. Since the atom structure of an atomic relation algebra is first-order interpretable in the algebra, it follows that At $\mathcal{A} \equiv$ At $\mathcal{B}$. So **CRAS** is not closed under elementary equivalence either, so is not elementary. □

See exercises 17.4(1) and 15.6(1) for the analogue of theorem 17.6 for n-square and infinitarily n-flat relativised representations, respectively, for finite $n \geq 5$.

17.3. There is no n-variable equational axiomatisation of **RRA**

Exercises

1. Recall from section 8.2.4 the Stebletsova–Venema n^2-ary operators Q_n^{ij}. It was shown in [SteVen98, Ste00] that if $\mathcal{A}$ is a relation algebra that has an expansion interpreting these operators in such a way as to satisfy certain equations, then the Q-operators are completely additive in the expansion, and $\mathcal{A}$ has a representation respecting their meaning as described in section 8.2.4.

    The following exercise shows that not all representable relation algebras have such an expansion. Let $\mathcal{B}$ be as in lemma 17.5.

    (a) Show that $\mathcal{B}$ is representable.

    (b) Fix an arbitrary representation M of $\mathcal{B}$. Define $\bar{a} = (a_{ij} : i, j < 4)$ by: $a_{01} = \mathsf{w}$, $a_{02} = \mathsf{g}_0$, $a_{03} = \mathsf{g}_1$, $a_{12} = a_{13} = \mathsf{y}$, and all other a_{ij} are 1. Show that the binary relation $Q_4^{01}(\bar{a})$ on M is equal to w^M.

    (c) Show that for any $\bar{b}$ that is the same as $\bar{a}$ except that $b_{23} \in \operatorname{At}\mathcal{B}$, the binary relation $Q_4^{01}(\bar{b})$ on M is $\emptyset$.

    (d) Deduce that $\mathcal{B}$ has no expansion to interpret the Q-operators so as to satisfy the equations of [SteVen98, Ste00]. [If $\mathcal{B}^*$ were one, show that $\mathcal{B}^* \models Q_4^{01}(\bar{a}) = \mathsf{w} \wedge Q_4^{01}(\bar{b}) = 0$ for all $\bar{a}, \bar{b}$ as above, contradicting complete additivity.]

17.3 There is no n-variable equational axiomatisation of RRA

We have seen that **RRA** cannot be axiomatised finitely. Now we go on to prove a stronger result, due to Jónsson — that **RRA** cannot be axiomatised by any set of equations using a fixed, finite number of variables. (To see that it is stronger, see exercise 11.5(6).) We will use a modified rainbow construction in which the red atoms have single subscripts instead of the standard two. For each finite n, we will build two algebras with one representable and the other not, and show that they cannot be distinguished by any n-variable equation.

DEFINITION 17.8 (Cf. section 16.2.) Let α and β be ordinals. Define the rainbow algebra $\mathcal{J}_{\alpha,\beta}$ to be the complex algebra over the following atom structure. The atoms are 1' together with

$$\{\mathsf{g}_i : i < \alpha\} \cup \{\mathsf{w}, \mathsf{w}_{ij} : i, j < \alpha\} \cup \{\mathsf{y}, \mathsf{b}\} \cup \{\mathsf{r}_i : i < \beta\}.$$

All atoms are self-converse. As before, we consider atoms g_i (any $i < \alpha$) to be *green*, atoms r_i (for $i < \beta$) are *red*, etc. The forbidden triples consist of all permutations of

$(1',s,t)$ unless $s = t$
$(g,g',g''),(g,g',W)$ any green atoms g,g',g'' and any white atom W
$(y,y,y),(y,y,b)$
(g_i,y,w_{jk}) unless $i \in \{j,k\}$
(r_i,r_i,r_j) any $i,j < \beta$.

Provided $\beta > 0$, the algebra $\mathcal{J}_{\alpha,\beta}$ will in fact be a relation algebra.

Fix $n < \omega$. Let $\alpha = 3 \cdot 2^n$ and $\beta = (\alpha+1)(\alpha+2)/2$. β is the number of subsets of $\alpha + 2$ of size 2. To show that **RRA** cannot be defined by any set of n-variable equations consider the algebras $\mathcal{A} = \mathcal{J}_{\alpha+2,\beta}$ and $\mathcal{B} = \mathcal{J}_{\alpha+2,\alpha}$.

LEMMA 17.9 $\mathcal{A} \in $ **RRA**, but $\mathcal{B} \notin $ **RRA**.

Proof (sketch). Note that both $\mathcal{A}$ and $\mathcal{B}$ are finite. Consider the game $G^{\alpha+4}_{\alpha+3}(\mathcal{B})$. As in the proof of proposition 16.8, $\forall$ can use the $\alpha + 2$ green atoms to force the construction of $\alpha + 2$ nodes $n_0, n_1, \ldots, n_{\alpha+1}$ in the last network N, such that $N(n_i, n_j)$ must be a red atom, for distinct $i, j < \alpha + 2$. But there are only α red atoms to choose so there must be $0 < i \neq j < \alpha+2$ such that $N(n_0,n_i) = N(n_0,n_j)$. But then, the triangle (n_0,n_i,n_j) is inconsistent, so $\forall$ has won the game. So by theorem 11.10(2), $\mathcal{B}$ is not representable.

To show that $\mathcal{A} \in $ **RRA** we show that $\exists$ has a winning strategy in $G^a_\omega(\mathcal{A})$, as in the proof of proposition 16.13. The key point in the argument is that with only $\alpha + 2$ green atoms, $\forall$ can only force a red clique of size $\alpha + 2$, not bigger. So $\exists$'s strategy, within red cliques, is to label each edge using a red atom and to ensure that each edge within the clique has a label unique to that edge (within the clique). Since the size of the clique is at most $\alpha + 2$ and there are β red atoms to choose, this can certainly be done. This strategy ensures that the suffixes of the three red atoms labelling the three edges of any triangle within a red clique are distinct. Such a triangle is consistent, by definition of the rainbow construction $\mathcal{A}$. For other edges, $\exists$'s strategy is the same as in proposition 16.13. So $\exists$ has a winning strategy in $G^a_\omega(\mathcal{A})$, and by theorem 11.10(2), $\mathcal{A}$ is representable. □

LEMMA 17.10 *Let $s = t$ be any L_{RA}-equation written with only the variables $\{x_0,\ldots,x_{n-1}\}$. Then*

$$\mathcal{A} \models (s=t) \iff \mathcal{B} \models (s=t).$$

Proof. Suppose that $\mathcal{A} \not\models (s=t)$. Then there is an assignment $h : \{x_0,\ldots,x_{n-1}\} \to \mathcal{A}$ such that $\mathcal{A}, h \models (s \neq t)$. We specify an assignment h' into $\mathcal{B}$ falsifying $s = t$ in $\mathcal{B}$, as follows. Note first that $\mathcal{A}$ and $\mathcal{B}$ have identical non-red atoms. So for any non-red atom a of $\mathcal{B}$ and any $i < n$, let $a \leq h'(x_i)$ iff $a \leq h(x_i)$. To determine h', it remains to identify which red atoms of $\mathcal{B}$ lie under $h'(x_i)$.

17.3. There is no n-variable equational axiomatisation of **RRA**

Let R be the set of all red atoms of $\mathcal{A}$, and let R' be the set of all red atoms of $\mathcal{B}$. By definition of $\mathcal{A}, \mathcal{B}$, we have

$$|R| = \beta \geq |R'| = \alpha = 3 \cdot 2^n.$$

The assignment h induces a partition of R into 2^n parts R_S, for $S \subseteq n = \{0, \ldots, n-1\}$, some of which may be empty, by:

$$R_S = \{r \in R : r \leq h(x_i) \text{ for } i \in S, \text{ and } r \cdot h(x_i) = 0 \text{ for } i \in n \setminus S\}.$$

Partition R' into 2^n parts $R'_S : S \subseteq n$ in such a way that

$$\begin{array}{ll} |R'| = |R| & \text{if } |R| < 3, \\ |R'| \geq 3 & \text{if } |R| \geq 3. \end{array}$$

Since $|R|, |R'| \geq 3 \cdot 2^n$, this can be done. Now, for each $i < n$ and each red atom r' of R', we complete the definition of $h'(x_i) \in \mathcal{B}$ by:

$$r' \leq h'(x_i) \iff r' \in R'_S \text{ for some } S \ni i.$$

It is now a simple matter to show, by term induction, that for any term u using only the variables $\{x_0, \ldots, x_{n-1}\}$, we have

$$\begin{array}{rcll} h(u) \supseteq R_S & \iff & h'(u) \supseteq R'_S & \text{for any } S \subseteq n, \\ h(u) \setminus R & = & h'(u) \setminus R' & \\ |h(u) \cap R| & = & |h'(u) \cap R'| & \text{if } |h(u) \cap R| < 3 \\ |h(u) \cap R| \geq 3 & \iff & |h'(u) \cap R'| \geq 3. & \end{array}$$

It follows that $\mathcal{B}, h' \models (s \neq t)$. Hence $\mathcal{B} \not\models (s = t)$. The reverse argument, that $\mathcal{B} \not\models (s = t) \Rightarrow \mathcal{A} \not\models (s = t)$, is entirely similar. □

Hence

THEOREM 17.11 (Jónsson, [Jón91]) *There is no equational theory Σ using only finitely many variables such that for all relation algebras $\mathcal{A}$ we have $\mathcal{A} \models \Sigma \iff \mathcal{A} \in$ **RRA**.*

Proof. If Σ is any n-variable equational theory then the two rainbow algebras of the preceding lemma either both validate Σ or neither do. Since one algebra is representable while the other is not, it follows that Σ does not define **RRA**. □

REMARK 17.12 The algebras $\mathcal{A}, \mathcal{B}$ are integral. By exercise 5.1(4), they are simple, and hence (by proposition 2.54) subdirectly irreducible. In a discriminator variety such as **RRA**, every universal prenex formula is equivalent in subdirectly irreducible algebras to an equation using the same variables. We conclude by the preceding theorem that **RRA** is not axiomatisable by any set of prenex universal sentences that uses a finite number of variables.

It is also possible to find an equivalent universal prenex formula for any universal formula, by standard first-order techniques. However, the prenex equivalent may have extra variable symbols. For example, consider the 2-variable universal sentence

$$\neg \exists x \exists y (x > y \land (\exists y (y > x \land \exists x (x > y)))).$$

The sentence means that there is no >-chain of four elements. A prenex universal equivalent is

$$\forall x \forall y \forall u \forall v \neg (x > y \land y > u \land u > v).$$

But in general it is not possible to find a prenex equivalent using the same variables.

PROBLEM 17.13 *Is there a universal axiomatisation of* **RRA** *using a finite number of variables?*

Going further,

PROBLEM 17.14 *Is there any first-order axiomatisation of* **RRA** *using a finite number of variables?*

Exercises

1. Show that the algebras $\mathcal{A}, \mathcal{B}$ can be distinguished by a 2-variable universal sentence.

17.4 $\mathbf{RA}_{n+1}$ is not finitely based over $\mathbf{RA}_n$

We saw in chapter 13 that the classes $\mathbf{RA}_n$ ($3 \leq n < \omega$) formed n-dimensional approximations or analogues of **RRA**. $\mathbf{RA}_n$ is a canonical variety, and by proposition 13.48,

$$\mathbf{RA}_3 \supseteq \mathbf{RA}_4 \supseteq \cdots \supseteq \mathbf{RRA}, \text{ and } \bigcap_{3 \leq n < \omega} \mathbf{RA}_n = \mathbf{RRA}.$$

We now investigate the relative strength of these approximations, for varying n. The inclusions above were shown to be strict in [Madd92]. Since $\mathbf{RA}_3 = \mathbf{SA}$ and $\mathbf{RA}_4 = \mathbf{RA}$, the inclusion $\mathbf{RA}_3 \supset \mathbf{RA}_4$ is finitely axiomatisable. Maddux conjectured [Madd83, page 90] that if $4 \leq n < l \leq \omega$ then $\mathbf{RA}_l$ is not finitely based

17.4. $\mathbf{RA}_{n+1}$ is not finitely based over $\mathbf{RA}_n$

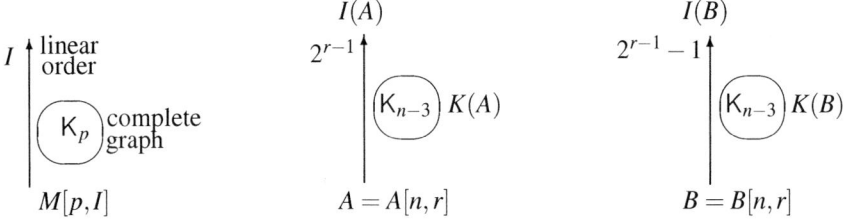

Figure 17.1: The structures $M[p,I]$, $A[n,r]$, $B[n,r]$

relative to $\mathbf{RA}_n$: that is, there is no first-order sentence σ such that $\mathcal{A} \in \mathbf{RA}_l$ iff $\mathcal{A} \in \mathbf{RA}_n$ and $\mathcal{A} \models \sigma$. Confirming the conjecture would provide one more piece of evidence that the $\mathbf{RA}_n$ behave similarly to $\mathbf{RRA}$, which as we just saw is not finitely axiomatisable.

Here, we confirm Maddux's conjecture. By exercise 1 of the introduction to part III, it suffices to show that $\mathbf{RA}_{n+1}$ is not finitely axiomatisable over $\mathbf{RA}_n$, for all finite $n \geq 4$. In chapter 15, we proved that $\mathbf{S\mathfrak{R}aCA}_{n+1}$ is not finitely axiomatisable over $\mathbf{S\mathfrak{R}aCA}_n$, by defining relation algebras $\mathfrak{A}(n,r)$ for $0 < r < \omega$, and showing that $\mathfrak{A}(n,r) \in \mathbf{S\mathfrak{R}aCA}_n \setminus \mathbf{S\mathfrak{R}aCA}_{n+1}$ and yet, although $\mathfrak{A}(n,r) \notin \mathbf{S\mathfrak{R}aCA}_{n+1}$, $\exists$ has a winning strategy in the game $G_r^{m,n+1}(\mathfrak{A}(n,r))$. Now we will prove the corresponding non-finite axiomatisability result for $\mathbf{RA}_n$. The proof works in a similar way, but here, the relation algebras we require are obtained from the rainbow construction.

Let L be the signature consisting of a single binary relation symbol, '$<$'. For any irreflexive linear order I, and any $p \leq \omega$, let the L-structure $M[p,I]$ be the disjoint union of I and the complete graph K_p of notation 17.1. The relation symbol $<$ is interpreted in $M[p,I]$ as

$$<^{M[p,I]} = <^I \cup <^{\mathsf{K}_p} \cup (I \times \mathsf{K}_p) \cup (\mathsf{K}_p \times I).$$

Warning: $<^{\mathsf{K}_p}$ is the edge relation on K_p, and so is symmetric.

Now fix finite $n, r \geq 4$. Regarding 2^{r-1} and $2^{r-1} - 1$ as ordered ordinals, define

$$\begin{aligned} A &= A[n,r] = M[n-3, 2^{r-1}], \\ B &= B[n,r] = M[n-3, 2^{r-1}-1], \\ \mathcal{A}_r^n &= \mathcal{A}_{A[n,r],B[n,r]} \quad \text{(as in section 16.2).} \end{aligned}$$

Write $I(A)$ for the linear order part of A, and $K(A)$ for the graph part. Define $I(B), K(B)$ similarly. See figure 17.1.

So far, we have been using the game $G^a(\mathcal{A})$. Here, we will need to use the game $G_r^n(\mathcal{A})$ as in the rainbow theorem 16.5. This is because by proposition 12.25, $\exists$ has a winning strategy in $G_\omega^n(\mathcal{A})$ if and only if $\mathcal{A}$ has an n-dimensional relational

522 Chapter 17. Applying the rainbow construction

basis. This is always a sufficient condition for $\mathcal{A} \in \mathbf{RA}_n$, and for finite $\mathcal{A}$, it is necessary. See definition 12.24 for the rules of G_r^n: there are r rounds played on atomic $\mathcal{A}$-networks with at most n nodes, in which $\forall$ can make a challenge by fitting a triangle onto an edge of the current network.

LEMMA 17.15 $\exists$ *has a winning strategy in the game* $G_\omega^n(\mathcal{A}_r^n)$.

Proof. By theorem 16.5, it suffices to show that $\exists$ has a winning strategy in the game $EF_\omega^{n-2}(A,B)$. $\exists$ plays in this game according to the following strategy. She can apply it in the combined rounds 0 and 1 by regarding them as separate rounds. If $\forall$ ever forfeits the game by leaving fewer than two elements of A pebbled, then $\exists$ certainly wins, so we can assume not.

Let $\forall$ pick up a spare pebble pair and place the first pebble of it on $a \in A$. By the rules of $EF_\omega^{n-2}(A,B)$, a is not currently occupied by a pebble. $\exists$ has to choose which element $b \in B$ to put the other pebble on.

1. If possible, $\exists$ chooses b to be an unoccupied element in $K(B)$.

2. If this is not possible, because all elements of $K(B)$ are occupied by pebbles, she chooses b to be an arbitrary element $x \in I(B)$.

Because there are only $n-2$ pebble pairs, $\exists$ can always implement this strategy. Because the substructure S of B with domain $K(B) \cup \{x\}$ is isomorphic to K_{n-2}, so that $B \models y < y'$ for any distinct $y, y' \in S$, it follows that the pebble positions always define a homomorphism from A to B, and she wins. $\square$

LEMMA 17.16 $\exists$ *has a winning strategy in the game* $G_r^\omega(\mathcal{A}_r^n)$.

Proof. By theorem 16.5, it is enough to provide a winning strategy for $\exists$ in $EF_{r-1}^\omega(A,B)$.

The rounds of this game are numbered $t = 0, 1, \ldots, r-2$. Notation: at each round, write l for the number of rounds remaining: so $l = r-1-t$. If $X \subseteq I(B)$, an *interval* of $I(B)$ generated by X is a maximal convex subset of $I(B)$ disjoint from X.

$\exists$'s strategy is simply to keep the following conditions:

1. If one pebble of a pair is on an element of $K(A)$, the other is on an element of $K(B)$.

2. Similarly for $I(A), I(B)$.

3. The partial map $: A \to B$ induced by the pebble positions is well-defined and one-to-one.

4. The partial map $: I(A) \to I(B)$ given by the positions of pebbles in the ordered part of A, B is order-preserving.

17.4. $\mathbf{RA}_{n+1}$ is not finitely based over $\mathbf{RA}_n$

5. Every interval of $I(B)$ generated by her current pebbling contains at least $2^l - 1$ points.

These conditions are true initially, since there are no pebbles in play, $l = r - 1$, and the unique interval of $I(B)$ generated by the current pebbling, namely $I(B)$, contains $2^{r-1} - 1$ points. They clearly remain true if $\forall$ chooses to remove some pairs of pebbles.

Suppose they hold just before round t ($0 \leq t \leq r - 2$). (If $t = 0$, $\forall$ will make two moves together; $\exists$ regards them as separate moves and calculates her responses as for normal rounds.) In round t, $\forall$ may place a pebble on an (unoccupied) element of $K(A)$. $\exists$ responds by placing the other pebble of the pair on an unoccupied element of $K(B)$. By the conditions, there will always be such an element. Or he may place a pebble in some interval of $I(A)$ generated by the other pebbles in play in $I(A)$. The corresponding interval of $I(B)$ (this notion is well-defined by condition 4) has at least $2^l - 1 = (2^{l-1} - 1) + 1 + (2^{l-1} - 1)$ points, so $\exists$ may find a point in the interval, 'near the middle', such that if $l > 1$, the two new intervals it generates have at least $2^{l-1} - 1$ points. $\exists$ places the other pebble of the pair on such a point. This preserves the listed conditions into the next round. Clearly, this strategy wins $EF_{r-1}^{\omega}(A,B)$ for $\exists$. □

LEMMA 17.17 $\forall$ *has a winning strategy in the game* $G_{\omega}^{n+1}(\mathcal{A}_r^n)$.

Proof. It suffices to show that $\forall$ has a winning strategy in $EF_{\omega}^{n-1}(A,B)$. $\forall$ will always place his pebbles on distinct elements of A. Recall that A is the union of a complete graph $K(A)$ of size $n - 3$ and an ordinal $I(A) = l = 2^{r-1}$, regarded as a linear order. $\forall$ uses rounds $0, 1, \ldots, n - 2$ to cover $K(A)$ and the elements $l - 1, l - 2 \in I(A)$ (i.e., the top two elements of $I(A)$) with $n - 1$ pebbles. Because at least two out of every three distinct elements here are related by $<$ both ways, if $\exists$ is not to lose she must respond by pebbling $K(B) \cup \{e, e'\}$, for some $e, e' \in I(B)$. (All her choices must be distinct, as $<^B$ is irreflexive, and she cannot use three distinct elements of $I(B)$ as it is not true that two of these are related by $<$ both ways.)

Consider now the pebbles that $\exists$ placed on $e, e' \in I(B)$. Let the other pebbles of the pairs that these came from be placed on $a, a' \in A$, respectively. If $\exists$ has not yet lost, we have $a, a' \in I(A)$. For if not, $A \models a < a' < a$, but $B \not\models b < b' < b$, so that $\{(a,b),(a',b')\}$ is not a partial homomorphism from A to B. So, assuming that he has not yet won, we see that $\forall$ has arranged that two elements of $I(A)$ are pebbled, the corresponding pebbles in B being in $I(B)$.

He will keep this condition from now on. Say that just after round t (for $t \geq n - 2$), $a_t < a_t'$ are pebbled in $I(A)$. If the corresponding elements of B are b_t, b_t', then $b_t < b_t'$ in $I(B)$, or $\exists$ has lost.

$\forall$ can now force $\exists$ to play a two-pebble game of length ω on $I(A), I(B)$, which he can win because $I(A)$ is longer than $I(B)$. He simply preserves the condition that if he has not won yet, then

$$b_t < a_t \text{ as ordinals, for each } t \geq n-2. \qquad (*)_t$$

This is true initially because $a_{n-2} = l - 2$ (by ∀'s choice) and $b_{n-2} = \min(e, e') \leq l - 3$ (because $|I(B)| = l - 1$). In each further round t, assuming that $(*)_{t-1}$ holds, he picks up the pebble on a'_{t-1}, and moves it to $a_{t-1} - 1$. (This will not forfeit him the game because $n \geq 4$, so at least two distinct elements of A remain pebbled at all times.) So $a_t = a_{t-1} - 1$, $a'_t = a_{t-1}$. ∃ must pick up the other pebble of the pair, and move it to some element b_t of $I(B)$ below b_{t-1}. By $(*)_{t-1}$,

$$b_t \leq b_{t-1} - 1 < a_{t-1} - 1 = a_t,$$

so $(*)_t$ holds.

Thus, eventually, b_t will be zero, a_t will not be, and ∀ will win after one more round. □

We can now read off our non-finite axiomatisability result.

THEOREM 17.18 *For each n with $4 \leq n < \omega$, the variety $\mathbf{RA}_{n+1}$ is not finitely based relative to $\mathbf{RA}_n$.*

Proof. Fix finite $n \geq 4$. Recall from proposition 12.25 that for any atomic non-associative algebra $\mathcal{A}$ — even uncountable — ∃ has a winning strategy in $G^n_\omega(\mathcal{A})$ iff $\mathcal{A}$ has an n-dimensional relational basis. This implies by definition 12.30 that $\mathcal{A} \in \mathbf{RA}_n$. For finite $\mathcal{A}$, the converse holds (by theorem 13.46(5 ⇒ 4), since $\mathcal{A} = \mathcal{A}^+$). So by lemmas 17.15 and 17.17, $\mathcal{A}^n_r \in \mathbf{RA}_n \setminus \mathbf{RA}_{n+1}$ for all $4 \leq r < \omega$.

By lemma 17.16, ∃ has a winning strategy in $G^{n+1}_r(\mathcal{A}^n_r)$ for all finite r. The fact that G^n_ω has an associated game tree was left as exercise 12.4(4). It follows from this and theorem 10.12 that ∃ has a winning strategy in $G^{n+1}_\omega(\prod_{4 \leq r < \omega} \mathcal{A}^n_r / D)$, for any non-principal ultrafilter D over $\omega \setminus 4$. $\prod_{4 \leq r < \omega} \mathcal{A}^n_r / D$ is certainly an atomic non-associative algebra, so $\prod_{4 \leq r < \omega} \mathcal{A}^n_r / D \in \mathbf{RA}_{n+1}$. Łoś' theorem now implies in the usual way that $\mathbf{RA}_{n+1}$ is not finitely axiomatisable over $\mathbf{RA}_n$. □

Exercises

1. For $3 \leq n < \omega$, let $\mathbf{RB}_n$ be the class of atomic non-associative algebras with an n-dimensional relational basis (definition 12.9).

    (a) Prove that for $n = 3, 4$, $\mathbf{RB}_n$ is elementary. [Cf. exercises 12.3(11, 12).]

    (b) Prove that for $5 \leq n < \omega$, $\mathbf{RB}_n$ is not elementary. [Let $A = M[n-4, \mathbb{Z}]$ and $B = M[n-4, \mathbb{N}]$. Show that $\mathcal{A}_{A,B} \notin \mathbf{RB}_n$ but $\mathcal{A}_{A,B}$ is elementarily equivalent to some $\mathcal{B} \in \mathbf{RB}_n$.]

17.5 Infinite-dimensional bases and relativised representations

So far we have restricted our attention to n-dimensional bases, n-square relativised representations, etc., for finite n. Of course the definitions do generalise to infinitely many dimensions, though there is more than one way of doing this. For this section only, we consider the infinite-dimensional case.

Recall definition 12.9, now generalised to infinitely many dimensions.

DEFINITION 17.19 Let $\mathcal{A}$ be an atomic non-associative algebra, and $\alpha \geq 2$ an ordinal. An α-*dimensional relational basis* (over $\mathcal{A}$) is a set $\mathcal{R}$ of atomic networks with set of nodes α, such that

- if $a \in \operatorname{At}\mathcal{A}$ then $N(0,1) = a$ for some $N \in \mathcal{R}$,
- if $a, b \in \operatorname{At}\mathcal{A}$, $N \in \mathcal{R}$, $i, j, k < \alpha$, $k \neq i, j$, and $N(i,j) \leq a;b$, then there is $M \in \mathcal{R}$ with $M \equiv_k N$ and $M(i,k) = a$, $M(k,j) = b$.

$\mathbf{RB}_\alpha$ denotes the class of all atomic non-associative algebras with α-dimensional relational bases.

We generalise definition 12.6.

DEFINITION 17.20 Let α, β be ordinals, let N be an α-dimensional atomic network, and let $\sigma : \beta \to \alpha$ be any map. We define $N\sigma$ to be the labelled graph with nodes β and edge labelling defined by

$$(N\sigma)(i,j) = N(\sigma(i), \sigma(j)),$$

for $i, j < \beta$. It is easy to check that $N\sigma$ is an atomic network (see the proof of lemma 12.7).

LEMMA 17.21

1. For infinite ordinals α, β, $\mathbf{RB}_\alpha = \mathbf{RB}_\beta$.

2. If $\mathcal{R}$ is an α-dimensional relational basis for the atomic relation algebra $\mathcal{A}$, then $\{N\sigma : N \in \mathcal{R},\ \sigma : \alpha \to \alpha,\ \operatorname{rng}(\sigma) \text{ is finite}\}$ is also a relational basis for $\mathcal{A}$.

Proof. For the first part, suppose $\mathcal{A} \in \mathbf{RB}_\alpha$ and let $\mathcal{R}$ be an α-dimensional relational basis for $\mathcal{A}$. Consider the set of β-dimensional atomic networks

$$S = \{N\sigma : N \in \mathcal{R},\ \sigma : \beta \to \alpha,\ \operatorname{rng}(\sigma) \text{ is finite}\}.$$

We claim that S is a β-dimensional relational basis for $\mathcal{A}$. We check the two conditions for a relational basis. Firstly, if $a \in \mathrm{At}\,\mathcal{A}$, then since $\mathcal{R}$ is a relational basis, there is $N \in \mathcal{R}$ with $N(0,1) = a$. For any map $\sigma : \beta \to \alpha$ with finite range, fixing 0 and 1, we have $(N\sigma)(0,1) = N(0,1) = a$ and $N\sigma \in S$.

Secondly, let $N\sigma \in S$, $a,b \in \mathrm{At}\,\mathcal{A}$, $i,j,k < \beta$, and $k \neq i,j$ be such that $(N\sigma)(i,j) \leq a;b$. Let σ' be the map that is identical to σ except that $\sigma'(k) = n$ for some $n < \alpha$ with $n \notin \mathrm{rng}(\sigma)$. Clearly, $N\sigma' \equiv_k N\sigma$. Note that $\mathrm{rng}(\sigma')$ is still finite, $N \in \mathcal{R}$, $n \neq \sigma'(i), \sigma'(j)$, and $N(\sigma'(i), \sigma'(j)) = (N\sigma)(i,j) \leq a;b$. Since $\mathcal{R}$ is a relational basis, there exists $M \in \mathcal{R}$ such that $M \equiv_n N$ and $M(\sigma'(i),n) = a$, $M(n,\sigma'(j)) = b$. By definition of σ', we obtain $M\sigma' \equiv_k N\sigma' \equiv_k N\sigma$ and $(M\sigma')(i,k) = M(\sigma'(i),\sigma'(k)) = M(\sigma'(i),n) = a$; similarly, $(M\sigma')(k,j) = b$. Since $M\sigma' \in S$, we see that S is a β-dimensional relational basis for $\mathcal{A}$.

Hence, $\mathbf{RB}_\alpha \subseteq \mathbf{RB}_\beta$. By symmetry of the above argument, the two classes are equal.

For the second part of the lemma, the class S defined above provides the solution, when $\beta = \alpha$. □

Recall definition 5.7.

DEFINITION 17.22 Let κ be a cardinal and let $\mathcal{A}$ be a non-associative algebra. A κ-*square relativised representation* M of $\mathcal{A}$ is a relativised representation such that for all cliques C of M with $|C| < \kappa$ and all $x,y \in C$, $a,b \in \mathcal{A}$, if $M \models (a;b)(x,y)$ then there is $z \in M$ with $M \models a(x,z) \wedge b(z,y)$ and such that $C \cup \{z\}$ is a clique.

$\mathbf{RA}_\kappa$ denotes the class of all non-associative algebras possessing κ-square relativised representations, and $\mathbf{CRA}_\kappa$ denotes the class of non-associative algebras with complete κ-square relativised representations.

By theorems 13.46 and 13.49, for finite $\kappa \geq 3$, this definition of $\mathbf{RA}_\kappa$ agrees with the earlier one given in definition 12.30. Also, the definition of ω-square is equivalent to the one in the proof of proposition 13.48.

The following extends proposition 13.48.

PROPOSITION 17.23 *For a cardinal* $\kappa \geq \omega$, *we have* $\mathbf{RA}_\kappa = \mathbf{RRA}$.

Proof. First note that for $\kappa < \lambda$ we have $\mathbf{RA}_\kappa \supseteq \mathbf{RA}_\lambda$. Recall from proposition 13.48 that $\bigcap_{n<\omega} \mathbf{RA}_n = \mathbf{RRA}$. It follows that $\mathbf{RRA} \supseteq \mathbf{RA}_\kappa$ for $\kappa \geq \omega$. Conversely, any classical representation of a relation algebra is easily seen to be κ-square for any κ, so $\mathbf{RRA} \subseteq \mathbf{RA}_\kappa$ as well. □

So κ-dimensional relational bases and κ-square relativised representations are not very interesting for infinite κ. But what about complete κ-square relativised representations?

LEMMA 17.24 *Let $\mathcal{A}$ be an atomic non-associative algebra. Then $\exists$ has a winning strategy in $G_\omega^\omega(\mathcal{A})$ iff $\mathcal{A} \in \mathbf{RB}_\omega$ iff $\mathcal{A} \in \mathbf{CRA}_\omega$.*

17.5. Infinite-dimensional bases and relativised representations

Proof (sketch). Suppose ∃ has a winning strategy σ in $G_\omega^\omega(\mathcal{A})$. The atomic networks in a play of this game are finite, so to obtain an ω-dimensional relational basis, we must convert them into ω-dimensional atomic networks. For a network N and map $v : \omega \to N$, define Nv as in definition 17.20. It is easy to check that

$$\{Nv : N \text{ occurs in some play of } G_\omega^\omega(\mathcal{A}) \text{ in which } \exists \text{ uses } \sigma, \text{ and } v : \omega \to N\}$$

is an ω-dimensional relational basis for $\mathcal{A}$. So $\mathcal{A} \in \mathbf{RB}_\omega$.

If $\mathcal{A}$ has an ω-dimensional relational basis, then by lemma 17.21 part 2 we can assume this basis consists of ω-dimensional atomic networks such that for each network N in the basis there is a finite bound on the size of any strict subnetwork of N. Then a complete ω-square relativised representation of $\mathcal{A}$ can be built step-by-step as in the proof of proposition 13.37. But here, instead of property 4 of proposition 13.37, we build a structure M so that for any *finite* clique C of M there is a network N in the relational basis and an embedding $v : N \to M$ such that $\mathrm{rng}(v) \supseteq C$. Properties 5, 7, and 8 of the proposition now refer to finite sequences of arbitrary length, rather than fixed-length n-tuples. Otherwise, there is no significant change to the proof of proposition 13.37. This modification to the proposition is sufficient to get lemma 13.40 to work, as an ω-square relativised representation only requires composition witnesses over finite-sized cliques, and hence we can construct a complete ω-square relativised representation of $\mathcal{A}$, as in theorem 13.38.

If $\mathcal{A}$ has a complete ω-square relativised representation, then ∃ can maintain an embedding from the current atomic network in a play of $G_\omega^\omega(\mathcal{A})$ into the representation so that edge labels are preserved, and thereby win the game. □

THEOREM 17.25 $\mathbf{CRA} \subset \mathbf{CRA}_\omega = \mathbf{RB}_\omega \subset \bigcap_{3 \leq n < \omega} \mathbf{RB}_n = \bigcap_{3 \leq n < \omega} \mathbf{CRA}_n$, where $\subset$ denotes strict inclusion.

Proof (sketch). Clearly, any complete classical representation is a complete κ-square relativised representation for any cardinal κ. Hence $\mathbf{CRA} \subseteq \mathbf{CRA}_\omega$. To show that the inclusion is proper, consider K_{ω_1} and K_ω — the undirected, loop-free graphs on ω_1 and ω nodes respectively, where ω_1 is the first uncountable ordinal. Define the non-associative algebra $\mathcal{A}$ to be $\mathcal{A}_{\mathrm{K}_{\omega_1}, \mathrm{K}_\omega}$. It is clear that ∃ has a winning strategy in $EF_\omega^\omega(\mathrm{K}_{\omega_1}, \mathrm{K}_\omega)$ — at each round there will only be finitely many pairs of pebbles in play and ∃ can always find an unpebbled node in K_ω for her moves. By theorem 16.5, ∃ has a winning strategy in $G_\omega^\omega(\mathcal{A})$, so by lemma 17.24, $\mathcal{A} \in \mathbf{CRA}_\omega$.

But now, one may show using the method of proposition 16.8 that in any complete representation of $\mathcal{A}$ there will be an uncountable red clique (indexed by $\{g_i : i < \omega_1\}$) and any edge within the clique is labelled by r_{ij} for some distinct $i, j < \omega$, and these indices must match in a triangle within the clique. Thus the nodes from the clique must have distinct induced indices but because ω is countable, it is easily seen that an uncountable red clique is impossible. We deduce that $\mathcal{A}$ has no complete representation, so $\mathcal{A} \notin \mathbf{CRA}$. Cf. exercise 16.4(5).

$\mathbf{CRA}_\omega = \mathbf{RB}_\omega$ is from lemma 17.24.

Evidently, $\mathbf{RB}_\omega \subseteq \bigcap_{n<\omega} \mathbf{RB}_n$ (by the argument of lemma 12.20, if $3 \leq \alpha < \beta$ and $\mathcal{R}$ is a β-dimensional relational basis then $\mathcal{R}\restriction_\alpha$ is an α-dimensional relational basis). To see that the inclusion is proper, we use the same example as in lemma 17.5: $\mathcal{A}_{K_\omega,K}$, where K is the disjoint union of graphs K_n for $0 < n < \omega$. Evidently, $\exists$ has a winning strategy in $EF_\omega^n(K_\omega,K)$ for any finite n, as she may always respond to $\forall$'s move by placing a pebble in K_n. But as we saw in lemma 17.5, $\forall$ has a winning strategy in $EF_\omega^\omega(K_\omega,K)$. By theorem 16.5, $\exists$ has a winning strategy in $G_\omega^n(\mathcal{A}_{K_\omega,K})$ for finite n, but not in $G_\omega^\omega(\mathcal{A}_{K_\omega,K})$. So by proposition 12.25 and lemma 17.24, $\mathcal{A}_{K_\omega,K} \in \bigcap_{n<\omega} \mathbf{RB}_n \setminus \mathbf{RB}_\omega$.

For finite $n \geq 3$, we have $\mathbf{RB}_n \subseteq \mathbf{CRA}_n$ by theorem 13.38. Conversely, if $\mathcal{A}$ has a complete n-square relativised representation then the set of all n-dimensional atomic networks that embed in it forms an n-dimensional relational basis, so $\mathcal{A} \in \mathbf{RB}_n$. Thus, $\mathbf{RB}_n = \mathbf{CRA}_n$. We deduce that $\bigcap_{3 \leq n < \omega} \mathbf{RB}_n = \bigcap_{3 \leq n < \omega} \mathbf{CRA}_n$. □

REMARK 17.26 So for $3 \leq n \leq \omega$ we have $\mathbf{CRA}_n = \mathbf{RB}_n$. But this equality can fail with uncountably many dimensions. To see this, note that the proof of theorem 17.25 can be used to show that $\mathcal{A}_{K_{\omega_1},K_\omega} \in \mathbf{CRA}_\omega \setminus \mathbf{CRA}_{\omega_1}$. Thus, $\mathbf{CRA}_{\omega_1} \subset \mathbf{CRA}_\omega = \mathbf{RB}_\omega = \mathbf{RB}_{\omega_1}$ (lemma 17.21). See exercise 17.5(3) below for a generalisation of this.

Exercises

1. Let $\mathcal{A}$ be an atomic relation algebra and let $\kappa \geq |\mathrm{At}\mathcal{A}| + \omega$. Show that $\mathcal{A} \in \mathbf{CRA}_\kappa \iff \mathcal{A} \in \mathbf{CRA}$.

2. Strengthen the notion of relational basis to obtain results of the form $\mathbf{RB}_\kappa = \mathbf{CRA}_\kappa$ for infinite κ.

3. Use the results of exercise 16.5(6) to show, for all infinite cardinals $\kappa < \lambda$, that $\mathbf{CRA}_\kappa \supset \mathbf{CRA}_\lambda$.

17.6 Weakly representable relation algebras

Recall from definition 5.14 that a *weak representation* of a relation algebra is one that respects the operations $\cdot$, 0, 1', ˘, and ;. A relation algebra $\mathcal{A}$ is said to be *weakly representable* if it has such a representation. We saw in example 9.2(5) that the class **wRRA** of weakly representable relation algebras is pseudo-universal, and hence can be axiomatised by games. As we said in section 5.2, Jónsson [Jón59] axiomatised **wRRA** by quasi-equations, using a step-by-step proof, and showed that $\mathbf{RA} \supset \mathbf{wRRA}$. Andréka proved in [And94] that $\mathbf{wRRA} \supset \mathbf{RRA}$, and indeed that **RRA** is not finitely axiomatisable over **wRRA**.

17.6. Weakly representable relation algebras

Here, we will prove the (implicit) result of Haiman [Hai87, Hai91] that **wRRA** is not finitely axiomatisable. [HodMik00] proved this explicitly, for any class K consisting of all subalgebras of generalised reducts of representable relation algebras, such that $\cdot$, ;, and $\breve{\ }$ are term-definable in K. For more recent results in this direction, see [Mik00].

The algebras we use are 'rainbow' algebras defined in the following way. For $n < \omega$, we let I_n be the structure in a language consisting of a single binary relation, say Suc, with domain n; for $i,j < n$, we let $I_n \models \mathrm{Suc}(i,j)$ iff $j = i+1$. So I_n is a directed graph consisting of an n-node chain. See figure 17.2.

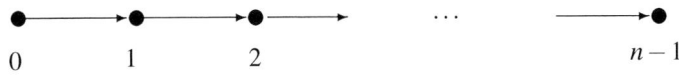

Figure 17.2: The structure I_n

We let $\mathcal{B}_n$ be the algebra $\mathcal{A}_{I_{2^n}, I_{2^n-1}}$, as defined in section 16.2.

LEMMA 17.27 *Let $0 < n < \omega$. The algebra $\mathcal{B}_n$ is not weakly representable.*

Proof. First, we attempt some calculations in $\mathcal{B}_n$. Recall that it is the complex algebra over its atom structure. We define $\mathsf{R} = \{r_{ij} : i,j < 2^n - 1\} \in \mathcal{B}_n$. For each $i < \omega$, we let $\mathsf{R}_i = \{r_{k,k+i} : 0 \leq k < 2^n - 1 - i\} \in \mathcal{B}_n$; note that this is zero if $i \geq 2^n - 1$. Then by definition of $\mathcal{B}_n$, we have

$$\begin{aligned}
(\mathsf{g}_i ; \mathsf{g}_j) \cdot (\mathsf{y} ; \mathsf{y}) &\leq \mathsf{R} &&\text{for all } i,j < 2^n, \\
(\mathsf{g}_i ; \mathsf{g}_{i+1}) \cdot (\mathsf{y} ; \mathsf{y}) &= \mathsf{R}_1 &&\text{for all } i < 2^n - 1, \\
\mathsf{R} \cdot (\mathsf{R}_i ; \mathsf{R}_j) &= \mathsf{R}_{i+j} &&\text{for all } i,j < \omega.
\end{aligned}$$

We now show that $\mathcal{B}_n \notin \mathbf{wRRA}$. In fact, we prove that $\mathcal{B}_n$ has no representation respecting the operations $\cdot$, $\breve{\ }$, and ;.

Assume for contradiction that there is such a representation — an injective map h from $\mathcal{B}_n$ into a proper relation algebra, preserving $\cdot$, $\breve{\ }$, and ;. It follows that $a \leq b$ in $\mathcal{B}_n$ iff $h(a) \subseteq h(b)$.

Since $\mathsf{w} \not\leq 0$, we have $h(\mathsf{w}) \not\subseteq h(0)$, so we may take $(x,y) \in h(\mathsf{w}) \setminus h(0)$. Now, $\mathsf{w} \leq \mathsf{g}_i ; \mathsf{y}$ for all $i < 2^n$, so there are points z_i for $i < 2^n$ with $(x, z_i) \in h(\mathsf{g}_i)$ and $(z_i, y) \in h(\mathsf{y})$. Since h preserves conversion, $(z_i, x) \in h(\mathsf{g}_i)$ and $(y, z_i) \in h(\mathsf{y})$, too.

We claim that $(z_i, z_j) \in h(\mathsf{R}_{j-i})$ for all $i < j < 2^n$. The proof is by induction on $j - i$. By the calculations above, for all $i < \omega$ with $i + 1 < 2^n$, we have $(z_i, z_{i+1}) \in h(\mathsf{g}_i ; \mathsf{g}_{i+1} \cdot \mathsf{y} ; \mathsf{y}) = h(\mathsf{R}_1)$, proving the claim when $j - i = 1$. Let $j > i + 1$, and assume the result for smaller $j - i$. By considering the paths z_i, x, z_j, z_i, y, z_j, and z_i, z_{j-1}, z_j, and using the inductive hypothesis, we see that

$$(z_i, z_j) \in h(\mathsf{g}_i ; \mathsf{g}_j \cdot \mathsf{y} ; \mathsf{y} \cdot \mathsf{R}_{j-1-i} ; \mathsf{R}_1) \subseteq h(\mathsf{R} \cdot (\mathsf{R}_{j-1-i} ; \mathsf{R}_1)) = h(\mathsf{R}_{j-i}),$$

completing the induction and proving the claim. In particular, it follows that $(z_0, z_{2^n-1}) \in h(\mathsf{R}_{2^n-1}) = h(0)$.

We are not yet finished, since h may not preserve zero, so that $h(0) \neq \emptyset$ is possible. Nonetheless, following the path x, z_0, z_{2^n-1}, y, we see that $(x,y) \in h((g_0;0);y) = h(0)$, which contradicts our original assumption on x,y. □

LEMMA 17.28 $\exists$ has a winning strategy in $G_{n+1}^\omega(\mathcal{B}_n)$ for each finite $n > 0$.

Proof. By theorem 16.5, we only need prove that $\exists$ has a winning strategy in $EF_n^\omega(I_{2^n}, I_{2^n-1})$. This is easily done, by the same argument as in lemma 17.16. □

We now obtain:

THEOREM 17.29 *The class* **wRRA** *of weakly representable relation algebras is not finitely axiomatisable.*

Proof. Let $\mathcal{B}$ be a non-principal ultraproduct over ω of the $\mathcal{B}_n$ ($0 < n < \omega$). By lemma 17.28 and theorem 10.12, $\exists$ has a winning strategy in $G_\omega^\omega(\mathcal{B})$; this is the same game as $G_\omega^a(\mathcal{B})$, so by theorem 11.10, there is $C \equiv \mathcal{B}$ with a complete (relation algebra) representation. Hence, $\mathcal{B} \in$ **RRA**. (We could equally say that $\mathcal{B}$ satisfies the Lyndon conditions and so is representable.) So certainly, $\mathcal{B} \in$ **wRRA**. But $\mathcal{B}_n \notin$ **wRRA** for all n, by lemma 17.27. It now follows by Łoś's theorem that **wRRA** is not finitely axiomatisable. □

In corollary 18.24, we will prove the stronger result that the problem of whether a finite relation algebra is weakly representable is undecidable.

The following problem and exercises are related to a question raised informally by Roger Maddux, namely whether **wRRA** = **RA**$_5$. (Maddux's question is resolved negatively by corollary 18.25, which shows that **RA**$_5 \not\subseteq$ **wRRA**. But if problem 17.30 has an affirmative answer, no finite rainbow algebra can distinguish the two classes.)

PROBLEM 17.30 *Is it the case that for any finite relational structures A,B in a given binary signature, if $\mathcal{A}_{A,B} \in$ **RA**$_5$ then $\mathcal{A}_{A,B} \in$ **wRRA**?*

Exercises

1. Show that $\mathcal{B}_n \in$ **RA** \ **RA**$_5$ for all $n > 0$.

2. Show that for any relational structures A,B in the same binary signature, if B is finite and $\mathcal{A}_{A,B} \in$ **wRRA** then $\mathcal{A}_{A,B} \in$ **RA**$_5$. [Use a weak representation of $\mathcal{A}_{A,B}$ to guide $\exists$ in $EF_\omega^3(A,B)$.]

17.7 Completions

Recall from section 2.5.4 that the *completion* of a relation algebra $\mathcal{A}$ (or more generally any completely additive BAO) is a complete extension of $\mathcal{A}$ in which $\mathcal{A}$ is dense; such an algebra always exists, and this condition defines it uniquely up to isomorphism over $\mathcal{A}$ (theorem 2.41). If $\mathcal{A}$ is atomic, its completion is therefore isomorphic to the full complex algebra over the atom structure of $\mathcal{A}$, since this algebra is complete and $\mathcal{A}$, having the same atoms, is dense in it. We saw this already, in remark 2.67.

RA$_4$ (**RA**) is closed under completions, by a result of Monk [Mon70] (see theorem 3.16); so is **RA**$_3$ (**SA**), as Maddux showed [Madd91b, theorem 36]. Indeed, both these varieties are conjugated and are defined by Sahlqvist equations, and so their closure under completions follows from theorem 2.96.

But a consequence of the fact that there are weakly but not strongly representable atom structures is that the class **RRA** is not closed under completions (corollary 14.15). Here we extend that result to the varieties **RA**$_n$ and $\mathbf{S\mathfrak{R}aCA}_n$, for $n \geq 6$. We will use the rainbow construction to construct an atomic relation algebra $\mathcal{A} \in \mathbf{RRA}$ whose completion is not even in **RA**$_6$. It follows immediately from the inclusions between the classes (see proposition 13.48) that $\mathbf{RRA}, \mathbf{S\mathfrak{R}aCA}_n, \mathbf{RA}_n$ are not closed under completions for any $n \geq 6$. The case $n = 5$, for both **RA**$_5$ and $\mathbf{S\mathfrak{R}aCA}_5$, is open.

17.7.1 The example

DEFINITION 17.31 Take complete irreflexive symmetric graphs K_4, K_3 of size 4 and 3, respectively. (Cf. figure 1.1, p. 19.) We take L to be the signature of graphs, with a single binary relation symbol, and regard K_4 and K_3 as L-structures by interpreting the binary relation symbol as the graph edge relation. We first consider the rainbow algebra $\mathcal{A}_{K_4,K_3}$.

LEMMA 17.32 $\mathcal{A}_{K_4,K_3} \notin \mathbf{RA}_6$.

Proof. $\forall$ clearly has a winning strategy in $EF_4^4(K_4, K_3)$. By theorem 16.5, he has a winning strategy in $G_5^6(\mathcal{A}_{K_4,K_3})$. But as $\mathcal{A}_{K_4,K_3}$ is finite, $\mathcal{A}_{K_4,K_3} \in \mathbf{RA}_6$ iff $\exists$ has a winning strategy in $G_\omega^6(\mathcal{A}_{K_4,K_3})$ (see proposition 12.25 and theorem 13.46(4 $\Leftrightarrow$ 5)). □

DEFINITION 17.33

1. Let S be the atom structure obtained from the atom structure of $\mathcal{A}_{K_4,K_3}$ by splitting each of the red atoms into ω copies.

   So S contains all non-red atoms of $\mathcal{A}_{K_4,K_3}$, and for each red atom r_{lm}, for $l,m \in K_3$, we put red atoms r_{lm}^n ($n < \omega$) into S.

A triple $(r^n_{lm}, r^{n'}_{l'm'}, r^{n^*}_{l^*m^*})$ of these new red atoms is consistent in S iff $n = n' = n^*$ and the corresponding triple $(r_{lm}, r_{l'm'}, r_{l^*m^*})$ of atoms of $\mathcal{A}_{K_4,K_3}$ is consistent in $\mathcal{A}_{K_4,K_3}$. A triple of atoms of S, not all of which are red, is consistent in S iff the corresponding triple of atoms of $\mathcal{A}_{K_4,K_3}$ is consistent in $\mathcal{A}_{K_4,K_3}$. This determines the consistent triples of atoms of S. The identity of S is the identity of $\mathcal{A}_{K_4,K_3}$. This determines the conversion relation on S in the way that one would expect.

2. Let C be the full complex algebra $\mathfrak{Cm}\,S$ over S. We identify each element $s \in S$ with the atom $\{s\}$ of C.

3. Let T be the subalgebra of C generated by S (i.e., T is the term algebra of S: cf. lemma 14.2).

By remark 2.67, C is the completion of T. We will show that $T \in \mathbf{RRA}$, while $C \notin \mathbf{RA}_6$.

LEMMA 17.34 $C \notin \mathbf{RA}_6$.

Proof. Notice that $\mathcal{A}_{K_4,K_3}$ embeds into C, via

- any non-red atom maps to itself,
- $r_{lm} \mapsto \{r^n_{lm} : n < \omega\}$,
- sums (unions) of atoms map to the corresponding sum in C.

This is a relation algebra embedding. Since $\mathcal{A}_{K_4,K_3} \notin \mathbf{RA}_6$, $\mathcal{A}_{K_4,K_3} \subseteq C$ up to isomorphism, and $\mathbf{RA}_6$ is a variety, we see that $C \notin \mathbf{RA}_6$. □

The 'reason' that $T \in \mathbf{RRA}$ is that splitting the red atoms gives ∃ the opportunity to prevaricate over which exact r_{ij} she is using to label an edge during a play of $G^\omega_\omega(T^+)$. Because there are infinitely many red atoms now, there is a non-principal filter of red atoms, namely the set of all cofinite sets of red atoms with distinct indices. This turns out to generate an ultrafilter ρ of T, and it is self-consistent: (ρ, ρ, ρ) is a consistent triple of atoms of T^+. So ∃ can use ρ to label all her red edges, and so win the game $G^a_\omega(T^+)$. Hence, T^+ and so T are representable.

This is the underlying idea, but to avoid repeating the proof of the rainbow theorem, we will actually prove it using the modified rainbow algebras of section 16.5. Let

$$\begin{aligned} \mathsf{D} &= \{r^n_{ll} : n < \omega,\, l \in K_3\} \in C, \\ \mathsf{R} &= \{r^n_{lm} : n < \omega,\, l,m \in K_3,\, l \neq m\} \in C. \end{aligned}$$

LEMMA 17.35 *Let $X \subseteq \mathsf{R}$. Then $X \in T$ if and only if X is finite or cofinite in R. A similar result holds for subsets X of D.*

17.7. Completions

Proof. First, we show that $\mathsf{R} \in \mathcal{T}$. Fix any distinct $i, j \in \mathsf{K}_4$. Observe that an atom x of $\mathcal{A}_{\mathsf{K}_4,\mathsf{K}_3}$ is red with distinct indices iff the triples of atoms (x, y, y) and (x, g_i, g_j) are both consistent. By definition of $\mathcal{S}$, the same holds in C. So $C \models \mathsf{R} = y ; y \cdot g_i ; g_j$. Since clearly, $y ; y \cdot g_i ; g_j \in \mathcal{T}$, we are done.

Given that $\mathsf{R} \in \mathcal{T}$ and that every red atom is in $\mathcal{T}$, the boolean operations ensure that any finite or cofinite subset of R is in $\mathcal{T}$. For the converse, let $X \in \mathcal{T}$ with $X \subseteq \mathsf{R}$; so there are a relation algebra term $\tau(\bar{x})$ and atoms $\bar{s} \in \mathcal{S}$ with $C \models \tau(\bar{s}) = X$. Let $n_0 < \omega$ be such that any red atom r^n_{lm} in $\bar{s}$ satisfies $n < n_0$. Assume that X is infinite; we prove it cofinite in R by showing that if $n > n_0$ and $l, m \in \mathsf{K}_3$ are distinct then $r^n_{lm} \in X$.

So fix such n, l, m. As X is infinite, we may pick $n' > n_0$ and distinct $l', m' \in \mathsf{K}_3$ such that $r^{n'}_{l'm'} \in X$. We show in two steps that some automorphism of C fixes X and takes $r^{n'}_{l'm'}$ to r^n_{lm}.

First, pick an automorphism θ of K_3 with $\theta(l') = l$ and $\theta(m') = m$. Let η be the permutation of $\mathcal{S}$ defined by: $\eta(r^{n'}_{ij}) = r^{n'}_{\theta(i)\theta(j)}$ (for all $i, j \in \mathsf{K}_3$), and $\eta(a) = a$ for all $a \in \mathcal{S}$ not of this form. This is easily seen to take consistent triples of $\mathcal{S}$ to consistent triples, and vice versa, and so it induces an automorphism $\hat{\eta}$ of C via: $\hat{\eta}(Y) = \{\eta(a) : a \in Y\}$, for $Y \subseteq \mathcal{S}$. Given our identification of elements of $\mathcal{S}$ with atoms of C specified in definition 17.33(2), it is clear that $\hat{\eta}$ takes $r^{n'}_{l'm'}$ to $r^{n'}_{lm}$. Also, $\hat{\eta}(\bar{s}) = \bar{s}$, and a standard induction on the term τ defining X now shows that $\hat{\eta}(X) = X$. Since $r^{n'}_{l'm'} \in X$, we have $C \models r^{n'}_{l'm'} \leq X$. As $\hat{\eta}$ is an automorphism of C, we have $C \models \hat{\eta}(r^{n'}_{l'm'}) \leq \hat{\eta}(X)$ — that is, $C \models r^{n'}_{lm} \leq X$. So $r^{n'}_{lm} \in X$.

Second, pick a permutation σ of ω that fixes every $p < n_0$ and takes n' to n. Let ψ be the permutation (automorphism) of $\mathcal{S}$ given by: $\psi(r^p_{ij}) = r^{\sigma(p)}_{ij}$ for all $p < \omega$ and $i, j \in \mathsf{K}_3$, and $\psi(a) = a$ for all non-red $a \in \mathcal{S}$. Again, ψ induces an automorphism of C, fixing $\bar{s}$ and hence X, and taking $r^{n'}_{lm}$ to r^n_{lm}. So $r^n_{lm} \in X$, as required.

The proof for subsets of D is similar, observing that for any $i \in \mathsf{K}_4$, we have $C \models \mathsf{D} = (y ; y \cdot g_i ; g_i) - 1' \in \mathcal{T}$. □

We now see that the embedding of $\mathcal{A}_{\mathsf{K}_4,\mathsf{K}_3}$ into C given in lemma 17.34 does not serve to embed it into $\mathcal{T}$ — for any $l, m \in \mathsf{K}_3$, the set of atoms $\{r^n_{lm} : n < \omega\}$ is not in $\mathcal{T}$.

Thus, there are unique non-principal ultrafilters δ, ρ of $\mathcal{T}$ containing D, R, respectively, namely

$$\delta = \{X \in \mathcal{T} : X \cap \mathsf{D} \text{ is cofinite in } \mathsf{D}\},$$
$$\rho = \{X \in \mathcal{T} : X \cap \mathsf{R} \text{ is cofinite in } \mathsf{R}\}.$$

Because $\mathsf{D} \cup \mathsf{R}$ is cofinite in $\mathcal{S}$, it follows that the set of atoms of the canonical extension $\mathcal{T}^+$ is, modulo the natural identification of principal ultrafilters with their generators, $\mathcal{S} \cup \{\delta, \rho\}$. The identity of $\mathcal{T}^+$ is 1', the converse in $\mathcal{T}^+$ of an atom in S is its converse in T, and δ, ρ are self-converse. By evaluating composition in T^+, the consistent triples of atoms in T^+ are easily seen to be precisely:

1. the consistent triples of atoms of S,

2. Peircean transforms of (δ, x, y), for any $x, y \in S \setminus \{1'\}$ that are not both red and are not distinct green atoms g_i, g_j,

3. Peircean transforms of (ρ, x, y), for any $x, y \in S \setminus \{1'\}$ that are not both red and are not both the same green atom g_i,

4. Peircean transforms of (δ, δ, x) and of (ρ, ρ, x), for any non-red $x \in S$,

5. (ρ, ρ, ρ). For, let $X, Y, Z \in \rho$; by lemma 3.26, we see that we must find $x \in X$, $y \in Y$, $z \in Z$ such that (x, y, z) is consistent in S. Now X, Y, Z all have cofinite intersection with R, so there is $n < \omega$ such that $\{r_{ij}^n : i \neq j \text{ in } K_3\} \subseteq X \cap Y \cap Z$. Let i, j, k be (the) distinct elements of K_3. Then $r_{ij}^n \in X$, $r_{jk}^n \in Y$, and $r_{ki}^n \in Z$, and $(r_{ij}^n, r_{jk}^n, r_{ki}^n)$ is S-consistent.

6. (ρ, ρ, δ) and (δ, δ, δ). This is proved similarly to 5.

LEMMA 17.36 $T \in \mathbf{RRA}$. *Hence,* $T \in \mathbf{RA}_n$ *and* $T \in \mathbf{S\Re aCA}_n$ *for all finite* $n \geq 3$.

Proof. Consider the graph Δ consisting of ω pairwise disjoint copies of K_3, plus a single copy of K_4. As notation, we identify the domain of Δ with $(K_3 \times \omega) \cup (K_4 \times \{\omega\})$ in the obvious way. Let $\mathcal{B}$ be the full complex algebra over the atom structure obtained from $\mathrm{At}\,\mathcal{A}_{K_4, \Delta}$ by deleting all red atoms r_{ij} where i, j lie in different connected components of Δ. So the red atoms remaining in $\mathcal{B}$ are $r_{(i,n),(j,n)}$ for $i, j \in K_3$ and $n < \omega$, plus $r_{(i,\omega),(j,\omega)}$ for $i, j \in K_4$.

As $\exists$ obviously has a winning strategy in the game $EF_\omega^\omega(K_4, K_4)$, by the rainbow theorem 16.5 she has a winning strategy in $G_\omega^\omega(\mathcal{A}_{K_4, K_4})$. Clearly, $\mathrm{At}\,\mathcal{A}_{K_4, K_4} \subseteq \mathrm{At}\,\mathcal{B} \subseteq \mathrm{At}\,\mathcal{A}_{K_4, \Delta}$. So by proposition 16.14, $\exists$ has a winning strategy in $G_\omega^\omega(\mathcal{B})$. Now $\mathcal{B}$ has countably many atoms, so by theorem 11.7, $\mathcal{B} \in \mathbf{RRA}$.

Now let $\theta : \mathrm{At}\,\mathcal{B} \to \mathrm{At}(T^+)$ be the map defined by

- $\theta(r_{(i,n),(j,n)}) = r_{ij}^n$ for all $i, j \in K_3$ and $n < \omega$,

- $\theta(r_{(i,\omega),(i,\omega)}) = \delta$ for $i \in K_4$,

- $\theta(r_{(i,\omega),(j,\omega)}) = \rho$ for distinct $i, j \in K_4$,

- θ is the identity on all other atoms.

17.7. *Completions*

It can be checked that θ is a surjective bounded morphism (see exercise 2.7(6)). It follows by that exercise that $\mathcal{T}^+$ embeds in $\mathcal{B}$, so that $\mathcal{T}^+ \in \mathbf{RRA}$. Hence, $\mathcal{T} \in \mathbf{RRA}$ also. The remainder of the theorem now follows from proposition 13.48. □

We obtain:

THEOREM 17.37 *The varieties* $\mathbf{RRA}$, $\mathbf{RA}_n$, *and* $\mathbf{S\mathfrak{R}aCA}_n$, *for all finite* $n \geq 6$, *are not closed under completions, and are not atom-canonical.*

Proof. By lemmas 17.36 and 17.34. □

Using a 'cylindric algebra' version of the rainbow construction, [Hodk97] shows that $\mathbf{RCA}_n$ is not closed under completions for any finite $n \geq 3$. Hence, it is not atom-canonical.

17.7.2 Corollaries and problems

Some corollaries include that for $V \in \{\mathbf{RRA}, \mathbf{RA}_n, \mathbf{S\mathfrak{R}aCA}_n : 6 \leq n < \omega\}$, there exists an algebra $\mathcal{A} \notin V$ such that $\mathcal{A}$ has a dense subalgebra in V. For example, a relation algebra with a dense representable subalgebra need not be representable itself (cf. corollary 14.15).

The $\mathbf{RRA}$-part of the following corollary is from [Ven97b]. Venema also proves that $\mathbf{RCA}_n$ is not a Sahlqvist variety for any finite $n \geq 3$, using the result of [Hodk97] that it is not closed under completions.

COROLLARY 17.38 *The varieties* $\mathbf{RRA}, \mathbf{RA}_n$, *and* $\mathbf{S\mathfrak{R}aCA}_n$ *(for all finite* $n \geq 6$*) are not axiomatisable by Sahlqvist equations.*

Proof. By theorem 2.96, any conjugated Sahlqvist variety is closed under completions. Since the above varieties are contained in $\mathbf{RA}$, they are conjugated. The corollary now follows from theorem 17.37. □

The following is not settled by corollary 17.38.

PROBLEM 17.39 *For which (finite)* $n \geq 5$ *do* $\mathbf{RA}_n$ *and* $\mathbf{S\mathfrak{R}aCA}_n$ *have a canonical equational axiomatisation? Does* $\mathbf{RRA}$ *have such an axiomatisation?*

A canonical axiomatisation is one consisting of canonical equations; an equation ε is canonical if $\mathcal{A} \models \varepsilon \Rightarrow \mathcal{A}^+ \models \varepsilon$. Of course, $\mathbf{RA}_3$ and $\mathbf{RA}_4$ are Sahlqvist-axiomatised, and Sahlqvist equations are canonical.

PROBLEM 17.40 *Are* $\mathbf{RA}_5$ *and* $\mathbf{S\mathfrak{R}aCA}_5$ *closed under completions?*

PROBLEM 17.41 *Is* **wRRA** *closed under completions?*

Exercises

1. Check that the map θ in the proof of lemma 17.36 is indeed a surjective bounded morphism.

2. Where does the proof of theorem 17.37 break down for $\mathbf{RA}_5$, when we work with $\mathcal{A}_{K_3,K_2}$ instead of $\mathcal{A}_{K_4,K_3}$? [Hint: consider the consistency of (ρ,ρ,ρ).]

Part V

Decidability

Chapter 18

Undecidability of the representation problem for finite algebras

18.1 Introduction

In this chapter we address an old question in algebraic logic: is there an algorithm that tells us whether or not a finite relation algebra is representable? We have not been able to pin down the origin of this problem precisely, but in all probability it originated with Roger Maddux. Maddux and McKenzie discussed it in the early 1980s, Maddux suggesting a solution by tiling (our approach here). It was raised again by McKenzie at a conference on universal algebra and lattice theory in 1996. The problem is listed in [AndMon+91, page 730, open problem 3] (credited to Maddux). There is a discussion of the question in [Madd94a, problems 13 and 14, page 463], where it is observed that the finite relation algebras can be partitioned into three classes:

(a) the non-representable ones,

(b) those that are representable over some finite set,

(c) the finite representable relation algebras with no representation over a finite set.

It is not hard to show that (a) and (b) are recursively enumerable.[1] The (isomorphism types of) finite relation algebras is clearly a recursive set. Consequently, (a),

---
[1] Strictly, we have to consider the isomorphism type of members of these classes, so that we are dealing with sets and not classes.

(b) and (c) are all recursive if and only if (c) is recursively enumerable. Maddux conjectured that both sides of this equivalence are false. In this chapter, we show that (a) is not recursive, thus confirming the conjecture. Hence, (c) is not recursively enumerable. One problem remains open: is (b) recursive — i.e., given a finite relation algebra, is it decidable whether it has a representation over a finite set?

We will reduce a certain tiling problem to the question of whether a finite relation algebra is representable. That will show that the question is undecidable. (This is utterly unsurprising, but our proof is rather complicated.) Incidentally, the same tiling construction shows that the problem of whether a finite algebra belongs to $\mathbf{S\Re aCA}_n$ (any n with $5 \leq n \leq \omega$) is also undecidable. This contrasts with the result of Maddux (corollary 12.32) that it is decidable whether a finite algebra belongs to $\mathbf{RA}_n$ for $3 \leq n < \omega$.

It is interesting to consider alternatives to the tiling reduction. For example, we have seen that the representable relation algebras cannot be defined by a finite number of axioms (theorem 17.3) [Mon64], and this might suggest that for finite relation algebras, the representability problem is undecidable. However, finite axiomatisability and decidability are not the same. If a class is finitely axiomatisable then this does give us a decision method to test whether a finite object belongs to the class or not. But the converse is false: Németi showed that the class $\mathbf{ICrs}_n$ of isomorphism types of relativised cylindric set algebras of any finite dimension $n \geq 3$ is not finitely axiomatisable (see [HenMon+85, 5.5.12, credited to Németi], and [Mon93, Mon00]), yet it is decidable whether a finite algebra is in this class [Ném96, theorem 4.1(ii, v)]. He further showed that the equational and indeed universal theory of this class is decidable. See fact 5.38 for more information. On the other hand, the undecidability of whether a finite relation algebra is representable does yield as a corollary that the equational theory of $\mathbf{RRA}$ is undecidable (see corollary 18.27).

One of the main motivations for Tarski's study of relation algebras was to define an alternative foundation for set theory. In [TarGiv87] it is shown that relation algebra can act as a vehicle for set theory and hence all of mathematics. It would seem, then, that undecidability results for relation algebra should be obtainable by this result. However, we have not been able to obtain the required result that way.

The tiling construction we give here is a variant of the rainbow construction given earlier. It looks rather different, at least at first, because the relation algebras are not integral: the identity contains three distinct atoms. So any representation of one of these relation algebras has its domain partitioned into three. One part is something like the x-axis of a tiling of the plane, another part is like the y-axis, and the third part is auxiliary. A perhaps more significant difference is that here, the analogues of the 'red cliques' of definition 16.10 are based on a single element, and not on an edge (c,c') as before. This has the effect that $\forall$ is able to label some

18.2. The tiling problem

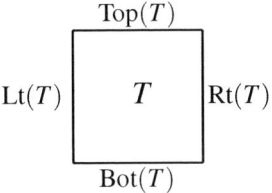

Figure 18.1: A tile T

of the edges in a red clique.

Applications In section 18.9, we apply the tiling technique to show that it is undecidable whether a finite relation algebra is weakly representable. In section 18.10, we use the undecidability of finite membership of **RRA** and related varieties to derive results on undecidability of their equational theories.

Recall from section 6.4 our brief discussion of the finitisation problem. The undecidability of representability of finite relation algebras has ramifications for this problem: see exercise 18.10(1), showing that **RRA** is not finitely axiomatisable in nth-order logic for any finite n. The undecidability result was also used in [HirHod$^+$02b] to show that for finite $n \geq 3$, every modal logic lying between K^n and $S5^n$ is undecidable and not finitely axiomatisable, and in [HodWol$^+$02] to show that the one-variable fragment of predicate CTL^* is undecidable. For further results, see [GabKur$^+$].

Exercises

1. Recall by exercise 3.4(9) that the set A of isomorphism types of finite non-representable relation algebras is recursively enumerable. Show that the set B of isomorphism types of finite relation algebras that are representable over some finite set is also recursively enumerable.

2. Deduce that the above two sets A, B, and also the set C of isomorphism types of finite representable relation algebras with no finite representation, are recursive if and only if C is recursively enumerable.

18.2 The tiling problem

An instance τ of the *tiling problem* is a finite set of square tiles $\tau = \{T_0, \ldots, T_{k-1}\}$. Each tile has a colour on each of its four edges: the four colours on the tile T_i are $\text{Top}(T_i), \text{Bot}(T_i), \text{Lt}(T_i)$ and $\text{Rt}(T_i)$. See figure 18.1. Note that the tiles have a fixed orientation and cannot be rotated or flipped over.

Such an instance τ is said to be a *yes-instance* if it is possible to tile the plane $\mathbb{Z}\times\mathbb{Z}$ using copies of tiles in τ. That is, there is a function $f:(\mathbb{Z}\times\mathbb{Z})\to\{0,\ldots,k-1\}$ such that for all $x,y\in\mathbb{Z}$ we have $\text{Lt}(T_{f(x+1,y)})=\text{Rt}(T_{f(x,y)})$ and $\text{Bot}(T_{f(x,y+1)})=\text{Top}(T_{f(x,y)})$. (Here and below, we abbreviate $f((x,y))$ to $f(x,y)$.) We call such an f a *tiling* (for τ). If there is no such tiling then we have a *no-instance*.

The tiling problem (given an instance, is it a yes-instance or a no-instance?) is known to be undecidable [Ber66]. The rough idea of the proof of this is to find a set of tiles that encodes the run of a given Turing machine whose input tape is initially blank. The tth row of a tiling of $\mathbb{Z}\times\mathbb{N}$ by such a tile set, where the 0th row (the 'x-axis') encodes the initial configuration of the machine, can be shown by induction on t to correspond to the machine's configuration at time t. Therefore, subject to this assumption on the 0th row of the tiling, the tiles will tile the upper half-plane $\mathbb{Z}\times\mathbb{N}$ iff the Turing machine does not halt. (Here, a tiling of $\mathbb{Z}\times\mathbb{N}$ is defined in the obvious way.) There is of course no Turing machine that decides whether an arbitrary Turing machine halts or not when started on the empty tape. Since the tiles are constructed effectively from a description of the given Turing machine, it follows that it is undecidable whether there is a tiling of $\mathbb{Z}\times\mathbb{N}$ with 0th row of a certain specified form. The assumption on the 0th row can be eliminated by using aperiodic tilings: this is in fact the hardest part of the argument. The conversion to $\mathbb{Z}\times\mathbb{Z}$ is straightforward.

It is not hard to show from this (see exercise 2 below) that the following problem is also undecidable. Given a finite set of tiles $\{T_0,\ldots,T_{k-1}\}$ as above, is it the case that for each $i<k$ there is a tiling f^i of the plane with T_i placed at $(0,0)$ (formally $f^i(0,0)=i$)? Indeed, establishing the undecidability of this problem does not need aperiodic tilings and is substantially easier than for the 'standard' tiling problem above. Exercise 5 below asks for a direct proof.

We lose no generality if we assume that one tile, T_0, is a special 'white' tile such that all of its four edges have the same colour but this colour is not used on any edge of any other tile. Thus, there is certainly a tiling f^0 with $f^0(0,0)=0$ given by $f^0(x,y)=0$ (all $x,y\in\mathbb{Z}$). T_0 can tile the plane on its own but not in combination with any of the other tiles. So given an arbitrary set of tiles $\{T_1,\ldots,T_{k-1}\}$, if we add the special tile T_0 then $\{T_0,T_1,\ldots,T_{k-1}\}$ is a yes-instance if and only if the original set $\{T_1,\ldots,T_{k-1}\}$ is a yes-instance.

It is this version of the tiling problem that we use here.

Now, given an instance τ of this tiling problem, we construct (by an algorithm) a finite weakly associative algebra $RA(\tau)$ such that τ is a yes-instance if and only if $RA(\tau)$ is a representable relation algebra. Because it is decidable whether a finite weakly associative algebra is a relation algebra, this suffices to prove the undecidability of the representation problem for finite relation algebras. See theorem 18.13.

18.3. The definition of $RA(\tau)$

Exercises

1. Assuming that the problem of whether a given finite set of tiles will tile $\mathbb{Z} \times \mathbb{N}$ is undecidable, show that the corresponding problem for $\mathbb{Z} \times \mathbb{Z}$ is undecidable too.

2. Consider the following tiling problem: given a finite set $\{T_0, \ldots, T_{k-1}\}$ of tiles, is it the case that for each $i < k$ there is a tiling f^i of $\mathbb{Z} \times \mathbb{Z}$ with $f^i(0,0) = i$? Assuming that the standard tiling problem (in which 'each $i < k$' is replaced by 'some $i < k$') is undecidable, show that this problem is also undecidable.

3. Let $\tau = \{T_0, \ldots, T_{k-1}\}$ be a set of tiles, and for $n < \omega$ let $R_n = \{(x,y) \in \mathbb{Z} \times \mathbb{Z} : -n \le x, y \le n\}$. We say that τ tiles R_n if there is a map $f : R_n \to k$ such that for all $(x,y) \in R_n$, if $x < n$ then $\mathrm{Rt}(T_{f(x,y)}) = \mathrm{Lt}(T_{f(x+1,y)})$, and if $y < n$ then $\mathrm{Top}(T_{f(x,y)}) = \mathrm{Bot}(T_{f(x,y+1)})$. Prove that τ tiles the plane $\mathbb{Z} \times \mathbb{Z}$ iff it tiles R_n for every $n < \omega$.

4. Sometimes it is more useful to specify explicitly which tiles may occur next to which. Formally, given two binary relations X, Y on $k = \{0, \ldots, k-1\}$, a tiling of (k, X, Y) is a function $f : \mathbb{Z} \times \mathbb{Z} \to k$ such that for all $x, y \in \mathbb{Z}$ we have $X(f(x,y), f(x+1,y))$ and $Y(f(x,y), f(x,y+1))$. Assuming that the 'standard' tiling problem for $\mathbb{Z} \times \mathbb{Z}$ is undecidable, prove that it is undecidable whether there is a tiling of a given (k, X, Y), and vice versa.

5. Given a Turing machine M, show how to construct a pair (τ, t_0), where τ is a finite set of tiles and $t_0 \in \tau$, such that there is a tiling of $\mathbb{Z} \times \mathbb{N}$ with t_0 placed at $(0,0)$ if and only if M does not halt when run on the empty (blank) tape. (Exercise 4 may help.)

18.3 The definition of $RA(\tau)$

In this section, we define the weakly associative 'tiling algebra' $RA(\tau)$, for a given set τ of tiles. $RA(\tau)$ is not integral[2] — in fact, the identity 1' is the sum of three identity atoms, e_0, e_1, e_2. If $RA(\tau)$ is representable, then (by exercise 5.1(12)) in any representation h, the domain D of the representation will be the disjoint union of three subsets — $D = D_0 \cup D_1 \cup D_2$ — and for any point $d \in D$ and $i < 3$ we have

$$(d, d) \in h(e_i) \iff d \in D_i.$$

As is standard for weakly associative algebras (see lemma 5.10), every atom a has a start identity atom $\mathrm{st}(a) = 1' \cdot (a; \breve{a})$ and an end identity atom $\mathrm{end}(a) = 1' \cdot (\breve{a}; a)$.

---

[2] A suitable integral algebra $RA(\tau)$ can also be constructed, but we will not do so here.

— these are atoms. Let st(a) = e_i and end(a) = e_j (some $i, j < 3$). If $RA(\tau)$ has a representation h with domain D, this tells us that for any pair of points $(e,d) \in h(a)$ we have $e \in D_i$ and $d \in D_j$. We'll call such an atom an $(i-j)$-*atom*.

We will give the atoms of $RA(\tau)$ subscripts to indicate their start and end atoms. If the subscripts are equal, we generally write just one of them, e.g., $a_{01}, e_{22} = e_2$, etc.

If n is any node in an atomic $RA(\tau)$-network N, there is a unique identity atom e_i such that $N(n,n) = e_i$ (some $i < 3$). We'll call such a node an *i-node*. If n is an *i*-node and m is a *j*-node of a network N, then the label $N(n,m)$ must be an $(i-j)$-atom. (n,m) is called an $(i-j)$-*edge*.

The atoms

If τ is a tiling instance with the k tiles $T_0, \ldots, T_{k-1}$, then $RA(\tau)$ has $2k+28$ atoms. They are:

start	end	Atoms
0	0	e_0, w_0
0	1	$g_{01}, u_{01}, v_{01}, w_{01}$
0	2	$g_{02}, u_{02}, v_{02}, w_{02}$
1	1	$e_1, +1_1, -1_1, w_1$
2	2	$e_2, +1_2, -1_2, w_2$
1	2	$t^i_{12}\,(i<k), w_{12}$

plus the converses of the $(0-1), (0-2)$ and $(1-2)$ atoms. If $i, j < 3$, $i \neq j$, and a_{ij} is any $(i-j)$ atom, we write a_{ji} for $\breve{a}_{ij}$. Thus, the converse of g_{02} is g_{20}. We consider some of the atoms to be coloured: the atoms $g_{01}, g_{10}, g_{02},$ and g_{20} are green, and the atoms $w_0, w_1, w_2, w_{01}, w_{02}, w_{12}$ and their converses are white.

The atom structure

To define $RA(\tau)$, it remains to define the operations of conversion and composition on the atoms. For conversion, we have already defined the converse of atoms with distinct subscripts. All the rest are self-converse except the following: the converse of $+1_1$ is -1_1 and the converse of $+1_2$ is -1_2, and vice versa.

Now we define composition. We do this by listing the inconsistent (or forbidden) triples (a,b,c) of atoms. As usual (cf. definition 3.25), this is defined to mean that $a\,;b\cdot\breve{c} = 0$. Recall that the *Peircean transforms* of the triple (a,b,c) are (b,c,a), (c,a,b), $(\breve{a},\breve{c},\breve{b})$, $(\breve{c},\breve{b},\breve{a})$, and $(\breve{b},\breve{a},\breve{c})$. By the Peircean law in weakly associative algebras, it follows from the inconsistency of (a,b,c) that its Peircean transforms must also be inconsistent. The following triples, plus all Peircean transforms of them, are defined to be inconsistent. Firstly, any triple where the indices

18.3. The definition of $RA(\tau)$

do not match is inconsistent: e.g., (x_{ij}, y_{kl}, a) and (x_j, y_{kl}, a) are inconsistent if $j \neq k$, for any atom a. Secondly, a triple (e_i, x, y) is inconsistent unless $x = \breve{y}$. Thirdly, the following are all inconsistent (the last column but one indicates the start and end atoms of the atoms in the triple).

$$(g_{10}, g_{02}, w_{21}) \qquad\qquad\qquad\qquad\qquad 0,1,2 \qquad (18.1)$$
$$(t^i_{12}, t^j_{21}, +1_1) \quad \text{any } i, j < k, \text{ unless } \mathrm{Lt}(T_i) = \mathrm{Rt}(T_j) \quad 1,1,2 \qquad (18.2)$$
$$(u_{10}, g_{02}, t^i_{21}) \quad \text{any } i \text{ with } 1 \leq i < k \qquad\qquad 0,1,2 \qquad (18.3)$$
$$(v_{10}, g_{01}, \pm 1_1) \qquad\qquad\qquad\qquad\qquad 0,1,1 \qquad (18.4)$$

Here and below, we write $\pm 1_1$ as an abbreviation for 'either $+1_1$ or -1_1'. There are also three dual rules for inconsistent triples, obtained from rules 18.2, 18.3 and 18.4 by swapping the subscripts 1 and 2 throughout and replacing Lt, Rt by Bot, Top, respectively. We will refer to these inconsistent triples by 'rules 18.1 to 18.4'.

All other triples are defined to be consistent. This suffices to define composition, though the resulting operation may not be associative. Note that $1' = e_0 + e_1 + e_2$ follows from this definition of consistency.

Clearly, we can obtain $RA(\tau)$ from τ effectively (by an algorithm).

LEMMA 18.1 *For any instance of the tiling problem τ, $RA(\tau)$ is a simple weakly associative algebra.*

Proof. Let C be the set of consistent triples of atoms of $RA(\tau)$. Using exercise 5.1(14), to show that $RA(\tau)$ is weakly associative it suffices to show that (i) C is closed under Peircean transforms, (ii) $x = y$ iff $(x, e_i, \breve{y}) \in C$ for some $i < 3$, and (iii) for any $i < 3$ and atoms x, y, z of $RA(\tau)$, if $(e_i, x, \breve{x}), (x, y, z) \in C$ then $(e_i, \breve{z}, z) \in C$. This is all rather easy to verify from the definition of the atom structure of $RA(\tau)$. To check (iii), for example, we note that if $(e_i, x, \breve{x}), (x, y, z) \in C$ then $\mathrm{st}(x) = \mathrm{end}(z) = i$, so that $(e_i, \breve{z}, z) \in C$.

To check simplicity, observe that for each atom x of $RA(\tau)$ we have $1; x; 1 = 1$. Hence, $1; x; 1$ is a discriminator term for $RA(\tau)$, which by theorem 2.55 is therefore simple. $\square$

It is helpful to think of the atoms t^i_{12} as corresponding to the tiles T_i ($i < k$). Because of this correspondence, we call an edge (n_1, n_2) of a network N a *tile edge* if $N(n_1, n_2) = t^i_{12}$ (some $i < k$). The atoms t^i_{12} ($i < k$) are called *tile atoms*.

Rule 18.2 (and its dual) force the tile edges to form a tiling pattern, as we'll see in theorem 18.2. The restriction to $i \geq 1$ in rule 18.3 is to do with our assumption that T_0 is a special tile that can tile the plane all by itself.

Exercises

1. Let τ be a set of tiles and let $RA(\tau)$ be the weakly associative algebra defined above. Prove that the following are equivalent. (Exercise 11.3(1) may help.)

   (a) $RA(\tau)$ is associative (and hence a relation algebra).

   (b) For each tile $T_i \in \tau$ there is a tile $T_j \in \tau$ such that $\text{Rt}(T_j) = \text{Lt}(T_i)$, along with three other conditions for the other sides of T_i.

2. Prove, for any set of tiles τ, that $RA(\tau) \in \mathbf{SA}$.

18.4 Games

To test representability of $RA(\tau)$, we will need to use games. Recall from definition 11.3 the game $G_\omega^a(\mathcal{A})$, played on atomic networks over an atomic relation algebra $\mathcal{A}$. A play of $G_\omega^a(\mathcal{A})$ consists of an ω-sequence of atomic $\mathcal{A}$-pre-networks $N_0 \subseteq N_1 \subseteq \cdots$. In round 0, $\forall$ chooses an atom a of $\mathcal{A}$ and $\exists$ chooses N_0 with a among its edge labels. In round $t > 0$, $\forall$ picks $x, y \in N_t$ and atoms a, b with $N(x,y) \leq a \mathbin{;} b$, and $\exists$ chooses $N_{t+1} \supseteq N_t$ with a node z such that $N_{t+1}(x,z) = a$ and $N_{t+1}(z,y) = b$. $\exists$ wins iff she always manages to find a suitable network N_t.

Theorem 11.7 states that for an atomic relation algebra $\mathcal{A}$ with countably many atoms, a winning strategy for player $\exists$ in $G_\omega^a(\mathcal{A})$ is equivalent to the existence of a complete representation of $\mathcal{A}$. By theorem 11.10(2), for a finite relation algebra $\mathcal{A}$, $\exists$ has a winning strategy in $G_\omega^a(\mathcal{A})$ iff $\mathcal{A}$ is representable.

More generally, the same holds for non-associative algebras, so certainly for weakly associative algebras. $G_\omega^a(\mathcal{A})$ is well-defined for any atomic non-associative algebra $\mathcal{A}$ — it is essentially the same as the game $G_\omega^\omega(\mathcal{A})$ of definition 12.24. We have:

- If $\mathcal{A}$ is a finite non-associative algebra, then $\mathcal{A} \in \mathbf{RRA}$ implies that $\exists$ has a winning strategy in $G_\omega^a(\mathcal{A})$ (because $\mathcal{A}$ is then certainly a relation algebra, so theorem 11.10(2) applies).

- If $\mathcal{A}$ is a finite non-associative algebra and $\exists$ has a winning strategy in $G_\omega^a(\mathcal{A})$, then $\mathcal{A} \in \mathbf{RRA}$. For by exercise 11.3(1), $\mathcal{A}$ is a relation algebra, so again theorem 11.10(2) applies.

18.5 Winning $\exists$-strategy implies tiling

Here, we prove the easy part of the reduction of tiling to representability, showing that if $RA(\tau)$ is representable then $\exists$ has a winning strategy in $G_\omega^a(RA(\tau))$. We will use games to prove this, though one could just as easily work in a representation.

18.5. Winning ∃-strategy implies tiling

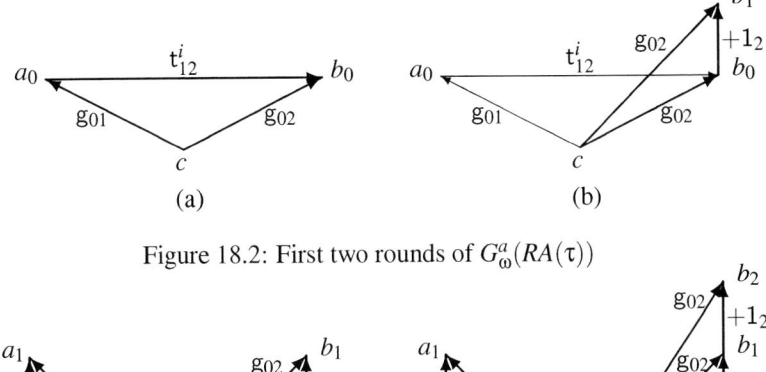

Figure 18.2: First two rounds of $G_\omega^a(RA(\tau))$

Figure 18.3: Next two rounds of $G_\omega^a(RA(\tau))$

THEOREM 18.2 *Let τ be a set of tiles such that T_0 is a special tile with all four of its edges equal to each other but distinct from the colours used by any other tile. If $\exists$ has a winning strategy for $G_\omega^a(RA(\tau))$, then for each $i < k$ there is a tiling of the plane with T_i at $(0,0)$, so τ is a yes-instance.*

Proof. Suppose $\exists$ has a winning strategy and let $\forall$ make the following moves. First he plays the atom t_{12}^i, and in the next round he picks atoms g_{01}, g_{02}, so forcing the triangle in figure 18.2(a). Then he picks the edge labelled g_{02} and two atoms $g_{02}, -1_2$, as in figure 18.2(b).

$\exists$ has to fill in and label the missing edge (a_0, b_1). She can do so because she has a winning strategy. The label must lie under $(g_{10}; g_{02}) \cdot (t_{12}^i; +1_2)$. By rule 18.1, she cannot choose w_{12}, so the label must be t_{12}^j for some $j < k$. By rule 18.2, she must choose j so that $\text{Bot}(T_j) = \text{Top}(T_i)$.

$\forall$ continues with the two moves shown in figure 18.3.

In this way $\forall$ constructs two sequences of nodes $a_0, a_1, a_2, \ldots$ and $b_0, b_1, b_2, \ldots$. He can also extend the sequences backwards $\ldots, a_{-1}, a_0, a_1, \ldots$ by playing suitable moves. $\exists$ has to label each edge (a_x, b_y) for $x, y \in \mathbb{Z}$, and in each case the only consistent choices are t_{12}^j for some $j < k$.

It is easy to see (by rule 18.2) that this defines a tiling f^i of the plane with T_i at $(0,0)$. Formally, if the edge (a_x, b_y) is labelled by t_{12}^j then let $f^i(x,y) = j$. □

18.6 $RA(\tau) \in \mathbf{S\Re aCA}_5$ implies tiling

By the preceding theorem, if $RA(\tau)$ is representable then τ is a yes-instance of the tiling problem. In fact, if $RA(\tau) \in \mathbf{S\Re aCA}_n$ for any $n \geq 5$ then this is already enough to ensure that τ is a yes-instance. This will allow us to prove that the finite algebra membership problem for $\mathbf{S\Re aCA}_n$ is also undecidable, for all finite $n \geq 5$.

THEOREM 18.3 *Let τ be a set of tiles as in theorem 18.2. If $RA(\tau) \in \mathbf{S\Re aCA}_5$, then τ is a yes-instance of the tiling problem defined in section 18.2.*

Proof. Assume that $RA(\tau) \in \mathbf{S\Re aCA}_5$. Since $RA(\tau)$ is finite, $RA(\tau)^+ \cong RA(\tau)$. The second half of theorem 13.46 now tells us that $RA(\tau)$ has a 5-dimensional hyperbasis, say $\mathcal{H}$. (See definition 12.11 for hyperbases.) We will extract from this hyperbasis a tiling of the plane with the tile T_i (some given $i < k$) at the origin.

First, we show that there exist hypernetworks $N_{xx'}^{yy'} \in \mathcal{H}$, for all even $x, y \in \mathbb{Z}$ and odd $x', y' \in \mathbb{Z}$ with $|x - x'| = |y - y'| = 1$, having the following features:

1. $N_{01}^{01}(1,2) = \mathsf{t}_{12}^i$,

2. $N_{xx'}^{yy'}(0,1) = N_{xx'}^{yy'}(0,3) = \mathsf{g}_{01}$,
   $N_{xx'}^{yy'}(0,2) = N_{xx'}^{yy'}(0,4) = \mathsf{g}_{02}$.

3. $N_{xx'}^{yy'}(1,3) = \begin{cases} +1_1 & \text{if } x < x' \\ -1_1 & \text{if } x > x' \end{cases}$
   $N_{xx'}^{yy'}(2,4) = \begin{cases} +1_2 & \text{if } y < y' \\ -1_2 & \text{if } y > y' \end{cases}$.

4. if $x < x'$ then $N_{xx'}^{yy'} \equiv_1 N_{x+2,x'}^{yy'}$,
   if $y < y'$ then $N_{xx'}^{yy'} \equiv_2 N_{xx'}^{y+2,y'}$,
   if $x > x'$ then $N_{xx'}^{yy'} \equiv_3 N_{x,x'+2}^{yy'}$,
   if $y > y'$ then $N_{xx'}^{yy'} \equiv_4 N_{x,x'}^{y,y'+2}$.

5. By rule 18.2 and (2) above, the edges $(1,2), (1,4), (3,2), (3,4)$ of $N_{xx'}^{yy'}$ are tile edges.

Recall that $N \equiv_i M$ means that the hypernetworks N, M agree off of i: see definition 12.5.

It may help to think of nodes 1 and 3 of the hypernetwork $N_{xx'}^{yy'}$ as corresponding to the points x, x' on the 'x-axis' of the plane $\mathbb{Z} \times \mathbb{Z}$, and nodes 2 and 4 of $N_{xx'}^{yy'}$ as

18.6. $RA(\tau) \in \mathbf{S\mathfrak{R}aCA}_5$ implies tiling 549

corresponding to y, y' on the y-axis. Nodes 1 and 3 are 1-nodes and will be used to 'walk along the x-axis of the plane'. Nodes 2 and 4 are 2-nodes and will be used to walk up and down the y-axis. Node 0 is a 0-node, like the node c of theorem 18.2. Although on this view there are four hypernetworks containing nodes corresponding to any given location (x, y) in the plane, they all 'fit together nicely', and the tiling of the plane will easily be obtained from their labels.

Now we proceed more formally. We will associate each hypernetwork $N_{xx'}^{yy'}$ with the point $(\min\{x, x'\}, \min\{y, y'\}) \in \mathbb{Z} \times \mathbb{Z}$. The relation between the hypernetworks is now as shown in figure 18.4.

$$
\begin{array}{cccccccccc}
 & \vdots & & \vdots & & \vdots & & \vdots & \\
 & & & \|_2 & & \|_2 & & \|_2 & \\
y=2 & \cdots & \equiv_3 & N_{01}^{23} & \equiv_1 & N_{21}^{23} & \equiv_3 & N_{23}^{23} & \equiv_1 & \cdots \\
 & & & \|_4 & & \|_4 & & \|_4 & \\
y=1 & \cdots & \equiv_3 & N_{01}^{21} & \equiv_1 & N_{21}^{21} & \equiv_3 & N_{23}^{21} & \equiv_1 & \cdots \\
 & & & \|_2 & & \|_2 & & \|_2 & \\
y=0 & \cdots & \equiv_3 & N_{01}^{01} & \equiv_1 & N_{21}^{01} & \equiv_3 & N_{23}^{01} & \equiv_1 & \cdots \\
 & & & \|_4 & & \|_4 & & \|_4 & \\
 & \vdots & & \vdots & & \vdots & & \vdots & \\
 & \cdots & & x=0 & & x=1 & & x=2 & & \cdots \\
\end{array}
$$

Figure 18.4: The arrangement of the hypernetworks in $\mathbb{Z} \times \mathbb{Z}$

The hypernetworks $N_{xx'}^{yy'} \in \mathcal{H}$ are defined by induction. To begin, by repeated use of the triangle addition property in $\mathcal{H}$ exactly as in the proof of the preceding theorem (18.2), we can find a hypernetwork $N_{01}^{01} \in \mathcal{H}$ with edge labelling given by

$$
\begin{aligned}
N_{01}^{01}(0, 1) &= N_{01}^{01}(0, 3) = g_{01}, \\
N_{01}^{01}(0, 2) &= N_{01}^{01}(0, 4) = g_{02}, \\
N_{01}^{01}(1, 2) &= t_{12}^i, \\
N_{01}^{01}(1, 3) &= +1_1, \quad N_{01}^{01}(2, 4) = +1_2.
\end{aligned}
$$

N_{01}^{01} is pretty much as shown in figure 18.3(a). Node 0 corresponds to c in the figure, nodes 1 and 3 to a_0, a_1, and nodes 2 and 4 to b_0, b_1, respectively.

Let $n < \omega$ and assume inductively that $N_{xx'}^{yy'}$, for each even $x, y \in \mathbb{Z}$ and odd $x', y' \in \mathbb{Z}$ with $-n \leq x, x', y, y' \leq n+1$, have been defined. These hypernetworks are associated with the points of the plane $\mathbb{Z} \times \mathbb{Z}$ lying in the region $R_n = \{(x, y) \in \mathbb{Z} \times \mathbb{Z} : -n \leq x, y \leq n\}$. We will define the hypernetworks for R_{n+1}.

Let $N_{xx'}^{yy'}$ be the hypernetwork associated with the top left-hand point $(-n, n)$ of R_n. Assume that $x < x'$, $y < y'$; the other cases are handled similarly. We know that $N_{xx'}^{yy'}(0,4) = g_{02} \leq g_{02}; -1_2$. So by the triangle addition property of $\mathcal{H}$, there is $N_{xx'}^{y+2,y'} \in \mathcal{H}$ with $N_{xx'}^{y+2,y'} \equiv_2 N_{xx'}^{yy'}$, $N_{xx'}^{y+2,y'}(0,2) = g_{02}$, and $N_{xx'}^{y+2,y'}(2,4) = -1_2$. Since also $N_{xx'}^{y+2,y'} \equiv_2 N_{xx'}^{yy'}$, it follows that $N_{xx'}^{y+2,y'}$ meets the above specifications. It is plainly associated with $(-n, n+1)$.

Now, we define the hypernetwork $N_{x+2,x'}^{y+2,y'}$ on the right of $N_{xx'}^{y+2,y'}$. We have

$$N_{xx'}^{y+2,y'}$$

$$\|\|_2$$

$$N_{xx'}^{yy'} \equiv_1 N_{x+2,x'}^{yy'}$$

So $N_{xx'}^{y+2,y'} \equiv_{12} N_{x+2,x'}^{yy'}$. By the amalgamation property of $\mathcal{H}$, there is $N_{x+2,x'}^{y+2,y'} \in \mathcal{H}$ satisfying:

$$N_{xx'}^{y+2,y'} \equiv_1 N_{x+2,x'}^{y+2,y'}$$

$$\|\|_2 \qquad \|\|_2$$

$$N_{xx'}^{yy'} \equiv_1 N_{x+2,x'}^{yy'}$$

It can be checked that $N_{x+2,x'}^{y+2,y'}$ satisfies the requirements listed above. For example, $N_{x+2,x'}^{y+2,y'}(1,3) = N_{x+2,x'}^{yy'}(1,3) = -1_1$, and $N_{x+2,x'}^{y+2,y'}(2,4) = N_{xx'}^{y+2,y'}(2,4) = -1_2$. Now, we have $N_{x+2,x'}^{y+2,y'} \equiv_{32} N_{x+2,x'+2}^{yy'}$, so again there is $N_{x+2,x'+2}^{y+2,y'} \in \mathcal{H}$ with $N_{x+2,x'}^{y+2,y'} \equiv_3 N_{x+2,x'+2}^{y+2,y'} \equiv_2 N_{x+2,x'+2}^{y,y'}$. Continuing in this way, we can define the hypernetworks associated with the points in the row $\{(m, n+1) : -n \leq m \leq n\}$ above R_n, moving from left to right along the row (see figure 18.5). We get round the top right-hand corner $(n+1, n+1)$ using the triangle addition property, down the right-hand side using the amalgamation property, and so on. The end result is that we have defined hypernetworks associated with all points in R_{n+1}. This completes the induction and the definition of the $N_{xx'}^{yy'}$.

18.7. Tiling implies winning ∃-strategy

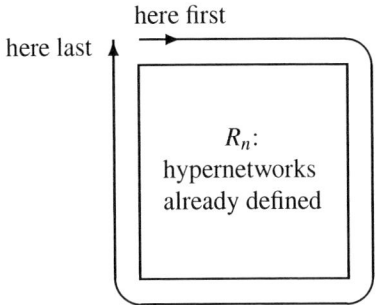

Figure 18.5: Defining the next round of hypernetworks

Now we define a tiling $f : \mathbb{Z} \times \mathbb{Z} \to k$ as follows. Let $(i,j) \in \mathbb{Z} \times \mathbb{Z}$, and let $N^{yy'}_{xx'}$ be the hypernetwork associated with (i,j), so $\min\{x,x'\} = i$ and $\min\{y,y'\} = j$. Then $f(i,j)$ is defined to be $l < k$, where l is given by:

$$t^l_{12} = \begin{cases} N^{yy'}_{xx'}(1,2), & \text{if } x < x', y < y', \\ N^{yy'}_{xx'}(1,4), & \text{if } x < x', y > y', \\ N^{yy'}_{xx'}(3,2), & \text{if } x > x', y < y', \\ N^{yy'}_{xx'}(3,4), & \text{if } x > x', y > y'. \end{cases}$$

We check that this is a valid tiling for τ. Let i,j,x,x',y,y',l be as above. Assume that $x < x'$ and $y < y'$ (the other cases are again similar). Thus, $N^{yy'}_{xx'}(1,2) = t^l_{12}$. Let $f(i+1,j) = m$, say, so that $N^{yy'}_{x+2,x'}(3,2) = t^m_{12}$. We have $N^{yy'}_{xx'} \equiv_1 N^{yy'}_{x+2,x'}$, so $N^{yy'}_{xx'}(3,2) = t^m_{12}$ as well. But $N^{yy'}_{xx'}(1,3) = +1_1$, and it follows from rule 18.2 of the definition of $RA(\tau)$ that $\mathrm{Lt}(T_m) = \mathrm{Rt}(T_l)$. Checking $f(i,j+1)$ is similar. So f tiles the plane, as required. We conclude that τ is a yes-instance of the tiling problem. □

18.7 Tiling implies winning ∃-strategy

To complete the reduction of the tiling problem to the representation problem, we need to prove the converse to theorem 18.2, by showing that ∃ has a winning strategy in $G^a_\omega(RA(\tau))$ whenever τ is a yes-instance of our tiling problem. This will be done in theorem 18.5 below.

It will simplify matters to adopt the following harmless conventions for the game $G^a_\omega(RA(\tau))$.

552   Chapter 18. Undecidability of the representation problem

CONVENTION 18.4

1. We will provide a winning strategy for $\exists$ in $G^a_\omega(RA(\tau))$ in which *she never adds more than a single node to the current network in each round of the play,* so that $|N_t| \leq |N_{t-1}| + 1$ for all $t > 0$.

2. If, in round t of the game, $\forall$ picks nodes $m, n \in N_{t-1}$ and atoms $a, b \in RA(\tau)$, and if there is already a node $p \in N_{t-1}$ such that $N_{t-1}(m, p) = a$ and $N_{t-1}(p, n) = b$, then $\exists$ does not need to make a proper extension but can let $N_t = N_{t-1}$. As $\forall$ is clearly wasting his time by making such a move, *we will assume throughout that he never makes a move of this kind.* With this assumption, $\exists$ is always forced to add a new point, and so $|N_t| = |N_{t-1}| + 1$ for all $t > 0$. If $\forall$ cannot make a move in some round then $\exists$ wins straight away.

3. Suppose, in some round of $G^a_n(RA(\tau))$, the current network is N_{t-1} and $\forall$ picks the nodes $m, n \in N_{t-1}$ and the new node added by $\exists$ is p. We regard $\forall$ as choosing the labels on the edges $(m, p), (p, m), (p, p), (p, n)$, and (n, p) of the new network N_t. (Of course, (m, p) determines (p, m), etc.) All other labels on edges of N_t involving p are regarded as having been chosen by $\exists$. This is as in theorem 16.13.

THEOREM 18.5 *Let $\tau = \{T_0, \ldots, T_{k-1}\}$ be a yes-instance of the tiling problem: each tile is part of a tiling of the plane, and T_0 is a special tile with all four edges the same colour, a colour not used by any other tile. Then $\exists$ has a winning strategy in the game $G^a_\omega(RA(\tau))$.*

Proof. Assume the hypotheses of the theorem statement. Let f^i be a tiling function where $f^i(0, 0) = i$ (each $i < k$). We will provide a winning strategy for $\exists$ in the game $G^a_\omega(RA(\tau))$. We will often use subscripts to denote implicitly the kind of network node we are discussing. For example, $n_i \in N$ is implicitly stated to be an i-node, m_1 and m'_1 are 1-nodes, and so on.

Initially, suppose that $\forall$ plays the atom a_{ij} of $RA(\tau)$. Recall from section 18.3 that i, j are the start and end indices of a_{ij}, respectively. $\exists$ responds with a network N_0 consisting only of the nodes n_i, n_j, equal if a_{ij} is an identity atom and distinct if not, with $N_0(n_i, n_j) = a_{ij}$, $N_0(n_j, n_i) = a_{ji}$, $N_0(n_i, n_i) = e_i$, and $N_0(n_j, n_j) = e_j$. By the second part of the definition of consistency of triples, this is clearly a well-defined network (in particular, if $a_{ij} \leq 1'$ then $a_{ij} = e_i = e_j$).

Suppose, at some stage in the continuing play of $G^a_\omega(RA(\tau))$, that the current network is N. (We assume inductively that it *is* a network.) $\forall$ picks two nodes $n_i, n_j \in N$ and two atoms $a_{il}, b_{l'j}$ such that $a_{il}; b_{l'j} \geq N(n_i, n_j)$. Necessarily, $l = l'$. An $\forall$-move of this kind is called an *l-move*, and, bearing in mind convention 18.4(2), it forces $\exists$ to add an *l*-node to N. She has to find a network M extending N with a node $n_l \in M$ such that $M(n_i, n_l) = a_{il}$ and $M(n_l, n_j) = b_{lj}$.

18.7. Tiling implies winning ∃-strategy

CONVENTION 18.6 Throughout, whenever we define the labelling of an edge $M(p,q) = c \in \text{At}(RA(\tau))$, the labelling on the converse edge is implicitly defined by $M(q,p) = \breve{c}$.

∃ first builds a labelled graph $N^\forall$ whose nodes are $N \cup \{n_l\}$, where n_l is a new l-node, not in N, and whose edges are those of N together with (n_i, n_l) and (n_l, n_i), (n_j, n_l) and (n_l, n_j), and (n_l, n_l). $N^\forall$ extends N, so that if $p, q \in N$ then $N^\forall(p,q) = N(p,q)$. Also, $N^\forall(n_i, n_l) = a_{il}$, $N^\forall(n_l, n_j) = b_{lj}$, and $N^\forall(n_l, n_l) = e_l$. By convention 18.6, labels on the reverse edges (n_l, n_i) and (n_j, n_l) are now specified as well. (These are the edges we regard as being labelled by ∀ — recall convention 18.4). It can be checked that this is well-defined if $n_i = n_j$. Clearly, $N^\forall$ is consistent — for any three nodes $p, q, r \in N^\forall$, if all three edges of the triangle (p, q, r) are in $N^\forall$ then $N^\forall(p,r) \leq N^\forall(p,q) ; N^\forall(q,r)$, else ∀ has made an illegal move.

We now define a strategy for ∃ to choose an atom to label each edge (x,y) not in $N^\forall$, where x, y are nodes of $N^\forall$. Such edges have the form (n_l, m) or (m, n_l) for $m \in N \setminus \{n_i, n_j\}$. Employing this strategy will result in a complete labelled graph, say M.

We show that the strategy is winning for ∃ by showing that the graph M is in fact an atomic $RA(\tau)$-network. In order to do this, we have to show for any three nodes $p, q, r \in M$ that the triangle (p, q, r) is consistent, i.e., that $M(p,r) \leq M(p,q) ; M(q,r)$. If all three edges $(p,r), (p,q)$ and (q,r) are in $N^\forall$ then we already know that the triangle is consistent. If not, then if two of p, q, r are equal, consistency is assured by convention 18.6 and our definition of $M(n_l, n_l) = e_l$, so long as ∃ always uses an l, m-atom to label an $(l - m)$-edge (and she will). So it suffices to check consistency of the triangles with three distinct nodes and an edge labelled by ∃ in the current round of the game. We will do this as we define the strategy: for each edge not in $N^\forall$, we will explain which atom ∃ chooses to label it, and check that any triangle containing it conforms with rules 18.1–18.4 of the definition of $RA(\tau)$.

REMARK 18.7 Since all this takes up most of what remains of the chapter, it may help the reader if we discuss the underlying idea a little, before plunging in. The critical part of the strategy is where ∃ is forced to choose a $(1 - 2)$-atom — a tile atom or w_{12} — to label a $(1 - 2)$ edge. The atom w_{12} is a sort of 'wild card' which may be adjacent to any tile as it is not mentioned in rule 18.2. So where possible, she chooses the atom w_{12} to label such edges.

However, rule 18.1 prohibits the use of w_{12} in some circumstances. When rule 18.1 applies, ∃ is forced to choose a tile atom. To help decide which one, we will assume that each tile edge in N is associated with a genuine tiling of the plane, in the same way as happened in theorem 18.2. This means that every tile edge (n_1, n_2) in N has an associated *tiling function* $f_{(n_1, n_2)} \in \{f^0, \ldots, f^{k-1}\}$, and sometimes also a pair of *coordinates* $\text{Co}(n_1, n_2) \in \mathbb{Z} \times \mathbb{Z}$, so that the tile atom labelling (n_1, n_2) is given by the tile $T_{f_{(n_1,n_2)}}(\text{Co}(n_1, n_2))$ if $f_{(n_1,n_2)} \neq f^0$, and by T_0, otherwise. Except in ∀'s current triangle, which is in $N^\forall$ so is assumed to be consistent, the tilings and coordinates will have to fit together in a coherent way rather as in theorem 18.2.

These tilings and coordinates will be assumed (inductively) to be given for N, and one of ∃'s tasks will be to extend them to M. (The u- and v-atoms, which are somewhat similar to the w_S-atoms in the rainbow construction, play an important role here. The way ∃'s strategy deployed v_{01} and v_{02} in earlier rounds of the game will ensure that she can define tiling functions for new tile edges coherently, and u_{01}, u_{02} do the same job for the coordinates.) When she has done this, it will be easy for her to decide which tile atom should label each new tile edge — she will just choose the one given by its tiling function and coordinates. Consistency of her choices will follow from the fact that the tiling function is a genuine tiling of the plane.

We will discuss the tiling and coordinate functions fully in sections 18.7.2–18.7.8 below, prior to describing ∃'s strategy for choosing tile atoms.

18.7.1 ∃'s strategy for non-tile edges

Here is ∃'s strategy for labelling edges of M that are not edges of $N^\forall$. We will define it so that:

(α) ∃ never chooses a green atom, $\pm 1_1$, or $\pm 1_2$ to label an edge.

(β) she chooses a tile atom $M(n_1, n_2) = t^i_{12}$ if and only if there is a node $n_0 \in N$ such that $N^\forall(n_0, n_1) = g_{01}$ and $N^\forall(n_0, n_2) = g_{02}$. In particular, she never chooses a tile atom unless one of the atoms chosen by ∀ in the current round of the game is green.

REMARK 18.8 Clearly, (β) completely determines which edges of M will be tile edges and which will not: an edge of M will be a tile edge iff either it is a tile edge of $N^\forall$, or it forms one side of a triangle in M whose other two edges are in $N^\forall$ and are g_{01}, g_{02}.

I: $(0-0), (1-1)$, **and** $(2-2)$ **edges.** For each of these three types of edges, ∃ chooses the label w_i for suitable i. Since w_i is not mentioned in any of rules 18.1 to 18.4, it follows that any triangle containing such an edge must be consistent.

II: $(0-1)$ **and** $(0-2)$ **edges.** We define the strategy for a $(0-1)$ edge (n_0, n_1); the strategy for $(0-2)$ edges is similar. ∃ always chooses either u_{01}, v_{01} or w_{01} for a $(0-1)$ edge.

18.7. Tiling implies winning ∃-strategy

1. Suppose there is a cycle $\gamma = \langle g^0, g^1, \ldots, g^{l-1} \rangle$ of distinct 1-nodes of $N^{\forall}$ (for some $l \geq 3$), such that

    (a) $n_1 \in \gamma$,
    (b) $N^{\forall}(g^i, g^{i+1}) = \pm 1_1$ (each $i < l - 1$), and $N^{\forall}(g^{l-1}, g^0) = \pm 1_1$,
    (c) for all $g^i \in \gamma$, if $g^i \neq n_1$ then $N^{\forall}(n_0, g^i) = \mathsf{g}_{01}$.

    Then $\exists$ lets $M(n_0, n_1) = \mathsf{u}_{01}$.

2. If there is a chain of distinct 2-nodes $C = \langle c_2, \ldots, d_2 \rangle$ in N, with $c_2 \neq d_2$, such that

    (a) for each $\alpha_2 \in C$, $N^{\forall}(n_0, \alpha_2) = \mathsf{g}_{02}$,
    (b) each edge between two consecutive nodes in the chain C is labelled by $\pm 1_2$ in N,
    (c) (n_1, c_2) and (n_1, d_2) will be tile edges of M (cf. remark 18.8),

    then she lets $M(n_0, n_1) = \mathsf{v}_{01}$.

3. Otherwise she lets $M(n_0, n_1) = \mathsf{w}_{01}$.

See figure 18.6.

We should check first that this strategy is well-defined — i.e., cases 1 and 2 do not apply simultaneously. So suppose both cases apply to the edge (n_0, n_1). Since case 1 applies, there is a cycle γ as described above; and since case 2 applies, there is a chain C as above with endpoints c_2 and d_2. As the edge (n_0, n_1) is just being labelled by $\exists$, either n_0 or n_1 is currently being added to the network. Now the new node, whichever it is, is incident with at most two edges in $N^{\forall}$, the ones labelled by $\forall$ in the current round. But n_0 is incident with four such edges, because $N^{\forall}(n_0, g) = \mathsf{g}_{01}$ ($g \in \gamma \setminus \{n_1\}$) and $N^{\forall}(n_0, \alpha_2) = \mathsf{g}_{02}$ ($\alpha_2 \in C$). So the node being currently added must be n_1. The two edges connecting n_1 to its neighbours in γ are in $N^{\forall}$, since they are labelled with $\pm 1_1$ in $N^{\forall}$, so these must be the two edges chosen by $\forall$ in the current round — in particular, he chose no green atoms. By (β), $\exists$ does not choose any tile atoms in this round, so the tile edges $(n_1, c_2), (n_1, d_2)$ must also be labelled by $\forall$ in the current round. This contradicts the fact that n_1 is incident with only two labelled edges in $N^{\forall}$, and shows that the strategy is well-defined.

Next we check that it is consistent: that when the labelling of all edges of M has been completed, no triangle containing an edge labelled by case II of the strategy is inconsistent. Though we have not yet described the rest of the strategy, so we don't know exactly how edges of M are labelled, we promise that conditions (α) and (β) will in the end be met, and these are all we need.

First, we check that if $\exists$ chooses the atom u_{01} for the edge (n_0, n_1), there is no 2-node n_2 such that the triangle (n_0, n_1, n_2) violates rule 18.3 in M. Suppose

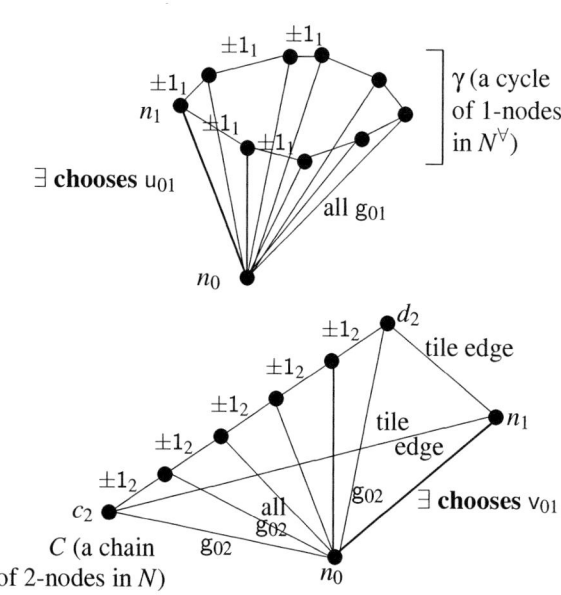

Figure 18.6: $\exists$'s strategy, case II(1,2)

this happened: so $M(n_0,n_1) = u_{01}$, $M(n_0,n_2) = g_{02}$, and $M(n_1,n_2) = t_{12}^i$ for some i with $1 \leq i < k$. Now, much as before, n_0 is incident with at least three green edges — two labelled g_{01} into the cycle γ that caused the use of u_{01}, and one labelled g_{02} to n_2 — so by (α), n_1 must be the node added in the current round. n_1 is incident with two edges labelled $\pm 1_1$ in the cycle γ, so by (α), these are the two edges chosen by $\forall$ in the current round. But then, as neither of $\forall$'s atoms is green, by (β) $\exists$ would not have chosen a tile atom for $M(n_1,n_2)$. Thus we have a contradiction. So if $\exists$ chooses $M(n_0,n_1) = u_{01}$ then any triangle involving this edge is consistent.

The second possible inconsistency that we have to check for is when $\exists$ chooses v_{01} for the edge (n_0,n_1), and there is a 1-node m_1 such that the triangle (n_0,n_1,m_1) violates rule 18.4 in M. So suppose there is a chain C with endpoints $c_2 \neq d_2$, with (n_1,c_2) and (n_1,d_2) both tile edges of M, as in case 2 of the strategy, and with $N^\forall(n_0,m_1) = g_{01}$ and $N^\forall(m_1,n_1) = \pm 1_1$ (these are $N^\forall$-edges, by (α)). As before, because the labels g_{02} on the edges (n_0,c_2) and (n_0,d_2) must have been chosen by $\forall$, and $c_2 \neq d_2$, this could only happen if n_1 is the node currently added.

18.7. Tiling implies winning ∃-strategy

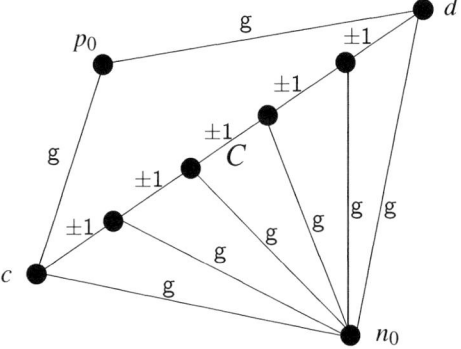

Figure 18.7: Non-existent configuration in N

Since the edge (n_1, m_1) is in $N^\forall$, at most one of the two tile edges (n_1, c_2), (n_1, d_2) can also be in $N^\forall$, so $\exists$ is going to label at least one of them with a tile atom. So by (β), $\forall$ must have chosen a green atom in the current round. His choices of atoms and nodes were therefore $\pm 1_1, g_{01}$, and m_1, p_0, say, for some $p_0 \in N$: we have $N^\forall(p_0, n_1) = g_{01}$. Hence, $\exists$ labelled *both* tile edges $(n_1, c_2), (n_1, d_2)$. By (α), (p_0, n_1) is the only green edge incident with n_1; so by (β), we must have $N(p_0, c_2) = N(p_0, d_2) = g_{02}$. Note that $p_0 \neq n_0$, since $M(n_1, n_0)$ is not green. This contradicts the following lemma.

LEMMA 18.9 *There do not exist distinct nodes $n_0, p_0 \in N$ and a chain of 1- or 2-nodes $C \subseteq N$ with distinct endpoints c, d, each edge between two consecutive nodes in C labelled by ± 1, and the edges $(p_0, c), (p_0, d), (n_0, x)$ (all $x \in C$) labelled green. See figure 18.7.*

Proof. Assume otherwise. Consider which node from $C \cup \{n_0, p_0\}$ was most recently added, as the game progressed through earlier rounds. Whichever it is, it can be connected to the rest of $C \cup \{n_0, p_0\}$ by at most two edges labelled by $\forall$. By (α), nodes in $C \setminus \{c, d\}$ (if any) are incident with three edges chosen by $\forall$ — two edges within C labelled ± 1, and one to n_0, labelled green — so it was not any of those. Nor was it c or d, as they are each incident with a ± 1 edge (within C) and two green edges, to p_0 and n_0. So the most recently added node must have been either n_0 or p_0. These are both connected to c and d by green edges. Because $c \neq d$, we see by (α) that in the round when the last of n_0, p_0 was built, $\forall$ must have selected the nodes c, d and two green atoms. But this is in contravention to convention 18.4, since there was a suitable node (namely whichever of n_0, p_0 already existed) in the network already, so $\forall$ should not have made such a move. $\square$

This contradiction shows that if $\exists$ chooses $M(n_0, n_1) = \mathsf{v}_{01}$ then any triangle involving this edge is consistent.

Hence, if this strategy is used, any triangle in M involving a $(0-1)$ edge labelled by $\exists$ must be consistent. Similarly, triangles involving $(0-2)$ edges chosen by $\exists$ are consistent.

Having dealt with all $(0-1), (0-2), (1-1)$ and $(2-2)$ edges now, we can see that (α) is indeed a true property of $\exists$'s strategy.

III: $(1-2)$ **non-tile edges** $\exists$ now proceeds to deal with the new $(1-2)$-edges not in $N^\forall$. Let (n_1, n_2) be such an edge. If it is consistent with $N^\forall$, she lets $M(n_1, n_2) = \mathsf{w}_{12}$. In more detail, if there is no node $m_0 \in N$ such that $N^\forall(m_0, n_1) = \mathsf{g}_{01}$ and $N^\forall(m_0, n_2) = \mathsf{g}_{02}$, then she lets $M(n_1, n_2) = \mathsf{w}_{12}$. Such an edge cannot be part of an inconsistent triangle in M, because the only inconsistent triples of atoms involving the atom w_{12} are those mentioned in rule 18.1 and because $\exists$ never chooses green atoms. The edges so labelled are not tile edges.

18.7.2 Tile edges

Finally, $\exists$ has to label the remaining $(1-2)$-edges of M, if any. These are the edges of the form (n_1, n_2) such that for some node $m_0 \in N$, $N^\forall(m_0, n_1) = \mathsf{g}_{01}$ and $N^\forall(m_0, n_2) = \mathsf{g}_{02}$. $\exists$ is not allowed to choose the atom w_{12} for (n_1, n_2) because of rule 18.1 — she must and will use a genuine tile atom here. This means that (β) is also a true property of $\exists$'s strategy.

To choose tile atoms for these edges, $\exists$ will take advantage of certain tilings and (possibly) coordinates associated with existing tile edges; and to continue winning later on, she will also have to extend these tilings and coordinates to the new tile edges constructed in the current round. This includes any new tile edges labelled by $\forall$, so that $\exists$ may have to define tilings/coordinates even if there are no new tile edges for her to label.

These tilings and coordinates will be assumed inductively to comply with the conditions T1, T2 and T3 below.

18.7.3 Attached and linked tile edges

To specify T1–T3, we need to define some terms.

DEFINITION 18.10 Let $X \in \{N, M\}$.

18.7. Tiling implies winning ∃-strategy

1. Let p,q,r be distinct nodes of X. The triangle (p,q,r) is said to be an ∀-*triangle* if it was constructed by ∀ in some round of the game. More formally, suppose (without loss of generality) that node r was the most recently constructed node out of p,q,r as the game progressed. Then triangle (p,q,r) is an ∀-triangle if in the round when r was added, ∀ chose p,q as his nodes (and $X(p,r), X(r,q)$ as his atoms). The order of the nodes p,q,r is not significant here, so if (p,q,r) is an ∀-triangle then so are (q,r,p) and (q,p,r).

2. Recall that a tile edge is one labelled with t_{12}^i for some $i < k$. Two edges (m_1, m_2) and (m_1, m'_2) of X are said to be *attached* to each other if they are both tile edges, $N(m_2, m'_2) = \pm 1_2$, and the triangle (m_1, m_2, m'_2) is *not* an ∀-triangle. Similarly, if (m_1, m_2) and (m'_1, m_2) are both tile edges, $N(m_1, m'_1) = \pm 1_1$, and the triangle (m_1, m'_1, m_2) is not an ∀-triangle, then the two edges (m_1, m_2) and (m'_1, m_2) are attached.

   In a nutshell, two tile edges are attached if they form two sides of a non-∀-triangle, the third side of which is labelled by a ± 1 atom.

   In this definition, we are not concerned with the orientation of the edges — we regard them as undirected edges.

3. Two tile edges (m_1, m_2) and (m'_1, m'_2) of X are said to be *linked in X* if and only if they are equal or there is a chain of tile edges in X from (m_1, m_2) to (m'_1, m'_2) with each edge in the chain attached to the next one. Thus, 'linked' is the reflexive, transitive closure in X of the 'attachment' relation. It is an equivalence relation on tile edges of X.

At this stage, ∃ has labelled all the edges of M except the $(1-2)$ tile edges. Although she has not yet labelled these, she knows which edges are going to be tile edges (remark 18.8), and she knows the labels on all $(1-1)$ and $(2-2)$ edges. She also knows which triangles ∀ has picked during the game, of course. Therefore it makes sense to say that two tile edges of M are attached or linked to each other in M.

18.7.4 Inductive conditions T1, T2, T3 on N

We require (inductively) that each tile edge (n_1, n_2) of N — whether its label was chosen by ∃ or ∀ — is associated with a tiling $f = f_{(n_1,n_2)} \in \{f^i : i < k\}$ in such a way that

T1 if the tile edges $(n_1, n_2), (n'_1, n'_2)$ are linked in N then $f_{(n_1,n_2)} = f_{(n'_1,n'_2)}$.

If $f = f^0$ we do not need coordinates. (Recall that all four edges of the tile T_0 have the same colour.) If $f \neq f^0$, we assume that associated with (n_1, n_2) there are the coordinates $\mathrm{Co}(n_1, n_2) = (x, y) \in \mathbb{Z} \times \mathbb{Z}$, say, such that:

560 Chapter 18. Undecidability of the representation problem

T2 If (n_1, n_2') is attached to (n_1, n_2), $f_{(n_1,n_2)} = f_{(n_1,n_2')} \neq f^0$, $N(n_2, n_2') = +1_2$, and $\mathrm{Co}(n_1, n_2) = (x, y)$, then $\mathrm{Co}(n_1, n_2') = (x, y+1)$.

Similarly, if (n_1', n_2) is attached to (n_1, n_2), the associated tiling of these edges is not f^0, $N(n_1, n_1') = +1_1$, and $\mathrm{Co}(n_1, n_2) = (x, y)$, then $\mathrm{Co}(n_1', n_2) = (x+1, y)$.

T3 For each tile edge (n_1, n_2), if $f_{(n_1,n_2)} = f^0$ then $N(n_1, n_2) = t_{12}^0$. If $f_{(n_1,n_2)} = f^i$ ($i > 0$) and $\mathrm{Co}(n_1, n_2) = (x, y)$ then $N(n_1, n_2) = t_{12}^{f^i(x,y)}$.

It is easy to arrange that the requirements T1, T2, and T3 hold in the initial network N_0. It will involve at most one tile edge (at most one edge of any kind), and if its label is t_{12}^i, we let the associated tiling be f^i, and if $i > 0$ we assign the coordinates $(0, 0)$.

Now assume inductively that tilings and coordinates are defined for N, satisfying T1–T3. First, we will describe how to extend them to M. Then, when all tilings and coordinates have been defined, we will check that T1 and T2 are satisfied. Finally, T3 will tell $\exists$ which tile to use for each new tile edge, and then we can check consistency of triangles involving them.

18.7.5 Tiling functions and coordinates for $\forall$'s tile edges

The first step is to define tiling functions and coordinates for any tile edges chosen by $\forall$ in the current round. If $\forall$ chooses $N^\forall(n_1, n_2) = t_{12}^j$, say, then let $f_{(n_1,n_2)} = f^j$, and if $j > 0$ let $\mathrm{Co}(n_1, n_2) = (0, 0)$, in agreement with T3.

18.7.6 Tiling functions for $\exists$'s new tile edges

$\exists$ must now find tilings and perhaps coordinates for the new tile edges in M, if any. Suppose that there are some; by (β), this implies that at least one of the atoms picked by $\forall$ is green. *Suppose also that the new node is a 2-node: i.e., the nodes of M are those of N plus a new node $n_2 \notin N$. The other case, where the new node is a 1-node, is dealt with similarly, using the $(1-2)$-symmetry of the rules 18.1–18.4 defining the atom structure. This n_2 will be fixed in the notation from now on.* The new tile edges are precisely those that are not edges of $N^\forall$ and have the form (n_1, n_2), where $n_1 \in N$ and $N(m_0, n_1) = g_{01}$, $N^\forall(m_0, n_2) = g_{02}$ for some $m_0 \in N$.

$\exists$ has to associate a tiling function $f_{(n_1,n_2)}$ with each such edge. She does this as follows:

- If (n_1, n_2) is linked (in M) to a tile edge t of $N^\forall$, then she associates with (n_1, n_2) the same tiling function that is associated with t. That is, she sets $f_{(n_1,n_2)} \stackrel{\text{def}}{=} f_{(t)}$.

18.7. Tiling implies winning ∃-strategy

- Otherwise, she lets $f_{(n_1,n_2)} = f^0$.

Of course, we have to show that this is well-defined. This is done by the following lemma.

LEMMA 18.11 *Under the above assumptions (in particular, that n_2 is the new node and (n_1, n_2) a tile edge to be labelled by $\exists$, so that $N^{\forall}(m_0, n_2) = g_{02}$ and $N(m_0, n_1) = g_{01}$ for some $m_0 \in N$):*

1. *If (n_1, n_2) is linked in M to a tile edge of $N^{\forall}$ that is not in N, then (n_1, n_2) is not linked in M to any other tile edge of $N^{\forall}$.*

2. *Let t, t' be tile edges of N. If t and t' are linked in M, then they are linked in N.*

For suppose that (n_1, n_2) is linked in M to two distinct tile edges t, t' of $N^{\forall}$. By (1) of the lemma, t, t' are edges of N. Because 'linked' is an equivalence relation on tile edges, t, t' are linked in M. By (2) of the lemma, they are linked in N, so by T1 for N, $f_{(t)} = f_{(t')}$. Thus, $\exists$'s definition of $f_{(n_1,n_2)} := f_{(t)}$ is well-defined.

Proof. (1) Assume the hypotheses. Then $\forall$'s chosen atoms in the current round must have been g_{02} and a tile atom. Neither of them was $\pm 1_2$.

Suppose that (n_1, n_2) is linked to a tile edge t in N. Then there is a chain of attached tile edges proceeding from (n_1, n_2) to t. At some stage, this chain crosses into N. This is impossible unless some edge of M incident with n_2 is labelled $\pm 1_2$. But $\forall$ did not choose $\pm 1_2$, and $\exists$ never does (by (α)). So (n_1, n_2) is not linked to any tile edge of N. As there is a unique tile edge of $N^{\forall}$ that is not in N, this proves part (1).

(2) Assume not. Take a counter-example pair of tile edges t, t' of N, linked in M but not in N, with shortest possible chain of attached tile edges e_i ($1 \le i \le s$) of M linking t with t': so t is attached to e_1, e_1 to e_2, ..., e_{s-1} to e_s, and e_s to t', and $s \ge 1$ is least possible. Clearly, no e_i is an edge of N, or a counter-example with a shorter chain would be possible. Also, the e_i are all distinct. So the e_i have the form (n_1^i, n_2) for distinct nodes $n_1^i \in N$ ($1 \le i \le s$). Since t, t' are edges of N, we must have $t = (n_1^1, m_2)$ and $t' = (n_1^s, m_2')$ for some 2-nodes $m_2, m_2' \in N$ with $M(n_2, m_2) = \pm 1_2$ and $M(n_2, m_2') = \pm 1_2$. Hence $(n_2, m_2), (n_2, m_2')$ are edges of $N^{\forall}$, chosen by $\forall$ in the current round. Now by the assumption of the lemma, $\forall$ also chose a green atom: $N^{\forall}(m_0, n_2) = g_{02}$. Because he only chose two edges, we must have $m_2 = m_2'$. Because $t \ne t'$, we see that $n_1^1 \ne n_1^s$, so $s \ge 2$. As $\forall$ did not choose a tile atom this round, the tile edges (n_1^i, n_2) are going to be labelled by $\exists$, so by (β) we must have $N(m_0, n_1^i) = g_{01}$ for all i. By considering the different types (0, 1, 2) of the nodes, we see that m_0, m_2, and the n_1^i are all distinct. See figure 18.8.

We claim that $N(m_0, m_2) = g_{02}$. Given the claim, we have:

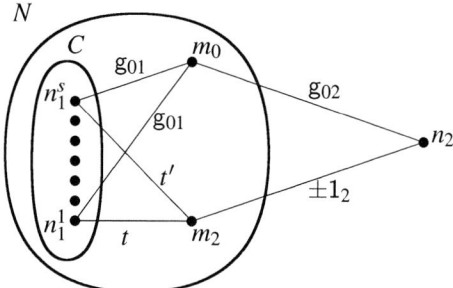

Figure 18.8: The chain linking t to t'

1. The edges (n_1^i, m_2) are all tile edges of N (by rule 18.1, since $N(m_0, n_1^i) = g_{01}$ for each i).

2. For each $1 \leq i < s$, (n_1^i, m_2) is attached to (n_1^{i+1}, m_2).

   For if not, (n_1^i, n_1^{i+1}, m_2) must be an $\forall$-triangle: two sides of it were chosen by $\forall$ when its last node was constructed. So if m_0 was already in existence when this triangle was completed, the label on the edge from the last node of the triangle to m_0 must have been chosen by $\exists$. But this last node is connected to m_0 by a green edge (we use the claim again here in the case where the last node was m_2), and $\exists$ never chooses green labels. So m_0 must have been constructed *after* the triangle (n_1^i, n_1^{i+1}, m_2). This is also impossible, since it is connected to all three nodes of the triangle by green edges and at least one of these edges must have been labelled by $\exists$. See figure 18.9.

3. So t is linked to t' in N, via the chain $t = (n_1^1, m_2), (n_1^2, m_2), \ldots, (n_1^s, m_2) = t'$. This is what we wanted to show.

To prove the claim, suppose for a contradiction that $N(m_0, m_2) \neq g_{02}$. We have a chain of 1-nodes

$$C = \langle n_1^1, \ldots, n_1^s \rangle \subseteq N$$

such that $N(m_0, n_1^i) = g_{01}$ for all $n_1^i \in C$ and consecutive nodes in the chain are connected by $\pm 1_1$ (figure 18.8). We know that the chain C has distinct endpoints: $n_1^1 \neq n_1^s \in C$. Which node out of $C \cup \{m_0, m_2\}$ was most recently added? The interior nodes $n_1^2, \ldots, n_1^{s-1}$ of C, if any, are incident with at least three edges within $C \cup \{m_0, m_2\}$ chosen by $\forall$ (consider green atoms and ± 1, as usual) and therefore none of these could be the most recently added node. The endpoints of C, n_1^1 and n_1^s, are each incident with one edge labelled $\pm 1_1$ and another labelled g_{01}. If n_1^1

18.7. Tiling implies winning ∃-strategy

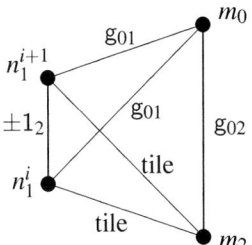

Figure 18.9: Successive tile edges are attached

(or n_1^s) was the most recently added then, as $N(m_0, m_2) \neq g_{02}$ by assumption, we cannot account for the fact that $N(n_1^1, m_2)$ (respectively, $N(n_1^s, m_2)$) is a tile atom. So this can't happen. Thus, the most recently added node must be m_0 or m_2.

If $\exists$ chose the edge (m_0, m_2) then, by her strategy for $(0-2)$ edges and because of the prior existence of the chain C, she would have chosen $N(m_0, m_2) = v_{02}$. But then, in the current round, $\forall$'s move must be illegal, as $g_{02}; \pm 1_2 \not\geq v_{02}$ (rule 18.4). So this can't happen, and it must be $\forall$ who chose the edge (m_0, m_2). Hence, the most recently added node was in fact m_2, for m_0 is incident with at least three edges, those to m_2, n_1^1, n_1^s, chosen by $\forall$.

We are supposing that (m_0, m_2) is not green and we know now that this edge was labelled by $\forall$. Therefore, at least one of the two tile edges $(n_1^1, m_2), (n_1^s, m_2)$ was labelled by $\exists$. So by (β), $\forall$ must have chosen the edge $N(p_0, m_2) = g_{02}$, for some $p_0 \in N$. Since (p_0, m_2) is green and (m_0, m_2) is not, we have $p_0 \neq m_0$. Hence $\forall$'s two edges were (m_0, m_2) and (p_0, m_2). He chose neither of the tile edges $(n_1^1, m_2), (n_1^s, m_2)$: both were chosen by $\exists$, and for her to do this we must have $N(p_0, n_1^1) = N(p_0, n_1^s) = g_{01}$. See figure 18.10. But this contradicts lemma 18.9, and completes the proof of the claim. □

Now we have a well-defined association of tiling functions to all new tile edges not in N.

18.7.7 Coordinates for $\exists$'s new tile edges

Next, we have to assign coordinates $Co(n_1, n_2)$ to each new tile edge (n_1, n_2) of M which is not an edge of $N^\forall$ and with $f_{(n_1, n_2)} = f^i$ for some $i > 0$.

If there are no such edges, there is nothing to do. So assume that there is at least one; let (n_1, n_2) be such an edge. Now, $i > 0$ means that (n_1, n_2) is linked to a tile edge of $N^\forall$ — either an edge of N, in which case $\forall$'s second atom must be $\pm 1_2$, or a tile edge chosen by $\forall$ in the current round, in which case $\forall$'s second

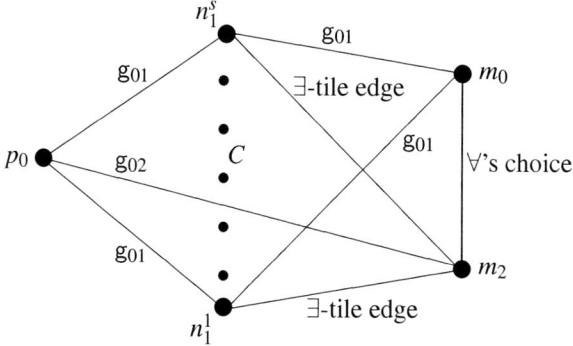

Figure 18.10: The node p_0

atom is a tile atom (cf. the proof of lemma 18.11(1)). Either way, we see that in the current round, $\forall$ chose exactly one green atom — say, $N^\forall(m_0, n_2) = g_{02}$, for some unique $m_0 \in N$. Like n_2, this m_0 will be fixed in our notation from now on.

Let
$$\Gamma = \{n_1 \in N : N(m_0, n_1) = g_{01}\}.$$

$\exists$ has to give coordinates to each of the edges (n_1, n_2), for $n_1 \in \Gamma$, if its tiling function is not f^0. Define a set of 1-nodes Γ^+ by

- $\Gamma^+ = \Gamma$ if $\forall$'s second atom is $\pm 1_2$,

- $\Gamma^+ = \Gamma \cup \{m_1\}$ if $\forall$'s second atom is $N^\forall(m_1, n_2) = t_{12}^j$ (some necessarily unique $m_1 \in N$ and $j < k$).

The tile edges incident with n_2 in M are $\{(n_1, n_2) : n_1 \in \Gamma^+\}$. Define an undirected graph G with nodes Γ^+ and with an edge connecting two nodes a, b of G if and only if $N(a, b) = \pm 1_1$.

LEMMA 18.12 *The graph G is acyclic (i.e., it has no cycles).*

Proof. First, suppose that $\forall$'s second atom is $\pm 1_2$, so that $\Gamma^+ = \Gamma$. Assume for contradiction that there is a cycle in G — say $\gamma = \langle g_0, \ldots g_{t-1} \rangle \subseteq G$ (some $t \geq 3$) with graph edges of G connecting g_0 to g_1, g_1 to g_2, $\ldots$, g_{t-1} to g_0. We ask the question: in the course of the game, which node of $\gamma \cup \{m_0\}$ was most recently added to N? Whichever it is, it can be connected to the rest of $\gamma \cup \{m_0\}$ by at most two edges labelled by $\forall$. Now each node in γ is incident with two edges labelled $\pm 1_1$ (because it is in the cycle γ) and one green edge connecting it to m_0 (because

18.7. Tiling implies winning ∃-strategy

it is in Γ), and all of these edges are only chosen by $\forall$. Thus, none of these can have been the most recently added node, and the last-added node must be m_0. But $|\gamma| \geq 3$, and m_0 is connected to every node of γ by a green edge. As before, this is impossible, since $\exists$ never chooses green labels. So G is acyclic in this case.

Now suppose instead that $\forall$ chooses $N^\forall(m_1,n_2) = t_{12}^j$ (some j), so $\Gamma^+ = \Gamma \cup \{m_1\}$. We claim that $j > 0$. To see this, recall that we are supposing that there is at least one edge (n_1,n_2) with $n_1 \in \Gamma$ and $f_{(n_1,n_2)} = f^i$ for some $i > 0$. By definition of $f_{(n_1,n_2)}$, this means that (n_1,n_2) is linked to a tile edge e in $N^\forall$ with tiling function f^i. However, as in lemma 18.11(1), when the two edges chosen by $\forall$ in the current round are $N^\forall(m_0,n_2) = g_{02}$ and $N^\forall(m_1,n_2) = t_{12}^j$, (n_1,n_2) cannot be linked to an edge in N, so it must be linked to the edge (m_1,n_2) and to no other edge of $N^\forall$. The tiling function for this edge is $f_{(m_1,n_2)} = f^j$. So $j = i > 0$, as claimed.

Suppose, for contradiction, that there is a cycle $\gamma \subseteq G$ ($|\gamma| \geq 3$). Now, if $m_1 \notin \gamma$ then we revert to the situation in the previous case: each node in $\gamma \cup \{m_0\}$ is incident with more than two $\forall$-edges within $\gamma \cup \{m_0\}$, which is impossible. So assume $m_1 \in \gamma$. Which node of $\gamma \cup \{m_0\}$ was most recently added to the network? The most recent node should be incident with at most two edges in $\gamma \cup \{m_0\}$ chosen by $\forall$. As above, each node in $\Gamma \cap \gamma$ is incident with two edges within γ labelled $\pm 1_1$ and one green edge connecting it to m_0, and all of these edges are only chosen by $\forall$. Therefore, either m_0 or m_1 was most recently added, and the edge (m_0,m_1) must have been chosen by $\exists$ — otherwise m_0 and m_1 are also incident with three edges chosen by $\forall$, since $m_1 \in \gamma$.

So which atom would $\exists$ have chosen to label the edge (m_0,m_1)? Her strategy for $(0-1)$ edges tells her to choose u_{01} because of the existence of the cycle γ. But then, the current $\forall$ move is illegal, as $j > 0$ and $g_{02}; t_{21}^j \not\geq u_{01}$ by rule 18.3, so that the triangle (m_0,m_1,n_2) is inconsistent. This gives us a contradiction and proves the lemma. □

This lemma allows us to define an integer *rank* $r(n_1)$ for each node n_1 of Γ^+ such that

$$\text{if } n_1, n_1' \in \Gamma^+ \text{ and } N(n_1,n_1') = +1_1 \text{ then } r(n_1') = r(n_1) + 1.$$

Now we can define coordinates for each new tile edge (n_1,n_2) with $n_1 \in \Gamma$ and $f_{(n_1,n_2)} = f^i$ for $i > 0$.

Case A Suppose that the two atoms chosen by $\forall$ in the current round of the game are $N^\forall(m_0,n_2) = g_{02}$ and $N^\forall(m_1,n_2) = t_{12}^j$ (for some $m_1 \in N$ and $j < k$). Here, $m_1 \in \Gamma^+$; obviously, it is unique. We define coordinates for (n_1,n_2) by

$$\text{Co}(n_1,n_2) = (r(n_1) - r(m_1), 0).$$

Case B Suppose that in the current round, $\forall$ chooses the two edges $N^\forall(m_0,n_2) = g_{02}$ and $N^\forall(m_2,n_2) = +1_2$ (the case where he chooses the atom -1_2 is similar). Again, m_2 is uniquely defined.

As $f(n_1,n_2) \neq f^0$, (n_1,n_2) must be linked to a tile edge of N. In fact, consideration of a shortest linking chain shows that *it must be linked to one of the form (n'_1,m_2), for some $n'_1 \in \Gamma$, by a chain of attached tile edges of M that are not edges of N*. We have $f_{(n'_1,m_2)} = f_{(n_1,n_2)} \neq f^0$, so (n'_1,m_2) has coordinates in N. Let $\mathrm{Co}(n'_1,m_2) = (x,y)$, say. The coordinates of the new edge (n_1,n_2) in M are now defined to be

$$\mathrm{Co}(n_1,n_2) = (x + r(n_1) - r(n'_1), y+1).$$

(If $\forall$'s second atom is -1_2 then replace $y+1$ by $y-1$.)

We have to show that this is well-defined. Suppose that (n_1,n_2) is linked to two distinct such tile edges (n'_1,m_2) and (n^*_1,m_2) by chains of the form stated. Then (n'_1,m_2) and (n^*_1,m_2) are themselves linked in M by such a chain — there is a chain of new tile edges

$$(n'_1,n_2) = (n^1_1,n_2),\ (n^2_1,n_2),\ \ldots,\ (n^s_1,n_2) = (n^*_1,n_2)$$

of M, each successive pair being attached, with (n'_1,m_2) attached to (n'_1,n_2) and (n^*_1,n_2) attached to (n^*_1,m_2). We have $n^1_1,\ldots,n^s_1 \in \Gamma$. The proof of the claim in lemma 18.11(2) now applies, to show that $N(m_0,m_2)$ is green, the (n^i_1,m_2) ($1 \leq i \leq s$) are all tile edges of N, and that (n^i_1,m_2) is attached to (n^{i+1}_1,m_2) for each $i < s$. So $(n'_1,m_2) = (n^1_1,m_2),\ (n^2_1,m_2),\ \ldots,\ (n^s_1,m_2) = (n^*_1,m_2)$ is a chain of attached tile edges of N.

Clearly, $n^1_1,n^2_1,\ldots,n^s_1$ is a path in G. So by T2 (in N) and the definition of the rank r, we have, for any $x,y \in \mathbb{Z}$, $\mathrm{Co}(n^i_1,m_2) = (x+r(n^i_1),y) \iff \mathrm{Co}(n^{i+1}_1,m_2) = (x+r(n^{i+1}_1),y)$ for each $i < s$. We now obtain

if $\mathrm{Co}(n'_1,m_2) = (x,y)$ then $\mathrm{Co}(n^*_1,m_2) = (x+r(n^*_1)-r(n'_1),y)$,

by induction on s. Thus, the coordinate function is well-defined.

We have now defined tilings for all tile edges of M that are not in N, and coordinates for those whose associated tiling is not f^0.

18.7.8 Conditions T1, T2 hold for M

Let us now check that M satisfies conditions T1 and T2. It is sufficient to check that if the tile edges e, e' of M are attached then they share a tiling function and, if appropriate, their coordinates match according to T2. Since T1 and T2 hold for N, we can assume that e, e' are not both edges of N.

18.7. Tiling implies winning ∃-strategy

It follows that e, e' are not both edges of $N^\forall$. For if they were, then being attached, they form two sides of a triangle Δ which is not a $\forall$-triangle, the third side of Δ being labelled by a ± 1 atom. Because $\exists$ never chose a ± 1 label in this (or any) round, the third side of Δ is also in $N^\forall$. So all three sides of Δ are in $N^\forall$. But by the preceding paragraph, Δ does not lie within N, so Δ must in fact be $\forall$'s triangle in the current round. This is a contradiction.

So we may assume that e is not an edge of $N^\forall$. This means that e is a tile edge labelled by $\exists$ in the current round. Say, $e = (n_1, n_2)$ for some $n_1 \in N$.

Case 1 Assume that e' is an edge of N. So e' must have the form (n_1, m_2) where $m_2 \in N$ and $M(m_2, n_2) = \pm 1_2$. Then as e, e' are certainly linked in M, by the well-definedness of the tiling function we have $f_{(e)} = f_{(e')}$, so T1 holds for e, e'. Moreover, if $f_{(e)} \neq f^0$, then by the well-definedness of the coordinates, if $Co(e') = (x, y)$ and $M(m_2, n_2) = +1_2$, say, we have $Co(e) = (x, y+1)$. Thus, the condition in T2 is met. The case where $M(m_2, n_2) = -1_2$ is similar.

Case 2 Assume that e' is an edge of $N^\forall$ but not an edge of N. Thus, it was chosen by $\forall$ in the current round, and has the form (m_1, n_2), where $N(m_1, n_1) = \pm 1_1$. Its coordinates, if any, are $(0, 0)$. As in the previous case, $f_{(e)} = f_{(e')}$, so T1 is satisfied. If $f_{(e)} \neq f^0$ and $N(m_1, n_1) = +1_1$, say, then again by definition of the coordinates (case A) we have $Co(e) = (1, 0)$. If $N(m_1, n_1) = -1_1$, $Co(e) = (-1, 0)$. Hence, T2 is met.

Case 3 Assume finally that e, e' are both edges of M that are not in $N^\forall$. Then they are linked in M. As 'linked' is an equivalence relation, if e is linked to a tile edge e^* of $N^\forall$ then so is e', and we have $f_{(e)} = f_{(e')} = f_{(e^*)}$. If e is not linked to any tile edge of $N^\forall$, then neither is e' and we have $f_{(e)} = f_{(e')} = f^0$. Hence $f_{(e)} = f_{(e')}$ in any case, and T1 is met.

Let $e' = (n_1', n_2)$, and assume that $N(n_1, n_1') = +1_1$ (the case $N(n_1, n_1') = -1_1$ is similar). If $f_{(e)} \neq f^0$ then e, e' are assigned coordinates as above, and there are two ways this can happen.

Case 3a Suppose first that e is linked in M to a tile edge $e^* = (m_1, n_2)$ chosen by $\forall$ in the current round. (So $\forall$'s atoms were $N^\forall(e^*)$ and the green atom forcing the existence of the tile edges e, e'.) Then e' is also linked to e^*, and we have

$$\begin{aligned} Co(e) &= (r(n_1) - r(m_1), 0) \\ Co(e') &= (r(n_1') - r(m_1), 0). \end{aligned}$$

As $N(n_1, n_1') = +1_1$, the rank r satisfies $r(n_1') = r(n_1) + 1$, and this yields that $Co(e)$ and $Co(e')$ are in accordance with T2.

Case 3b Suppose now that e is linked in M to a tile edge e^* of N by a chain of attached tile edges of M that are not in N. Then so is e' (by extending the chain by the extra link (e',e)). We saw that e^* can be taken to be of the form (n_1^*, m_2), where $N^\forall(m_2, n_2) = \pm 1_2$ — say, -1_2. Let $\text{Co}(e^*) = (x,y)$. Then the definition of the new coordinates (above) gives

$$\begin{aligned} \text{Co}(e) &= (x + r(n_1) - r(n_1^*), y - 1) \\ \text{Co}(e') &= (x + r(n_1') - r(n_1^*), y - 1). \end{aligned}$$

From the definition of the rank r, $r(n_1') = r(n_1) + 1$, so that if $\text{Co}(e) = (z,t)$ then $\text{Co}(e') = (z+1,t)$, in accordance with T2.

18.7.9 $\exists$'s strategy for tile edges, T3, and consistency

Now we have a tiling function and (if necessary) coordinates for each new tile edge to be labelled by $\exists$. We can now specify the remainder of $\exists$'s strategy.

IV: $(1-2)$ tile edges For each such edge e, $\exists$ lets

$$M(e) = \begin{cases} t_{12}^0, & \text{if } f_{(e)} = f^0, \\ t_{12}^{f_{(e)}(\text{Co}(e))}, & \text{otherwise,} \end{cases}$$

in accordance with T3. It remains to check that triangles involving these edges are consistent. There are two rules involving tile edges: rules 18.3 and 18.2.

For rule 18.3, suppose for a contradiction that the new tile edge (n_1, n_2) lies in a triangle Δ in M whose other sides are labelled by g_{01} and u_{02}, or alternatively g_{02} and u_{01}. There can only be a problem with rule 18.3 if $M(n_1, n_2) = t_{12}^i$ for some i with $i > 0$. Now $\exists$ only chooses t_{12}^i with $i > 0$ if (n_1, n_2) is linked to a tile edge of $N^\forall$. That means that in the current round, $\forall$ must have chosen one green atom g_{02} and either $\pm 1_2$ or a tile atom t_{12}^j (some j). All other edges incident with n_2 were chosen by $\exists$. The edge (m_0, n_2) (say) of $N^\forall$ labelled g_{02} is not in Δ, as we know that $N(m_0, n_1) = g_{01}$ (else $\exists$ would not choose a tile to label (n_1, n_2)). So it follows that $\exists$ must have chosen a second side of the triangle Δ as well as (n_1, n_2). $\exists$ never chooses a green atom, so this second side must be labelled with the u-atom. But we have already checked (when we considered her strategy for $(0-1)$ and $(0-2)$ edges) that no triangles involving u-labelled edges chosen by $\exists$ are inconsistent. Thus, rule 18.3 cannot be violated.

That leaves the crucial rule 18.2. Suppose then that $\exists$ chooses the label $M(n_1, n_2) = t_{12}^i$, and suppose that this edge is part of a triangle (n_1, n_2, n_1')

18.8. Conclusion 569

with $M(n'_1, n_2) = t^j_{12}$ and $M(n'_1, n_1) = +1_1$, say (the other three cases, in which $+1$ is replaced by -1 and/or n'_1 is replaced by a 2-node n'_2, are entirely similar).

This triangle (n_1, n_2, n'_1) is not an $\forall$-triangle, because if it were it would be in $N^\forall$, whereas in fact one of its edges is a tile edge labelled by $\exists$. Hence the two tile edges in it are attached. By T1, they share a tiling function, say f^l. If $l = 0$ then by T3 we have $i = j = 0$, so there is no problem with rule 18.2. If $l > 0$, by T2 we have $\text{Co}(n'_1, n_2) = (x, y)$ and $\text{Co}(n_1, n_2) = (x+1, y)$ for some $x, y \in \mathbb{Z}$. By T3, $f^l(x, y) = j$ and $f^l(x+1, y) = i$. Because f^l is a valid tiling, $\text{Lt}(T_i) = \text{Rt}(T_j)$. Hence the labels on the triangle (n_1, n_2, n'_1), which are $(t^i_{12}, t^j_{21}, +1_1)$, are consistent with rule 18.2.

Thus, triangles involving new edges are consistent with all the rules.

This completes $\exists$'s strategy. In each case she is able to choose an atom without creating any inconsistent triangles and she is able to define coherent tilings and coordinates where appropriate. In this way she can continue forever and win the game $G^a_\omega(RA(\tau))$. □

18.8 Conclusion

We can now prove the theorem we wanted.

THEOREM 18.13 *The following problems are undecidable.*

1. *Instance: a finite, simple relation algebra $\mathcal{A}$.*
   Yes-instance: if $\mathcal{A}$ is representable; no-instance otherwise.

2. *Instance: a finite relation algebra $\mathcal{A}$.*
   Yes-instance: if $\mathcal{A}$ is representable; no-instance otherwise.

3. *Let $5 \leq n \leq \omega$.*
   Instance: a finite, simple relation algebra $\mathcal{A}$.
   Yes-instance: if $\mathcal{A} \in \mathbf{S\Re aCA}_n$; no-instance otherwise.

4. *Let $5 \leq n \leq \omega$.*
   Instance: a finite relation algebra $\mathcal{A}$.
   Yes-instance: if $\mathcal{A} \in \mathbf{S\Re aCA}_n$; no-instance otherwise.

Proof. We recall from section 18.4 that a finite non-associative algebra $\mathcal{A}$ is a representable relation algebra iff $\exists$ has a winning strategy in $G^a_\omega(\mathcal{A})$. So by theorems 18.2 and 18.5,

$$\tau \text{ is a yes-instance} \iff RA(\tau) \in \mathbf{RRA}. \tag{18.5}$$

1. Loosely speaking, we reduce the tiling problem to our representability problem. If the representation problem for finite simple relation algebras were decidable by some algorithm, then we could decide whether a given tiling instance τ is a yes-instance of our undecidable tiling problem, as follows. We construct $RA(\tau)$, which by lemma 18.1 is a finite simple weakly associative algebra. Now the problem of whether a finite weakly associative algebra is a relation algebra (i.e., with associative composition) is certainly decidable. So we can decide algorithmically whether $RA(\tau)$ is a relation algebra. If it is not, then certainly $RA(\tau) \notin \mathbf{RRA}$, so by (18.5), τ is a no-instance. If $RA(\tau)$ is a relation algebra, then as it is also simple, our hypothetical algorithm will decide whether it is representable. By (18.5), this also decides whether τ is a yes-instance. Hence, the tiling problem would be decidable, which, since it is not, is a contradiction.

2. This problem extends the one in the previous part, so it is undecidable.

3. This is similar to (1). We assume for contradiction that for finite simple relation algebras, membership of $\mathbf{S\Re aCA}_n$ is decidable. Given a tiling instance τ, we construct $RA(\tau)$. If it is not a relation algebra, then as before, τ is a no-instance. If it is, then since it is also simple, we can apply our hypothetical algorithm to determine whether $RA(\tau) \in \mathbf{S\Re aCA}_n$. If the answer is yes, then by exercise 5.6(1), $RA(\tau) \in \mathbf{S\Re aCA}_n \subseteq \mathbf{S\Re aCA}_5$, so by theorem 18.3, τ is a yes-instance. If the answer is no, then by exercise 5.6(2), $RA(\tau) \notin \mathbf{RRA}$, so by (18.5), τ is a no-instance. Thus, we have decided the tiling problem algorithmically, which is impossible. This contradiction completes the proof.

4. Covered by previous case.

$\square$

The following corollaries are straightforward. (We will see further corollaries in the exercises and in chapter 19.)

COROLLARY 18.14 *For any finite $n \geq 5$, the problem of whether a finite (simple) relation algebra has an n-dimensional hyperbasis is undecidable.*

For a proof, use theorems 18.13 and 13.46.

COROLLARY 18.15 *For any class K such that $\mathbf{RRA} \subseteq \mathsf{K} \subseteq \mathbf{S\Re aCA}_5$, the problem of whether a finite relation algebra is in K is undecidable. Hence, K is not finitely axiomatisable.*

For a proof, note that a set τ of tiles is a yes-instance of our tiling problem iff $RA(\tau) \in \mathsf{K}$, and that a finite axiomatisation of K would yield a decision procedure for it.

COROLLARY 18.16 *The set of isomorphism types of representable finite relation algebras with only infinite representations is not recursively enumerable.*

For a proof, see the discussion in section 18.1, and exercise 18.1(2).

Problems

PROBLEM 18.17 *For finite $n \geq 5$, is it decidable whether a finite relation algebra has a finite n-dimensional hyperbasis? Is it decidable whether a finite algebra that is known to be in $\mathbf{S\Re aCA}_n$ has such a hyperbasis?*

PROBLEM 18.18 (Maddux) *Is it decidable whether a finite relation algebra has a finite representation?*

We conjecture that the answer is 'no' in both cases.

Exercises

1. Show that no class lying between **RRA** and $\mathbf{S\Re aCA}_5$ is finitely axiomatisable in second-order logic, or indeed n^{th}-order logic for any finite n.

18.9 Weak representability is undecidable

In this section, we apply the earlier work of this chapter to the class **wRRA** of weakly representable relation algebras. We first saw these algebras in section 5.2, and in theorem 17.29 we used simple rainbow algebras to prove that **wRRA** is not finitely axiomatisable. Here, we will strengthen this result by showing that it is undecidable whether a finite relation algebra is weakly representable — the analogue of theorem 18.13 for **wRRA**.

Deterministic tilings The proof again involves tilings. The kind of tiling problem we need is defined as follows.

DEFINITION 18.19 *Let τ be a finite set of tiles.*

1. *For $S \subseteq \mathbb{Z} \times \mathbb{Z}$, a map $f : S \to \tau$ is said to be a partial tiling if*
   - *whenever $(i,j), (i+1,j) \in S$ then $\mathrm{Rt}(f(i,j)) = \mathrm{Lt}(f(i+1,j))$,*
   - *whenever $(i,j), (i,j+1) \in S$ then $\mathrm{Top}(f(i,j)) = \mathrm{Bot}(f(i,j+1))$.*

2. *Let $t_0 \in \tau$. The pair (τ, t_0) is said to be deterministic if there is a sequence of finite sets $S_n \subseteq \mathbb{Z} \times \mathbb{Z}$ $(n < \omega)$ satisfying:*

(a) $S_0 = \{(0,0)\}$,
(b) for each $n < \omega$, $S_{n+1} = S_n \cup \{(i,j)\}$ for some $(i,j) \in \mathbb{Z} \times \mathbb{Z} \setminus S_n$ such that $S_n \cap \{(i+1,j),(i-1,j),(i,j+1),(i,j-1)\} \neq \emptyset$,
(c) $\bigcup_{n<\omega} S_n = \mathbb{Z} \times \mathbb{Z}$,
(d) for any $n < \omega$ and partial tiling $f : S_n \to \tau$ with $f(0,0) = t_0$, there is at most one partial tiling $f' : S_{n+1} \to \tau$ extending f (i.e., with $f' \restriction S_n = f$).

Evidently, if (τ, t_0) is deterministic then there is at most one tiling of the plane $\mathbb{Z} \times \mathbb{Z}$ using τ and with t_0 at the origin; but the conditions above are stronger than this, since they say that any such tiling can be constructed step-by-step, at each step there being at most one tile that fits onto the partial tiling constructed so far.

DEFINITION 18.20 The *deterministic tiling problem* is the following:
Instance: a pair (τ, t_0), where τ is a finite set of tiles, $t_0 \in \tau$ is arbitrary, and (τ, t_0) is deterministic.
Yes-instance: if there is a tiling of $\mathbb{Z} \times \mathbb{Z}$ with t_0 at the origin; no-instance otherwise.

Note that any instance (τ, t_0) of the deterministic tiling problem is deterministic: it is not part of the problem to decide whether (τ, t_0) is deterministic or not.

FACT 18.21 *The deterministic tiling problem is undecidable.*

This follows from the known undecidability of whether a deterministic Turing machine halts when started on the empty tape. We leave the details to the exercises below.

Weak representability of tiling algebras A set τ of tiles is said to *have a white tile* if there is a tile $t_w \in \tau$ like the tile T_0 in section 18.2: the four edges of t_w all have the same colour, and this colour does not occur as the colour of an edge of any other tile in τ. Now, given an instance (τ, t_0) of the deterministic tiling problem such that τ has a white tile t_w, we define the following finite set of finite weakly associative tiling algebras:

$$S(\tau, t_0) = \{RA(\sigma) : \{t_0, t_w\} \subseteq \sigma \subseteq \tau\}.$$

Here, $RA(\sigma)$ is as defined in section 18.3. Note that $S(\tau, t_0)$ is effectively constructible from (τ, t_0).

LEMMA 18.22 *Let τ be a finite set of tiles with a white tile t_w, let $t_0 \in \tau$, and suppose that (τ, t_0) is deterministic. The following are equivalent:*

1. (τ, t_0) *is a yes-instance of the deterministic tiling problem,*

18.9. Weak representability is undecidable

2. $S(\tau, t_0) \cap \mathbf{RRA} \neq \emptyset$,

3. $S(\tau, t_0) \cap \mathbf{wRRA} \neq \emptyset$.

Proof. The proof of $(1 \Rightarrow 2)$ is straightforward. If (τ, t_0) is a yes-instance of the deterministic tiling problem then let σ be the set of tiles used in the tiling of $\mathbb{Z} \times \mathbb{Z}$ by τ-tiles with t_0 at the origin, together with t_w. It is clear that $RA(\sigma) \in S(\tau, t_0)$. Moreover, every tile in σ occurs in a tiling of $\mathbb{Z} \times \mathbb{Z}$ using tiles from σ. By theorem 18.5, $\exists$ has a winning strategy in $G_\omega^a(RA(\sigma))$, and as in section 18.4 we obtain $RA(\sigma) \in \mathbf{RRA}$. So $S(\tau, t_0) \cap \mathbf{RRA} \neq \emptyset$.

Since $\mathbf{RRA} \subseteq \mathbf{wRRA}$, $(2 \Rightarrow 3)$ is trivial. It remains to show $(3 \Rightarrow 1)$. The proof is similar to that of theorem 18.2, with the complication that 'not every arc in a weak representation is labelled by an atom'. Let $RA(\sigma) \in S(\tau, t_0) \cap \mathbf{wRRA}$, where $\{t_0, t_w\} \subseteq \sigma \subseteq \tau$. Recall from section 5.2 that a weak representation of $RA(\sigma)$ can be regarded as a model of the following theory $T = W_{RA(\sigma)}$, where R, S range over $RA(\sigma)$:

$$\forall x, y\, [1'(x,y) \leftrightarrow (x = y)]$$
$$\forall x, y\, [R \cdot S(x,y) \leftrightarrow R(x,y) \wedge S(x,y)]$$
$$\forall x, y\, [\check{R}(x,y) \leftrightarrow R(y,x)]$$
$$\forall x, y\, [R;S(x,y) \leftrightarrow \exists z (R(x,z) \wedge S(z,y))]$$
$$\forall x, y\, \neg 0(x,y)$$
$$\exists x, y\, [R(x,y) \wedge \neg S(x,y)] \qquad \text{if } R \not\leq S.$$

Therefore, there exists a model, say M, of the above theory T. Note that for $x, y \in M$ with $M \models 1(x,y)$, $\{a \in RA(\sigma) : M \models a(x,y)\}$ is only a filter of $RA(\sigma)$, not necessarily an ultrafilter. There may be no atom holding on (x,y).

Now we refer to the definition of $RA(\sigma)$ given in section 18.3. Since $g_{01} \not\leq 0$, by the last axiom schema of T we may choose $r, x_0 \in M$ such that $M \models g_{01}(r, x_0) \wedge \neg 0(r, x_0)$. Since $(g_{01}, (t_0)_{12}, g_{20})$ is a consistent triple of atoms of $RA(\sigma)$, we have $g_{01} \leq g_{02}; (t_0)_{21}$. (Here and below, given any tile $t \in \sigma$, for simplicity of notation we write t_{12}, t_{21} for the corresponding tile atoms of $RA(\sigma)$.) Since $M \models g_{01}(r, x_0)$, the relational product and conversion axioms of T ensure that there exists $y_0 \in M$ with $M \models g_{02}(r, y_0) \wedge (t_0)_{12}(x_0, y_0)$. The axioms for $\cdot$ and 0 ensure that $(t_0)_{12}$ is the unique atom holding on (x_0, y_0).

Now we define $x_i, y_i \in M$ for $i, j \in \mathbb{Z}$ by induction, satisfying

- $M \models g_{01}(r, x_i) \wedge g_{02}(r, y_j)$,

- $M \models +1_1(x_i, x_{i+1}) \wedge +1_2(y_j, y_{j+1})$,

for all $i, j \in \mathbb{Z}$. x_0, y_0 are already defined. Assume we have defined $x_{-i}, \ldots, x_i$ and $y_{-i}, \ldots, y_i$. Define x_{i+1} as follows: since $g_{01} \leq g_{01}; -1_1$, the relational product axioms and conversion axioms of T ensure that there is some $x_{i+1} \in M$ such that

$M \models g_{01}(r, x_{i+1}) \wedge +1_1(x_i, x_{i+1})$. Define x_{-i-1} from x_{-i} similarly, using the fact that $g_{01} \leq g_{01}; +1_1$. Define y_{i+1} and y_{-i-1} similarly, using $g_{02} \leq g_{02}; \pm 1_2$.

Claim. For each $i, j \in \mathbb{Z}$, there is a unique tile $t^{ij} \in \sigma$ such that $M \models t_{12}^{ij}(x_i, y_j)$. Moreover, the map $f : \mathbb{Z} \times \mathbb{Z} \to \sigma$ defined by $f(i, j) = t^{ij}$ is a tiling of $\mathbb{Z} \times \mathbb{Z}$ with $f(0, 0) = t_0$.

Proof of claim. Since (τ, t_0) is deterministic, we may choose finite subsets $S_n \subseteq \mathbb{Z} \times \mathbb{Z}$ ($n < \omega$) as in definition 18.19. It suffices to show that for all $n < \omega$ and $(i, j) \in S_n$, there is a unique tile $t^{ij} \in \sigma$ such that $M \models t_{12}^{ij}(x_i, y_j)$, and such that the map $f : S_n \to \sigma$ given by $f(i, j) = t^{ij}$ satisfies $f(0, 0) = t_0$ and is a partial tiling.

This is done by induction on n. We have it already for $n = 0$, by choice of x_0, y_0. Assume the result for n. Let $f : S_n \to \sigma$ be the partial tiling given by: $f(k, l)$ is the unique tile $t \in \sigma$ such that $M \models t_{12}(x_k, y_l)$. Let (i, j) be the unique element of $S_{n+1} \setminus S_n$. By definition of x_i, y_j, we have $M \models g_{01}(r, x_i) \wedge g_{02}(r, y_j)$. By the conversion axioms of T, $M \models g_{10}(x_i, r) \wedge g_{02}(r, y_j)$. But by definition of $RA(\sigma)$, $g_{10}; g_{02} = \sum_{t \in \sigma} t_{12}$. Therefore, by the relational product axioms of T, we have $M \models \left(\sum_{t \in \sigma} t_{12}\right)(x_i, y_j)$.

Now let

$$\alpha = \prod \{a \in RA(\sigma) : M \models a(x_i, y_j)\}.$$

So $\alpha \leq \sum_{t \in \sigma} t_{12}$. It follows by the boolean product axioms of T and the finiteness of $RA(\sigma)$ that $M \models \alpha(x_i, y_j)$. So by the zero axiom of T, $\alpha \neq 0$.

α is thus a non-zero sum of tile atoms arising from tiles in σ. We want to show that it is in fact a tile atom t_{12} for some $t \in \sigma$ that can be used to extend the partial tiling f to S_{n+1}.

Pick any tile $t \in \sigma$ with $t_{12} \leq \alpha$, and define $f_t : S_{n+1} \to \sigma$ by $f_t \restriction_{S_n} = f$ and $f_t(i, j) = t$. We first show that f_t is a partial tiling. Because $f : S_n \to \sigma$ is inductively a partial tiling, the tiles associated by f_t with any two adjacent pairs in S_n fit together. So it suffices to show that for each $(k, l) \in S_n \cap \{(i+1, j), (i-1, j), (i, j+1), (i, j-1)\}$, the tiles $f_t(i, j)$ and $f_t(k, l)$ fit together. We only consider the case $(k, l) = (i+1, j)$; the others are handled similarly. Assume then that $(i+1, j) \in S_n$. By definition of the x_i, y_j, and the inductive assumption on f, we have $M \models +1_1(x_i, x_{i+1}) \wedge f_t(i+1, j)_{12}(x_{i+1}, y_j)$. The relational product axioms of T yield $M \models (+1_1; f_t(i+1, j)_{12})(x_i, y_j)$. So by definition of α, we have

$$t_{12} \leq \alpha \leq +1_1; f_t(i+1, j)_{12}.$$

So $(+1_1, f_t(i+1, j)_{12}, t_{21})$ is a consistent triple of atoms of $RA(\sigma)$. By definition of $RA(\sigma)$ and of f_t, we obtain $\mathrm{Rt}(f_t(i, j)) = \mathrm{Rt}(t) = \mathrm{Lt}(f_t(i+1, j))$, as required.

So f_t is a partial tiling extending f. Since (τ, t_0) is deterministic and $f(0, 0) = t_0$, there is at most one partial tiling $f : S_{n+1} \to \tau$ extending f. So if $t, u \in \sigma$ are tiles such that $t_{12}, u_{12} \leq \alpha$, then $f_t = f_u$, so $t = f_t(i, j) = f_u(i, j) = u$. Hence, α is a

18.9. Weak representability is undecidable

tile atom: say, $\alpha = t_{12}$ for some tile $t \in \sigma$. We already showed that $M \models t_{12}^{ij}(x_i, y_j)$ and that $f_t : S_{n+1} \to \sigma$ is a partial tiling. This completes the induction, and proves the claim.

It is immediate from the claim that (τ, t_0) is a yes-instance of the deterministic tiling problem. So we have established (1). $\square$

We can now derive the results we want.

THEOREM 18.23 *For any class* K *satisfying* **RRA** $\subseteq$ K $\subseteq$ **wRRA**, *the problem of whether an arbitrary finite relation algebra is in* K *is undecidable. Hence,* K *is not finitely axiomatisable in n^{th}-order logic, for any finite n.*

Proof. Assume for contradiction that **RRA** $\subseteq$ K $\subseteq$ **wRRA** and there is an algorithm to decide whether a given finite relation algebra is in K. We show how to use this algorithm to decide whether an arbitrary instance (τ, t_0) of the deterministic tiling problem is a yes-instance.

Let τ^+ denote τ augmented by a new, 'white' tile t_w. It is clear that (τ^+, t_0) is deterministic. Given (τ, t_0), we construct $S(\tau^+, t_0)$; this can be done effectively. By assumption, for each *relation algebra* $\mathcal{A} \in S(\tau^+, t_0)$, we can decide whether $\mathcal{A} \in$ K. As is easily seen, **wRRA** $\subseteq$ **RA**, so no *non-relation algebra* in $S(\tau^+, t_0)$ can be in K. Since whether a finite weakly associative algebra is a relation algebra is certainly decidable, it follows that we can decide whether $S(\tau^+, t_0) \cap$ K $\neq \emptyset$. If yes, then as K $\subseteq$ **wRRA**, we have $S(\tau^+, t_0) \cap$ **wRRA** $\neq \emptyset$, so by lemma 18.22, (τ^+, t_0) is a yes-instance. If no, then since **RRA** $\subseteq$ K, we have $S(\tau^+, t_0) \cap$ **RRA** $= \emptyset$. By lemma 18.22 again, (τ^+, t_0) is a no-instance.

Since (τ^+, t_0) is clearly a yes-instance of the deterministic tiling problem iff (τ, t_0) is, this procedure decides effectively whether or not (τ, t_0) is a yes-instance. This contradicts fact 18.21, and proves the first part of the theorem. For the second part, note that any such axiomatisation of K would immediately yield a decision procedure for the class of finite algebras in K, contradicting the first part. $\square$

COROLLARY 18.24 *The problem of whether a finite relation algebra is weakly representable is undecidable. The class* **wRRA** *is not finitely axiomatisable in n^{th}-order logic, for any finite n.*

Proof. Immediate from theorem 18.23, by taking K = **wRRA**. $\square$

The second part strengthens theorem 17.29.

Jónsson showed in [Jón59] that **wRRA** $\subset$ **RA**. Hence, **RA**$_n \nsubseteq$ **wRRA** for $n = 3, 4$. We can strengthen this as follows:

COROLLARY 18.25 *For no finite $n \geq 3$ do we have* **RA**$_n \subseteq$ **wRRA**.

Proof. Assume for contradiction that $\mathbf{RA}_n \subseteq \mathbf{wRRA}$ for some $n \geq 3$. By proposition 13.48, $\mathbf{RRA} \subseteq \mathbf{RA}_n \subseteq \mathbf{wRRA}$. But by corollary 12.32, we can decide whether a finite relation algebra is in $\mathbf{RA}_n$. This contradicts theorem 18.23. □

This answers negatively a question of Maddux: namely, whether $\mathbf{wRRA} = \mathbf{RA}_5$. The following remains open:

PROBLEM 18.26 *Do we have $\mathbf{wRRA} \subseteq \mathbf{RA}_n$ for some $n \geq 5$? What inclusions hold between $\mathbf{wRRA}$ and the $\mathbf{S\Re aCA}_n$ $(n \geq 5)$?*

Exercises

Let $M = (Q, \Sigma, q_0, \delta, F)$ be a deterministic Turing machine, where Q is a finite set of states, $q_0 \in Q$ is the initial state and $F \subseteq Q$ is the set of halting states, Σ is a finite alphabet containing a 'blank' character '$\wedge$', and $\delta : (Q \setminus F) \times \Sigma \to Q \times \Sigma \times \{0, 1, -1\}$ is the 'instruction table'. M has a single two-way-infinite tape, with squares indexed by the integers $\mathbb{Z}$. Each square holds a single character from Σ; initially, all squares are blank. M starts off in state q_0 with its read-write head over square 0. At each step, if M is in state $q \in Q$ and its head is in square $n \in \mathbb{Z}$ which contains character $a \in \Sigma$, then if $q \in F$, M halts; otherwise, if $\delta(q,a) = (q', a', d)$, M overwrites the a in square n with a', moves its head to square $n + d$, and goes to state q' ready for the next step.

Consider the set τ_M of tiles shown in figure 18.11. In the figure, q, a denote

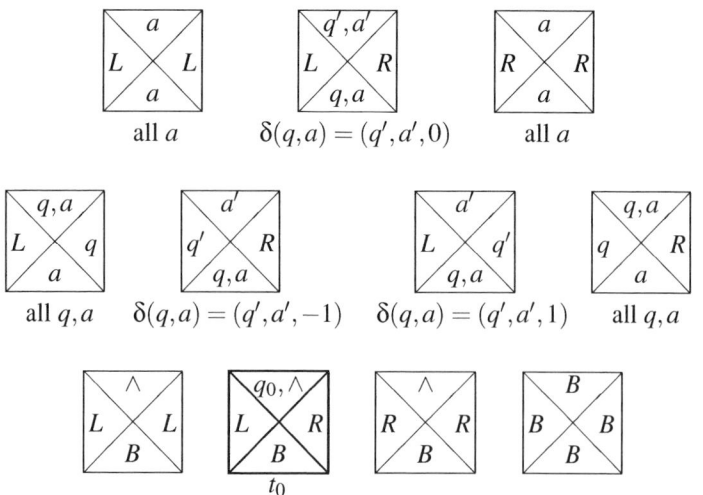

Figure 18.11: Tiles for the Turing machine M

elements of Q, Σ, respectively, and L, R, B are new symbols. So, e.g., for each $a \in \Sigma$

18.10. Undecidability of equational theories

there is a tile with top and bottom labelled by a and left and right labelled by L; for each $q \in Q \setminus F$ and $a \in \Sigma$ with $\delta(q,a) = (q',a',1)$, there is a tile with bottom labelled q,a, top a', left L, and right q'; etc. The colours $\text{Top}(t)$, etc., of a tile t are given by the symbols labelling the sides of t. So two tiles fit together iff their touching sides carry the same symbols.

Let t_0 be the tile with heavy lines (as indicated in the figure). Show that:

1. Any infinite (non-halting) run of M starting on a blank tape determines a tiling of the plane with t_0 at the origin. (For $y \geq 0$, the tile t at (x,y) should encode square x of M's tape at time y. If $\text{Bot}(t)$ involves a state of M, M should be in that state and with its head in square x.)

2. Any tiling of the plane with t_0 at the origin determines an infinite run of M starting on a blank tape.

3. (τ_M, t_0) is a deterministic tiling. (You will need to define sets S_n ($n < \omega$) as in definition 18.19.)

4. The deterministic tiling problem is undecidable.

18.10 Undecidability of equational theories

For a functional signature L, the *equational theory* of a class of L-algebras is the set of L-equations that are valid in every algebra in the class. Another easy corollary of theorem 18.13 is:

COROLLARY 18.27 *The equational theory of* **RRA** *is undecidable. For $5 \leq n < \omega$, the equational theory of* $S\Re aCA_n$ *is undecidable.*

Proof. We reduce the problem of telling if a finite, simple relation algebra is not in **RRA** (or $S\Re aCA_n$) to the problem of telling if an equation in the language L_{RA} of relation algebras is valid in **RRA** (or in $S\Re aCA_n$, respectively).

Let $\mathcal{A}$ be a finite, simple relation algebra. We define a conjunction $\Delta(\mathcal{A})$ of L_{RA}-equations as follows. $\Delta(\mathcal{A})$ is similar to the diagram of $\mathcal{A}$, familiar from model theory, but it uses variables rather than constants. For each $a \in \mathcal{A}$, let x_a be a variable symbol (distinct from all others). $\Delta(\mathcal{A})$ includes the conjuncts $x_0 = 0$, $x_1 = 1$, and $x_{1'} = 1'$. For all $a,b,c \in \mathcal{A}$ with $a;b = c$, $\Delta(\mathcal{A})$ includes a conjunct $x_a; x_b = x_c$. $\Delta(\mathcal{A})$ also includes a conjunct $x_a + x_b = x_c$ for all $a,b,c \in \mathcal{A}$ with $a + b = c$, and similarly a finite number of conjuncts corresponding to each of the other operators $-, \smile$.

Consider the existential sentence $\exists_{a \in \mathcal{A}} x_a \Delta(\mathcal{A})$, where $\exists_{a \in \mathcal{A}} x_a$ stands for the string of quantifiers $\exists x_{a_0} \ldots \exists x_{a_l}$, for some arbitrary enumeration $a_0, \ldots a_l$ of $\mathcal{A}$. For any relation-type algebra $\mathcal{B}$, let us temporarily write $\mathcal{A} \sqsubseteq \mathcal{B}$ to denote that there is an L_{RA}-embedding from $\mathcal{A}$ into $\mathcal{B}$. Then it is easily seen that $\mathcal{B} \models \exists_{a \in \mathcal{A}} x_a \Delta(\mathcal{A})$ iff $\mathcal{A} \subseteq \mathcal{B}$. Since **RA** is a discriminator variety, we know from theorem 2.55 and lemma 2.59 that the quantifier-free formula $\neg \Delta(\mathcal{A})$ is equivalent, over simple relation algebras, to an equation $\varepsilon(\mathcal{A})$. This equation is effectively constructible from $\mathcal{A}$. So for any simple $\mathcal{B} \in \mathbf{RA}$, we have

$$\mathcal{B} \models \varepsilon(\mathcal{A}) \iff \mathcal{A} \not\subseteq \mathcal{B}.$$

So $\varepsilon(\mathcal{A})$ is valid *over simple representable* relation algebras if and only if for all simple $\mathcal{B} \in \mathbf{RRA}$ we have $\mathcal{A} \not\subseteq \mathcal{B}$. Since $\mathcal{A}$ is simple and **RRA** is closed under subalgebras, this is if and only if $\mathcal{A} \notin \mathbf{RRA}$. Similarly, $\varepsilon(\mathcal{A})$ is valid over simple algebras in $\mathbf{S\mathfrak{R}aCA}_n$ if and only if $\mathcal{A} \notin \mathbf{S\mathfrak{R}aCA}_n$, for $n \geq 5$.

But $\mathbf{RRA}, \mathbf{S\mathfrak{R}aCA}_n$ are varieties (theorem 3.37, proposition 5.48); being contained in **RA**, they are discriminator varieties. An equation is valid over the simple members of a discriminator variety if and only if it is valid over the whole variety (corollary 2.50 and theorem 2.55). Thus, $\varepsilon(\mathcal{A})$ is valid over **RRA** (or $\mathbf{S\mathfrak{R}aCA}_n$) if and only if $\mathcal{A} \notin \mathbf{RRA}$ (respectively $\mathcal{A} \notin \mathbf{S\mathfrak{R}aCA}_n$). So if we could algorithmically decide validity of equations over **RRA** or over $\mathbf{S\mathfrak{R}aCA}_n$ ($n \geq 5$), we could decide non-membership of these classes for finite simple relation algebras, contradicting theorem 18.13. □

The undecidability of the equational theories of **RA** and **RRA** was proved a long time ago by Tarski [Tar41, p. 8]. [TarGiv87, xii, p. 255] shows that the equational theory of any variety between **RA** and **RRA** is undecidable. This also follows from the next theorem, from [AndGiv+97]:

THEOREM 18.28 (Andréka, Givant, Németi) *If K is a class of relation algebras satisfying one of the two conditions below, then K has an undecidable equational theory.*

1. *For each $n < \omega$, there is a simple algebra in K with at least n elements below the identity.*

2. *For each $n < \omega$, there is an algebra in K containing a set G with at least n elements such that G is a group under composition and conversion, and the elements of G are pairwise disjoint.*

Observe that the undecidability of the equational theories of **RA**, **RRA**, **GRA**, **wRRA**, **RA**$_n$, and $\mathbf{S\mathfrak{R}aCA}_n$ ($n \geq 4$) is given by their result. The undecidability of the equational theory of the class **SA** of semi-associative algebras was proved in [Madd78b, theorem 2, page 222] and in [Madd94b].

Exercises

1. Check that the equational theories of **RA**, **GRA**, and **RA**$_n$ (for finite $n \geq 4$) are undecidable.

18.10. Undecidability of equational theories

2. Let **RRA** $\subseteq$ K $\subseteq$ **wRRA**. Show that the equational theory of SK is undecidable.

3. Construct a finite relation algebra $\mathcal{A}$ such that the network satisfaction problem (see §7.9) over $\mathcal{A}$ is undecidable [Hir99]. For this, consider the following tiling problem. Given a fixed, finite set τ of tiles, an instance of $P(\tau)$ is a non-empty finite sequence $S(0,0), S(1,0), \ldots, S(n,0) \in \tau$ such that $\text{Rt}(S(i,0)) = \text{Lt}(S(i+1,0))$, for $i = 0, \ldots, n-1$. The sequence is a yes-instance of $P(\tau)$ if it is possible to extend it to a tiling of the whole plane, and a no-instance otherwise.

    (a) Prove that there exists a finite set of tiles τ such that $P(\tau)$ is undecidable.

    (b) Find a set of tiles τ' such that (i) $P(\tau')$ is still undecidable, (ii) for each $T \in \tau'$ there is a tiling of the plane with T at $(0,0)$, and (iii) τ' contains one tile which can tile the plane on its own but cannot be adjacent to any other tile.

    (c) Reduce $P(\tau')$ to the network satisfaction problem over the relation algebra $RA(\tau')$. Theorem 18.5 should be helpful.

Chapter 19

Finite base property

19.1 Introduction

This chapter complements the undecidability results of section 18.10, in the sense that we prove decidability of the validity of equations or universal first-order sentences in certain classes of algebras — that is to say, the decidability of the equational or universal theories of the classes. However, our treatment of decidability is not very conventional, since we obtain our results via the so-called 'finite base property'. In fact, this property is the main topic of the chapter.

What is the finite base property? The finite base property is actually a clutch of several related properties. The simplest kind of 'finite base' question is probably this: given a class K of representable algebras, *does any finite algebra in* K *have a finite representation?*[1] We imagine that the notion of representation provides a concrete algebra of relations on some base set, isomorphic to any given algebra in K; a *finite* representation is simply one whose base set is finite. Other kinds of finite base property involve whether non-validity of equations (etc.) in K can always be witnessed by finite algebras, or by algebras with finite representations. This is where the decidability connection lies.

Take the classical notion of representation for relation algebras. A little thought shows that **RRA** does not have the finite base property. The point algebra (see section 4.4) is finite, with only eight elements, but all its classical representations are (unions of) dense linear orders and so are infinite. Many of the tiling algebras of chapter 18 also fit this pattern. On the other hand, the proof of Stone's theorem (2.10) shows that any finite boolean algebra can be represented as a field of

---

[1] In section 19.3, we call this the 'finite algebra on finite base property'; for now, we will stick with the briefer 'finite base property'.

subsets of a finite set (namely, the set of ultrafilters of the algebra): so the finite base property is true for boolean algebras.

Well, we have seen this kind of phenomenon before, and the long-suffering reader will now be expecting a chapterful of negative results on the finite base property for relation-type algebras. In fact, *all the results of this chapter are broadly positive!* This is because we concentrate on relativised representations. For **WA**, there is a straight positive result, and the method used to prove it works for the cylindric-type algebras **Crs, D, G** (see chapter 5 for the definitions), and several other classes characterised by relativised representations too. We get the finite base property in all its pertinent forms, and it follows that the equational (and indeed universal) theories of these classes are decidable.

We can also obtain the simple form of the finite base property — that any finite algebra has a finite representation — for some classes whose equational theory is undecidable, namely $\mathbf{RA}_n$ and a certain subclass of $\mathbf{S\Re aCA}_n$, for all finite $n \geq 4$. Using the representation theory developed in chapter 13, we will show that any finite algebra in $\mathbf{RA}_n$ has a finite n-square relativised representation. We will also show that any subalgebra of the relation algebra reduct of a finite n-dimensional cylindric algebra has a finite n-flat relativised representation. So, for example, although the point algebra just mentioned has no finite classical representation, it does have a finite n-flat relativised representation for all finite n (exercise 19.6(3) below). We will see that for $n \geq 5$, it is not true that every finite algebra in $\mathbf{S\Re aCA}_n$ has a finite n-flat relativised representation. Although any finite algebra in $\mathbf{S\Re aCA}_n$ is a subalgebra of the relation algebra reduct of some n-dimensional cylindric algebra C, it is not always possible to find a *finite C*, and in such cases the algebra has no finite n-flat relativised representation. However, every finite relation algebra has a finite 4-flat relativised representation; this adds to the potential utility of flat relativised representations in studying arbitrary relation algebras.

Why the finite base property? We have chosen to concentrate on the finite base property instead of on decidability for several reasons. First, there already exist excellent references for decidability in algebraic logic, such as [AndGiv+97], while work on the finite base property is rarer. Second, the finite base property is more in tune with the broadly model-theoretic approach of this book, and indeed, our treatment uses some fairly sophisticated results from modern model theory. Third, while of course we have no objection to games and step-by-step arguments, the techniques we use here provide an interesting finitistic alternative. Fourth, they give interesting additional information beyond plain decidability: for instance, a rough complexity bound. The finite base property can also be useful in obtaining other theorems. For example, it follows from the finite base property for cylindric relativised set algebras ($\mathbf{Crs}_n$) that any finite $\mathbf{Crs}_n$ can be obtained by relativising

19.1. Introduction

a finite cylindric set algebra ($\mathbf{Cs}_n$). This was used by Simon [Sim97] to prove that every finite cylindric algebra of dimension 3 can be obtained by twisting and relativisation from a finite $\mathbf{Cs}_3$. And fifth, classes such as $\mathbf{RA}_n$ and $\mathbf{S\Re aCA}_n$ have a form of the finite base property but have undecidable equational theories. Here, we have no option to study decidability at all.

The disadvantages of the finite base property approach to decidability include that the resulting complexity bounds are not usually optimal, and that some classes of cylindric-type algebras fail to have the finite base property but still have decidable equational theories — see [MikMar99], for example. This last will not come up here in our work with relation algebras, though.

Why decidability of equational theories? Most of the book has been concerned with representations of various kinds, and any readers who have come this far will perhaps understand our interest in representations with finite base. It may be less clear why decidability of the equational theory of a class of algebras is of interest. (For example, why not consider the full first-order theory?) One important reason is the connection that classes of algebras of relations bear to conventional logical systems. We already saw this at work in fact 15.16 and theorem 15.17. Taking modal logic as another example, it is now well known that Kripke's semantics for modal logic was anticipated by Jónsson and Tarski's work on boolean algebras with operators [JónTar51, JónTar52]. To take advantage of this, one translates the modal connective $\Diamond$ to a unary operator symbol $f_\Diamond$, and the boolean operators $\wedge, \neg$ to $\cdot, -$. One may now translate any modal formula φ to a term t_φ of the signature $L = \{+, -, 0, 1, f_\Diamond\}$, a signature of BAOs. For example, the formula $\varphi = \Box(p \to \Diamond q)$ translates to the term $t_\varphi = -f_\Diamond-(-p + f_\Diamond(q))$; here, p, q are variables. A formula φ is valid in the basic modal logic K iff the equation $t_\varphi = 1$ is valid in the class C of all L-BAOs. So decidability of the *equational* theory of C entails (in fact is equivalent to) decidability of the modal logic K. Many other results of this kind exist. The connection between modal and algebraic logic was of great interest in the 1990s — see, e.g., [Ven92, Ben96] — and the view of many today is that one should pass back and forth between the two subjects as and when appropriate, just as the 1990s workers travelled between Amsterdam and Budapest.

Guarded fragment of first-order logic We will establish the finite base property by taking advantage of another way to translate modal formulas: by the 'standard translation' into *first-order logic* (see, e.g., [Ben76, Ben85]). For example, $\Box(p \to \Diamond q)$ is translated to

$$\forall y(R(x, y) \to [P(y) \to \exists z(R(y, z) \wedge Q(z))]). \tag{19.1}$$

The translation mimics the Kripke semantics for modal logic: R is a binary relation symbol to be interpreted as the accessibility relation on the underlying frame of the Kripke model, and P,Q are unary relation symbols intended to be true at the worlds where p,q are true. Not every first-order formula (with one free variable in the appropriate signature) is the standard translation of a modal formula; so the formulas that are form a proper *fragment* of first-order logic, and one that inherits the nice properties of the 'basic' modal logic K, such as decidability with reasonable complexity, interpolation, and in particular the *finite model property:* every satisfiable formula is satisfied in some finite model. This often remains true when we vary the modal similarity type: various multimodal, temporal, and dynamic logics also correspond by standard translation to well behaved modal-style fragments of classical logic.

Finding 'modal fragments' of first-order logic is an old problem in modal correspondence theory. The aim is to identify syntactically defined fragments of first-order logic that contain the standard translations of various modal logics and share their nice properties 'for the same reasons'. One such (hopefully) modal fragment, the *guarded fragment,* was put forward by Andréka, van Benthem, and Németi in [AndBen+98]. Their idea was to look at quantification patterns. Only relativised quantification (along the accessibility relation of the Kripke frame) is allowed in modal formulas; so in the guarded fragment, all quantification must be relativised to some atomic formula. Thus, if $\varphi(x,y,z)$ is a formula of the guarded fragment, then so are $\exists yz(R(x,y,z) \wedge \varphi(x,y,z))$ and $\forall yz(R(x,y,z) \to \varphi(x,y,z))$, because x and the quantified y,z are 'guarded' by the relation symbol R. All free variables of φ must occur in the guard, which must be an atomic formula (here, $R(x,y,z)$). The plain $\exists yz\varphi(x,y,z)$ would not be acceptable. Observe that (19.1) is in the guarded fragment.

The definition of this fragment was suggested by the classes $\mathbf{Crs}_n$ (see chapter 5), in which cylindrification is always relativised to the unit. The equational theories of these classes were proved decidable by Németi in [Ném86], using 'mosaics'. And sure enough, the guarded fragment does have the hoped-for nice properties. Its decidability, and other results such as a Łoś–Tarski theorem, were proved in [AndBen+98, Ben96, Ben97]. Complexity results are established in [Grä99b, Mar97]: deciding validity for sentences of the guarded fragment, with or without equality, is complete for double-exponential time, but n-variable fragments of the guarded fragment (for finite $n \geq 2$) are EXPTIME-complete, and some 2-variable guarded fragments are even in PSPACE. It was proved in [Grä99b] that the guarded fragment has the finite model property — any guarded sentence with a model has a finite model. (For further discussion of surrounding issues, see, e.g., [AreMon+99, Mar01b, Var97] as well as the citations already given.)

Because of results like these, the guarded fragment and various extensions of it (e.g., by fixed-point operators) have become rather popular. But the guarded fragment was objected to on the ground that the standard translations of some quite respectable modal-style formulas, such as temporal formulas involving Since and Until, fall outside the fragment. The temporal formula $U(p,q)$ expresses that q will

19.1. Introduction

be true until such a time as p is true. Its standard translation is

$$\exists y(x < y \wedge P(y) \wedge \forall z(x < z \wedge z < y \to Q(z))) \tag{19.2}$$

— this is not in the guarded fragment because $\forall z(x < z \wedge z < y \to Q(z))$ is not. However, the $\forall z$ is clearly guarded to some extent in (19.2): z doesn't occur with x, y in a single atomic formula, but each pair of variables from x, y, z do (x and y become guarded in this way higher up the formula, by $x < y$). So van Benthem [Ben97] proposed the *loosely guarded fragment*, which he also calls the *pairwise guarded fragment*.

Loosely guarded fragment Roughly speaking, in the loosely guarded fragment, quantified variables must be pairwise guarded by atomic formulas. For example, if $\varphi(x, y, z)$ is a formula of the loosely guarded fragment then so is $\exists yz(R(x, y) \wedge R(y, z) \wedge S(x, z) \wedge \varphi(x, y, z))$, where R, S are binary relation symbols. (See definition 19.1 for details.) We are not aware of any explicit algebraic origin of this idea, but it is clearly reminiscent of the definition of composition in relativised representations of weakly associative algebras, the Jónsson Q-operators (see section 8.2.4), and the n-square and n-flat relativised representations of chapter 13: we introduce it here precisely because it is able to express the quantification patterns in these examples. The loosely guarded fragment does contain (19.2), and of course (19.1). It is much more powerful than the guarded fragment, but still has many nice properties, such as decidability and reasonable complexity [AndBen+98, Grä99b, Mar97, Ben97] — identical complexity results to those already cited for the guarded fragment hold for the loosely guarded fragment. The same goes for the 'packed fragment' (definition 19.2), in which the guards themselves may involve existential quantifiers. This fragment was defined by Marx [Mar01b] in order to characterise the loosely guarded fragment in terms of back-and-forth systems of partial isomorphisms defined on packed sets. Grädel [Grä99a] defined the 'clique-guarded fragment', a syntactic variant of the packed fragment, and proved similar results. See, e.g., [HodOtt01] for connections between these fragments.

Why the loosely guarded fragment? The loosely guarded fragment also has the finite model property (as does the packed fragment). We can use this to prove many finite base property results, essentially by translating the algebraic problem into modal logic and then, by the standard translation, into the loosely guarded fragment. This two-step translation (but done in a single step) of algebra into classical first-order logic has been central to the book. We saw it explicitly in definition 2.75 when we were axiomatising the atom structures of a variety, but there are many other examples: to mention just two, the theories $T_{\mathcal{A}}$ and $R_{\mathcal{A}}$ defining (relativised) representations of the relation algebra $\mathcal{A}$ (definitions 3.30 and 5.1) arise by viewing algebraic terms as first-order formulas. Here, we take advantage for one last time of the model-theoretic insights that it offers: $R_\mathcal{A}$ and related theories are loosely guarded, so the finite model property of the loosely guarded fragment directly yields the finite base property results that we are looking for.

Outline of chapter

In section 19.2 we formally define the loosely guarded and packed fragments, which have the finite model property. Section 19.3 will define the various finite base properties more precisely, and explain the connections between them. We then prove our finite base property results: section 19.4 covers **WA** and (in the exercises) several other classes of algebras with relativised representations, section 19.5 covers $\mathbf{RA}_n$, and section 19.6, $\mathbf{S\mathfrak{R}aCA}_n$.

19.2 Guarded fragments

Here we define the loosely guarded and packed fragments of first-order logic, and state the finite model property for them. We also briefly discuss the clique-guarded fragment.

19.2.1 Loosely guarded fragment

We begin by recalling from [Ben97, p. 9] the definition of the loosely guarded fragment.

DEFINITION 19.1 Let L be a signature without function symbols. The *loosely guarded fragment* LGF(L) over L consists of (just) the following L-formulas:

- Any atomic L-formula is in LGF(L).

- LGF(L) is closed under boolean combinations.

- If

    - γ (the 'guard') is a conjunction of atomic L-formulas,[2]
    - $\varphi \in \mathsf{LGF}(L)$,
    - every free variable of φ is free in γ,
    - $\bar{y}$ is a tuple of free variables of γ,

---
[2] The original definition [Ben97] did not allow equalities in guards, but their presence does not affect the expressive power or the finite model property so we include them.

19.2. Guarded fragments

- if x is a free variable of γ, y is a variable from $\bar{y}$, and $x \neq y$,[3] then there is a conjunct of γ in which x, y both occur,

then $\exists \bar{y}(\gamma \wedge \varphi) \in \mathsf{LGF}(L)$.

Note that γ may have more free variables than φ, and more than two variables may be 'guarded' by a single conjunct of γ. The reader may check that the standard translation of $U(p,q)$ (formula (19.2) of section 19.1) is loosely guarded.

19.2.2 Packed fragment

We will also state the finite model property for the 'packed fragment', which was defined in [Mar01b].

DEFINITION 19.2 Let L be a signature without function symbols. An L-formula γ is said to be a *packing guard* if γ is a conjunction of atomic and existentially quantified atomic formulas such that for any distinct free variables x, y of γ, there is a conjunct of γ in which x, y both occur free.

The *packed fragment* $\mathsf{PF}(L)$ consists of (just) the following formulas:

- Any atomic L-formula is in $\mathsf{PF}(L)$.

- $\mathsf{PF}(L)$ is closed under boolean combinations.

- If the L-formula γ is a packing guard, $\varphi \in \mathsf{PF}(L)$, every free variable of φ is free in γ, and $\bar{y}$ is a tuple of variables, then $\exists \bar{y}(\gamma \wedge \varphi) \in \mathsf{PF}(L)$.[4]

$\mathsf{LGF}(L)$ is not a subfragment of $\mathsf{PF}(L)$ because guards γ of $\mathsf{PF}(L)$ must bind *every* pair of free variables of φ. For example, the standard translation of $U(p,q)$ (formula (19.2) of section 19.1) is loosely guarded but not packed. However, every $\mathsf{LGF}(L)$-sentence is equivalent to a sentence of $\mathsf{PF}(L)$, and conversely for signatures with only binary relation symbols. An example, due to Marx, of a packed sentence which is not equivalent to a loosely guarded sentence is

$$\exists xyz(\exists w R(x,y,w) \wedge \exists w R(x,z,w) \wedge \exists w R(z,y,w) \wedge \neg R(x,y,z)).$$

---

[3] [Grä99b, definition 2.2] omits the restriction $x \neq y$; this does not reduce the expressive power if equalities are allowed as conjuncts of γ.

[4] [Mar01b] requires that γ and φ have exactly the same free variables. We do not need this restriction, but adding it does not reduce the expressive power since we may add conjuncts $x = x$ to φ until its free variables are the same as γ's. Also, [Mar01b] does not allow existentially quantified equalities in guards; it is plain that this does not change the expressive power either.

19.2.3 Clique-guarded fragment

We discuss this (briefly) because of its connection to the packed fragment and to clique-relativised semantics. The clique-guarded fragment was introduced by Grädel in [Grä99a] and can be viewed[5] as an alternative formulation of the packed fragment. We will not define it formally here, but it can be obtained by 'moving existential quantifiers of packing guards outside'. For example, if φ is quantifier-free, the formula $\exists xy (\exists z R(x,y,z) \land \varphi(x,y))$ of the packed fragment is equivalent to the formula $\exists xyz (R(x,y,z) \land \varphi(x,y))$ of the clique-guarded fragment. Similarly, $\forall xy (\exists z R(x,y,z) \to \varphi(x,y))$ is equivalent to $\forall xyz (R(x,y,z) \to \varphi(x,y))$.

There is also a semantic version of the clique-guarded fragment — what we call the 'clique-guarded semantics'.

DEFINITION 19.3 Let L be a relational signature and M an L-structure.

1. The *Gaifman graph* $\mathfrak{G}(M)$ of M is the undirected graph $(\mathrm{dom}(M), E)$, where for distinct $a, b \in M$, $E(a,b)$ holds iff there are n-ary $R \in L$ and $a_1, \ldots, a_n \in M$ with $M \models R(a_1, \ldots, a_n)$ and $a, b \in \{a_1, \ldots, a_n\}$.

2. As usual, a *clique* in $\mathfrak{G}(M)$ is a set $C \subseteq M$ such that $E(a,b)$ for all distinct $a, b \in C$.

3. The 'clique-guarded semantics' $M \models_G \varphi(\bar{a})$, where $\varphi(\bar{x})$ is an L-formula, $\bar{a} \in M$, and $\mathrm{rng}(\bar{a})$ is a clique in $\mathfrak{G}(M)$, are defined as follows.

   - For atomic φ, we let $M \models_G \varphi(\bar{a})$ iff $M \models \varphi(\bar{a})$.
   - The semantics of the boolean connectives are defined in the usual way.
   - $M \models_G \exists x \varphi(x, \bar{a})$ iff $M \models_G \varphi(b, \bar{a})$ for some $b \in M$ such that $\mathrm{rng}(b\bar{a})$ is a clique in $\mathfrak{G}(M)$.

The connection of the clique-guarded semantics to the packed fragment is given by the following proposition.

PROPOSITION 19.4 *Let L be a finite relational signature.*

1. *For any L-formula $\varphi(\bar{x})$, there is a formula $\varphi^p(\bar{x})$ of $\mathrm{PF}(L)$ such that for any L-structure M and any $\bar{a} \in M$ where $\mathrm{rng}(\bar{a})$ is a clique in $\mathfrak{G}(M)$, we have $M \models_G \varphi(\bar{a})$ iff $M \models \varphi^p(\bar{a})$.*

2. *Conversely, if $\varphi(\bar{x})$ is a formula of $\mathrm{PF}(L)$, then for any L-structure M and any $\bar{a} \in M$ such that $\mathrm{rng}(\bar{a})$ is a clique in $\mathfrak{G}(M)$, we have $M \models \varphi(\bar{a})$ iff $M \models_G \varphi(\bar{a})$.*

---
[5]This was observed by C. Hirsch and S. Tobies (private communication, 2000).

19.2. Guarded fragments

Proof (sketch). For the first part, φ^p is defined by induction on φ. The atomic and boolean cases are straightforward. Consider the case $\varphi = \exists y \psi(\bar{x}, y)$, and assume that ψ^p has been defined. Because L is finite, the reflexive closure of the edge relation in the Gaifman graph $\mathfrak{G}(M)$ is definable in M by a finite disjunction of existentially quantified atomic formulas. So we can find a formula expressing that $\bar{x}y$ forms a clique in $\mathfrak{G}(M)$. Such a formula can be written as a disjunction $\bigvee_i \gamma_i$ of packing guards $\gamma_i(\bar{x}, y)$. The quantifier $\exists y$ distributes over this disjunction, so relativising to $\bigvee_i \gamma_i$ yields the formula $\varphi^p(\bar{x}) = \bigvee_i \exists y (\gamma_i \wedge \psi^p) \in \mathsf{PF}(L)$. Then φ^p is as required. The second part is easy, since for $\varphi \in \mathsf{PF}(L)$, φ^p is logically equivalent to φ. We leave the details to exercise 2 below. □

For a relativised representation M of a relation algebra, the clique-guarded semantics is identical to the clique-relativised semantics $\models_C$ of definition 13.3. The edge relation of the Gaifman graph $\mathfrak{G}(M)$ is just 1^M, and a clique in $\mathfrak{G}(M)$ is just a clique in the sense of definition 13.2. So proposition 19.4 can be proved without the finite-signature restriction:

PROPOSITION 19.5 *Let M be a relativised representation of a relation algebra or non-associative algebra $\mathcal{A}$, let $3 \leq n < \omega$, and let $\varphi(x_0, \ldots, x_{n-1})$ be an $\mathcal{L}(\mathcal{A})^n$-formula.*

1. *For any $\bar{a} \in C^n(M)$ (as in definition 13.2), we have $M \models_G \varphi(\bar{a})$ iff $M \models_C \varphi(\bar{a})$.*

2. *We can construct a formula φ^p of $\mathsf{PF}(L)$ such that $M \models_C \varphi(\bar{a})$ iff $M \models \varphi^p(\bar{a})$ for any $\bar{a} \in C^n(M)$.*

19.2.4 Finite model property

We will use the following results:

THEOREM 19.6 *For any relational signature L, the packed fragment $\mathsf{PF}(L)$ has the finite model property.*

COROLLARY 19.7 *For any relational signature L, the loosely guarded fragment $\mathsf{LGF}(L)$ has the finite model property.*

Remarks on the proofs Corollary 19.7 follows quite easily from theorem 19.6. Theorem 19.6 is proved using techniques of Herwig. Herwig's theorems [Her95, Her98] give a way of extending a finite structure to a larger one, still finite and inheriting some properties of the original structure, in such a way that all partial isomorphisms of the smaller structure extend to automorphisms of the larger one. Earlier results in this direction include [Tru92], and in particular, those of Hrushovski

[Hru92], on which the techniques used by Herwig are based. The method uses permutation groups; a combinatorial alternative is given in [HerLas00], which also established striking equivalences between partial isomorphism extension results and known theorems in free groups, and proved a very strong extension theorem. An equivalent version was proved in [Alm99, AlmDel99], using free groups.

All this work was motivated by pure model-theoretic (or group-theoretic) considerations. Using Herwig's theorems to prove the finite base property originated in joint work with several people [HirHod+98, AndHod+99], and in [Grä99b] the finite model property for the guarded fragment was proved this way. For applications in other areas, see, e.g., [Gro96]. Theorem 19.6 and corollary 19.7 were proved in [Hodk02], using a modification of [Her98]. A more recent combinatorial construction of [HodOtt01] derived theorem 19.6 much more simply and directly from the finite model property for the guarded fragment [Grä99b], and also from Herwig's earlier result of [Her95].

REMARK 19.8 It is not hard to see that we can add constants to L and keep the finite model property for $\mathsf{PF}(L)$ and $\mathsf{LGF}(L)$. (Note that guards need not guard constants, so, e.g., $\forall x(x = x \to R(c, x))$ is a loosely guarded sentence.) However, the loosely guarded fragment in signatures with function symbols is undecidable (this follows from results in [MikMar99]) and therefore does not have the finite model property.

Exercises

1. Prove that the loosely guarded fragment and the packed fragment are decidable: there is an algorithm that decides whether or not a given sentence of these fragments has a model. (See [Grä99b, Mar97] for a better algorithm!)

2. Check the details of the proof of proposition 19.4.

3. Show that every $\mathsf{LGF}(L)$-sentence is equivalent to a $\mathsf{PF}(L)$-sentence. Use this to deduce corollary 19.7 from theorem 19.6.

4. Show that if L is a signature consisting of only unary and binary relation symbols, every $\mathsf{PF}(L)$-sentence is equivalent to a $\mathsf{LGF}(L)$-sentence.

5. Show that adding constants to L keeps the finite model property for $\mathsf{PF}(L)$ and $\mathsf{LGF}(L)$.

19.3 The finite base property

We begin our work on the finite base property proper by defining somewhat more formally the various forms it takes. We thank Hajnal Andréka and Istvan Németi for help in preparing parts of this section.

19.3. The finite base property

DEFINITION 19.9 Let K be a class of algebras of (functional) signature L.

1. K is said to have the *finite algebra property* if any universal first-order L-sentence that is true in all finite algebras in K is true in every algebra in K.

    In other words, the universal first-order theory of K is the same as the universal first-order theory of the class of all finite members of K.

Assume now that K is a class of representable algebras (for some notion of representation of the algebras as set algebras: see below).

2. K is said to have the *finite base property* if any universal first-order L-sentence that is true in all algebras in K having a representation over a finite base set is valid in K.

3. K is said to have the *finite algebra on finite base property* if any finite algebra in K has a representation with finite base.

4. We say that K has the finite algebra (or base) property *for equations* if the relevant property holds for K when we replace 'universal first-order sentence' by 'equation' above. The finite algebra/base property for other kinds of sentence is defined similarly.

This definition is deliberately informal to the extent that the notion of representation in parts 2–4 is not specified. All specific representations of algebras seen in this book are interpretations of the elements of the algebra as relations on some base set, and the usual meaning of the above definition is with respect to these kinds of representation. More generally, K may be a subclass of a pseudo-elementary or pseudo-universal class in the sense of definition 9.1: $K \subseteq \{\mathcal{A} : \mathcal{A} = M^{\mathbf{a}}\!\restriction_L \text{ for some } M \models U\}$ for some two-sorted defining theory U with sorts $\mathbf{a}, \mathbf{r}$. In that case, a representation of $\mathcal{A} \in K$ is a two-sorted structure M with $M^{\mathbf{a}}\!\restriction_L = \mathcal{A}$. But the important part of the representation is the 'new' part outside $\mathcal{A}$, so we define the base set of the representation to be the domain of $M^{\mathbf{r}}$. However, definition 19.9 only needs that any representation has a base, so it may apply to much more general notions of representation than these — this is why the precise definition of representation is left open.

Some of the results below will need mild restrictions on the notion of representation, such as that any subalgebra of a representable algebra is representable on the same base (this is true for 'pseudo-universal representations'), and that no infinite algebra has a representation with finite base.

Crudely, the following lemma shows that 'finite base property = finite algebra property + finite algebra on finite base property'.

LEMMA 19.10 [ANDHOD+99] *Let L, K be as above.*

1. *If K has the finite algebra (respectively, finite base) property, then it has the finite algebra (respectively, finite base) property for equations.*

2. *If K has both the finite algebra property and the finite algebra on finite base property, then it has the finite base property. The same holds for equations.*

3. *Suppose that the notion of representation is such that any representation of an infinite algebra has infinite base. If K has the finite base property, then it has the finite algebra property. The same holds for equations.*

4. *Suppose that L is finite, and that the notion of representation is such that any subalgebra of a representable algebra in K is representable and on the same base. If K has the finite base property, then it has the finite algebra on finite base property.*

Proof.

1. — is trivial, since any equation is a universal sentence.

2. Let σ be a universal sentence that is true in any algebra in K that has a representation with finite base. Then by the finite algebra on finite base property, σ is true in every *finite* algebra in K. So by the finite algebra property, σ is true in *every* algebra in K. The proof for equations is the same.

3. If σ is any sentence true in every finite algebra in K, then σ is true in every algebra in K that has a representation with finite base, since such an algebra must by assumption be finite. So the finite base property implies the finite algebra property, for universal sentences and for equations.

4. To derive the finite algebra on finite base property from the finite base property, let $\mathcal{A} \in K$ be finite, with domain $\{a_0, \ldots, a_{n-1}\}$, say. Let $\delta(x_0, \ldots, x_{n-1})$ be a quantifier-free formula as in corollary 18.27, describing the diagram of $\mathcal{A}$. So δ is the conjunction of all L-formulas of the form $x_{i_0} = f(x_{i_1}, \ldots, x_{i_r})$, where $f \in L$ is an r-ary function symbol, $\{i_0, \ldots, i_r\} \subseteq n$, and $\mathcal{A} \models a_{i_0} = f(a_{i_1}, \ldots, a_{i_r})$, together with $\bigwedge_{i<j<n} x_i \neq x_j$. (We use that L is finite here, to obtain δ; L is functional because we are dealing with algebras.) Then $\forall \bar{x} \neg \delta(\bar{x})$ is not valid in K (it fails in $\mathcal{A}$). So by the finite base property, it must be false in some algebra $\mathcal{B} \in K$ that has a representation with finite base. So $\mathcal{B} \models \exists \bar{x} \delta(\bar{x})$. If $b_0, \ldots, b_{n-1} \in \mathcal{B}$ are such that $\mathcal{B} \models \delta(b_0, \ldots, b_{n-1})$, then $(a_i \mapsto b_i)$ is an embedding of $\mathcal{A}$ into $\mathcal{B}$. By the assumption on the notion of representation, $\mathcal{A}$ must have a representation with finite base, too. □

19.3. The finite base property

We list some of the known facts about the finite base property and its variants.[6]

FACT 19.11

1. For $n < \omega$, **Crs**$_n$, **D**$_n$, and **G**$_n$ (see section 5.5) have the finite base property. This was first proved by Andréka and Németi (unpublished manuscript); a short proof was published in [AndHod+99]. See exercise 19.4(3). Németi proved in [Ném96, theorem 4.1] that these classes have the finite algebra property.

2. **RA** and **SA** do not have the finite algebra property for equations. **RRA** does not have the finite algebra on finite base property for classical representations. The proofs for these are adaptations of ones for the cylindric cases (see (4) below).

3. It was proved in [Ném87a] that **WA** has the finite algebra property. We will prove in theorem 19.15 below that **WA** has the finite base property (a result of [AndHod+99], where it was shown that certain counting quantifiers can also be included).

4. A *Cartesian space* is a set of the form ^nU for some set U. Recall from definition 5.16 that $\mathbf{Cs}_n = \{\mathcal{A} \in \mathbf{Crs}_n : 1^{\mathcal{A}} \text{ is a Cartesian space}\}$. Then $\mathbf{Cs}_2, \mathbf{RCA}_2$ have the finite base property by [HenMon+85, 4.2.8, 3.2.66], while for $n \geq 3$, $\mathbf{Cs}_n$ has neither the finite algebra property with respect to equations, nor the finite algebra on finite base property. The same holds for $\mathbf{RCA}_n$: cf. [Ném84] and [Mon91, status of Problem 4.15]. Of course, the notion of representation for $\mathbf{RCA}_n$ here is the classical one.

5. **Crs**$_n$ for $n \geq 2$ does not have the finite algebra property with respect to existential sentences. This follows from work of Mikulás [Mik96a], in which an equation ε with free variable x is given such that $\mathcal{A} \models \varepsilon$ for some $\mathcal{A} \in \mathbf{Crs}_n$ while for all $\mathcal{A} \in \mathbf{Crs}_n$, $\mathcal{A} \models \varepsilon$ implies that $\mathcal{A}$ is infinite. That is, the existential sentence $\exists x \neg \varepsilon$ fails in **Crs**$_n$ and it can fail only in infinite algebras in **Crs**$_n$. Roughly, the idea in [Mik96a] is to express with $\forall x \varepsilon$ that the unit 'contains' an injective but non-surjective function on (part of) the base.

The link to decidability is given in the following proposition; as one would expect, it mostly involves the more syntactic finite base properties. It is not really a formal result, as the precise definition of 'representation' in parts 3 and 4 is again left open, for the same reasons as before. However, it certainly covers the kinds of representation mentioned above.

---
[6] Some of this information appeared in [AndHod+99]; we thank the copyright owners Association for Symbolic Logic for granting permission for this republication.

PROPOSITION 19.12 *Let* K *be a class of algebras of finite signature L. Assume any of the following:*

1. *The isomorphism types of finite algebras in* K *are recursively enumerable, and* K *is recursively axiomatisable and has the finite algebra property.*

2. K *is finitely axiomatisable and has the finite algebra property.*

3. *All algebras in* K *are representable, the isomorphism types of algebras in* K *that are representable over a finite set are recursively enumerable, and* K *is recursively axiomatisable and has the finite base property.*

4. K *is pseudo-universal with finite defining theory, and has the finite base property.*

5. K *has the finite base property in the strong sense that there is a total recursive function f from universal L-sentences to natural numbers such that if* σ *is a universal sentence that is not valid in* K*, there is an algebra in* K *witnessing non-validity of* σ *with a representation with base of size at most* $f(\sigma)$. *Also, the isomorphism types of algebras in* K *that are representable over a finite set of a given size n are recursively enumerable, uniformly in n.*

Then the universal theory of K *is decidable.*

Proof.

1. This is analogous to the well known decidability argument in modal logic, using the 'finite model property'.

   Clearly, we can recursively enumerate the validities of K. We can also recursively enumerate all finite algebras in K and all universal *L*-sentences. Hence, we can evaluate the universal sentences in the finite algebras, in ω stages: in stage *n*, we evaluate the first *n* sentences in the first *n* algebras in the given enumeration of finite algebras in K. All sentences found not to be true in some finite algebra are 'printed out'. By the finite algebra property, this recursively enumerates all of the non-K-valid universal sentences. Since any universal *L*-sentence will appear in precisely one of these enumerations, this provides an effective decision procedure for the universal theory of K.

2. — a special case of (1).

3. Similar to (1).

4. By theorem 9.28, K is recursively axiomatisable, and we can recursively enumerate the isomorphism types of algebras in K that are representable over a finite set by enumerating all finite models of the finite defining theory of K. Decidability now follows from (3).

19.4. Finite base property for **WA**

5. Given σ, enumerate all algebras in K with a representation having base of size $\leq f(\sigma)$. If one is found invalidating σ, declare σ not to be valid in K. Otherwise, declare it valid. This gives the right answer, in finite time.

$\square$

Not many classes of relation-type algebras can be dealt with in this way, for the good reason that even the equational theory of **SA** and its usual subclasses is already undecidable. However, we can deal with **WA**, and the classes **Crs**, **D**, **G** of cylindric-type algebras. We will also establish the finite algebra on finite base property for **RA**$_n$ and a subclass of **S$\Re\mathfrak{a}$CA**$_n$ in terms of the representation theory discussed in chapter 13, even though the equational theories of these classes are undecidable. Further results are contained in the exercises.

Exercises

1. Show that the variety of boolean algebras has the finite base property.

2. Let K be a class of algebras of finite signature L. Assume that the isomorphism types of finite algebras in K are recursively enumerable, and that K has the finite base property in the strong sense of proposition 19.12(5). Show that the universal theory of K is decidable.

19.4 Finite base property for WA

We begin our application of the finite model property for the loosely guarded fragment (corollary 19.7) by proving that **WA** has the finite base property. We also prove the finite algebra on finite base property: any finite weakly associative algebra has a finite relativised representation. Indeed, this follows from the finite base property by proposition 19.12. The same proposition shows that the universal theory of **WA** is decidable.

The finite base property for **WA** was first proved in joint work with Andréka and Németi [AndHod+99]. Marx has shown that the problem of deciding validity of an equation in **WA** is EXPTIME-complete [Mar97].

Let $\mathcal{A}$ be a non-associative algebra. We regard the elements of $\mathcal{A}$ as binary relation symbols of a new signature, $L(\mathcal{A})$. Recall from definition 5.1 that $R_\mathcal{A}$ is the first-order $\mathcal{L}(\mathcal{A})$-theory consisting of the following axioms:

$$\forall xy[1'^{\mathcal{A}}(x,y) \leftrightarrow (x=y)]$$
$$\forall xy[r(x,y) \leftrightarrow s(x,y) \vee t(x,y)] \quad \text{for all } r,s,t \in \mathcal{A} \text{ with } r = s+t$$
$$\forall xy[1^{\mathcal{A}}(x,y) \rightarrow (r(x,y) \leftrightarrow \neg s(x,y))] \quad \text{for all } r,s \in \mathcal{A} \text{ with } r = -s$$
$$\forall xy[r(x,y) \leftrightarrow s(y,x)] \quad \text{for all } r,s \in \mathcal{A} \text{ with } r = \breve{s}$$
$$\forall xy[1^{\mathcal{A}}(x,y) \rightarrow (r(x,y) \leftrightarrow \exists z(s(x,z) \wedge t(z,y)))] \quad \text{for all } r,s,t \in \mathcal{A} \text{ with } r = s;t$$
$$\exists xy\, r(x,y) \quad \text{for all } r \in \mathcal{A} \text{ with } r \neq 0.$$

As before, we will generally drop the superfixes $\mathcal{A}$, but sometimes we have to include them for clarity. A *relativised representation* of $\mathcal{A}$ is a model of $R_{\mathcal{A}}$.

It follows from the axioms that in any relativised representation of $\mathcal{A}$, the unit 1 is interpreted as a reflexive relation (since 1' is, and $1' \leq 1$ in $\mathcal{A}$). Also, because we did not relativise the 'converse' axioms to the unit, and $\breve{1} = 1$, the interpretation of the unit must also be symmetric. Thus, for us, *relativised representations automatically have reflexive and symmetric unit.* See the discussion in section 5.1.

By theorem 7.5, due to Maddux, any weakly associative algebra has a relativised representation. So for any $\mathcal{A} \in \mathbf{WA}$, there is a model $M \models R_{\mathcal{A}}$. However, this relativised representation is built in a step-by-step fashion. The construction does not of itself yield a finite representation even if $\mathcal{A}$ is finite. But this is now easily remedied:

THEOREM 19.13 *Any finite weakly associative algebra has a finite relativised representation. That is,* **WA** *has the finite algebra on finite base property.*

Proof. We observe that each axiom in $R_{\mathcal{A}}$ is logically equivalent to a sentence of the loosely guarded fragment:

- $\forall xy[1'(x,y) \leftrightarrow (x=y)]$ is equivalent to
  $\forall xy[1'(x,y) \rightarrow x = y] \wedge \forall x[x = x \rightarrow 1'(x,x)]$,

- $\forall xy[[r+s](x,y) \leftrightarrow r(x,y) \vee s(x,y)]$ is equivalent to the conjunction of
  - $\forall xy[r(x,y) \rightarrow [r+s](x,y)]$,
  - $\forall xy[s(x,y) \rightarrow [r+s](x,y)]$,
  - $\forall xy[[r+s](x,y) \rightarrow r(x,y) \vee s(x,y)]$,

- $\forall xy[1(x,y) \rightarrow (r(x,y) \leftrightarrow \neg[-r](x,y))]$ is already in the loosely guarded fragment,

- $\forall xy[r(x,y) \leftrightarrow \breve{r}(y,x)]$ is equivalent to
  $\forall xy[r(x,y) \rightarrow \breve{r}(y,x)] \wedge \forall xy[\breve{r}(x,y) \rightarrow r(y,x)]$,

- $\forall xy[1(x,y) \rightarrow ([r;s](x,y) \leftrightarrow \exists z(r(x,z) \wedge s(z,y)))]$ is equivalent to the conjunction of

19.4. Finite base property for WA

- $\forall xy[1(x,y) \wedge [r;s](x,y) \to \exists z(r(x,z) \wedge s(z,y) \wedge \top)]$,
- $\forall xyz[1(x,y) \wedge r(x,z) \wedge s(z,y) \to [r;s](x,y)]$,

- $\exists xy\, r(x,y)$ is equivalent to $\exists xy(r(x,y) \wedge \top)$.

Hence, if $\mathcal{A}$ is any finite weakly associative algebra, $\bigwedge R_{\mathcal{A}}$ is logically equivalent to a sentence σ of the loosely guarded fragment. By theorem 7.5, σ has a model, so by corollary 19.7, it has a finite model. This model is a finite relativised representation of $\mathcal{A}$. □

We can also prove that **WA**— has the stronger 'finite base property'. We need a preliminary lemma, related to the 'standard translation' seen in section 19.1.

LEMMA 19.14 *Let $\mathcal{A} \in$ **WA**, let $\psi(\bar{x})$ be a quantifier-free formula of the signature of **WA**, and let $\bar{a} \in \mathcal{A}$ with $|\bar{a}| = |\bar{x}|$. Then there is a loosely guarded $L(\mathcal{A})$-sentence $\tau_{\mathcal{A}}(\psi(\bar{a}))$, whose relation symbols are among $0^{\mathcal{A}}, 1^{\mathcal{A}}, 1'^{\mathcal{A}}$ and the relation symbols in $\bar{a}$, such that for any relativised representation $M \models R_{\mathcal{A}}$, we have $\mathcal{A} \models \psi(\bar{a})$ iff $M \models \tau_{\mathcal{A}}(\psi(\bar{a}))$.*

Proof. Let $\bar{a} \in \mathcal{A} \in$ **WA** be given. First let $t(\bar{x})$ be a **WA**-term. For any distinct variables u, v, we translate $t(\bar{a})$ into a loosely guarded formula $\tau_{\mathcal{A}}^{uv}(t(\bar{a}))$ of $L(\mathcal{A})$, with free variables u, v. The translation is defined by induction on t.

- If t is a variable, then $t(\bar{a})$ is a for some $a \in \mathrm{rng}(\bar{a})$, and we let $\tau_{\mathcal{A}}^{uv}(t(\bar{a}))$ be $a(u, v)$.

- If t is a constant, say 1', we let $\tau_{\mathcal{A}}^{uv}(t)$ be $1'^{\mathcal{A}}(u, v)$. 0 and 1 are handled similarly.

Assuming inductively that $t(\bar{a}), t'(\bar{a})$ are translated, we translate terms made from them as follows. We suppress the $\bar{a}$ as it plays no role here.

- $\tau_{\mathcal{A}}^{uv}(-t) = 1(u, v) \wedge \neg \tau_{\mathcal{A}}^{uv}(t)$,
- $\tau_{\mathcal{A}}^{uv}(t + t') = \tau_{\mathcal{A}}^{uv}(t) \vee \tau_{\mathcal{A}}^{uv}(t')$,
- $\tau_{\mathcal{A}}^{uv}(\breve{t}) = \tau_{\mathcal{A}}^{vu}(t)$,
- $\tau_{\mathcal{A}}^{uv}(t ; t') = 1(u, v) \wedge \exists w(1(u, w) \wedge 1(w, v) \wedge \tau_{\mathcal{A}}^{uw}(t) \wedge \tau_{\mathcal{A}}^{wv}(t'))$, where w is a new variable.

Let $M \models R_{\mathcal{A}}$ be a relativised representation of $\mathcal{A}$. It is easily seen that for any $t(\bar{a}), t'(\bar{a})$, we have

$$\mathcal{A} \models t(\bar{a}) = t'(\bar{a}) \iff M \models \forall uv\bigl(\tau_{\mathcal{A}}^{uv}(t(\bar{a})) \leftrightarrow \tau_{\mathcal{A}}^{uv}(t'(\bar{a}))\bigr).$$

Therefore, we define the translation $\tau_{\mathcal{A}}$ on quantifier-free formulas $\psi(\bar{a})$ by stipulating that

- for terms $t(\bar{x}), t'(\bar{x})$, and $\bar{a} \in \mathcal{A}$, we choose arbitrary distinct variables u, v and define

$$\tau_{\mathcal{A}}(t(\bar{a}) = t'(\bar{a})) \stackrel{\text{def}}{=} \forall uv \big(1(u,v) \to \big(\tau_{\mathcal{A}}^{uv}(t(\bar{a})) \leftrightarrow \tau_{\mathcal{A}}^{uv}(t'(\bar{a}))\big)\big),$$

and if $\tau_{\mathcal{A}}(\psi), \tau_{\mathcal{A}}(\psi')$ are defined, then

- $\tau_{\mathcal{A}}(\psi \wedge \psi') \stackrel{\text{def}}{=} \tau_{\mathcal{A}}(\psi) \wedge \tau_{\mathcal{A}}(\psi')$,

- $\tau_{\mathcal{A}}(\neg \psi) \stackrel{\text{def}}{=} \neg \tau_{\mathcal{A}}(\psi)$.

Clearly, for any quantifier-free $\psi(\bar{x})$ and $\bar{a} \in \mathcal{A}$, $\tau_{\mathcal{A}}(\psi(\bar{a}))$ is a loosely guarded $L(\mathcal{A})$-sentence, and $\mathcal{A} \models \psi(\bar{a})$ iff $M \models \tau_{\mathcal{A}}(\psi(\bar{a}))$. □

THEOREM 19.15 (Andréka et. al., [AndHod$^+$99]) *Let $\forall \bar{x} \psi$ be a universal first-order sentence of the language of weakly associative algebras, where $\psi(\bar{x})$ is quantifier-free, and suppose that $\forall \bar{x} \psi$ is valid in every weakly associative algebra $\mathcal{A}$ that has a finite relativised representation (i.e., $R_{\mathcal{A}}$ has a finite model). Then $\forall \bar{x} \psi$ is valid in* **WA**.
In short, **WA** *has the finite base property.*

Proof. Let $\forall \bar{x} \psi(\bar{x})$ be a prenex universal sentence of the signature $\{+, -, 0, 1, 1', \smile, ;\}$ of **WA**, where ψ is quantifier-free. Assume that $\mathcal{A} \models \neg \forall \bar{x} \psi(\bar{x})$ for some $\mathcal{A} \in \mathbf{WA}$. Thus, $\mathcal{A} \models \neg \psi(\bar{a})$ for some $\bar{a} \in \mathcal{A}$. We may assume without loss of generality that $0^{\mathcal{A}}, 1^{\mathcal{A}}, 1'^{\mathcal{A}} \in \text{rng}(\bar{a})$. Let $M \models R_{\mathcal{A}}$ be a relativised representation of $\mathcal{A}$ (see theorem 7.5 for existence). By lemma 19.14, $M \models \tau_{\mathcal{A}}(\neg \psi(\bar{a}))$, and all relation symbols in $\tau_{\mathcal{A}}(\neg \psi(\bar{a}))$ are from $\bar{a}$.

Let ρ be the conjunction of

- $\forall u (u = u \to 1^{\mathcal{A}}(u,u))$ and $\forall uv (1^{\mathcal{A}}(u,v) \to 1^{\mathcal{A}}(v,u))$,

- $\forall uv (a(u,v) \to 1^{\mathcal{A}}(u,v))$ for every relation symbol a in $\bar{a}$,

- $\forall uv (1'^{\mathcal{A}}(u,v) \leftrightarrow u = v)$ and $\neg \exists uv 0^{\mathcal{A}}(u,v)$.

Then ρ is logically equivalent to a loosely guarded $L(\mathcal{A})$-sentence, and $M \models \rho$.

Now $\rho \wedge \tau_{\mathcal{A}}(\neg \psi(\bar{a}))$ is equivalent to a loosely guarded $L(\mathcal{A})$-sentence, and $M \models \rho \wedge \tau_{\mathcal{A}}(\neg \psi(\bar{a}))$. By corollary 19.7, there is a finite model $N \models \rho \wedge \tau_{\mathcal{A}}(\neg \psi(\bar{a}))$.

Define the relation-type algebra

$$\mathcal{B} = (\wp((1^{\mathcal{A}})^N), \cup, \setminus, \emptyset, (1^{\mathcal{A}})^N, \text{Id}_N, -^{-1}, |),$$

where $\text{Id}_N = \{(x,x) : x \in N\}$, the unit $1^{\mathcal{B}}$ of $\mathcal{B}$ is $(1^{\mathcal{A}})^N$, and the interpretations $\setminus, |$ of the function symbols $-, ;$ in $\mathcal{B}$ are relativised to $(1^{\mathcal{A}})^N$. Hence, $\mathcal{B}$ is closed under these operations. Now the interpretations $\text{Id}_N, -^{-1}$ of $1', \smile$ are not so relativised; but

19.4. Finite base property for WA

because $N \models \rho$, the unit $1^{\mathcal{B}} = (1^{\mathcal{A}})^N$ of $\mathcal{B}$ is reflexive and symmetric, so $\mathrm{Id}_N \subseteq 1^{\mathcal{B}}$ and $a^{-1} = \{(x,y) : (y,x) \in a\} \subseteq 1^{\mathcal{B}}$ for all $a \subseteq 1^{\mathcal{B}}$. So $\mathcal{B}$ is closed under all the operations and is indeed a (finite relation-type) algebra.

By regarding each element of $\mathcal{B}$ as a binary relation symbol interpreted as itself, we see that $N \models R_{\mathcal{B}}$, so N is a relativised representation of $\mathcal{B}$. So by proposition 5.5, $\mathcal{B}$ is a weakly associative algebra.

Now by the second conjuncts of ρ, each relation symbol a in $\bar{a}$ is interpreted in N as a subset of $(1^{\mathcal{A}})^N$, and hence as an element of $\mathcal{B}$. We identify a with the corresponding element a^N of $\mathcal{B}$, so $\bar{a}$ becomes a tuple of elements of $\mathcal{B}$. Because all relation symbols in $\tau_{\mathcal{A}}(\neg\psi(\bar{a}))$ are from $\bar{a}$, $\tau_{\mathcal{A}}(\neg\psi(\bar{a}))$ is a sentence of $\mathcal{L}(\mathcal{B})$ under this identification.

We claim that $\tau_{\mathcal{A}}(\neg\psi(\bar{a})) = \tau_{\mathcal{B}}(\neg\psi(\bar{a}))$. The proof is a straightforward induction on the construction of $\neg\psi$; the only delicate case is that of the constant symbols. Consider 1'. Since $N \models \rho$, we have $(1'^{\mathcal{A}})^N = \mathrm{Id}_N$, and hence, under the identification above, $1'^{\mathcal{A}} = 1'^{\mathcal{B}} \in \mathcal{B}$. So, referring to lemma 19.14, we have

$$\tau_{\mathcal{A}}^{uv}(1') = 1'^{\mathcal{A}}(u,v) = 1'^{\mathcal{B}}(u,v) = \tau_{\mathcal{B}}^{uv}(1').$$

The cases of 0, 1 are similar. This completes our proof of the claim.

As $N \models \tau_{\mathcal{A}}(\neg\psi(\bar{a}))$, the claim tells us that $N \models \tau_{\mathcal{B}}(\neg\psi(\bar{a}))$. By lemma 19.14, we obtain $\mathcal{B} \models \neg\psi(\bar{a})$, so that $\mathcal{B} \models \neg\forall\bar{x}\psi$. We have found a finite weakly associative algebra ($\mathcal{B}$) with finite base (N) in which $\forall\bar{x}\psi$ is false, completing the proof. □

We can deduce from this the decidability of the universal (and hence the equational) theory of **WA**:

COROLLARY 19.16 (Németi, [Ném86]) *The universal first-order theory of* **WA** *is decidable.*

Proof. By theorem 19.15 and proposition 19.12, since **WA** is finitely axiomatised. □

We remark here that the equational and universal theories of **WA** and **NA** were proved decidable in [Ném87a].

Exercises

1. Let $X \subseteq \{\text{reflexive, symmetric}\}$. Prove that the class of non-associative algebras having a 'general' relativised representation in which the unit has the properties in X has the finite base property.

2. [Jónsson Q-algebras; [Jón91, p. 248]] Expand the signature of **WA** by an m^2-ary function symbol Q_m, for each $m < \omega$. For any algebra $\mathcal{A}$ of this signature, let $R_{\mathcal{A}}^Q$ be the theory consisting of $R_{\mathcal{A}}$ plus the axioms

$$\forall x_0 x_1 (r(x_0, x_1) \leftrightarrow \exists x_2 \ldots x_{m-1} \bigwedge_{i,j<m} r_{ij}(x_i, x_j)),$$

for each $m < \omega$ and $r, r_{ij} \in \mathcal{A}$ $(i, j < m)$ such that $\mathcal{A} \models r = Q_m(r_{ij} : i, j < m)$. Let Q be the class of all algebras $\mathcal{A}$ such that $R_{\mathcal{A}}^Q$ has a model. Prove that Q has the finite base property.

3. [Andréka–Németi, [AndHod$^+$99]] Prove that $\mathbf{Crs}_n$, $\mathbf{D}_n$, $\mathbf{G}_n$ have the finite base property for any finite n.

4. Polyadic $\mathbf{Crs}_n$ is the class of all subalgebras of algebras of the form

$$\langle \wp(W), \cdot, -, 0, 1, c_i, s_{[i,j]}, s_{[i/j]} \rangle_{i,j<n},$$

where $W \subseteq {}^n U$ for some set U. Here, all operations are relativised to W, and for a map $\sigma : n \to n$ and $r \subseteq W$,

$$s_\sigma(r) = \{(a_0, \ldots, a_{n-1}) \in W : (a_{\sigma(0)}, \ldots, a_{\sigma(n-1)}) \in r\}.$$

We write $[i, j]$ for the transposition map $\sigma : n \to n$ given by $\sigma(i) = j$, $\sigma(j) = i$, and $\sigma(k) = k$ for all $k \in n \setminus \{i, j\}$. We write $[i/j]$ for the substitution map $: n \to n$ given by: $[i/j](i) = j$, and $[i/j](k) = k$ for all $k < n$ with $k \neq i$.

Prove that polyadic $\mathbf{Crs}_n$ has the finite base property (for all finite n).

5. We obtain $\mathbf{Crs}_n^+$ from $\mathbf{Crs}_n$ by adding the operations $s_{\sigma, \Gamma}$ for all $\sigma : n \to n$ and $\Gamma \subseteq n$. These operations combine substitution and cylindrification. Take an algebra $\mathcal{A}$ in $\mathbf{Crs}_n$, let its unit be W, and let $a \in \mathcal{A}$. Then

$$s_{\sigma, \Gamma}(a) = \{\bar{s} \in W : \bar{s} \circ \sigma \equiv_\Gamma \bar{t} \text{ for some } \bar{t} \in a\}.$$

Note that in the above definition, $\bar{s} \circ \sigma$ does not necessarily belong to W, and indeed, $s_{\sigma, \Gamma}$ is not expressible from s_σ and the generalised cylindrification $c_{(\Gamma)}$ for arbitrary W.

Show that for all n, $\mathbf{Crs}_n^+$ has the finite base property.

6. Prove that each of the above classes of algebras has decidable universal theory.

7. Show that **WA** has the finite base property in the 'strong sense' of proposition 19.12(5). Indeed, there is an algorithm that, given a universal sentence that is not valid in **WA**, constructs a finite **WA** with a finite relativised representation in which the sentence is false.

8. Prove that arrow logic in the relativised semantics (see [MarPól+96]) has the finite model property.

PROBLEM 19.17 Let $\mathcal{L}$ be the class of all algebras $(A, ;, \backslash, /, \leq)$, where A is a non-empty set of binary relations on some set U, $W = \bigcup A$ is transitive and $\{x, y : (x, y) \in W\} = U$, and for $r, s \in A$,

- $r ; s = \{(x, y) : \exists z ((x, z) \in r \land (z, y) \in s)\}$,
- $r \backslash s = \{(x, y) \in W : \forall z ((z, x) \in r \Rightarrow (z, y) \in s)\}$,
- $r / s = \{(x, y) \in W : \forall z ((y, z) \in s \Rightarrow (x, z) \in r)\}$,
- $r \leq s$ iff $r \subseteq s$.

The operations $/$, $\backslash$ are called *residuals* (cf. exercise 3.2(11)); U is called the *base* of the algebra.

Is any finite algebra in $\mathcal{L}$ isomorphic to an algebra in $\mathcal{L}$ with finite base? [If so, it would follow by results of Andréka and Mikulás [AndMik94] that the Lambek calculus has the finite model property. The finite model property for the loosely guarded fragment can be used to prove it if the condition that W is transitive is dropped.]

19.5 Finite algebra on finite base property for RA$_n$

Here, we prove that **RA**$_n$ (for finite $n \geq 3$) has the finite algebra on finite base property with respect to the n-square relativised semantics introduced in chapter 13. Note that the finite base property fails here, as the equational theories of these classes are undecidable. The argument of theorem 19.15 does not adapt to prove the finite base property for square (or flat) relativised representations: the algebra $\mathcal{B}$ of that proof is only a weakly associative algebra, and not necessarily in **RA**$_n$ or **SℜaCA**$_n$.

THEOREM 19.18 *Let $n \geq 3$ and $\mathcal{A} \in \mathbf{RA}_n$ be a finite algebra. Then $\mathcal{A}$ has a finite n-square relativised representation. Thus, $\mathbf{RA}_n$ has the finite algebra on finite base property for n-square relativised representations.*

The converse also holds: for this, and further equivalent conditions, see theorem 13.51.

Proof. By proposition 13.17, an n-square relativised representation of a relation-type algebra $\mathcal{A}$ is simply a model of the theory Sq$^n(\mathcal{A})$ of definition 13.16, consisting of the axioms $R_\mathcal{A}$ defining a relativised representation, plus the further axioms

$$\forall \bar{x} \bigl(\mathrm{Clique}(\bar{x}) \land (r;s)(x_i, x_j) \bigr) \rightarrow \exists x_k \bigl(\mathrm{Clique}(\bar{x}) \land r(x_i, x_k) \land s(x_k, x_j) \bigr),$$

for each $r,s \in \mathcal{A}$ and $i,j,k < n$ with $k \neq i,j$, where $\bar{x} = (x_0,\ldots,x_{n-1})$ is a tuple of distinct variables and Clique($\bar{x}$) abbreviates the formula $\bigwedge_{l,m<n} 1(x_l,x_m)$. As seen in theorem 19.13, the sentences of $R_{\mathcal{A}}$ are logically equivalent to sentences of the loosely guarded fragment, and the extra axioms are clearly already in the loosely guarded fragment. Thus, for finite $\mathcal{A}$, $\bigwedge \mathrm{Sq}^n(\mathcal{A})$ is a loosely guarded sentence.

The proof is now concluded as before. We know (theorems 13.46 and 13.49) that $\mathcal{A}$ has an n-square relativised representation, so that $\bigwedge \mathrm{Sq}^n(\mathcal{A})$ has a model. By corollary 19.7, it has a finite model — a finite n-square relativised representation of $\mathcal{A}$. □

19.6 The finite algebra on finite base property for $\mathbf{S\Re aCA}_n$?

We now try to extend theorem 19.18 to n-flat, infinitarily n-flat, and n-smooth relativised representations, and $\mathbf{S\Re aCA}_n$. By exercise 13.7(8), any finite n-flat relativised representation is infinitarily n-flat and n-smooth.

Of course, the finite base property fails for $\mathbf{S\Re aCA}_n$ for any $n \geq 4$, as the equational theories of these classes are undecidable (corollary 18.27). However, there is a problem even with the finite algebra on finite base property: while any finite algebra in $\mathbf{RA}_n$ has a finite n-dimensional relational basis, simply because there are only finitely many n-dimensional networks, the corresponding result may not hold for $\mathbf{S\Re aCA}_n$, using hyperbases. This is because even over a finite algebra, there are infinitely many n-dimensional hypernetworks — the hyperlabels are not restricted to come from a finite set.

We begin by showing that this problem is a real one:

PROPOSITION 19.19 *For every $n \geq 5$, there exist finite algebras in $\mathbf{S\Re aCA}_n$ that do not have a finite n-dimensional hyperbasis.*

Proof. Assume for contradiction that every finite algebra in $\mathbf{S\Re aCA}_n$ has a finite n-dimensional hyperbasis. The following algorithm will now decide membership of $\mathbf{S\Re aCA}_n$, for finite algebras:

- Using the recursive axiomatisation of $\mathbf{S\Re aCA}_n$ given in theorem 13.55, recursively enumerate all (isomorphism types of) finite relation algebras that are not in $\mathbf{S\Re aCA}_n$.

- Recursively enumerate all (isomorphism types of) finite relation algebras $\mathcal{A}$. For each $\mathcal{A}$, enumerate all finite sets of n-dimensional hypernetworks over $\mathcal{A}$, using natural numbers as hyperlabels, and check each one to see if it is a hyperbasis. (All to be done in parallel.) When a hyperbasis is found, 'print out' $\mathcal{A}$. This recursively enumerates all and only the finite algebras

19.6. The finite algebra on finite base property for $S\mathfrak{R}a\mathbf{CA}_n$?

in $S\mathfrak{R}a\mathbf{CA}_n$, by assumption and the fact (theorem 13.46) that any relation algebra with an n-dimensional hyperbasis is in $S\mathfrak{R}a\mathbf{CA}_n$.

Since any finite relation algebra will occur, up to isomorphism, in exactly one of these two enumerations, the process will decide whether or not it is in $S\mathfrak{R}a\mathbf{CA}_n$, in finite time. But now, we have contradicted the result of theorem 18.13 that this problem is undecidable. □

Certainly, if $\mathcal{A} \in S\mathfrak{R}a\mathbf{CA}_n$ has no finite n-dimensional hyperbasis then there is not much hope of finding a finite n-flat relativised representation of it. (Actually, theorem 19.20 below shows that there is no hope, so that the fails for $S\mathfrak{R}a\mathbf{CA}_n$ with respect to n-flat relativised representations.) It may be a hard problem to tell whether a given finite relation algebra has a finite n-dimensional hyperbasis — see problem 18.17.

The 'solution' is to restrict to algebras that do have a finite hyperbasis. That is, we consider the subclass $S\mathfrak{R}a\{C \in \mathbf{CA}_n : C \text{ is finite}\}$ of $S\mathfrak{R}a\mathbf{CA}_n$. This class does have the finite algebra on finite base property.

THEOREM 19.20 *Let $\mathcal{A}$ be a finite relation algebra and let $n \geq 4$. The following are equivalent:*

1. *$\mathcal{A}$ has a finite n-flat, infinitarily n-flat, and n-smooth relativised representation.*

2. *$\mathcal{A} \subseteq \mathfrak{R}aC$ for some finite $C \in \mathbf{CA}_n$.*

3. *$\mathcal{A}$ has a finite n-dimensional hyperbasis.*

Proof. This is again similar to theorem 19.13. Inspection of the proof of theorems 13.20 and 13.27 shows that $(1) \Rightarrow (2) \Rightarrow (3)$ holds. So we only have to prove $(3) \Rightarrow (1)$. For this, we will use proposition 13.37.

Let $\mathcal{H}$ be a finite n-dimensional hyperbasis for $\mathcal{A}$. By lemma 12.15, we may assume that $\mathcal{H}$ is symmetric; inspection of the proof of the lemma shows that this does not affect the finiteness of $\mathcal{H}$. Let L be the finite relational signature consisting of $L(\mathcal{A})$ (a binary relation symbol for each $a \in \mathcal{A}$), together with an n-ary relation symbol R_N for each $N \in \mathcal{H}$. Take the structure M constructed from $\mathcal{H}$ in proposition 13.37, and make it an L-structure by defining

- $M \models a(x,y)$ iff (x,y) is an edge of M and $M(x,y) \leq a$, for all $x,y \in M$,

- $M \models R_N(x_0, \ldots, x_{n-1})$ iff $M(x_0, \ldots, x_{n-1}) = N$, for all $x_0, \ldots, x_{n-1} \in M$,

for all $a \in \mathcal{A}$ and $N \in \mathcal{H}$. Observe that (x,y) is an edge of M iff $M \models 1(x,y)$. One may check that M satisfies the following finite set Σ of axioms, corresponding to properties 1–8 of proposition 13.37. In the axioms, a,b range over atoms of $\mathcal{A}$, N,N' range over $\mathcal{H}$, $\bar{x}$ denotes an n-tuple $(x_0, \ldots, x_{n-1})$ of distinct variables, and Clique$(\bar{x})$ abbreviates the formula $\bigwedge_{l,m<n} 1(x_l, x_m)$.

1. $\forall x\, 1(x,x)$, and $\forall xy(1(x,y) \to 1(y,x))$

2. $\forall xy\left(1(x,y) \leftrightarrow \bigvee_a a(x,y)\right)$, and $\bigwedge_a \forall xy\left(a(x,y) \to \bigwedge_{b \neq a} \neg b(x,y)\right)$

3. $\bigwedge_{a \leq 1'} \forall xy\bigl(a(x,y) \to x = y\bigr)$, plus $\forall x\left(1(x,x) \to \bigvee_{a \leq 1'} a(x,x)\right)$

4. $\forall \bar{x}\left(\mathrm{Clique}(\bar{x}) \to \bigvee_N \left(R_N(\bar{x}) \wedge \bigwedge_{N' \neq N} \neg R_{N'}(\bar{x})\right)\right)$

5. $\bigwedge_N \forall \bar{x}\left(R_N(\bar{x}) \to \bigwedge_{i,j<n} [N(i,j)](x_i, x_j)\right)$

6. $\bigwedge_N \forall \bar{x}\left(R_N(\bar{x}) \to \bigwedge_{\sigma: n \to n} R_{N\sigma}(\bar{x} \circ \sigma)\right)$

7. $\bigwedge_N \forall \bar{x}\left(\mathrm{Clique}(\bar{x}) \to \left(\exists x_k R_N(\bar{x}) \leftrightarrow \bigvee_{N' \equiv_k N} R_{N'}(\bar{x})\right)\right)$, for all $k < n$

8. $\bigwedge_N \exists \bar{x}\, R_N(\bar{x})$.

It is clear that these sentences are, up to logical equivalence, loosely guarded. Since Σ is finite and $M \models \Sigma$, we see by corollary 19.7 that Σ has a finite model. Such a model satisfies the properties 1–8 of proposition 13.37, and since theorem 13.38 only used these properties (cf. remark 13.43), we see that there is a finite relativised representation of $\mathcal{A}$ which is n-flat, infinitarily n-flat, and n-smooth. (The special case where $\mathcal{H}$ is only a relational basis gives another proof of theorem 19.18.) □

COROLLARY 19.21 *Any finite relation algebra is a subalgebra of $\mathfrak{Ra}C$ for some finite $C \in \mathbf{CA}_4$, and has a finite relativised representation that is 4-flat, infinitarily 4-flat, and 4-smooth.*

Proof. By theorem 19.20 and exercise 12.3(12). □

To summarise: with respect to n-flat relativised representations, $\mathbf{S}\mathfrak{Ra}\mathbf{CA}_n$ for $n \geq 5$ does not have the finite algebra on finite base property, but a finite algebra in $\mathbf{S}\mathfrak{Ra}\mathbf{CA}_n$ has a finite n-flat relativised representation iff it is in $\mathbf{S}\mathfrak{Ra}\{C \in \mathbf{CA}_n : C$ is finite$\}$, iff it has a finite n-dimensional hyperbasis.

19.6. The finite algebra on finite base property for $S\Re aCA_n$?

Exercises

1. Check that all the sentences in the proof of theorem 19.20 are logically equivalent to loosely guarded sentences.

2. Let $n \geq 5$.

   (a) Show that the set of isomorphism types of finite algebras in $S\Re aCA_n$ that have no finite n-dimensional hyperbasis is not recursively enumerable.

   (b) Let K be the class of finite algebras in $S\Re aCA_n$ that have an infinite n-dimensional hyperbasis. Show that K is pseudo-elementary, but the set of isomorphism types of algebras in K is not recursively enumerable.

3. Show that any finite relation algebra with a homogeneous (classical) representation, such as the point algebra of section 4.4, has a finite n-flat relativised representation for all finite $n \geq 4$. (Cf. theorem 12.41.)

4. More generally, let $\mathcal{A}$ be a finite weakly associative algebra. Show that:

   (a) for all finite $n \geq 3$, $\mathcal{A}$ has an n-dimensional cylindric basis iff it has a finite n-homogeneous relativised representation (cf. exercise 13.6(1)),

   (b) $\mathcal{A}$ has a homogeneous (classical) representation iff it has a finite n-homogeneous relativised representation for all finite $n \geq 3$ (cf. theorem 12.41).

5. State and prove a version of corollary 19.21 for semi-associative algebras.

6. Show that the equational theory of $S\Re a\{C \in CA_n : C \text{ is finite}\}$ is undecidable for any finite $n \geq 4$. [See theorem 18.28.]

7. Show that there exists a recursive bound on the size of a smallest n-square relativised representation of a finite $\mathcal{A} \in RA_n$, in terms of n and $|\mathcal{A}|$. Indeed, there is an algorithm that constructs such a representation from $\mathcal{A}$ and n.

PROBLEM 19.22 For $n \geq 4$, let $S\Re aCA_n^f$ denote the class $S\Re a\{C \in CA_n : C \text{ is finite}\}$. Exercise 15.4(1) showed that $S\Re aCA_n^f \supset S\Re aCA_{n+1}^f$ for all finite $n \geq 3$. Study the class $\bigcap_{n<\omega} S\Re aCA_n^f$.

This is the class of all finite relation algebras with a finite n-dimensional hyperbasis for every finite $n \geq 3$. Clearly it is a class of finite representable relation algebras. It contains a finite relation algebra with no finite representation — for example, the point algebra. Is it the class of finite relation algebras with ω-categorical representations? (It contains all of these.) Is it the class of finite relation algebras that have a representation M with finitely many $\mathcal{L}(\mathcal{A})_n$-definable n-ary relations for all finite n? Is membership of it decidable? Is it the case that for all finite $n \geq 4$ there is a finite *representable* relation algebra $\mathcal{A} \in \mathbf{S}\mathfrak{R}\mathbf{a}\mathbf{CA}_n^f \setminus \mathbf{S}\mathfrak{R}\mathbf{a}\mathbf{CA}_{n+1}^f$? (This is true if the condition that $\mathcal{A} \in \mathbf{RRA}$ is dropped.)

Part VI

Epilogue

Chapter 20

Brief summary

Here follows an attempt to summarise the key definitions and results in the book. Unattributed definitions and theorems may be due to several people, or the authors: more details may be found in the main text.

20.1 Basic definitions

DEFINITION 20.1 [Relation algebras, definitions 3.1 and 3.8, Tarski 1948]

- A *relation algebra* is an algebra $\mathcal{A} = \langle A, 0, 1, -, +, 1', \smile, ; \rangle$ satisfying the axioms:

  R0. The equations defining a boolean algebra
  R1. $a;(b;c) = (a;b);c$
  R2. $(a+b);c = a;c+b;c$
  R3. $a;1' = a$
  R4. $\breve{\breve{a}} = a$
  R5. $(a+b)\smile = \breve{a} + \breve{b}$
  R6. $(a;b)\smile = \breve{b};\breve{a}$
  R7. $\breve{a};(-(a;b)) \leq -b$.

- **RA** denotes the class of all relation algebras.

- A *proper relation algebra* $\mathcal{P} = \langle P, \emptyset, U, \setminus, \cup, \mathrm{Id}_B, ^{-1}, | \rangle$ is a set P of binary relations over some base set B, containing the binary relations $\emptyset$, U (the 'unit', an equivalence relation on B), $\mathrm{Id}_B = \{(x,x) : x \in B\}$, and closed under the formation of finite unions, complement relative to U, conversion and composition of relations.

It is easily checked that any proper relation algebra is a relation algebra. **RA** is a discriminator variety (theorem 3.19); it is conjugated (lemma 3.13), so completely additive (theorem 2.40); and it is defined by Sahlqvist equations, so is canonical (theorem 2.95) and closed under completions (theorem 2.96).

DEFINITION 20.2 [Representations, definitions 3.28 and 3.30] Let $\mathcal{A}$ be a relation algebra.

- A *representation* of $\mathcal{A}$ is an isomorphism mapping $\mathcal{A}$ to a proper relation algebra.

- Alternatively, a representation can be defined as a model of the following first-order theory $T_\mathcal{A}$. The signature $\mathcal{L}(\mathcal{A})$ of $T_\mathcal{A}$ has one binary relation symbol for each element of $\mathcal{A}$. Below, we write R, S, T for elements of $\mathcal{A}$.

$$\begin{array}{rcll}
\sigma_{1'} & = & \forall x, y\, [1'(x,y) \leftrightarrow (x=y)] & \\
\sigma_+(R,S,T) & = & \forall x, y\, [R(x,y) \leftrightarrow S(x,y) \vee T(x,y)] & : \; R = S+T \\
\sigma_-(R,S) & = & \forall x, y\, [1(x,y) \rightarrow (R(x,y) \leftrightarrow \neg S(x,y))] & : \; R = -S \\
\sigma_\smile(R,S) & = & \forall x, y\, [R(x,y) \leftrightarrow S(y,x)] & : \; R = \check{S} \\
\sigma_;(R,S,T) & = & \forall x, y\, [R(x,y) \leftrightarrow \exists z(S(x,z) \wedge T(z,y))] & : \; R = S;T \\
\sigma_{\neq 0}(R) & = & \exists x, y\, R(x,y) & : \; R \neq 0.
\end{array}$$

Essentially, a model of $T_\mathcal{A}$ is a proper relation algebra isomorphic to $\mathcal{A}$.

- **RRA** denotes the class of all representable relation algebras.

RRA is a canonical variety (Monk, Tarski; theorems 3.36, 3.37).

Complete representations

DEFINITION 20.3 Let $\mathcal{A}$ be a relation algebra.

- A representation $h : \mathcal{A} \to \mathcal{P}$ is *complete* if arbitrary suprema (and infima) are preserved, wherever they exist in the algebra. That is, whenever $S \subseteq \mathcal{A}$ and the least upper bound $\sum S$ of S exists in $\mathcal{A}$, then $h(\sum S) = \bigcup_{s \in S} h(s)$.

- A representation $h : \mathcal{A} \to \mathcal{P}$ is *atomic* if $h(1) = \bigcup \{h(a) : a \text{ an atom of } \mathcal{A}\}$. (An atom is a minimal, non-zero element.)

A representation is complete iff it is atomic (theorem 2.21), and in this case, $\mathcal{A}$ is an atomic relation algebra (corollary 2.22).

20.2 Games for representability

Not all relation algebras are representable. We characterise and axiomatise those that are representable using networks and games, as follows.

DEFINITION 20.4 [Networks, definitions 7.1, 11.1 and 12.1] Let $\mathcal{A}$ be a relation algebra.

- A *network* over $\mathcal{A}$ is a finite, complete directed graph N with each edge (x,y) labelled by an element $N(x,y)$ of $\mathcal{A}$, and satisfying the consistency conditions $N(x,x) \leq 1$' and $(N(x,y) ; N(y,z)) \cdot N(x,z) \neq 0$, for all nodes $x,y,z \in N$.

- An *atomic network* has the additional property that each edge is labelled by an atom of $\mathcal{A}$.

- Let Λ be a non-empty set, and let $3 \leq n < \omega$. Recall that $n = \{0,\ldots,n-1\}$. An *n-dimensional Λ-hypernetwork* over $\mathcal{A}$ is a labelled hypergraph N with nodes $\{0,1,\ldots,n-1\}$, hyperedges $^{\leq n}n$, such that edges (elements of $^2 n$) are labelled by atoms of $\mathcal{A}$, all other sequences are labelled by elements of Λ, and satisfying:

  1. the edge labellings of N form an atomic network,
  2. if $\bar{x}, \bar{y} \in {^{\leq n}n}$ are labelled hyperedges of the same length l, say, and $N(x_i, y_i) \leq 1$' for each $i < l$, then $N(\bar{x}) = N(\bar{y})$.

- For n-dimensional hypernetworks M, N, and $i < n$, we write $M \equiv_i N$ if M and N agree on all labellings of hyperedges not involving the node i. That is, $M(\bar{x}) = N(\bar{x})$ whenever $\bar{x} \in {^{\leq n}(n \setminus \{i\})}$. Similarly, $M \equiv_{ij} N$ means that the two hypernetworks agree on all sequences not involving i or j.

We can determine whether a relation algebra $\mathcal{A}$ is representable by defining a game $G_\omega(\mathcal{A})$ (definition 7.12) in which two players, $\forall$ and $\exists$, construct a countable sequence of complete directed graphs

$$N_0 \subseteq N_1 \subseteq \cdots$$

with edges labelled by elements of $\mathcal{A}$. In round 0, $\forall$ picks an arbitrary non-zero element $a \in \mathcal{A}$ and $\exists$ must respond with a network N_0 with nodes x,y such that $N(x,y) = a$. In round $i > 0$, $\forall$ picks a directed edge e from N_{i-1} and a pair of elements $a, b \in \mathcal{A}$. $\exists$ responds in one of two ways. She may *reject,* by letting N_i be the same as N_{i-1} except that $N_i(e) = N_{i-1}(e) \cdot -(a;b)$. Or she may *accept,* by letting N_i have one node z in addition to those of N_{i-1} and labelling the edges of N_i as follows:

- $N_i(x,z) = a$, $N_i(z,z) = 1$', $N_i(z,y) = b$

- $N_i(e) = N_{i-1}(e) \cdot (a;b)$
- for any edge f of N_{i-1} other than e, $N_i(f) = N_{i-1}(f)$
- all other edges of N_i are labeled by 1.

If in any round $i < \omega$, $\exists$ chooses a labelled graph N_i that is not a network, then she loses the play; if all the N_i are networks then she wins.

We also define an r-round game $G_r(\mathcal{A})$, for $r < \omega$, which is defined similarly, but play stops after the construction of N_r.

THEOREM 20.5 *Let $\mathcal{A}$ be a relation algebra.*

1. *(Theorem 7.19) $\exists$ has a winning strategy in $G_\omega(\mathcal{A})$ if and only if $\mathcal{A}$ is representable.*

2. *(Proposition 7.24) $\exists$ has a winning strategy in each game $G_r(\mathcal{A})$ ($r < \omega$) if and only if she has a winning strategy in $G_\omega(\mathcal{A})$.*

3. *(Proposition 8.2) For each $r < \omega$, there is a universal first-order sentence σ_r of the signature of relation algebras such that $\mathcal{A} \models \sigma_r$ if and only if $\exists$ has a winning strategy in $G_r(\mathcal{A})$. The sentence σ_r does not depend on $\mathcal{A}$.*

Hence, the universal theory $\{\sigma_r : r < \omega\}$, together with the axioms defining relation algebras, defines the class **RRA** of representable relation algebras. Using the theory of discriminators from universal algebra, it is possible to replace these universal sentences by equations defining **RRA**. A rather simpler axiomatisation of **RRA** can be given in an expanded signature with function symbols analogous to the Q-operators of Jónsson (theorem 8.5).

Axiomatising pseudo-elementary classes A class of structures is said to be *pseudo-elementary* if it is the class of all structures arising as the first sort of a model of a fixed two-sorted, first-order theory (definition 9.1). Generalising the above, a method is given for finding explicit universal first-order axioms for any pseudo-elementary class that is closed under substructures (i.e., a 'pseudo-universal class' — see theorem 9.28), and also explicit first-order axioms for the elementary closure of any pseudo-elementary class (corollary 9.38). Axiomatisations using generalised Q-operators can also be given (theorem 9.39).

Game trees The game G_ω, above, generalises to 'game trees' (definition 10.2), based on Hintikka games. These games can be used to define many classes of algebras. For example, if we insist that in G_r, $\forall$'s moves are restricted so that he can only choose atoms and $\exists$ is required to select atomic networks for her moves (so she does not simply accept or reject, she must actually choose atoms to label

20.3. Relativised representations, bases, reducts

new edges of the network) we define an 'atomic game' G_r^a (definition 11.3). We obtain:

THEOREM 20.6 *Let $\mathcal{A}$ be an atomic relation algebra.*

1. *(Theorem 11.10) $\mathcal{A}$ is elementarily equivalent to a relation algebra with a complete representation iff $\exists$ has a winning strategy in $G_r^a(\mathcal{A})$ for all $r < \omega$.*

2. *(Theorem 11.7) If $\mathcal{A}$ has countably many atoms, then $\mathcal{A}$ has a complete representation if and only if $\exists$ has a winning strategy in $G_\omega^a(\mathcal{A})$.*

3. *(Theorem 11.10) If $\mathcal{A}$ is finite, the following are equivalent:*

   (a) $\mathcal{A}$ is representable,

   (b) $\mathcal{A}$ has a complete representation,

   (c) $\exists$ has a winning strategy in $G_r^a(\mathcal{A})$ for all $r < \omega$,

   (d) $\exists$ has a winning strategy in $G_\omega^a(\mathcal{A})$.

For any game defined by a game tree, we can find an axiom that determines whether $\exists$ has a winning strategy in the game curtailed to r rounds (r finite), though in general the axioms need not be universal. (The axioms for the games $G_r^a : r < \omega$ are collectively equivalent to the *Lyndon conditions* of [Lyn50].) But part 2 of theorem 20.5 does not hold for arbitrary games. In general, we can deduce that $\exists$ has a winning strategy in the full-length game from a winning strategy for her in all the finite-length games, only in the case where she has a finite number of choices for each of her moves. (This follows from König's tree lemma.) In the case of the game G_ω, she has always two choices. This fails for the game G_ω^a.

20.3 Relativised representations, bases, reducts

20.3.1 Relativised representations

Not every relation algebra is representable. But every relation algebra has a relativised representation (with reflexive and symmetric unit).

DEFINITION 20.7 [Definition 5.1] Let $\mathcal{A}$ be an algebra of the signature of relation algebras. A *relativised representation* of $\mathcal{A}$ is a model M of the following theory $R_\mathcal{A}$. $R_\mathcal{A}$ is rather like the theory $T_\mathcal{A}$, but the composition axioms are *relativised* to 1^M, the biggest binary relation. 1^M must be reflexive and symmetric but might not be transitive. So, for $R = S;T$, the $T_\mathcal{A}$-axioms $\sigma_;(R,S,T)$ (for all $R = S;T$ in $\mathcal{A}$) are replaced in $R_\mathcal{A}$ by axioms

$$\forall x,y[1(x,y) \to (R(x,y) \leftrightarrow \exists z(S(x,z) \wedge T(z,y)))].$$

Maddux (1982) showed that the class of algebras that have relativised representations is precisely the class **WA** of *weakly associative algebras*. **WA** is defined as follows [Madd82]:

- **WA** is defined by the axioms for relation algebras but with the weak associativity law $(1' \cdot x); 1 = ((1' \cdot x); 1); 1$ in place of the associativity axiom (R1).

Maddux's proof showed that the class of weakly associative algebras with complete representations is the class of atomic weakly associative algebras, so is elementary. It is known that the loosely guarded fragment of first-order logic has the finite model property (corollary 19.7, from [Hodk02, HodOtt01]). It follows from this that:

THEOREM 20.8 (Finite base property for WA)

1. *Every finite weakly associative algebra has a finite relativised representation ([AndHod+99]; see theorem 19.13).*

2. *It can be similarly shown that every equation and universal sentence valid over finite weakly associative algebras is valid over all weakly associative algebras ([AndHod+99]; see theorem 19.15).*

3. *It follows that the equational and universal theories of* **WA** *are decidable [Ném86].*

Maddux also defined the classes **SA** and **NA**:

- **SA** (semi-associative algebras) is defined by the axioms for relation algebras, but with the associativity axiom (R1) replaced by the semi-associative law $x; 1 = (x; 1); 1$,

- **NA** (non-associative algebras): similar, but (R1) is dropped altogether.

All three are canonical conjugated Sahlqvist varieties, closed under completions, and **NA** $\supset$ **WA** $\supset$ **SA** $\supset$ **RA**. Also, **SA** is a discriminator variety (see [Madd91b]). Section 5.1 has more details.

To find varieties between **RA** and **RRA**, we can refine the notion of relativised representation, use 'bases', or consider relation algebra reducts. Each of these can be done in any finite dimension $n \geq 3$, and all three approaches are equivalent. There are two variants of each approach; for $n = 3, 4$ the variants are equivalent, but not for larger dimensions. We discuss this next.

20.3. Relativised representations, bases, reducts

Square and flat relativised representations

DEFINITION 20.9 Let M be a relativised representation of $\mathcal{A}$: i.e., $M \models R_{\mathcal{A}}$. Let $3 \leq n < \omega$.

- A *clique* C of M is a set of points of the domain of M such that $M \models 1(x,y)$ for every $x, y \in C$.

- (Definitions 5.7 and 13.4) M is said to be *n-square* if for every clique C with $|C| < n$, every $x, y \in C$, and every $r, s \in \mathcal{A}$, if $M \models [r;s](x,y)$ then there is $z \in M$ such that $C \cup \{z\}$ is a clique, $M \models r(x,z)$, and $M \models s(z,y)$.

- (Definition 13.6) M is said to be *n-flat* if for every n-variable $L(\mathcal{A})$-formula φ, the formulas $\exists x \exists y \varphi$ and $\exists y \exists x \varphi$ are equivalent when evaluated in M with all quantifiers relativised to n-sized cliques in M.

- (Definition 13.6) M is said to be *infinitarily n-flat* if for every n-variable infinitary formula $\varphi \in L(\mathcal{A})^n_{\infty\omega}$, $\exists x \exists y \varphi$ and $\exists y \exists x \varphi$ are equivalent in M when evaluated with quantifiers relativised to n-sized cliques.

Any infinitarily n-flat relativised representation is trivially n-flat, and any n-flat relativised representation is n-square (lemma 13.10), but for $n \geq 5$ there are relation algebras with n-square relativised representations but not n-flat ones.

20.3.2 Relational bases and hyperbases

DEFINITION 20.10 [Bases, definitions 12.9 and 12.11] Let $3 \leq n < \omega$, let Λ be a non-empty set, and let $\mathcal{A}$ be an atomic non-associative algebra. Let $\mathcal{S}$ be a set of n-dimensional Λ-hypernetworks.

- [Madd78b] $\mathcal{S}$ is called an *n-dimensional relational basis* for $\mathcal{A}$ if
    1. for every atom $a \in \mathcal{A}$ there is $N \in \mathcal{S}$ with $N(0,1) = a$,
    2. if $N \in \mathcal{S}$, $i, j < n$, $a, b \in \text{At}(\mathcal{A})$ are such that $a;b \geq N(i,j)$, and $k < n$ satisfies $k \notin \{i,j\}$, then there is $M \in \mathcal{S}$ with $M \equiv_k N$ and $M(i,k) = a$, $M(k,j) = b$.

- $\mathcal{S}$ is called an *n-dimensional Λ-hyperbasis* if $\mathcal{S}$ is an n-dimensional relational basis, and
    3. whenever $M, N \in \mathcal{S}$, $i, j < n$, and $M \equiv_{ij} N$, there is $L \in \mathcal{S}$ with $M \equiv_i L \equiv_j N$.

DEFINITION 20.11 [Madd83] For $3 \leq n < \omega$, $\mathbf{RA}_n$ is the class of all subalgebras of (complete) atomic non-associative algebras with an n-dimensional relational basis.

[Madd83] proved among other things that **RA**$_n$ is a canonical variety, **RA**$_3$ = **SA**, and **RA**$_4$ = **RA**.

Games and bases

There are games associated with these bases. Using game trees, we can define an n-pebble game G_r^n, similar to G_r^a, but where the number of nodes in a network is limited to n. So it may be necessary for $\forall$ to delete a node before making his move, in order to stay within this bound. For $r \leq n \leq \omega$, G_r^n is essentially the same as G_r^a. See definition 12.24 for details.

THEOREM 20.12 *Let $\mathcal{A}$ be an atomic non-associative algebra and let $2 \leq n < \omega$.*

- *(Proposition 12.25) $\mathcal{A}$ has an n-dimensional relational basis iff $\exists$ has a winning strategy in the game $G_\omega^n(\mathcal{A})$.*

- *(Exercise 12.4(3)) If $\mathcal{A}$ is finite, $\mathcal{A}$ has an n-dimensional relational basis iff $\exists$ has a winning strategy in $G_r^n(\mathcal{A})$ for all finite r.*

There is also a game that characterises hyperbases, similar to G_ω^n but played on n-dimensional hypernetworks whose nodes are always $\{0, 1, \ldots, n-1\}$, and in which $\forall$ is also allowed to make 'amalgamation moves' by picking $i, j < n$ and any two previously played hypernetworks M, N such that $M \equiv_{ij} N$. $\exists$ must respond with a hypernetwork L with $M \equiv_i L \equiv_j N$. See definition 12.26.

These games will be used below.

20.3.3 Relation algebra reducts

DEFINITION 20.13 [Cylindric algebra, definition 5.16] Let α be an ordinal. An α-dimensional *cylindric algebra* is an algebra $C = (C, +, -, 1, 0, c_i, d_{ij})_{i,j<\alpha}$ (where the c_i are unary functions and the d_{ij} constants), satisfying, for every $x, y \in C$, $i, j, k < \alpha$:

C0. $(C, +, -, 0, 1)$ is a boolean algebra

C1. $c_i 0 = 0$

C2. $x \leq c_i x$

C3. $c_i(x \cdot c_i y) = c_i x \cdot c_i y$

C4. $c_i c_j x = c_j c_i x$

C5. $d_{ii} = 1$

C6. if $k \neq i, j$, then $d_{ij} = c_k(d_{ik} \cdot d_{kj})$

20.3. Relativised representations, bases, reducts

C7. if $i \neq j$, then $c_i(d_{ij} \cdot x) \cdot c_i(d_{ij} \cdot -x) = 0$.

$\mathbf{CA}_\alpha$ denotes the class of all α-dimensional cylindric algebras. For $i, j < \alpha$ and $a \in C$, we define the *i-for-j* substitution operator

$$s^i_j a = \begin{cases} a, & \text{if } i = j; \\ c_i(d_{ij} \cdot a), & \text{otherwise.} \end{cases}$$

DEFINITION 20.14 [Relation algebra reduct, definition 5.40] For $\alpha \geq 3$ and an α-dimensional cylindric algebra C, we define the *relation algebra reduct of C*, $\mathfrak{Ra}C$, to be the relation-type algebra with domain $\{a \in C : c_i a = a \text{ whenever } 2 \leq i < \alpha\}$, the operations $+, -, 0, 1$ being defined as in C, 1' being defined as d_{01}, $\breve{a}$ being defined as $s^2_0 s^0_1 s^1_2 a$, and $a;b$ being defined as $c_2(s^1_2 a \cdot s^0_2 b)$.

If $\alpha \geq 4$, $\mathfrak{Ra}C$ is a relation algebra. $\mathbf{S\mathfrak{Ra}CA}_\alpha$ denotes the class of all subalgebras of relation algebra reducts of α-dimensional cylindric algebras. It is a canonical variety (proposition 5.48).

DEFINITION 20.15 [**Crs**, **D**, **G**, definitions 5.34, 5.35] Let α be an ordinal.

- A *cylindric relativised set algebra of dimension* α consists of a set S of α-ary relations over some base set U, equipped with constants 0 and 1 (here, 1 is any α-ary relation over U) and for $i, j < \alpha$, the operations $\cup, \setminus$ (complement relative to 1), diagonal elements $D_{ij} = \{\bar{x} \in 1 : x_i = x_j\}$, and the cylindrifications C_i where $C_i(X) = \{\bar{u} \in 1 : \exists \bar{x} \in X \, \forall j < \alpha (j \neq i \to x_j = u_j)\}$.

- The class of all α-dimensional cylindric relativised set algebras is denoted $\mathbf{Crs}_\alpha$.

- $\mathbf{D}_\alpha$ is the class of all algebras in $\mathbf{Crs}_\alpha$ such that if $\bar{x} \in 1$ and $i, j < \alpha$, then $(x_0, \ldots, x_{i-1}, x_j, x_{i+1}, \ldots) \in 1$.

- $\mathbf{G}_\alpha$ is the class of all algebras in $\mathbf{Crs}_\alpha$ (and of $\mathbf{D}_\alpha$) such that if $\pi : \alpha \to \alpha$ is any map, and $\bar{x} \in 1$, then $(x_{\pi(0)}, x_{\pi(1)}, \ldots) \in 1$.

DEFINITION 20.16 [Relation algebra reduct embedding, definition 5.49] Let $\alpha \geq 3$ be an ordinal.

- For a non-associative algebra $\mathcal{A}$ and $C \in \mathbf{Crs}_\alpha$, a map $\theta : \mathcal{A} \to C$ is said to be a *relation algebra reduct embedding* if $C_i(\theta(a)) = \theta(a)$ for all $a \in \mathcal{A}$ and $2 \leq i < \alpha$, θ is one-one and preserves $+, -, 0,$ and 1, $\theta(1') = D_{01}$, $\theta(\breve{a}) = s^2_0 s^0_1 s^1_2 \theta(a)$, for all $a \in \mathcal{A}$, and $\theta(a;b) = C_2(s^1_2 a \cap s^0_2 b)$ for all $a, b \in \mathcal{A}$, where s^i_j is defined using the **Crs** operations as above.

- $\mathbf{S\mathfrak{Ra}D}_\alpha$ denotes the class of all non-associative algebras $\mathcal{A}$ such that there exists $C \in \mathbf{D}_\alpha$ and a relation algebra reduct embedding from $\mathcal{A}$ into C. $\mathbf{S\mathfrak{Ra}G}_\alpha$ is defined similarly.

20.3.4 Equivalences between the notions

We can connect relational bases with square relativised representations:

THEOREM 20.17 *Let $4 \leq n < \omega$.*

- *An atomic non-associative algebra has a complete n-square relativised representation iff it has an n-dimensional relational basis (theorem 13.45).*

- *An arbitrary non-associative algebra has an n-square relativised representation iff it is in $\mathbf{RA}_n$, iff it is in $\mathbf{S\mathfrak{R}aD}_n$, iff it is in $\mathbf{S\mathfrak{R}aG}_n$ (theorem 13.46).*

- *A finite non-associative algebra is in $\mathbf{RA}_n$ iff it has an n-dimensional relational basis, iff there exists a relation algebra reduct embedding from it into some finite $C \in \mathbf{D}_n$ (or $\mathbf{G}_n$), iff it has a finite n-square relativised representation (theorems 13.51, 19.18).*

A similar result can be proved for $n = 3$ (see theorem 13.49). We can also relate flat relativised representations, hyperbases, and relation algebra reducts:

THEOREM 20.18 *Let $4 \leq n < \omega$.*

- *An atomic non-associative algebra has a complete infinitarily n-flat relativised representation iff it has an n-dimensional Λ-hyperbasis, for some Λ (theorem 13.45).*

- *An arbitrary non-associative algebra has an (infinitarily) n-flat relativised representation iff it is in $\mathbf{S\mathfrak{R}aCA}_n$, iff its canonical extension has an n-dimensional Λ-hyperbasis, for some Λ (theorem 13.46).*

- *A finite non-associative algebra $\mathcal{A}$ is a subalgebra of $\mathfrak{R}aC$ for some finite $C \in \mathbf{CA}_n$ iff $\mathcal{A}$ has a finite (infinitarily) n-flat relativised representation, iff $\mathcal{A}$ has a finite n-dimensional Λ-hyperbasis, for some finite Λ (theorem 19.20).*

However, for $5 \leq n < \omega$, not every finite algebra in $\mathbf{S\mathfrak{R}aCA}_n$ is the relation algebra reduct of a *finite n*-dimensional cylindric algebra, or has a finite *n*-flat relativised representation or a finite hyperbasis (proposition 19.19).

20.4 The rainbow construction

The *rainbow construction* is a method of building a relation algebra $\mathcal{A}_{A,B}$ from two binary relational structures A, B in such a way that $\exists$ has a winning strategy in the game characterising relational bases over $\mathcal{A}_{A,B}$ if and only if she has a winning strategy in an Ehrenfeucht–Fraïssé-style 'forth' game $EF_r^p(A,B)$ from A to B with p pairs of pebbles and r rounds. For $\exists$ to win, the pebble positions must always define a partial homomorphism from A to B.

20.4. The rainbow construction

THEOREM 20.19 (Theorem 16.5) *Let A, B be structures in a relational signature L consisting of only binary relation symbols. Let $p, r \leq \omega$. Then $\exists$ has a winning strategy in the game $G_{1+r}^{2+p}(\mathcal{A}_{A,B})$ if and only if she has a winning strategy in the game $EF_r^p(A, B)$.*

This construction allows us to find relation algebras in which $\exists$ can win games of certain lengths but not others. The rainbow construction can be used to construct an atomic relation algebra $\mathcal{A}$ for which $\exists$ has a winning strategy in any finite-length game $G_r^a(\mathcal{A}) = G_r^r(\mathcal{A})$, for $r < \omega$, but where $\forall$ has a winning strategy in the full-length game $G_\omega^a(\mathcal{A}) = G_\omega^\omega(\mathcal{A})$. Because $\exists$'s having a winning strategy in G_r^r for finite r is an elementary property, we can use theorem 20.6 and some model theory to conclude that:

THEOREM 20.20 (Theorem 17.6) *The class of relation algebras with complete representations is not closed under elementary equivalence, so is not elementary.*

Using the rainbow construction again, for each $r < \omega$ we construct a finite relation algebra $\mathcal{A}_r$ for which $\exists$ has a winning strategy in $G_r^a(\mathcal{A}_r)$ but not in the full-length game $G_\omega^a(\mathcal{A}_r)$. Using theorem 20.6(3), we obtain:

THEOREM 20.21 (Theorem 17.3, Monk [Mon64]) *The class **RRA** of all representable relation algebras is not finitely axiomatisable.*

RRA is not even axiomatisable by equations using a bounded (finite) number of variables ([Jón91], exercise 11.5(4), theorem 17.11). Furthermore,

THEOREM 20.22 (Theorem 17.18)

- For $4 \leq n < \omega$ and $r < \omega$ there is an algebra $\mathcal{A}(n, r)$ such that (i) $\exists$ can win $G_\omega^n(\mathcal{A}(n, r))$, (ii) $\forall$ can win $G_\omega^{n+1}(\mathcal{A}(n, r))$, but (iii) $\exists$ can win $G_r^{n+1}(\mathcal{A}(n, r))$.

- Hence $\mathbf{RA}_{n+1}$ cannot be defined by finitely many axioms over $\mathbf{RA}_n$.

Exercise 17.4(1) extended the construction to show that for finite $n \geq 5$, the class of atomic non-associative algebras with an n-dimensional relational basis (and hence, by theorem 13.45, with a complete n-square relativised representation) is not elementary. The corresponding classes for $n = 3, 4$ are elementary, being the atomic algebras in $\mathbf{RA}_3, \mathbf{RA}_4$, respectively (by theorem 13.45 and exercises 12.3(11, 12)).

A slightly easier construction using 'Monk algebras' instead of the rainbow construction is used to show that for all finite $n \geq 4$, $\mathbf{S\mathfrak{R}aCA}_{n+1}$ is not finitely axiomatisable over $\mathbf{S\mathfrak{R}aCA}_n$ (theorem 15.1), nor over $\mathbf{RA}_{n+1}$ (remark 15.13). This uses the 'game for hyperbases' mentioned after theorem 20.12. Exercise 15.6(1) showed similarly that for finite $n \geq 5$, the class of atomic non-associative algebras with an n-dimensional hyperbasis (and hence, by theorem 13.45, with a complete infinitarily n-flat or n-smooth relativised representation) is not elementary. (For $n = 3,4$, the same theorem plus exercises 12.3(11, 12) showed that the corresponding classes are just the atomic algebras in $S\mathfrak{Ra}CA_3, S\mathfrak{Ra}CA_4$, respectively, and so are elementary.) Similarly, for $3 \leq m < n < \omega$, the class $S\mathfrak{Nr}_m CA_{n+1}$, of subalgebras of m-dimensional neat reducts of $(n+1)$-dimensional cylindric algebras (definition 5.40), is not finitely based over $S\mathfrak{Nr}_m CA_n$ (theorem 15.1). Consequently:

THEOREM 20.23 (theorem 15.17) *For $3 \leq m < n < \omega$, there is no finite set Σ of m-variable axiom schemas such that*

$$\Sigma \vdash_{m,n} \varphi \iff \vdash_{m,n+1} \varphi$$

where φ is an arbitrary m-variable formula in a relational signature with m-ary relation symbols and $\vdash_{m,n}$ is a certain Hilbert system for proving m-variable formulas with n-variable proofs.

20.5 Atom structures

The following was defined by Lyndon in [Lyn50], though very similar ideas can be found in [JónTar52, section 5].

DEFINITION 20.24 [Relation algebra atom structure, definition 3.21] Let $\mathcal{A}$ be an atomic relation algebra. Its *atom structure* is defined to be

$$\mathrm{At}(\mathcal{A}) = \langle \{\text{atoms of } \mathcal{A}\}, \{a \in \mathrm{At}(\mathcal{A}) : a \leq 1'\},$$
$$(a \mapsto \breve{a})_{a \in \mathrm{At}(\mathcal{A})}, \{(a,b,c) \in \mathrm{At}(\mathcal{A}) : c \leq a;b\}\rangle.$$

Since the relation algebra operations are completely additive, the boolean structure plus the atom structure of a relation algebra determine the algebra uniquely, up to isomorphism. An atom structure is *weakly representable* if *some* atomic representable relation algebra has that atom structure, and it is *strongly representable* if all atomic relation algebras with that atom structure are representable (definition 14.1). A result of Venema (theorem 2.84) shows that the class of weakly representable atom structures forms an elementary class. But a probabilistic graph construction of Erdös can be used to prove that the class of strongly representable atom structures is not elementary. Erdös proves that there exist graphs where both the chromatic number and minimal cycle size are arbitrarily large. This implies that there exist a sequence of graphs none of which can be coloured with finitely many colours and yet a non-principal ultraproduct of them can be coloured with just two colours. From an infinite graph Γ we construct an atom structure $\alpha(\Gamma)$. We show that $\alpha(\Gamma)$ is strongly representable iff Γ cannot be coloured finitely. It follows that the class of strongly representable atom structures is not closed under ultraproducts and hence is not elementary (see theorem 14.3 for details). The ultraproduct of the $\alpha(\Gamma)$s is a weakly but not strongly representable atom structure. It can be shown from this that (a) **RRA** is not closed under completions, and (b) **RRA** cannot be defined using Sahlqvist equations (see corollary 14.15). The rainbow construction shows the same for $\mathbf{RA}_n$ and $\mathbf{S\mathcal{R}aCA}_n$ for finite $n \geq 6$ (theorem 17.37 and corollary 17.38; the case $n = 5$ is open).

20.6 Decidability

A certain undecidable tiling problem can be reduced to the problem of deciding whether an arbitrary finite relation algebra is representable or not. This shows that the latter problem is also undecidable (theorem 18.13). The reduction constructs from a yes-instance of the tiling problem a finite representable relation algebra, and from a no-instance a weakly associative algebra that is not even in $\mathbf{S\mathcal{R}aCA}_5$. So the proof shows that it is undecidable whether a finite relation algebra is in $\mathbf{S\mathcal{R}aCA}_n$ for any $n \geq 5$ (also theorem 18.13), or indeed, in any class K with $\mathbf{RRA} \subseteq \mathsf{K} \subseteq \mathbf{S\mathcal{R}aCA}_5$ (corollary 18.15). Hence, these classes are not finitely axiomatisable in m^{th}-order logic for any finite m. On the other hand, for any finite n, it is decidable whether a finite non-associative algebra is in $\mathbf{RA}_n$ (corollary 12.32). Nonetheless, the equational theories of **RA**, **RRA**, **SA**, $\mathbf{RA}_n$, $\mathbf{S\mathcal{R}aCA}_n$ (for finite $n \geq 4$) are undecidable (theorem 18.28). (The equational theories of **NA**, **WA** are decidable [Ném87a].)

A relation algebra is *weakly representable* if it has a representation respecting the operations $0, 1, 1', \breve{\ }$, and ; (but not necessarily $+, -$). The class of weakly representable relation algebras is denoted by **wRRA** (definition 5.14). The proof just mentioned shows that it is undecidable whether a finite relation algebra is weakly representable (corollary 18.24). Indeed, for every class K with $\mathbf{RRA} \subseteq \mathsf{K} \subseteq \mathbf{wRRA}$, the class of isomorphism types of finite algebras in K is undecidable, and K is not finitely axiomatisable in m^{th}-order logic for any finite m (theorem 18.23). **wRRA** can also be proved non-finitely axiomatisable by the rainbow construction (theorem 17.29).

20.7 Summary of relations between the classes

$$\mathbf{NA} \supset \mathbf{WA} \supset \mathbf{SA} = \mathbf{RA}_3 \supset \mathbf{RA} = \mathbf{RA}_4 = \mathbf{S\mathcal{R}aCA}_4$$
$$\mathbf{RRA} = \bigcap_{3 \leq n < \omega} \mathbf{RA}_n = \bigcap_{3 \leq n < \omega} \mathbf{S\mathcal{R}aCA}_n$$

WA	=	$\mathbf{S\mathcal{R}aG}_3$			
SA	=	$\mathbf{S\mathcal{R}a(CA}_3 \cap \mathbf{G}_3)$			
$\mathbf{RA}_n$	=	$\mathbf{S\mathcal{R}aD}_n$	=	$\mathbf{S\mathcal{R}aG}_n$	for all finite $n \geq 4$
$\mathbf{S\mathcal{R}aCA}_n$	=	$\mathbf{S\mathcal{R}a(CA}_n \cap \mathbf{D}_n)$	=	$\mathbf{S\mathcal{R}a(CA}_n \cap \mathbf{G}_n)$	for all finite $n \geq 4$

Also, we have

$$\mathbf{RA}_4 \supset \mathbf{RA}_5 \supset \ldots \supset \mathbf{RRA}, \qquad \mathbf{S\Re aCA}_4 \supset \mathbf{S\Re aCA}_5 \supset \ldots \supset \mathbf{RRA}.$$

'$\supset$' denotes strict inclusion and each of the inclusions is not finitely axiomatisable.

$$\mathbf{RA}_n \supset \mathbf{S\Re aCA}_n, \quad \text{for each } n \text{ with } 5 \leq n < \omega.$$

This inclusion is not finitely axiomatisable either. Further, for $n \geq 5$ there is no $m \geq 5$ such that $\mathbf{RA}_m \subseteq \mathbf{S\Re aCA}_n$ (remark 15.13).

20.8 Summary of properties of classes

The following tables summarise some of the information given here.

	canonical variety	finitely axiomatisable	closed under completions	Sahlqvist variety	decidable equational theory
NA	✓	✓	✓	✓	✓
WA	✓	✓	✓	✓	✓
SA	✓	✓	✓	✓	×
RA	✓	✓	✓	✓	×
RA$_5$	✓	×	?	?	×
RA$_n$ ($6 \leq n < \omega$)	✓	×	×	×	×
S$\Re$aCA$_5$	✓	×	?	?	×
S$\Re$aCA$_n$ ($6 \leq n < \omega$)	✓	×	×	×	×
RRA	✓	×	×	×	×

	finite membership is decidable	finite algebra on finite base property*	CK† is elementary	CK† = {atomic algebras in K}
NA	✓	n/a	n/a	n/a
WA	✓	✓	✓	✓
SA	✓	✓	✓	✓
RA	✓	✓	✓	✓
RA$_5$	✓	✓	×	×
RA$_n$ ($6 \leq n < \omega$)	✓	✓	×	×
S$\Re$aCA$_5$	×	×	×	×
S$\Re$aCA$_n$ ($6 \leq n < \omega$)	×	×	×	×
RRA	×	×	×	×

20.8. Summary of properties of classes

Notes for the tables

? Open problem.

n/a not applicable (no suitable notion of representation).

* with respect to a suitable notion of representation: i.e., classical representations for **RRA**; relativised representations for **WA**, 3- and 4-square relativised representations for **SA** and **RA**, respectively; n-square relativised representations for **RA**$_n$; and n-flat, infinitarily n-flat, or n-smooth relativised representations for **S$\Re$aCA**$_n$.

† CK denotes the class of all algebras in K that have a complete representation (of the appropriate kind, as in *).

Here is a vague question, partly arising from the table. Consider the following evidence:

- The equational theories of boolean algebras and weakly associative algebras are decidable.

  Like that of **RRA**, the equational theory of **RA**$_n$ and **S$\Re$aCA**$_n$ for $n \geq 3$ is undecidable.

- **RA**$_3$ and **RA**$_4$ are finitely axiomatisable, as are boolean algebras.

  But for $n \geq 5$, **RA**$_n$ and **S$\Re$aCA**$_n$ are not finitely axiomatisable; nor is **RRA**.

- **CRA**$_3$ and **CRA**$_4$ are elementary, consisting of the atomic algebras in **RA**$_n$. (**CRA**$_n$ is the class of non-associative algebras with a complete n-square relativised representation.) A similar result holds for boolean algebras.

  But for $n \geq 5$, **CRA**$_n$ is not elementary; a similar result holds for **RRA**.

  Analogous results hold for flat relativised representations (exercise 15.6(1)).

- **RA**$_3$ and **RA**$_4$ are closed under completions and are axiomatised by Sahlqvist equations, as are boolean algebras.

  But for $n \geq 6$, **RA**$_n$ and **S$\Re$aCA**$_n$ are not closed under completions and are not Sahlqvist-axiomatisable, and neither is **RRA**.

For $n = 3$, it seems that the algebras in **RA**$_n$ and **S$\Re$aCA**$_n$ behave like boolean algebras, while for $n \geq 5$, or maybe 6, they become more like full-blooded representable relation algebras. The exceptions are decidability of equational theories, where $n = 3$ is the threshold, but mostly it is 5.

PROBLEM 20.25 *Find other evidence bearing on this. Is it really true? Why?*

Chapter 21

Problems

We have included exercises in the text. Some of them are simply there for practising the techniques, and some of them are more significant problems. Here we highlight some of the more important problems whose solution, we think, would make a significant contribution to the research in this field.

1. **[Problem 17.14]** Prove that **RRA** cannot be axiomatised with a set of first-order sentences using only finitely many variables.

    A related problem is from graph theory. Let $c \geq 2$. Find two finite, non-isomorphic graphs G, H such that $\exists$ has a winning strategy in the c-colour game defined as follows. In round 1, $\forall$ takes one of the c colours and places it on a *set* of nodes of one of the two graphs. $\exists$ takes the same colour and places it on a set of nodes of the other graph. In any subsequent round $\forall$ either takes a colour not already in play and places it on a set of nodes of one of the two graphs, in which case $\exists$ also takes that colour and places it on a set of nodes of the other graph, or he picks up a colour which is already in play ($\exists$ picks up the same colour which is placed on the other graph) and he places that colour on a set of nodes of either graph (and $\exists$ places the same colour on a set of nodes of the other graph).

    At any stage the two graphs have $k \leq c$ colours placed on them; this partitions the nodes of each graph into 2^k corresponding parts. Let us call these parts $G_0, \ldots, G_{2^k-1}$ in G and $H_0, \ldots, H_{2^k-1}$ in H. If for some $i < 2^k$ we have G_i is empty but H_i is not, or the other way round, then $\forall$ wins at this stage. Also if for some $i, j < 2^k$ there is an edge connecting a node in G_i to a node in G_j but there is no edge from a node in H_i to one in H_j (or the other way round) then $\forall$ wins. The play continues for ω rounds. If $\forall$ never wins then $\exists$ wins.

Essentially, we seek two finite non-isomorphic graphs G, H which cannot be distinguished in a second-order language with c monadic predicate symbols, one binary predicate symbol denoting the edge relation and equality, in which second-order quantification is only allowed over the monadic predicate symbols. If such graphs can be found (say G does not embed into H) then it is possible to use the rainbow construction to construct two relation algebras $\mathcal{A}_{G,H}, \mathcal{A}_{H,H}$ such that the second is representable but the first is not, yet the two algebras cannot be distinguished by any c-variable formula. If this could be done for all finite c then we could solve the previous problem. Currently we have not been able to solve this problem even for $c = 2$.

2. **[Problem 17.39, Venema]** Find (or show non-existence of) a set of equations $\{e_i : i < \omega\}$ with the following properties:

    - $\mathcal{A} \models \{e_i : i < \omega\} \iff \mathcal{A} \in \mathbf{RRA}$.
    - Each e_i is canonical, i.e. $\mathcal{A} \models e_i \Rightarrow \mathcal{A}^+ \models e_i$.

    Repeat, replacing **RRA** by **S$\mathfrak{R}$aCA**$_n$ for $n \geq 5$. Repeat for **RA**$_n$ as well.

3. **[Problem 3.27, Jónsson]** Find all simple relation algebras with no subalgebras other than the whole algebra and the degenerate relation algebra with just one element in its domain.

    This is [Madd94a, problem P2]. Maddux stated there that 22 simple relation algebras with no non-trivial subalgebras had been found.

4. **[Problem 5.18]** If $C \in \mathbf{RCA}_\omega$ does C^+ necessarily have a complete, ω-dimensional representation?

5. **[Problem 9.3]** Is the class of relation algebras with a homogeneous representation an elementary class?

6. **[Problem 9.4, Németi, [Ném96, AndGol$^+$98, And01]]** Is the class **IG**$_\omega$ (the isomorphism-closure of the ω-dimensional cylindric relativised set algebras in which the unit is closed under substitutions and permutations) a variety, or even a pseudo-elementary class? Is it closed under ultraproducts?

7. [Old problem, stated in [AndTho88, p. 681] as well as other sources.] For infinite α, is the equational theory of $\mathbf{D}_\alpha$ decidable?

8. **[Problem 18.26, partly Maddux]** A relation algebra is said to be *weakly representable* if it has a representation respecting the relation algebra operations $(1', \check{\,}, ;)$ and $0, 1, \cdot$, but not necessarily $+, -$. The class of weakly representable relation algebras is denoted by **wRRA**. Do we have **wRRA** $\subseteq$ **RA**$_n$ for some $n \geq 5$? What inclusions hold between **wRRA** and the **S$\mathfrak{R}$aCA**$_n$ $(n \geq 5)$?

Chapter 21. Problems 627

9. Is **wRRA** a variety? Is it canonical?

10. [**Problem 17.41**] Is the class **wRRA** of weakly representable relation algebras closed under completions?

11. [**Problem 17.40**] Are $\mathbf{RA}_5$ and $\mathbf{S\mathfrak{R}aCA}_5$ closed under completions?

12. [**Sayed Ahmed**] Which $\mathbf{S\mathfrak{N}r}_m\mathbf{CA}_n$ for $m < n < \omega$ are closed under completions?

13. [**Cf. problem 5.47, Németi–Sayed Ahmed**] For finite $n \geq 4$, is $\mathfrak{R}a\mathbf{CA}_n$ closed under subalgebras? Is it elementary?

14. [**Problem 14.20**] For which $\alpha \geq 3$ is $\operatorname{Str}\mathbf{RCA}_\alpha$ elementary?

15. [**Cf. problem 12.38**] For finite $n \geq 4$, is every algebra in $\mathbf{S\mathfrak{R}aCA}_n$ embeddable in a relation algebra with an n-dimensional cylindric basis?

16. [**Problem 18.18, Maddux**] Is it decidable whether an arbitrary finite relation algebra has a finite representation?

17. [**Problem 18.17**] For fixed finite $n \geq 5$, is it decidable whether an arbitrary finite relation algebra has a finite n-dimensional hyperbasis?

18. [**Problem 15.18**] Let $\vdash_{m,n}$ be the n-variable proof relation of m-variable formulas of [AndNém$^+$01] (see §15.7). Let $3 \leq m < \omega$. Is there a finite set of m-schemata whose set Σ of m-instances satisfies

$$\Sigma \vdash_{m,m} \phi \iff \vdash_{m,m+1} \phi$$

for all m-variable formulas ϕ?

19. [**Problem 9.16, Venema**] If a class of structures is closed under ultraproducts, must it be pseudo-elementary?

20. [**Madd94a, problems 15, 16**] Following on from exercise 12.5(13), is it true that *almost all* finite relation algebras are representable? More precisely, if $RA(n), RRA(n)$ are the numbers of isomorphism types of relation algebras and representable relation algebras (respectively) with no more than n elements, is it the case that

$$\lim_{n \to \infty} \frac{RRA(n)}{RA(n)} = 1?$$

21. [**Madd94a, problem 9**] Let $\mathcal{A}$ be a finite relation algebra with a flexible atom. Does $\mathcal{A}$ necessarily have a finite representation?

Bibliography

Numbers in brackets at the end of each reference
refer to the pages on which it is cited.
The bibliography thus serves as an author index.

[AddHen$^+$65]   J. Addison, L. Henkin, and A. Tarski, editors. *The theory of models*. Proc. 1963 Internat. Symposium at Berkeley. North-Holland, Amsterdam, 1965. [274]

[All81]   J. Allen. An interval-based representation of temporal knowledge. In P. J. Hayes, editor, *Proc. 7th International Joint Conference on Artificial Intelligence*, pages 221–226. William Kaufmann, 1981. [252]

[All83]   J. Allen. Maintaining knowledge about temporal intervals. *Communications of the ACM* 26 no. 11 (1983) 832–843. [142]

[All84]   J. Allen. Towards a general theory of action and time. *Artificial Intelligence* 23 no. 2 (1984) 123–154. [99, 252, 253]

[AllHay85a]   J. Allen and P. Hayes. A commonsense theory of time. In A. Joshi, editor, *Proc. 9th International Joint Conference on Artificial Intelligence*, pages 528–531. Morgan Kaufmann, 1985. [99]

[AllHay85b]   J. Allen and P. Hayes. Moments and points in an interval-based temporal logic. Technical Report TR180, Department of Computer Science, University of Rochester, 1985. [252]

[AllKau83]   J. Allen and H. Kautz. A model of naïve temporal reasoning. In J. Hobbs and R. Moore, editors, *Contributions to Artificial Intelligence, Volume 1*. Ablex, Norwood, N.J., 1983. [99, 252]

[AllKoo83]   J. Allen and J. Koomen. Planning using a temporal world model. In *Proc. 8th International Joint Conference on Artificial Intelligence*, pages 711–714. Morgan Kaufmann, 1983. [99, 252]

[Alm99]   J. Almeida. Hyperdecidable pseudovarieties and the calculation of semidirect products. *Int. J. Algebra and Computation* 9 (1999) 241–261. Available at http://www.fc.up.pt/cmup/home/jalmeida/. [590]

[AlmDel99]   J. Almeida and M. Delgado. Sur certains systèmes d'équations avec contraintes dans un groupe libre. *Port. Math* 56 (1999) 409–417. Available at http://www.fc.up.pt/cmup/home/jalmeida/. [590]

[And87]   H. Andréka. Non-representable PEA_3 with representable CA-reduct. Preprint, 1987. [204]

[And88]   H. Andréka. On taking subalgebras of relativized relation algebras. *Algebra Universalis* 25 (1988) 96–110. [154]

[And89]   H. Andréka. On the 'union-relation composition' reducts of relation algebras. *Abstracts Amer. Math. Soc.* 10 no. 2 (1989) 174. Full, unpublished manuscript called 'On the representation problem of distributive semilattice-ordered semigroups', Mathematical Institute, Hungarian Academy of Sciences, Budapest. [205]

[And90a]   H. Andréka. Finite axiomatizability of $S\mathfrak{N}\mathfrak{r}_n CA_{n+1}$ and non-finite axiomatizability of $S\mathfrak{N}\mathfrak{r}_n CA_{n+2}$. Lecture notes for a series of lectures given at algebraic logic meeting, Oakland, CA, 1990. [192, 465, 488]

[And90b]   H. Andréka. The equational theories of representable positive cylindric and relation algebras are decidable, 1990. Preprint. [206]

[And91]   H. Andréka. Representation of distributive lattice-ordered semigroups with binary relations. *Algebra Universalis* 28 (1991) 12–25. [205]

[And94]   H. Andréka. Weakly representable but not representable relation algebras. *Algebra Universalis* 32 (1994) 31–43. [17, 164, 165, 206, 528]

[And97a]   H. Andréka. Complexity of equations valid in algebras of relations, Part I: Strong non-finitizability. *Ann. Pure. Appl. Logic* 89 (1997) 149–209. [169, 185, 204, 205, 215, 465]

[And97b] H. Andréka. Complexity of equations valid in algebras of relations, Part II: Finite axiomatizations. *Ann. Pure. Appl. Logic* 89 (1997) 211–229. [206]

[And01] H. Andréka. A finite axiomatization of locally square cylindric-relativized set algebras. *Studia Sci. Math. Hungar.* 38 (2001) 1–11. Preprint (1995) available at http://www.math-inst.hu/pub/algebraic-logic. [12, 184, 185, 274, 283, 626]

[AndBen+98] H. Andréka, J. van Benthem, and I. Németi. Modal logics and bounded fragments of predicate logic. *J. Philosophical Logic* 27 (1998) 217–274. [16, 182, 212, 584, 585]

[AndDün+92] H. Andréka, I. Düntsch, and I. Németi. A nonpermutational integral relation algebra. *Michigan Mathematics Journal* 39 (1992) 371–384. [281]

[AndGiv+94a] H. Andréka, S. Givant, and I. Németi. Decision problems for equational theories of relation algebras. *Bulletin of Section of Logic* 23 no. 2 (1994) 47–52. [142]

[AndGiv+94b] H. Andréka, S. Givant, and I. Németi. The lattice of varieties of representable relation algebras. *J. Symbolic Logic* 59 no. 2 (1994) 631–661. [139, 149]

[AndGiv+95] H. Andréka, S. Givant, and I. Németi. Perfect extensions of boolean algebras with operators, and derived algebras. *J. Symbolic Logic* 60 no. 3 (1995) 775–795. [59, 96]

[AndGiv+97] H. Andréka, S. Givant, and I. Németi. *Decision problems for equational theories of relation algebras*, volume 126 of *Memoirs*. Amer. Math. Soc., Providence, Rhode Island, 1997. [21, 117, 136, 145, 210, 578, 582]

[AndGiv+98] H. Andréka, S. Givant, S. Mikulás, I. Németi, and A. Simon. Notions of density that imply representability in algebraic logic. *Ann. Pure. Appl. Logic* 91 (1998) 93–190. [72, 210]

[AndGol+98] H. Andréka, R. Goldblatt, and I. Németi. Relativised quantification: Some canonical varieties of sequence-set algebras. *J. Symbolic Logic* 63 (1998) 163–184. [184, 185, 283, 626]

[AndHod+99] H. Andréka, I. Hodkinson, and I. Németi. Finite algebras of relations are representable on finite sets. *J. Symbolic Logic* 64 (1999) 243–267. [vi, 590, 592, 593, 595, 598, 600, 614]

[AndMad+91] H. Andréka, R. Maddux, and I. Németi. Splitting in relation algebras. *Proc. Amer. Math. Soc.* 111 (1991) 1085–1093. [22, 211]

[AndMad94] H. Andréka and R. Maddux. Representations for small relation algebras. *Notre Dame J. Formal Logic* 35 no. 4 (1994) 550–562. [137, 140, 144, 206]

[AndMik94] H. Andréka and S. Mikulás. Lambek calculus and its relational semantics: completeness and incompleteness. *J. Logic, Language and Information* 3 (1994) 1–37. [206, 601]

[AndMon+91] H. Andréka, J. Monk, and I. Németi, editors. *Algebraic logic*, volume 54 of *Colloq. Math. Soc. J. Bolyai*. North-Holland, Amsterdam, 1991. [130, 204, 539]

[AndNém84a] H. Andréka and I. Németi. On a problem of J. S. Johnson about representability of polyadic algebras. Preprint, 1984. [204]

[AndNém84b] H. Andréka and I. Németi. Term definability of substitutions in Gs's. Preprint, 1984. [203]

[AndNém+01] H. Andréka, I. Németi, and I. Sain. Algebraic logic. In D. Gabbay and F. Guenthner, editors, *Handbook of philosophical logic*, volume 2, pages 133–247. Kluwer Academic Publishers, 2nd edition, 2001. [9, 276, 465, 486, 487, 627]

[AndTho88] H. Andréka and R. Thompson. A Stone type representation theorem for algebras of relations of higher rank. *Trans. Amer. Math. Soc.* 309 no. 2 (1988) 671–682. [12, 183–185, 274, 626]

[AndTuz88] H. Andréka and Z. Tuza. Nonfinite axiomatizability of the polyadic operations in algebraic logic. *Abstracts Amer. Math. Soc.* 9 (1988) 500. [204]

[AneHou91] I. Anellis and N. Houser. Nineteenth century roots of algebraic logic and universal algebra. In Andréka et al. [AndMon+91], pages 1–36. [4]

[AreMon+99] C. Areces, C. Monz, H. de Nivelle, and M. de Rijke. The guarded fragment: ins and outs. In J. Gerbrandy, M. Marx, M. de Rijke, and Y. Venema, editors, *JFAK. Essays dedicated to J. van Benthem on the occasion of his 50th birthday*, Vossiuspers. Amsterdam University Press, Amsterdam, 1999. CD-ROM. ISBN 90 5629 104 1. [182, 584]

Bibliography

[Ari63]   Aristotle. *Categories*. Clarendon Aristotle series. Clarendon Press, 1963. [2]

[Bar77]   J. Barwise. On Moschovakis closure ordinals. *J. Symbolic Logic* 42 (1977) 292–296. [274, 494]

[BauFri[+]96]   G. Baum, M. Frias, A. Haeberer, and P. López. From specifications to programs: a fork-algebraic approach to bridge the gap. In W. Penczek and A. Szalas, editors, *Mathematical Foundations of Computer Science*, volume 1113 of *Lecture notes in computer science*, pages 180–191. Springer, 1996. [210]

[Ben76]   J. van Benthem. *Modal correspondence theory*. PhD thesis, Mathematical Institute, University of Amsterdam, 1976. [275, 583]

[Ben80]   J. van Benthem. Some kinds of modal completeness. *Studia Logica* 39 (1980) 125–158. [89, 130]

[Ben85]   J. van Benthem. *Modal logic and classical logic*. Bibliopolis, Naples, 1985. [85, 94, 275, 583]

[Ben96]   J. van Benthem. *Exploring logical dynamics*. Studies in Logic, Language and Information. CSLI Publications & FoLLI, Stanford, 1996. [182, 208, 212, 583, 584]

[Ben97]   J. van Benthem. Dynamic bits and pieces. Technical Report LP-97-01, ILLC, Universiteit van Amsterdam, 1997. [16, 182, 584–586]

[Ber66]   R. Berger. *The undecidability of the domino problem*, volume 66 of *Memoirs*. Amer. Math. Soc., Providence, Rhode Island, 1966. [542]

[Bir35]   G. Birkhoff. On the structure of abstract algebras. *Proc. Cambr. Philos. Soc.* 31 (1935) 433–454. [68]

[Bir44]   G. Birkhoff. Subdirect unions in universal algebra. *Bull. Amer. Math. Soc.* 50 (1944) 764–768. [69]

[Birò92]   B. Birò. Non-finite axiomatizability results in algebraic logic. *J. Symbolic Logic* 57 (1992) 832–843. [206]

[BlaRij[+]01]   P. Blackburn, M. de Rijke, and Y. Venema. *Modal logic*. Tracts in Theoretical Computer Science. Cambridge University Press, Cambridge, UK, 2001. [85, 94–96, 210]

[BloPig89]	W. Blok and D. Pigozzi. *Algebraizable logics*, volume 396 of *Memoirs*. Amer. Math. Soc., Providence, Rhode Island, 1989. [212]
[Boo51]	G. Boole. *The mathematical analysis of logic, being an essay towards a calculus of deductive reasoning*. Oxford, Basil Blackwell, 1951. Original work published in 1847 in Cambridge by MacMillan, Barclay and MacMillan and in London by George Bell. [2, 38]
[Bre77]	D. Bredikhin. Abstract characteristic of some classes of algebras of relations. *Algebra and theory of numbers* 2 (1977) 3–19. In Russian. [205]
[Bre93]	D. Bredikhin. The equational theory of relation algebras with positive operations. *Izv. Vyash. Uchebn. Zaved. Math.* 3 (1993) 23–30. In Russian. [206]
[BreSch78]	D. Bredikhin and B. Schein. Representation of ordered semigroups and lattices by binary relations. *Colloquium Mathematicum* 39 (1978) 1–12. [206]
[BriKah+97]	C. Brink, W. Kahl, and G. Schmidt, editors. *Relational methods in computer science*. Advances in computing sciences. Springer-Verlag, Wien, New York, 1997. [212]
[BruRys49]	R. Bruck and H. Ryser. The nonexistence of certain finite projective planes. *Canad. J. Math.* 1 (1949) 88–93. [146]
[Bur82]	J. Burgess. Axioms for tense logic I: "since" and "until". *Notre Dame J. Formal Logic* 23 no. 2 (1982) 367–374. [274]
[BurSan81]	S. Burris and H. Sankappanavar. *A course in universal algebra*, volume 78 of *Graduate texts in mathematics*. Springer-Verlag, New York, 1981. Available at http://www.thoralf.uwaterloo.ca/htdocs/ualg.html. [25, 69, 71]
[Cam91]	P. J. Cameron. Projective and polar spaces. Maths Notes 13, Queen Mary, University of London, 1991. [146]
[ChagZak97]	A. Chagrov and M. Zakharyaschev. *Modal logic*, volume 35 of *Oxford Logic Guides*. Clarendon Press, Oxford, 1997. [95]
[ChaKei90]	C. Chang and H. Keisler. *Model theory*. North-Holland, Amsterdam, 3rd edition, 1990. [25, 31, 35, 37, 42, 51, 58, 123, 130, 284, 288, 290, 304, 329, 330, 333]

Bibliography

[ChiTar51]   L. Chin and A. Tarski. Distributive and modular laws in the arithmetic of relation algebras. *University of California Publications in Mathematics, New Series* 1 (1951) 341–384. [4, 105, 117, 158]

[Com69]   S. Comer. Classes without the amalgamation property. *Pacific J. Math.* 28 (1969) 309–318. [210]

[Com83]   S. Comer. A remark on chromatic polygroups. *Congr. Numer.* 38 (1983) 85–95. [142]

[Com84]   S. Comer. Combinatorial aspects of relations. *Algebra Universalis* 18 (1984) 77–94. [141]

[Com91]   S. Comer. A remark on representable positive cylindric algebras. *Algebra Universalis* 28 (1991) 150–151. [206]

[CsiGab+95]   L. Csirmaz, D. Gabbay, and M. de Rijke, editors. *Logic colloquium '92*. Studies in logic, language and computation. CSLI Publications & FoLLI, Stanford, 1995. [212]

[DaiMon63]   A. Daigneault and J. Monk. Representation theory for polyadic algebras. *Fundamenta Math.* 52 (1963) 151–176. [203]

[DawLin+95]   A. Dawar, S. Lindell, and S. Weinstein. Infinitary logic and inductive definability over finite structures. *Information and computation* 119 (1995) 160–175. [274, 404]

[DeaMcD87]   T. Dean and D. McDermott. Temporal database management. *Artificial Intelligence* 32 (1987) 1–55. [99, 252]

[DecMei+91]   R. Dechter, I. Meiri, and J. Pearl. Temporal constraint networks. *Artificial Intelligence* 49 (1991) 61–95. [99, 143, 252]

[Die97]   R. Diestel. *Graph theory*. Number 173 in Graduate texts in mathematics. Springer-Verlag, 1997. [448]

[Dün91]   I. Düntsch. Small integral relation algebras generated by a partial order. *Period. Math. Hungar.* 23 (1991) 129–138. [142]

[Dün93]   I. Düntsch. A note on cylindric lattices. In Rauszer [Rau93], pages 231–238. [206, 212]

[EbbFlu99]   H.-D. Ebbinghaus and J. Flum. *Finite model theory*. Perspectives in mathematical logic. Springer-Verlag, New York, 2nd edition, 1999. [387, 404]

[Ehr61]	A. Ehrenfeucht. An application of games to the completeness problem for formalized theories. *Fund. Math.* 49 (1961) 128–141. [274]
[Erd59]	P. Erdös. Graph theory and probability. *Canadian Journal of Mathematics* 11 (1959) 34–38. [447, 448]
[Fag74]	R. Fagin. Generalized first-order spectra and polynomial-time recognizable sets. In R. M. Karp, editor, *Complexity of computation*, number 7 in SIAM–AMS Proceedings, pages 43–73, 1974. [59, 287, 291]
[Fin75]	K. Fine. Some connections between elementary and modal logic. In Kanger [Kan75], pages 15–31. [89, 123, 130]
[Fra54]	R. Fraïssé. Sur l'extension aux relations de quelques propriétés des ordres. *Ann. Sci. École Norm. Sup.* 71 (1954) 363–388. [13, 395]
[Fri02]	M. Frias. *Fork algebras in algebra, logic and computer science.* Advances in Logic. World Scientific Publishing Co., 2002. [210]
[FriBau+95]	M. Frias, G. Baum, A. Haeberer, and P. Veloso. Fork algebras are representable. *Bulletin of the section of logic* 24 no. 2 (1995) 64–75. [210]
[FriBau+96]	M. Frias, G. Baum, and A. Haeberer. Adding design strategies to fork algebras. In D. Bjrner, M. Broy, and I. Pottosin, editors, *2nd International Andrei Ershov Memorial Conference, Akademgorodok, Novosibirsk*, volume 1181 of *Lecture notes in computer science*, pages 214–226. Springer, 1996. [210]
[FriBau+97]	M. Frias, G. Baum, and A. Haeberer. Fork algebras in algebra, logic and computer science. *Fundamenta Informaticae* 32 no. 1 (1997) 1–25. [210]
[Gab81]	D. Gabbay. An irreflexivity lemma with applications to axiomatizations of conditions on tense frames. In U. Monnich, editor, *Aspects of Philosophical Logic*, pages 67–89. Reidel, Dordrecht, 1981. [274]
[GabHod+94]	D. Gabbay, I. Hodkinson, and M. Reynolds. *Temporal logic: mathematical foundations and computational aspects, Vol. 1.* Clarendon Press, Oxford, 1994. [13, 290]

[GabKur+] D. Gabbay, A. Kurucz, F. Wolter, and M. Zakharyaschev. *Many-dimensional modal logics: theory and applications*. Studies in Logic. North-Holland, Amsterdam. In preparation. [541]

[GalSte53] D. Gale and F. Stewart. Infinite games with perfect information. *Contributions to the theory of games II, Annals of mathematical studies* 28 (1953) 291–296. [238, 318, 498]

[Giv91] S. Givant. Tarski's development of logic and mathematics based on the calculus of relations. In Andréka et al. [AndMon+91], pages 189–215. [486]

[Giv99] S. Givant. Universal classes of simple relation algebras. *J. Symbolic Logic* 64 (1999) 575–589. [71, 90, 129, 136, 145]

[Giv01] S. Givant. Inequivalent representations of geometric relation algebras. (2001). Submitted. [148]

[GivAnd02] S. Givant and H. Andréka. Groups and algebras of binary relations. *Bull. Symbolic Logic* 8 (2002) 38–64. [209, 210]

[GivVen99] S. Givant and Y. Venema. The preservation of Sahlqvist equations in completions of Boolean algebras with operators. *Algebra Universalis* 41 (1999) 47–84. [94, 95]

[Gol89] R. Goldblatt. Varieties of complex algebras. *Ann. Pure. Appl. Logic* 44 (1989) 173–242. [90, 91, 93, 96, 97, 130, 455]

[Gol91] R. Goldblatt. On closure under canonical embedding algebras. In Andréka et al. [AndMon+91], pages 217–229. [90, 93, 131]

[Gol95] R. Goldblatt. Elementary generation and canonicity for varieties of boolean algebras with operators. *Algebra Universalis* 34 (1995) 551–607. [90, 92, 93, 97, 184]

[Gol01] R. Goldblatt. Persistence and atomic generation for varieties of boolean algebras with operators. *Studia Logica* 68 no. 2 (2001) 155–171. [91, 93, 97, 98]

[Gor00] L. Gordeev. Combinatorial principles relevant to finite variable logic. In J. Desharnais, editor, *RelMiCS 2000*, pages 95–111. Département d'Informatique, Université Laval, 2000. [464, 488]

[Grä99a] E. Grädel. Decision procedures for guarded logics. In *Automated Deduction - CADE16*, volume 1632 of *Lecture notes in computer science*, pages 31–51. Springer-Verlag, 1999. [182, 585, 588]

[Grä99b]	E. Grädel. On the restraining power of guards. *J. Symbolic Logic* 64 (1999) 1719–1742. [182, 584, 585, 587, 590]
[Grät79]	G. Grätzer. *Universal algebra*. Springer-Verlag, New York, 2nd edition, 1979. [25]
[Gro96]	M. Grohe. Arity hierarchies. *Ann. Pure. Appl. Logic* 82 (1996) 103–163. [590]
[GurHar82]	Y. Gurevich and L. Harrington. Trees, automata, and games. In *Theory of computing (proc. 14th annual A.C.M. symposium, San Francisco)*, pages 60–65. Association for Computing Machinery, 1982. [10]
[HaeFri$^+$97]	A. Haeberer, M. Frias, G. Baum, and P. Veloso. Fork algebras. In Brink et al. [BriKah$^+$97], chapter 4. [210]
[Hai87]	M. Haiman. Arguesian lattices which are not linear. *Bull. Amer. Math. Soc.* 16 (1987) 121–123. [165, 513, 529]
[Hai91]	M. Haiman. Arguesian lattices which are not type I. *Algebra Universalis* 28 (1991) 128–137. [165, 513, 529]
[Hal57]	P. Halmos. Algebraic logic, IV. *Trans. Amer. Math. Soc.* 86 (1957) 1–27. [200]
[Hal62]	P. Halmos. *Algebraic logic*. Chelsea Publishing Co., New York, 1962. [167, 201, 203]
[Han95]	B. Hansen. Finitizability questions for some reducts of cylindric algebras. In Csirmaz et al. [CsiGab$^+$95], pages 115–134. [206]
[Hen68]	L. Henkin. Relativization with respect to formulas and its use in proofs of independence. *Compositio Math.* 20 (1968) 88–106. [180]
[HenMon$^+$71]	L. Henkin, J. Monk, and A. Tarski. *Cylindric algebras part I*. North-Holland, 1971. [9, 11, 78, 82, 96, 166, 167, 169–171, 185, 187, 188, 191–193, 215, 261]
[HenMon$^+$81]	L. Henkin, D. Monk, A. Tarski, H. Andréka, and I. Németi. *Cylindric set algebras*, volume 883 of *Lecture Notes in Math*. Springer, Berlin, 1981. [185]
[HenMon$^+$85]	L. Henkin, J. Monk, and A. Tarski. *Cylindric algebras part II*. North-Holland, 1985. [16, 22, 166, 168–170, 174, 177, 183, 185, 186, 189, 191–193, 200, 201, 204, 211, 212, 215, 261, 271, 282, 432, 465, 486, 487, 540, 593]

[Her95]	B. Herwig. Extending partial isomorphisms on finite structures. *Combinatorica* 15 (1995) 365–371. [589, 590]
[Her98]	B. Herwig. Extending partial isomorphisms for the small index property of many ω–categorical structures. *Israel J. Math.* 107 (1998) 93–124. [589, 590]
[HerLas00]	B. Herwig and D. Lascar. Extending partial isomorphisms and the profinite topology on the free groups. *Trans. Amer. Math. Soc.* 352 (2000) 1985–2021. [590]
[Hin73]	J. Hintikka. Quantifiers vs. quantification theory. *Dialectica* 27 (1973) 329–358. [274]
[Hir95]	R. Hirsch. Intractability in the Allen & Koomen planner. *Computational Intelligence Journal* 11 no. 4 (1995) 553–564. [99]
[Hir96]	R. Hirsch. Relation algebras of intervals. *Artificial Intelligence Journal* 83 (1996) 1–29. [99, 143]
[Hir99]	R. Hirsch. A relation algebra with undecidable network satisfaction problem. *Bulletin of the Interest Group in Pure and Applied Logics* 7 no. 4 (1999) 547–554. [255, 579]
[Hir00]	R. Hirsch. Tractable approximations for temporal constraint handling. *Artificial Intelligence Journal* 116 (2000) 287–295. [vi, 253]
[HirHod97a]	R. Hirsch and I. Hodkinson. Complete representations in algebraic logic. *J. Symbolic Logic* 62 no. 3 (1997) 816–847. [vi, 22, 47, 181, 302, 443]
[HirHod97b]	R. Hirsch and I. Hodkinson. Step by step — building representations in algebraic logic. *J. Symbolic Logic* 62 no. 1 (1997) 225–279. [vi, 123, 148, 269, 282]
[HirHod$^+$98]	R. Hirsch, I. Hodkinson, M. Marx, S. Mikulás, and M. Reynolds. Mosaics and step-by-step. In Orlowska [Orl98], pages 158–167. Appendix to [VenMar98]. [590]
[HirHod00]	R. Hirsch and I. Hodkinson. Relation algebras with n-dimensional relational bases. *Ann. Pure. Appl. Logic* 101 (2000) 227–274. [vi, 361]
[HirHod01a]	R. Hirsch and I. Hodkinson. Relation algebras from cylindric algebras, II. *Ann. Pure. Appl. Logic* 112 (2001) 267–297. [vi, 361, 373, 467]

[HirHod01b] R. Hirsch and I. Hodkinson. Representability is not decidable for finite relation algebras. *Trans. Amer. Math. Soc.* 353 (2001) 1403–1425. [vi]

[HirHod02a] R. Hirsch and I. Hodkinson. Strongly representable atom structures of relation algebras. *Proc. Amer. Math. Soc.* 130 (2002) 1819–1831. [vi]

[HirHod⁺02b] R. Hirsch, I. Hodkinson, and A. Kurucz. On modal logics between $K \times K \times K$ and $S5 \times S5 \times S5$. *J. Symbolic Logic* 67 no. 1 (2002) 221–234. [200, 541]

[HirHod⁺02c] R. Hirsch, I. Hodkinson, and R. Maddux. Provability with finitely many variables. *Bull. Symbolic Logic* (2002). To appear. [361, 464, 466]

[HirHod⁺02d] R. Hirsch, I. Hodkinson, and R. Maddux. Relation algebra reducts of cylindric algebras and an application to proof theory. *J. Symbolic Logic* 67 no. 1 (2002) 197–213. [464]

[Hod85] W. Hodges. *Building Models by Games.* Number 2 in London Mathematical Society Student Texts. Cambridge University Press, 1985. [10, 11, 215, 226, 232, 238, 274, 276]

[Hod93] W. Hodges. *Model theory,* volume 42 of *Encyclopedia of mathematics and its applications.* Cambridge University Press, 1993. [vi, 25, 31, 35, 37, 38, 42, 58, 60, 123, 130, 148, 215, 244, 274, 284, 287, 289, 290, 297, 304, 310, 326, 329, 330, 333, 425, 456, 458]

[Hod97] W. Hodges. Games in logic. In P. Dekker, M. Stokhof, and Y. Venema, editors, *Proc. 11th Amsterdam Colloquium,* pages 13–18. Institute for Logic, Language and Computation/Department of Philosophy, Universiteit van Amsterdam, 1997. [10]

[Hodk97] I. Hodkinson. Atom structures of cylindric algebras and relation algebras. *Ann. Pure. Appl. Logic* 89 (1997) 117–148. [22, 169, 443, 535]

[Hodk02] I. Hodkinson. Loosely guarded fragment of first-order logic has the finite model property. *Studia Logica* 70 (2002) 205–240. [vi, 590, 614]

[HodMik00] I. Hodkinson and S. Mikulás. Non-finite axiomatizability of reducts of algebras of relations. *Algebra Universalis* 43 (2000) 127–156. [22, 165, 166, 200, 206, 443, 513, 529]

[HodMik+01]	I. Hodkinson, S. Mikulás, and Y. Venema. Axiomatizing complex algebras by games. *Algebra Universalis* 46 (2001) 455–478. [136, 275]
[HodOtt01]	I. Hodkinson and M. Otto. Finite conformal hypergraph covers and Gaifman cliques in finite structures. (2001). Preprint. [182, 585, 590, 614]
[HodWol+02]	I. Hodkinson, F. Wolter, and M. Zakharyaschev. Decidable and undecidable fragments of first-order branching temporal logics, 2002. Proc. Logic in Computer Science (LICS'02), accepted. Available at http://www.dcs.kcl.ac.uk/staff/mz. [541]
[Hoo01]	E. Hoogland. *Definability and interpolation – model-theoretic investigations*. PhD thesis, Institute for Logic, Language and Computation, Amsterdam, 2001. DS-2001-05. [210]
[How78]	J. Howie. Idempotent generators in finite full transformation semigroups. *Proc. Royal Soc. Edinburgh* 81A (1978) 317–323. [369]
[Hru92]	E. Hrushovski. Extending partial isomorphisms of graphs. *Combinatorica* 12 (1992) 411–416. [590]
[HugPip73]	R. Hughes and F. Piper. *Projective planes*. Graduate texts in mathematics. Springer-Verlag, 1973. [146]
[Hun04]	E. Huntington. Sets of independent postulates for the algebra of logic. *Trans. Amer. Math. Soc.* 5 (1904) 288–309. [5]
[ImmKoz87]	N. Immerman and D. Kozen. Definability with bounded number of bound variables. In *LICS87, Proceedings of the Symposium on Logic in Computer Science, Ithaca, New York*, pages 236–244, Washington, 1987. Computer Society Press. [274]
[Jip93]	P. Jipsen. Discriminator varieties of boolean algebras with residuated operators. In Rauszer [Rau93], pages 239–252. [70, 72, 76]
[JipMad97]	P. Jipsen and R. Maddux. Nonrepresentable sequential algebras. *Logic J. IGPL* 5 no. 4 (1997) 565–574. [206]
[Joh69]	J. Johnson. Nonfinitizability of classes of representable polyadic algebras. *J. Symbolic Logic* 34 (1969) 344–352. [200, 204]
[Jón59]	B. Jónsson. Representation of modular lattices and relation algebras. *Trans. Amer. Math. Soc.* 92 (1959) 449–464. [112, 144, 146, 163–166, 247, 388, 528, 575]

[Jón62]	B. Jónsson. Defining relations for full semigroups of finite transformations. *Michigan Mathematics Journal* 9 (1962) 77–85. [170, 369]
[Jón82]	B. Jónsson. Varieties of relation algebras. *Algebra Universalis* 15 (1982) 273–298. [120]
[Jón91]	B. Jónsson. The theory of binary relations. In Andréka et al. [AndMon+91], pages 245–292. [12, 129, 145, 205, 207, 215, 264, 267, 351, 362, 513, 519, 600, 619]
[Jón95]	B. Jónsson. On the canonicity of Sahlqvist identities. *Studia Logica* 53 (1995) 473–491. [94, 95]
[JónTar48]	B. Jónsson and A. Tarski. Representation problems for relation algebras. *Bull. Amer. Math. Soc.* 54 (1948) 80, 1192. [134, 136, 208]
[JónTar51]	B. Jónsson and A. Tarski. Boolean algebras with operators I. *American Journal of Mathematics* 73 (1951) 891–939. [60, 61, 63, 64, 77–80, 83, 94, 95, 117, 583]
[JónTar52]	B. Jónsson and A. Tarski. Boolean algebras with operators II. *American Journal of Mathematics* 74 (1952) 127–162. [70, 71, 105, 110, 112, 117, 158, 162, 166, 208, 209, 583, 620]
[Kan75]	S. Kanger, editor. *Proc. 3rd Scandinavian logic symposium, Uppsala, 1973*. North Holland, Amsterdam, 1975.
[Kant92]	I. Kant. *Kant's introduction to logic and his essay on the mistaken subtilty of the four figures*. Thoemmes Press, 1992. [1]
[Kar94]	B. von Karger. The class of representable sequential algebras is not finitely axiomatizable. Draft, 1994. [206]
[Karp65]	C. Karp. Finite-quantifier equivalence. In Addison et al. [AddHen+65], pages 407–412. [274]
[KauLad91]	H. Kautz and P. Ladkin. Integrating metric and qualitative temporal reasoning. In *Proceedings of the ninth national conference on artificial intelligence, Anaheim, California*, pages 241–246. AAAI Press/MIT Press, 1991. [99, 252]
[Kay91]	R. Kaye. *Models of Peano Arithmetic*, volume 15 of *Oxford Logic Guides*. Clarendon Press, Oxford, 1991. [304]

Bibliography

[Kec95] A. Kechris. *Classical descriptive set theory*. Springer-Verlag, New York, 1995. [238]

[Kei65] H. Keisler. Finite approximations of infinitely long formulas. In Addison et al. [AddHen+65], pages 158–169. [9, 274, 275]

[Kra99] M. Kracht. *Tools and techniques in modal logic*, volume 142 of *Studies in Logic and the Foundations of Mathematics*. North-Holland, Amsterdam, 1999. [94, 95]

[Kram91] R. Kramer. Relativized relation algebras. In Andréka et al. [AndMon+91], pages 293–349. [154, 211]

[Kur97] A. Kurucz. *Decision problems in algebraic logic*. PhD thesis, Hungarian Academy of Sciences, Budapest, 1997. Available at http://www.math-inst.hu/pub/algebraic-logic/kuagthes.dvi. [208, 210]

[Kur00a] A. Kurucz. Arrow logic and infinite counting. *Studia Logica* 65 (2000) 199–222. [208]

[Kur00b] A. Kurucz. Weakly associative relation algebras and projection elements. Preprint, 2000. [208]

[KurNém00] A. Kurucz and I. Németi. Representability of pairing relation algebras depends on your ontology. *Fundamenta Informaticae* 44 no. 4 (2000) 397–420. [208]

[LadMad94] P. Ladkin and R. Maddux. On binary constraint problems. *Journal of the Association of Computing Machinery* 41 (1994) 435–469. [143, 144, 253, 254]

[LadRei93] P. Ladkin and A. Reinefeld. A symbolic approach to interval constraint problems. *Artificial Intelligence and Symbolic Mathematical Computing* 737 (1993) 65–84. [252]

[Lig90] G. Ligozat. Weak representation of interval algebras. In *Proceedings of eighth national conference on artificial intelligence, Mass.*, pages 715–720, 1990. [99]

[Lig94] G. Ligozat. Tractable relations in temporal reasoning: pre-convex relations. In B. Neumann, editor, *ECAI-94, Workshop on Spatial and Temporal Reasoning*, Amsterdam, August 1994. John Wiley and Sons Ltd. [99]

[Lyn50] R. Lyndon. The representation of relational algebras. *Annals of Mathematics* 51 no. 3 (1950) 707–729. [8, 11, 14, 112, 115, 119, 165, 215, 217, 275, 280, 337, 341, 343, 346, 350, 355, 459, 613, 620]

[Lyn56] R. Lyndon. The representation of relation algebras, II. *Annals of Mathematics* 63 no. 2 (1956) 294–307. [9, 215, 246, 261, 275, 343, 516]

[Lyn61] R. Lyndon. Relation algebras and projective geometries. *Michigan Mathematics Journal* 8 (1961) 207–210. [8, 118, 145, 147, 165, 392]

[Madaa] J. Madarász. Hereditarily non-finite axiomatizability of relation algebras and their variants. Manuscript consists of 3 TEX pages and 18 handwritten pages. [206, 207]

[Madab] J. Madarász. Non-finite axiomatizability of kinds of algebras of relations. Manuscript. [206, 207]

[MadaNém+97] J. Madarász, I. Németi, and G. Sági. On the finitization problem of relation algebras. *Bulletin of the section of logic* 26 no. 3 (1997) 140–145. [206, 207]

[MadaSay01] J. Madarász and T. Sayed Ahmed. Amalgamation, interpolation and epimorphisms, solutions to all problems of Pigozzi's paper, and some more. Preprint, Alfréd Rényi Institute of Mathematics, Hungarian Academy of Sciences, 2001. [211]

[Madd] R. Maddux. *Relation algebras*. In preparation. [137, 144]

[Madd78a] R. Maddux. Some sufficient conditions for the representability of relation algebras. *Algebra Universalis* 8 (1978) 162–172. [209]

[Madd78b] R. Maddux. *Topics in relation algebra*. PhD thesis, University of California, Berkeley, 1978. [210, 217, 225, 338, 348, 356, 357, 389, 432, 459, 578, 615]

[Madd80] R. Maddux. The equational theory of CA_3 is undecidable. *J. Symbolic Logic* 45 no. 2 (1980) 311–316. [200, 210]

[Madd82] R. Maddux. Some varieties containing relation algebras. *Trans. Amer. Math. Soc.* 272 no. 2 (1982) 501–526. [12, 107, 141, 155, 157–159, 163, 182, 198, 215–217, 225, 274, 614]

[Madd83]	R. Maddux. A sequent calculus for relation algebras. *Ann. Pure. Appl. Logic* 25 (1983) 73–101. [13, 15, 16, 123, 195, 338, 356, 376, 384, 385, 387, 389, 432, 486, 520, 615, 616]
[Madd85]	R. Maddux. Finite integral relation algebras. In S. Comer, editor, *Universal algebra and lattice theory*, number 1149 in Lecture Notes in Mathematics, pages 175–197. Springer-Verlag, 1985. Proc. Southeastern Conference in Universal Algebra and Lattice Theory, July 1984. [144, 162]
[Madd89a]	R. Maddux. Canonical relativized cylindric set algebras. *Proc. Amer. Math. Soc.* 107 (1989) 465–478. [421]
[Madd89b]	R. Maddux. Non-finite axiomatizability results for cylindric and relation algebras. *J. Symbolic Logic* 54 no. 3 (1989) 951–974. [13, 206, 338, 356, 357, 361, 372, 389, 390, 402, 486]
[Madd90a]	R. Maddux. A relation algebra which is not a cylindric reduct. *Algebra Universalis* 27 (1990) 279–288. [192]
[Madd90b]	R. Maddux. Necessary subalgebras of simple nonintegral semiassociative relation algebras. *Algebra Universalis* 27 (1990) 544–558. [161, 162]
[Madd91a]	R. Maddux. Introductory course on relation algebras, finite-dimensional cylindric algebras, and their interconnections. In Andréka et al. [AndMon⁺91], pages 361–392. [139, 168, 464]
[Madd91b]	R. Maddux. Pair-dense relation algebras. *Trans. Amer. Math. Soc.* 328 no. 1 (1991) 83–131. [158, 161–163, 209, 360, 375, 384, 387, 389, 390, 433, 531, 614]
[Madd91c]	R. Maddux. The neat embedding property and the number of variables required in proofs. *Proc. Amer. Math. Soc.* 112 (1991) 195–202. [17, 464]
[Madd91d]	R. Maddux. The origin of relation algebras in the development and axiomatization of the calculus of relations. *Studia Logica* 3/4 (1991) 421–455. [3]
[Madd92]	R. Maddux. Relation algebras of every dimension. *J. Symbolic Logic* 57 no. 4 (1992) 1213–1229. [361, 520]
[Madd93]	R. Maddux. Finitary axiomatizations of the true relational equations. In Rauszer [Rau93], pages 201–208. [210]

[Madd94a] R. Maddux. A perspective on the theory of relation algebras. *Algebra Universalis* 31 (1994) 456–465. [118, 539, 626, 627]

[Madd94b] R. Maddux. Undecidable semiassociative relation algebras. *J. Symbolic Logic* 59 (1994) 398–418. [578]

[Madd95] R. Maddux. On the derivation of identities involving projection functions. In Csirmaz et al. [CsiGab+95], pages 145–163. [388]

[Madd96] R. Maddux. Relation-algebraic semantics. *Theoretical Computer Science* 160 (1996) 1–85. [210, 212]

[MaddTar76] R. Maddux and A. Tarski. A sufficient condition for the representability of relation algebras. *Notices Amer. Math. Soc.* 23 (1976) A–447. [209, 210]

[Mak64] M. Makkai. On PC_Δ-classes in the theory of models. *Matematikai Kutató Intézetének Közleményei* 9 (1964) 159–194. [287]

[Mal71a] A. Mal'cev. A few remarks on quasi-varieties of algebraic systems. In Wells [Mal71c], pages 416–421. [69]

[Mal71b] A. Mal'cev. A general method for obtaining local theorems in group theory. In Wells [Mal71c], pages 15–21. [287]

[Mal71c] B. Wells, III, editor. *The metamathematics of algebraic systems: collected papers, 1936–1967, Anatolii Mal'cev*. North-Holland, 1971.

[Mar95] M. Marx. *Algebraic relativization and arrow logic*. PhD thesis, University of Amsterdam, 1995. ILLC Dissertation Series 95-3. [185, 208]

[Mar97] M. Marx. Complexity of modal logics of relations. Technical Report ML–97–02, Institute for Logic, Language and Computation, University of Amsterdam, May 1997. [584, 585, 590, 595]

[Mar98] M. Marx. Amalgamation in relation algebras. *J. Symbolic Logic* 63 no. 2 (1998) 479–484. [211]

[Mar99a] M. Marx. Relation algebras can tile. *Information Sciences* 119 (1999) 173–191. [210]

[Mar99b] M. Marx. Relativized relation algebras. *Algebra Universalis* 41 (1999) 23–45. [154, 158, 211]

Bibliography

[Mar01a] M. Marx. Relation algebra with binders. *J. Logic Computat.* 11 (2001) 691–700. [210]

[Mar01b] M. Marx. Tolerance logic. *J. Logic, Language and Information* 10 (2001) 353–373. [584, 585, 587]

[MarPól+96] M. Marx, L. Pólos, and M. Masuch, editors. *Arrow logic and multi-modal logic*. Studies in Logic, Language and Information. CSLI Publications & FoLLI, Stanford, 1996. [210, 212, 601]

[MarVen97] M. Marx and Y. Venema. *Multi-dimensional modal logic*, volume 4 of *Applied logic series*. Kluwer, 1997. [185, 208, 210]

[McKe66] R. McKenzie. *The representation of relation algebras*. PhD thesis, University of Colorado at Boulder, 1966. [119, 123, 136]

[McKe70] R. McKenzie. Representations of integral relation algebras. *Michigan Mathematics Journal* 17 (1970) 279–287. [119, 136, 140, 281]

[McKe75] R. McKenzie. On spectra, and the negative solution of the decision problem for identities having a finite nontrivial model. *J. Symbolic Logic* 40 (1975) 186–196. [72]

[McKeMcN+87] R. McKenzie, G. McNulty, and W. Taylor. *Algebras, lattices, varieties*. Cole mathematics series. Wadsworth & Brooks, Monterey, Calif., 1987. [25]

[McKi43] J. McKinsey. The decision problem for some classes of sentences without quantifiers. *J. Symbolic Logic* 8 (1943) 61–76. [69]

[Mei91] I. Meiri. Combining qualitative and quantitative constraints in temporal reasoning. Technical Report R-160, UCLA, December 1991. [99]

[Mik93a] S. Mikulás. Gabbay-style calculi. *Bulletin of the Section of Logic* 22 no. 2 (1993) 50–66. [210]

[Mik93b] S. Mikulás. Strong completeness of the Lambek calculus with respect to relational semantics. In Rauszer [Rau93], pages 209–217. [206]

[Mik95] S. Mikulás. *Taming logics*. PhD thesis, University of Amsterdam, 1995. ILLC Dissertation Series 95–12. [208]

[Mik96a] S. Mikulás. A note on expressing infinity in cylindric-relativized set algebras. Preprint, King's College London, 1996. [593]

[Mik96b]   S. Mikulás. Gabbay-style calculi. In H. Wansing, editor, *Proof theory of modal logic*, pages 243–252. Kluwer, 1996. [210]

[Mik00]   S. Mikulás. Axiomatizability of algebras of binary relations — a survey. In B. Löwe, W. Malzkorn, and T. Räsch, editors, *Foundations of the formal sciences II*, Nov 2000. Submitted. [529]

[MikMar99] S. Mikulás and M. Marx. Undecidable relativizations of algebras of relations. *J. Symbolic Logic* 64 (1999) 747–760. [583, 590]

[Mon61a]   J. Monk. On the representation theory for cylindric algebras. *Pacific J. Math.* 11 (1961) 1447–1457. [192, 488]

[Mon61b]   J. Monk. *Studies in cylindric algebra*. PhD thesis, University of California, Berkeley, 1961. [191, 192, 375, 432, 433]

[Mon64]   J. Monk. On representable relation algebras. *Michigan Mathematics Journal* 11 (1964) 207–210. [8, 12, 14, 119, 136, 145, 148, 165, 215, 351, 355, 454, 479, 513, 514, 540, 619]

[Mon69]   J. Monk. Nonfinitizability of classes of representable cylindric algebras. *J. Symbolic Logic* 34 (1969) 331–343. [9, 12, 169, 211, 215, 355, 441, 465]

[Mon70]   J. Monk. Completions of boolean algebras with operators. *Mathematische Nachrichten* 46 (1970) 47–55. [65, 94, 531]

[Mon91]   J. Monk. Remarks on the problems in the books Cylindric Algebras, part I and part II and Cylindric Set Algebras. In Andréka et al. [AndMon+91], pages 719–722. [593]

[Mon93]   J. Monk. Lectures on cylindric set algebras. In Rauszer [Rau93], pages 253–290. [184, 540]

[Mon00]   J. Monk. An introduction to cylindric set algebras. *Logic J. IGPL* 8 no. 4 (2000) 451–496. [170, 184, 185, 540]

[Mor60]   A. de Morgan. On the syllogism, no. iv, and on the logic of relations. *Transactions of the Cambridge Philosophical Society* 10 (1860) 331–358. Republished in [Mor66]. [1]

[Mor66]   A. de Morgan. *On the syllogism and other logical writings*. Rare masterpieces of philosophy and science. Routledge and Kegan Paul, 1966. W Stark, ed. [2]

Bibliography

[NebBür94] B. Nebel and H.-J. Bürckert. Reasoning about temporal relations: A maximal tractable subclass of Allen's interval algebra. In *Proceedings of the 12th National Conference of the American Association for Artificial Intelligence*, pages 356–361, Seattle, WA, July 1994. MIT Press. [99]

[Ném81] I. Németi. Connections between cylindric algebras and initial algebra semantics of CF languages. In B. Domolki and T. Gergely, editors, *Mathematical logic in computer science*, volume 26 of *Colloq. Math. Soc. J. Bolyai*, pages 561–605. North-Holland, 1981. [183]

[Ném83] I. Németi. The class of neat reducts is not a variety but is closed w.r.t HP. *Notre Dame J. Formal Logic* 24 (1983) 399–409. [191]

[Ném84] I. Németi. Neither the variety of cylindric algebras nor the variety of representable cylindric algebras is generated by its finite members for dimensions greater than 2. Preprint, Mathematical Institute, Budapest, 1984. [593]

[Ném85] I. Németi. Cylindric-relativized set algebras have strong amalgamation. *J. Symbolic Logic* 50 (1985) 689–700. [185, 211]

[Ném86] I. Németi. *Free algebras and decidability in algebraic logic*. PhD thesis, Hungarian Academy of Sciences, 1986. [13, 184, 185, 192, 217, 584, 599, 614]

[Ném87a] I. Németi. Decidability of relation algebras with weakened associativity. *Proc. Amer. Math. Soc.* 100 no. 2 (1987) 340–344. [593, 599, 621]

[Ném87b] I. Németi. On varieties of cylindric algebras with applications to logic. *Ann. Pure. Appl. Logic* 36 (1987) 235–277. [210]

[Ném91] I. Németi. Algebraisations of quantifier logics, an introductory overview. *Studia Logica* 50 no. 3/4 (1991) 485–570. An extended version of this paper is available at http://math-inst.hu/pub/algebraic-logic/survey.dvi. [185, 204–206, 212]

[Ném95] I. Németi. Decidable versions of first-order logic and cylindric-relativized set algebras. In Csirmaz et al. [CsiGab+95], pages 177–241. [13, 184, 185, 421]

[Ném96] I. Németi. A fine-structure analysis of first-order logic. In Marx et al. [MarPól+96], pages 221–247. [13, 184–186, 283, 540, 593, 626]

[NémAnd91] I. Németi and H. Andréka. On Jónsson's clones of operations on binary relations. In Andréka et al. [AndMon+91], pages 431–442. [208]

[NémSág00] I. Németi and G. Sági. On the equational theory of representable polyadic algebras. *J. Symbolic Logic* 65 (2000) 1143–1167. [203]

[NémSim97] I. Németi and A. Simon. Relation algebras from cylindric and polyadic algebras. *Logic J. IGPL* 5 (1997) 575–588. [191]

[Orl98] E. Orlowska, editor. *Logic at work. Essays dedicated to the memory of Elena Rasiowa*, volume 24 of *Studies in fuzziness and soft computing*. Springer-Verlag, Berlin/Heidelberg, 1998.

[Oxt71] J. Oxtoby. *Measure and category*. Springer-Verlag, New York, 1971. [215]

[Pei33] C. Peirce. *Collected papers*. Harvard University Press, Cambridge, Mass, 1933. C. Hartshorne and P. Weiss, eds. [4]

[Pel88] R. Pelavin. *A formal approach to planning with concurrent actions and external events*. PhD thesis, University of Rochester, 1988. [99, 252]

[Pig71] D. Pigozzi. Amalgamation, congruence-extension, and amalgamation properties in algebras. *Algebra Universalis* 1 (1971) 269–349. [211]

[Pin73] C. Pinter. A simpler set of axioms for polyadic algebras. *Fundamenta Math.* 79 (1973) 223–232. [205]

[Pra79] V. Pratt. Models of program logics. In *Proc. 20th IEEE Symposium on Foundations of Computer Science, San Juan*, pages 115–122, 1979. [386]

[Rau93] C. Rauszer, editor. *Algebraic methods in logic and in computer science*, volume 28 of *Banach Center publications*. Institute of Mathematics, Polish Academy of Sciences, 1993.

[Res75] D. Resek. *Some results on relativized cylindric algebras*. PhD thesis, University of California, Berkeley, 1975. [170, 183]

[ResTho91] D. Resek and R. Thompson. Characterizing relativized cylindric algebras. In Andréka et al. [AndMon+91], pages 519–538. [173, 183, 185]

Bibliography

[Rij95a]   M. de Rijke. A Lindström theorem for modal logic. In A. Ponse, M. de Rijke, and Y. Venema, editors, *Modal logic and process algebra*, volume 53 of *Lecture notes*, pages 217–230. CSLI, Stanford, CA, 1995. ISBN 1-881526-95-X. [275]

[Rij95b]   M. de Rijke. Modal model theory. Technical Report CS-R9517, CWI, Amsterdam, 1995. [275]

[RijVen95]   M. de Rijke and Y. Venema. Sahlqvist's theorem for boolean algebras with operators. *Studia Logica* 54 (1995) 61–78. [94, 95]

[Ság99]   G. Sági. *On the finitization problem of algebraic logic*. PhD thesis, Hungarian Academy of Sciences, Budapest, 1999. [203, 207, 208]

[Ság00]   G. Sági. A completeness theorem for higher order logics. *J. Symbolic Logic* 65 (2000) 857–884. [208]

[Sah75]   H. Sahlqvist. Completeness and correspondence in the first and second order semantics for modal logic. In Kanger [Kan75], pages 110–143. [94, 95]

[Sai82]   I. Sain. Amalgamation and epimorphisms of cylindric algebras and boolean algebras with operators. Preprint, Math. Inst. of Hungarian Academy of Sciences, Budapest, 1982. [62]

[Sai90]   I. Sain. Beth's and Craig's properties via epimorphisms and amalgamation in algebraic logic. In C. H. Bergman, R. D. Maddux, and D. L. Pigozzi, editors, *Algebraic logic and universal algebra in computer science*, volume 425 of *Lecture Notes in Computer Science*, pages 209–226. Springer-Verlag, Berlin, 1990. [210]

[Sai95]   I. Sain. On the problem of finitizing first-order logic and its algebraic counterpart (a survey of results and methods). In Csirmaz et al. [CsiGab[+]95], pages 243–292. [181, 206]

[Sai00]   I. Sain. On the search for a finitizable algebraization of first-order logic. *Logic J. IGPL* 8 no. 4 (2000) 497–591. [206]

[SaiGyu97]   I. Sain and V. Gyuris. Finite schematizable algebraic logic. *Logic J. IGPL* 5 no. 5 (1997) 699–751. [206]

[SaiTho91]   I. Sain and R. Thompson. Strictly finite schema axiomatization of quasi-polyadic algebras. In Andréka et al. [AndMon[+]91], pages 539–571. [204]

[SamVac89]	G. Sambin and V. Vaccaro. A new proof of Sahlqvist's theorem on modal definability and completeness. *J. Symbolic Logic* 54 (1989) 992–999. [94, 95]
[Say]	T. Sayed Ahmed. On amalgmation of reducts of polyadic algebras. *Algebra Universalis*. To appear. [206]
[Say01]	T. Sayed Ahmed. The class of neat reducts is not elementary. *Logic J. IGPL* 9 (2001) 625–660. [191, 465]
[SayNém01]	T. Sayed Ahmed and I. Németi. On neat reducts of algebras of logics. *Studia Logica* 62 no. 2 (2001) 229–262. [191]
[Sch91]	B. Schein. Representation of subreducts of Tarski relation algebras. In Andréka et al. [AndMon$^+$91], pages 621–635. [205]
[Sco65]	D. Scott. Logic with denumerably long formulas and finite strings of quantifiers. In Addison et al. [AddHen$^+$65], pages 329–341. [274]
[She71]	S. Shelah. Every two elementarily equivalent models have isomorphic ultrapowers. *Israel J. Math.* 10 (1971) 224–233. [58]
[Sik64]	R. Sikorski. *Boolean algebras*, volume 25 of *Ergebnisse der Mathematik und ihre Grenzgebiete*. Springer Verlag, Berlin, 2nd edition, 1964. [38, 48, 50]
[Sim93]	A. Simon. What the finitization problem is not. In Rauszer [Rau93], pages 95–116. [205]
[Sim97]	A. Simon. *Nonrepresentable algebras of relations*. PhD thesis, Math. Inst. Hungar. Acad. Sci., Budapest, 1997. http://www.renyi.hu/pub/algebraic-logic/simthes.html. [191, 192, 203, 211, 212, 583]
[Ste00]	V. Stebletsova. *Algebras, relations and geometries*. PhD thesis, Zeno, Leiden–Utrecht Research Institute of Philosophy, Heidelberglaan 8, 3584 CS Utretcht, Netherlands, 2000. Number XXXII in Quæstiones Infinitæ. [145, 200, 207, 210, 268, 275, 362, 517]
[SteVen98]	V. Stebletsova and Y. Venema. Q-algebras. *Algebra Universalis* 40 (1998) 19–49. [207, 268, 275, 362, 388, 517]
[Stevens72]	F. W. Stevenson. *Projective planes*. Freeman and Co., San Franciso, CA, 1972. [146]

Bibliography 653

[Sto36] M. Stone. The theory of representations for boolean algebras. *Trans. Amer. Math. Soc.* 40 (1936) 37–111. [42]

[Tar35] A. Tarski. Zur Grundlegung der Bool'schen algebra. *Fundamenta Math.* 24 (1935) 177–198. [48]

[Tar41] A. Tarski. On the calculus of relations. *J. Symbolic Logic* 6 (1941) 73–89. [4, 5, 16, 70, 119, 578]

[Tar46] A. Tarski. A remark on functionally free algebras. *Ann. Math.* 47 (1946) 163–165. [68]

[Tar53] A. Tarski. Some metalogical results concerning the calculus of relations. *J. Symbolic Logic* 18 (1953) 188–189. [209]

[Tar54] A. Tarski. Contributions to the theory of models, I, II. In *Proc. Konink. Nederl. Akad. van Wetensch.*, volume 57 (= Indag. Math. 16) of *A*, pages 572–581 and 582–588 resp., 1954. [287]

[Tar55] A. Tarski. Contributions to the theory of models, III. *Koninkl. Nederl. Akad. Wetensch Proc.* 58 (= Indag. Math. 17) (1955) 56–64. [126, 136]

[Tar86] S. Givant and R. McKenzie, editors. *Collected papers of Alfred Tarski*. Birkhäuser, Basel, Boston, 1986. 4 volumes. [4]

[TarGiv87] A. Tarski and S. Givant. *A formalization of set theory without variables*. Number 41 in Colloquium Publications. Amer. Math. Soc., Providence, Rhode Island, 1987. [16, 22, 120, 130, 144, 209, 212, 464, 486, 540, 578]

[Tho93] R. Thompson. Complete description of substitutions in cylindric algebras and other algebraic logics. In Rauszer [Rau93], pages 327–342. [170, 173, 174, 369]

[Tru92] J. Truss. Generic automorphisms of homogeneous structures. *Proc. London Math. Soc.* 64 (1992) 121–141. [589]

[Var97] M. Vardi. Why is modal logic so robustly decidable? In N. Immerman and P. Kolaitis, editors, *Descriptive complexity and finite models*, volume 31 of *DIMACS Series in Discrete Mathematics and Theoretical Computer Science*, pages 149–184. Amer. Math. Soc., Providence, RI, 1997. [584]

[Ven91] Y. Venema. Relational games. In Andréka et al. [AndMon+91], pages 695–718. [208]

[Ven92]	Y. Venema. *Many-dimensional modal logic*. PhD thesis, University of Amsterdam, 1992. [16, 210, 583]
[Ven93]	Y. Venema. Derivation rules as anti-axioms. *J. Symbolic Logic* 58 (1993) 1003–1034. [94, 210]
[Ven97a]	Y. Venema. Atom structures. In M. Kracht, M. de Rijke, H. Wansing, and M. Zakharyaschev, editors, *Advances in Modal Logic '96*, pages 291–305. CSLI Publications, Stanford, 1997. [84, 88, 457]
[Ven97b]	Y. Venema. Atom structures and Sahlqvist equations. *Algebra Universalis* 38 (1997) 185–199. [12, 89, 92, 94, 98, 169, 205, 215, 455, 535]
[Ven98]	Y. Venema. Rectangular games. *J. Symbolic Logic* 63 (1998) 1549–1564. [200, 210, 274]
[VenMar98]	Y. Venema and M. Marx. A modal logic of relations. In Orlowska [Orl98], pages 124–167. Also available as tech. report IR-396, Faculteit der Wiskunde en Informatica, Vrije Universiteit Amsterdam, 1995. [13, 208, 210, 274, 421]
[VilKau86a]	M. Vilain and H. Kautz. Constraint propagation algorithms for temporal reasoning. In *Proc. fifth national conference on artificial intelligence*, pages 377–382. Morgan Kaufmann, 1986. [252]
[VilKau86b]	M. Vilain and H. Kautz. Constraint propagation algorithms for temporal reasoning. In *Proceedings of the fifth AAAI*, pages 377–382, 1986. [252, 253]
[VilKau$^+$89]	M. Vilain, H. Kautz, and P. van Beek. Constraint propagation algorithms for temporal reasoning — a revised report. In D. S. Weld and J. de Kleer, editors, *Readings in Qualitative Reasoning about Physical Systems*, pages 373–381. Morgan Kaufmann, 1989. [252]
[Wer78]	H. Werner. *Discriminator algebras*, volume 6 of *Stud. Algebra Anwendungen*. Akademie-Verlag, Berlin, 1978. [70]

Symbol index

General mathematics		
Symbol	Meaning	Page
$\vdash_{m,n}$	n-variable proof system with m-variable atomic formulas and theorems	487
$^{\alpha}U$	the set of functions from α to U	167
$^{<\alpha}U$	$\bigcup_{\beta<\alpha} {}^{\beta}U$	29
$^{\leq\alpha}U$	$^{<\alpha+1}U$	29
$\chi(\Gamma)$	chromatic number of graph Γ	448
$(\bar{x} \mapsto \bar{y})$	the relation $\{(x_i, y_i) : i < |\bar{x}|\}$	404
$f\restriction_X$	the restriction of the function f to the set X	28
$f \circ g$	composition of functions f and g: $f \circ g(x) = f(g(x))$	29
$g(\Gamma)$	girth (length of shortest cycle) of graph Γ	448
$\mathrm{ht}_T(t), \mathrm{ht}(t)$	height of node t in tree T	311
$\mathrm{ht}(T)$	height of tree T	311
Id_X	identity relation over (or map on) the set X	29
$[i/j]$	i-for-j substitution map	173
$[i,j]$	i-for-j transposition map	600
K_κ	complete, undirected, loop-free graph on κ nodes	514
$\mathbb{N}, \mathbb{Z}, \mathbb{Q}, \mathbb{R}$	natural numbers, integers, rationals, reals	27
$\wp(S)$	the power set (set of subsets) of S	26
$\mathrm{rng}(f), \mathrm{dom}(f)$	the range and domain of the function f	28
$\mathrm{rng}(\bar{x})$	$\{x_0, \dots, x_{|\bar{x}|-1}\}$	29
$(\bar{x}_s : s \leq t)$	concatenation of tuples $\bar{x}_s$, for $s \leq t$ in a tree	311
$\mathrm{Suc}(t,u)$	u is an immediate successor of t in a tree	311

$\mathrm{Sym}(X)$	symmetric group of all permutations on the set X	30
$\hat{t}$	set of predecessors of node t in a tree	310
$T\restriction_\alpha$	restriction of tree T to nodes of height $< \alpha$	311
$\bar{x} \in M$	$\bar{x}$ is a tuple of elements of the structure M	33
$\bar{x} \in S$	$\bar{x}$ is a tuple of elements of the set S	29
$\bar{x}^i_j$	tuple $(x_0, \ldots, x_{i-1}, x_j, x_{i+1}, \ldots)$	195
$\bar{x}\bar{y}$	concatenation of sequences $\bar{x}, \bar{y}$	404
$\bar{x} \equiv_i \bar{y}$	the sequences $\bar{x}, \bar{y}$ agree off of i	180
$\bar{x} \equiv_{\bar{a}} \bar{y}$	the sequences $\bar{x}, \bar{y}$ agree off of $\mathrm{rng}(\bar{a})$	400
$\bar{x} \equiv_{ij} \bar{y}$	the sequences $\bar{x}, \bar{y}$ agree off of i and j	400
$\bar{x}$	the tuple $(x_0, x_1, \ldots)$	29
$|\bar{x}|$	the length (number of entries) of the tuple $\bar{x}$	29
$|X|$	the cardinality of the set X	28
X/E	set of equivalence classes of equivalence relation E on X	27
X^n	$X \times X \times \cdots \times X$ (n times)	26
$[X]^n$	set of all subsets of X of size n	31
$^Y X$	set of all maps from Y into X	29
ω (ω_1)	the first infinite (uncountable) ordinal and cardinal	28

Symbol index

\multicolumn{3}{c}{General model theory}		
Symbol	Meaning	Page
$\cong$	isomorphism of structures	36
$\equiv$	elementary equivalence	35
$\subseteq$	substructure or subalgebra	36
$\preceq$	elementary substructure	36
$\models$	classical Tarskian satisfaction relation	34
$\models_G$	satisfaction relation in clique-guarded semantics	588
a	'algebra' sort of two-sorted signature L^s	277
$\text{Aut}(M)$	automorphism group of structure M	36
$\text{dom}(M)$	domain of structure M (usually identified with M)	33
$\mathfrak{G}(M)$	Gaifman graph of structure M	588
$\mathsf{K} \models \Phi$	$M \models \phi$ for all $M \in \mathsf{K}$, $\phi \in \Phi$	35
$\text{LGF}(L)$	loosely guarded fragment of first-order language of signature L	586
L^n	n-variable fragment of first-order language L	32
$L_{\infty\omega}, L^n_{\infty\omega}, L^\omega_{\infty\omega}$	infinitary languages	32
L^s	two-sorted signature for pseudo-elementary class	277
$|M|$	cardinality of domain of structure M	33
$M{\upharpoonright}_L$	reduct of structure M to signature L	34, 284
$M^{\mathbf{a}}$	**a**-sorted part of L^s-structure M	277
$\text{Mod}(\sigma)$	class of models of sentence σ	34
$\text{Mod}(T)$	class of models of theory T	34
M_P	P-part of structure M	285
$PC^{(\prime)}, PC^{(\prime)}_\Delta$	pseudo-elementary classes (single-sorted definition)	285
$\text{PF}(L)$	packed fragment of first-order language L	587
$\varphi(\bar{x})$	the free variables of formula φ are among $\bar{x}$	32, 284
φ^χ	relativisation of formula φ to formula χ	37
r	'representation' sort of two-sorted signature L^s	277
r^M	interpretation of the symbol r in the structure M	33
$\text{Th}(\mathsf{K}), \text{Th}(M)$	theory of structures in class K, or of structure M	34

Model theory of algebras		
Symbol	Meaning	Page
$\models_C$	satisfaction relation in clique-relativised semantics	401
$\mathrm{Clique}(\bar{x})$	formula expressing that $\mathrm{rng}(\bar{x})$ is a clique in a relativised representation	405
$C^n(M)$	set of all n-tuples with 'clique range' in the relativised representation M	400
$\mathrm{Fl}(\mathcal{A})$	first-order theory defining an n-flat relativised representation of the non-associative algebra $\mathcal{A}$	402
L^a	signature of atom structures of L-BAOs	78
$\mathcal{L}(\mathcal{A})$	relational first-order signature corresponding to the relation algebra $\mathcal{A}$	119
$\mathcal{L}(\mathcal{A})^n$	the n-variable $\mathcal{L}(\mathcal{A})$-formulas	400
$\mathcal{L}(\mathcal{A})^n_{\infty\omega}$	n-variable infinitary language with signature $\mathcal{L}(\mathcal{A})$	400
L_{BA}	signature of boolean algebras	38
L_c^d	signature of d-dimensional cylindric algebras	269
LC	the set of Lyndon conditions	341
L_{RA}	signature of relation algebras	106
PL	Peircean law	107, 159
φ^C	set of tuples satisfying φ in clique-relativised semantics	409
$R_{\mathcal{A}}$	first-order theory defining a relativised representation of the non-associative algebra $\mathcal{A}$	152
$\mathrm{Sm}^n(\mathcal{A})$	first-order theory defining an n-smooth relativised representation of the non-associative algebra $\mathcal{A}$	406
$\mathrm{Sq}^n(\mathcal{A})$	first-order theory defining an n-square relativised representation of the non-associative algebra $\mathcal{A}$	405
$T_{\mathcal{A}}$	first-order theory defining a representation of the relation algebra $\mathcal{A}$	119
$W_{\mathcal{A}}$	first-order theory defining a weak representation of the relation algebra $\mathcal{A}$	165
WL	weak associativity law	157

Symbol index 659

\multicolumn{3}{c}{Universal algebra}		
Symbol	Meaning	Page
$\prod_{\lambda \in \Lambda} M_\lambda / D$, $\prod_D M_\lambda$	ultraproduct of the structures M_λ ($\lambda \in \Lambda$) by the ultrafilter D over the set Λ	56
$\prod_D M$, M^Λ/D	ultrapower of the structure M by ultrafilter D over Λ	57
$\subseteq^c$	$\mathcal{A} \subseteq^c \mathcal{B}$ iff $\mathcal{A}$ is a subalgebra of $\mathcal{B}$ and inclusion is a complete embedding from $\mathcal{A}$ into $\mathcal{B}$	45, 61
a/D	element of $\prod_D M_\lambda$ corresponding to $a \in \prod_{\lambda \in \Lambda} M_\lambda$	57
$\mathcal{A}^+$	the canonical extension of the BAO $\mathcal{A}$	81
$\mathcal{A}\!\upharpoonright_\alpha$	relativisation of BAO $\mathcal{A}$ to α	72
At $\mathcal{A}$	atom structure of the atomic BAO $\mathcal{A}$	78
At $\mathcal{B}$	set of atoms of the atomic boolean algebra $\mathcal{B}$	40
At V	the class of atom structures of atomic algebras in V	84
$\mathcal{B}^+$	the canonical extension of the boolean algebra $\mathcal{B}$	44
BAO	boolean algebra with operators	60
bool(C)	the boolean part (boolean reduct) of the BAO C	60
BPI	boolean prime ideal theorem	42
$\mathfrak{Cm}\,\alpha$	the complex algebra of the relational structure α	79
Cst V	$\{\text{At}\,\mathcal{A}^+ : \mathcal{A} \in \text{V}\}$	92
$\text{FO}^k(\text{V})$	class of atom structures with first-order algebra in V	98
HK	closure of the class K under homomorphic images	68
IK	closure of the class K under isomorphism	68
ker(g)	$\{(i,j) \in {}^2\text{dom}(g) : g(i) = g(j)\}$ – kernel of function g	173
ker(h)	$\{b : h(b) = 0\}$, the kernel of homomorphism h	61
PK	closure of the class K under products	68
π_α	projection $\mathcal{A} \to \mathcal{A}\!\upharpoonright_\alpha$	72
SK	closure of the class K under subalgebras and isomorphism (see also $\mathbf{S\mathfrak{R}aD}_\alpha, \mathbf{S\mathfrak{R}aG}_\alpha$)	68
Sir K	class subdirectly irreducible algebras in K	69
Str V	class of atom structures with complex algebra in V	91
Up K	closure of the class K under ultraproducts	68
Uf $\mathcal{A}$	atom structure from the ultrafilters of the BAO $\mathcal{A}$	81
Wst V	class of atom structures with term algebra in V	98

Algebras and atom structures		
Symbol	Meaning	Page
$\mathcal{A}_{A,B}$	rainbow algebra based on structures A, B	494
$\mathcal{A}_n$	Lyndon algebras	146
$\mathfrak{A}(n, r)$	variant of Monk algebra	466
$\alpha(\Gamma)$	atom structure constructed from graph Γ	449
C	containment algebra	143
$\mathfrak{Ca}(\mathcal{H})$	cylindric-type algebra with atom structure from the set $\mathcal{H}$ of hypernetworks	371, 373
$\mathfrak{Cm}(\mathcal{G})$	complex algebra of the group $\mathcal{G}$	134
C_r	complex algebra over hyperbasis for $\mathfrak{A}(n, r)$	467
$\mathcal{D}(M), \mathcal{D}^0(M)$	algebras of definable sets in a relativised representation M	409
$\Delta^k(A)$	k-variable first-order algebra over atom structure A	98
$\mathcal{F}(B)$	the full proper relation algebra on set B	104, 133
I	Allen interval algebra	142
$\mathcal{K}$	McKenzie's algebra (the smallest non-representable relation algebra)	140
$\mathcal{M}$	metric point algebra	143
$\mathcal{M}_n$	Monk algebras	142
$\mathcal{N}_1$	left-linear point algebra	142
$\mathfrak{Nr}_\beta C$	neat reduct to β dimensions of cylindric algebra C	186
$\mathcal{P}$	point algebra	139
$\mathcal{PA}$	pentagonal algebra	139
$\mathfrak{Ra}C$	relation algebra reduct of the cylindric algebra C	186
$RA(\tau)$	relation algebra constructed from tiling instance τ	544
$\mathcal{T}, \mathcal{T}'$	two-atom relation algebras	138, 139
$\tau(S)$	term algebra over atom structure S	446
$\mathcal{X}_q$	strongly but not completely representable relation algebra atom structure of Maddux	459

Symbol index

Operations in algebras		
Symbol	Meaning	Page
$\sum S, \prod S$	sum (supremum), product (infimum) of subset S of a boolean algebra	45
$\sim, \simeq$	congruences on s-c-words	174
0'	the relation algebra 'diversity' operation	109
1'	the relation algebra 'identity' operation	106
$\smile$	the relation algebra 'converse' operation	106
;	the relation algebra 'composition' operation	106
c_i	the ith cylindrifier of a cylindric algebra	167
d_{ij}	the ijth diagonal of a cylindric algebra	167
end(a)	$1' \cdot \breve{a}; a$, the 'end' of atom a	159
$\widehat{\varphi}$	cylindric algebra term obtained from formula φ	487
$\mathrm{Id}_B, {}^{-1}, \mid$	the concrete relational operations: identity on B, inverse, composition	101
${}_k\mathsf{s}(i,j)$	transposition operator	189
$\mathsf{R}_X, \mathsf{B}_X, \mathsf{G}_X$	sets of red, blue, green atoms of $\mathfrak{Cm}(\alpha(\Gamma))$ with indices in X	450
ρ_Γ	string of cylindrifications/substitutions for set Γ	177
$\rho_{(i_0,\ldots,i_{n-1})}$	string of cylindrifications/substitutions for the sequence $i_0, \ldots, i_{n-1}$	176
$\mathsf{s}_{\bar{a}}$	s-c word w with $\widehat{w} = \bar{a}$	417
s^i_j	substitution operator (*see also* ${}_k\mathsf{s}(i,j)$)	170
s_σ	polyadic substitution operator	201, 600
$\mathsf{s}_{\sigma,\Gamma}$	combined cylindrification–substitution operator	600
s_{ijk}	s-c word w with $\widehat{w} = (0 \mapsto i, 1 \mapsto j, 2 \mapsto k)$	417
st(a)	$1' \cdot a; \breve{a}$, the 'start' of atom a	159
$\widehat{w}$	map: $\alpha \to \alpha$ associated with an s-c word w of $\mathbf{CA}_\alpha$	174
w^*	$\bigcup_{w=uv} \ker(\widehat{v})$, for an s-c word w	174

Algebraic logic classes		
Symbol	Meaning	Page
CA_α	class of α-dimensional cylindric algebras	168
CRA	class of completely representable relation algebras	515
CRA_κ	class of non-associative algebras with complete κ-square relativised representations	526
$CRAS$	class of atom structures of completely representable relation algebras	455
Crs_α	class of α-dimensional relativised set algebras	180
Crs_n^+	class of generalised α-dimensional relativised set algebras	600
Cs_α	class of α-dimensional cylindric set algebras	167
D_α	class of Crs_αs with unit closed under substitutions	180
Df_α	class of α-dimensional diagonal-free algebras	200
DS	class of distributive lattice ordered semigroups	207
$FOAS$	class of relation algebra atom structures with representable first-order algebra	456
G_α	class of Crs_αs with unit closed under all maps	180
GRA	class of group relation algebras	134
HRA	class of relation algebras with homogeneous representations	281
LC	class of atomic relation algebras satisfying the Lyndon conditions	341
$LCAS$	class of relation algebra atom structures satisfying the Lyndon conditions	337
NA	class of non-associative algebras	155
$\mathfrak{Nr}_\beta K$	neat β-dimensional reducts of algebras in the class K of cylindric algebras	190
PA_α	class of α-dimensional polyadic algebras	201
PEA_α	class of α-dimensional polyadic equality algebras	201
QPA_α	class of α-dimensional quasi-polyadic algebras	202
$QPEA_\alpha$	class of α-dimensional quasi-polyadic equality algebras	202
RA	class of relation algebras	106
$\mathfrak{Ra}K$	relation algebra reducts of algebras in the class K of cylindric algebras	190

Symbol index

Symbol	Meaning	Page
$\mathbf{RA}_\kappa$	class of non-associative algebras with κ-square relativised representations	526
$\mathbf{RA}_n$	class of subalgebras of algebras in $\mathbf{RB}_n$	384
$\mathbf{RB}_n$	class of atomic non-associative algebras with n-dimensional relational bases	367, 525
$\mathbf{RCA}_\alpha$	class of α-dimensional representable cylindric algebras	168
$\mathbf{Rdf}_\alpha$	class of α-dimensional representable diagonal-free algebras	200
$\mathbf{RPA}_\alpha$	class of α-dimensional representable polyadic algebras	202
$\mathbf{RPEA}_\alpha$	class of α-dimensional representable polyadic equality algebras	202
$\mathbf{RRA}$	class of representable relation algebras	118
$\mathbf{SA}$	class of semi-associative algebras	157
$\mathbf{S\mathfrak{N}r}_\beta\mathbf{CA}_\alpha$	class of subalgebras of neat β-dimensional reducts of α-dimensional al cylindric algebras	190
$\mathbf{S\mathfrak{R}aCA}_\alpha$	class of subalgebras of relation algebra reducts of α-dimensional cylindric algebras	190
$\mathbf{S\mathfrak{R}aD}_\alpha$	class of algebras having a relation algebra reduct embedding into an algebra in $\mathbf{D}_\alpha$	194
$\mathbf{S\mathfrak{R}aG}_\alpha$	class of algebras having a relation algebra reduct embedding into an algebra in $\mathbf{G}_\alpha$	194
$\mathbf{SRAS}$	class of strongly representable atom structures ($=\mathrm{Str}\,\mathbf{RRA}$)	445
$\mathbf{WA}$	class of weakly associative algebras	157
$\mathbf{WRAS}$	class of weakly representable atom structures ($=\mathrm{At}\,\mathbf{RRA}$)	445
$\mathbf{wRRA}$	class of weakly representable relation algebras	163

Sets of hypernetworks, hyperbases, and restrictions of them		
Symbol	Meaning	Page
$H_n(\mathcal{A}, \Lambda)$	set of all n-dimensional Λ-hypernetworks over the atomic relation algebra $\mathcal{A}$	367
$H_m^n(\mathcal{A}, \Lambda)$	set of all n-wide m-dimensional Λ-hypernetworks over the atomic relation algebra $\mathcal{A}$	373
$\mathcal{H}\!\upharpoonright_m$	restriction of hyperbasis $\mathcal{H}$ to m dimensions	372
$\mathcal{H}\!\upharpoonright_m^n$	restriction of hyperbasis $\mathcal{H}$ to m dimensions but keeping hyperlabels of length up to n	373

Networks		
Symbol	Meaning	Page
$\mathsf{Acc}(N,x,y,u,v)$	term network for accepting move by $\exists$ (RA case)	262
$\beta(n,R)$	index of node n in red clique R ($\beta(n)$ for short)	496
$\mathsf{CNet}^d(N)$	formula from d-dimensional cylindric term network N	270
$\mathsf{edges}(N)$	set of edges of a relativised pre-network	226
I_a	network with nodes $0, 1$, with $I_a(0,1) = a$, etc.	239
$\mathsf{In}(N,i,\bar{x},\tau)$	term network for accepting move by $\exists$ (CA case)	269
$\mathsf{Ind}(i,I,J)$	index of node i with respect to maps I, J	471
$\iota(F)$	instantiation of free variables of formulas in F (from syntactic network) by assignment ι	298
$M \equiv_i N$	the hypernetworks M, N agree on all labels not involving i	365
$M \equiv_{i_0,\ldots,i_{k-1}} N$	the hypernetworks M, N agree on all labels not involving any of $i_0, \ldots, i_{k-1}$	365, 373
$|N|$	cardinality of set of nodes of N	220
$\mathsf{nodes}(N)$	set of nodes of N	220
$N \subseteq N'$	N is a subnetwork of N'; N' refines N	222, 226
N^ι	pre-network obtained from term network N under assignment ι	262
$N\!\restriction_m$	restriction of hypernetwork N to m dimensions	372
$N\!\restriction_m^n$	restriction of hypernetwork N to m dimensions but keeping hyperlabels of length up to n	373
$N\sigma$	the hypernetwork that embeds by σ into the hypernetwork N	365, 525
$\mathsf{Out}(N,i,\bar{x},\tau)$	term network for rejecting move by $\exists$ (CA case)	269
$\mathsf{PreNet}(\mathcal{A})$	class of all pre-networks over $\mathcal{A}$	236
$r(n_1)$	rank of node n_1 in graph Γ^+	565
$R_N(c,c')$	red clique based on (c,c') in network N	499
$\mathsf{rk}(I,J)$	rank of maps I, J	471
$\mathsf{rk}(N)$	rank of the network N	224
$\mathsf{Rej}(N,x,y,u,v)$	term network for rejecting move by $\exists$ (RA case)	262
$\mathsf{Var}(F)$	set of free variables of formulas in term network F	298
$\bar{x} \sim_N \bar{y}$	l-tuples $\bar{x}, \bar{y}$ are logically equal in hypernetwork N	467

Symbol index

\multicolumn{3}{c}{Games, game trees}		
Symbol	Meaning	Page
$\forall, \exists$	players in games	226
$\mathrm{form}_{\leq}(t)$, $\mathrm{form}_{<}(t)$	formulas expressing winning strategy in game-tree game	323
$G(T, \mathcal{A}, v_r)$	game defined by game tree T, structure $\mathcal{A}$, valuation v_r	312
$G_n(N, \mathcal{A})$	n-round representation game over relation algebra $\mathcal{A}$	233
$G_n(N, C)$	n-round representation game over cylindric algebra C	249
$G_r^a(\mathcal{A})$	r-round atomic game over atomic relation algebra $\mathcal{A}$	339
$G_r^a(\mathcal{A}, N)$	as $G_r^a(\mathcal{A})$ but starts from atomic network N	339
$G_r^{m,n}(\mathcal{A})$	r-round game played over n-node hypernetworks with m-node $\forall$-moves	378
$G_r^p(\mathcal{A})$	r-round game played over p-node atomic networks over $\mathcal{A}$	376
$G_\omega^{\mathrm{rel}}(\mathcal{A})$	ω-round game played on relativised networks over weakly associative algebra $\mathcal{A}$	226
$\Gamma(U, \mathcal{A})$	game to determine if $\mathcal{A}$ is in the pseudo-universal class defined by U	293
$\Gamma^+(U, \mathcal{A})$	game to determine if $\mathcal{A}$ is in elementary closure of pseudo-elementary class defined by U	303
$\Gamma_n(U, \mathcal{A}, N, S)$	n-round game like $\Gamma(U, \mathcal{A})$ but starting from (N, S)	296
$\mathbf{EF}_r^n(A, B)$	standard Ehrenfeucht–Fraïssé game from A to B with n pebble pairs, r rounds	492
$EF_r^n(A, B)$	modified Ehrenfeucht–Fraïssé game from A to B with n pebble pairs, r rounds	493
$\mathcal{J}_n(\mathcal{A})$	Jónsson-style game characterising **wRRA**	247
$\Lambda_n^m(\alpha, \beta, \mathcal{A})$	Lyndon-style game characterising **RRA**	246
$T(m, n)$	game tree for $G_\omega^{m,n}$	382
$T \!\upharpoonright_n$	restriction of game tree T to nodes of height $< n$	312
$T_{\geq t}$	game tree based on $\{u \in T : u \geq t\}$	313
$T \subseteq U$	T is a sub-game tree of game tree U	313
θ_T	formula expressing winning strategy for $\exists$ in game over finite tree T	323
Θ_T	theory $\{\theta_{T \upharpoonright_n} : n < \omega\}$	324
$\bigcup_{n<\omega} T_n$	union of chain of game trees $T_n : n < \omega$	313

Subject index

Boldface page numbers refer to definitions. *Italic* page numbers refer to material in the summary chapter 20.

The symbol ∼ denotes the current heading. The symbol ? indicates an open problem.

Terms beginning with a symbol are listed under that symbol. For example, *Q-operator* is listed under Q. Where the symbol is a variable, the choice of variable is governed by the usual usage in the text, and is almost always '*n*'. For example, *n-type, n-homogeneous representation,* etc., are listed under N.

3-consistent pre-network **253**
∀-triangle **559**
α-ary relation **167**
ω-big structure 130
ω-categoricity 143, 605

abelian group **30**
abstract algebraic logic 212
accepting move by ∃ **234**, 250, 319, *611*
additive function **60**
algebra 5, **34**, 37 *see also individual types of algebra*
algebra of relations **164**
algebraic logic 1, 4, 99, 212
algebraisation 99, 201, 204, 212
algorithm *see also* decidability; undecidability
   for membership of $\mathbf{RA}_n$ 386, 387
   propagation ∼ **253**, 252–258
Allen interval algebra **142**, 143, 252, 253, 254, 258
amalgamation *see also* game: for hyperbasis ($G_r^{m,n}$)
   for boolean algebras 53
   homogeneity and 13, 395–398

amalgamation (*cont.*)
   in algebraic logic 206, 210–211
   in cylindric basis 356, **389**, 392, 395–397
   in game G_ω^{rel} 229
   in hyperbasis **368**, 443, 467, 469, 472, 550
amalgamation move in game $G_r^{m,n}$ **379**, 383, 480–483
anti-Monk algebra **348**
antisymmetric binary relation **26**
approximations to **RRA** 10, 15–17, 208, 215, 355–362, 399 *see also* basis; network; $\mathbf{RA}_n$; relation algebra reduct; relativised representation
arc *see* edge
arity **31**, **60**
arrow logic 208, 210, 212, 601
assignment **34**
associative function **28**, 30, 39 *see also* composition
   composition in relation algebra reducts 190, 197
   composition in relation algebras **106**, 113
   game characterisation 345–346, 383, 496
   in tiling algebra 545, 546, 570

667

associative function (*cont.*)
    weakened associativity 154–155, 157–158 *see also* semi-associative algebra; weakly associative algebra
atom **40** *see also* flexible atom
atom-canonical variety **89**, 89–92, 98
    RA 110
    $\mathbf{RA}_n, \mathbf{S\mathfrak{R}aCA}_n$ ($n \geq 6$) are not 535
    $\mathbf{RCA}_1, \mathbf{RCA}_2$ 169
    $\mathbf{RCA}_n$ ($3 \leq n < \omega$) are not 535
    RRA is not 91, 455, 535
atom structure **78**, 77–98 *see also* complex algebra; first-order algebra; term algebra
    ∼s of atomic algebras in a variety (AtV) **84**, 84–90, 97, 283, 335–336
    ∼s of canonical extensions of algebras in a variety (CstV) **92**, 98, 336
    ∼s of complex algebras in a variety (StrV) **91**, 91–93, 97–98, 336
    ∼s with term algebra in a variety (WstV) **98**
    of canonical extension **81**, 116, 130, 458, 461–462, 533–534
    of cylindric algebra 169, 371, 458, 627
    of relation algebra **112**, 111–117, 335–338, *620 see also* relation algebra ∼
    of weakly associative algebra **163**, 545
    variety generated by class of ∼s **89**, 89–93, 97, 131, 336
atomic boolean algebra **40**, 51, 53, 54
    complete embedding and 45–46 *see also* complete embedding
    complete representation and 48 *see also* complete representation
    completion 50, 55
atomic formula **32**
atomic game *see* game: atomic (G_n^a)
atomic network 221, 222, **339**, *611* see also game: atomic (G_n^a); hypernetwork
    from other ∼ by transformation **525**
    hypernetwork and 364
    induced index of node **460, 496, 544**
    partial isomorphism **396**, 396–398
atomic relation algebra **109**, 111–116, 335–338 *see also* relation algebra atom structure
atomic relativised (pre-)network **225–226**, 227

atomic representation **47**, **124**, 279, 343, *610 see also* complete representation
atomless boolean algebra **41**, 51, 54
attached tile edges **559**, 560, 561, 566
automorphism **36**, 108, 533
    extending partial isomorphism 130, 395–398, 589
    of representation 130, 281, 395–398 *see also* homogeneous representation
axiom of choice 26, 42, 314 *see also* boolean prime ideal theorem
axiom of foundation 208
axiomatisation **35** *see also* finitely axiomatisable class; formula; non-finite axiomatisability; recursive ∼; Sahlqvist variety
    atom structures of atomic algebras in a variety (AtV) 84–89, 90
    atom structures of complex algebras in a variety (StrV) 92
    atom structures with first-order algebra in a variety (FO^{k}V) 98
    boolean algebras 5, **39**
    $\mathbf{CA}_\alpha \cap \mathbf{ICrs}_\alpha, \mathbf{ICrs}_\alpha, \mathbf{ID}_\alpha, \mathbf{IG}_\alpha$ 182–184
    complex algebras over a variety 275
    CRA (elementary closure) **349** *see also* Lyndon conditions
    cylindric algebras **168**
    finite-variable logic 486
    games and 9–13, 261, 273–277, 292, *611–612*
    general relativised representations 154
    Lyndon algebras (atom structures, equational theory, subalgebras) 145
    non-orthodox ∼ 145, 210
    pseudo-elementary class 276, 289
        elementary closure 290, **305**, 302–305, *612*
        expansion by Q-operators 306–308, *612*
        pseudo-universal class 289, **301**, 292–302, *612*
    $\mathbf{RA}_n$ 360, 387, **438**
        by existential second-order sentence 387
        expansion by Q-operators 208, 362, 388
    $\mathbf{RCA}_1, \mathbf{RCA}_2$ 168
    $\mathbf{RCA}_\alpha$ 261, **270**, 269–272
    subreducts 206

Subject index

axiomatisation (*cont.*)
  relation algebras 4, 7, **106** *see also* relation algebra: axioms
    RRA 9, 215, **264**, 261–264, *612* *see also* Lyndon conditions
      canonical $\sim$? 264, 535, 626
      expansion by Q-operators 264–268, 275, 517, *612*
      finite-variable $\sim$? 520, 625
      Lyndon game (Λ_n^m) 246
      subreducts 206
    S$\mathfrak{R}$aCA$_n$ 360, **438**
    wRRA 165, 247, 388, 528
  axioms 4, 5–8, 99

back-and-forth system 13, 274, 404, 407, 425, 585
Banach–Mazur theorem 215
BAO *see* boolean algebra with operators
base 591 *see also* base-isomorphism; finite base property
  of cylindric set algebra 167
  of field of sets 38
  of proper relation algebra **101**, 102, *609*
    with residuals **601**
  of relativised representation 153
  of representation
    of boolean algebra 42
    of relation algebra 118
base-isomorphism **119**, 120
  between representations of
    Allen interval algebra 143
    full proper relation algebras 129
    group relation algebras 135
    Lyndon algebras 147–148
    pentagonal algebra 139
    point algebra 144
    two-atom algebra $\mathcal{T}'$ 144
basis 13, 356 *see also* cylindric $\sim$; hyper$\sim$; **RA**$_n$; relational $\sim$; symmetric $\sim$
  encodes winning strategy 356, 377–378, 380, 397
Beth property 206, 210
big structure 130
bijective function **29**
binary relations 2, **26**, 101
  laws governing 4
  operations on 3–4, 100, 101, 105, 268, 600, 601

binary symbol **31**
binding conventions for relation algebra operations 106
Birkhoff's theorem **68**, 71, 126, 193, 385
bisimulation 13, 274
black atom **495**
boolean algebra 5–7, **39**, 38–55 *see also* atomic $\sim$; $\sim$ with operators; complete embedding; dense subalgebra
  amalgamation 53
  atom **40**
  canonical extension **44**, 53, 59
  complete representation **46**, 46–48, 55
  completion **50**, 48–50, 55
  distributive law 53, 54
  filter and ultrafilter **41**, 42, 44, 51–52, 54
  finite base property 581–582, 595
  ideal **41**
  infinite sums and products 44–46, 53–55
  over a graph 454
  perfect extension **83**
  representation *see* complete representation: of $\sim$; representation: of $\sim$
boolean algebra with operators **60**, 60–98 *see also* atom structure; completely additive operator; complex algebra; dense subalgebra; simple algebra; subdirectly irreducible algebra
  atomic, complete, etc. **61**
  boolean reduct *see* boolean reduct: of $\sim$
  canonical extension *see* canonical extension: of $\sim$
  completion *see* completion: of $\sim$
  filter and ultrafilter **61** *see also* ultrafilter
  ideal **62**, 61–63, 192–193
  (proper) relation algebra is 104, 109
boolean combination **32**
boolean functions
  complement 3, **39**
  product, sum 3, **39**, **45**, 44–46, 53–55
boolean homomorphism **40**, 51, 52
boolean part *see* boolean reduct
boolean prime ideal theorem **42**, 43, 44, 52, 82, 83, 128, 387
boolean reduct **61**
  induced representation of 123, 155–156, 400
  neat reducts and 187, 416

boolean reduct (*cont.*)
  of boolean algebra with operators 61, 62, 65, 78, 83
  of relation algebra 108, 351–352, 451
bound variable 32
bounded morphism 96, 535
boxed variable 93
BPI *see* boolean prime ideal theorem
branch of tree 313
Bruck–Ryser theorem 146, 148

c-word 172 *see also* s-c-word
canonical axiomatisation 264, **535**, 626
canonical class 81
canonical embedding algebra *see* canonical extension
canonical equation 94, **264**, 535, 626 *see also* Sahlqvist equation
canonical extension 44, 80 *see also* canonical variety
  atom structure of **81**, 116, 130, 458, 461–462, 533–534
  complete embedding and 53, 82, 188, 431
  completion versus 48, 80
  of boolean algebra **44**, 53, 59
  of boolean algebra with operators **81**, 80–83, 95, 96
    perfect extension and 83, 95
    preserves Sahlqvist equations 94, 188
  of cylindric algebra 169, 170, 188, 431, 626
  of relation algebra 116
    game played on 264, 350–351, 454, 461, 532, 534
    Lyndon conditions and 337, 338, 351
    preserves relativised representability 360, 406, 408, 432
    preserves representability 124–126, 130, 350–351
  of weakly representable relation algebra 627
  products and 59, 96
  relation algebra reducts and 188, 193, 195, 431
  saturation and 124–126, 130, 387, 406, 408

canonical variety **81** *see also* canonical extension
  atom structures and 89–93, 97, 98, 130–131, 336
  CA_α 168, 188, 431
  $CA_\alpha \cap ICrs_\alpha, ICrs_\alpha, ID_\alpha, IG_\alpha$ 183, 184
  NA, SA, WA 158, 159, 361
  RA 110, 361
  RA_n 360, 361, 385, 387, 432
  RCA_1, RCA_2 169
  RRA 128, 129, 351, 361
  Sahlqvist variety 94
  $S\mathfrak{N}r_\beta CA_\alpha$ 192
  $S\mathfrak{R}aCA_\alpha$ 192, 361, 432
  wRRA? 627
cardinal 28
Cartesian space **593**
categoricity 143, 605
Cayley representation **134**, 135
Cayley's theorem 30
chromatic number 447, **448**, 450–453
classical representation **118**, 119, 152 *see also* representation
clique *see also* red ∼
  in graph **30**, **421**, **588**
  in relativised representation **158**, 358–359, **400**, 400–407, 433, *615*
clique-guarded fragment 401, 585, 588
clique-guarded semantics **588**, 589
clique-relativised semantics 359, **401**, 405, 407, 589
  algebra of definable sets 409
  in n-flat relativised representation 402–404, 407, 425
clopen set **30**, 43–44
closed game 238
closed game tree **313**, 320
closed set **30**, 143
closed term 33
closure
  of unit under maps **180–181**
  reflexive 26
  topological 49
  transitive 26
  under functions **29**, 34
  under homomorphic images, isomorphism, products, subalgebras 68 *see also* Birkhoff's theorem; homomorphic image
cofinite set **28**, 51, 52

collineation 148
colouring *see* graph: ~
common part of two hypernetworks 368
commutative function **28**, 30, 39
commutative relation algebra **117**
compact topology **31**, 43–44, 83
compactness **37**, 60, 122, 289
  boolean prime ideal theorem and 42
  Hilbert systems 488
  non-finite axiomatisability by 332, 454
  representations and 129, 231, 433
compatible tuple 266
complete boolean algebra **49**, 49–50, 55
  see also completion
complete embedding **45**, 54–55, **61**
  atomic boolean algebra and 45–46
  bases and 375
  bounded morphism and 96
  canonical extension and 53, 82, 188, 431
  reducts and 187–188, 409, 411, 412, 416, 429
complete graph **30**, 514
complete homomorphism **96**, 416
complete relativised representation **153**, 359, 400
  basis and 423, 428–429, 526–528, *618*
  is atomic 163, 400
  of canonical extension 406, 408
  of weakly associative algebra 232
  relation algebra reduct and 409, 428–429, *618*
  relativised representation and 527, 528
complete representation *see also* completely representable structure
  is atomic 47, 124, 279, *610*
  of boolean algebra **46**, 46–48, 55
  of cylindric algebra 170, 626
  of cylindric relativised set algebra 181–182, 412
  of relation algebra **123**, 129, 458, *610*
    complete relativised representation and 527, 528
    game and 347–348, 349–350, *613*
    hyperbasis from 375
    of canonical extension 124–126, 337
    of complex group 134
    of Lyndon algebra 145
    of metric point algebra 143, 144
    of rainbow algebra 511
complete theory **33**

completely additive boolean algebra with operators 63
completely additive operator **63** *see also* boolean algebra with operators
  CA operators 169, 416, 420, 485
  completion and 65–66
  complex algebra and 79, 82, 89, 90
  conjugated function is 64
  Crs operators 181, 182
  dense subalgebra and 66, 89, 90
  determined by values on atoms 77–78, 85, 87, 88
  finitisation problem and 206
  first-order correspondent and 86–89, 97
  Q-operators 268, 517
  RA operators 109, 446
  substitutions 170, 181, 416
completely representable structure *see also* complete representation
  atom structure 447, **455**, 456, 458, 516
  boolean algebra **46**
  relation algebra **123**
    Lyndon conditions 280, 337, 343, 349–350, 456, 459
    no elementary characterisation 336–337, 516, *619*
    point-dense, pair-dense 209
    pseudo-elementary characterisation 279–280
completeness 5–8, 43, 100, 119, 252
completion
  $\mathbf{CA}_\alpha$ closed under 168
  canonical extension versus 48, 80
  of boolean algebra **50**, 48–50, 55
  of boolean algebra with operators 65–66, **66**, 98
    atomic case 80, 531
    Sahlqvist equations and 94, 455, 535
  open problems 535, 627
  RA closed under 110
  $\mathbf{RA}_3, \mathbf{RA}_4$ closed under 531
  $\mathbf{RA}_n, \mathbf{S\mathfrak{R}aCA}_n$ ($n \geq 6$) not closed under 361, 535, *621*
  $\mathbf{RCA}_1, \mathbf{RCA}_2$ closed under 169
  $\mathbf{RCA}_n$ ($3 \leq n < \omega$) not closed under 535
  RRA not closed under 336, 455, 535, *621*
complex algebra **79**, 80, 89–93 *see also* atom structure; canonical extension
  completion and 80, 531

complex algebra (*cont.*)
  ∼s over a variety 275
  cylindric, over hyperbasis 371–374, 375
  of group *see* group relation algebra
  of relation algebra atom structure **112**, 461–462, 529, 532, 534
    strongly representable atom structure and 446, 450, 455, 459
  weakly associative 163
complex group **134**, 398 *see also* group relation algebra
complexity
  cylindric basis problem 395
  equational theory of **WA** 595
  guarded fragments 584, 585
  logic and 59, 287
  membership of RA_n 385–386, 387
  network satisfaction problem 252–258
  relativisation and 211
composition *see also* associative function
  in relation algebra reducts **187**, 188–190, 194, 196–198, 417
  in relation algebras **106**, 138
  of binary relations 3, **101**
  of partial functions 29
  of ultrafilters 116, 125, 462, 533–534
computer programs for relation algebras 144
congruence
  on boolean algebra with operators **62**
  on network **222**, 235, 243, 251, 398
  on s-c-words **175**, 175–180, 417, 420
conjugate **64**
conjugated boolean algebra with operators **64**, 64–65
conjugated class of algebras **64**
conjugated operator **64**, 64–65
  CA operators 169
  Crs operators 181
  Q-operators 268
  RA operators 109
  substitutions 181
conjugated Sahlqvist variety
  atom-canonical 92, 98
  closed under completions 94, 455, 535
conjugated variety **64** *see also* variety
  CA_α 168
  $ICrs_\alpha$ 181, 183
  NA 162
  RA 109
  RA_n 361, 385

conjugated variety (*cont.*)
  RCA_n 169
  RRA 361
  $S\mathfrak{R}aCA_n$ 361
connected graph **30**
consistent theory **33**
consistent triple of atoms **115**, 138
consistent type **122**
constant **31**
  lemma on ∼s 289
containment algebra **143**
converse relation 3, **101**
conversion operation **106**
  determined by composition 117
  (hyper)network labels and 221, 339, 365
  in relation algebra reducts **186**, 198
coordinates 554, 558, 559, 560, **565**, **566**, 563–568
correspondent *see* Sahlqvist ∼
countable set **28**
counting modality 208, 593
Craig's trick 290
cycle **448**, 454, 564–565
cylindric algebra **168**, 166–170, *616 see also* cylindric relativised set algebra; neat reduct; relation algebra reduct; representable ∼
  amalgamation 211
  atom structure 169, 371, 458, 627
  canonical extension *see* canonical extension: of ∼
  cylindrification *see* cylindrification operation
  diagonal **167**, 170
  diagonal-free reduct 200, 251
  from cylindric set algebra 212, 583
  from hyperbasis 371–374, 375, 467
  generated by lower-dimensional elements 200, 203
  lattice reduct 206
  polyadic algebra and 202–204
  rainbow construction for 443, 535
  splitting, twisting 211–212, 583
  substitution **170** *see also* substitution operation
cylindric basis 356, 357, **389** *see also* hyperbasis; relational basis
  3-, 4-dimensional, and **SA**, **RA** 394
  complexity of detecting 395
  homogeneity and 356, 395–398, 428, 605

Subject index 673

cylindric basis (cont.)
  hyperbasis and 390–395
  relation algebra reduct and 357, 627
  relational basis and 394–395
  representability and 394
  restriction of 394, 396
  weak associativity and 390, 394
cylindric reduct of polyadic equality algebra **202**, 204
cylindric relativised set algebra **180**, 180–186, *617* see also decidability
  amalgamation 211
  axiomatisation 182–184
  finite base property 593, 600
  from cylindric set algebra 582–583
  guarded fragment and 182, 584
  IG_ω a variety? 184, 283, 626
  not finitely axiomatisable 183, 540
  polyadic algebra and 203, 204
  polyadic $\sim$ **600**
  relation algebra reducts and 194–198, 409–411, 412–415 see also relation algebra reduct
cylindric set algebra **167**, 212, 583, 593
cylindric term network **269**
cylindric-type algebra, BAO **167**
cylindrification operation **167**, 170, **180**
  completely additive
    in **CA** 169, 416, 420, 485
    in **Crs** 181, 182
  polyadic $\sim$ **202**
cylindrification-substitution operation **600**

database 99, 206, 212
De Morgan laws **54**
De Morgan's Theorem K 107 see also Peircean law
decidability 33 see also complexity; undecidability
  arrow logic 208
  equational theory 583
    $\mathbf{Crs}_\alpha, \mathbf{D}_\alpha, \mathbf{G}_\alpha$ 183, 184, 198, 540, 584, 626
    Lyndon algebras 145
    NA 599, *621*
    positive reducts 206
    WA 599, *614*, *621*
  finite axiomatisability and 540, 594
  finite base property and 581, 593–595

decidability (cont.)
  finite hyperbasis problem 571, 603, 627
  finite representation problem 540, 571, 627
  guarded fragment 584
  loosely guarded fragment 585, 590
  membership of $\mathbf{ICrs}_n$ 186, 540
  membership of $\mathbf{RA}_n$ 362, 385, 540, *621*
  packed fragment 585, 590
  recursive axiomatisability and 594, 602
  recursive enumerability and 540, 594, 595
  relativised representation properties 437
  universal theory 186, 594, 600
    $\mathbf{Crs}_\alpha, \mathbf{D}_\alpha, \mathbf{G}_\alpha$ 183, 184, 540, 600
    NA 599, *621*
    WA 599, *614*, *621*
decidable theory **33** see also decidability
default labelling **476**, 478, 480, 481–483
defect 223, 231, 235, 433
definable set **37**, 85
  algebra of $\sim$s **409** see also first-order algebra
defining theory 275, **277** see also pseudo-elementary class
degenerate algebra 40, 51, **61**
  simple 71
  subdirectly irreducible 69
degenerate relation algebra 138
degree of node in graph **30**, 457
dense order 122, 139, 142, 144
dense subalgebra **41**, **61**
  atomic case 51
  complete additivity preserved in 66, 89, 90
  complete embedding and 46
  completion and 49–50, 65–66, 80, 531
  infinite sums preserved in 54, 66
  of relation algebra reduct 193
  representable $\sim$ 55, 283
    of non-representable algebra 455, 535
dense subset **41**, 51
desarguesian projective plane 146, 165
determined game 10, **238**, 239, 318, 451, 498
deterministic pair (τ, t_0) **571**
deterministic strategy 237, 238, 314, **315**, 329, 330
deterministic tiling problem **572**, 572–575, 577

diagonal element **167**, 170, **180**
diagonal-free algebra **200**, 199–200, 206
 polyadic algebra and 202, 203
diagonal-free reduct of cylindric algebra 200, 251
diagonal-free set algebra **199**
diagram 289, 577, 592
dilation 211
Diophantine equation 208, 210
directed cylindric algebra 208
directed graph **29**
directed labelled graph 220, 262, 421 *see also* network
directed labelled hypergraph 249, 364, 421 *see also* hypernetwork
discrete topology **31**, 49–50
discriminator class **129**
discriminator term **70**, 70–76, 96
discriminator variety **70**, 70–76
 boolean algebras 76
 CA_α ($\alpha < \omega$) 168
 CA_α ($\alpha \geq \omega$) is not 170, 202
 equations characterising 72, 76
 equations from universal sentences 74–75, 76
  axiomatisation with 264, 270, 302, 307, 438, 519
  non-embeddability with 578
 PEA_α 202
 RA 110
 RA_n 360, 385
 SA 162
 WA is not 163
disjoint elements of boolean algebra **52**
disjoint union
 of atom structures 97, 131, 385
 of relational structures **34**, 448, 516, 528, 534
 of representations 58, 103, 158, 398
 of sets **26**
distributive lattice-ordered semigroup 207
distributive law in boolean algebras 53, 54
diversity atom **109**
domain *see also* base
 of binary relation **26**
 of partial function **29**
 of structure **33**
duality
 atom structures and algebras 77–83, 85–87, 94–98

duality (*cont.*)
 atom structures and varieties 84–93, 97–98, 335–338
 filters and ideals 42
 logics and quasi-varieties 212
dynamic logic 208, 283

edge
 of graph **30**
 of hypernetwork **364**
 of network **220**
 of relativised (pre-)network **226**
Ehrenfeucht–Fraïssé game 18, 274, 442, 618 *see also* rainbow construction; winning strategy: in ~
 determined 498
 modified ~ **493**
 standard ~ **492**
 transfinite variant 511
elementary chain 303, 326, 329–330
 ~ theorem 330, 333
elementary class **35** *see also* non-~; pseudo-~; pseudo-universal class
 atom structures of a variety 84–93, 97, 98, 336
 closed under ultraproducts and ultraroots 58, 290, 449, 453
 HRA? 282, 626
 pseudo-elementary class and 289, 290, 291
 $\Re\mathfrak{a}CA_n$? 192, 627
 RB_3, RB_4 524, *619*
 $Str RCA_\alpha$? 169, 458, 627
 weakly representable atom structures 446, *620*
elementary closure **35**, 290, 305, 349 *see also* elementary equivalence
elementary diagram 289
elementary embedding **36**
 into ultrapower **58**
elementary equivalence **35**, 58 *see also* elementary closure
 class closed under 58, 91, 449, 454
elementary extension **36**, 122, 123, 130, 275
elementary map **36**, 130
elementary substructure **36**, 291
 game played on 329–331, 485, 516

Subject index

embedding 36 *see also* relation algebra reduct ∼
  into ultrapower 58
  of hypernetwork 365
    into structure 421–423
  of network 483
    into representation 246, 252, 268, 395
empty network 293
empty s-c-word 172
end index of atom 459
end of atom 159, 159–161, 163, 228, 543
equality 32
equation 32, 51 *see also* merry-go-round ∼; Sahlqvist ∼; simple ∼
  atom structure translation 88
  variety and 67, 68, 70 *see also* discriminator variety; variety
equational axiomatisation 35 *see also* axiomatisation
equational class 35
equational theory 33 *see also* axiomatisation; decidability: ∼; undecidability: ∼ of
  of a class 35, 583
equivalence class 27
equivalence relation 27 *see also* congruence; kernel
  from ultrafilter 57
  in n-smooth relativised representations 404
  lattice of commuting ∼s 164
  unit of proper relation algebra 102, 134, 152, *609*
  unit of relativised representation 158
equivalent formulas 35
equivalent game trees 325
Erdös construction 17, 447–449, *620*
`eval` predicate 281
existential formula 33, 494, 593
existential second-order sentence 38, 59, 287, 387
expansion of structure 34 *see also* Q-operator
  finitisation problem 206–208
extension of a structure 36

faithfulness of representation 122, 227, 234, 242
field of relations 100, 101

field of sets 38, 38–39, 51
  boolean algebra and 42–44, 46–48, 49, 53, 58
  proper relation algebra is 101
filter 41, 42, 44, 51–52, 61 *see also* ultra∼
  but not ultra∼ 165, 573
final segment 29
finished play 315
finitary map 173, 202
  product of substitutions 369, 375, 390, 412, 416 *see also* substitution map
finite algebra on finite base property 581, **591**, 591–593
  Cs_n, RCA_n 593
  RA 604
  RA_n 601, *618*
  RRA lacks 593
  SA 605
  $S\mathfrak{R}aCA_n$ lacks 603, *618*
    but subclass has 603–604
  WA 596, *614*
finite algebra property **591**, 591–594
  Crs_n, D_n, G_n 593
  Crs_n^+ 600
  RA, SA lack 593
  WA 593
finite base property 16, 581–586, **591**, 590–593
  boolean algebras 581–582, 595
  Crs_n, D_n, G_n 593, 600
  Cs_n, RCA_n 593
  decidability and 581, 593–595
  Q-algebras 600
  WA 597–599, 600, *614*
finite hyperbasis problem 571, 603, 627
finite intersection property 42, 83, 128
finite model property 16, **584**, 587
  arrow logic 208
  guarded fragments 584, 589–590, 597, 598, 602, 604, *614*
finite-variable logic 32, **400** *see also* clique-relativised semantics; non-finite axiomatisability
  axiomatising **RRA** 520, 625
  finitisation problem 207
  pairing axiom 130
  proof theory 15, 212, 356, 465, 486–489, *620*, 627
  relation algebra from formulas 136–137

finitely axiomatisable class **35**, 58, 75, 290
  see also non-finite axiomatisability
  $CA_n \cap ICrs_n, ID_n, IG_n$ 183
  decidability and 540, 594
  Q_n-algebras 208, 362
  RCA_1, RCA_2 168
  reducts of polyadic algebras 206
  relation algebras with representations respecting $\{0,1,+,1',\breve{},;\}$ 166
  RRA
    expansions 206–208
    subclasses 208–210
    subreducts 206
  $S\mathfrak{Nr}_m CA_{m+1}$ 192, 465, 488
finitely axiomatisable theory 33
finitely based variety **75** *see also* finitely axiomatisable class
finitely branching tree **311**, 328
finitely generated structure 37, 68, 128, 239, 246
finitely satisfiable type **122**, 303, 305
finitisation problem 152, 205–210, 541
first-order algebra 98, 130, **456**, 457, 515
first-order correspondent *see* Sahlqvist correspondent
first-order language **31**
  recursive $\sim$ **288**
fixed-point logic 387, 404
flat relativised representation *see* n-$\sim$
flexible atom **141**, **348**, 627
forbidden triple of atoms **115**, 138
forcing 11, 274, 276, 295–296
fork algebra 210
formula **31**, 31–33, 34, 37–38
  cylindric algebra term from 487
  expressing networkhood 262, 270, 299, 341–342, 346
    by Q-operators 265, 307
  expressing winning strategy 11, 273–277
    see also winning strategy
  $EF_r^p(A,B)$ 494
  $G(T,\mathcal{A})$ 323–325
  $G_n^a(\mathcal{A})$ 342, 346 *see also* Lyndon conditions
  $G_n(I_a, \mathcal{A})$ 262–263, 265–267
  $G_n(N^\iota, C)$ 270
  $\Gamma(U, \mathcal{A})$ 292, 298, 306–308
  $\Gamma^+(U, \mathcal{A})$ 303, 306–308
  logically equivalent $\sim$s **35**
  preserved in ultraproducts 57

formula (*cont.*)
  relation algebra from $\sim$s 136–137
  valid $\sim$ **35**
'forth' game *see* Ehrenfeucht–Fraïssé game
free variable **32**, 402
frozen red clique 499
Fréchet filter **52**
full complex algebra *see* complex algebra
full proper relation algebra **104**, 129, **133**, 241 *see also* proper relation algebra
function **28**, 33
function symbol **31**
functional element **208**
functional signature **31**

Gabbay-style axiomatisation 145, 210
Gaifman graph **588**
Gale–Stewart theorem 238, 318, 451, 498
game 9–15, 17–18, 273–277, 309–310, 441–443, 491, *611–613* *see also* axiomatisation; determined $\sim$; Ehrenfeucht–Fraïssé $\sim$; formula: expressing winning strategy; $\sim$ tree; strategy; winning strategy
  atomic (G_n^a) **339**, 546, *613*
    associativity and 345–346
    complete representations and 347–348, 349–350, *613*
    finite approximation 337, 349–351
    G_n and 340, 346
    G_r^n and 377, 514
    Lyndon conditions and 337–338, 341–343, 349–351
    on canonical extension 350–351, 454, 461, 532, 534
    transfinite variant 348, 349
  axiomatisation and 9–13, 261, 273–277, 292, *611–612*
  defined by game tree ($G(T, \mathcal{A}, v)$) **312**
    see also game tree
  determined 318
  finite approximation 321, 322, 326, 332
  non-finite axiomatisability and 325–326, 331
  on elementary substructure 329–331, 485, 516
  on ultraproduct 326–328

Subject index 677

game (*cont.*)
  elementary substructure and 329–331, 485, 516
  for homogeneous representation 397
  for hyperbasis ($G_r^{m,n}$) **378**, 378–384, 464, 476–483, *616*
  for pseudo-elementary class (Γ^+) 276, **303**, 302–308
  for pseudo-universal class (Γ) 292, **293**
    finite approximation 296–298
    pseudo-universal class and 294–296, 301
  for relational basis (G_r^n) **376**, *616*
    finite approximation 491, 495
    G_n^a and 377, 514
    relational basis and 356, 377, 384, 526
    (semi-)associativity and 383, 496
    transfinite variant 511
  for relativised representation (G_ω^{rel}) **226**, 226–232
  for representation of cylindric algebra ($G_n(N,C)$) **249**, 250, 269–270
  for representation of relation algebra (G_n) 215, **233**, 233–236, *611–612*
    canonical axiomatisation and 264
    determined 238, 239, 451
    finite approximation 244
    G_n^a and 340, 346
    rank and 224–225, 239
    representation and 239, 245, 246
  for representation of relation algebra (Λ_n^m) 246
  for weak representation ($\mathcal{J}_n$) 247
  n-pebble 18, 274, 309, 377, 443, *616*
    see also Ehrenfeucht–Fraïssé ∼; ∼: for relational basis (G_r^n)
  non-elementary class and 326, 331, 441
  non-finite axiomatisability and 13, 325–326, 331, 441
  play 227, **315**
  rules 227
  scheduling of moves 232, 245, 295, 347, 380, 397
game tree 311, 311–313, *612 see also* game: defined by ∼ ($G(T, \mathcal{A}, v)$)
  class defined by 310, **331**, 333
  equivalent ∼s 325
  for G_n^a 343–345, 349, 515, 516
  for $G_r^{m,n}$ 380–383, 384, 484
  for G_r^n 384, 524

game tree (*cont.*)
  for G_n 319–321
  restrictions on formulas in 325
  subtree **313**
general relativised representation **153**, 153–155, 162, 599
generalised cylindric set algebra **168**
generalised diagonal-free set algebra **199**
generalised subreduct 105, 205–206, 529
generating set **37**, 200, 203
girth 447, **448**
graph **29** *see also* clique
  chromatic number 447, **448**, 450–453
  colouring 59, 290, **447**, 454, 457, *620*
    non-finitely colourable ∼ 449, 451, 454
  of edges 457, 458
  cycle **448**, 454, 564–565
  Erdös construction 17, 447–449, *620*
  for tiling coordinates **564**
  girth 447, **448**
  in rainbow construction 499, 514, 516, 521, 527, 529, 531
  independent set **447**
  isomorphism problem 625
  relation algebra from **140**, **450**
green atom **495**, 544
group **30**, 99, 578, 590 *see also* group relation algebra
  of automorphisms **36**, 130, 281
  of collineations 148
  of permutations **30**
group relation algebra 90, **134**, 134–136, 148, 280, 398
guard 584, **586**
guarded fragment 16, 182, 583, 584 *see also* loosely ∼

halting problem 542, 572
height of (node in) tree **311**
Henkin's two equations for $\mathbf{RCA_2}$ 169
hereditarily refinable network 223, 227, 235, **237**, 239
Herwig's theorem 589
higher-order logic 38, 571, 575, *621 see also* second-order logic
Hilbert system 465, 486, 488–489, *620*
Hintikka game 310
holds predicate 276, 279, 437

homogeneous representation **281**, 284, **395**, 605
   cylindric basis and 356, 395–398
   elementary property? 282, 626
   Lyndon algebra without 441
   n-homogeneous **284** *see also* n-homogeneous relativised representation
homomorphic image **36**, 68
   class closed under $\sim$s
      RA_n 385
      RRA 126–128
      $S\mathfrak{R}a\mathbf{CA}_\alpha$ 192–193
      variety 68
homomorphism **35**, 40, 72–74, 511
   ideal and 61–63, 193
hybrid logic 210
hyperbasis 357, 359, **368**, *615*, *618* *see also* cylindric basis; hypernetwork; wide $\sim$; relational basis; symmetric basis
   3-, 4-dimensional, and **SA**, **RA** 375–376, 387, 432, 433–434
   complete embedding and 375
   completely representable relation algebra has 375
   cylindric algebra from 371–374, 375, 467
   cylindric basis and 390–395
   game to test **378**, 378–384, 464, 476–483, *616*
   interpolants in 370
   Monk algebra $\mathfrak{A}(n,r)$ and 467–475
   neat reduct and 373, 465, 469
   non-elementary property 486, *620*
   relation algebra reduct and 359, 372, 428–432, 434, 464, *618*
      finite 435–436, 602–605, *618*
      hyperbasis from 416–420
   relational basis and 368, 376, 394, 478
   relativised representation and 359, 428–432, 434, *618*
      finite 435–436, 603–604, *618*
      from hyperbasis 421–426
      hyperbasis from 427
   restriction of **372**, 373–375
   structure encoding 380–382, 437, 484–485
   tiling from 548–551
   undecidability of having $\sim$ 570
      finite $\sim$? 571, 603, 627
hyperedge **30**, **364**

hypergraph **30**
hyperlabel **364**
hypernetwork 357, **363**, *611* *see also* atomic network; hyperbasis; wide $\sim$
   agrees off of x with other $\sim$ **365**, *611*
   atomic network and 364
   embedding *see* embedding: of $\sim$
   from other $\sim$ by transformation 365–366
   from relation algebra reduct 418–420
   maximal strict labelled hypergraph in **422**
   restriction of **372**

ideal **41**, **62**
   homomorphism and 61–63, 192–193
idempotent function **28**, 39
identity atom **109**, **163**, 394, 543
identity map **29**
identity operation (1') **106**
   determined by composition 117
   in relation algebra reducts **186**, 198
   is atom in integral algebra 161
identity relation 3, **101**, 164
iff **26**
illegal position **314**
immediate predecessor, successor **311**
inclusion map **29**
inclusion relation **26**
inconsistent triple of atoms **115**, 138
independent set **447**
index of atom in Maddux algebra **459**
index of node
   in network **460**, **544**
   in red clique **496**
   with respect to functions I,J **471**
infimum *see* boolean functions
infinitarily n-flat relativised representation 359, **402**, 402–403, 407–408, *615* *see also* relativised representation
   is n-smooth 405, 426
   n-flat and 405, 412, 436
infinitary language **32**, 207, 359, **400**, 458
initial segment **29**
injective function **29**
instance of schema **487**
integers **27**
integral algebra
   non-associative algebra **161**, 161–162
   relation algebra **117**, 130, 136, 143
interior **49**

Subject index

interpolation 206, 210
  in hyperbasis 370
interpretation
  of a structure in another 37, 351, 516
  of symbols in a structure 33
inverse 29
involution 29
irreflexive binary relation 26
irreflexive partial order 27
isomorphism 36, 68

join *see* boolean functions
Jónsson Q-algebra *see* Q-algebra
Jónsson Q-operator 267–268, 585, 600

Keisler game 310
Keisler–Shelah theorem 58
kernel 61, 126, 173
Kleene star operator 283
König's tree lemma 244, 297, 313

labelled (hyper)graph 30, 220, 262, 421
  see also default labelling; network
  maximal strict $\sim$ 422
Lambek calculus 206, 601
language 31
lattice reduct of cylindric algebra 206
leaf of tree 311
left linear point algebra 142
legal position 314
lemma on constants 289
length
  of cycle 448
  of path 30
  of s-c-word 172
  of tuple 29
limit ordinal 28
linear order 27, 42, 52, 122, 139, 144
linked tile edges 559, 560–562, 565, 566
local isomorphism *see* partial isomorphism
locally cubic or square unit 181
logically equivalent formulas 35
loop-free graph 30

loosely guarded fragment 16, 585, **586**, 590
  see also guarded fragment
  axioms for relativised representation 596–597, 602
  finite model property 589–590, 597, 598, 602, 604, *614*
  packed fragment and 587, 590
Łoś' theorem 57, 60, 327, 328, 449
  non-finite axiomatisability by 331, 485, 515, 524, 530
Łoś–Tarski theorem 35, 584
Löwenheim–Skolem–Tarski theorem 37, 120, 232, 329
Lyndon algebra 146, 144–149, 200, 207, 351–352, 392–393, 441
Lyndon conditions 8, 337–338, 341, 346, *613*
  atom structures satisfying 456, 456–457, 515
  atomic game and 337–338, 341–343, 349–351
  complete representability and 280, 337, 343, 349–350, 456, 459
  representations and 349–351, 456, 459–462
  strongly representable atom structures and 456–457, 459–462

m-dimensional move in game $G_r^{m,n}$ 379, 382, 477, 478, 482
Maddux algebra 459, 459–462
many-sorted logic 31
  defining pseudo-elementary class 275, 277, 281, 285, 437
  for game $G_r^{m,n}$ 380, 485
map 28
maximal ideal 41
McKenzie's algebra 140, 143, 144, 166, 386
meet *see* boolean functions
merry-go-round equation 182, 183, 184, 191, 203, 204
metric point algebra 143, 144, 252, 258
middle index of atom 459
modal logic 72, 99, 123, 200, 212, 541, 583
model 34
model theory 19, 31–38, 284–291
modest s-c-word 179

modus ponens 487
Monk algebra 17, **141**, 211, 441–442, *619*
  anti-∼ 348
  for reducts ($\mathfrak{A}(n,r)$) 464, **466**
    hyperbasis and 467–475
    hyperbasis game and 476–483
    non-finite axiomatisability and 484–485
    uncountable variant 485
  from graph ($\mathfrak{Cm}\,\alpha(\Gamma)$) **450**, 450–454, 457, 458
  rainbow construction and 18, 443
Monk's theorem *see* canonical extension: of relation algebra: preserves representability; non-finite axiomatisability: **RRA**
monochromatic element **450**, 451
monochromatic triangle 17, 142, 349, 441, 450, 457, 458
monoid 118
monotonic operator **61**
monotonic term 86
mosaic 13, 217, 421, 584

n-ary
  function **28**, 33
  operator **60**
  relation **29**, 33
  symbol **31**
n-colourable graph **59**
n-dense relation algebra **209**
n-flat relativised representation 358–360, **402**, 402–403, 408, *615* *see also* relativised representation
  finite 582, 603–605, *618*
  infinitarily n-flat and 405, 412, 436
  is n-square 403, 405
  model of Fl^n($\mathcal{A}$) 406
n-generated structure **37**
n-homogeneous relativised representation **428**, 605
n-homogeneous representation **284**
n-max element **209**
n-pebble game *see* game: n-pebble
n-smooth relativised representation **404**, 407, 408 *see also* relativised representation
  infinitarily n-flat implies 405, 426
  model of Sm^n($\mathcal{A}$) 406
  n-homogeneous implies 428

$(n-\frac{1}{2})$-square relativised representation **411**
n-square relativised representation **158**, 358–360, **401**, 407–408, **526**, *615* *see also* relativised representation
  classical representations and 433, 526–528
  finite 582, 601, 604, 605, *618*
  model of Sq^n($\mathcal{A}$) 406
  n-flat implies 403, 405
  RA$_n$ and 359–360, 430–431, 433–434, 526, 582, 601, *618*
n-tuple **29**
n-type **122**
n-variable logic *see* finite-variable logic
n-wide hyperbasis **373**, 373–375, 467, 469 *see also* hyperbasis
n-wide hypernetwork **373**, 379, 380–382, 477 *see also* hypernetwork
natural numbers **27**
neat embedding theorem 123, 192
  for relation algebras 432
neat reduct **186** *see also* finitely axiomatisable class: **S**$\mathfrak{Nr}_m$**CA**$_{m+1}$; non-finite axiomatisability: **S**$\mathfrak{Nr}_m$**CA**$_n$; relation algebra reduct
  atomic case 187, 193, 418
  class of subalgebras of ∼s **190**, 191
    variety 192, 193, 488
  complete embedding and 187–188
  complete homomorphism and 416
  completion of 627
  hierarchy 192, 442, 463–465, 475, 484, *620*
  hyperbasis and 373, 465, 469
  non-elementary property 191, 465
  proof theory and 487–488
negative term **86**, 93, 97
network *see also* atomic ∼; hyper∼; ∼ satisfaction problem; pre-∼; relativised ∼; term ∼
  approximation to representation 217–220, 221
  refining 222–225, 227 *see also* game; hereditarily refinable ∼
  embedding *see* embedding: of ∼
  first-order characterisation *see* formula: expressing ∼hood
  for cylindric algebra 248, **249**
  for pseudo-universal class **292**
  for relation algebra **220**, *611*

Subject index 681

network (*cont.*)
  rank **224**, 224–225, 239
  weak associativity and 391, 394
network addition property **483**
network satisfaction problem **252**
  algorithms for 252–255
  approximations 255–258
  undecidable 223, 255, 579
network-universal relation algebra **254**, 256
no-instance **542**
node *see also* index of $\sim$
  of graph **29**
  of hypernetwork **364**
  of network **220**
  of pre-network **233**
  of relativised (pre-)network **225**
  of tree **310**
non-associative algebra **155**, 156, 158–159, 162, 363, *614*
  universal theory decidable 599, *621*
  with cylindric but no relational basis 394–395
non-desarguesian projective plane 146, 165
non-elementary class *see also* elementary class
  atomic relation algebras with
    hyperbasis 486, *620*
    relational basis (RB_n) 524, *619*
  CRA 336–337, 516, *619*
  CRAS 456, 516
  games and 326, 331, 441
  non-finitely colourable graphs 449
  $\mathfrak{Nr}_m CA_n$ ($1 < m < n$) 191, 465
  pseudo-elementary 291
  SRAS (= Str**RRA**) 92, 442, 447, 457, *620*
  WRAS \ **SRAS** 458
non-finite axiomatisability 13, 17 *see also* finitisation problem
  chains 362
  classes between
    RRA and $S\mathfrak{Ra}CA_5$ 570, 571, *621*
    RRA and **wRRA** 575, *621*
  compactness and 332, 454
  FOAS 515
  games and 13, 325–326, 331, 441
  group relation algebras 136
  higher-order logic 571, 575, *621*
  ICrs$_\alpha$ 183, 540
  LCAS 515

non-finite axiomatisability (*cont.*)
  modal logics 200
  no finite-variable axiomatisation
    equational 205, 215, 352, 517–519, *619*
    universal 204, 207, 352, 465, 520
  proof theory 487, *620*
  RA$_n$ over **RA**$_{n-1}$ 361, 524, *619*
  RCA$_n$ 169, 215
    over $S\mathfrak{Nr}_n CA_m$ 484
  Rdf$_\alpha$ 200
  representable polyadic algebras 204
  representable sequential algebras 206
  RRA 215, 351, 454, 479, 515, *619*
    expansions 206–207
    Monk's method 8, 145, 165, 355
    over **RA**$_n$, $S\mathfrak{Ra}CA_n$ 361, 479
    over **wRRA** 165, 528
    subreducts 205–206
  $S\mathfrak{Nr}_m CA_n$ ($n \geq m+2$) 465
    over $S\mathfrak{Nr}_m CA_{n-1}$ 442, 464–465, 484, 485, *620*
  $S\mathfrak{Ra}CA_n$
    over **RA**$_n \cap S\mathfrak{Ra}CA_{n-1}$ 479
    over $S\mathfrak{Ra}CA_{n-1}$ 361, 442, 464, 479, 485, *619*
  SRAS = Str**RRA** 148
  ultraproducts and 326, 331 *see also* Łoś' theorem
  WRAS = At**RRA** 148, 515
  wRRA 165, 529–530, 575, *621*
non-finitely colourable graph 449, 451, 454
non-orthodox axiomatisation 145, 210
non-principal filter **41**
non-principal ultraproduct **57**
non-recursively enumerable class *see also* recursively enumerable class
  finite relation algebras with
    infinite hyperbasis 605
    no finite hyperbasis 605
    only infinite representations 540, 571
non-well-founded set theory 208
normal operator **60**, 64, 265
  pseudo-normality axiom 307
NP (complexity) 59, 253, 254, 287, 395
nullary operator 61
nullary symbol **31**

one-one function **29**

onto function 29
open cover **31**, 44
open set **30**, 143
operator **60**, 61 *see also* completely additive $\sim$; conjugated $\sim$; Q-$\sim$
operator symbol **61**
orbit **30**
order-isomorphism **27**
order of projective plane **145**
order type **27**
ordered pair **26**
ordinal **27**
outcome of play **294**
owner of node in game tree **311**

P-part of structure **285**
packed fragment 585, **587**, 587–590
packing guard **587**
pair-dense relation algebra **209**
pair (in relation algebra) **209**
pair (ordered) **26**
pairing axiom 130
pairwise guarded fragment *see* loosely guarded fragment
parameters
  definable with **37**, 85, 98, 456
  of type 122
partial elementary map 130
partial function **29**
partial homomorphism **492**, 510
partial isomorphism **36**, 510
  extendable 284, 395–398, 428, 589
  in n-smooth relativised representation 404
  of atomic network **396**, 396–398
partial isomorphism move in game **397**
partial order **27**, 40, 51, 142
partial tiling **571**
partition
  of element of boolean algebra **54**
  of set **27** *see also* graph: colouring
path-consistency 222, 254
path in graph **30**
$PC, PC', PC_\Delta, PC'_\Delta$ classes **285**, 284–291, 387 *see also* pseudo-elementary class
PC^s class **277**, 287, 291
PDL (propositional dynamic logic) 283
pebble game *see* game: n-pebble

Peircean law **107**, 107–108
  equivalent equations 117
  extended version 159–161
  Q-operators and 268
Peircean transform **115**, 544
pentagonal algebra **139**, 255
perfect extension **83**, 95 *see also* canonical extension
permutation **30**
permutation group **30**
permutational representation **130**, 136, 281
permutationally invariant operation 206–207
Pinter's substitution algebras 191, **204**
planning 99, 138, 252
play of game 227, **315**
players ∀ and ∃ 9, 226, 292, 312
  labels chosen by **477**, **502**, **552**
point algebra **139**, 252
  Allen interval algebra from 143
  networks over 221
  representations of 139, 144, 348, 581, 605
point-dense relation algebra **209**
point (in relation algebra) **209**
pole position **245**, **297**, **304**
polyadic (equality) algebra 191, **201**, 201–204, 206, 283, **600**
polyadic (equality) set algebra **202**
position in game **312**, **314**, 333
positive boolean combination 32
positive reduct 206
positive term **86**
  strictly $\sim$ **93**
power set **26** *see also* complex algebra; field of sets
pre-network *see also* network
  for cylindric algebra 249
  for pseudo-elementary class 302
  for pseudo-universal class 292
  for relation algebra **233** *see also* network satisfaction problem
  rank **239**
pre-order **27**
prenex formula **33**, 75, 520
prenex normal form **33**
primary triangle **506**
prime ideal theorem *see* boolean $\sim$
primitive recursive function 302
principal filter **41**

Subject index 683

principal ultraproduct 57
private game 232, 327, 496, 499, 502–509
probability 388, 448, *620*, 627
product
   boolean *see* boolean functions
   cardinal and ordinal 28
   of sets 26
   of structures 56, 68–70, 74 *see also* subdirect $\sim$; ultra$\sim$
   canonical extension and 59, 96
program view of s-c-words 172–174, 180
projection 56, 74, 208
projective geometry 144–148, 200 *see also* Lyndon algebra
projective plane **145**, 145–148, 165
proof theory 15, 212, 356, 465, 486–489, *620*, 627
propagation algorithm **253**, 252–258
proper filter **41**
proper inclusion **26**
proper relation algebra **101**, 101–104, 152, *609* *see also* full $\sim$; representable relation algebra
   examples 138
   generated by subset **134**
   is relation algebra 116
propositional dynamic logic 283
pseudo-elementary class 11, 273, **277**, *612* *see also* axiomatisation: $\sim$
   as notion of representation 211, 276, 591
   defining theory 275, **277**
   examples 278–283, 348, 458, 605
   IG_ω? 283, 626
   game for 276, **303**, 302–308
   game tree defines 310, 333
   model theory of 284–291
   single-sorted view 284–287
   ultraproducts and 290, 291, 627
pseudo-normality axiom 307
pseudo-universal class **277**, *612* *see also* axiomatisation: pseudo-elementary class: $\sim$
   examples 278–283, 360, 437
   game for *see* game: for $\sim$ (Γ)
   is universal 289, 290
   model theory of 285–290
   single-sorted view 285–287

Q-algebra 264, 283, 362, 600

Q-operator
   axiomatising expansion of
     RA_n 208, 362, 388
     RRA 264–268, 275, 517, *612*
   generalised $\sim$ 306–308, *612*
   Jónsson $\sim$ **267–268**, 585, 600
   Stebletsova–Venema $\sim$ **268**, 517
quantifier elimination 120–121, 407
quantifier-free formula **32**, 74–75
quasi-equation **67**
quasi-polyadic (equality) algebra 191, **202**
quasi-product 210
quasi-projection **209**, 388
quasi-variety **67**, 67–70, 76
   wRRA 165, 166

rainbow algebra **494**
   complete representability 511
   semi-associativity 495, 510
rainbow construction 17–19, 442–443, 491, **494–495**, 499–500, *618* *see also* tiling algebra
   applications 513
     completions 531–535
     CRA 516
     RA_n 522–524
     relational bases 491, 524, 527–528
     RRA 514, 518
     wRRA 529–530
   for cylindric algebras 443, 535
   modifications 509, 510, 517, 534, 540
   Monk algebras and 18, 443
   rainbow theorem 495–509, 511, *619*
   w_S atoms **495**, 499–500, 504, 505, 506, 508, 554 *see also* tiling algebra: u and v atoms
Ramsey's theorem **31**
   graph colourings 457
   Monk algebras 17, 142, 441, 443, 450
RA_n 15–16, **384**, 386–388, **526** *see also* completion
   axiomatisation *see* axiomatisation: $\sim$
   canonical variety 360, 361, 385, 387, 432
   equational theory undecidable 578, *621*
   finite algebra on finite base property 601, *618*
   hierarchy 361, 386, 443, 524, *619*

RA_n (cont.)
  membership problem decidable 362, 385, 540, *621*
    complexity 385, 387
    probability of membership 388
    pseudo-elementary class 387, 437
  relation algebra reduct characterisation 208, 210, 359, 362, 388, 430–432, 433–435, *618*
  relativised representation characterisation 359–360, 430–431, 433–434, 526, 582, 601, *618*
  RRA and 360, 432, 479
  SA, RA and 360, 361, 387, 390, 432, 433–435
  $S\mathfrak{R}aCA_n$ and 361–362, 432, 478
  weakly representable relation algebras and 530, 575, 576, 626
range
  of binary relation 26
  of function 28
rank
  of atom in Maddux algebra **459**
  of functions I, J **471**, 476
  of network **224**, 224–225, 239
  of node in tiling graph **565**
  of operator **60**
  of position in game 333
  of pre-network **239**
rational numbers **27**, 122, 139, 142
real numbers **27**, 143, 291
rectangular density 200, 210, 274
recursive axiomatisation **288**
  decidability and 594, 602
  discriminator variety 75, 302, 438
  pseudo-elementary class 290, 305
  pseudo-universal class 301, 302
  RRA 263–264
recursive game tree **313**, 325
recursive language **288**
recursive saturation **303**
recursive type 303
recursively enumerable axiomatisation 264, 276, **288**, 289, 290
recursively enumerable class *see also* non-$\sim$ co-PC' class 291
  decidability and 540, 594, 595
  finite relation algebras with
    finite representation 539, 541
    no representation 130, 539, 541

red atom **495**
red clique **496**, 496–506, 508, 518
  based on edge **499**
  based on node 540
  frozen 499
  uncountable 527
reduct of structure **34**, **284** *see also* neat reduct; relation algebra reduct
reducts of relation algebras 163–164, 205–206 *see also* relation algebra reduct
refinement *see* network
reflexive binary relation **26**
reflexive closure **26**
reflexive unit 102, 154–156, 599
regular open set **49**
rejecting move by $\exists$ **234**, **250**, 319, *611*
relation **29**, 99, **167**
relation algebra 7, **106**, *609* *see also* atomic $\sim$; $\sim$ reduct; representable $\sim$; simple $\sim$
  alternative definition 117
  atomic, complete, etc. **109**
  axioms 4, **106**, 107–118, 154–156, 189–190, 191, 195–197 *see also* associative function; Peircean law
  basis characterisation 376, 387, 389, 394, 432
  binding conventions for operations 106
  canonical extension *see* canonical extension: of $\sim$
  classification 144
  constructions of $\sim$s 441–443 *see also* Maddux algebra; Monk algebra; rainbow construction; tiling algebra
  equational theory undecidable 578, *621*
  examples 133–147 *see also individual algebras*
  finite algebra on finite base property 604
  finite algebra property fails 593
  game characterisation 345–346, 383, 496
  history 1–9
  RA_4 and 360, 361, 387, 390, 432
  relation algebra reduct is 189, 195, 464
    and conversely 191, 195, 388, 432
  Sahlqvist variety of $\sim$s 110, 335
  subdirectly irreducible $\sim$ **109**, 111
relation algebra atom structure **112**, 111–117, 335–338, *620 see also* strongly representable atom structure; weakly representable atom structure

Subject index 685

relation algebra atom structure (*cont.*)
  algebras over *see* complex algebra; first-order algebra; term algebra
  game G_n^a played on 339
  of anti-Monk algebra 349
  of canonical extension 116, 458, 461–462, 533–534
  of Lyndon algebras 145
  of Monk algebra $\mathfrak{A}(n,r)$ **466**
  of Monk algebra $(\alpha(\Gamma))$ 447, **449**
  of rainbow algebra **494–495**
    variants 509, 517, 534
  of tiling algebra **544–545**
  representability not determined by 336, 455
  satisfying Lyndon conditions **456**, 456–457, 515
relation algebra reduct 15–16, 357, *617 see also* composition: in ∼s; completion; neat reduct
  atomic case 187, 417
  axiomatisation 360, **438**
  basis and 357, 359, 372, 376, 412–420, 428–434, 464, *618*, 627
    finite 376, 435–436, 602–605, *618*
  canonical extension and 188, 193, 195, 431
  closed under subalgebras? 627
  complete embedding and 188, 409, 411, 412, 416, 429
  elementary property? 192, 627
  finitisation problem and 206–207
  hierarchy 361, 442, 463–464, 475, 479, 485, *619*
  is relation algebra 189, 195, 464
    and conversely 191, 195, 388, 432
  of cylindric algebra **186**, 186–193, 605
  of cylindric relativised set algebra 194–198, 409–411, 412–415
  of Q_n-algebra 208, 362, 388
  pseudo-universal view 282, 437
  $\mathbf{RA}_n$ and 208, 359, 362, 388, 430–435, *618*
  relativised representation and 359–360, 409, 428–434, *618*
    finite 435–436, 582, 603–604, *618*
  semi-associative algebra and 191, 198, 203, 433–435, 605
  weakly associative algebra and 197–198, 361, 435

relation algebra reduct embedding **194**, 194–198, 376, 409–411, 429, 431, *617*
relation algebra term 106
relation symbol **31**
relation-type algebra 7, **106**
relation-type BAO **152**
relational basis 356, **367**, *615 see also* cylindric basis; game: for ∼ (G_r^n); hyperbasis; $\mathbf{RA}_n$; symmetric basis; triangle addition property
  $2\frac{1}{2}$-dimensional 436
  3-, 4-dimensional
    is cylindric basis 394
    is hyperbasis 376, 394
    SA, **RA** and 375–376, 387, 389–390, 432, 433
  algorithm to find 385, 387
  canonical extension and 384, 385, 387, 432
  closed under substitutions 369
  complete embedding and 375
  cylindric basis and 394–395
  finite 385–386, 435, 602
  from other ∼ by transformation 369, 525
  hyperbasis and 368, 376, 394, 478
  infinite-dimensional 384, **525**, 525–528
  Maddux's definition **389**
  non-elementary property 524, *619*
  probability of having 388
  rainbow construction and 491, 524, 527–528
  relation algebra reduct and 359, 376, 412–415, 428–435, *618*
  relativised representation and 359, 421–425, 428–434, 526–528, *618*
    finite 435, 604
  restriction of **372**, 372, 375
  structure encoding 437
relational signature **31**
relational structure **34**
relative product **3**
relative sum **105**
relativisation
  of algebra 72, 211, 583
  of formula 37, 284, 405, 589
relativised cylindric set algebra *see* cylindric relativised set algebra
relativised network **226**, 226–231
relativised pre-network **225**, 225–227

relativised representation 15, 16, **153**, 152–158, 357–360, 399–400, *613 see also* complete ∼; general ∼; infinitarily n-flat ∼; n-flat ∼; n-smooth ∼; n-square ∼
  basis and 359, 421–434, 526–528, *618*
    finite 435–436, 603–604, *618*
  canonical extension has 360, 406, 408, 432
  clique **158**, 358–359, **400**, 400–407, 433, *615*
  decidable 437
  finite 437, 582, 595–597, 601–605, *618*
  finitisation problem and 152, 208
  game to build **226**, 226–232
  loosely guarded definition 596–597, 602
  model of $R_{\mathcal{A}}$ 153, 229, 399–400, 596
  packed fragment and 589
  RA_n and 430–431, 433–434 *see also* n-square ∼: RA_n and
  relation algebra reduct and 359–360, 409, 428–434, *618*
    finite 435–436, 582, 603–604, *618*
  semi-associative algebra and 433–434, 605
  weakly associative algebra and 157, 225–232, 596, *614*
representable boolean algebra **42**
representable cylindric algebra **168**
  axiomatisation *see* axiomatisation; non-finite axiomatisability: RCA_n; Sahlqvist variety
  canonical extension of 169, 170, 626
  completion of 169, 535
  cylindric algebra from 212, 583
  diagonal-free reduct 200
  finite base property 593
  from hyperbasis 375
  game to test **249**, 250, 269–270
  neat embedding theorem 123, 192
  polyadic algebras and 204
  pseudo-universal view 283
  relation algebra reduct 193
  Str RCA_α 169, 458, 627
  (sub)reducts 206
  variety of ∼s 168
representable diagonal-free algebra **200**
representable polyadic (equality) algebra **202**, 203–204

representable relation algebra **118**, *610 see also* approximations to **RRA**; proper relation algebra; relation algebra atom structure; representation; weakly ∼
  axiomatisation *see* axiomatisation: **RRA**; finitely axiomatisable class; finitisation problem; non-finite axiomatisability: **RRA**; Sahlqvist variety: **RRA** is not
  canonical extension of 124–126, 130, 350–351
  canonical variety of ∼s 128, 129, 351, 361
  completely ∼ *see* completely representable structure: relation algebra
  completion of 336, 455, 535, *621*
  cylindric basis and 394
  finite algebra on finite base property fails 593
  finite ∼
    undecidability 205, 569, *621*
    with finite representation 539–540, 541, 571, 627
    without finite representation 139, 144, 348, 539–540, 541, 571
  from formulas 137
  game to test *see* game: for representation of relation algebra (G_n)
  Lyndon's characterisation 246, 343 *see also* Lyndon conditions
  neat embedding theorem for 432
  probability of representability 627
  pseudo-universal view 279, 282
  Q-operator and 264–268, 275, 517, *612*
  RA_n and 360, 432, 479
  smallest counter-example *see* McKenzie's algebra
  sufficient conditions 208–210 *see also* Lyndon conditions
  variety of ∼s 126, 129
  with infinite square representation 280
representable sequential algebra 206
representation 6–13 *see also* atomic ∼; base; complete ∼; finite base property; game; homogeneous ∼; relativised ∼; *representable algebras*
  of boolean algebra 6–7, **42**, 42–43, 53, 58, 581–582
  of cylindric algebra **168**

Subject index

representation (*cont.*)
 of relation algebra 8, 100, **118**, *610*
  arbitrarily large $\sim$ 129
  canonical extension has 124–126, 130, 350–351
  finite $\sim$ problem 540, 571, 627
  game to build *see* game: for $\sim$ of relation algebra (G_n)
  isomorphism *see* base-isomorphism
  Lyndon conditions and 349–351, 456, 459–462
  model of $T_{\mathfrak{A}}$ 120, 279, *610*
  pseudo-universal view 279
  relativised $\sim$ and 433, 526–528
  respecting $\{0, 1, +, 1', \breve{\,}, ;\}$ 166
 of relation-type algebra 7
 pseudo-elementary view 211, 276, 591
 ultrafilter induced by 47
residual operator 105, 601
restricted proper relation algebra **103**, 103–104
restriction
 of basis **372**, 372–375, 394, 396
 of function 28
root of tree **311**
rooted tree **311**
rules of game 227

s-word **172** *see also* s-c-word
s-c-word **172**, 172–180, 416–420
Sahlqvist correspondent
 of equations defining **RA** 113
 of equations defining **SA**, **WA** 163
 of Henkin's two equations 169
 of Sahlqvist equation 84–88, 92, **94**, 97
 of simple equation **87**, 87–88, 96
Sahlqvist equation 92, **93**, 93–95, 188, 535
  see also canonical equation; simple equation
Sahlqvist variety **93** *see also* conjugated $\sim$
 atom structures 92
 CA$_\alpha$ 168
 canonicity 94
 ID$_\alpha$, **IG**$_\alpha$ 184
 NA, **WA**, **SA** 159
 RA 110, 335
 RA$_n$, $\mathsf{S}\mathfrak{R}\mathfrak{a}$**CA**$_n$ ($n \geq 6$) are not 361, 535, *621*
 RCA$_1$, **RCA**$_2$ 169

Sahlqvist variety (*cont.*)
 RCA$_n$ ($n \geq 3$) is not 205, 215, 535
 RRA is not 205, 215, 336, 455, 535, *621*
satisfiable pre-network **252**
saturation **122**, 122–123, 303
 canonicity via 124–126, 130, 387, 406, 408
 games and 303, 326
scheduling of moves in game 232, 245, 295, 347, 380, 397
schema **487**
second-order logic 38, 59, 205, 287, 387, 571, 626 *see also* higher-order logic
secondary triangle **506**
semi-associative algebra **157**, 157–159, 161–163, *614*
 basis characterisation 375–376, 387, 389–390, 394, 433–434
 equational theory undecidable 578, *621*
 finite algebra (on finite base) property 593, 605
 game characterisation 383
 polyadic equality algebra and 203
 RA$_3$ and 360, 361, 387, 390, 433–435
 rainbow algebra is 495, 510
 relation algebra reducts and 191, 198, 203, 433–435, 605
 relativised representation and 433–434, 605
 tiling algebra is 546
 triangle addition property and 375, 415
semi-associative law **157**, 161, 163, 415, *614*
semigroup 206, 207, 210
semilattice-ordered semigroup 206
sentence **33**
sequential algebra 105, 206, 283
set theory 26, 42, 208, 212, 408, 540
sibling of node in tree **311**
signature **31**
similar structures **34**, 61
similarity type *see* signature
simple algebra **71**, 76, 433 *see also* simple relation algebra
simple equation **86**, 86–88, 93, 96, 98, 163, 168 *see also* Sahlqvist equation
simple relation algebra 104, **109**, 111, 129, 162, 433, 569
 Jónsson's problem 118, 626
simple term **86**

smooth relativised representation *see n-∼*
solitaire game   247
sort   **31**   *see also* many-sorted logic
sorts **a, r**   277
splitting   211, 394, 531–535
square proper relation algebra   **102**, 104, 129, 134   *see also* proper relation algebra
square representation   **118**, 120–121, 129   *see also* n-square relativised representation
standard translation   **85**, 85–88, 583, 585, 597   *see also* Sahlqvist correspondent
start index of atom   **459**
start of atom   **159**, 159–161, 163, 228, 543
Stebletsova–Venema Q-operator   **268**, 517
step-by-step construction   *see also* game
  by games   9, 215, 235, 273–276
  for axiomatisation   8, 13, 184, 343, 528
  of relation algebra   145
  of representation   8, 223, 225, 231, 433, 527, 596
  of tiling   572
Stone space topology   **43**
Stone's theorem   7, **42**, 48, 49, 80, 581
strategy   10, **236–237**, 239, **294**, 314, **315**   *see also* deterministic $\sim$; winning $\sim$
  used in play   **237**, **315**, 316
strict network   **220**, 222, 234–235
strictly positive term   **93**
strongly balanced formula   **207**
strongly representable atom structure   441, **446**, *620*
  chromatic number and   450–453
  class of $\sim$s is
    closed under ultraroots   458
    not closed under elementary equivalence, ultrapowers   454
    not closed under ultraproducts   283, 447, 453
    not elementary   92, 442, 447, 457, *620*
    not finitely axiomatisable   148
    not pseudo-elementary   283
  complex algebra of   446, 450, 455, 459
  Lyndon conditions and   456–457, 459–462
  related notions   455–458
structure   **33**
  generated by a set   **37**

subalgebra   **36**, 68–69
subdirect decomposition   **69**, 75, 104
subdirect product   **69**, 71, 111, 168, 199, 202
subdirect representation   **69**, 74
subdirectly irreducible algebra   **69**, 69–71, 74, 76
subdirectly irreducible component   **69**
subdirectly irreducible relation algebra   **109**, 111
subformula   **32**
subgraph   **30**
subgroup   **30**
subnetwork   **222**
subreduct   205–207, 208, 210, 362
substitution algebra   191, **204**
substitution-cylindrification operation   **600**
substitution map   **173**, 366, 369, 600   *see also* finitary map
substitution operation   **170**, 170–180, *617*
  completely additive
    in **CA**   170, 416
    in **Crs**   181
  polyadic   600
  polyadic $\sim$ **202**
  relation algebra reducts and   189–190, 194, 196–198, 416–420
substructure   **36**   *see also* elementary $\sim$
subtree of game tree   **313**
successor ordinal   **28**
sum (boolean)   *see* boolean functions
sum (cardinal and ordinal)   **28**
supremum   *see* boolean functions
surjective function   **29**
syllogism   1–2
symmetric basis   **369**, 369–370, 396, 414, 421, 423
symmetric binary relation   **26**
symmetric closure of basis   **369–370**
symmetric relation algebra   112, **117**, 348
symmetric unit   103, 154–156, 599
syntactic network   **298**   *see also* term network

tabular relation algebra   **209**
temporal constraints   252–258
temporal reasoning   99, 138, 252, 541, 584–585

Subject index

term **31**
  cylindric $\sim$ from formula 487
  standard translation of 85–88 *see also* Sahlqvist correspondent
term algebra **89**, 98, **446**, 457, 458, 532
term network **262**, 265, **269**, 270, 319, 343–345
ternary symbol **31**
theory **33**
  of a class **34**
Thompson's theorem 174–179
tile atom **545**
tile edge **545**
  attached or linked to other $\sim$ **559**, 560–562, 565, 566
tiling **542** *see also* $\sim$ problem
  hyperbasis yields 548–551
  partial **571**
  weak representation yields 573–575
  winning strategy from *see* winning strategy: in atomic game on tiling algebra
  winning strategy yields 547
tiling algebra **543–545**, 581 *see also* rainbow construction
  representability and tiling 569–570 *see also* winning strategy: in atomic game on $\sim$
  (semi-)associativity 545, 546, 570
  u and v atoms **544–545**, 554, 563, 565 *see also* rainbow construction: w_S atoms
tiling problem 210, 540, **541**, 541–543, *621* *see also* tiling
  deterministic $\sim$ **572**, 572–575, 577
  undecidable 542–543, 570, 572, 577
timestamp 477
top element *see* unit
topology **30**, 43–44, 49–50, 83
total function **29**
totally disconnected topology **31**, 43–44, 83
transitive binary relation **26**
transitive closure **26**
transitive group action **30**, 130, 281
transitive unit 103, 158
transposition map **600**
tree **310**, 311, 313–314 *see also* game tree

triangle addition property 367 *see also* triangle move
  in game G_r^n 377, 378
  in hyperbasis 467, 468, 550
  network addition property and 484
  restricted to nodes 0–2: 375, 413, 415
  semi-associative algebras and 375, 415
triangle-consistency 220
triangle move *see also* triangle addition property
  in game for homogeneous representation **397**
  in game $G_r^{m,n}$ **379**, 383, 477–479, 482
tuple **29**, 33
Turing machine 276, 288, 313, 542, 543, 572, 576
twisting 212, 583
two-sorted language *see* many-sorted logic
type **122**, 130, 303, 305

ultrafilter *see also* Stone's theorem; ultraproduct
  atom structure of $\sim$s **81**, 116, 130, 458, 461–462, 533–534
  composition 116, 125, 462, 533–534
  of boolean algebra **41**, 52, 54
  of boolean algebra with operators **61**
  over a graph 454
  over a set **56**
  representation induces 47
ultrapower **57**, 58, 91, 449, 454
ultraproduct **57**, 56–60, 68 *see also* Łoś' theorem
  atom structures and 91, 97, 98, 131, 283, 447, 458
  canonicity and 90, 98, 131
  elementary class and 58, 290, 449, 453
  game played on 326–328
  IG_α and 184, 283, 626
  iteration 59
  non-finite axiomatisability and 326, 331 *see also* Łoś' theorem
  pseudo-elementary class and 290, 291, 627
  quasi-variety and 69
  RRA closed under 129
  saturation 123, 130
ultraroot **57**, 58, 91, 288, 290, 449, 458
unary relation 2

unary symbol **31**
uncountable set **28**
undecidability *see also* decidability
  Diophantine equations and 208, 210
  equational theory of
    Df_n, Rdf_n 200
    GRA, RA, RRA, SA, RA_n, $S\mathfrak{R}aCA_n$
      198, 577–578, *621*
    Lyndon algebras 145, 200
    relation algebra reducts of finite cylindric algebras 605
  halting problem 542, 572
  hyperbasis problem 570
  membership of classes between
    RRA and **$S\mathfrak{R}aCA_5$** 570, *621*
    RRA and **wRRA** 575, *621*
  membership of **$S\mathfrak{R}aCA_n$** 386, 569, 603, *621*
  modal logics 200
  network satisfaction problem 223, 255, 579
  representation problem 205, 569, *621*
  semigroups and 210
  tiling and 210, 540, 542–543, 569–570, 572, 575, 577
  weak representation problem 575, *621*
undirected graph **30**
unfinished play **315**
unit
  of cylindric relativised set algebra **180**
  of proper relation algebra **101**, 102, 134, 152, *609*
  of relativised representation **153**, 154–158
  of representation **118**, 217
  reflexive and symmetric 102–103, 154–156, 599
universal axiomatisation **35** *see also* axiomatisation
universal class **35**, 289, 290
universal formula **33**
universal game tree **325**
universal generalisation **487**
universal relation 3
universal second-order sentence **38**
universal theory **33**
  of a class **35**
universally axiomatisable class **35**
unsharpness problem 388

'until' connective 584–585, 587

valid formula **35**
variable **31**
variety **67**, 67–76 *see also* atom-canonical $\sim$; atom structure; Birkhoff's theorem; canonical $\sim$; conjugated $\sim$; discriminator $\sim$; Sahlqvist $\sim$
  CA_α 168
  $CA_\alpha \cap ICrs_\alpha$, $ICrs_\alpha$, ID_α, IG_α 182–184
  complex algebras over 275
  generated by class of atom structures **89**, 89–93, 97, 131, 336
  IG_ω? 184, 283, 626
  NA, SA, WA 158
  RA 109
  RA_n 385, 387
  RCA_α 168
  RRA 126, 129
  $S\mathfrak{Nr}_\beta CA_\alpha$ 192, 193, 488
  $S\mathfrak{R}aCA_\alpha$ 192
  subalgebras of Lyndon algebras 145
  wRRA? 627
vertex of graph **29**
vocabulary **31**

weak associativity law **157**, 160, 163, 197–198, *614* *see also* weakly associative algebra
weak representation **164**, 165, 529, 573–575 *see also* weakly representable relation algebra
weakly associative algebra **157**, 157–163, *614*
  arithmetic of 159–161
  complexity of equational theory 595
  cylindric and hyperbases of 390
  finite algebra on finite base property 596, *614*
  finite algebra property 593
  finite base property 597–599, 600, *614*
  networks and 228, 391, 394
  relativised representation of 157, 225–232, 596, *614*
  $S\mathfrak{R}aG_3$ and 197–198, 361, 435
  tiling algebra is 545
  universal theory decidable 599, *614*, *621*

Subject index

weakly representable atom structure 441, **446**, *620*
  but not strongly representable 442, 455, 458, 531, *621*
  elementary class 446, *620*
  not finitely axiomatisable 148, 515
  pseudo-elementary view 283
  related notions 455–458
weakly representable relation algebra **164**, 163–166, *621*
  canonical extension 627
  completion 535, 627
  game to test 247
  Jónsson's axiomatisation 165, 247, 388, 528
  non-finite axiomatisability 165, 529–530, 575, *621*
  pseudo-universal view 281
  RA$_n$ and 530, 575, 576, 626
  RRA not finitely axiomatisable over 165, 528
  undecidable for finite algebras 575, *621*
  variety of $\sim$s? 627
well-founded order **27**, 314, 323
well-order **27**, 42
white atom **495**, **544** *see also* rainbow construction: w$_S$ atoms
white tile **542**, 572
wide hyperbasis **373**, 373–375, 467, 469 *see also* hyperbasis
wide hypernetwork **373**, 379, 380–382, 467, 468, 477 *see also* hypernetwork
winning position **316**
winning strategy 10, 274 *see also* formula: expressing $\sim$; game
  basis encodes $\sim$ 356, 377–378, 380, 397
  countable elementary substructures and 329–331, 485, 516
  in atomic game on rainbow algebra 514–516
  in atomic game on tiling algebra **554**, **558**, **568**, 569
    consistent 555–558, 568–569
    conventions 552, 553
    coordinates 563–568, 569
    discussion 553–554, 558
    tiling from $\sim$ 547
    tiling functions 560–563, 566–567, 569
    well-defined 555, 561–563, 566

winning strategy (*cont.*)
  in Ehrenfeucht–Fraïssé game 492, 493–494, 496–509, 510, 511
  for **CRA** 516
  for **RA**$_n$ 522–524, 531
  for **RB**$_\omega$ 528
  for **RRA** 514, 534
  for **wRRA** 530
  in game tree game 314, **316**, 316–318, 322, 324, 326, 332
  in hyperbasis game on $\mathfrak{A}(n,r)$ 442, 476–483
  in pseudo-elementary game 303, 304
  in pseudo-universal game **294**, 294–296, 297–298
  in relational basis game on $\mathfrak{A}(n,r)$ 479
  in relational basis game on rainbow algebra 442–443, 491, 499
  for $\forall$ 496–498, 499–500, 509
  for $\exists$ 500–509, 510
  in relativised representation game 227
  in representation game on cylindric algebra 250
  in representation game on relation algebra 235, **237**, 238–244, *612*
    Lyndon-style 247
    Monk algebra $\mathfrak{Cm}\alpha(\Gamma)$ 451–453
    rank and 239
    ultraproducts and 326–328
word *see* s-c-word

yellow atom **495**
yes-instance **542**

zero relation 3